STRUCTURAL DYNAMICS
THEORY AND APPLICATIONS

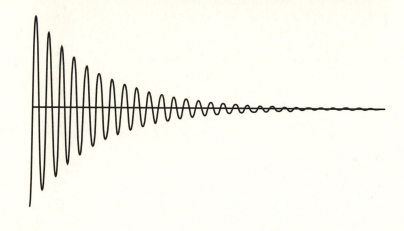

JOSEPH W. TEDESCO
Department of Civil Engineering
Auburn University

WILLIAM G. McDOUGAL
Department of Civil Engineering
Oregon State University

C. ALLEN ROSS
Graduate Engineering Research Center
University of Florida

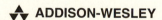 **ADDISON-WESLEY**

An imprint of Addison Wesley Longman, Inc.

Menlo Park, California • Reading, Massachusetts • Harlow, England
Berkeley, California • Don Mills, Ontario • Sydney • Bonn • Amsterdam • Tokyo • Mexico City

Senior Acquisitions Editor: Michael Slaughter
Associate Editor: Susan Slater
Production Manager: Pattie Myers
Senior Production Editor: Teri Hyde
Art and Design Supervisor: Kevin Berry
Composition: Lachina Publishing Services
Illustrations: Scientific Illustrators
Cover Design: Juan Vargas
Cover Image: Joseph W. Tedesco
Text Design: R. Kharibian & Associates
Text Printer and Binder: World Color Book Services
Cover Printer: Phoenix Color Corp.

Library of Congress Cataloging-in-Publication Data
Tedesco, Joseph W.
    Structural dyamics : theory and applications / Joseph W. Tedesco,
  William G. McDougal, C. Allen Ross.
      p. cm.
    Includes bibliographical references and index.
    ISBN 0-673-98052-9
    1. Structural dynamics.   I. McDougal, William G.   II. Ross, C.
  Allen.   III. Title.
  TA654.T43   1998
  624.1'7--dc21   98-28805
                                                         CIP

Instructional Material Disclaimer
The programs presented in this book have been included for their instructional value. They have been tested with care but are not guaranteed for any particular purpose. Neither the publisher or the authors offer any warranties or representations, nor do they accept any liabilities with respect to the programs.

The full complement of supplemental teaching materials is available to qualified instructors.

ISBN 0–673–98052–9

1 2 3 4 5 6 7 8 9 10—RNT—02 01 00 99 98

Addison Wesley Longman, Inc.
2725 Sand Hill Road
Menlo Park, California 94025

Contents

Preface

The dynamic analysis of complex structures has experienced impressive progress since the 1970s. Among the reasons for this trend are the advent of digital computers and the development of sophisticated numerical analysis tools, particularly the finite element method. As technologies in these areas continue to advance, practical dynamic analyses, both linear and nonlinear, of extremely complicated systems are becoming more commonplace. Therefore, it is imperative that engineers familiarize themselves with these modern numerical solution techniques and their implementation on digital computers.

The motivation for this book is to provide engineers with an understanding of the dynamic response of structures and of the common analysis techniques employed to evaluate these responses. Although the book emphasizes numerical solution techniques for a range of applications in structural dynamics, a comprehensive treatment of the classical analytical methods is also included. Among the special topics addressed in the book are the response of structures to earthquake excitation, the analysis of blast loading, wave forces on structures, wave propagation in elastic media, and nonlinear dynamic response. Moreover, the solution techniques demonstrated throughout the text are versatile and not limited to these topics, and are appropriate for many other applications in civil, mechanical, and aerospace engineering.

The book contains material for several courses on structural dynamics. The material includes a wide range of subjects, from very elementary to advanced, arranged in increasing order of difficulty. To systematize presentation of the material, the book is organized into five parts: I. Single-Degree-of-Freedom (SDOF) Systems; II. Multi-Degree-of-Freedom (MDOF) Systems; III. Continuous Systems; IV. Nonlinear Dynamic Response; and V. Practical Applications. The material in Part I is suitable for an elementary introductory course in structural dynamics at the junior or senior level. A more comprehensive course in introductory structural dynamics, taught to advanced seniors and first-year graduate students, can be offered from the material in Parts I and II. An advanced graduate level course in structural dynamics can include the material in Parts III and IV, and several selected topics from Part V.

Throughout the book, detailed derivations and implementation of numerical solution techniques are presented. Indeed, many of the end-of-chapter homework problems require a PC computer solution. Depending on a student's level of sophistication, they may write their own computer routines or use commercially available software packages such as MATLAB, MATHCAD, and MAPLE to solve the problems. As a convenience, a suite of computer programs written in FORTRAN for a PC that may be employed for the problem solutions are available on the authors' website at www.Structural-Dynamics.com.

This book has been written to serve not only as a textbook for college and university students, but also as a reference book for practicing engineers. The analytical formulations and numerical solution techniques presented throughout the book underlie most computer programs used by engineers in analyzing and designing structures subject to dynamic loadings.

The contents of this book are the result of teaching courses in structural dynamics and wave mechanics at Auburn University, Oregon State University and the University of Florida. The content was strongly influenced by our research experience. Organizations that have supported our research include the Air Force Office of Scientific Research, U.S. Army Corps of Engineers Waterways Experiment Station, Wright Laboratory Armament Directorate, Wright Laboratory Air Base Survivability Section, Office of Naval Research, SeaGrant, the Federal Highway Administration, and the Alabama Department of Transportation. We are indebted to the colleagues with whom we worked at these organizations.

We are very appreciative to the following individuals for their careful reviews of the manuscript and for their constructive suggestions: Thomas Baker, University of Virginia; James F. Doyle, Purdue University; Faoud Fanos, Iowa State University; Winfred A. Foster, Auburn University; Ronald B. Guenther, Oregon State University; Robert T. Hudspeth, Oregon State University; Barry T. Rosson, University of Nebraska; Parthe Sakar, Texas Tech University; Avi Singhal, Arizona State University; Bozidar Stojadinovic, University of Michigan; Theodore Toridis, George Washington University; Penny Vann, Texas Tech University; A. Neil Williams, University of Houston; Solomon C.S. Yim, Oregon State University; and Norimi Mitzutani, Nagoya University. We are also thankful to many former students who assisted in the solutions of the in-text examples and the end-of-chapter homework exercises, especially Mahmoud El-Mihilmy, Sanjoy Chakraborty, Prabhakar Marur, Dennis Tow, Johnathan Powell, Molly Hughes, Nathan Porter, and Robert Williams.

Joseph W. Tedesco
William G. McDougal
C. Allen Ross

1 Basic Concepts

1.1 INTRODUCTION TO STRUCTURAL DYNAMICS

This text is concerned with the analysis of structures subjected to dynamic loads. In this context, dynamic means time varying. That is, the application and/or removal of the loads necessarily varies with time [1].[†] Moreover, the response (i.e., resulting deflections, internal stresses, etc.) of a structure resisting such loads is also time dependent or dynamic in nature.

In reality, no loads that are applied to a structure are truly static. Since all loads must be applied to a structure in some particular sequence, a time variation of the force is inherently involved. However, whether or not a load should be considered dynamic is a relative matter. The most significant parameter influencing the extent of the dynamic effect a load has upon a structure is the *natural period of vibration* of the structure, *T*. Briefly stated, the natural period of vibration is the time required for the structure to go through one complete cycle of *free vibration*. If the application time for the load is large compared to the natural period of the structure, then there will be no dynamic effect, and the load can be considered static. If, on the other hand, the application time for the load is in close proximity to the natural period of the structure, it will induce a dynamic response.

Situations in which dynamic loading must be considered are quite numerous. Examples include: the response of bridges to moving vehicles; the action of wind gusts, ocean waves, or blast pressures upon a structure; the effect of landing impact upon aircraft; the effect on a building structure whose foundation is subjected to earthquake excitation; and the response of structures subjected to alternating forces caused by oscillating machinery [2]. Under these types of loading conditions, either the entire structure or certain components of the structure are set in motion (i.e., caused to vibrate). Therefore, it is necessary to apply the principles of dynamics rather than those of statics to evaluate the structural response. It will be demonstrated throughout this text that the maximum deflections, stresses, strains, and various other response quantities exhibited by a structure are generally more severe when loads of a given amplitude are applied dynamically rather than statically.

---

† Numbers in brackets refer to end-of-chapter references.

1.2 TYPES OF DYNAMIC LOADS

The response of a structure to dynamic loads may be categorized as either *deterministic* or *nondeterministic* [3]. If the magnitude, point of application, and time variation of the loading are completely known, the loading is said to be *prescribed,* and the analysis of the structural response to this prescribed loading is defined as a deterministic analysis. However, if the time variation and other characteristics of the loading are not completely known, but can be defined only in a statistical sense, the loading is referred to as *random,* and the corresponding analysis of the structural response is termed nondeterministic. This text emphasizes the deterministic response of structures to prescribed dynamic loading.

To expedite the dynamic analysis of structures, it is convenient to classify dynamic loads as either *periodic* or *nonperiodic.* Periodic loadings repeat themselves at equal time intervals. A single time interval is called the period T_0. The simplest form of periodic loading can be represented by a sine function as shown in Figure 1.1a. This type of periodic loading is referred to as *simple harmonic.* Another form of periodic load is illustrated in Figure 1.1b. This loading is termed *periodic, nonharmonic.* Most periodic loads may be accurately represented by summing a sufficient number of harmonic terms in a *Fourier series.* Any loading that cannot be characterized as periodic is nonperiodic. Nonperiodic loads range from short-duration impulsive types, such as a wind gust or a blast pressure (Figure 1.1c), to fairly long duration loads, such as an earthquake ground motion (Figure 1.1d).

1.3 SOURCES OF DYNAMIC LOADS

Sources of dynamic loads on structures are many and varied. However, the origin of the majority of significant dynamic loads can be attributed to one of the following sources: (1) environmental, (2) machine induced, (3) vehicular induced, and (4) blast induced. Undoubtedly, the single most common source of dynamic loads on structures is environmental in nature. Some typical examples of environmentally induced dynamic loads on structures include wind loads, earthquake loads, and wave loads.

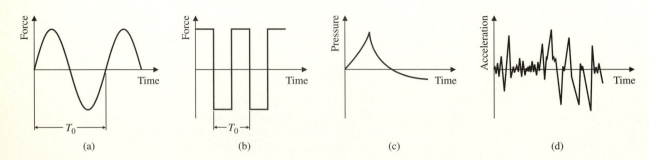

Figure 1.1

Types of dynamic loadings: (a) simple harmonic; (b) periodic, nonharmonic; (c) nonperiodic, short duration; (d) nonperiodic, long duration.

Wind loads on structures are a function of wind velocity, height of the structure, and shape and stiffness characteristics of the structure. Aerodynamic forces in the form of drag and lift forces can be computed from a mean wind velocity record similar to that shown in Figure 1.2. Earthquakes are another environmental phenomenon that can induce significant dynamic loads on structures. Earthquake forces that develop in structures result from the acceleration of the structure's base by a highly irregular and complex earthquake ground motion as illustrated in Figure 1.1d. In the design of coastal and offshore structures, wave-induced loads represent the primary design criterion. Wave forces include components due to drag, inertia, lift, and buoyancy to estimate the total hydrodynamic loads acting on the structure.

Another significant source of dynamic loads on structures, commonplace in industrial installations, is that attributed to equipment or machinery such as reciprocating and rotating engines, turbines, and conveyor mechanisms. The nature of machine-induced dynamic loading is usually periodic. Vehicular-induced vibrations represent still another source of dynamic loading, and they can be categorized as either internal or external. A common example of an externally induced vehicular dynamic load is that caused to a highway bridge from speeding trucks traveling across it. Similarly, the same truck traffic may induce vibrations that cause cracks in plaster and other minor damage to buildings located close to the highway. Oftentimes, equipment located within a vehicle must be isolated from forces generated by internal vehicular vibration. For instance, sensitive navigational equipment mounted within aircraft must be isolated from dynamic forces induced by take-offs and landings or in-flight turbulence.

Another major source of dynamic loads is attributed to blasts, either from explosive devices or accidental chemical explosions. Blast-induced loads on structures have been a longtime concern in the design of military installations. In recent years, however, there has been a heightened awareness within the civilian sector of the susceptibility of government and institutional facilities to terrorist bombings. The dynamic loads induced to a structure from blasts manifest themselves primarily in the form of an overpressure, as illustrated in Figure 1.1c, but in some instances the loads may instigate significant ground shaking, as illustrated in Figure 1.3. The blast overpressure is most devastating to structures in close proximity to the explosion; however, the ground-shaking component resulting from a large underground detonation may affect structures many miles away.

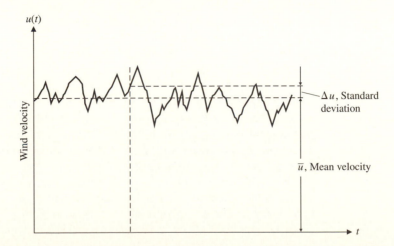

Figure 1.2
Typical wind velocity record.

Figure 1.3

Typical ground motion records produced by an underground explosion. *Blast Vibration Analysis* by G.A. Bollinger, Copyright © 1971. Reprinted by permission of Southern Illinois University Press, Carbondale, IL.

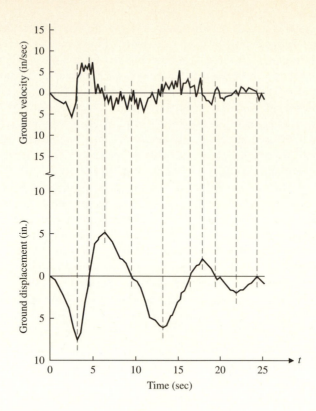

1.4 DISTINGUISHING FEATURES OF A DYNAMIC PROBLEM

A structural dynamics problem differs from its static counterpart in two essential aspects [4]. The first and most obvious difference is the time-varying nature of the excitation (applied loads) and the response (resulting deflections, stresses, etc.). That is, both are functions of time in a structural dynamics problem. This precludes the existence of a single solution. The analyst must investigate the solution over a specific interval of time to fully evaluate the structural response. Thus, a dynamic analysis is inherently more computationally intensive than a static analysis.

However, the most important feature differentiating a dynamic problem from the corresponding static problem is the occurrence of *inertia forces* when the loading is dynamically applied. Consider the vertical cantilever structure shown in Figure 1.4. If a force F is applied statically at the tip of the cantilever, as illustrated in Figure 1.4a, the resulting shear force V, bending moment M, and associated stresses and deflections in the structure can be computed from the basic static structural analysis principles, and are directly proportional to the force F. If, however, a time-varying force $F(t)$ is applied to the tip of the cantilever, as illustrated in Figure 1.4b, the structure is set in motion, i.e., vibrates and experiences accelerations. Inertia forces proportional to the mass then develop in the structure that resist these accelerations. The significance of the contribution made by inertia forces to the shear force $V(t)$, bending moment $M(t)$, and related stresses and deflections in the structure determines whether a dynamic analysis is warranted.

Figure 1.4

Cantilever structure subjected to (a) a static load; (b) a dynamic load.

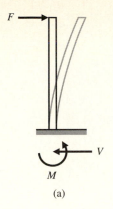

(a)

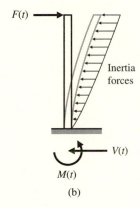

(b)

1.5 METHODOLOGY FOR DYNAMIC ANALYSIS

Typical of any problem in engineering mechanics, an appropriate methodology for conducting a dynamic structural analysis is essential to achieve a viable solution. One such methodology is summarized in Figure 1.5, which defines three basic phases of a dynamic analysis: (1) identification of the physical problem, (2) definition of the mechanical model, and (3) solution of the mechanical model.

Phase 1 entails recognition of the problem as it exists in nature. This includes accurately identifying and describing the physical structure, or structural component, and the source of the dynamic loading. Phase 2 requires an interpretation of the physical problem into a form conducive to available analysis techniques. This involves defining a mechanical model that accurately represents the dynamic behavior of the physical problem in terms of geometry, kinematics, loading, and boundary conditions. The idealization of the physical problem to a mechanical model conducive to available analysis techniques generally involves some simplifying assumptions, which influences the formulation of the differential equations governing the structural response. In Phase 3 the governing differential equations are solved to obtain the dynamic response. The solution is only as accurate as the representation provided by the mechanical model. Therefore, this step generally requires an assessment for accuracy. If the predefined accuracy criteria are met, the mechanical model has then been solved

Figure 1.5

Methodology for conducting a dynamic analysis.

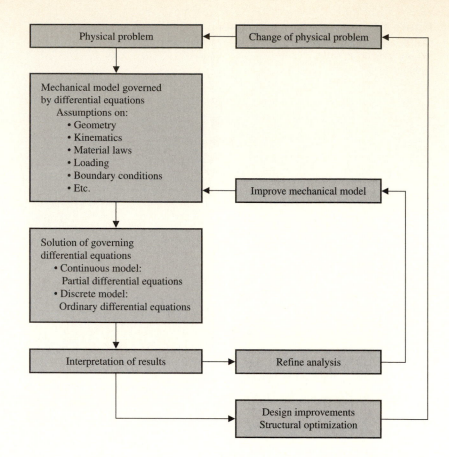

with a satisfactory level of confidence, and the analysis results can be interpreted in a meaningful manner. For complex structures, it may be necessary to refine the analysis by considering a more detailed mechanical model or to introduce design improvements for structural optimization, which leads to further analyses involving several iterations.

The complexity of the analysis depends largely on the physical problem under consideration and on the mechanical model that must be employed to obtain a sufficiently accurate response prediction. A linear analysis can be a routine task, although a fully three-dimensional solution may require a significant amount of human effort and computing resources. On the other hand, a nonlinear dynamic analysis can represent a major challenge to the ingenuity of the analyst and require very significant resources.

Indeed, the most important step in the dynamic analysis procedure is defining a mechanical model that accurately represents the physical problem. Theoretically, all structures possess an infinite number of *degrees of freedom* (DOF). In other words, an infinite number of independent spatial coordinates are required to completely specify the position of all points on the structure at any instant of time [6]. However, most practical analyses are conducted on mechanical models having a finite number of DOF. For each DOF exhibited by a structure, there exists a *natural frequency* (or natural period) of vibration. For each natural frequency, the structure vibrates in a particular mode of vibration.

For most large, complex structures, however, it is not necessary to determine all the system natural frequencies, since relatively few of these vibration modes contribute appreciably to the dynamic response. Therefore, the mechanical model should be defined in such a manner that only those vibration modes that significantly contribute to the dynamic response are accurately represented.

In general, the mechanical model can be categorized as either *continuous* or *discrete*. The type of mechanical model employed for an analysis affects the nature of the governing differential equations and their subsequent solution. For a continuous model, the mathematical formulation of the problem results in a system of partial differential equations. However, for a discrete system the mathematical formulation yields a set of ordinary differential equations, one for each DOF. Analytical solutions for partial differential equations and for large systems of ordinary differential equations are quite cumbersome, if not impossible in many cases. Therefore, in most practical applications numerical solution techniques must be employed.

Consider the transverse vibration of the multistory building structure illustrated in Figure 1.6a. In reality, the structure manifests distributed mass and stiffness characteristics along its height. The continuous model representation of the structure is shown in Figure 1.6b. The mathematical formulation of the continuous model incorporates the distributed mass, $m(y)$, and stiffness, $k(y)$, characteristics of the structure. Moreover, the independent displacement variable $x(y,t)$ is a function of both position y and time t. Therefore, the resulting equations of motion must be partial differential equations. Discrete model representations of the same structure are illustrated in Figure 1.6c and d. These models are commonly referred to as *lumped mass models* because the system is assumed to be represented by a small number of localized (or lumped) masses. The representation portrayed in Figure 1.6c is a single-degree-of-freedom (SDOF) system, in which the entire mass m of the structure is localized (lumped) at the top and the structure has constant stiffness k. The independent displacement variable $x(t)$ of the mass is a function of time alone. Thus the single resulting equation of motion is an ordinary differential equation. The lumped mass representation shown in Figure 1.6d is a three-DOF system in which the mass of the structure is localized at three locations. Each localized mass, m_1, m_2, and m_3, has its own displacement variable, $x_1(t)$, $x_2(t)$, and $x_3(t)$, respectively. This model also exhibits three discrete stiffness terms, k_1, k_2, and k_3. The resulting equations of motion for this model are a set of three simultaneous ordinary differential equations.

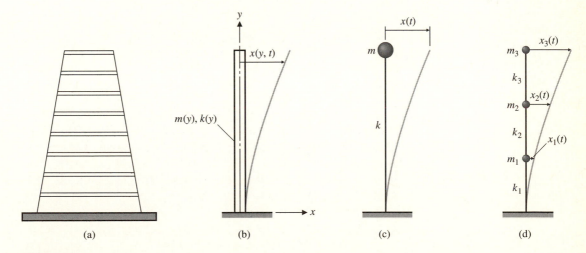

(a) (b) (c) (d)

Figure 1.6

Mechanical models for multistory building structure: (a) physical representation; (b) continuous model; (c) SDOF discrete model; (d) three-DOF discrete model.

Practical dynamic analysis of large, complicated multidegree-of-freedom (MDOF) structures is generally accomplished through a computer-implemented numerical analysis technique known as the *finite element method* (FEM). In FEM analyses, continuous systems are characterized as discrete MDOF systems. Many commercially available, general purpose FEM computer programs, such as ADINA [7], have been used successively in modeling very complex problems in various areas of engineering. FEM models possessing tens of thousands of DOF are not uncommon. The complexity and detail of the FEM model used in a particular analysis is highly dependent on the unique aspects of the response the analyst seeks.

Consider, for example, the multigirder steel highway bridge illustrated in Figure 1.7. The steel girders shown in the plan view (Figure 1.7a) serve as the primary load-

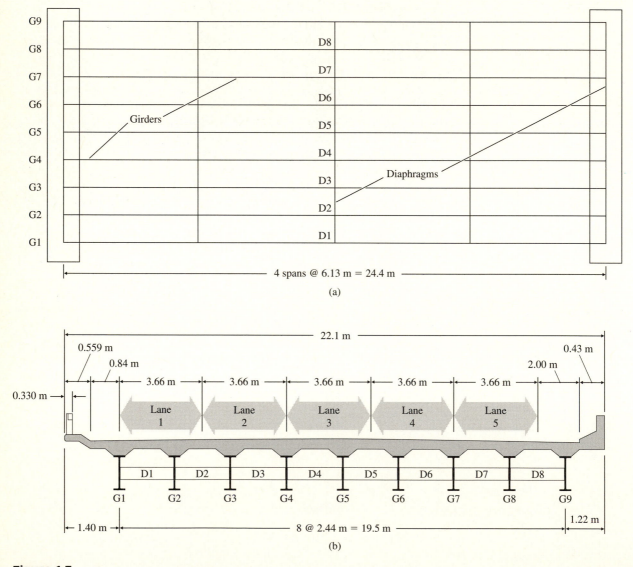

Figure 1.7
Multigirder steel highway bridge: (a) plan view; (b) typical cross section.

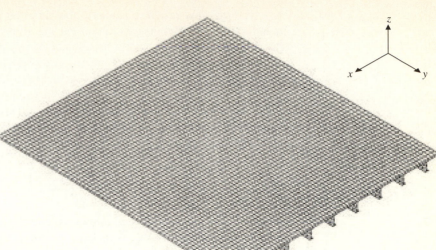

Figure 1.8
Isometric view of FEM model for bridge.

carrying members. The structural system is typically tied together by a reinforced concrete deck slab and transverse steel members, or diaphragms, that span transversely between the girders as shown in the bridge cross-section illustration (Figure 1.7b). To assess the effectiveness of the diaphragms in distributing truck traffic loads transversely among the girders, a detailed FEM model of the bridge is required. An isometric view of the FEM model for the bridge structure of Figure 1.7 is illustrated in Figure 1.8. This model exhibits over 50,000 active DOF.

1.6 TYPES OF STRUCTURAL VIBRATION

There are several types of structural vibration. A structure can be caused to vibrate by imposing upon it some *initial conditions,* or disturbances [8]. Initial conditions generally manifest themselves as an energy input such as a velocity imparted to the mass of the structure (kinetic energy) or a displacement configuration imposed upon the structure (potential energy) that is suddenly removed. The resulting structural vibration occurs in the absence of any externally applied forces, and is termed *free vibration.* Free vibration usually occurs at the fundamental natural frequency (lowest frequency) of the structure. Since there is no external excitation acting on the structure, the vibrations diminish with time as the energy input to the structure from the initial conditions eventually is dissipated, or damped out.

The vibrations of a structure under the influence of external excitation are called *forced vibrations.* If the source of the excitation is periodic, the vibration consists of a *steady-state response* and a *transient response.* The steady-state response transpires at the frequency of the excitation. When the excitation frequency coincides with one of the natural frequencies of the structure, a condition known as *resonance* exists. At resonance, the amplitudes of the vibrations become exceedingly large and are limited only by the damping in the structure.

The transient response is due to the initial energy stored in the structure and is manifested as oscillations generally occurring at the fundamental natural frequency of the structure. The transients are usually completely damped out after several cycles of vibration, are insignificant in comparison to the steady-state vibrations, and can be ignored. However, if the structure is excited by a suddenly applied nonperiodic force (such as a blast), the subsequent response is also a transient response, since steady-state oscillations are not produced. Similar to free vibration, the transient oscillations occur at the natural frequencies of the structure. The amplitude of the transient response varies in a manner dependent upon the nature of the excitation.

1.7 ORGANIZATION OF THE TEXT

This book was conceived as an introductory text for both undergraduate and graduate courses in civil engineering. It includes a variety of topics in the theory of structural dynamics as well as applications of this theory to the analysis and response of structures subject to earthquake excitation, blast loading, and wave forces. The book also addresses wave propagation phenomena in elastic media. Although the applications of structural dynamics in civil engineering are different from those encountered in mechanical engineering, engineering mechanics, and aerospace engineering, the principles and solution techniques are basically the same. Therefore, this text emphasizes these principles and solution techniques (especially numerical solution techniques), and illustrates them with over 135 worked-out examples, over 460 homework problems, and approximately 800 illustrations.

The book is organized in five parts: I. Single-Degree-of-Freedom (SDOF) Systems; II. Multidegree-of-Freedom (MDOF) Systems; III. Continuous Systems; IV. Nonlinear Dynamic Response; and V. Practical Applications. Although the emphasis of the book is directed toward linear problems in structural dynamics, techniques for solving a limited class of nonlinear structural dynamics problems are also introduced. Because of the wide variety of topical coverage presented in this text, many of the mathematical symbols and other notation have multiple representations. Therefore, to maintain consistency and clarity of presentation, a separate notation section is provided at the end of each chapter.

Part I, which includes Chapters 2 through 8, addresses the vibration of SDOF systems. SDOF systems are studied in great detail because many practical problems can be solved with this formulation and many of the solution techniques developed for SDOF systems can be easily modified and extended to the solution of MDOF systems. Chapter 2 discusses the formulation of the equations of motion for SDOF systems by application of Newton's second law, the energy method, and the principle of virtual displacement. Chapter 3 addresses the undamped free vibration of SDOF systems, and Chapter 4 discusses the free vibration of SDOF systems possessing viscous, hysteresis, or Coulomb damping. Chapter 5 examines the response of SDOF systems to harmonic excitation. Some of the most important fundamental concepts in structural dynamics are discussed in this chapter, such as resonance, force transmission, and vibration isolation. Chapter 6 discusses the response of SDOF systems to periodic (nonharmonic) excitation as well as the response to arbitrary dynamic excitation by implementation of the Duhamel integral method; this chapter also introduces the concept of response spectrum. Chapter 7 presents numerical techniques for evaluating the dynamic response of SDOF systems. Part I concludes with Chapter 8, a discussion of frequency domain solution techniques for SDOF systems.

Part II, which includes Chapters 9 through 13, addresses the dynamic response of discrete MDOF systems. These chapters examine general MDOF systems, whose dynamic response can be characterized by a finite number of DOF. Chapter 9 discusses some fundamental properties of MDOF vibrating systems, such as the mass and stiffness matrices, and formulation of the system eigenproblem. Chapter 10 examines the basic concepts and procedures for the free vibration analysis of MDOF systems, and also presents several approximate methods for estimating the fundamental frequency of MDOF systems. Chapter 11 presents several commonly employed numerical solution techniques for extracting the natural frequencies (eigenvalues) and mode shapes (eigenvectors) for MDOF systems. Chapter 12 discusses the evaluation of the dynamic response of MDOF systems by the mode superposition method, and Chapter 13 examines the dynamic response of MDOF systems by direct numerical integration, and presents a discussion of the relative advantages of the mode superposition and direct numerical integration methods.

Part III of the text, Chapter 14, concerns the vibrations of systems having distributed mass and stiffness properties, or continuous systems. Continuous systems possess an infinite number of DOF and their equations of motion must be expressed in the form of partial differential equations. Chapter 14 examines the free vibration of uniform rods, cables, and beams, as well as the undamped forced vibration of beams, and discusses approximate solution techniques that effectively transform continuous systems into equivalent discrete systems.

Part IV of the text, Chapter 15, is devoted to nonlinear dynamic response. Chapter 15 examines the various types of nonlinearities and discusses the incremental formulation of the equations of motion for systems possessing nonlinear characteristics. This chapter also presents several commonly employed numerical solution techniques for nonlinear equilibrium equations as well as a rigorous analysis of both SDOF and MDOF elastoplastic systems.

Part V, Chapters 16 through 21, deals with several practical applications of various aspects of basic structural dynamics theory discussed in Parts I through IV. Chapter 16 addresses one-dimensional wave propagation in elastic media and discusses applications to stress wave velocities in uniform rods and collinear impact of bars. Chapter 17 presents a brief seismological background on causes and characteristics of earthquakes, as well as a discussion of earthquake ground motions. Chapter 18 presents deterministic procedures for evaluating the response of structures to earthquake ground motions and addresses both the time-history and response spectrum methods of analysis. Chapter 19 discusses the basic concepts that define blast loads on structures and the corresponding structural response. Chapter 20 discusses the basic theories to describe water waves and methodologies to select design waves. Finally, Chapter 21 discusses the response of structures to wave forces, as well as formulations for wave forces on small bodies and large bodies, with applications to both fixed structures and moving structures.

Many of the worked out examples and end-of-chapter problems presented in Chapter 18 require the north-south ground motion component of the 1994 Northridge, California earthquake as input. Numerical values for this ground acceleration record (i.e., digitized accelerogram) are available on the author's web site: www.Structural-Dynamics.com.

A major emphasis of this text is the development of numerical solution techniques for a wide variety of structural dynamics problems. Therefore, listings for a number of computer algorithms are presented throughout the text in the solutions of various in-text examples. A suite of computer programs that may be employed for the solution of many of the end-of-chapter problems are described on the author's web site. Both the source and executable codes for these computer programs are available on the web site.

1.8 SYSTEMS OF UNITS

The problems in this text are written primarily using the English system of units. However, in recognition of the fact that there will eventually be a change from the English system of units to the International System of Units (SI), and realizing that the two systems will coexist for some years, a limited number of problems are presented in SI units. Table 1.1 shows some quantities typically used in structural dynamics in both English and SI units, as well as conversion factors for transforming from English units to SI units.

TABLE 1.1 Systems of Units

Quantity	English System	SI System	Conversion Factor
Length	foot (ft)	meter (m)	0.3048
	inch (in)	meter (m)	0.0254
Force	pound (lb)	newton (N)	4.4482
	kip (1000 lb)	newton (N)	4448.2
Mass	slug (lb-sec^2/ft)	kilogram (kg)	14.59
	pound-mass (lbm)	kilogram (kg)	0.045359
Mass density	lbm/ft^3	kg/m^3	16.02
	lbm/in^3	kg/m^3	27680.0
Stress or pressure	lb/ft^2 (psf)	N/m^2 (Pa)	47.88
	lb/in^2 (psi)	N/m^2 (Pa)	6894.8
Acceleration	ft/sec^2	m/sec^2	0.3048
	in/sec^2	m/sec^2	0.0254
Velocity	ft/sec	m/sec	0.3048
	in/sec	m/sec	0.0254
Volume	ft^3	m^3	0.028317
Moment or torque	in-lb	N-m	0.113
	ft-lb	N-m	1.356

REFERENCES

1 Irvine, H.M., *Structural Dynamics for the Practicing Engineer,* Allyn and Unwin, Boston, 1986.

2 Tauchert, T.R., *Energy Principles in Structural Mechanics,* McGraw-Hill, New York, 1974.

3 Clough, R.W. and Penzien, J., *Dynamics of Structures,* McGraw-Hill, New York, 1975.

4 Craig, R.R., *Structural Dynamics, An Introduction to Computer Methods,* Wiley, New York, 1981.

5 Bathe, K.J., "Some Advances in Finite Element Procedures for Nonlinear Structural and Thermal Problems," *Proceedings of the Symposium on Future Directions of Computational Mechanics,* Winter Annual Meeting, 1986.

6 Beards, C.F., *Structural Vibration Analysis: Modelling, Analysis and Damping of Vibrating Structures,* Wiley, New York, 1983.

7 ADINA, "A Finite Element Computer Program for Automatic Dynamic Incremental Nonlinear Analysis," *Report ARD 90-1,* ADINA R&D, Inc., Watertown, MA, 1990.

8 Tse, F.S., Morse, I.E., and Hinkle, R.T., *Mechanical Vibrations, Theory and Applications,* 2nd ed., Allyn and Bacon, Boston, 1978.

9 Stallings, J.M., Cousins, T.E., and Tedesco, J.W., "Fatigue of Diaphragm—Girder Connections," *Final Report RP 930-307,* Auburn University Highway Research Center, Auburn University, AL, 1996.

10 Tedesco, J.W., Stallings, J.M., and Tow, D.R., "Finite Element Method Analysis of Bridge Girder—Diaphragm Interaction," *Computers and Structures,* Vol. 56, No. 2, 1995, pp. 461–473.

PART I

Single-Degree-of-Freedom (SDOF) Systems

2 ▲ Equation of Motion and Natural Frequency

In reality, all structures possess an infinite number of degrees of freedom. That is, an infinite number of independent spatial coordinates are necessary to completely define a structure's *configuration*—the geometric location of all the masses of the structure or system [1]. A system for which only one spatial coordinate is required to define the configuration is a single-degree-of-freedom (SDOF) system. Although a SDOF model may not provide accurate representation for detailed analysis of many structures, information obtained from the investigation of a SDOF system may be adequate for a preliminary analysis of a complicated structure. Moreover, a thorough understanding of the basic principles of vibration of a SDOF system is essential before a dynamic analysis of a more complex multidegree-of-freedom (MDOF) system can be undertaken.

This chapter deals with SDOF systems, that is, systems described by a single, second-order ordinary differential equation. This mathematical expression, which defines the dynamic equilibrium of a system, is called the *equation of motion* of the structure. An important result from the solution of the equation of motion is the displacement-time history of a structure subjected to a prescribed time-varying load.

Quite often, the formulation of the equation of motion represents one of the most important phases of a dynamic analysis. In this chapter several methods for formulating that equation for SDOF systems are discussed, including d'Alembert's principle, the energy method, and the principle of virtual displacements. The concept of natural frequency is also introduced in this chapter.

2.1 FUNDAMENTAL COMPONENTS OF A VIBRATING SYSTEM

Before establishing methodologies for formulating the equations of motion for SDOF systems, it is important to define the basic components comprising the vibrating system. These include mass, stiffness, damping, and forcing. Damping is the energy loss mechanism, and forcing is the source of excitation.

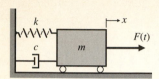

Figure 2.1

Mechanical model for a simple SDOF system.

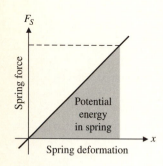

Figure 2.2

Force-deformation relationship for a linear spring.

The mechanical model for a simple SDOF vibrating system is depicted in Figure 2.1. It consists of a rigid body of mass m, constrained to move in only one translational direction, whose position is completely defined by the single displacement coordinate x. A spring of constant stiffness k, fixed at one end and attached at the other end to the mass, provides elastic resistance to displacement. The energy dissipation mechanism is represented by a damper or dashpot having a damping coefficient c. The external excitation to the system is provided by the time-varying force $F(t)$. Vibration in the absence of externally applied forces is also possible. Such vibration is referred to as *free vibration.*

The displacement of the mass is measured from its static equilibrium position and is defined as a function of time by the spatial coordinate $x(t)$. The motion of the mass is resisted by the force F_S that develops in the spring and is defined by

$$F_S = kx \tag{2.1}$$

where k is the spring constant. The units of k are generally defined as pounds per inch (lb/in) or newtons per meter (N/m). The relationship between the deformation in the spring and the force in the spring is illustrated in Figure 2.2. For a completely elastic system, the spring serves as an energy storage device. The energy stored in the spring is called the *strain energy,* or potential energy, of the system. The strain energy V is calculated as the area under the force-displacement curve of the spring and is given by

$$V = \frac{1}{2}kx^2 \tag{2.2}$$

A *conservative system* will continue to vibrate indefinitely even after the external excitation has ceased. However, all practical structures exhibit energy dissipation, or damping, that prevents this from happening. Damping is a very complex phenomenon for which numerous analytical models exist to describe its effect. The most commonly employed analytical damping model is the linear viscous dashpot model [2]. The damping force F_D is proportional to the velocity $\dot{x}$ of the mass and is given by

$$F_D = c\dot{x} \tag{2.3}$$

where c is the viscous damping coefficient having units of pound-seconds per inch (lb-sec/in) or newton-seconds per meter (N-sec/m).

It was mentioned in Chapter 1 that the primary feature distinguishing a dynamic problem from the corresponding static problem was the presence of inertia forces in a vibrating system. The inertia force F_I is the product of the mass and the acceleration of the mass $\ddot{x}$ and is given by

$$F_I = -m\ddot{x} \tag{2.4}$$

The negative sign indicates that the inertial force opposes the acceleration of the mass.

2.2 D'ALEMBERT'S PRINCIPLE OF DYNAMIC EQUILIBRIUM

D'Alembert's principle of dynamic equilibrium is a convenient method for establishing the equations of motion for simple SDOF and MDOF systems. It essentially involves invoking Newton's second law of motion to the system. Newton's second law states

Figure 2.3

Free-body diagram for SDOF system.

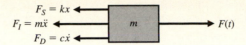

that the rate of change of momentum is proportional to the applied force and occurs along the line of action of the force. For a constant mass, the rate of change of momentum is equal to the product of the mass and its acceleration.

A free-body diagram of the SDOF system illustrated in Figure 2.1 is shown in Figure 2.3. The expression for dynamic equilibrium, using d'Alembert's principle, is given by

$$\sum (\text{forces})_x - m\ddot{x} = 0 \tag{2.5}$$

Thus, by introducing the appropriate inertia force, it can be reasoned that the applied force on the mass is in equilibrium with the inertia force. Therefore, the dynamic problem is reduced to an equivalent problem of statics. Applying Eq. (2.5) to the free-body diagram in Figure 2.3 results in the equation of motion for the system:

$$F(t) - kx - c\dot{x} - m\ddot{x} = 0 \tag{2.6}$$

or

$$m\ddot{x} + c\dot{x} + kx = F(t) \tag{2.7}$$

Dividing Eq. (2.7) through by m results in

$$\ddot{x} + \frac{c}{m}\dot{x} + \frac{k}{m}x = \frac{F(t)}{m} \tag{2.8}$$

or

$$\ddot{x} + \frac{c}{m}\dot{x} + \omega^2 x = \frac{F(t)}{m} \tag{2.9}$$

where the term ω is called the *natural circular frequency* of the system, with units of radians per second, and is given by

$$\omega = \sqrt{\frac{k}{m}} \tag{2.10}$$

As illustrated by Eq. (2.10), the natural frequency is defined solely by the system's mass and stiffness characteristics. Natural frequency plays a vital role in vibration analysis and will be referred to extensively throughout the text.

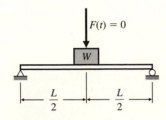

Figure 2.4

Simple beam with concentrated weight at midspan.

EXAMPLE 2.1 ▼

A simply supported beam having a concentrated weight W at its midspan is shown in Figure 2.4. Consider the mass of the beam to be negligible in comparison to the mass of the concentrated weight, and neglect the effects of damping in the system. Determine the equation of motion for the system and calculate its natural circular frequency. Let $W = 10$ kips, the moment of inertia of the beam $I = 200$ in⁴, Young's modulus for the beam material $E = 29,000$ ksi, and the beam length $L = 20$ ft.

Solution

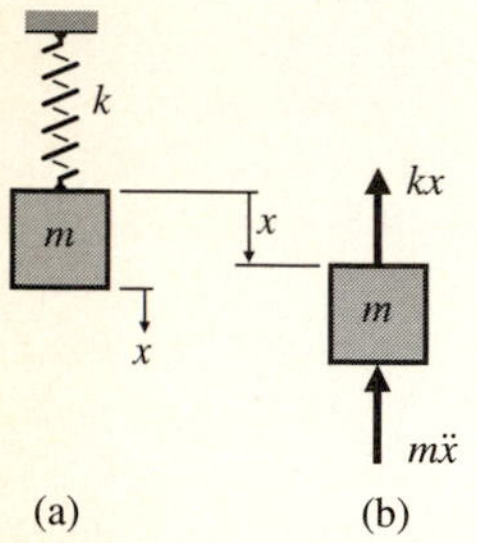

The mechanical model representing this system is illustrated in Figure 2.5a, and the free-body diagram is shown in Figure 2.5b. Application of d'Alembert's principle gives the equation of motion as

$$m\ddot{x} + kx = F(t) = 0 \tag{1}$$

or

$$\ddot{x} + \frac{k}{m}x = 0 \tag{2}$$

The natural frequency of the system can be determined directly from Eq. (2.10). However, it is first necessary to evaluate the mass and stiffness terms. With the acceleration of gravity $g = 386.4$ in/sec², the mass term is given by

$$m = \frac{W}{g} = \frac{10,000 \text{ lb}}{386.4 \text{ in/sec}^2} = 25.9 \text{ lb-sec}^2/\text{in}$$

and the stiffness is expressed as

$$k = \frac{48EI}{L^3} = \frac{48(29,000 \text{ ksi})(200 \text{ in}^4)}{(240 \text{ in})^3}$$
$$= 20.139 \text{ kips/in} = 20,139.0 \text{ lb/in}$$

Therefore, from Eq. (2.10),

$$\omega = \sqrt{\frac{k}{m}} = \sqrt{\frac{20,139.0}{25.9}} = 27.88 \text{ rad/sec}$$

Notice that the left-hand side of Eq. (1) contains no energy dissipation or damping term $c\dot{x}$, indicating that the vibration is *undamped*. Also note that the right-hand side of Eqs. (1) and (2) is zero, which indicates that there is no externally applied time-varying force. Vibration in the absence of externally applied forces is called *free vibration*, which is typically caused by imparting an initial displacement or velocity to the mass. Undamped free vibration of SDOF systems is the subject of Chapter 3, and Chapter 4 discusses damped free vibration of SDOF systems. ▲

Figure 2.5
(a) SDOF idealization of simple beam; (b) free-body diagram of SDOF idealization.

Example 2.1 is representative of a system exhibiting localized mass and localized stiffness. That is, the elastic restoring force F_S in the system is localized in the massless spring (the beam), and the inertia force F_I is assumed to be concentrated at the mass center of the block mass m. There exists a category of SDOF systems, however, for which localizing the mass would render an erroneous formulation of the equation of motion. This is typical of systems composed of rigid bars of finite length, for which the mass is distributed along the length, that exhibit a rotational degree of freedom. Such a system is illustrated in the following example.

EXAMPLE 2.2 ▼

A uniform rigid rod having mass m and length L is restrained to move vertically at one end by a translational spring having a stiffness k, and by a rotational spring having stiffness k_R at the other end as shown in Figure 2.6. Determine the equation of motion for this system and calculate its natural circular frequency.

Solution

Refer to the free-body diagram in Figure 2.7, noting that the system exhibits a rotational degree of freedom θ about point A, and apply the dynamic equilibrium equation of moment $\Sigma M - I_{0_A}\ddot{\theta} = 0$. Since there is no externally applied time-varying force, the equation of motion will again represent free vibration. The equation of motion is then expressed as

$$I_{0_A}\ddot{\theta} + kL\Delta + k_R\theta = 0 \tag{1}$$

where $\ddot{\theta}$ is the rotational acceleration and I_{0_A} is the mass moment of inertia of the rod with the axis of rotation at end A; I_{0_A} is equal to $mL^2/3$ (see Table 2.1). At point B the vertical deflection $\Delta = L \sin \theta$. For small angles of rotation $\sin \theta \cong \theta$; therefore, $\Delta = L\theta$. Making these substitutions into Eq. (1) yields

$$\frac{mL^2}{3}\ddot{\theta} + kL^2\theta + k_R\theta = 0 \tag{2}$$

Dividing Eq. (2) through by $mL^2/3$ gives the equation of motion as

$$\ddot{\theta} + 3\frac{(kL^2 + k_R)}{mL^2}\theta = 0 \tag{3}$$

The natural circular frequency of the system is then calculated as

$$\omega = \sqrt{\frac{3(kL^2 + k_R)}{mL^2}} \quad \text{rad/sec}$$

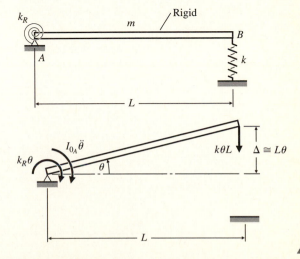

Figure 2.6
Elastically restrained rigid rod.

Figure 2.7
Free-body diagram of
elastically restrained rigid rod.

▲

Expressions for calculating the mass moment of inertia for several commonly encountered inertial elements are presented in Table 2.1.

TABLE 2.1 Expressions for Calculating Mass Moment of Inertia for Several Common Shapes

Shape	Illustration	Expression
Uniform slender bar of mass m		$I_{0_{cg}} = mL^2/12$ $I_{0_A} = mL^2/3$
Thin circular disk of uniform thickness and mass m		$I_{0_{cg}} = mR^2/4$
Rectangular plate of uniform thickness and mass m		$I_{0_{cg}} = m(a^2 + b^2)/12$
Triangular plate of uniform thickness and mass m		$I_{0_{cg}} = m(a^2 + b^2)/18$

Shape	Illustration	Expression
Elliptical plate of uniform thickness and mass m		$I_{0_{cg}} = m(a^2 + b^2)/16$
Sphere of mass m		$I_{0_{cg}} = 2/3mR^2$

2.3 THE ENERGY METHOD

The equation of motion for a conservative system can be established from energy considerations. If a conservative system such as a simple spring-mass system is set in motion, its total mechanical energy is the sum of the kinetic energy and the potential energy. The kinetic energy T is due to the velocity of the mass, and the potential energy V is due to the position of the mass or the strain energy of the spring as it undergoes deformation. For a conservative system, the total mechanical energy is constant at any instant of time, and therefore its time derivative is zero. It follows that

$$T + V = \text{constant} \tag{2.11}$$

and

$$\frac{d}{dt}(T + V) = 0 \tag{2.12}$$

Equation (2.12) yields the equation of motion for the system under consideration.

Figure 2.8
Simple pendulum.

EXAMPLE 2.3 ▼

A simple massless pendulum of length L has the mass m attached to its end as shown in Figure 2.8. Determine the equation of motion for the system and calculate the natural circular frequency. Assume the mass of the pendulum is negligible compared to the attached mass m.

Figure 2.9
Free-body diagram of
simple pendulum.

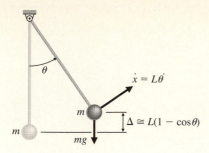

Solution

Assume small rotation of the mass about the pivot and refer to the free-body diagram in Figure 2.9; the kinetic energy is expressed as

$$T = \frac{1}{2}m\dot{x}^2 = \frac{1}{2}m(L\dot{\theta})^2 \tag{1}$$

where $\dot{x}$ is the tangential velocity of the mass and $\dot{\theta}$ is the rotational velocity of the pendulum. The potential energy is due to the position of the mass and is given by

$$V = mg\Delta = mgL(1 - \cos\theta) \tag{2}$$

where g is the acceleration due to gravity and Δ is the vertical displacement of the mass. Substituting Eqs. (1) and (2) into Eq. (2.12) yields

$$mL^2\dot{\theta}\ddot{\theta} + mgL(\sin\theta)\dot{\theta} = 0 \tag{3}$$

Dividing Eq. (3) through by $\dot{\theta}$ gives

$$mL^2\ddot{\theta} + mgL\sin\theta = 0 \tag{4}$$

For small rotations, $\sin\theta \cong \theta$ and dividing Eq. (4) through by mL^2 yields the equation of motion as

$$\ddot{\theta} + \left(\frac{g}{L}\right)\theta = 0 \tag{5}$$

The natural circular frequency of the system is given by

$$\omega = \sqrt{\frac{g}{L}} \quad \text{rad/sec}$$

▲

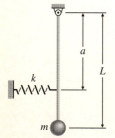

Figure 2.10
Elastically restrained
pendulum.

EXAMPLE 2.4 ▼

A simple pendulum is connected to a mass m and a translational spring having stiffness k as shown in Figure 2.10. Determine the equation of motion and calculate the natural frequency. Assume the mass of the pendulum is negligible with respect to the attached mass m.

Solution

Assume small rotations θ of the mass about the pivot and refer to the free-body diagram in Figure 2.11; the kinetic energy is expressed as

$$T = \frac{1}{2}m\dot{x}^2 = \frac{1}{2}m(L\dot{\theta})^2 \tag{1}$$

and the potential energy is given by

$$V = \frac{1}{2}k\Delta_h^2 + mg\Delta_v \tag{2}$$

Noting that $\Delta_h \cong a\theta$ and $\Delta_v = L(1-\cos\theta)$ in Eq. (2), then

$$V = \frac{1}{2}k(a\theta)^2 + mgL(1 - \cos\theta) \tag{3}$$

Substituting Eqs. (1) and (3) into Eq. (2.12) yields

$$mL^2\dot{\theta}\ddot{\theta} + mgL(\sin\theta)\dot{\theta} + ka^2\theta\dot{\theta} = 0 \tag{4}$$

Dividing through Eq. (4) by $\dot{\theta}$ results in

$$mL^2\ddot{\theta} + (mgL\ \sin\theta + ka^2\theta) = 0 \tag{5}$$

For small rotations, $\sin\theta \cong \theta$, thus Eq. (5) yields the equation of motion as

$$\ddot{\theta} + \left(\frac{mgL + ka^2}{mL^2}\right)\theta = 0 \tag{6}$$

The natural circular frequency is expressed as

$$\omega = \sqrt{\frac{mgL + ka^2}{mL^2}}\ \ \text{rad/sec}$$

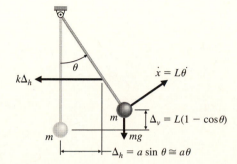

Figure 2.11
Free-body diagram of elastically restrained pendulum.

2.4 THE PRINCIPLE OF VIRTUAL DISPLACEMENTS

The analytical models for many engineering systems are often complex. For such systems, the formulation of the equation of motion by d'Alembert's principle may be a difficult task. In such instances, a more methodical procedure is required. The principle of virtual displacements (PVD) provides the basis for systematic procedures to establish the equation of motion. In this section the application of the PVD to formulate the equations of motion for systems exhibiting localized stiffness and distributed mass, and for systems with distributed stiffness and distributed mass, are presented.

2.4.1 Systems Having Localized Stiffness and Distributed Mass

As illustrated in Example 2.2, some engineering structures may be idealized as having localized stiffness and distributed mass. For this type of system the elastic deformations are limited entirely to localized spring elements, and displacements of only a single form or shape are permitted [3]. With this type of formulation it ʃ convenient to calculate the system inertia forces on the assumption that the mass and the mass moment of inertia of the system components are concentrated at their respective mass centers. It is also effective to represent distributed external loads applied to the system by their resultant forces.

A *virtual displacement* δv is an imaginary, infinitesimal change in configuration of the system that does not violate the kinematic constraints or geometric boundary conditions for that system. The *virtual work* δW is the work done by all forces acting on the system as it undergoes a virtual displacement. According to the *principle of virtual displacements* (PVD), if a system in dynamic equilibrium undergoes a virtual displacement, the virtual work done by all the forces (including the inertia forces) acting on the system must be zero [4]. Thus

$$\delta W = 0 \tag{2.13}$$

Equation (2.13) may be more conveniently expressed as

$$(F_I + F_S + F_D + F_E)\, \delta v = 0 \tag{2.14}$$

where F_E are the forces associated with the externally applied loads. Application of Eq. (2.14) results in the equation of motion for the system.

EXAMPLE 2.7 ▼

Use the PVD to derive the equation of motion for the system depicted in Figure 2.12. The bar is of uniform mass per length $\bar{m}$ with a concentrated mass m located at the free end B. Assume the bar to be rigid and let the angular rotation of the bar θ be the independent displacement coordinate. The transverse linearly distributed load has maximum amplitude $\bar{F}_0(t)$ that varies with time; however, the axial load N remains constant and horizontal.

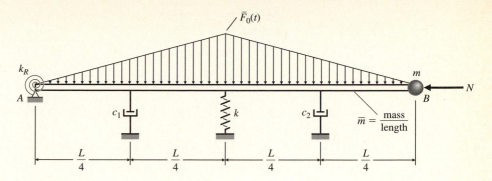

Figure 2.12
System having localized stiffness and distributed mass.

Solution

A free-body diagram of the system is shown in Figure 2.13. The bar is given a clockwise rotational virtual displacement $\delta\theta$. In the calculation of virtual work, a quantity is considered positive when the force acts in the same direction as the virtual displacement. The real vertical displacement at any location on the structure as a function of time $v(x, t)$ is expressed by

$$v(x, t) = x \sin \theta(t) \tag{1}$$

But for small θ, $\sin \theta \cong \theta$; therefore Eq. (1) becomes

$$v(x, t) = x \theta(t) \tag{2}$$

and the corresponding vertical virtual displacement $\delta v(x, t)$ is given by

$$\delta v(x, t) = x \delta \theta \tag{3}$$

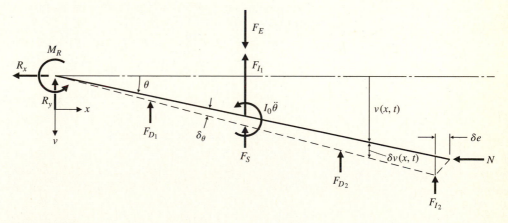

Figure 2.13
Free-body diagram of SDOF system shown in Figure 2.12.

Applying the virtual work equation (note that the reactions R_x and R_y do no work) given by

$$\delta W = 0 \tag{4}$$

to the system, when expressed in the form given by Eq. (2.14), results in

$$-F_S\left(\frac{L}{2}\delta\theta\right) - F_{D_1}\left(\frac{L}{4}\delta\theta\right) - F_{D_2}\left(\frac{3L}{4}\delta\theta\right) - M_R\,\delta\theta$$

$$-F_{I_1}\left(\frac{L}{2}\delta\theta\right) - I_0\ddot\theta\delta\theta - F_{I_2}(L\delta\theta) + F_E\left(\frac{L}{2}\delta\theta\right) + N\,\delta e = 0 \tag{5}$$

where M_R = torsional resisting moment developed in the rotational spring at support A

I_0 = mass moment of inertia of the bar about its center of mass (refer to Table 2.1)

δe = rotation induced virtual displacement in the horizontal direction experienced by the normal force N during application of $\delta\theta$.

From Figure 2.13 it is apparent that

$$\delta e = L\delta\theta\sin\theta \cong L\delta\theta \cdot \theta \tag{6}$$

Calculating the force components acting on the system yields

$$F_S = k\left(\frac{L}{2}\right)\theta \qquad F_{I_1} = (\bar m L)\left(\frac{L}{2}\right)\ddot\theta = \frac{\bar m L^2}{2}\ddot\theta$$

$$F_{D_1} = c_1\left(\frac{L}{4}\right)\dot\theta \qquad I_0 = (\bar m L)\left(\frac{L^2}{12}\right)\ddot\theta = \frac{\bar m L^3}{12}\ddot\theta \tag{7}$$

$$F_{D_2} = c_2\left(\frac{3L}{4}\right)\dot\theta \qquad F_{I_2} = mL\ddot\theta$$

$$M_R = k_R\theta \qquad\qquad F_E = \frac{\bar F_0(t)L}{2}$$

Substituting these force components into the virtual work equation given by Eq. (5) yields

$$-\frac{kL^2}{4}\delta\theta - \frac{c_1 L^2}{16}\dot\theta\delta\theta - \frac{9c_2 L^2}{16}\dot\theta\delta\theta - k_R\theta\delta\theta - \frac{\bar m L^3}{4}\ddot\theta\delta\theta$$

$$-\frac{\bar m L^3}{12}\ddot\theta\delta\theta - mL^2\ddot\theta\delta\theta + \frac{\bar F_0(t)L^2}{2\ 2}\delta\theta + NL\theta\delta\theta = 0 \tag{8}$$

Rearranging the terms in Eq. (8) gives

$$\left[\left(\frac{\bar m L^3}{3} + mL^2\right)\ddot\theta + \left(\frac{c_1 L^2 + 9c_2 L^2}{16}\right)\dot\theta + \left(\frac{kL^2}{4} + k_R - NL\right)\theta\right]\delta\theta$$

$$= \left(\frac{\bar F_0(t)L^2}{4}\delta\theta\right) \tag{9}$$

Since $\delta\theta \neq 0$ in Eq. (9) and dividing through by L, the equation of motion is

$$\left(\frac{\overline{m}L^2}{3} + mL\right)\ddot{\theta} + \left(\frac{c_1L + 9c_2L}{16}\right)\dot{\theta} + \left(\frac{kL}{4} + \frac{k_R}{L} - N\right)\theta$$

$$= \frac{\overline{F}_0(t)L}{4} \qquad (10)$$

Noting that the general expression for natural circular frequency is given by $\omega = \sqrt{k/m}$, then for this system

$$\omega = \sqrt{\frac{(kL/4) + (k_R/L) - N}{(\overline{m}L^2/3) + mL}} \quad \text{rad/sec}$$

▲

2.4.2 Systems Having Distributed Stiffness and Distributed Mass

For systems idealized with distributed mass and distributed stiffness characteristics, flexural deformations occur that allow the structure to exhibit an infinite number of degrees of freedom. However, a generalized SDOF analysis can be made if it is assumed that only a single deflection (vibration) pattern could occur. Therefore, if it is assumed that the system will vibrate in some predetermined shape, and that only the amplitude of vibration varies with time, then the motion of the system can be described by a single variable, or *generalized coordinate,* and only one DOF exists. Systems idealized in this manner are referred to as *generalized* SDOF systems.

To implement the procedure it is necessary to select a displacement configuration for the system. The displacement configuration or shape function selected must be statically admissible. That is, the shape function must satisfy the geometric (essential) boundary conditions of the system and possess derivatives of an order at least equal to that appearing in the strain energy expression for the system. Ideally, the shape function should satisfy the force (natural) boundary conditions as well [5].

Consider the cantilever beam shown in Figure 2.14. If it is assumed that the beam has distributed mass and distributed stiffness properties, then the vertical displacement v at any location on the beam is a function of both position and time, that is, $v(x, t)$. To approximate this system with a single DOF, it is necessary to assume a shape function that specifies the only configuration in which the structure may vibrate. The shape function is represented by $\psi(x)$ and the time-dependent amplitude of the motion is represented by the generalized coordinate $z(t)$. Therefore, the motion of the structure is described by

$$v(x, t) = \psi(x)z(t) \qquad (2.15)$$

The accuracy of the method is dependent upon how closely the shape function represents the true motion of the structure. The more accurately the shape function predicts the actual displacement of the structure, the more accurate is the analysis. In many instances the deflection configuration produced by applying the loads statically provides a reasonable estimate for the shape function. To expedite the analysis, the generalized coordinate should be selected as the displacement of a convenient reference point in the system. For the cantilever beam example illustrated in Figure 2.14, the free end or tip displacement of the cantilever provides a most convenient reference.

Figure 2.14

(a) System having distributed stiffness and distributed mass; (b) assumed shape function.

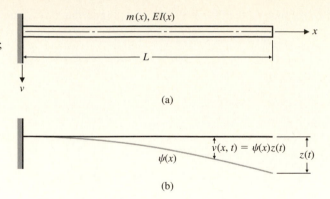

$m(x), EI(x)$

L

v

(a)

$v(x, t) = \psi(x)z(t)$

$\psi(x)$

$z(t)$

(b)

Based upon the displacement relation defined by Eq. (2.15), the equation of motion for a generalized SDOF system can be formulated by application of the principle of virtual displacements. However, in dealing with systems having distributed mass and stiffness characteristics, application of the virtual work equation can be tedious. Nevertheless, once the formulations of the generalized parameters for mass, stiffness, damping, etc., have been established, the procedure for constructing the equation of motion becomes quite direct.

To derive the expressions for the generalized parameters by the PVD, consider the system shown in Figure 2.15. The structure has distributed stiffness and distributed mass, designated by $EI(x)$ and $m(x)$, respectively. It also has a localized mass m, a localized translational spring k_c and distributed spring $k(x)$, and a localized damper c_d and distributed damping $c(x)$. The beam is subjected to a time-varying distributed load $F(x, t)$, a time-varying concentrated load $F(t)$, and an axial load N that remains constant and horizontal for all times t.

Before proceeding with the analysis, select an appropriate shape function $\psi(x)$. The minimum conditions imposed on $\psi(x)$ are that it satisfy the geometric boundary conditions and that it be of an order at least equal to that appearing in the strain energy expression. In this example, the geometric boundary conditions require that the dis-

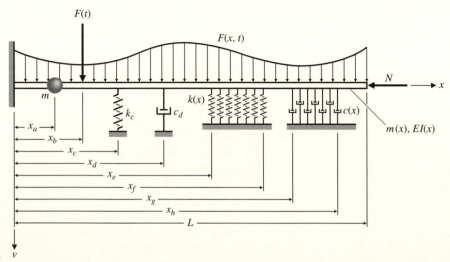

$F(t)$

$F(x, t)$

N

x

m

k_c

c_d

$k(x)$

$c(x)$

$m(x), EI(x)$

x_a

x_b

x_c

x_d

x_e

x_f

x_g

x_h

L

v

Figure 2.15

General system having both distributed stiffness and distributed mass characteristics.

placement and slope of the beam be equal to zero at the fixed end (at $x = 0$), or $v(0, t) = v'(0, t) = 0$. For this example, we choose the deflection at the tip of the cantilever as the generalized coordinate $z(t)$, and select the shape function such that $\psi(L) = 1$. The assumed deflected shape of the structure is indicated in Figure 2.16, along with the virtual displacement $\delta v(x, t)$.

To evaluate the virtual work equation, it is first necessary to determine the virtual work done by all forces acting on the system. The virtual work is considered positive if the originating force acts in the same direction as the virtual displacement. The following relationships will prove helpful in computing the virtual work expressions:

$$
\begin{aligned}
v'(x, t) &= \psi'(x)z(t) & \dot{v}(x, t) &= \psi(x)\dot{z}(t) \\
v''(x, t) &= \psi''(x)z(t) & \ddot{v}(x, t) &= \psi(x)\ddot{z}(t) \\
\delta v(x, t) &= \psi(x)\,\delta z(t) & \delta\dot{v}(x, t) &= \psi(x)\,\delta\dot{z}(t) \\
\delta v'(x, t) &= \psi'(x)\,\delta z(t) & \delta\ddot{v}(x, t) &= \psi(x)\,\delta\ddot{z}(t)
\end{aligned}
\tag{2.16}
$$

The terms in the virtual work equation are determined as follows:

1. Inertia, for both distributed mass $m(x)$ and localized mass m,

$$
\delta W_{\text{inertia}} = -\int_0^L [m(x)\ddot{v}(x, t)\delta v(x, t)\,dx] - m\ddot{v}(x_a, t)\delta v(x_a, t)
\tag{2.17}
$$

or

$$
\delta W_{\text{inertia}} = -\left[\int_0^L m(x)\psi^2(x)\,dx + m\psi^2(x_a)\right]\ddot{z}(t)\,\delta z(t)
\tag{2.18}
$$

2. Nonconservative forces, including damping and externally applied loads,

   a. Damping, for both distributed damping $c(x)$ and localized damper c_d,

$$
\delta W_{\text{damping}} = -\int_{x_g}^{x_h} c(x)\dot{v}(x, t)\delta v(x, t)\,dx - c_d\dot{v}(x_d, t)\delta v(x, t)
\tag{2.19}
$$

or

$$
\delta W_{\text{damping}} = -\left[\int_{x_g}^{x_h} c(x)\psi^2(x)\,dx + c_d\psi^2(x_d)\right]\dot{z}(t)\delta z(t)
\tag{2.20}
$$

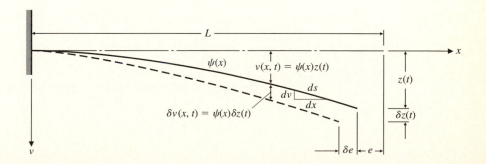

Figure 2.16

Assumed deflected shape for general system illustrated in Figure 2.15.

b. Transverse loads, including distributed forces $F(x, t)$ and concentrated forces $F(t)$,

$$\delta W_F = \int_0^L F(x, t)\,\delta v(x, t)\,dx + F(t)\,\delta v(x_b, t) \tag{2.21}$$

or

$$\delta W_F = \left[\int_0^L F(x, t)\psi(x)\,dx + F(t)\psi(x_b)\right]\delta z(t) \tag{2.22}$$

c. Axial load N,

$$\delta W_N = N\,\delta e \tag{2.23}$$

The axial displacement of the end of the beam e due to flexural deformation, which appears in Eq. (2.23), is determined assuming small structure displacements. The elemental length along the beam, ds, as shown in Figure 2.16, is expressed as

$$ds = (dx^2 + dv^2)^{1/2} = \left[1 + \left(\frac{dv}{dx}\right)^2\right]^{1/2} dx \tag{2.24}$$

If the displacements are small, then the slope squared is very small. For this case, the elemental length ds can be approximated as

$$ds = 1 + \frac{1}{2}\left(\frac{dv}{dx}\right)^2 \tag{2.25}$$

Equation (2.25) is an application of the binomial expansion. In general, if some quantity $\varepsilon \ll 1$, then we can write

$$(1 + \varepsilon)^\alpha = 1 + \alpha\varepsilon + O(\varepsilon^2) \tag{2.26}$$

where $O(\varepsilon^2)$ represents the binomial expansion of ε, and the truncation error is of order ε^2. The total length of the deflected beam can then be expressed as

$$L = \int_0^{L-e} ds = \int_0^{L-e}\left[1 + \frac{1}{2}\left(\frac{dv}{dx}\right)^2\right]dx = (L - e) + \frac{1}{2}\int_0^{L-e}\left(\frac{dv}{dx}\right)^2 dx \tag{2.27}$$

Therefore, the axial displacement at the end of the beam is given by

$$e = \frac{1}{2}\int_0^{L-e}\left(\frac{dv}{dx}\right)^2 dx \tag{2.28}$$

The limits of integration of Eq. (2.28) may be expressed as

$$e = \frac{1}{2}\int_0^L\left(\frac{dv}{dx}\right)^2 dx - \frac{1}{2}\int_{L-e}^L\left(\frac{dv}{dx}\right)^2 dx \tag{2.29}$$

Since $e/L <\!< 1$, the second integral in Eq. (2.29) is negligible; thus the axial displacement e is given as

$$e = \frac{1}{2}\int_0^L \left(\frac{dv}{dx}\right)^2 dx = \frac{1}{2}\int_0^L (v')^2\, dx \tag{2.30}$$

The determination of the horizontal virtual displacement at the end of the beam, δe, follows a similar derivation. The total beam length is expressed as

$$L = \int_0^{L-e-\delta e} \left[1 + \frac{1}{2}\left(\frac{d}{dx}(v + \delta v)\right)^2\right] dx$$

$$= \int_0^{L-e-\delta e} \left[1 + \frac{1}{2}\left(1 + \frac{(\delta v/dx)}{(dv/dx)}\right)^2 \left(\frac{dv}{dx}\right)^2\right] dx \tag{2.31}$$

Because the change in the virtual displacement is much smaller than the equilibrium displacement, the binomial expansion of Eq. (2.31) gives

$$L = \int_0^{L-e-\delta e} \left[1 + \frac{1}{2}\left(1 + 2\frac{(\delta v/dx)}{(dv/dx)}\right)\left(\frac{dv}{dx}\right)^2\right] dx \tag{2.32}$$

Expanding Eq. (2.32) yields

$$L = (L - e - \delta e) + \frac{1}{2}\int_0^{L-e-\delta e} \left(\frac{dv}{dx}\right)^2 dx + \int_0^{L-e-\delta e} \frac{d\delta v\, dv}{dx\, dx} dx \tag{2.33}$$

The first integral of Eq. (2.33) is equal to e because $\delta e/e <\!< 1$. As previously noted, $e/L <\!< 1$; therefore the integration limits on the second integral in Eq. (2.33) can be approximated as 0 to L. This yields the expression for δe as

$$\delta e = \int_0^{L-e-\delta e} \frac{dv\, d\delta v}{dx\, dx} dx = \int_0^L v'\, \delta v'\, dx \tag{2.34}$$

Thus, substituting the expression for δe given by Eq. (2.34) into Eq. (2.23) results in

$$\delta W_N = N \int_0^L [\psi'(x)]^2 dx\ z(t)\delta z(t) \tag{2.35}$$

3. Conservative forces, which include forces associated with the potential energy in springs and the strain energy due to bending; thus

$$\delta W_{\text{spring}} = \delta V_{\text{spring}}$$

$$= -\int_{x_e}^{x_f} k(x)v(x,t)\ \delta v(x,t)\ dx - k_c v(x_c,t)\ \delta v(x_c,t) \tag{2.36}$$

or

$$\delta W_{\text{spring}} = \delta V_{\text{spring}}$$

$$= -\left[\int_{x_e}^{x_f} k(x)\psi^2(x)\,dx + k_c\psi^2(x_c) \right] z(t)\delta z(t) \tag{2.37}$$

and

$$\delta W_{\text{bending}} = \delta V_{\text{bending}} = -\int_0^L EI(x)v''(x,t)\,\delta v''(x,t)dx \tag{2.38}$$

or

$$\delta W_{\text{bending}} = \delta V_{\text{bending}} = -\left\{ \int_0^L EI(x)[\psi''(x)]^2 dx \right\} z(t)\,\delta z(t) \tag{2.39}$$

and combining Eqs. (2.37) and (2.39) gives

$$\delta W_{\text{spring}} = \delta V_{\text{bending}}$$

$$= -\left\{ \int_{x_e}^{x_f} k(x)\psi^2(x)\,dx + k_c\psi^2(x_c) \right.$$

$$\left. + \int_0^L EI(x)[\psi''(x)]^2 dx \right\} z(t)\,\delta z(t) \tag{2.40}$$

The equation of motion can now be obtained by summing the contributing terms to the total virtual work, given by Eqs. (2.18), (2.20), (2.22), (2.35), and (2.40) into the general virtual work equation expressed by Eq. (2.13). Then, expanding Eq. (2.13) yields

$$-\left[\int_0^L m(x)\psi(x)\,dx + m\psi^2(x_a) \right] \ddot{z}(t)\,\delta z(t)$$

$$-\left[\int_{x_g}^{x_h} c(x)\psi^2(x)\,dx + c_d\psi^2(x_d) \right] \dot{z}(t)\,\delta z(t)$$

$$-\left\{ \int_{x_e}^{x_f} k(x)\psi^2(x)\,dx + k_c\psi^2(x_c) + \int_0^L EI(x)[\psi''(x)]^2\,dx \right\} z(t)\,\delta z(t) \tag{2.41}$$

$$+\left[\int_0^L F(x,t)\psi(x)dx + F(t)\psi(x_b) \right] \delta z(t)$$

$$+\left\{ N\int_0^L [\psi'(x)]^2 dx \right\} z(t)\,\delta z(t) = 0$$

Grouping similar terms in Eq. (2.41) and rearranging yields

$$\left[\int_0^L m(x)\psi(x)\,dx + m\psi^2(x_a)\right]\ddot{z}(t)\,\delta z(t)$$

$$+ \left[\int_{x_g}^{x_h} c(x)\psi^2(x)\,dx + c_d\psi^2(x_d)\right]\dot{z}(t)\,\delta z(t)$$

$$+ \left\{\int_{x_e}^{x_f} k(x)\psi^2(x)\,dx + k_c\psi^2(x_c) - N\int_0^L [\psi'(x)]^2 dx\right\}z(t)\,\delta z(t) \qquad (2.42)$$

$$- \left[\int_0^L F(x,t)\psi(x)\,dx + F(t)\psi(x_b)\right]\delta z(t) = 0$$

Simplifying the form of Eq. (2.42) results in the expression

$$[M^*\ddot{z} + C^*\dot{z} + K^*z - K_G^*z - P^*(t)]\delta z(t) = 0 \qquad (2.43)$$

Since $\delta z(t) \neq 0$ in Eq. (2.43), the generalized SDOF equation of motion written in the generalized coordinate $z(t)$ becomes

$$M^*\ddot{z} + C^*\dot{z} + (K^* - K_G^*)z = P^*(t) \qquad (2.44)$$

It should be noted that Eq. (2.44), with the exception of the K_G^* term, is identical in form to the equation of motion for a simple lumped mass SDOF system represented by Eq. (2.7). The equation of motion for any generalized SDOF, regardless of how complex, can always be reduced to the form given by Eq. (2.44). The terms in this equation for generalized mass M^*, generalized damping C^*, generalized stiffness K^*, generalized geometric stiffness K^*_G, and generalized force $P^*(t)$ can be readily determined once an appropriate shape function is selected. These generalized parameters can be expressed in a more general form:

$$M^* = \int m(x)\psi^2(x)\,dx + \sum_i m_i\psi^2(x_i) \qquad (2.45)$$

$$C^* = \int c(x)\psi^2(x)\,dx + \sum_i c_i\psi^2(x_i) \qquad (2.46)$$

$$K^* = \int k(x)\psi^2(x)\,dx + \int EI(x)[\psi''(x)]^2 dx + \sum_i k_i\psi^2(x_i) \qquad (2.47)$$

$$K_G^* = N\int [\psi'(x)]^2 dx \qquad (2.48)$$

$$P^*(t) = \int F(x,t)\psi(x)\,dx + \sum_i F_i\psi(x_i) \qquad (2.49)$$

Thus, for any system, regardless of its complexity, a generalized SDOF equation of motion can be formulated merely by implementing the expressions of the generalized parameters defined by Eqs. (2.45) through (2.49) into Eq. (2.44). In addition, the natural frequency for the generalized system can be determined from the expression

$$\omega = \sqrt{\frac{K^* - K_G^*}{M^*}} \tag{2.50}$$

The generalized geometric stiffness term K_G^*, given by Eq. (2.48) and appearing in Eqs. (2.43) and (2.44), represents the effect of an axial load on the transverse or lateral stiffness of a structural member. For a compressive axial force N, the transverse stiffness of the member is decreased, and therefore the combined generalized stiffness is equal to $K^* - K_G^*$. However, if the applied axial force N is tensile, the transverse stiffness is increased. In this case the combined generalized stiffness is equal to $K^* + K_G^*$. An interesting observation from this phenomenon is that if the applied axial force is compressive, the resulting natural frequency for the system, given by Eq. (2.50), is less than that for the corresponding system with no applied axial force, for which K_G^* would be equal to zero. In direct contrast, if the applied axial force is tensile, then the resulting natural frequency is determined from Eq. (2.50) by replacing $K^* - K_G^*$ with $K^* + K_G^*$. Therefore the resulting natural frequency would be greater than that for the corresponding system for which the axial force is zero.

Another interesting feature associated with the use of the generalized parameter approach is the ability to determine the *critical buckling load* N_{cr} for the system. N_{cr} may be evaluated by equating the combined generalized stiffness to zero. That is, by imposing the condition

$$K^* - K_G^* = 0 \tag{2.51}$$

or

$$\int k(x)\psi^2(x)dx + \int EI(x)[\psi''(x)]^2 dx$$
$$+ \sum_i k_i \psi^2(x_i) - N \int [\psi'(x)]^2 dx = 0 \tag{2.52}$$

solving for N yields the expression for the critical buckling load given by

$$N_{cr} = \frac{\int k(x)\psi^2(x)dx + \int EI(x)[\psi''(x)]^2 dx + \sum_i k_i \psi^2(x_i)}{\int [\psi'(x)]^2 dx} \tag{2.53}$$

EXAMPLE 2.8 ▼

Use the generalized parameter approach to determine the equation of motion and the natural frequency for transverse vibration of the beam shown in Figure 2.17. Assume a deflected shape given by

$$\psi(x) = 16 \left(\frac{x}{L}\right)^2 \left(1 - \frac{x}{L}\right)^2$$

Figure 2.17

System of Example 2.8.

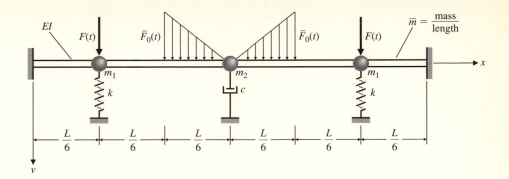

The beam has flexural rigidity EI, uniform mass per unit length $\bar{m}$, concentrated masses m_1 and m_2, linearly varying distributed loads having a maximum amplitude of $\bar{F}_0(t)$, and concentrated loads of magnitude $F(t)$. Use the deflection at midspan as the generalized coordinate.

Solution

(a) The first step in the analysis should always be a check that the assumed deflection shape satisfies the kinematic boundary conditions. Reference to Figure 2.18 shows the geometric boundary conditions of displacement and slope to be

$$v(0, t) = v(L, t) = v'(0, t) = v'\left(\frac{L}{2}, t\right) = v'(L, t) = 0 \qquad (1)$$

Noting that

$$\psi(x) = \frac{16x^2}{L^2} - \frac{32x^3}{L^3} + \frac{16x^4}{L^4} \qquad (2)$$

and

$$\psi'(x) = 32\frac{x}{L^2} - 96\frac{x^2}{L^3} + 64\frac{x^3}{L^4} \qquad (3)$$

Figure 2.18

Assumed deflected shape and imposed virtual displacement for system shown in Figure 2.17.

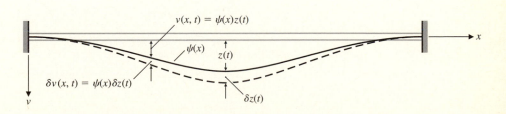

and then applying the boundary conditions of Eq. (1) to Eqs. (2) and (3) gives

$$\psi(0) = 0 \qquad \psi'(0) = 0 \quad \psi'(L) = 0$$

$$\psi(L) = 0 \qquad \psi'\left(\frac{L}{2}\right) = 0$$

(4)

for which we note that all kinematic boundary conditions are satisfied.

(b) Select the generalized coordinate $z(t)$ as the midspan deflection at $x = L/2$ and also note that

$$\psi\left(\frac{L}{2}\right) = 1.0$$

(5)

(c) Evaluate the generalized parameters:

(i) The generalized mass is evaluated from Eq. (2.45) as

$$M^* = \bar{m} \int_0^L \psi^2(x)\,dx + m_1 \psi^2\left(\frac{L}{6}\right) + m_1 \psi^2\left(\frac{5L}{6}\right) + m_2 \psi^2\left(\frac{L}{2}\right)$$

(6)

Evaluating the terms in Eq. (6) yields

$$M^* = \bar{m}\int_0^L \left[256\,\frac{x^4}{L^4} - 1024\,\frac{x^5}{L^5} + 1536\,\frac{x^6}{L^6} - 1024\,\frac{x^7}{L^7} + 256\,\frac{x^8}{L^8}\right]dx$$

$$+256m_1\left[\frac{\left(\frac{L}{6}\right)^4}{L^4} - \frac{4\left(\frac{L}{6}\right)^5}{L^5} + \frac{6\left(\frac{L}{6}\right)^6}{L^6} - \frac{4\left(\frac{L}{6}\right)^7}{L^7} + \frac{\left(\frac{L}{6}\right)^8}{L^8}\right]$$

$$+256m_1\left[\frac{\left(\frac{5L}{6}\right)^4}{L^4} - \frac{4\left(\frac{5L}{6}\right)^5}{L^5} + \frac{6\left(\frac{5L}{6}\right)^6}{L^6} - \frac{4\left(\frac{5L}{6}\right)^7}{L^7} + \frac{\left(\frac{5L}{6}\right)^8}{L^8}\right]$$

$$+256m_2\left[\frac{\left(\frac{L}{2}\right)^4}{L^4} - \frac{4\left(\frac{L}{2}\right)^5}{L^5} + \frac{6\left(\frac{L}{2}\right)^6}{L^6} - \frac{4\left(\frac{L}{2}\right)^7}{L^7} + \frac{\left(\frac{L}{2}\right)^8}{L^8}\right]$$

(7)

which after integrating and simplifying results in

$$M^* = 0.406\,\bar{m}L + 0.191\,m_1 + m_2$$

(8)

(ii) The generalized damping is calculated from Eq. (2.46) as

$$C^* = c\psi^2\left(\frac{L}{2}\right) = c(1.0)^2 = c$$

(9)

(iii) The generalized stiffness is determined from Eq. (2.47), yielding

$$K^* = \int_0^L EI(x)[\psi''(x)]^2 dx + k\psi^2\left(\frac{L}{6}\right) + k\psi^2\left(\frac{5L}{6}\right) \tag{10}$$

or

$$K^* = EI\int_0^L\left(\frac{32}{L^2} - \frac{192x}{L^3} + \frac{192x^2}{L^4}\right)dx + k\psi^2\left(\frac{L}{6}\right) + k\psi^2\left(\frac{5L}{6}\right) \tag{11}$$

Integrating Eq. (11) results in

$$K^* = EI\left[\frac{1024}{L^3} - \frac{12{,}288}{2L^3} + \frac{49{,}152}{3L^3} - \frac{73{,}728}{4L^3} + \frac{36{,}864}{5L^3}\right]$$

$$+ 256k\left[\frac{\left(\frac{L}{6}\right)^4}{L^4} - \frac{4\left(\frac{L}{6}\right)^5}{L^5} + \frac{6\left(\frac{L}{6}\right)^6}{L^6} - \frac{4\left(\frac{L}{6}\right)^7}{L^7} + \frac{\left(\frac{L}{6}\right)^8}{L^8}\right] \tag{12}$$

$$+ 256k\left[\frac{\left(\frac{5L}{6}\right)^4}{L^4} - \frac{4\left(\frac{5L}{6}\right)^5}{L^5} + \frac{6\left(\frac{5L}{6}\right)^6}{L^6} - \frac{4\left(\frac{5L}{6}\right)^7}{L^7} + \frac{\left(\frac{5L}{6}\right)^8}{L^8}\right]$$

and evaluating Eq. (12) yields

$$K^* = 204.8\,\frac{EI}{L^3} + 0.191k \tag{13}$$

Since $N = 0$, it follows that the generalized geometric stiffness given by Eq. (2.48) is

$$K_G^* = 0 \tag{14}$$

(iv) The generalized force is calculated from Eq. (2.49) as

$$P^*(t) = \int_{L/3}^{L/2}\left[-6\overline{F}_0(t)\frac{x}{L} + 3\overline{F}_0(t)\right]\psi(x)\,dx \tag{15}$$

$$+ \int_{L/2}^{2L/3}\left[6\overline{F}_0(t)\frac{x}{L} - 3\overline{F}_0(t)\right]\psi(x)\,dx + F(t)\psi\left(\frac{L}{6}\right) + F(t)\psi\left(\frac{5L}{6}\right)$$

or

$$P^*(t) = 48\overline{F}_0(t)\int_{L/3}^{L/2}\left(\frac{x^2}{L^2} - \frac{4x^3}{L^3} + \frac{5x^4}{L^4} - \frac{2x^5}{L^5}\right)dx$$

$$+ 48\overline{F}_0(t)\int_{L/2}^{2L/3}\left(-\frac{x^2}{L^2} + \frac{4x^3}{L^3} - \frac{5x^4}{L^4} + \frac{2x^5}{L^5}\right)dx \tag{16}$$

$$+ F(t)\psi\left(\frac{L}{6}\right) + F(t)\psi\left(\frac{5L}{6}\right)$$

Integrating Eq. (16) and collecting similar terms results in

$$P^*(t) = 48\overline{F}_0(t)\left[\frac{(L/2)^3}{3L^2} - \frac{(L/2)^4}{L^3} + \frac{(L/2)^5}{L^4} - \frac{(L/2)^6}{3L^5}\right.$$

$$-\frac{(L/3)^3}{3L^2} + \frac{(L/3)^4}{L^3} - \frac{(L/3)^5}{L^4} + \frac{(L/3)^6}{3L^5}\right]$$

$$+48\overline{F}_0(t)\left[-\frac{(2L/3)^3}{3L^2} + \frac{(2L/3)^4}{L^3} - \frac{(2L/3)^5}{L^4} + \frac{(2L/3)^6}{3L^5}\right.$$

$$+\frac{(L/2)^3}{3L^2} - \frac{(L/2)^4}{L^3} + \frac{(L/2)^5}{L^4} - \frac{(L/2)^6}{3L^5}\right]$$

$$+F(t)\left[16\frac{(L/6)^2}{L^2} - 32\frac{(L/6)^3}{L^3} + 16\frac{(L/6)^4}{L^4}\right.$$

$$+(16)\frac{(5L/6)^2}{L^2} - 32\frac{(5L/6)^3}{L^3} + 16\frac{(5L/6)^4}{L^4}\right] \quad (17)$$

and simplifying Eq. (17) yields

$$P^*(t) = 0.149\,\overline{F}_0(t)L + 0.617\,F(t) \quad (18)$$

(d) Substituting the generalized parameters given by Eqs. (8), (9), (13), and (18) into the generalized equation of motion represented by Eq. (2.44) yields

$$(0.406\,\overline{m}L + 0.191\,m_1 + m_2)\ddot{z} + c\dot{z} + \left(204.8\,\frac{EI}{L^3} + 0.191\,k\right)z$$

$$= 0.149\overline{F}_0(t)L + 0.617F(t) \quad (19)$$

(e) The natural circular frequency for the generalized SDOF system can be established from Eq. (2.50) as

$$\omega = \sqrt{\frac{204.8\,(EI/L^3) + 0.191\,k}{(0.406\,\overline{m}L + 0.191\,m_1 + m_2)}}\ \text{rad/sec} \quad (20)$$

▲

EXAMPLE 2.9 ▼

Determine the critical buckling load for the system shown in Figure 2.17. Assume an axial compressive load *N*, which does not vary with time and remains horizontal, acting at either end of the structure.

Solution

(a) Calculate the generalized geometric stiffness from Eq. (2.48) as

$$K_G^* = N \int_0^L [\psi'(x)]^2 dx \tag{1}$$

Integrating Eq. (1) yields

$$K_G^* = 1024N \int_0^L \left(\frac{x^2}{L^4} - \frac{6x^3}{L^5} + \frac{13x^4}{L^6} - \frac{12x^5}{L^7} + \frac{4x^6}{L^8} \right) dx \tag{2}$$

thus

$$K_G^* = 1024N \left(\frac{1}{3L} - \frac{6}{4L} + \frac{13}{5L} - \frac{12}{6L} + \frac{4}{7L} \right) \tag{3}$$

Simplifying Eq. (3) results in

$$K_G^* = 4.876 \frac{N}{L} \tag{4}$$

(b) Calculating the combined generalized stiffness results in

$$K^* - K_G^* = 204.8 \frac{EI}{L^3} + 0.191k - 4.876 \frac{N}{L} \tag{5}$$

(c) Equating the combined stiffness given by Eq. (5) to zero and solving for N_{cr} yields

$$204.8 \frac{EI}{L^3} + 0.191k - 4.876 \frac{N_{cr}}{L} = 0 \tag{6}$$

or

$$N_{cr} = 42 \frac{EI}{L^2} + 0.0391 \, kL \tag{7}$$

▲

Indeed, the degree of accuracy for the PVD method for systems having distributed mass and distributed stiffness characteristics depends upon how closely the assumed shape function $\psi(x)$ represents the actual mode of vibration (or buckled configuration). This method is generally employed to approximate the fundamental, or first, mode of vibration. However, estimates of the second or higher modes of vibration can be approximated by selecting the shape function $\psi(x)$ such that it is representative of one of the higher modes.

REFERENCES

1 Tse, F.S., Morse, I.E., and Hinkle, R.T., *Mechanical Vibrations, Theory and Applications,* 2nd ed., Allyn and Bacon, Boston, 1978.

2 Craig, Roy R., Jr., *Structural Dynamics,* Wiley, New York, 1981.

3 Clough, R.W. and Penzien, J., *Dynamics of Structures,* McGraw Hill, New York, 1975.

4 Berg, Glen V., *Elements of Structural Dynamics,* Prentice Hall, Englewood Cliffs, NJ, 1989.

5 Humar, J.L., *Dynamics of Structures,* Prentice Hall, Englewood Cliffs, NJ, 1990.

NOTATION

c	viscous damping coefficient	M_R	torsional resisting moment
C^*	generalized damping	M^*	generalized mass
ds	differential length of elemental beam segment	N	amplitude of externally applied axial force
dx	horizontal projection of elemental beam segment	N_{cr}	critical buckling load
dy	vertical projection of elemental beam segment	$P^*(t)$	generalized force
dv	differential displacement of elemental beam segment	t	time
e	beam axial displacement	T	kinetic energy; also natural period of vibration
E	Young's modulus	v	displacement coordinate
F_D	damping force	V	strain energy
F_E	forces associated with externally applied loads	x	translational displacement
F_I	inertia force	$\dot{x}$	translational velocity
F_S	restoring force in spring	$\ddot{x}$	translational acceleration
$F(t)$	time-varying externally applied concentrated force	$z(t)$	generalized coordinate
$F(x, t)$	time-varying externally applied distributed force	PVD	principle of virtual displacements
$\overline{F}_0(t)$	maximum amplitude of time-varying externally applied linearly distributed force	SDOF	single degree of freedom
		δe	axial virtual displacement
g	acceleration due to gravity	δv	translational virtual displacement
I	static moment of inertia	δW	virtual work
I_o	mass moment of inertia	$\delta z(t)$	virtual displacement of generalized coordinate
k	translational spring constant	$\delta\theta$	rotational virtual displacement
k_R	rotational spring constant	Δ	deflection
K^*	generalized stiffness	$\psi(x)$	shape function
K_G^*	generalized geometric stiffness	ω	natural circular frequency (rad/sec)
L	length	θ	angle of rotation
m	mass	$\dot{\theta}$	rotational velocity
$\overline{m}$	mass per unit length	$\ddot{\theta}$	rotational acceleration

PROBLEMS

2.1 A massless cantilever beam having flexural rigidity EI supports a weight having a mass of 1.5 m as shown in Figure P2.1. Determine the equation of motion by d'Alembert's principle, and calculate the natural frequency.

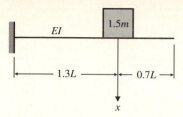

Figure P2.1

2.2 A simply supported beam supports a mass *m* as shown in Figure P2.2. Determine the equation of motion by d'Alembert's principle, and calculate the natural frequency. Assume the beam has flexural rigidity *EI* and neglect its mass.

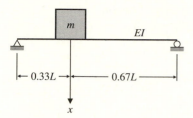

Figure P2.2

2.3 For the structural frame shown in Figure P2.3 determine the equation of motion by d'Alembert's principle, and calculate the natural frequency. Assume the horizontal member to be rigid and support a mass 2.5 *m*. Assume $E = 30 \times 10^6$ psi, $I = 150$ in^4, $L = 12.0$ ft, and $m = 1.0$ lb-sec^2/in.

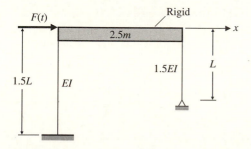

Figure P2.3

2.4 Use d'Alembert's principle to determine the equation of motion for the structural frame shown in Figure P2.4, and calculate the natural frequency. Assume that the horizontal mem-

ber is rigid and supports a total mass *m*. The vertical members have flexural rigidity *EI*.

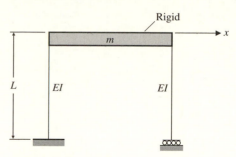

Figure P2.4

2.5 The rigid bar of mass *m* shown in Figure P2.5 is pinned at its left end, where it is also partially restrained from rotation by a rotational spring of stiffness k_R. It is also supported by a translational spring of stiffness $1.5k$. Use d'Alembert's principle to determine the equation of motion, and calculate the natural frequency. Assume small rotations.

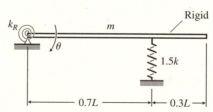

Figure P2.5

2.6 Use d'Alembert's principle to determine the equation of motion for the system shown in Figure P2.6, and calculate the natural frequency. The rod is rigid and has a total mass 1.5*m*. Assume small rotations.

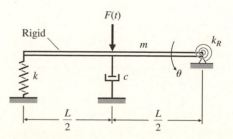

Figure P2.6

2.7 A rigid, massless pendulum of length L has a mass m attached at its free end as shown in Figure P2.7. It is partially restrained from rotation at its pivot point by a rotational spring k_R. Use d'Alembert's principle to determine the equation of motion, and calculate its natural frequency. Assume small oscillations.

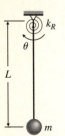

Figure P2.7

2.8 Repeat Problem 2.6 using the energy method.

2.9 The rigid, massless pendulum shown in Figure P2.8 is partially restrained from rotating at its pivot point by a rotational spring k_R and has a translational spring of stiffness k attached to its supported mass m. Use the energy method to determine the equation of motion, and calculate the natural frequency. Assume small oscillations.

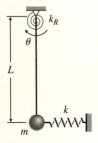

Figure P2.8

2.10 Repeat Problem 2.7 using the energy method.

2.11 The rigid pendulum shown in Figure P2.9 has a mass m_2 and supports a mass m_1 at its end. It is partially restrained at its pivot point by a rotational spring k_R and has a translational spring attached to the mass m_1. Use the energy method to determine the equation of motion, and calculate its natural frequency.

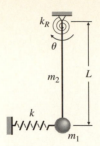

Figure P2.9

2.12 Use the PVD to derive the equation of motion for the rigid beam shown in Figure P2.10. The beam has a mass per unit length $\bar{m}$ and is subject to a time-varying uniformly distributed force of maximum amplitude $\bar{F}_0$ (t), but the end force N remains constant and horizontal. Assume small angles of rotation. Also calculate the natural frequency.

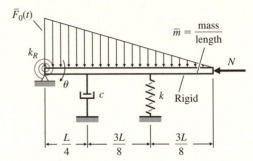

Figure P2.10

2.13 The rigid bar of mass m shown in Figure P2.11 is subject to a time-varying linearly distributed force of maximum amplitude $\bar{F}_0$ (t). Use the PVD to derive the equation of motion, and calculate the natural frequency. Assume small angles of rotation.

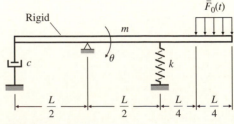

Figure P2.11

2.14 Repeat Problem 2.11 using the PVD.

2.15 The rigid beam shown in Figure P2.12 has uniform mass per unit length $\bar{m}$. Derive the equation of motion by the PVD, and calculate the natural frequency. Assume small angles of rotation.

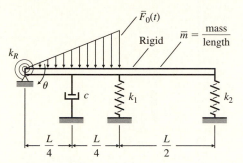

Figure P2.12

2.16 For the structure having distributed mass and stiffness shown in Figure P2.13 use the generalized parameter approach to derive the equation of motion for transverse vibration. Assume a deflected shape given by

$$\psi(x) = \left(\frac{x}{L}\right)^2\left(\frac{3}{2} - \frac{x}{2L}\right)$$

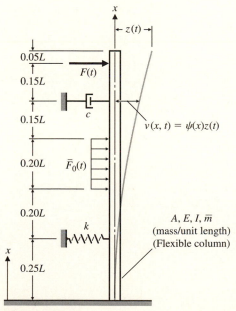

Figure P2.13

Use the deflection at the top of the structure as the generalized coordinate. Denoting the uniformly distributed mass per unit length as $\bar{m}$, the uniform stiffness by EI, and the maximum amplitude of the uniformly distributed force per length by $\bar{F}_0(t)$, evaluate the generalized physical properties M^*, K^*, and C^* and the generalized loading $P^*(t)$. Calculate the natural frequency of the system. If a downward axial load N is applied at the top of the structure, evaluate the generalized geometric stiffness K_G^* and the generalized combined stiffness $K^* - K_G^*$. Evaluate the critical buckling load N_{cr}.

2.17 Repeat the procedure described in Problem 2.16 for the structure shown in Figure P2.14. Assume a deflected shape given by

$$\psi(x) = 3\left(\frac{x}{L}\right)^2 - \left(\frac{x}{L}\right)^3$$

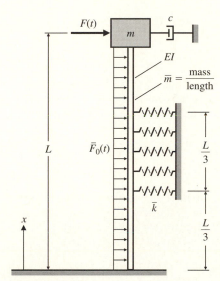

Figure P2.14

Use the transverse deflection at the top as the generalized coordinate. Note that the middle third of the structure is restrained by translational springs having uniform stiffness per length $\bar{k}$.

2.18 Repeat the procedure described in Problem 2.16 for the structure shown in Figure P2.15. Assume a deflected shape given by

$$\psi(x) = \left(\frac{x}{L}\right)^2$$

Use the transverse deflection at the top as the generalized coordinate.

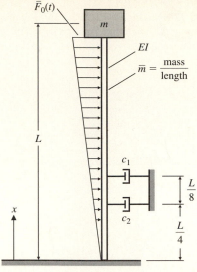

Figure P2.15

2.19 Determine the equation of motion for transverse vibration for the beam shown in Figure P2.16 by the generalized parameter approach. Assume a deflected shape given by

$$\psi(x) = 16\left(\frac{x}{L}\right)^2\left(1 - \frac{x}{L}\right)^2$$

Use the deflection at midspan as the generalized coordinate. Calculate the natural frequency as $\omega = \sqrt{K^*/M^*}$. If a compressive axial load N is applied to either end of the beam, calculate K_G^* and determine the critical buckling load N_{cr}.

2.20 Remove the right-hand support from the beam shown in Figure P2.16 and repeat the procedure described in Problem 2.19. Assume a deflected shape given by

$$\psi(x) = 3\left(\frac{x}{L}\right)^2 - \left(\frac{x}{L}\right)^3$$

Use the deflection at the free end as the generalized coordinate.

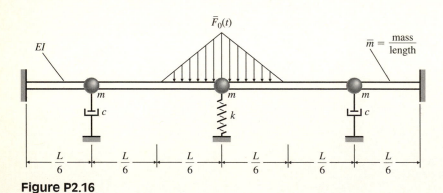

Figure P2.16

3 ▲ Undamped Free Vibration

Vibrations are generally classified as either *free vibrations* or *forced vibrations*. Free vibration occurs in the absence of externally applied forces. The impetus for the free vibration is usually an initial displacement and/or velocity imparted to the mass. A system undergoing free vibration will oscillate at one or more of its natural frequencies. A SDOF system has only one natural frequency.

Forced vibration occurs under the excitation of externally applied forces. If the excitation is transient (i.e., of short duration), the system response is at its natural frequency (once the disturbance terminates). However, if the excitation is oscillatory (periodically repetitive) and continues with time, the system vibrates at the excitation frequency. In situations where the excitation frequency coincides with the natural frequency of the system, a condition known as *resonance* occurs. At resonance, the amplitude of vibration becomes extremely large, and damage to the system is imminent if the vibration continues at the resonant frequency.

All structures exhibit some form of energy dissipation, or damping. Typically, the energy dissipation is due to frictional resistance or material hysteresis. In most practical engineering structures and mechanical systems, the damping is relatively small and, therefore, has very little influence on the natural frequency. However, even a small level of damping has a significant effect in limiting the amplitude of vibration, especially at resonance.

This chapter focuses on the free vibration of undamped, SDOF systems, examines the derivation of the governing differential equation, and presents an interpretation of its solution. The chapter also introduces the concept of equivalent stiffness and discusses calculation of the natural frequency by the Rayleigh method.

3.1 SIMPLE HARMONIC MOTION

The simplest mechanical model for a SDOF vibrating system is composed of a single mass element that is connected to a rigid support through a linear elastic spring and viscous dashpot as shown in Figure 3.1a. The equation of motion for this system was given by Eq. (2.7) as

$$m\ddot{x} + c\dot{x} + kx = F(t)$$

For the case of undamped, free vibration, the dashpot and external forces are removed as shown in Figure 3.1b. The equation of motion then becomes

$$m\ddot{x} + kx = 0 \tag{3.1}$$

Equation (3.1) describes mathematically the oscillations of the system. The system is set into motion by application of an initial displacement and/or an initial velocity to the mass. For example, if the mass is displaced from its static equilibrium position and then suddenly released, the potential energy stored in the spring is converted to kinetic energy as the mass moves back toward its equilibrium position. However, since the mass has acquired kinetic energy, it will pass through the equilibrium position, whereupon the kinetic energy of the mass will be transferred back to potential energy in the spring. The rate of energy transfer between the mass and the spring is the natural frequency of the system. Theoretically, this process will continue indefinitely (since no energy is being dissipated from the system), resulting in continuous vibration of the mass about its static equilibrium position. This type of oscillation is called *free vibration*. The essence of this motion is termed *simple harmonic* since its variation with time can be represented by a sine or cosine function.

Rearranging Eq. (3.1) yields

$$\ddot{x} + \frac{k}{m}x = 0 \tag{3.2}$$

which is recognized as a second-order, homogeneous, ordinary differential equation with constant coefficients. Equation (3.2) is more conveniently expressed as

$$\ddot{x} + \omega^2 x = 0 \tag{3.3}$$

where $\omega^2 = k/m$ and ω is the natural circular frequency, having units of radians per second.

The solution to Eq. (3.2) represents the motion of the mass as a continuous function of time. Therefore, the solution is a function $x(t)$ such that its second derivative plus a constant multiple of the function itself must be identically zero and independent of time. This implies that the function $x(t)$ must be such that its form is unchanged by

Figure 3.1

Mechanical models for SDOF systems: (a) viscously damped system subject to externally applied time-varying force; (b) undamped system in free vibration.

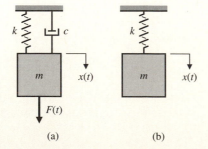

the process of differentiation [1]. This requirement is satisfied by the linear exponential function. Therefore, it is reasonable to assume a solution to Eq. (3.2) having the form

$$x(t) = Ce^{st} \tag{3.4}$$

where C and s are constants that must be determined. The time derivatives for the assumed solution are

$$\dot{x}(t) = Cse^{st} \tag{3.5}$$

and

$$\ddot{x}(t) = Cs^2e^{st} \tag{3.6}$$

Substitution of Eqs. (3.4) and (3.6) into Eq. (3.2) results in

$$Cs^2e^{st} + \frac{k}{m}Ce^{st} = 0 \tag{3.7}$$

or

$$\left(s^2 + \frac{k}{m}\right)Ce^{st} = 0 \tag{3.8}$$

Equation (3.8) must be satisfied for all values of time. Since the exponential e^{st} is non-zero for all time, the only solutions to Eq. (3.8) are $C = 0$ and $s^2 + k/m = 0$. For the case $C = 0$, the assumed solution for $x(t)$ itself is identically zero. This is termed a trivial solution and is worthless. Therefore the only plausible solution is determined from the algebraic equation

$$s^2 + \frac{k}{m} = 0 \tag{3.9}$$

which is called the auxiliary equation for the original differential equation given by Eq. (3.2).

Application of the quadratic formula to Eq. (3.9) results in

$$s = \frac{\pm\sqrt{-4\frac{k}{m}}}{2} \tag{3.10}$$

yielding two values for s:

$$s_1 = i\left(\frac{k}{m}\right)^{1/2} \tag{3.11}$$

and

$$s_2 = -i\left(\frac{k}{m}\right)^{1/2} \tag{3.12}$$

where $i = \sqrt{-1}$. Substitution of Eqs. (3.11) and (3.12) into Eq. (3.4) gives two independent solutions:

$$x_1(t) = C_1e^{s_1t} \tag{3.13}$$

and

$$x_2(t) = C_2 e^{s_2 t} \tag{3.14}$$

where C_1 and C_2 are arbitrary constants. Since the system under consideration is linear, the complete solution is the sum of the two independent solutions and is given by

$$x(t) = C_1 e^{s_1 t} + C_2 e^{s_2 t} \tag{3.15}$$

or

$$x(t) = C_1 e^{i(k/m)^{1/2}t} + C_2 e^{-i(k/m)^{1/2}t} \tag{3.16}$$

To convert the complex exponential given by Eqs. (3.15) and (3.16) to equivalent trigonometric functions, two relations known as *Euler's formulas* [2] can be utilized. Euler's formulas are given as

$$e^{i\alpha} = \cos\alpha + i\sin\alpha \tag{3.17}$$

and

$$e^{-i\alpha} = \cos\alpha - i\sin\alpha \tag{3.18}$$

where $\alpha = (k/m)^{1/2}t$ is a real function. Expanding Eqs. (3.17) and (3.18) yields

$$e^{i(k/m)^{1/2}t} = \cos\left(\frac{k}{m}\right)^{1/2}t + i\sin\left(\frac{k}{m}\right)^{1/2}t \tag{3.19}$$

and

$$e^{-i(k/m)^{1/2}t} = \cos\left(\frac{k}{m}\right)^{1/2}t - i\sin\left(\frac{k}{m}\right)^{1/2}t \tag{3.20}$$

Substitution of Eqs. (3.19) and (3.20) into Eq. (3.16) results in

$$x(t) = C_1\cos\left(\frac{k}{m}\right)^{1/2}t + iC_1\sin\left(\frac{k}{m}\right)^{1/2}t$$
$$+ C_2\cos\left(\frac{k}{m}\right)^{1/2}t - iC_2\sin\left(\frac{k}{m}\right)^{1/2}t \tag{3.21}$$

Rearranging the terms in Eq. (3.21) gives

$$x(t) = (C_1 + C_2)\cos\left(\frac{k}{m}\right)^{1/2}t + i(C_1 - C_2)\sin\left(\frac{k}{m}\right)^{1/2}t \tag{3.22}$$

Two new constants A and B are defined as

$$A = C_1 + C_2 \tag{3.23}$$

and

$$B = i(C_1 - C_2) \tag{3.24}$$

and the resulting solution is

$$x(t) = A\sin\left(\frac{k}{m}\right)^{1/2}t + B\cos\left(\frac{k}{m}\right)^{1/2}t \tag{3.25}$$

or

$$x(t) = A\sin\omega t + B\cos\omega t \tag{3.26}$$

where $\omega = (k/m)^{1/2}$. For all physical systems, C_1 and C_2 are complex conjugates; therefore A and B are real numbers [1].

Equation (3.26) represents the solution for the displacement of the mass as a function of time. The constants A and B are determined from the initial conditions. The most general set of initial conditions is a displacement and a velocity simultaneously imparted to the mass at time $t = 0$. The initial conditions are represented as $x(0) = x_0$ and $\dot{x}(0) = \dot{x}_0$.

To determine the constants A and B, Eq. (3.26) is evaluated at time $t = 0$; thus

$$x(0) = A(0) + B(1.0) = x_0 \tag{3.27}$$

and

$$B = x_0 \tag{3.28}$$

Next, the value of the constant A is determined by first differentiating Eq. (3.26) with respect to time to obtain

$$\dot{x}(t) = A\omega\cos\omega t - B\omega\sin\omega t \tag{3.29}$$

Equation (3.29) represents the velocity of the mass at any time. Evaluating Eq. (3.29) at time $t = 0$ yields

$$\dot{x}(0) = A\omega(1.0) - B\omega(0) + \dot{x}_0 \tag{3.30}$$

and

$$A = \frac{\dot{x}_0}{\omega} \tag{3.31}$$

Substituting the values of B and A, as determined by Eqs. (3.28) and (3.31), respectively, into Eq. (3.26) yields the general free vibration solution for the motion of the mass in terms of the prescribed initial conditions. Thus

$$x(t) = \frac{\dot{x}_0}{\omega}\sin\omega t + x_0\cos\omega t \tag{3.32}$$

Equation (3.32) reveals that the motion of the mass is the sum of two harmonic functions and will vary with time in a cyclic manner. The resulting motion, plotted as displacement $x(t)$ versus time t, is shown in Figure 3.2. Also indicated in the figure are the amplitude of free vibration X and the natural period of vibration T. The motion is periodic and repeats itself every T seconds, where

$$T = \frac{2\pi}{\omega} \tag{3.33}$$

Figure 3.2

Undamped free vibration of a SDOF system subject to initial displacement x_0 and initial velocity $\dot{x}_0$.

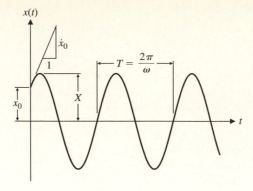

3.2 INTERPRETATION OF THE SOLUTION

To gain insight into the physical meaning of the solution, it is advantageous to express the solution in terms of a single harmonic component. Since A and B are arbitrary constants, they may be replaced by another set of arbitrary constants. For convenience, let $A = X \cos \phi$ and $B = X \sin \phi$, where X is the *amplitude of free vibration* and ϕ is the *phase angle*. Substituting these expressions for A and B into Eq. (3.26) gives

$$x(t) = X(\cos \phi \sin \omega t + \sin \phi \cos \omega t) \tag{3.34}$$

Making use of the trigonometric identity $\cos \phi \sin \omega t = \sin(\omega t + \phi)$ reduces Eq. (3.34) to

$$x(t) = X \sin(\omega t + \phi) \tag{3.35}$$

The new arbitrary constants X and ϕ can be evaluated from the initial conditions in the following manner:

$$A^2 + B^2 = X^2 \cos^2 \phi + X^2 \sin^2 \phi \tag{3.36}$$

Noting that $\sin^2 \phi + \cos^2 \phi = 1$ and that $A = \dot{x}_0/\omega$ and $B = x_0$ yields

$$X = \sqrt{x_0^2 + \left(\frac{\dot{x}_0}{\omega}\right)^2} \tag{3.37}$$

The phase angle ϕ can be determined from the ratio

$$\frac{B}{A} = \frac{X \sin \phi}{X \cos \phi} = \tan \phi \tag{3.38}$$

or

$$\tan \phi = \frac{B}{A} \tag{3.39}$$

Thus

$$\phi = \tan^{-1}\left(\frac{x_0\omega}{\dot{x}_0}\right) \tag{3.40}$$

Harmonic motion such as that described by Eq. (3.35) is often represented as the projection on a straight line of a point that is moving on a circle at constant speed [3]. Consider the rotating vector (OP) of radius X illustrated in Figure 3.3. With the angular speed of the radius designated by ω, and the phase angle ϕ measured counterclockwise from the horizontal, the displacement $x(t)$ is given by Eq. (3.35). At time zero, the position of the mass is found by projecting horizontally from point P to the displacement axis at $t = 0$. This indicates that the initial displacement of the mass x_0 is equal to $X \sin \phi$. If the radius is imagined to rotate in a counterclockwise direction, it will make one complete rotation in $2\pi/\omega$ sec, the natural period. The amplitude of the oscillations varies between $\pm X$ and the motion repeats itself every $2\pi/\omega$ sec as indicated in Figure 3.3. The maximum displacement occurs whenever $\sin(\omega t + \phi) = \pm 1$.

While the quantity ω is known as the natural circular frequency, there is another quantity, f, known merely as the natural frequency. The *natural frequency* is defined as the number of cycles of oscillation per unit of time and is generally given in cycles per second (cps) or hertz (Hz), defined as

$$f = \frac{1}{T} = \frac{\omega}{2\pi} \tag{3.41}$$

The velocity of the mass at any time is determined by taking the first time derivative of Eq. (3.35) and is expressed as

$$\dot{x}(t) = X\omega \cos(\omega t + \phi) \tag{3.42}$$

The maximum value of the velocity is $\pm X\omega$, which occurs whenever $\cos(\omega t + \phi) = \pm 1$. Similarly, the acceleration of the mass at any time is determined by taking the second time derivative of Eq. (3.35) and is given as

$$\ddot{x}(t) = -X\omega^2(\sin \omega t + \phi) \tag{3.43}$$

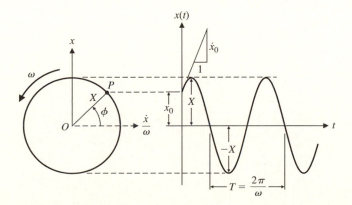

Figure 3.3

Simple harmonic motion illustrated by a rotating vector.

This expression indicates that the maximum acceleration experienced by the mass is $\pm X\omega^2$, which occurs whenever $\sin(\omega t + \phi) = \pm 1$.

Close scrutiny of Eqs. (3.35), (3.42), and (3.43) reveals some interesting characteristics of harmonic motion. From Eqs. (3.35) and (3.42) it is apparent that the maximum velocity occurs when the displacement is zero, and that the maximum displacement occurs when the velocity is zero. Also from Eqs. (3.35) and (3.43) it is apparent that the maximum acceleration and maximum displacement occur at the same time; however, they have opposite signs (directions). Therefore, the displacement, velocity, and acceleration are said to be out of phase with one another. The velocity leads the displacement by $\pi/2$ rad, and the acceleration leads the displacement by π rad. Both the velocity and the acceleration are harmonic and have the same frequency of oscillation as the displacement. The time variation between displacement $x(t)$, velocity $\dot{x}(t)$, and acceleration $\ddot{x}(t)$ in harmonic motion is illustrated in Figure 3.4. The quantity ϕ/ω represents the time by which the resultant motion lags behind the harmonic term in the response.

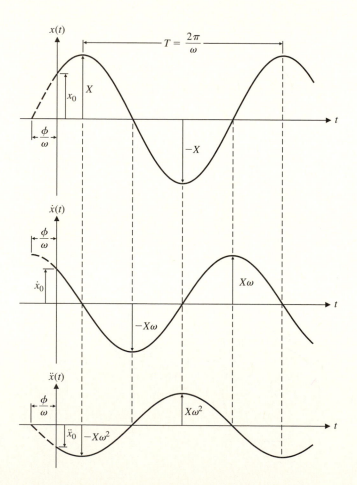

Figure 3.4

Phase difference between displacement $x(t)$, velocity $\dot{x}(t)$, and acceleration $\ddot{x}(t)$ for simple harmonic motion.

EXAMPLE 3.1 ▼

Consider the simply supported beam of Example 2.1 shown again in Figure 3.5. The beam supports a concentrated weight W at midspan, and its natural circular frequency is 27.88 rad/sec. If the midspan of the beam is displaced downward from its static equilibrium position through a distance of 2.0 in and suddenly released with an upward velocity of 3.0 in/sec, determine (a) the natural frequency, (b) the natural period, (c) the maximum midspan displacement, (d) the maximum midspan velocity, (e) the maximum midspan acceleration, and (f) the phase angle. Plot the resulting displacement, velocity, and acceleration as a function of time.

Solution

The mechanical model for this system is similar to that shown in Figure 3.1b. The differential equation of motion is given by Eq. (3.3) as

$$\ddot{x} + \frac{k}{m}x = 0 \tag{1}$$

with initial conditions $x(0) = x_0 = 2.0$ in, and $\dot{x}(0) = \dot{x}_0 = -3.0$ in/sec

(a) The natural frequency is given by Eq. (3.41),

$$f = \frac{\omega}{2\pi} = \frac{27.88 \text{ rad/sec}}{2\pi \text{ rad}} = 4.44 \text{ Hz}$$

(b) The natural period is determined from Eq. (3.33),

$$T = \frac{1}{f} = \frac{2\pi}{\omega} = 0.225 \text{ sec}$$

(c) The displacement as a function of time is given by Eq. (3.35) as

$$x(t) = X \sin(\omega t + \phi) \tag{2}$$

and the maximum midspan displacement is determined from Eq. (3.37),

$$x_{max} = X = \sqrt{(x_o)^2 + \left(\frac{\dot{x}_0}{\omega}\right)^2} = \sqrt{(2.0)^2 + \left(\frac{3.0}{27.88}\right)^2} = 2.005 \text{ in}$$

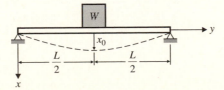

Figure 3.5
Simple supported beam of Example 3.1.

(d) The velocity as a function of time is given by Eq. (3.42) as

$$\dot{x}(t) = X\omega\cos(\omega t + \phi) \tag{3}$$

and the maximum midspan velocity is given by

$$\dot{x}_{max} = X\omega = 2.005(27.88) = 55.90 \text{ in/sec}$$

(e) The acceleration as a function of time is determined from Eq. (3.43) as

$$\ddot{x}(t) = -X\omega^2\sin(\omega t + \phi) \tag{4}$$

and the maximum midspan acceleration is

$$\ddot{x}_{max} = X\omega^2 = \dot{x}_{max}\omega = 55.9(27.88) = 1558.48 \text{ in/sec}^2$$

(f) From Eq. (3.40) the phase angle is calculated as

$$\phi = \tan^{-1}\left(\frac{x_0\omega}{\dot{x}_0}\right) = \tan^{-1}\left[\frac{(2)(27.88)}{(-3.0)}\right] = (-1.517 \text{ rad}) = 273°$$

Plots of the resulting displacement, velocity, and acceleration time histories are presented in Figure 3.6. These figures clearly indicate the phase difference between the displacement, velocity, and acceleration.

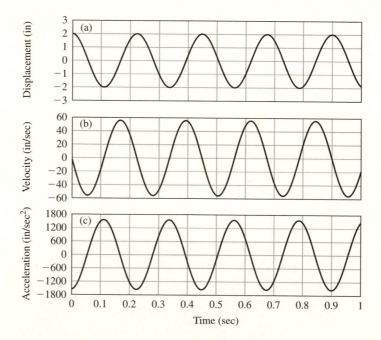

Figure 3.6

Free vibration response for beam of Example 3.1: (a) displacement response history $x(t)$; (b) velocity response history $\dot{x}(t)$; (c) acceleration response history $\ddot{x}(t)$.

EXAMPLE 3.2   ▼

A 6.8-kg mass is suspended by a linear spring. The mass receives an impact such that its motion begins with an initial velocity but no initial displacement. In the ensuing motion, the natural period of the system is measured to be 0.25 sec, and the displacement amplitude of the oscillations is 50 mm. Find (a) the spring constant for the system and (b) the initial velocity $\dot{x}_0$ imparted to the mass. (c) Also determine the displacement, velocity, and acceleration of the mass at time $t = 0.63$ sec.

Solution

The mechanical model for this system is similar to that shown in Figure 3.1b. The solution for displacement of the mass and its time derivatives are given by Eqs. (3.35), (3.42), and (3.43), respectively

$$x(t) = X \sin(\omega t + \phi) \tag{1}$$

$$\dot{x}(t) = X \omega \cos(\omega t + \phi) \tag{2}$$

$$\ddot{x}(t) = -X \omega^2 \sin(\omega t + \phi) \tag{3}$$

The vibration amplitude X is determined from Eq. (3.37),

$$X = \sqrt{x_0^2 + \left(\frac{\dot{x}_0}{\omega}\right)^2} = 50 \text{ mm} \tag{4}$$

and the natural circular frequency is determined from Eq. (3.33),

$$\omega = \frac{2\pi}{T} = \frac{2\pi}{0.25 \text{ sec}} = 25.13 \text{ rad/sec}$$

(a) The spring constant can be determined from the expression for natural circular frequency,

$$\omega = \sqrt{\frac{k}{m}} \tag{5}$$

so that

$$k = m\omega^2 = (6.8 \ N - \sec^2/m)(25.13 \text{ rad/sec})^2$$

and

$$k = 4242 \ N/m$$

(b) The vibration amplitude given by Eq. (4) can be used to determine the initial velocity $\dot{x}_0$. Since $x_0 = 0$, from Eq. (4) it follows that

$$\dot{x}_0 = (50 \text{ mm})(25.13 \text{ rad/sec}) = 1256.5 \text{ mm/sec}$$

(c) To determine the displacement, velocity, and acceleration at time $t = 0.63$ sec, Eqs. (1), (2), and (3) are evaluated at $t = 0.63$ sec:

$$x(0.63) = 50 \sin[(25.13)(0.63)] = -6.18 \text{ mm}$$

$$\dot{x}(0.63) = 1257 \cos[(25.13)(0.63)] = -1247.4 \text{ mm/sec}$$

$$\ddot{x}(0.63) = -31{,}576 \sin[(25.13)(0.63)] = 3903.4 \text{ mm/sec}^2$$

▲

3.3 EQUIVALENT STIFFNESS

Many undamped SDOF systems can be represented by one mass and one spring element. The single spring element is generally an equivalent characterization of the elastic stiffness of the structure. In many practical cases the elastic stiffness of the physical structure is described by several spring elements, which can be reduced to a single spring element to represent the equivalent stiffness of the structure.

The various spring elements in many SDOF vibrating systems can be envisioned to exist in parallel, in series, or in some combination of these [4]. A system with springs in parallel is depicted in Figure 3.7a. The linear springs have constants k_1 and k_2 and the mass is suspended from a rigid support. It is assumed that the mass does not rotate. To assess the equivalent stiffness of the springs, represented by k_e, as shown in Figure 3.7b, it is convenient to consider the static equilibrium of the system. The constraints imposed upon the system dictate that the deflection in each spring must be identical. If the deflection in each spring due to the weight of the mass $W = mg$ (where g is the

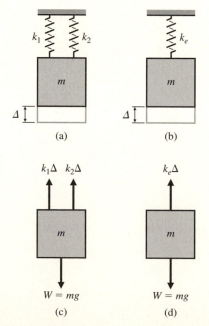

Figure 3.7

Springs in parallel:
(a) original system;
(b) equivalent system;
(c) free-body diagram of original system; (d) free-body diagram of equivalent system.

acceleration due to gravity) is assumed to be Δ, as shown in Figure 3.7c, then the deflection in the equivalent spring, k_e, must also be Δ, as shown in Figure 3.7d. Thus, from consideration of these free-body diagrams, the condition for equivalence can be stated as

$$k_e\Delta = (k_1 + k_2)\Delta \tag{3.44}$$

from which

$$k_e = k_1 + k_2 \tag{3.45}$$

In this case, the equivalent stiffness is equal to the sum of the constants of the two springs of the original system. For any system where the deformation of any elastic element is the same as all other elastic elements of the system, the elastic elements are said to be in parallel, and the equivalent stiffness is the sum of the individual elastic constants. The general expression for the equivalent stiffness for n springs in parallel is given by

$$k_e = \sum_{i=1}^{n} k_i \tag{3.46}$$

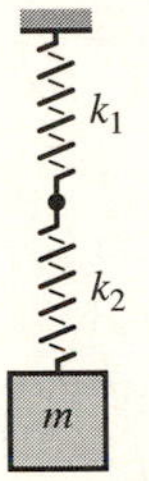

A system with springs in series is shown in Figure 3.8. From consideration of the static equilibrium of the system, it is apparent that the force in each spring must be equal to the weight of the mass ($W = mg$). Furthermore, the displacement of the mass Δ is given by

$$\Delta = \Delta_1 + \Delta_2 \tag{3.47}$$

where Δ_1 and Δ_2 are the elongations of springs 1 and 2, respectively. Therefore, for an equivalent spring having a constant k_e, it follows from Eq. (3.47) that

$$\frac{W}{k_e} = \frac{W}{k_1} + \frac{W}{k_2} \tag{3.48}$$

from which

$$\frac{1}{k_e} = \frac{1}{k_1} + \frac{1}{k_2} \tag{3.49}$$

In general, the reciprocal of the equivalent spring constant for n springs in series is equal to the sum of the reciprocals of the elastic constants of the individual springs and is given by

$$\frac{1}{k_e} = \sum_{i=1}^{n} \frac{1}{k_i} \tag{3.50}$$

Expressed in terms of k_e directly, Eq. (3.50) becomes

$$k_e = \frac{1}{\sum_{i=1}^{n} 1/k_i} \tag{3.51}$$

Figure 3.8

System with springs in series.

Figure 3.9
Cantilever beam with tip load.

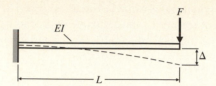

As a more practical example of the equivalent spring constant for an elastic element, consider the cantilever beam loaded at its free end as shown in Figure 3.9. The deflection of the beam at the point of application of the force F is given by

$$\Delta = \frac{FL^3}{3EI} \tag{3.52}$$

where E is Young's modulus for the beam material and I is the moment of inertia of the beam's cross-sectional area about the neutral axis. The equivalent stiffness for the beam is the ratio of the applied load to the deflection at the point of application of the load. Thus, from Eq. (3.52) the equivalent stiffness is

$$k_e = \frac{F}{\Delta} = \frac{3EI}{L^3} \tag{3.53}$$

The equivalent spring constants for beams with various other types of loading and/or boundary conditions can be obtained in a similar manner. Some of the more commonly encountered cases of equivalent stiffness are summarized in Table 3.1.

TABLE 3.1 **Equivalent Spring Constants**

Case	Equivalent Spring Constant
1. n axial springs in parallel	$k_1 + k_2 + k_3 + \cdots + k_n$
2. n axial springs in series	$\dfrac{1}{\dfrac{1}{k_1} + \dfrac{1}{k_2} + \ldots + \dfrac{1}{k_n}}$

(continued)

Case	Equivalent Spring Constant
3. Springs in parallel and in series 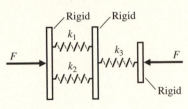	$$\dfrac{k_1 k_3 + k_2 k_3}{k_1 + k_2 + k_3}$$
4. Inclined axial spring 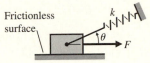	$$k \cos^2 \theta$$
5. Rotating bar with spring support 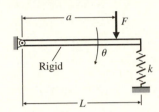	$$k\left(\dfrac{L}{a}\right)^2$$
6. Rigid bar supported on two springs 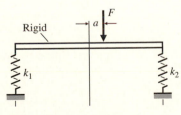	$$\dfrac{4 k_1 k_2}{k_1[1 + a/L]^2 + k_2[1 + a/L]^2}$$
7. Rigid bar supported on three springs 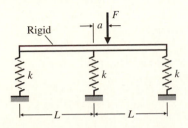	$$\dfrac{3k}{1 + \dfrac{3}{2}(a/L)^2}$$
8. Axially loaded bar 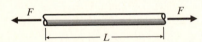	$$\dfrac{AE}{L}$$ A = cross-sectional area E = elastic modulus

(continued)

Case	Equivalent Spring Constant
9. Axially loaded tapered bar 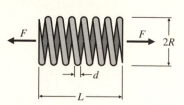	$$\frac{\pi E D_a D_b}{4L}$$
10. Axial helical spring 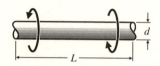	$$\frac{Gd^4}{G4nR^3}$$ n = active number of turns G = elastic shear modulus
11. Torsion of a uniform shaft 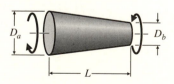	$$\frac{GJ}{L}$$ J = torsional constant of cross section ($\pi d^4/32$)
12. Torsion of tapered circular shaft	$$\frac{3\pi}{32} \frac{D_b^4 G}{L[D_b/D_a + (D_b/D_a)^2 + (D_b/D_a)^3]}$$
13. Spiral torsional spring 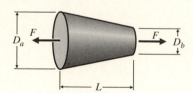	$$\frac{EI}{L}$$ E = Young's modulus I = moment of inertia of cross-sectional area L = total length of spiral
14. Cantilever beam, end load 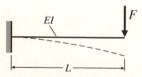	$$\frac{3EI}{L^3}$$

(continued)

Case	Equivalent Spring Constant
15. Simply supported beam, load at midspan 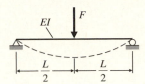	$\dfrac{48EI}{L^3}$
16. Simply supported beam, load anywhere between supports 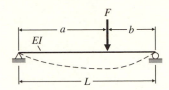	$\dfrac{3EIL}{a^2b^2}$
17. Fixed-fixed beam, load at midspan 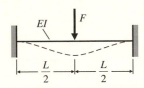	$\dfrac{192EI}{L^3}$
18. Fixed-fixed beam, off-center load 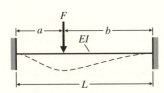	$\dfrac{3EI(a+b)^3}{a^3b^3}$
19. Propped cantilever, load at midspan 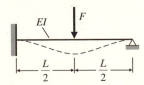	$\dfrac{768EI}{7L^3}$
20. Propped cantilever with overhang, load at free end 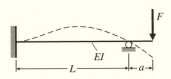	$\dfrac{24EI}{a^2(3L+8a)}$

(continued)

Case	Equivalent Spring Constant
21. Simple beam with overhang, load at free end 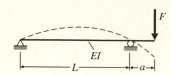	$\dfrac{3EI}{(L+a)a^2}$
22. Shear frame, fixed base 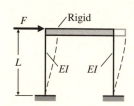	$\dfrac{12EI}{L^3}(2)$
23. Shear frame, pinned base 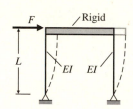	$\dfrac{3EI}{L^3}(2)$

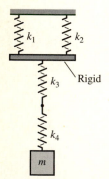

Figure 3.10

System with springs in parallel and in series.

For systems that have spring elements both in parallel and in series, the equivalent stiffness is determined by calculating the equivalent spring constant for the elastic elements in parallel $(k_e)_p$, and for the elastic elements in series $(k_e)_s$, and then appropriately combining the two. An example of a system having spring elements in parallel and in series is illustrated in Figure 3.10. The equivalent stiffness for the system is calculated by

$$k_e = (k_e)_p + (k_e)_s \tag{3.54}$$

or

$$k_e = \frac{1}{[1/(k_e)_p] + [1/(k_e)_s]} \tag{3.55}$$

where

$$(k_e)_p = k_1 + k_2 \tag{3.56}$$

and

$$(k_e)_s = \frac{1}{(1/k_3) + (1/k_4)} \tag{3.57}$$

At this point it should be recognized that $(k_e)_p$ and $(k_e)_s$ are in series; therefore, k_e is determined by Eq. (3.55) as follows:

$$k_e = \frac{1}{\left(\dfrac{1}{k_1 + k_2}\right) + \left(\dfrac{1}{k_3}\right) + \left(\dfrac{1}{k_4}\right)} \tag{3.58}$$

EXAMPLE 3.3 ▼

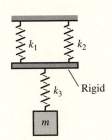

Figure 3.11
System of
Example 3.3.

For the system shown in Figure 3.11, determine (a) the differential equation of the motion of the mass m and (b) the natural circular frequency of vibration of the mass m.

Solution

Springs k_1 and k_2 are in parallel with one another, and spring k_3 is in series with k_1 and k_2. Let $k' = k_1 + k_2$; then from Eq. (3.49),

$$\frac{1}{k_e} = \frac{1}{k'} + \frac{1}{k_3} \tag{1}$$

or

$$k_e = \frac{k'k_3}{k' + k_3} = \frac{k_3(k_1 + k_2)}{k_1 + k_2 + k_3} \tag{2}$$

(a) The equation of motion is expressed as

$$\ddot{x} + \frac{k_e}{m}x = 0 \tag{3}$$

and substitution of Eq. (2) into Eq. (3) yields

$$\ddot{x} + \frac{k_3(k_1 + k_2)}{m(k_1 + k_2 + k_3)}x = 0 \tag{4}$$

(b) The natural circular frequency is calculated as

$$\omega = \sqrt{\frac{k_e}{m}} = \left[\frac{k_3(k_1 + k_2)}{m(k_1 + k_2 + k_3)}\right]^{1/2} \text{rad/sec} \tag{5}$$

▲

EXAMPLE 3.4 ▼

A spherical container of weight W is supported at the tip of a cantilever beam having flexural rigidity EI and length L as shown in Figure 3.12. The container is also

Figure 3.12

Cantilever beam of Example 3.4.

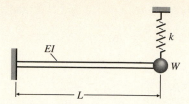

supported by a spring of stiffness k. Determine (a) the equivalent spring stiffness for this system and (b) the natural frequency of vibration. Neglect the mass of the beam and the spring.

Solution

(a) The spring having stiffness k and the cantilever beam represent two springs in parallel. Let k_b represent the stiffness for the beam. From Table 3.1

$$k_b = \frac{3EI}{L^3} \tag{1}$$

and thus

$$k_e = k + k_b = k + \frac{3EI}{L^3} = \frac{kL^3 + 3EI}{L^3} \tag{2}$$

(b) The natural frequency is expressed as

$$\omega = \sqrt{\frac{k_e}{m}} \tag{3}$$

where

$$m = \frac{W}{g} \tag{4}$$

and g is the acceleration due to gravity. Thus substituting Eqs. (2) and (4) into Eq. (3) yields the expression for natural circular frequency,

$$\omega = \left[\frac{kL^3 + 3EI}{L^3(W/g)} \right]^{1/2} \text{ rad/sec} \tag{5}$$

▲

EXAMPLE 3.5 ▼

For the steel shear frame structure shown in Figure 3.13, determine the natural circular frequency of vibration for oscillations in the horizontal direction. The horizontal girder is assumed to be infinitely rigid with respect to the columns, and it

Figure 3.13

Steel shear frame structure
of Example 3.5.

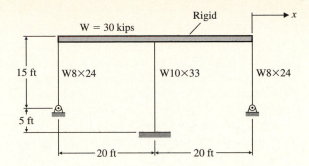

supports a total dead weight W of 30 kips uniformly distributed along its length. Neglect the mass of the columns that bend about their strong axis.

Solution

The three columns represent springs in parallel. For the W8×24 columns, I_{xx} = 82.8 in^4, and for the W10×33, I_{xx} = 170 in^4; E = 29,000 ksi for steel.

For the W8×24s the transverse stiffness is

$$k_1 = \frac{3EI}{L^3}(2) = \frac{(6)(29,000)(82.8)}{[(15 \text{ ft})(12 \text{ in})]^3} = 1.235 \text{ kips/in}$$

and for the W10×33,

$$k_2 = \frac{12EI}{L^3} = \frac{(12)(29,000)(170)}{[(20 \text{ ft})(12 \text{ in})]^3} = 4.28 \text{ kips/in}$$

Thus, the equivalent stiffness is calculated from Eq. (3.50) as

$$k_e = \sum_{i=1}^{2} k_i = 6.75 \text{ kips/in}$$

and the natural circular frequency is expressed by

$$\omega = \sqrt{\frac{k_e}{m}}$$

where

$$m = \frac{W}{g} = \frac{30 \text{ kips}}{(32.2 \text{ ft}/\text{sec}^2)(12 \text{ in/ft})}$$
$$= 0.0776 \text{ kip-sec}^2/\text{in}$$

Therefore

$$\omega = \left(\frac{6.75 \text{ kips/in}}{0.0776 \text{ kip-sec}^2/\text{in}}\right)^{1/2} = 9.3 \text{ rad/sec}$$

▲

3.4 RAYLEIGH METHOD

The equation of motion and the natural frequency of a conservative system (no damping) can be established by the principle of conservation of energy, since no energy is removed from the system. The kinetic energy T is stored in the mass and is proportional to the square of the velocity. The potential energy or strain energy V is stored in the spring and is proportional to its elastic deformation. The total energy in the system is composed of various amounts of kinetic energy and potential energy as the system vibrates; however, the sum is constant and therefore its time derivative is zero. This may be expressed as [5]

$$T + V = \text{(total mechanical energy)} = \text{constant} \tag{3.59}$$

and

$$\frac{d}{dt}(T + V) = 0 \tag{3.60}$$

The use of Eq. (3.60) to establish the equation of motion for a conservative system was illustrated in Chapter 2. The natural frequency of a conservative system can also be determined from consideration of Eq. (3.59). For this purpose let T_1 and V_1, and T_2 and V_2, represent the kinetic and potential energy in the system at two different instances of time. From the principle of conservation of energy, Eq. (3.59) can be written as

$$T_1 + V_1 = T_2 + V_2 = \text{constant} \tag{3.61}$$

The amounts of kinetic energy and potential energy in the system may vary with time; however, their sum is constant or invariant. If T_1 and V_1 are considered at the instant of time when the mass is passing through its static equilibrium position, then there is no elastic deformation or strain energy in the spring; consequently, $V_1 = 0$. Furthermore, if T_2 and V_2 are considered at the instant of time corresponding to the maximum displacement of the spring, the velocity of the mass at this time is zero; consequently, $T_2 = 0$. For these two instants of time Eq. (3.61) then becomes

$$T_1 + 0 = V_2 + 0 \tag{3.62}$$

For a system undergoing harmonic motion, T_1 and V_2 are maximum values; therefore from Eq. (3.62)

$$T_{\max} = V_{\max} \tag{3.63}$$

Application of Eq. (3.63) to a conservative system leads directly to the calculation of natural frequency. This procedure is referred to as the *Rayleigh method*.

EXAMPLE 3.6 ▼

A simply supported beam of length L and flexural rigidity EI supports a mass m as shown in Figure 3.14. The mass of the beam is negligible with respect to the mass m. Determine the natural circular frequency of vertical vibration of the beam-mass system by the Rayleigh method.

Figure 3.14
Simply supported beam
of Example 3.6.

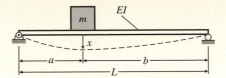

Solution

The maximum potential and kinetic energies, respectively, are expressed by

$$V_{max} = \frac{1}{2}k_e x_{max}^2 \tag{1}$$

and

$$T_{max} = \frac{1}{2}m\dot{x}_{max}^2 \tag{2}$$

From Table 3.1 the equivalent stiffness is

$$k_e = \frac{3EIL}{a^2 b^2} \tag{3}$$

For harmonic motion, the displacement and velocity are given by Eqs. (3.35) and (3.42), respectively, as

$$x(t) = X\sin(\omega t + \phi) \tag{4}$$

$$\dot{x}(t) = X\omega\cos(\omega t + \phi) \tag{5}$$

from which

$$x_{max} = X \quad \text{and} \quad \dot{x}_{max} = X\omega \tag{6}$$

Therefore, substituting Eq. (6) into Eqs. (1) and (2) results in

$$T_{max} = \frac{1}{2}mX^2\omega^2 \quad \text{and} \quad V_{max} = \frac{1}{2}k_e X^2 \tag{7}$$

Equating the expressions for T_{max} and V_{max} given by Eq. (7) and solving for ω yields

$$\omega = \sqrt{\frac{k_e}{m}} = \sqrt{\frac{3EIL}{a^2 b^2 m}} \tag{8}$$

▲

EXAMPLE 3.7 ▼

A uniform rigid rod of mass m and length L, shown in Figure 3.15, is restrained by two translational springs of stiffness k_1 and k_2 and by a rotational spring of

Figure 3.15
Uniform rigid rod of
Example 3.7.

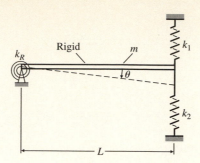

stiffness k_R. Determine the natural circular frequency of angular oscillation of the rod by the Rayleigh method. Assume small angles of rotation, θ.

Solution

The maximum potential and kinetic energies, respectively, are expressed by

$$V_{\text{max}} = \frac{1}{2}k_R\theta_{\text{max}}^2 + \frac{1}{2}k_1\theta_{\text{max}}^2 L^2 + \frac{1}{2}k_2\theta_{\text{max}}^2 L^2 \tag{1}$$

and

$$T_{\text{max}} = \frac{1}{2}\left(\frac{1}{3}m\right)(\dot{\theta}_{\text{max}}L)^2 \tag{2}$$

For harmonic motion, the equations for rotational displacement $\theta(t)$ and velocity $\dot{\theta}$ are

$$\theta(t) = \beta\sin(\omega t + \phi) \tag{3}$$

and

$$\dot{\theta}(t) = \beta\omega\cos(\omega t + \phi) \tag{4}$$

where β is the amplitude of rotational vibration and from which

$$\theta_{\text{max}} = \beta \quad \text{and} \quad \dot{\theta}_{\text{max}} = \beta\omega \tag{5}$$

Therefore, substituting Eq. (5) into Eqs. (1) and (2) yields

$$T_{\text{max}} = \frac{1}{6}\beta^2\omega^2 L^2 m \tag{6}$$

and

$$V_{\text{max}} = \frac{1}{2}k_R\beta^2 + \frac{1}{2}k_1\beta^2 L^2 + \frac{1}{2}k_2\beta^2 L^2 \tag{7}$$

Equating T_{max} and V_{max} given by Eqs. (6) and (7), respectively, and solving for ω yields

$$\omega = \sqrt{\frac{3(k_R + k_1 L^2 + k_2 L^2)}{mL^2}} \tag{8}$$

▲

Figure 3.16

Flexible beam with uniformly distributed mass.

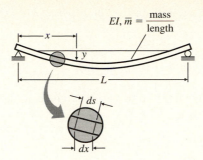

The Rayleigh method may be extended to structures with distributed mass and distributed stiffness properties. Consider the beam shown in Figure 3.16, having flexural rigidity EI and uniform mass per length equal to $\bar{m}$. If y is the amplitude of the assumed deflection curve, then

$$T_{max} = \frac{1}{2}\int_0^L \dot{y}_{max}^2 \, dm = \frac{1}{2}\omega^2 \int_0^L y^2 \, dm \qquad (3.64)$$

where dm is the mass of a differential length of beam, ds.

Since the beam resists load primarily through bending moment or flexure, the strain energy in the member is calculated as

$$V_{max} = \frac{1}{2}\int_0^L M \, d\theta \qquad (3.65)$$

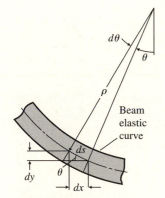

Figure 3.17

Differential segment of beam elastic curve.

where M is the bending moment and θ is the slope of the elastic curve as shown in Figure 3.17. For small deflections the differential length of the elastic curve, ds, is approximately equal to dx. Therefore, the following relationships can be assumed

$$\theta = \frac{dy}{ds} \cong \frac{dy}{dx} \qquad (3.66)$$

and

$$\rho \, d\theta = dx \qquad (3.67)$$

where ρ is the radius of curvature. It follows from Eqs. (3.66) and (3.67) that

$$\frac{1}{\rho} = \frac{d\theta}{dx} = \frac{d^2y}{dx^2} \qquad (3.68)$$

From conventional beam theory we have the relationship

$$\frac{1}{\rho} = \frac{M}{EI} \qquad (3.69)$$

and from Eq. (3.68) we note that

$$d\theta = \frac{dx}{\rho} \qquad (3.70)$$

Combining Eqs. (3.68), (3.69), and (3.70) into Eq. (3.65) results in

$$V_{max} = \frac{1}{2}\int_0^L \frac{M}{\rho}dx = \frac{1}{2}\int_0^L EI\left(\frac{d^2y}{dx^2}\right)^2 dx \tag{3.71}$$

Equating T_{max} given by Eq. (3.64) with V_{max} given by Eq. (3.71) and solving for ω^2 yields

$$\omega^2 = \frac{\displaystyle\int_0^L EI\left(\frac{d^2y}{dx^2}\right)^2 dx}{\displaystyle\int_0^L y^2 dm} \tag{3.72}$$

Equation (3.72) will yield the fundamental frequency of vibration of the system. However, to use this method the deflection y must be known as a function of x. Therefore, an appropriate displacement function must be selected. The displacement function must be compatible with the geometric boundary conditions of the system and should be differentiable to the second degree. The static deflection curve corresponding to a loading condition proportional to the mass distribution of the system is generally a reasonable selection for the displacement function.

EXAMPLE 3.8 ▼

Determine the natural frequency of the cantilever beam shown in Figure 3.18 by the Rayleigh method. The beam has uniform mass per length $\bar{m}$ and flexural rigidity EI.

Solution

The natural frequency may be determined by application of Eq. (3.72). However, before Eq. (3.72) can be implemented, a continuous displacement function for y must be selected that satisfies the boundary conditions. In this example, the displacement function representing the static deflection of a uniformly loaded cantilever beam has been selected and is given by

$$y(x) = \frac{A(x^4 - 4L^3x + 3L^4)}{3L^4} \tag{1}$$

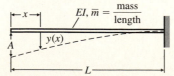

Figure 3.18
Cantilever beam of
Example 3.8.

where A represents the maximum deflection at the tip of the cantilever. The pertinent derivatives of y are

$$\frac{dy}{dx} = A\frac{4x^3 - 4L^3}{3L^4} \tag{2}$$

and

$$\frac{d^2y}{dx^2} = A\frac{4x^2}{L^4} \tag{3}$$

The *geometric boundary conditions* that the displacement function must satisfy are

1. The deflection (y) of the elastic curve at $x = L$ must equal zero.
2. The slope (dy/dx) of the elastic curve at $x = L$ must equal zero.

    The *force boundary condition* that the displacement function must satisfy is

3. The bending moment (d^2y/dx^2) at $x = 0$ must equal zero.

Substituting the expressions for y and d^2y/dx^2 given by Eqs. (1) and (3), respectively, into Eq. (3.72) yields

$$\omega^2 = \frac{EIA^2\int_0^L (4x^2/L^4)^2\,dx}{A^2\int_0^L [(x^4 - 4L^3x + 3L^4)/3L^4]^2\,\overline{m}\,dx} \tag{4}$$

Integrating and evaluating Eq. (4) over the interval from 0 to L results in

$$\omega^2 = \frac{3.2EIA^2/L^3}{11.6\overline{m}A^2} = \frac{0.276EI}{\overline{m}L^4} \tag{5}$$

or

$$\omega = 0.525\sqrt{\frac{EI}{\overline{m}L^4}} \tag{6}$$

▲

REFERENCES

1 Hutton, D.V., *Applied Mechanical Vibrations,* McGraw Hill, New York, 1981.

2 *Standard Mathematical Tables,* 23rd ed., CRC Press, Cleveland, OH, 1975.

3 Thompson, W.T., *Theory of Vibrations With Applications,* 2nd ed., Prentice Hall, Englewood Cliffs, NJ, 1981.

4 James, M.L., Smith, G.M., Wolford, J.C., and Whaley, P.W., *Vibrations of Mechanical and Structural Systems,* Harper and Row, New York, 1989.

5 Blevins, Robert D., *Formulas for Natural Frequency and Mode Shape,* Van Nostrand Reinhold, New York, 1979.

NOTATION

A	constant	s	constant in linear exponential function
B	constant	s_1, s_2	roots of the auxiliary equation
c	viscous damping coefficient	t	time
C	constant	T	natural period (sec); also represents kinetic energy
C_1	constant	V	potential energy
C_2	constant	W	weight
e^{st}	linear exponential function	x	translational displacement
$e^{i\alpha}, e^{-i\alpha}$	Euler equation exponentials	x_0	initial displacement
E	Young's modulus	$\dot{x}$	translational velocity
f	natural frequency (cps or Hz)	$\dot{x}_0$	initial velocity
$F(t)$	time-varying externally applied force	$\ddot{x}$	translational acceleration
g	acceleration due to gravity	X	displacement amplitude of free vibration
i	imaginary number ($\sqrt{-1}$)	SDOF	single degree of freedom
I	static moment of inertia	β	amplitude of rotational displacement
k	translational stiffness	Δ	translational deflection
k_e	equivalent stiffness	ϕ	phase angle (rad)
k_R	rotational stiffness	ω	natural circular frequency (rad/sec)
L	length	θ	rotational displacement
M	bending moment	$\dot{\theta}$	rotational velocity
m	mass	ρ	radius of curvature
$\overline{m}$	mass per unit length		

PROBLEMS

3.1 A 50-lb weight is suspended from a spring having an elastic constant of 10 lb/in. Determine the natural frequency for the system.

3.2 A mass of 10 kg is attached to a linear spring. The mass is impacted such that the motion is instigated with an initial velocity with no initial displacement. In the resulting motion the period is determined to be 0.2 sec and the amplitude of vibration is 60 mm. Determine (a) the spring constant and (b) the initial velocity.

3.3 A 10-lb weight attached to a linear spring causes a static equilibrium deflection of 0.25 in. Determine the natural frequency of the system if it is displaced from its equilibrium position and suddenly released.

3.4 A simple spring-mass system undergoing free vibration reaches a maximum acceleration of 50 in/sec² and exhibits a natural frequency of 100 Hz. Determine (a) the amplitude of vibration and (b) the maximum velocity of the mass.

3.5 A linear spring supports a body weighing 50 lb. In free vibration the body oscillates at a frequency of 400 cycles/min. Determine the spring constant k for the system.

3.6 A mass m is attached to a linear spring of stiffness k. The natural period of free vibration is observed to be 0.35 sec. When a mass weighing 2 lb is added to the system, the period increases by 10%. Determine the weight of the mass m and the stiffness k for the original system.

3.7 Determine the magnitude of the mass that must be attached to a linear spring of stiffness 200 N/m so that the system will have a natural frequency of 50 Hz.

3.8 The maximum velocity attained by the mass in a simple spring-mass system is 20 in/sec, and the natural period of free vibration is 1.5 sec. If the mass was released with an initial displacement of 3.0 in, determine (a) the amplitude of free vibration, (b) the maximum acceleration of the mass, (c) the initial velocity, and (d) the phase angle.

3.9 A spring-mass system has a natural frequency of 10 Hz. When the spring constant is increased by 5 lb/in, the frequency is changed by 25%. Determine (a) the spring constant for the original system and (b) the mass of the original system.

3.10 Determine the length required for a simple pendulum of mass m if the natural period of vibration is to be 0.75 sec. Assume small angle of oscillation.

3.11 The simple 20-ft-long beam shown in Figure P3.1 supports a weight of 1000 lb. If the midspan of the beam is displaced downward 1.5 in with an initial velocity of 4.5 in/sec in the same direction, determine (a) the natural period of vibration, (b) the maximum displacement of the mass, (c) the maximum velocity of the mass, (d) the maximum acceleration of the mass, and (e) the phase angle. Assume $E = 29,000$ ksi and $I = 1830$ in.[4]

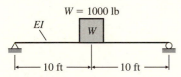

Figure P3.1

3.12 The cantilever beam shown in Figure P3.2 is given an initial displacement at its free end of 2.0 in with a velocity of 3.0 in/sec in the same direction. Determine (a) the natural frequency of free vibration; (b) the maximum displacement, velocity, and acceleration of the mass; and (c) the phase angle. Also plot the resulting displacement, velocity, and acceleration of the mass as a function of time between 0 and 5 sec. Assume $E = 30,000$ ksi, $I = 2500$ in[4], $L = 10$ ft, and $W = 750$ lb.

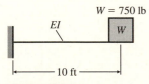

Figure P3.2

3.13 A 100-lb body is suspended by a linear spring. The body is subjected to an impact that results in motion with an initial

velocity but no initial displacement. In the ensuing motion the period of vibration is measured to be 0.15 sec and the amplitude of oscillation is 1.5 in. Determine (a) the spring constant, (b) the initial velocity, and (c) the velocity and acceleration of the mass at $t = 0.41$ sec.

3.14 A 50-lb weight is attached to a linear spring. The body is subjected to an initial velocity of 2 in/sec with no initial displacement. The natural period of vibration is observed to be 0.3 sec. Determine (a) the spring constant, (b) the amplitude of vibration, and (c) the velocity and acceleration of the mass at $t = 0.5$ sec.

3.15 For the systems shown in Figure P3.3, determine (a) the equivalent stiffness, (b) the equation of motion, and (c) the natural frequency of vibration.

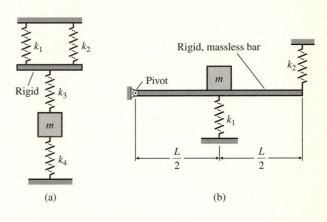

(a)   (b)

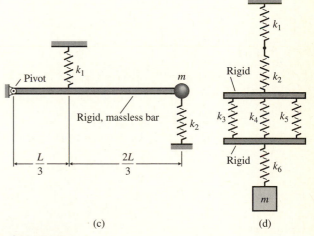

(c)   (d)

Figure P3.3

3.16 Determine (a) the equivalent stiffness, (b) the equation of motion, and (c) the natural frequency of vibration for the systems shown in Figure P3.4.

3.17 For the systems shown in Figure P3.5, determine (a) the natural frequency by the Rayleigh method and (b) the equation of motion. Assume small oscillations.

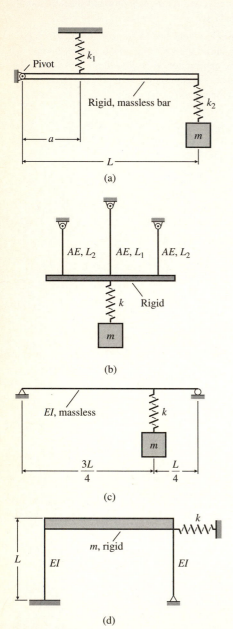

(a)

(b)

(c)

(d)

Figure P3.4

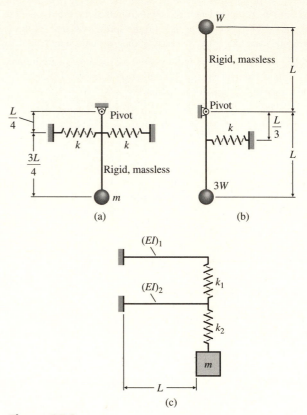

(a)   (b)

(c)

Figure P3.5

3.18 A massless, rigid bar *AB* pivots about point *A* as shown in Figure P3.6. Determine (a) the equation of motion and (b) the natural frequency for the system. Assume small oscillations.

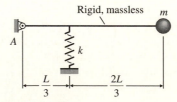

Figure P3.6

3.19 For the torsional disk-shaft systems shown in Figure P3.7, determine (a) the natural torsional frequency by the Rayleigh method and (b) the differential equation of motion.

Assume the disks have a mass moment of inertia about their mass center of I_0 and assume the shafts to be solid and massless.

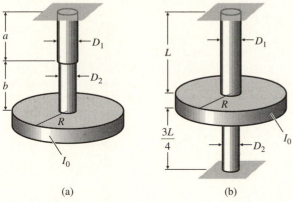

(a)   (b)   (c)

Figure P3.7

3.20 For the torsional disk-shaft system shown in Figure P3.8, the disk has a weight of 200 lb and a radius $R = 4.5$ in, and the solid shaft has a length $L = 40$ in and a diameter $D = 2$ in. If the natural frequency of torsional oscillation is 100 Hz, determine the shear modulus of the shaft material.

Figure P3.8

$W = 200$ lb

D

L

R

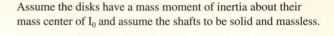

m

Rigid base k k

Ground surface

h

Figure P3.9

3.22 A cart having a weight $W_1 = 400$ lb rolls with a velocity of 5 mph and impacts a loading dock buffer as shown in Figure P3.10. The buffer has a weight of $W_2 = 200$ lb and a spring constant k of 200 lb/in. After impact the cart remains in contact with the buffer. Determine (a) the equation of motion and natural frequency for the cart-buffer system and (b) the maximum displacement for the car-buffer system after impact.

3.21 A body of mass m is mounted on a rigid base as shown in Figure P3.9 and dropped to the ground surface from a height h. Determine (a) the equation of motion for the system; (b) the natural frequency; and (c) the maximum displacement, velocity, and acceleration of the mass. Assume no rebound of the base.

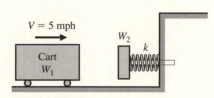

$V = 5$ mph

W_2

Cart W_1

k

Figure P3.10

3.23 A mass m_1 is supported by a linear spring of stiffness k and is in static equilibrium as shown in Figure P3.11. A second mass m_2 is dropped from a height h and sticks to mass m_1 without rebound. Determine (a) the equation of motion and natural frequency for this system and (b) the maximum displacement of the combined mass $(m_1 + m_2)$.

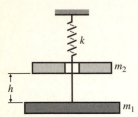

Figure P3.11

3.24 Determine the maximum displacement, velocity, and acceleration of the combined mass $(m_1 + m_2)$ described in Problem 3.23 if the weight of mass m_1 is 50 lb, the weight of mass m_2 is 10 lb, $h = 20$ in, and the spring stiffness $k = 250$ lb/in.

3.25 A mass m is attached to the tip of a cantilever beam of length L, also having mass m that is uniformly distributed over its entire length as shown in Figure P3.12. The flexural rigidity of the beam is EI. The static deflection curve for the beam is $y(x) = (A/2L^3)(3Lx^2 - x^3)$. Determine the natural frequency of the beam by the Rayleigh method.

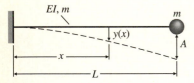

Figure P3.12

3.26 The fixed-ended beam shown in Figure P3.13 has flexural rigidity EI and a total mass m that is uniformly distributed over its length L. The static deflection curve for the beam is $y(x) = 16A(x/L)^2[1 - (x/L)]^2$. Determine the natural frequency of the beam by the Rayleigh method.

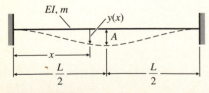

Figure P3.13

3.27 Repeat Problem 3.25 using a static deflection curve for the beam given by

$$y(x) = A[1 - \cos(\pi x / 2L)].$$

3.28 Repeat Problem 3.26 using a static deflection curve for the beam given by

$$y(x) = (4Ax^2/L^3)(3L - 4x), \quad x \le \frac{L}{2}.$$

3.29 A pin-ended column of length L is subject to an axial load P as shown in Figure P3.14. The mass of the column m is uniformly distributed along the length. By the Rayleigh method determine the natural frequency of the column and the maximum value of P for stability. Assume flexural rigidity EI and assume a static deflection curve $y = A \sin(\pi x/L)$.

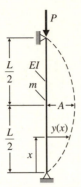

Figure P3.14

3.30 Repeat Problem 3.29 using a static deflection curve for the column given by

$$y(x) = \frac{A(x^4 - 4L^3 + 3L^4)}{3L^4}$$

3.31 A cylindrical buoy is weighted such that it has a low center of gravity as shown in Figure P3.15. If the buoy is displaced from its equilibrium position an amount x_0 and suddenly released, determine (a) the equation of motion for vertical oscillation of the buoy and (b) the natural frequency of oscillation. Assume the density of the fluid to be ρ and the diameter of the buoy to be D.

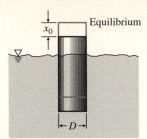

Figure P3.15

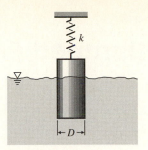

Figure P3.17

3.32 A spherical buoy shown in Figure P3.16 is in static equilibrium when half submerged in water. If the buoy is depressed and released at a vertical distance $x_0 = R/10$, derive (a) the equation of motion and (b) an expression for the natural frequency. If $R = 3.5$ ft and the density of the water is 62.4 lb/ft^3, determine the natural frequency and period of vibration.

3.35 A U-tube manometer open at both ends and containing a column of liquid mercury of length l and mass density ρ is shown in Figure P3.18. If the column of mercury is displaced a distance x_0 and suddenly released, determine the natural frequency of its oscillations by the Rayleigh method.

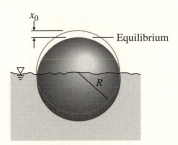

Figure P3.16

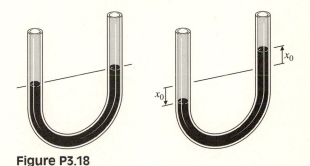

Figure P3.18

3.33 Determine the natural frequency and period of vibration for the cylindrical buoy described in Problem 3.31 if the weight of the buoy is 300 lb, $D = 3$ ft, and $\rho = 64$ lb/ft^3.

3.34 The cylindrical buoy described in Problems 3.31 and 3.33 is attached to a fixed structure by a linear spring $k = 200$ lb/in as shown in Figure P3.17. If the buoy is displaced from its equilibrium position 6 in and released, determine (a) the natural frequency of oscillations and (b) the amplitude of the oscillations.

4 ▲ Damped Free Vibration

In the previous discussion of undamped free vibration, the effects of energy dissipation forces, such as those due to friction or hysteresis, were not addressed. In such undamped systems, vibration would theoretically continue indefinitely, since they are void of any energy dissipation mechanisms. In reality, however, all structural and mechanical systems possess several inherent dissipative mechanisms of one form or another.

Damping is the process of energy dissipation in a vibrating system. Damping often manifests itself in most structural and mechanical systems as one, or a combination, of three forms: (1) *viscous damping,* (2) *Coulomb or dry-friction damping,* or (3) *hysteresis damping.* Generally, one of these three forms dominates, thus making a reasonable analysis possible in most cases. However, the viscous type of damping is the most conducive to mathematical formulation (since it is velocity dependent) and is therefore most frequently employed in vibration analysis. In fact, many systems for which the damping is known not to be viscous are often analyzed as systems with equivalent viscous damping. The equivalent viscous damping is determined in such a manner as to yield the same dissipation of energy per cycle as that produced by the actual damping mechanism. In this chapter, viscous damping is considered first, in detail, followed by a discussion of dry-friction and hysteresis damping.

4.1 FREE VIBRATION WITH VISCOUS DAMPING

Viscous damping is a commonly employed form of damping in vibration analysis in which the damping force is proportional to the velocity of the mass. The damping force always opposes the motion of the mass such that it is a continuous linear function of the velocity. Thus, the mass will be subjected to a damping force having a magnitude given by

$$F_D = c\dot{x} \tag{4.1}$$

where $\dot{x}$ is the velocity of the mass and c is the viscous damping coefficient.

Consider the viscously damped SDOF system shown in Figure 4.1a. The free-body diagram for the system, shown in Figure 4.1b, will aid in the derivation of the

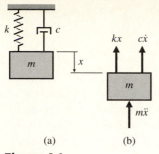

(a) **(b)**

Figure 4.1

Viscously damped
SDOF system.

differential equation of motion. Application of Newton's second law gives the differential equation of motion as

$$m\ddot{x} + c\dot{x} + kx = 0 \tag{4.2}$$

or

$$\ddot{x} + \frac{c}{m}\dot{x} + \frac{k}{m}x = 0 \tag{4.3}$$

Applying the solution procedures developed in Chapter 3 to Eq. (4.3), a solution for displacement of the mass in the form $x(t) = Ce^{st}$ is assumed to obtain the auxiliary equation

$$s^2 + \frac{c}{m}s + \frac{k}{m} = 0 \tag{4.4}$$

which has the roots

$$s_{1,2} = \frac{1}{2}\left[-\frac{c}{m} \pm \sqrt{\left(\frac{c}{m}\right)^2 - 4\frac{k}{m}}\right] \tag{4.5}$$

or

$$s_{1,2} = -\frac{c}{2m} \pm \sqrt{\left(\frac{c}{2m}\right)^2 - \frac{k}{m}} \tag{4.6}$$

Depending on whether the quantity within the radical, $(c/2m)^2 - k/m$ is zero, positive, or negative, the solution may take one of three forms. If this quantity is zero, then Eq. (4.6) yields

$$\frac{c}{2m} = \sqrt{\frac{k}{m}} = \omega \tag{4.7}$$

or

$$c = 2m\omega \tag{4.8}$$

which results in the repeated roots $s_1 = s_2 = -c/2m$, and the solution to Eq. (4.3) defining the displacement of the mass is

$$x(t) = (A + Bt)e^{-(c/2m)t} \tag{4.9}$$

Since the occurrence of repeated roots presents a case having special significance, the corresponding value of the damping coefficient shall be referred to as the *critical damping constant,* denoted by $C_c = 2m\omega$. Utilizing the notation for the critical damping constant allows the roots given by Eq. (4.6) to be rewritten as

$$s_{1,2} = \frac{c}{C_c}\omega \pm \omega\sqrt{\left(\frac{c}{C_c}\right)^2 - 1} \tag{4.10}$$

or

$$s_{1,2} = (-\zeta \pm \sqrt{\zeta^2 - 1})\omega \tag{4.11}$$

where $\omega = (k/m)^{1/2}$ is the natural circular frequency of the corresponding undamped system and

$$\zeta = \frac{c}{C_c} = \frac{c}{2m\omega} \tag{4.12}$$

is the *damping factor,* or the fraction of the critical damping. The damping factor is the ratio of the actual damping present in a system to the critical damping constant for the system. Equation (4.11) indicates that the nature of the roots of the auxiliary equation depends on whether the value of the damping factor ζ is less than, equal to, or greater than unity.

4.1.1 Case 1: Less than Critical Damping (Underdamped)

If damping in the system is less than critical, $\zeta < 1$ (or $c < 2m\omega$), then the roots of the auxiliary equation represented by Eq. (4.4) are both imaginary and are given by

$$s_{1,2} = \left(-\zeta \pm i\sqrt{1 - \zeta^2} \right)\omega \tag{4.13}$$

then the solution to Eq. (4.3) for the displacement of the mass becomes

$$x(t) = e^{-\zeta\omega t}\left(A \sin\sqrt{1 - \zeta^2}\,\omega t + B \cos\sqrt{1 - \zeta^2}\,\omega t \right) \tag{4.14}$$

The solution represented by Eq. (4.14) is more conveniently written and interpreted in the form

$$x(t) = Xe^{-\zeta\omega t}\sin(\omega_d t + \phi) \tag{4.15}$$

where X is the maximum free vibration amplitude evaluated from the initial conditions and

$$\omega_d = \sqrt{1 - \zeta^2}\,\omega \tag{4.16}$$

is the damped natural circular frequency; ϕ is the phase angle of the damped oscillations, also determined from the initial conditions. Equation (4.16), relating the damped and undamped natural circular frequencies, is plotted in Figure 4.2. The displacement of the mass represented by Eq. (4.14) or Eq. (4.15) is a harmonic function having an

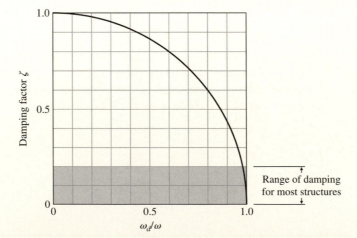

Figure 4.2

Damped natural frequency as a function of undamped natural frequency and damping factor.

Figure 4.3

Free vibration response of underdamped system ($\zeta < 1.0$).

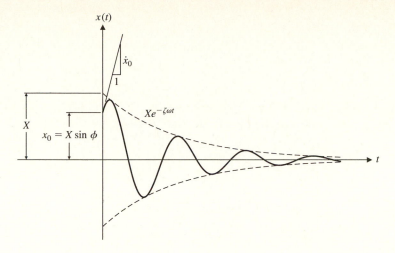

amplitude X that decays exponentially with time. With initial conditions of displacement $x(0) = x_0$ and velocity $\dot{x}(0) = \dot{x}_0$ specified, the constants A and B in Eq. (4.14) can be evaluated, and Eq. (4.14) becomes

$$x(t) = e^{-\zeta\omega t}\left[\frac{\dot{x}_0 + \zeta\omega x_0}{\omega_d}\sin\omega_d t + x_0\cos\omega_d t\right] \qquad (4.17)$$

which is also equivalent to Eq. (4.15). The general form of the free vibration motion is illustrated in Figure 4.3. Motion of this type is considered to be *underdamped,* and the damping is less than critical.

4.1.2   Case 2: Critical Damping

When the damping coefficient is equal to the critical damping constant, $\zeta = 1$ (or $c = C_c = 2m\omega$) and the system is said to be *critically damped.* The displacement of the mass is represented by Eq. (4.9), which can be expressed as

$$x(t) = (A + Bt)e^{-\omega t} \qquad (4.18)$$

The solution represented by Eq. (4.18) is the product of a linear function of time and a decaying exponential. Depending on the values of A and $B,$ many forms of motion are possible. The general solution for nonzero initial conditions of displacement $x(0)$ and velocity $\dot{x}(0)$ can be established from Eq. (4.17) by letting $\zeta \to 1$, resulting in

$$x(t) = e^{-\omega t}\{[\dot{x}(0) + \omega x(0)]t + x(0)\} \qquad (4.19)$$

Each form of the displacement solution represented by Eq. (4.19) is characterized by an amplitude that decays without oscillations. Three types of critically damped responses with an initial displacement $x(0) = x_0$ and several different values for initial velocity $\dot{x}(0)$ are shown in Figure 4.4.

4.1.3   Case 3 : Greater than Critical Damping (Overdamped)

Finally, for the case $\zeta > 1$ (or $c > 2m\omega$), the system is said to be *overdamped.* The roots of the auxiliary equation represented by Eq. (4.4) are both real and are given by

$$s_{1,2} = \left(-\zeta \pm i\sqrt{\zeta^2 - 1}\right)\omega \qquad (4.20)$$

Figure 4.4

Free vibration response for system with critical damping ($\zeta = 1.0$). *Theory of Vibration with Application*, 5/e, by Thompson/Dahleh, Copyright © 1993. Reprinted by permission of Prentice-Hall, Inc., Upper Saddle River, NJ.

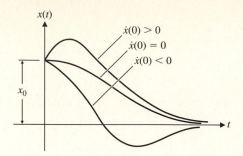

Since $\zeta > 1$, then $\sqrt{\zeta^2 - 1} < \zeta$ and the roots s_1 and s_2 of Eq. (4.20) are both negative. Therefore, the displacement solution of the mass is the sum of two decaying exponentials given by

$$x(t) = A e^{(-\zeta + \sqrt{\zeta^2 - 1})\omega t} + B e^{(-\zeta - \sqrt{\zeta^2 - 1})\omega t} \tag{4.21}$$

where the constants A and B, evaluated for initial conditions $x(0) = x_0$ and $\dot{x}(0) = \dot{x}_0$, are expressed as

$$A = \frac{\dot{x}_0 + \left(\zeta + \sqrt{\zeta^2 - 1}\right)\omega x_0}{2\omega\sqrt{\zeta^2 - 1}} \tag{4.22}$$

and

$$B = \frac{-\dot{x}_0 - \left(\zeta - \sqrt{\zeta^2 - 1}\right)\omega x_0}{2\omega\sqrt{\zeta^2 - 1}} \tag{4.23}$$

The resulting displacement of the mass is represented by an exponentially decreasing function of time, as illustrated in Figure 4.5, and is referred to as *nonoscillatory,* or *aperiodic.*

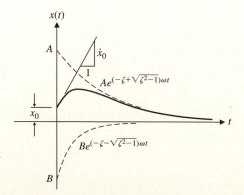

Figure 4.5

Free vibration response of overdamped system ($\zeta > 1.0$).

EXAMPLE 4.1 ▼

The damped spring-mass system shown in Figure 4.1 has a mass $m = 0.0052$ lb-sec^2/in, a stiffness $k = 12$ lb/in, and a damping coefficient $c = 0.1$ lb-sec/in. If the mass is subject to an initial displacement of $x(0) = 1.5$ in and suddenly released with zero initial velocity, determine the amplitude of free vibration of the mass after 5, 10, and 20 oscillations. Evaluate the response using Eq. (4.15).

Solution

The undamped natural circular frequency ω is calculated as

$$\omega = \sqrt{\frac{k}{m}} = \sqrt{\frac{12 \text{ lb/in}}{0.0052 \text{ lb-sec/in}}} = 48 \text{ rad/sec}$$

and the critical damping constant C_c for this system is determined by

$$C_c = 2\sqrt{mk} = 2\sqrt{(0.0052 \text{ lb-sec/in})(12 \text{ lb/in})} = 0.50 \text{ lb-sec/in}$$

The system damping factor ζ is then calculated from Eq. (4.12) as

$$\zeta = \frac{c}{C_c} = \frac{0.1}{0.5} = 0.2$$

and the damped natural circular frequency ω_d is evaluated from Eq. (4.16), yielding

$$\omega_d = \omega\sqrt{1 - \zeta^2} = 47.0 \text{ rad/sec}$$

The damped natural period for the system can then be determined from the expression

$$T_d = \frac{2\pi}{\omega_d} = 0.1337 \text{ sec}$$

Applying the initial condition for the displacement of $x(0) = 1.5$ in to Eq. (4.15) yields

$$x(0) = X \sin\phi = 1.5 \tag{1}$$

and then applying the initial condition for the velocity of $\dot{x}(0) = 0$ to the first time derivative of Eq. (4.15) gives

$$\dot{x}(0) = \omega_d X \cos\phi - \zeta\omega X \sin\phi = 0 \tag{2}$$

Solving Eqs. (1) and (2) simultaneously results in $X = 1.531$ in and $\phi = 1.369$ rad. Therefore, after n oscillations or at time $t = nT_d$, where T_d is the damped natural period, the vibration amplitude X_n can be expressed as

$$X_n(t = nT_d) = 1.5e^{-\zeta\omega nT_d} \tag{3}$$

or

$$X_n(t = nT_d) = 1.5e^{-\zeta\omega(2\pi n\zeta/\sqrt{1-\zeta^2}\omega)} = 1.5e^{-2\pi n\zeta/\sqrt{1-\zeta^2}} \tag{4}$$

Therefore, from Eq. (4), after 5 oscillations ($n = 5$), $X_5 = 2.5 \times 10^{-3}$ in, after 10 oscillations ($n = 10$), $X_{10} = 4.12 \times 10^{-6}$ in, and after 20 oscillations ($n=20$), $X_{20} = 1.1 \times 10^{-11}$ in. ▲

Most structural and mechanical systems exhibit damping factors ranging from approximately 0.01 to an extreme maximum of about 0.2. At these levels, the difference between the undamped and damped natural frequencies (or periods) is not significant, as illustrated in Figure 4.2 and by the previous example. However, only very small amounts of damping are necessary to quickly dissipate the free vibration oscillations of the same systems, as was clearly illustrated by this example. Figure 4.6 illustrates the general effect of damping on the amplitude and natural period of free vibration of an underdamped system.

Figure 4.6

General effect of damping on free vibration response.

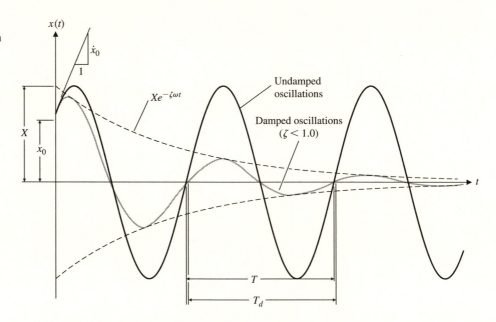

EXAMPLE 4.2 ▼

A weight of 10 lb is suspended from a spring having a stiffness of 30 lb/in, and the system is critically damped. The weight is given an initial displacement of 2 in from its equilibrium position, held there momentarily, and suddenly released with zero initial velocity. Determine the damping coefficient c for the system and the displacement of the mass $x(t)$ from its equilibrium position after 0.01, 0.1, and 1.0 sec.

Solution

(a) For a critically damped system, note that $c = C_c = 2m\omega$. The mass of the system is

$$m = \frac{W}{g} = \frac{10 \text{ lb}}{(386.4 \text{ in/sec}^2)} = 0.0259 \text{ lb-sec}^2/\text{in}$$

and the undamped natural circular frequency is

$$\omega = \sqrt{\frac{k}{m}} = \sqrt{\frac{30}{0.0259}} = 34.03 \text{ rad/sec}$$

Therefore the system damping coefficient is

$$c = C_c = 2m\omega = 2(0.0259 \text{ lb-sec}^2/\text{in})(34.03 \text{ rad/sec})$$
$$= 1.763 \text{ lb-sec/in}$$

(b) For $\zeta = 1$, the solution for $x(t)$ determined from Eq. (4.18) is

$$x(t) = (A + Bt)e^{-\omega t} = Ae^{-\omega t} + Bte^{-\omega t} \tag{1}$$

and the velocity $\dot{x}(t)$ as determined from the first time derivative of Eq. (4.15) is

$$\dot{x}(t) = -A\omega e^{-\omega t} + Be^{-\omega t} - B\omega e^{-\omega t} \tag{2}$$

Applying the initial condition for displacement of $x(0) = 2$ to Eq. (1) yields

$$A = 2 \text{ in}$$

Applying the initial condition for velocity of $\dot{x}(0) = 0$ to Eq. (2) yields

$$0 = -(2)(34.03) + B - 0 \tag{3}$$

and therefore from Eq. (3),

$$B = 68.06 \text{ in/sec}$$

Substituting the values for A and B into Eq. (1) yields the displacement solution given by

$$x(t) = (2 + 68.06 \, t)e^{-34.06t} \tag{4}$$

Then, from Eq. (4), the displacement of the mass at the specified times is determined as

$$x(0.01) = 1.9 \text{ in}$$
$$x(0.1) = 0.28 \text{ in}$$
$$x(1.0) = 6.4 \times 10^{-14} \text{ in}$$

▲

4.2 LOGARITHMIC DECREMENT

Because it is not possible to analytically ascertain the amount of damping in most structural and mechanical systems, it must be determined experimentally. The degree of damping in an underdamped system ($\zeta < 1$) may be defined in terms of the rate of decay of the free oscillations. The larger the amount of damping in the system, the greater the rate of decay.

The *logarithmic decrement* δ is the natural logarithm of the ratio of any two successive displacement amplitudes in the same direction. If a damped vibration displacement solution, expressed by Eq. (4.15) and illustrated graphically in Figure 4.7, is considered, then the logarithmic decrement of two successive amplitudes X_1 and X_2 is given as

$$\delta = \ln\left(\frac{X_1}{X_2}\right) = \ln\left[\frac{e^{-\zeta\omega t_1}\sin(\omega t_1 + \phi)}{e^{-\zeta\omega(t_1+T_d)}\sin[\omega_d(t_1 + T_d) + \phi]}\right] \tag{4.24}$$

where t_1 is the time corresponding to X_1. Moreover, since the values of the sine terms appearing in the right-hand side of the equation are equal when the time is increased by the damped period T_d, Eq. (4.24) reduces to

$$\delta = \ln\frac{e^{-\zeta\omega t_1}}{e^{-\zeta\omega(t_1+T_d)}} = \ln e^{\zeta\omega T_d} = \zeta\omega T_d \tag{4.25}$$

Since the damped natural period may be expressed as

$$T_d = \frac{2\pi}{\omega_d} = \frac{2\pi}{\omega\sqrt{1 - \zeta^2}} \tag{4.26}$$

then by substituting Eq. (4.26) for T_d into Eq. (4.25), the expression for logarithmic decrement becomes

$$\delta = \frac{2\pi\zeta}{\sqrt{1 - \zeta^2}} \tag{4.27}$$

which is an exact equation.

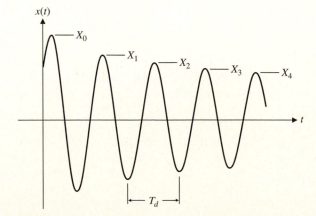

Figure 4.7

Rate of decay of damped oscillations measured by logarithmic decrement.

For small values of ζ, that is, when $\sqrt{1 - \zeta^2} \cong 1$ (which would be true of most physical systems having a $\zeta \leq 0.2$), an approximation for the logarithmic decrement is given as

$$\delta = 2\pi\zeta \tag{4.28}$$

which may be used in most practical applications. An illustration of a comparison between the exact and approximate values of δ, given by Eqs. (4.27) and (4.28), respectively, as a function of ζ is presented in Figure 4.8. This figure clearly illustrates that there is virtually no difference in δ calculated by the two expressions for $\zeta \leq 0.2$.

It is also possible to obtain the logarithmic decrement from two nonsuccessive cycles of damped oscillations. Consider the ratio of two nonconsecutive amplitudes X_i and X_{i+n}, where n is any integer. The logarithmic decrement for this situation is expressed as

$$\ln \frac{X_i}{X_{i+n}} = \ln \frac{e^{-\zeta\omega t_i}}{e^{-\zeta\omega(t_i + nT_d)}} = n\zeta\omega T_d = n\delta \tag{4.29}$$

where t_1 is the time corresponding to X_i. This expression is particularly useful for lightly damped systems where successive oscillation peaks would have very similar ordinates. Equation (4.29) would provide better results in such instances by considering two peaks that are several cycles apart.

The useful purpose of the logarithmic decrement is to estimate the amount of damping in the system in a practical manner. Therefore, to this end, with δ having been determined by use of either Eq. (4.25) or Eq. (4.29) as appropriate, Eq. (4.27) may be used to estimate the damping factor ζ by

$$\zeta = \frac{\delta}{\sqrt{(2\pi)^2 + \delta^2}} \tag{4.30}$$

Figure 4.8

Logarithmic decrement as a function of damping factor. *Theory of Vibration with Application*, 2/e, by Thompson, Copyright © 1981. Reprinted by permission of Prentice-Hall, Inc., Upper Saddle River, NJ.

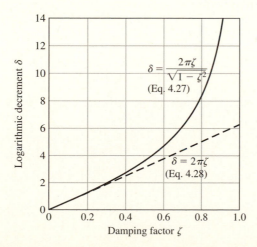

EXAMPLE 4.3 ▼

For a viscously damped system, a particular vibration displacement amplitude in a free vibration trace is measured as 75% of the immediately preceding amplitude. Determine the damping factor for the system.

Solution

The ratio of two successive vibration amplitudes X_i and X_{i+1} yields

$$\frac{X_i}{X_{i+1}} = \frac{1}{0.75} = 1.333$$

The logarithmic decrement can then be determined from Eq. (4.25) as

$$\delta = \ln\left(\frac{X_i}{X_{i+1}}\right) = \ln(1.333) = 0.288$$

and then from Eq. (4.30) the damping factor is established by

$$\zeta = \frac{\delta}{\sqrt{(2\pi)^2 + \delta^2}} = 0.046$$

▲

EXAMPLE 4.4 ▼

For a certain viscously damped system, measurements of a free vibration trace show that the vibration amplitude reduction is 80% in 15 cycles. The critical damping constant C_c for the system is known to be 60 N-sec/m. Determine the damping factor ζ and the damping coefficient c for the system.

Solution

From Eq. (4.29),

$$n\delta = \ln\frac{X_i}{X_{i+n}} = \ln\frac{X_i}{0.2X_i} = 1.609$$

from which the logarithmic decrement for $n = 15$ is established by

$$\delta = \frac{1}{n}(1.609) = \frac{1.609}{15} = 0.1073$$

Applying Eq. (4.30) yields the damping factor

$$\zeta = \frac{\delta}{\sqrt{(2\pi)^2 + \delta^2}} = 0.017$$

and the system damping coefficient is determined from Eq. (4.12) as

$$c = \zeta C_c = 0.017(60 \text{ N-sec/m}) = 1.02 \text{ N-sec/m}$$

▲

EXAMPLE 4.5 ▼

A bridge girder is deflected at midspan (by winching the bridge down with a crane) and suddenly released. After the initial disturbance, the oscillations were found to decay exponentially from an amplitude of 4 in to 2.7 in after 5 cycles of free vibration. Determine the damping factor for the bridge girder.

Solution

Implementing Eq. (4.29) yields

$$n\delta = \ln \frac{X_i}{X_{i+n}} = \ln\left(\frac{4}{2.7}\right) = 0.393$$

from which the logarithmic decrement is calculated for $n = 5$ as

$$\delta = \frac{1}{n}(0.393) = \frac{0.393}{5} = 0.0786$$

The damping factor for the bridge girder can then be determined from Eq. (4.30) as

$$\zeta = \frac{\delta}{\sqrt{(2\pi)^2 + \delta^2}} = 0.0125$$

▲

EXAMPLE 4.6 ▼

To determine the dynamic characteristics of a large-scale model concrete dolos armor unit, a drop test is conducted as illustrated in Figure 4.9. In the drop test the vertical fluke of the unit is raised a specific height h from its seated equilibrium and suddenly released. The subsequent impact of the bottom of the fluke with the concrete base induces vibrations in the dolos. The dolos armor unit is instrumented with strain gages at various locations on its surface. A typical strain gage trace of the vibration oscillations after the unit impacts the concrete base is presented in Figure 4.10. If the concrete armor unit weighs approximately 60 lb, determine (a) the natural damped period of vibration, (b) the natural damped frequency of vibration, and (c) the damping factor and damping coefficient.

Figure 4.9

Schematic of dolos drop test.

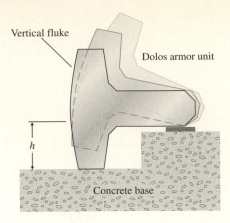

Vertical fluke

Dolos armor unit

h

Concrete base

Solution

(a) From Figure 4.10, the damped period T_d is estimated as

$$T_d \cong \frac{0.005}{4} = 0.00125 \text{ sec}$$

(b) Therefore, the damped natural frequency is given by

$$f_d = \frac{1}{T_d} = 800 \text{ Hz}$$

and the damped natural circular frequency is

$$\omega_d = 2\pi f_d = 5026 \text{ rad/sec}$$

(c) From Figure 4.10, the vibration amplitudes after 7 and 16 cycles are estimated as

$$X_7 = 40 \quad \text{and} \quad X_{16} = 20$$

and from Eq. (4.29), $n\delta$ is estimated as

$$n\delta = \ln\left(\frac{40}{20}\right) = 0.693$$

Then from Eq. (4.28) the logarithmic decrement (noting that $n = 9$) is approximated as

$$\delta = 0.077 \cong 2\pi\zeta$$

from which the damping factor is calculated as $\zeta = 0.0123$. The critical damping constant is given by

$$C_c = 2m\omega \tag{1}$$

where the undamped natural circular frequency ω is expressed as

$$\omega = \frac{\omega_d}{\sqrt{1 - \zeta^2}} = \frac{5026 \text{ rad/sec}}{\sqrt{1 - (0.0123)^2}} = 5026.4 \text{ rad/sec}$$

Figure 4.10

Strain gage trace recorded on surface of dolos after impact.

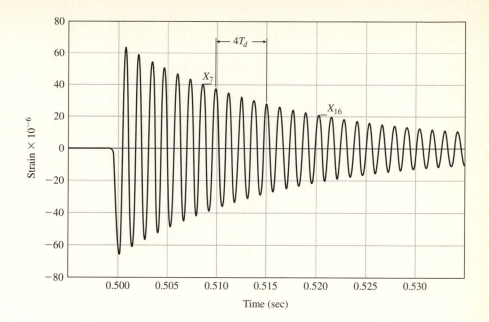

Then the critical damping constant is determined from Eq. (1) as

$$C_c = 2\left(\frac{60 \text{ lb}}{386.4 \text{ in/sec}^2}\right)(5026.4 \text{ rad/sec}) = 1561 \text{ lb-sec/in}$$

and the system damping coefficient c is determined from Eq. (4.12) as

$$c = \zeta C_c = 19.2 \text{ lb-sec/in}$$

▲

4.3   HYSTERESIS DAMPING

In reality, most structural and mechanical systems do not exhibit the highly idealized form of viscous damping considered thus far. When materials are cyclically stressed, energy is dissipated within the material itself due primarily to internal friction caused by the slipping and sliding of particles at internal planes during deformation. Such internal damping is generally referred to as *hysteresis damping,* or *structural damping.* This form of damping results in a phase lag between the damping force and deformation as illustrated in Figure 4.11. This curve is generally referred to as a *hysteresis loop,* thus spawning the term hysteresis damping. The area enclosed within the loop represents the energy loss or dissipated energy per loading cycle.

If ΔU represents the energy loss per cycle, as illustrated in Figure 4.11, then

$$\Delta U = \int F_D \, dx \qquad (4.31)$$

Experiments [1, 2] conducted on the internal damping that occurs in solid materials and structures subjected to cyclic stressing have shown that the energy dissipated per cycle is independent of frequency and proportional to the square of the amplitude of vibration. Thus, the energy loss per cycle may be expressed as

$$\Delta U = \pi \eta k X^2 \tag{4.32}$$

where η = a dimensionless structural damping coefficient for the material
k = the equivalent stiffness of the system
X = the displacement amplitude
π = a convenient proportionality constant

Because hysteresis damping is defined in terms of energy loss per cycle and is a nonlinear function of displacement amplitude, it does not readily lend itself to analytical solution. However, a viscous-type damping formulation is quite conducive to mathematical analysis. Therefore, for convenience of analysis, hysteresis damping can be defined as an equivalent viscous damping such that the energy loss per cycle is equal to that predicted by Eq. (4.32). The energy dissipated per cycle by the viscous damping force $c\dot{x}$ is given by

$$\Delta U = 4 \int_0^X c\dot{x} \; dx \tag{4.33}$$

assuming harmonic motion and $x(t) = X \sin \omega t$ for the complete cycle. Noting that

$$\dot{x} = \omega X \cos \omega t \tag{4.34}$$

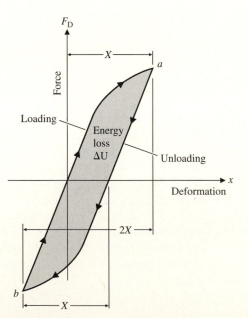

Figure 4.11
Hysteresis loop.

then substituting Eq. (4.34) into Eq. (4.33) results in

$$\Delta U = 4 \int_0^{\pi/2} cX^2 \omega^2 \cos\omega t \, dt = \pi c \omega X^2 \tag{4.35}$$

An equivalent viscous damping coefficient c_e can be determined by equating the expressions for ΔU, given by Eqs. (4.30) and (4.35), as follows:

$$\pi c_e \omega X^2 = \pi \eta k X^2 \tag{4.36}$$

Solving Eq. (4.36) for c_e yields the expression

$$c_e = \frac{\eta k}{\omega} \tag{4.37}$$

For linear vibrations, ΔU is relatively small and the motion is approximately harmonic. Thus the *structural damping coefficient* η can be evaluated experimentally by determining the logarithmic decrement in a procedure similar to that described in Section 4.2. Reference to the oscillations illustrated in Figure 4.12 shows that the energy equation for a half cycle of vibration from point a to point b is

$$\frac{kX_1^2}{2} - \frac{\pi k \eta X_1^2}{4} - \frac{\pi k \eta X_{1.5}^2}{4} = \frac{kX_{1.5}^2}{4} \tag{4.38}$$

which can be reduced to

$$\frac{X_1^2}{X_{1.5}^2} = \frac{1 + (\pi\eta/2)}{1 - (\pi\eta/2)} \tag{4.39}$$

Similar consideration for the next half cycle of motion from point b to point c yields

$$\frac{X_{1.5}^2}{X_2^2} = \frac{1 + (\pi\eta/2)}{1 - (\pi\eta/2)} \tag{4.40}$$

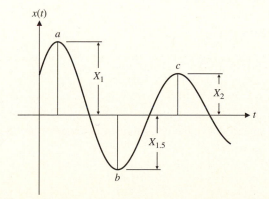

Figure 4.12

Free vibration motion for system with structural damping.

Multiplying Eqs. (4.39) and (4.40) results in

$$\frac{X_1^2}{X_2^2} = \left[\frac{1 + (\pi\eta/2)}{1 - (\pi\eta/2)}\right]^2 \tag{4.41}$$

or

$$\frac{X_1}{X_2} = \frac{1 + (\pi\eta/2)}{1 - (\pi\eta/2)} \tag{4.42}$$

For very small values of η, Eq. (4.42) can be approximated by

$$\frac{X_1}{X_2} \cong 1 + \pi\eta \tag{4.43}$$

The logarithmic decrement is then given as

$$\delta = \ln\left(\frac{X_1}{X_2}\right) = \ln(1 + \pi\eta) \cong \pi\eta \tag{4.44}$$

Therefore, for a structure considered to exhibit hysteretic or structural damping characteristics, the coefficient η can be determined by measuring successive amplitudes of the oscillation and then applying either Eq. (4.43) or Eq. (4.44). Then the structure can be analyzed as an equivalent viscously damped system by calculating the equivalent viscous damping coefficient from Eq. (4.37).

EXAMPLE 4.7 ▼

The main span of a bridge structure is considered as a SDOF system for calculation of its fundamental frequency. From preliminary vibration tests, the effective mass of the structure was determined to be 375×10^3 kg and the effective stiffness to be 38,850 kN/m. The ratio of successive displacement amplitudes from a free vibration trace was measured to be 1.05. Calculate the values of the structural damping coefficient η and the equivalent viscous damping coefficient c_e.

Solution

The ratio of successive amplitudes is defined as

$$\frac{X_1}{X_2} = 1.05$$

From Eq. (4.44), the logarithmic decrement is expressed as

$$\delta = \ln\left(\frac{X_1}{X_2}\right) = 0.049 \cong \pi\eta \tag{1}$$

Therefore, the structural damping coefficient is calculated from Eq. (1) as

$$\eta = \frac{0.049}{\pi} = 0.0156$$

The equivalent viscous damping coefficient determined from Eq. (4.37) is

$$c_e = \frac{\eta k}{\omega} \qquad (2)$$

where

$$\omega = \sqrt{\frac{k}{m}} = \sqrt{\frac{38,850 \times 10^3 \text{N/m}}{375,000 \text{ kg}}} = 10.18 \text{ rad/sec}$$

Thus, from Eq. (2) the equivalent viscous damping coefficient is

$$c_e = \frac{0.0156(38,850)}{10.18} = 59.5 \text{ kN-sec/m}$$

▲

EXAMPLE 4.8 ▼

A simply supported W24 × 55 steel beam having a span of 40 ft supports a concentrated weight at its midspan of $W = 20$ kips. Free oscillations of the system exhibit an amplitude decay of 0.75% per cycle. If the mass of the beam is negligible with respect to the concentrated weight, determine the structural damping coefficient η and the equivalent viscous damping coefficient c_e. Assume a Young's modulus $E = 29 \times 10^3$ ksi for the steel beam.

Solution

The ratio of successive amplitudes is expressed as

$$\frac{X}{(1 - 0.0075)X} = 1.00756$$

From Eq. (4.43),

$$1 + \pi\eta = 1.00756$$

from which the structural damping coefficient is

$$\eta = \frac{0.00756}{\pi} = 0.0024$$

The calculated natural circular frequency of the system ω is determined as

$$\omega = \sqrt{\frac{k}{m}}$$

where

$$k = \frac{48EI}{L^3} = \frac{48(29,000 \text{ ksi})(1350 \text{ in}^4)}{(480 \text{ in})^3} = 17 \text{ kips/in}$$

and the mass m is

$$m = \frac{W}{g} = \frac{20,000 \text{ lb}}{386.4 \text{ in/sec}^2} = 51.76 \text{ lb-sec}^2/\text{in}$$

Thus

$$\omega = \sqrt{\frac{17 \times 10^3 \text{ lb/in}}{51.76 \text{ lb-sec}^2/\text{in}}} = 18.12 \text{ rad/sec}$$

and from Eq. (4.37), the equivalent viscous damping coefficient is

$$c_e = \frac{0.0024(17 \times 10^3 \text{ lb/in})}{18.12 \text{ rad/sec}} = 2.25 \text{ lb-sec/in}$$

▲

4.4 COULOMB DAMPING

In many structures, damping occurs when relative motion takes place at interfaces or joints between adjacent members. This form of damping is referred to as *Coulomb,* or *dry-friction, damping.* During free vibration, friction forces develop that are independent of vibration amplitude and frequency. These forces always oppose the motion of the mass, and their magnitude is essentially constant.

Coulomb damping in a SDOF system is illustrated in Figure 4.13. The friction damping force F_d is constant and is given by

$$F_d = \mu N \tag{4.45}$$

where μ is the coefficient of friction and N is the normal force on the contact surface. The friction force always acts in a direction opposite to that of the velocity or motion of the body, as in the case of a dashpot damper. However, the frictional resistance is assumed to be constant regardless of the velocity as shown in Figure 4.13.

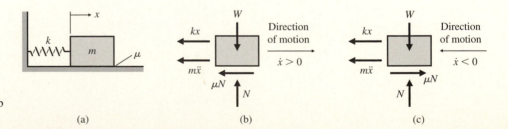

Figure 4.13
SDOF system with Coulomb damping.

(a) (b) (c)

Because of the directional nature of the friction force, the motion of the mass is not a continuous function, and the system is said to exhibit *piecewise-linear* characteristics. That is, the system exhibits linear characteristics for motion in a particular direction. For clarity of solution, two differential equations of motion are required to describe the motion, and are given as (referring to Figure 4.13)

$$m\ddot{x} + kx + F_d = 0 \qquad (4.46)$$

for motion from left to right ($\dot{x} > 0$) and

$$m\ddot{x} + kx - F_d = 0 \qquad (4.47)$$

for motion from right to left ($\dot{x} < 0$). Equations (4.46) and (4.47) may be combined and more succinctly written as

$$\ddot{x} + \frac{k}{m}x = \pm\frac{F_d}{m} \qquad (4.48)$$

Due to the presence of the constant term F_d/m on the right-hand side of Eq. (4.48), they are nonhomogeneous. Therefore, their solutions consist of both a complementary and a particular part. The complementary solution is very similar to that for an undamped system undergoing free vibration, and the particular solution is merely a constant term that is obtained by direct substitution into Eq. (4.48). Thus, the solutions are

$$x(t) = A_1 \sin \omega t + B_1 \cos \omega t - \frac{F_d}{k} \qquad (4.49)$$

for motion from left to right, and

$$x(t) = A_2 \sin \omega t + B_2 \cos \omega t + \frac{F_d}{k} \qquad (4.50)$$

for motion from right to left. The constants A_1 and B_1 in Eq. (4.49) are evaluated from the initial conditions for motion to the right, and the constants A_2 and B_2 in Eq. (4.50) are evaluated from the initial conditions for motion to the left.

For the case at hand, consider an initial displacement to the right, or $x(0) = x_0$, with no initial velocity $\dot{x}(0) = 0$. The initial motion is therefore from right to left. Thus, applying the prescribed initial condition of displacement to Eq. (4.50) and the initial condition of velocity to the first time derivative of Eq. (4.50) results in values for $A_2 = 0$ and $B_2 = x_0 - (F_d/k)$, and the solution for displacement of the mass from right to left is given by

$$x(t) = \left(x_0 - \frac{F_d}{k}\right)\cos \omega t + \frac{F_d}{k} \qquad (4.51)$$

This solution is valid only until motion to the left ceases, or when the velocity is equal to zero. This time is established by setting the first time derivative of Eq. (4.51) equal to zero and solving for time t as follows:

$$\dot{x}(t) = -\omega\left(x_0 - \frac{F_d}{k}\right)\sin \omega t = 0 \qquad (4.52)$$

from which it is found that $t = \pi/\omega$, which corresponds to one-half the natural period, or $T/2$. The displacement at this time as determined by Eq. (4.51) is

$$x\left(\frac{\pi}{\omega}\right) = -\left(x_0 - \frac{2F_d}{k}\right) \tag{4.53}$$

which indicates that the maximum displacement amplitude for motion to the left is an amount $2F_d/k$ less than the initial displacement amplitude to the right x_0.

The next half cycle of motion to the right is described by Eq. (4.49). The constants A_1 and B_1 in Eq. (4.49) must be evaluated from the conditions of displacement and velocity at time $t = \pi/\omega$. Hence,

$$x\left(\frac{\pi}{\omega}\right) = -\left(x_0 - \frac{2F_d}{k}\right) = -B_1 - \frac{F_d}{k} \tag{4.54}$$

from which $B_1 = x_0 - (3F_d/k)$. The velocity at time $t = \pi/\omega$ is evaluated from the first time derivative of Eq. (4.49). This gives

$$\dot{x}\left(\frac{\pi}{\omega}\right) = 0 = -\omega_1 A_1 \tag{4.55}$$

from which $A_1 = 0$. Equation (4.49) then becomes

$$x(t) = \left(x_0 - \frac{3F_d}{k}\right)\cos\omega t - \frac{F_d}{k} \tag{4.56}$$

Equation (4.56) is valid until the velocity once again equals zero. To find the time at which this occurs, the first time derivative of Eq. (4.56) is set equal to zero. Thus,

$$\dot{x}(t) = -\omega\left(x_0 - \frac{3F_d}{k}\right)\sin\omega t = 0 \tag{4.57}$$

and $t = 2\pi/\omega$, which is equal to the natural period T. The corresponding displacement is determined from Eq. (4.56) as

$$x\left(\frac{2\pi}{\omega}\right) = x_0 - \frac{4F_d}{k} \tag{4.58}$$

Therefore, in each half cycle of motion the amplitude loss is $2F_d/k$, and in each full cycle of motion the amplitude loss is $4F_d/k$, as illustrated in Figure 4.14.

Some very important observations can be made from the motion demonstrated in Figure 4.14. First, it is noticed that the decay of the oscillations is linear. That is, the curve connecting the successive peak displacement amplitudes is a straight line. A second observation is that the magnitude of the natural period (frequency) is unaffected by Coulomb damping. One other point of interest is that the mass need not return to its original rest equilibrium position once the vibrations have ceased. Motion will end at some amplitude X_s when the restoring force in the spring kX_s cannot overcome the frictional damping force F_d.

Figure 4.14

Free vibration motion of
system with Coulomb
damping.

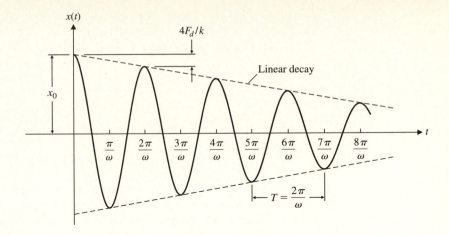

EXAMPLE 4.9 ▼

For the system shown in Figure 4.13, $W = 1$ kip, $k = 2$ kips/in, $\mu = 0.15$, and the
initial conditions are $x_0 = 6$ in, for displacement, and $\dot{x}_0 = 0$, for velocity. Determine the vibration displacement amplitude after 4 cycles and the number of cycles
of motion completed before the mass comes to rest.

Solution

The Coulomb friction force is

$$F_d = \mu N = 0.15(1000 \text{ lb}) = 150 \text{ lb}$$

and the natural circular frequency is

$$\omega = \sqrt{\frac{k}{m}} = \sqrt{\frac{2000 \text{ kips/in}}{\dfrac{1000 \text{ lb}}{386.4 \text{ in/sec}^2}}} = 27.8 \text{ rad/sec}$$

Then the natural period is

$$T = \frac{2\pi}{\omega} = 0.226 \text{ sec}$$

At the end of n cycles, the amplitude X_n is expressed by Eq. (4.58) as

$$X_n = x_0 - 4n\frac{F_d}{k} \tag{1}$$

Therefore, after 4 cycles of motion Eq. (1) yields

$$X_4 = 6 - 4\frac{(4)(150)}{2000} = 4.8 \text{ in}$$

Motion will cease when the amplitude of the nth cycle is such that $kX_n \leq F_d$ or $X_n \leq 0.075$ in. Noting that the amplitude loss per cycle is $4F_d/k = 4(150)/2000 = 0.3$ in, then the number of cycles completed until motion ceases is represented as

$$X_n = x_0 - n\left(\frac{4F_d}{k}\right) \leq 0.075 \tag{2}$$

from which $n = 19.75$. This indicates that motion will terminate after 19 3/4 cycles.   ▲

Due to the directional nature of the Coulomb damping force and its independence of velocity, it poses some cumbersome analytical obstacles. However, in a manner similar to that used for hysterectic damping, an equivalent viscous damping coefficient can be derived by equating the energy loss per cycle for a viscously damped system with a system possessing Coulomb damping. For a viscously damped system, the energy dissipated per cycle is given by Eq. (4.35) as

$$\Delta U = \pi c \omega X^2$$

The energy dissipated per cycle of Coulomb damping is approximately represented by

$$\Delta U = 4F_d X \tag{4.59}$$

By equating Eqs. (4.35) and (4.59) an equivalent viscous damping coefficient can be defined by

$$c_e = \frac{4F_d}{\pi \omega X} \tag{4.60}$$

which may be applied in much the same fashion as the equivalent viscous damping coefficient for a system exhibiting structural damping.

REFERENCES

1   Vierck, Robert K., *Vibration Analysis,* 2nd ed., Harper and Row, New York, 1979.

2   Beards, C.F., *Structural Vibration Analysis: Modelling, Analysis and Damping of Vibrating Structures,* Halsted Press, New York, 1983.

3   James, M.L., Smith, G.M., Wolford, J.C., and Whaley, P.W., *Vibration of Mechanical and Structural Systems,* Harper and Row, New York, 1989.

4   Haberman, Charles M., *Vibration Analysis,* Charles E. Merrill, Columbus, OH, 1960.

5   Steidel, Robert F., *An Introduction to Mechanical Vibrations,* Wiley, New York, 1971.

NOTATION

A, A_1, A_2	constants	c_e	equivalent viscous damping coefficient
B, B_1, B_2	constants	C_c	critical damping constant
c	viscous damping coefficient	C_1, C_2	constants

e^{st}	linear exponential function	W	weight
E	Young's modulus	x	translational displacement
f	natural frequency (cps)	x_0	initial displacement
f_d	damped natural frequency (cps)	$\dot{x}$	translational velocity
F_d	friction damping force	$\dot{x}_0$	initial velocity
F_D	viscous damping force	$\ddot{x}$	translational acceleration
g	acceleration due to gravity	X	amplitude of free vibration
i	imaginary number ($\sqrt{-1}$)	X_i	amplitude of free vibration for ith oscillation
k	translational stiffness	SDOF	single degree of freedom
m	mass	δ	logarithmic decrement
n	an integer	ΔU	energy loss per cycle
N	normal force	ϕ	phase angle (rad)
s	constant in linear exponential function	η	structural damping coefficient
s_1, s_2	roots of the auxiliary equation	μ	coefficient of friction
t	time	ω	natural circular frequency (rad/sec)
T	undamped natural period (sec)	ω_d	damped natural circular frequency (rad/sec)
T_d	damped natural period (sec)	ζ	damping factor

PROBLEMS

4.1 A viscously damped SDOF system exhibits a static deflection of 0.75 in due to its own weight of 50 lb. Determine the value of the critical damping constant for the system.

4.2 A viscously damped system has a total weight of 50 lb, a spring stiffness $k = 40$ lb/in, and a damping coefficient $c = 0.40$ lb-sec/in. Determine the displacement, velocity, and acceleration of the mass as a function of time if it is disturbed from its equilibrium position with an initial velocity of 10 in/sec with no initial displacement. Calculate the displacement, velocity, and acceleration of the mass at $t = 1.75$ sec.

4.3 The water level meter shown in Figure P4.1 consists of a rigid, massless arm of length $L = 20$ in and a floating cylinder of diameter $d = 3$ in. The weight of the cylinder is 1.5 lb and the specific weight of the water $\gamma_w = 64.0$ lb/ft³. Determine the value of the damping coefficient c of the dashpot required for critical damping.

4.4 Write the equation of motion for small oscillations, and determine expressions for critical damping and damped natural frequency for the system shown in Figure P4.2. Assume the rod is rigid and weightless and pivots about its left end.

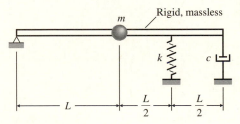

Figure P4.2

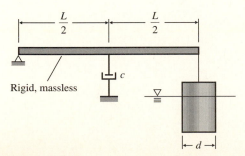

Figure P4.1

4.5 A vehicle weighing 5000 lb is supported by four identical springs and four identical viscous dampers. The static deflection of the vehicle under its own weight is 12 in. Determine the required damping coefficient for each of the dampers to achieve critical damping in the vehicle.

4.6 An overdamped system has a weight of 25 lb, a spring constant $k = 100$ lb/in, and a damping coefficient $c = 10$ lb-sec/in. The system is set in motion with an initial velocity of $\dot{x}_0 = 1200$ in/sec. Determine (a) an expression for the displacement of the mass as a function of time, (b) the maximum displacement of the mass from its static equilibrium position, and (c) the time at which the maximum displacement occurs.

4.7 An underdamped system has a weight of 5 lb, a spring constant $k = 20$ lb/in, and a damping coefficient $c = 0.2$ lb-sec/in. The system is displaced from its equilibrium position by 2 in and suddenly released. Determine the vibration amplitude after 10 oscillations and after 20 oscillations. Determine the time at which the system comes to rest.

4.8 Determine the ratio of successive amplitudes of vibration for a viscously damped system whose damping factor ζ is known to be 0.3.

4.9 In a certain viscously damped system, a particular vibration amplitude was measured to be 80% of the amplitude immediately preceding it. Determine the damping factor ζ for the system.

4.10 A vibrating system is connected to an adjustable viscous damping apparatus. At a particular setting, the ratio of successive amplitudes of vibration is 1 to 5. If the amount of damping is doubled, determine the resulting ratio of successive amplitudes.

4.11 A viscously damped system having a weight of 40 lb has a spring constant $k = 100$ lb/in and a damping coefficient $c = 0.08$ lb-sec/in. The system is set into free vibration by sudden release from an initial displacement of 4 in. Determine the amplitude of vibration after 10 cycles, 15 cycles, and 20 cycles.

4.12 The design for a shock absorber must be limited to 15% overshoot (displacement beyond the static equilibrium posi-

tion) when displaced from equilibrium and released. Calculate the required damping factor ζ.

4.13 Measurements indicate that the amplitude reduction for a particular viscously damped system is 75% in 15 cycles of free vibration. Determine the damping factor ζ for the system.

4.14 A viscously damped system weighing 100 lb exhibits a static deflection of 0.5 in due to its own weight. The mass is given an initial displacement $x_0 = 0.25$ in and suddenly released. After 3 cycles of free vibration the amplitude is measured to be 0.1 in. Determine (a) the logarithmic decrement, (b) the damping factor ζ, (c) the damping coefficient c, and (d) the frequency of damped vibration.

4.15 A viscously damped system has a mass of 10 kg and a stiffness $k = 60$ N/m. When the mass is given an initial displacement and released, the overshoot is 20%. Determine the damping factor and the damping coefficient for the system.

4.16 Given the following values of logarithmic decrement for four different materials, determine the corresponding damping factor ζ: material $A = 0.004$, material $B = 0.25$, material $C = 0.05$, and material $D = 0.10$.

4.17 A viscously damped structure is set into free vibration with an initial velocity. The resulting damped oscillations are shown in Figure P4.3. Determine (a) the natural period of vibration, (b) the logarithmic decrement, and (c) the damping factor ζ.

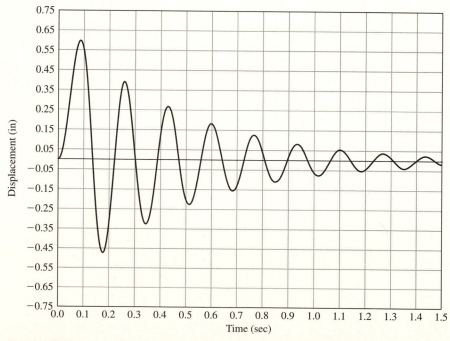

Figure P4.3

4.18 A projectile weighing 1.5 lb and traveling horizontally at a velocity $v_0 = 500$ in/sec, as shown in Figure P4.4, strikes a body initially at rest and remains imbedded in it after impact. The body weighs 60 lb, $k = 20$ lb/in, and $c = 0.7$ lb-sec/in. Determine the equation for the displacement of the object body as a function of time after impact.

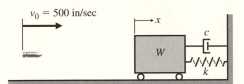

Figure P4.4

4.19 A body of mass m is mounted as shown in Figure P4.5. The assembly is dropped from a height h to a rigid surface. Assume the assembly does not bounce after contact. Determine (a) the equation of motion of the mass m after contact and (b) the expression for the acceleration of the mass after contact as a function of time. If $m = 40$ kg, $k = 4$ kN/m, $c = 100$ N-sec/m, and $h = 2$ m, determine the maximum acceleration experienced by the mass.

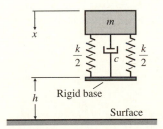

Figure P4.5

4.20 The assembly shown in Figure P4.6 impacts the surface at a velocity $v_0 = 75$ ft/sec. An instrument of mass $m = 0.05$ lb-sec^2/in is secured on a mount having stiffness $k = 75$ lb/in and a viscous damping coefficient $c = 2$ lb-sec/in. Determine the maximum acceleration of the mass after impact. Assume the assembly does not bounce.

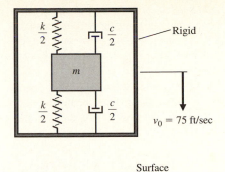

Figure P4.6

4.21 A projectile having mass $m = 10$ kg is fired into a shock tube with a velocity $v_0 = 20$ m/sec as shown in Figure P4.7. At the end of the tube the projectile impacts a stopper with a stiffness $k = 40$ kN/m and damping coefficient $c = 0.5$ kN-sec/m. Determine (a) the maximum displacement of the projectile after impacting the stopper and (b) the time after impact at which the maximum displacement is attained.

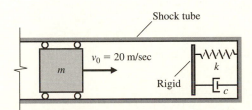

Figure P4.7

4.22 A weight $W = 5$ lb is supported by a spring with a stiffness $k = 0.75$ lb/in. The weight is given an initial displacement of 2.5 in from its equilibrium position and then released. After 150 cycles of oscillation, the vibration amplitude was measured as 1.25 in. Determine the hysteresis damping coefficient and the equivalent viscous damping coefficient.

4.23 A weight $W = 100$ lb is positioned at midspan of a simply supported beam having length $L = 60$ in. The amplitude decay of free oscillations of the system is 1% per cycle. Determine the hysteresis damping coefficient and the equivalent viscous damping coefficient. Assume $E = 29,000$ ksi and $I = 4$ in^4 for the beam.

4.24 A simply supported beam deflects 1 in from a 1000-lb load positioned at midspan. The resulting deflected shape closely resembles the vibration *mode shape* of the beam for which the natural frequency is 15 Hz. Determine the hysteresis damping coefficient.

4.25 A structure exhibiting hysteresis damping has a period of vibration of 0.5 sec. When set into free vibration it is observed that the amplitude of the tenth cycle is 90% of the amplitude of the first cycle. Determine the hysteresis damping coefficient.

4.26 For the system shown in Figure P4.8, $W = 25$ lb, $k = 50$ lb/in, and the coefficient of friction between the mass and the surface is $\mu = 0.2$. If the mass is given an initial displacement of 6 in and is suddenly released, determine (a) the maximum velocity of the mass, (b) the displacement amplitude decrease per cycle, and (c) the displacement when the mass comes to rest.

4.27 For the system shown in Figure P4.8, $W = 1.5$ kips, $k = 3.5$ kips/in, and the coefficient of friction between the mass and the surface is $\mu = 0.15$. If the mass is set in motion with initial conditions of displacement $x_0 = 7.5$ in and velocity $\dot{x}_0 = 4$ in/sec, determine (a) the vibration displacement amplitude after 5 cycles and (b) the number of cycles of motion completed before the mass comes to rest.

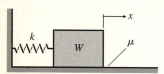

Figure P4.8

4.28 The system shown in Figure P4.9 has a mass $m = 8$ kg, a stiffness $k = 3.5$ N/mm, a viscous damping coefficient $c = 0.12$ N-sec/mm, and a coefficient of friction $\mu = 0.13$. The mass is displaced from its equilibrium position a distance of 70 mm and released. Determine (a) the differential equation of motion, (b) the displacement of the mass as a function of time, and (c) the number of oscillations completed before the mass comes to rest.

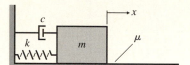

Figure P4.9

4.29 A particular system consisting of a mass $m = 60$ kg and a spring of stiffness $k = 10$ kN/m is subject to a Coulomb damping force of 10 N. If the initial displacement amplitude is 8 cm, determine (a) the amplitude at the end of 6 cycles, (b) the amplitude at the end of 8 cycles, and (c) the number of cycles completed before the mass comes to rest.

4.30 A projectile of mass m impacts a body of mass m_1 at a velocity v_0 as shown in Figure P4.10. The system has a stiffness k and the coefficient of friction between the body and the surface is μ. Determine (a) the expression for displacement of the mass as a function of time and (b) the natural frequency of the system.

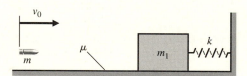

Figure P4.10

5 ▲ Response to Harmonic Excitation

As discussed in Chapter 1, prescribed or deterministic dynamic loads can be conveniently classified as either periodic or nonperiodic. Periodic loads may be harmonic or nonharmonic. In either case, however, periodic loads are repetitive and exhibit the same amplitude variation successively over many cycles. Nonperiodic loads, on the other hand, may be either short-duration impulsive events such as a blast or long-duration occurrences such as an earthquake excitation.

Harmonic excitation is frequently encountered in industrial and power installations. It is commonly produced by the imbalance of forces in rotating machinery such as fans, blowers, or compressors. The harmonic excitation continues as long as the machinery is in operation, and causes the support structure to undergo sustained vibrations. These vibrations have two components, a transient component and a steady-state component. The transient vibrations are instigated by initial conditions induced at start-up. They occur at the natural frequency of the system or structure, but are quickly damped out and are, therefore, generally neglected. The steady-state vibrations, however, continue long after the transient component has died out. They occur at the frequency of the exciting force and can result in a highly undesirable condition known as *resonance* if the excitation frequency is in close proximity to the natural frequency of the structure. At resonance, the steady-state response may build up very large displacement amplitudes, regardless of damping, and cause severe overstressing or failure in the system.

5.1 FORCED HARMONIC RESPONSE OF UNDAMPED SYSTEMS

Harmonic excitation is frequently encountered in many types of structural and mechanical systems. It is usually attributed to the standard operating conditions of various types of machinery. There are three broad categories of machines in present day practice that cause harmonic excitation: (1) reciprocating and rotating engines, motors, and compressors; (2) turbines; and (3) special machines such as radar towers and conveyor mechanisms. The harmonic excitation usually manifests itself as either a force or a displacement.

Figure 5.1a depicts a simple spring-mass system that is subjected to a time-varying external harmonic force given by $F(t) = F_0 \sin \Omega t$, where F_0 is the force amplitude and Ω is the circular frequency of the forcing function. To obtain the differential equation of motion, Newton's second law is written for the free-body diagram of the mass in some displaced position as shown in Figure 5.1b, giving

$$\sum F_x = m\ddot{x} = -kx + F(t) \tag{5.1}$$

which may be written in standard form as

$$\ddot{x} + \frac{k}{m}x = \frac{F(t)}{m} \tag{5.2}$$

or

$$\ddot{x} + \frac{k}{m}x = \frac{F_0}{m} \sin \Omega t \tag{5.3}$$

Equation (5.3) is a nonhomogeneous, second-order differential equation with constant coefficients. The complete solution is the sum of a complementary or homogeneous solution $x_h(t)$ and a particular solution $x_p(t)$ such that

$$x(t) = x_h(t) + x_p(t) \tag{5.4}$$

where

$$\ddot{x}_h + \frac{k}{m}x_h = 0 \tag{5.5}$$

and

$$\ddot{x}_p + \frac{k}{m}x_p = \frac{F_0}{m} \sin \Omega t \tag{5.6}$$

Equation (5.5) is the differential equation of motion for free vibration of a simple SDOF system, for which the solution has previously been determined in Section 3.1 as

$$x_h(t) = A \sin \omega t + B \cos \omega t \tag{5.7}$$

where $\omega = (k/m)^{1/2}$ is the natural circular frequency of the system.

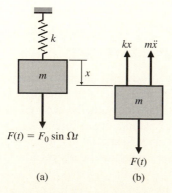

Figure 5.1

Harmonic excitation of an undamped SDOF system.

To obtain the particular solution corresponding to Eq. (5.6), it is noted that the differential equation involves $x_p(t)$, its second derivative, and a sine function. Since the second derivative of a sine function is itself a sine function, a solution for $x_p(t)$ can be assumed as

$$x_p(t) = X_f \sin \Omega t \qquad (5.8)$$

where X_f is an unknown constant representing the displacement amplitude of the forced response and must be determined such that the assumed solution satisfies the differential equation. Substituting this assumed solution for $x_p(t)$ into Eq. (5.6) yields

$$-X_f\Omega^2 \sin \Omega t + X_f\frac{k}{m} \sin \Omega t = \frac{F_0}{m} \sin \Omega t \qquad (5.9)$$

from which the resulting value of X_f is

$$X_f = \frac{F_0/m}{(k/m) - \Omega^2} \qquad (5.10)$$

to satisfy the differential equation. Multiplying both the numerator and the denominator of Eq. (5.10) by m/k yields

$$X_f = \frac{F_0/k}{1 - (m/k)\Omega^2} = \frac{F_0/k}{1 - (\Omega/\omega)^2} \qquad (5.11)$$

since $m/k = 1/\omega^2$. The resulting particular solution is

$$x_p(t) = \frac{F_0/k}{1 - (\Omega/\omega)^2} \sin \Omega t \qquad (5.12)$$

and the complete solution for the motion of the mass, the sum of the homogeneous and particular solutions, is

$$x(t) = A \sin \omega t + B \cos \omega t + \frac{F_0/k}{1 - (\Omega/\omega)^2} \sin \Omega t \qquad (5.13)$$

The constants A and B that result from the homogeneous solution are evaluated by invoking the initial conditions $x(0) = x_0$ and $\dot{x}(0) = \dot{x}_0$ to Eq. (5.13) and its first derivative, respectively.

To lend physical insight to the complete solution, Eq. (5.13) can be expressed in the alternative form

$$x(t) = X \sin(\omega t + \phi) + \frac{F_0/k}{1 - (\Omega/\omega)^2} \sin \Omega t \qquad (5.14)$$

where X and ϕ are the amplitude and phase angle of the free vibration response, respectively. The quantity F_0/k is the equivalent static deflection that would result from applying a force of magnitude F_0 to the spring-mass system. The quantity Ω/ω is termed the *frequency ratio* and represents the ratio of the frequency of the forcing function to the natural circular frequency of the spring-mass system. Letting $X_0 = F_0/k$ and $r = \Omega/\omega$, the amplitude of the forced response is expressed as

$$X_f = \frac{X_0}{1 - r^2} \tag{5.15}$$

In Eq. (5.15), the amplitude of the forced response X_f is readily noted to be the equivalent static deflection X_0 multiplied by the *dynamic magnification factor* (DMF), which is equal to $1/(1 - r^2)$. Therefore, the value of X_f may be less than, equal to, or greater than the equivalent static deflection, depending on the value of the frequency ratio r.

Introducing the foregoing expressions for X_f and r into Eq. (5.14) gives the expression for the complete solution,

$$x(t) = X \sin(\omega t + \phi) + \frac{X_0}{1 - r^2} \sin \Omega t \tag{5.16}$$

which indicates that the motion of the mass is the sum of a free vibration component (transient) and a forced response component (steady state), each of which varies sinusoidally with time. Three distinct types of motion are possible, depending on whether the value of the frequency ratio r is less than, greater than, or equal to one.

If the frequency ratio r is less than one, then $\Omega < \omega$, indicating that the natural frequency of the free response (transient) is greater than the frequency of the forced response (steady state). The resulting motion is represented in Figure 5.2. The free vibration portion of the motion completes several cycles in the time required for one cycle of the forced response. The total motion (depicted by the solid line in Figure 5.2) exhibits a sinusoidal variation about a lower frequency base curve (represented by the dashed line in Figure 5.2). Moreover, since $r < 1$, the DMF or $1/(1 - r^2)$ is greater than one, thus causing the forced response to be greater than the equivalent static deflection X_0.

When the frequency ratio is greater than one, then $\Omega > \omega$ and the total motion is characterized as the forced response oscillating about the free vibration portion of the response as indicated in Figure 5.3. Also, if $r > \sqrt{2}$, the amplitude of the forced response will be less than the equivalent static deflection X_0.

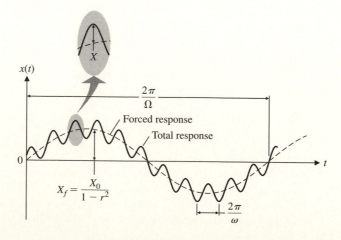

Figure 5.2

Response of an undamped SDOF system to harmonic excitation.

Figure 5.3

Response of an undamped SDOF system to harmonic excitation.

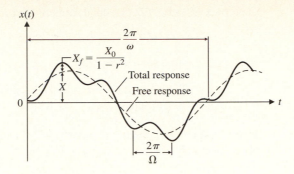

EXAMPLE 5.1 ▼

The undamped spring-mass system shown in Figure 5.1 has a mass of 4.5 kg and a spring stiffness of 3.5 N/mm. It is excited by a harmonic force having an amplitude $F_0 = 100$ N and an excitation frequency $\Omega = 18$ rad/sec. The initial conditions are $x(0) = 15$ mm and $\dot{x}(0) = 150$ mm/sec. Determine (a) the frequency ratio, (b) the amplitude of the forced response, (c) the displacement of the mass at time $t = 2$ sec, and (d) the velocity of the mass at time $t = 4$ sec.

Solution

The natural frequency of the system is calculated as

$$\omega = \left(\frac{k}{m}\right)^{1/2} = \sqrt{\frac{3500}{4.5}} = 27.89 \text{ rad/sec}$$

(a) The frequency ratio is expressed as

$$r = \left(\frac{\Omega}{\omega}\right) = \frac{18}{27.89} = 0.645$$

(b) The frequency ratio is used in Eq. (5.15) to determine the amplitude of the forced response as

$$|X_f| = \left|\frac{F_0/k}{1 - r^2}\right| = \left|\frac{100/3.5}{1 - (0.645)^2}\right| = 48.92 \text{ mm}$$

(c) To determine the displacement at $t = 2$ sec, the constants in the complete solution must be evaluated by applying the initial conditions. The complete solution is given by Eq. (5.14) as

$$x(t) = X \sin(\omega t + \phi) + \frac{F_0/k}{1 - r^2} \sin \Omega t \tag{1}$$

or by Eq. (5.13) as

$$x(t) = A \sin \omega t + B \cos \omega t + \frac{F_0/k}{1 - r^2} \sin \Omega t \tag{2}$$

Therefore, from Eq. (2), the displacement solution is

$$x(t) = A \sin(27.89t) + B \cos(27.89t) + 48.92 \sin(18t) \tag{3}$$

and taking the first time derivative of Eq. (3) yields the velocity solution

$$\dot{x}(t) = 27.89A \cos(27.89t) - 27.89B \sin(27.89t)$$
$$+ 881.44 \cos(18t) \tag{4}$$

Applying the initial conditions of $x(0) = 15$ and $\dot{x}(0) = 150$ to Eqs. (3) and (4), respectively yields

$$x(0) = B = 15 \tag{5}$$

$$\dot{x}(0) = 27.89A + 881.44 = 150 \tag{6}$$

Solving Eqs. (5) and (6) simultaneously gives

$$A = -26.23 \text{ mm} \quad \text{and} \quad B = 15 \text{ mm/sec}$$

The complete solutions for displacement and velocity are given by Eqs. (3) and (4), respectively, as

$$x(t) = -26.23 \sin(27.89t) + 15 \cos(27.89t) + 48.92 \sin(18t) \tag{7}$$

and

$$\dot{x}(t) = -731.44 \cos(27.89t) - 418.35 \sin(27.89t)$$
$$+ 881.44 \cos(18t) \tag{8}$$

(c) The displacement at $t = 2$ sec is determined from Eq. (7) as

$$x(2) = 18.23 + 10.78 - 48.57 = -19.56 \text{ mm}$$

(d) The velocity at $t = 4$ sec is determined from Eq. (8) as

$$\dot{x}(4) = -24.47 + 418.12 - 852.58 = -458.93 \text{ mm/sec}$$

▲

5.2 BEATING AND RESONANCE

5.2.1 Beating

Two very important phenomena occur when the frequency of the forcing function Ω approaches the natural circular frequency of the system ω, which in turn causes the frequency ratio to approach unity. First consider the case in which Ω and ω are *nearly* the same, or ω is slightly greater than Ω; thus, $(\omega - \Omega)$ is very small. Recalling the general solution given by Eq. (5.13) in the form

$$x(t) = A \sin \omega t + B \cos \omega t + \frac{X_0}{1 - r^2} \sin \Omega t \tag{5.17}$$

and assuming nonzero initial conditions $x(0) = x_0$ and $\dot{x}(0) = \dot{x}_0$, then $A = (\dot{x}_0/\omega) - X_0 r/(1 - r^2)$ and $B = x_0$, and Eq. (5.17) becomes

$$x(t) = \frac{\dot{x}_0}{\omega}\sin \omega t + x_0 \cos \omega t + \frac{X_0}{1 - r^2}(\sin \Omega t - r\sin \omega t) \qquad (5.18)$$

To illustrate the *beating phenomenon,* let initial conditions $x_0 = \dot{x}_0$ in Eq. (5.18), and since $r \cong 1$, Eq. (5.18) then reduces to

$$x(t) = \frac{X_0}{1 - r^2}(\sin \Omega t - \sin \omega t) \qquad (5.19)$$

Next, letting

$$\frac{X_0}{1 - r^2} = \frac{X_0\omega^2}{\omega^2 - \Omega^2} \qquad (5.20)$$

Eq. (5.19) becomes

$$x(t) = \frac{X_0\omega^2}{\omega^2 - \Omega^2}(\sin \Omega t - \sin \omega t) \qquad (5.21)$$

Introducing the trigonometric identity

$$\sin \alpha - \sin \beta = 2\cos\left(\frac{\alpha + \beta}{2}\right)\sin\left(\frac{\alpha - \beta}{2}\right) \qquad (5.22)$$

in which $\alpha = \Omega t$ and $\beta = \omega t$, allows Eq. (5.21) to be written as

$$x(t) = \frac{2X_0\omega^2}{\omega^2 - \Omega^2}\cos\left(\frac{\Omega + \omega}{2}t\right)\sin\left(\frac{\Omega - \omega}{2}t\right) \qquad (5.23)$$

Since Ω is just slightly less than ω in Eq. (5.23), let

$$\frac{\omega - \Omega}{2} = \varepsilon \qquad (5.24)$$

in which ε is a relatively small quantity. Then the denominator in Eq. (5.23) can be expressed as

$$\omega^2 - \Omega^2 = (\omega + \Omega)(\omega - \Omega) \cong -4\Omega\varepsilon \qquad (5.25)$$

Finally, substituting Eq. (5.25) into Eq. (5.23) and letting $\omega = \Omega$, the resulting expression for displacement is

$$x(t) = -\frac{X_0\Omega}{2\varepsilon}\cos \Omega t \sin \varepsilon t \qquad (5.26)$$

Since Ω is much larger than ε, the term $\sin \varepsilon t$ oscillates with a much larger period than does $\cos \Omega t$. The resulting motion, depicted in Figure 5.4, is a rapid oscillation with a slowly varying amplitude and is referred to as a *beat*. Sometimes the two sinusoids add to each other, and at other times they cancel each other out, resulting in a *beating phenomenon*. The beating phenomenon often manifests itself in mechanical equipment as an emitted sound having a similar cyclically varying magnitude.

Figure 5.4

Beating phenomenon for an undamped SDOF system.

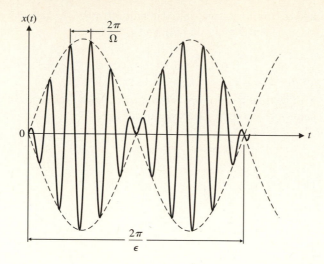

EXAMPLE 5.2 ▼

An undamped system is harmonically forced, resulting in a beating condition. The natural frequency of the system is 1500 cycles/min and the excitation frequency is 1450 cycles/min. Determine (a) the beat period of the resulting motion, (b) the number of oscillations within each beat, and (c) the maximum amplitude of the oscillations if the weight of the system is 10,000 lb and the amplitude of the steady-state force is 5000 lb.

Solution

(a) The frequency components for this system are

$$\omega = 2\pi\frac{(1500)}{60} = 157 \text{ rad/sec}$$

$$\Omega = 2\pi\frac{(1450)}{60} = 151.8 \text{ rad/sec}$$

$$\varepsilon = \frac{\omega - \Omega}{2} = 2.6 \text{ rad/sec}$$

and the beat period is determined as

$$\frac{2\pi}{\varepsilon} = 2.417 \text{ sec.}$$

(b) The number of oscillations within each beat is given by

$$\frac{2\pi/\varepsilon}{2\pi/\Omega} = 58.4$$

(c) The maximum amplitude of oscillations is determined as

$$\left|\frac{X_0\Omega}{2\varepsilon}\right| = \left|\frac{(F_0/k)\Omega}{2\varepsilon}\right| \tag{1}$$

where

$$k = m\omega^2 = \frac{10,000}{386.4}(157)^2 = 637,914 \text{ lb/in}$$

Thus, from Eq. (1) the maximum amplitude of the oscillations is

$$\left|\frac{(F_0/k)\Omega}{2\varepsilon}\right| = \left|\frac{(5000/637,914)(151.8)}{2(2.6)}\right| = 0.239 \text{ in}$$

▲

5.2.2 Resonance

Next, consider the case for which the frequency of the forcing function, Ω, becomes exactly equal to the systems's natural circular frequency ω. This condition is known as *resonance,* and it is very important in the vibration of structural and mechanical systems. At resonance, the frequency ratio r is equal to unity and the DMF is equal to infinity, suggesting that the amplitude of vibration increases without bound. Indeed, both the plot of DMF versus frequency ratio, illustrated in Figure 5.5, and the solution given by Eq. (5.16) misleadingly indicate that the amplitude of the forced response is infinite for all values of time for $r = 1$.

This interpretation of the resonant condition is neither physically possible nor mathematically correct. To lend further insight to the resonance phenomenon, a reevaluation of the solution provided by Eq. (5.16) is warranted. Obviously, the error lies in the assumed solution for the forced response given by Eq. (5.8),

$$x_p(t) = X_f \sin \Omega t$$

It has been recommended [1, 2] that a more appropriate particular solution is provided by the expression

$$x_p(t) = X_f t \sin \omega t \tag{5.27}$$

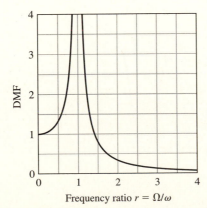

Figure 5.5

Dynamic magnification factor versus frequency ratio for an undamped SDOF system excited by harmonic forcing function.

in which it is noted that $\Omega = \omega$ at resonance. Restricting focus to the differential equation for the forced response, the governing differential equation for the particular solution originally stated by Eq. (5.6) can be written as

$$\ddot{x}_p + \omega^2 x_p = \frac{F_0}{m} \sin \omega t \tag{5.28}$$

Substituting the assumed solution given by Eq. (5.27) into Eq. (5.28) yields

$$X_f(2\omega \cos \omega t - \omega^2 t \sin \omega t) + X_f \omega^2 t \sin \omega t = \frac{F_0}{m} \sin \omega t \tag{5.29}$$

from which (neglecting the harmonic terms)

$$X_f = \frac{F_0}{2m\omega} = \frac{X_0 \omega}{2} \tag{5.30}$$

Substituting the value of X_f given by Eq. (5.30) into Eq. (5.27), and then adding the transient response, the complete solution for the resonant case then becomes

$$x(t) = X \sin(\omega t + \phi) + \frac{X_0 \omega}{2} t \sin \omega t \tag{5.31}$$

Equation (5.31) indicates that for a system operating at resonance, the amplitude of the forced response increases linearly with time by πX_0 per cycle as illustrated in Figure 5.6. Theoretically, the amplitude will eventually approach infinity. In reality, however, the system will break down once the amplitude becomes intolerably large for the spring mechanism. Fortunately, since the steady-state amplitude varies directly with time, the system would have to operate at resonance for an extended period before the amplitude becomes destructively large. Therefore, it is acceptable for a system or machine in route to its operating frequency to quickly pass through the resonance condition.

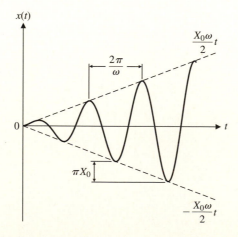

Figure 5.6

Forced response of an undamped SDOF system at resonance, $\Omega = \omega$.

EXAMPLE 5.3 ▼

A SDOF system similar to that depicted in Figure 5.1 has a total weight of 1000 lb and a spring stiffness of 2500 lb/in. The system is excited at resonance by a harmonic force of amplitude $F_0 = 600$ lb. Determine the displacement amplitude of the forced response after (a) 1 1/4 cycles, (b) 10 1/4 cycles, and (c) 20 1/4 cycles.

Solution

At resonance, $r = 1$ and the forced response is given by Eq. (5.27) as

$$x(t) = \frac{X_0 \omega}{2} t \sin \omega t \tag{1}$$

in which

$$X_0 = \frac{F_0}{k} = \frac{600}{2500} = 0.24 \text{ in}$$

and

$$\omega = \left(\frac{k}{m}\right)^{1/2} = \left[\frac{(2500)(386.4)}{1000}\right]^{1/2} = 31.1 \text{ rad/sec}$$

Therefore, Eq. (1) becomes

$$x(t) = 3.732t \sin \omega t \tag{2}$$

(a) Noting that the natural period of the system T is equal to $2\pi/\omega$, then after 1 1/4 cycles,

$$t = 1.25T = 1.25\left(\frac{2\pi}{\omega}\right) = \frac{2.5\pi}{\omega}$$

and Eq. (2) yields

$$x(t) = 3.732\left(\frac{2.5\pi}{31.1}\right) \sin 2.5\pi = 0.943 \text{ in}$$

(b) After 10 1/4 cycles,

$$t = 10.25T = 10.25\left(\frac{2\pi}{\omega}\right) = \frac{20.5\pi}{\omega}$$

and Eq. (2) yields

$$x(t) = 3.732\left(\frac{20.5\pi}{31.1}\right) \sin 20.5\pi = 7.73 \text{ in}$$

(c) After 20 1/4 cycles,

$$t = 20.25T = 20.25\left(\frac{2\pi}{\omega}\right) = \frac{40.5\pi}{\omega}$$

and Eq. (2) yields

$$x(t) = 3.732\left(\frac{40.5\pi}{31.1}\right)\sin 40.5\pi = 15.22 \text{ in}$$

▲

5.3 FORCED HARMONIC VIBRATIONS WITH VISCOUS DAMPING

Consider the viscously damped system shown in Figure 5.7a, in which the mass is subjected to a harmonic, external exciting force $F_0 \sin \Omega t$. Applying Newton's second law to the free-body diagram in Figure 5.7b yields

$$m\ddot{x} = -kx - c\dot{x} + F_0 \sin \Omega t \tag{5.32a}$$

or

$$\ddot{x} + \frac{c}{m}\dot{x} + \frac{k}{m}x = \frac{F_0}{m}\sin \Omega t \tag{5.32b}$$

as the differential equation of motion, where c is the viscous damping coefficient and $c\dot{x}$ is the viscous damping force. Equation (5.32b) is nonhomogeneous, for which the complete solution will be the sum of a homogeneous and a particular solution. The homogeneous solution x_h is given by Eq. (4.14), Eq. (4.18), or Eq. (4.21), depending upon whether the damping factor ζ is less than, equal to, or greater than unity, respectively. The homogenous solution x_h, or the *transient response,* dies out with time for all nonzero values of ζ. Therefore, in vibration analysis, engineers are generally interested

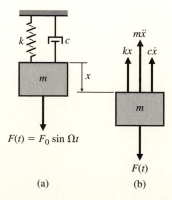

Figure 5.7

Viscously damped SDOF system subject to harmonic excitation.

in only the particular solution x_p or the *steady-state* response, which remains with time as long as the exciting force persists. This phenomenon is demonstrated in Figure 5.8, in which the transient (free) response depicted by Figure 5.8a combines with the steady-state response shown in Figure 5.8b in the early stages of the vibration. However, after a relatively short period of time the transient response is completely damped out as illustrated in Figure 5.8c.

The particular solution to Eq. (5.32b) will be of the form

$$x_p(t) = C \sin \Omega t + D \cos \Omega t \tag{5.33}$$

where C and D are constants. Substitution of this assumed solution and its time derivatives into Eq. (5.32) yields

$$-C\Omega^2 \sin \Omega t - D\Omega^2 \cos \Omega t + \frac{c}{m}\Omega C \cos \Omega t - \frac{c}{m}\Omega D \sin \Omega t$$

$$+\frac{k}{m}C \sin \Omega t + \frac{k}{m}D \cos \Omega t = \frac{F_0}{m} \sin \Omega t \tag{5.34}$$

Equating coefficients of similar harmonic terms of time in Eq. (5.34) yields

$$\left(\frac{k}{m} - \Omega^2\right)C - \frac{c}{m}\Omega D = \frac{F_0}{m} \tag{5.35}$$

and

$$\frac{c}{m}\Omega C + \left(\frac{k}{m} - \Omega^2\right)D = 0 \tag{5.36}$$

Simultaneous solution of Eqs. (5.35) and (5.36) results in expressions for C and D as

$$C = \frac{[(k/m) - \Omega^2](F_0/m)}{[(k/m) - \Omega^2]^2 + [(c/m)\Omega]^2} \tag{5.37}$$

$$D = \frac{-(c/m)\Omega(F_0/m)}{[(k/m) - \Omega^2]^2 + [(c/m)\Omega]^2} \tag{5.38}$$

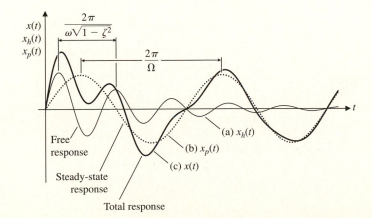

Figure 5.8

Response of viscously damped SDOF system to harmonic excitation:
(a) transient response;
(b) steady-state response;
(c) total response.

The use of the relations

$$\left(\frac{k}{m} - \Omega^2\right)^2 = (\omega^2 - \Omega^2)^2 = \omega^4(1 - r^2)^2 \tag{5.39}$$

$$\left(\frac{c}{m}\Omega\right)^2 = \left(\frac{2c\omega^2}{2m\omega}\frac{\Omega}{\omega}\right)^2 = \omega^4(2\zeta r)^2 \tag{5.40}$$

and

$$\frac{F_0}{m} = \frac{F_0}{m}\frac{m}{k}\omega^2 = \frac{F_0}{k}\omega^2 = X_0\omega^2 \tag{5.41}$$

may simplify the expressions for C and D. Substituting Eqs. (5.39), (5.40), and (5.41) into Eqs. (5.37) and (5.38) yields

$$C = \frac{(1 - r^2)X_0}{(1 - r^2)^2 + (2\zeta r)^2} \tag{5.42}$$

and

$$D = \frac{-2\zeta r X_0}{(1 - r^2)^2 + (2\zeta r)^2} \tag{5.43}$$

where X_0 is the equivalent static deflection equal to F_0/k. Equation (5.33), for the particular solution x_p, can be expressed in the more useful form

$$x_p(t) = X_f \sin(\Omega t - \psi) \tag{5.44}$$

in which the steady-state amplitude X_f is given by

$$X_f = \sqrt{C^2 + D^2} = \frac{X_0}{\sqrt{(1 - r^2)^2 + (2\zeta r)^2}} \tag{5.45}$$

Substituting Eq. (5.45) into Eq. (5.44), the particular or steady-state solution is

$$x_p(t) = \frac{X_0}{\sqrt{(1 - r^2)^2 + (2\zeta r)^2}} \sin(\Omega t - \psi) \tag{5.46}$$

in which the phase angle of the steady-state solution, ψ, is expressed as

$$\psi = \tan^{-1}\left(-\frac{D}{C}\right) = \tan^{-1}\left(\frac{2\zeta r}{1 - r^2}\right) \tag{5.47}$$

The phase angle represents the *lag* of the steady-state displacement response behind the harmonic excitation. Equation (5.46) reveals that the amplitude of the steady-state response, X_f, is equal to the equivalent static deflection X_0 multiplied by the *dynamic magnification factor* (DMF) given by

$$\text{DMF} = \frac{1}{\sqrt{(1 - r^2)^2 + (2\zeta r)^2}} \tag{5.48}$$

The *complete solution* (including the transient term) can be expressed as

$$x(t) = e^{-\zeta \omega t}(A\sin \omega_d t + B\cos \omega_d t)$$

$$+ \frac{X_0}{\sqrt{(1 - r^2)^2 + (2\zeta r)^2}}\sin(\Omega t - \psi) \qquad (5.49)$$

or

$$x(t) = Xe^{-\zeta \omega t}(\sin \omega_d t + \phi) + \frac{X_0}{\sqrt{(1 - r^2)^2 + (2\zeta r)^2}}\sin(\Omega t - \psi) \quad (5.50)$$

in which the damped natural circular frequency $\omega_d = \omega\sqrt{1 - \zeta^2}$ and the constants A and B in Eq. (5.49), or X and ϕ in Eq. (5.50), are evaluated from the initial conditions of displacement and velocity.

EXAMPLE 5.4 ▼

A structure having a mass of 100 kg and a translational stiffness of 40,000 N/m shown in Figure 5.9 is subject to a harmonic force having an amplitude of 500 N and an operating frequency of 15 rad/sec. The damping factor for the structure is $\zeta = 0.10$. For the sustained or steady-state vibration, determine (a) the amplitude of the steady-state displacement, (b) its phase with respect to the exciting force, and (c) the maximum velocity of the response. For initial conditions $x(0) = 6$ cm and $\dot{x}(0) = 0$, determine the equation that describes the total displacement of the structure as a function of time.

Solution

From the given data it is known that

$$F_0 = 500 \text{ N} \qquad m = 100 \text{ kg}$$

$$\Omega = 15 \text{ rad/sec} \qquad \zeta = 0.1$$

$$\omega = \sqrt{\frac{k}{m}} = \sqrt{\frac{40,000}{100}} = 20 \text{ rad/sec}$$

$$r = \frac{\Omega}{\omega} = \frac{15}{20} = 0.75$$

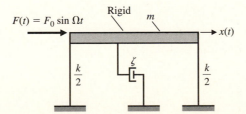

Figure 5.9

Structure of Example 5.4.

(a) The amplitude of the steady-state displacement is given by

$$X_f = X_0(\text{DMF}) \tag{1}$$

where

$$X_0 = \frac{F_0}{k} = \frac{500}{40{,}000} = 0.0125 \text{ m} = 12.5 \text{ mm}$$

and the DMF is determined from Eq. (5.48) as

$$\text{DMF} = \frac{1}{\sqrt{[1 - (0.75)^2]^2 + [2(0.1)(0.75)]^2}} = 2.162$$

and thus from Eq. (1) we have

$$X_f = 12.5(2.162) = 27.025 \text{ mm}$$

(b) The phase angle ψ is determined from Eq. (5.47) as

$$\psi = \tan^{-1}\left[\frac{2(0.1)(0.75)}{1 - (0.75)^2}\right] = 0.33 \text{ rad}$$

(c) The steady-state-velocity expression is given by the first time derivative of Eq. (5.46), or

$$\dot{x}(t) = \frac{X_0 \Omega}{\sqrt{(1 - r^2)^2 + (2\zeta r)^2}} \cos(\Omega t - \psi) \tag{2}$$

the velocity is maximum when $\cos(\Omega t - \psi) = 1.0$; therefore from Eq. (2),

$$\dot{x}_{\max} = X_0 \Omega(\text{DMF}) = (0.0125)(15)(2.162) = 0.41 \text{ m/sec}$$

The homogeneous solution representing the transient response is given by Eq. (4.14) for $\zeta < 1$,

$$x_h(t) = (A \sin \omega_d t + B \cos \omega_d t)e^{-\zeta \omega t} \tag{3}$$

where $\omega_d = \omega\sqrt{1 - \zeta^2} = 20(0.995) = 19.9$ rad/sec and the particular solution representing the steady-state response is determined from Eq. (5.44) as

$$x_p(t) = 27.025 \sin(15t - 0.33) \text{ mm} \tag{4}$$

The complete solution for the displacement is obtained by combining Eqs. (3) and (4); thus

$$x(t) = (A \sin 19.9t + B \cos 19.9t)e^{-2t} + 27.025 \sin(15t - 0.33) \tag{5}$$

in which the constants A and B are evaluated by applying the initial conditions to Eq. (5); thus

$$x(0) = 60 = B - 8.76 \tag{6}$$

$$\dot{x}(0) = 0 = -2B + 19.9A + 383.5 \tag{7}$$

Simultaneous solution of Eqs. (6) and (7) yields $A = -12.36$ and $B = 68.75$. The total displacement is then determined from Eq. (5) as

$$x(t) = (-12.36 \sin 19.9t + 68.75 \cos 19.9t)e^{-2t}$$
$$+ 27.025 \sin(15t - 0.33) \text{ mm} \tag{8}$$

▲

5.4 EFFECT OF DAMPING FACTOR ON STEADY-STATE RESPONSE AND PHASE ANGLE

In Section 5.3, expressions for the dynamic magnification factor and phase angle were derived and are given by Eqs. (5.48) and (5.47), respectively. For the case of zero damping, the expression for the dynamic magnification factor reduces to

$$\text{DMF} = \frac{1}{1 - r^2} \tag{5.51}$$

and it is noted from Eq. (5.48) that

$$\tan \psi = 0 \tag{5.52}$$

It is not readily apparent from Eq. (5.52) whether the phase angle for zero damping should be 0° or 180°. Further examination of the effect of damping on Eq. (5.48) will lend valuable insight to this dilemma.

The effect of the damping factor on the steady-state amplitude X_f may be demonstrated by plotting the dynamic magnification factor as a function of the frequency ratio r for several values of damping factor ζ as illustrated in Figure 5.10. The maximum value for the dynamic magnification factor occurs at resonance ($r = 1$) for zero damping, which was previously illustrated in Figure 5.5. For low excitation frequencies, represented by $r \ll 1.0$, Figure 5.10 exhibits dynamic magnification factors close to unity. This is the expected result, since a disturbing force having a frequency close to zero approximates a statically applied force. For high excitation frequencies where $r \gg 1$, the dynamic magnification factor becomes quite small. In fact, as r approaches infinity, the dynamic magnification factor approaches zero, which is representative of the case where the excitation frequency is so large in comparison to the natural frequency of the system that there is insufficient time for the system to respond.

For a given damping factor, the frequency ratio corresponding to the maximum dynamic magnification factor can be determined by setting the derivative of the dynamic magnification factor with respect to r equal to zero. Thus

$$\frac{d(\text{DMF})}{dr} = \frac{r(1 - r^2 - 2\zeta^2)}{[(1 - r^2)^2 + (2\zeta r)^2]^{3/2}} = 0 \tag{5.53}$$

The instances for $r = 0$ and $r = \infty$, which satisfy Eq. (5.53), have been previously discussed. The remaining condition satisfying Eq. (5.53) is $1 - r^2 - 2\zeta^2 = 0$, which yields a maximum dynamic magnification factor at

$$r = \sqrt{1 - 2\zeta^2} \tag{5.54}$$

Figure 5.10

Dynamic magnification factor (DMF) versus frequency ratio for various levels of damping. *Mechanical Vibrations, Theory and Applications,* 2/e, by Tse/Morse/Hinkle. Reprinted by permission of authors.

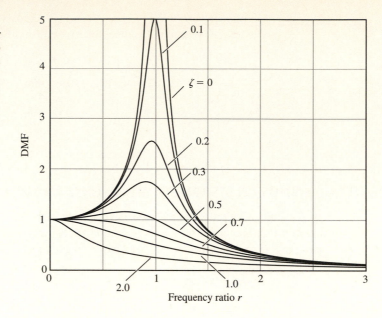

for damping factors $\zeta \le 1/\sqrt{2}$. Therefore, the maximum dynamic magnification factor is obtained by substituting Eq. (5.54) into Eq. (5.48), yielding

$$(\text{DMF})_{\text{max}} = \frac{1}{2\zeta\sqrt{1 - 2\zeta^2}} \tag{5.55}$$

For damping factors $\zeta \ge 1/\sqrt{2}$, the maximum dynamic magnification factor occurs at $r = 0$ and is equal to unity. By contrast, at resonance, Eq. (5.54) is equal to unity and Eq. (5.55) becomes

$$(\text{DMF})_{\text{res}} = \frac{1}{2\zeta} \tag{5.56}$$

For most practical applications, however, since ζ is a relatively small quantity, $(\text{DMF})_{\text{res}} \cong (\text{DMF})_{\text{max}}$.

The phase angle ψ given by Eq. (5.47) is plotted in Figure 5.11 as a function of the frequency ratio r for selected values of damping factor ζ. For low excitation frequencies, $r \ll 1$, the phase angle is close to zero and the displacement is nearly in phase with the disturbing force. However, for high excitation frequencies, $r \gg 1$, the phase angle approaches 180° and is nearly out of phase with the disturbing force. At resonance, Eq. (5.47) must equal zero; therefore

$$\psi_{\text{res}} = 180° \tag{5.57}$$

For small values of damping, the change in phase angle occurs rather abruptly near the resonance condition as illustrated in Figure 5.11. If the damping factor is zero, then the phase angle is either 0° or 180°, depending upon whether the excitation frequency is less than or greater than the resonant frequency.

Figure 5.11

Phase angle versus frequency ratio for various levels of damping. *Mechanical Vibrations, Theory and Applications,* 2/e, by Tse/Morse/Hinkle. Reprinted by permission of authors.

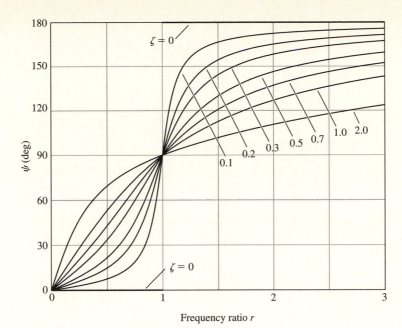

EXAMPLE 5.5 ▼

A structure weighing 1200 lb, having a stiffness $k = 800$ lb/in, and a damping coefficient $c = 6.0$ lb-sec/in is subject to a harmonic force of amplitude $F_0 = 400$ lb. Determine the resonant and maximum amplitudes of steady-state vibration.

Solution

The pertinent system parameters are

$$\omega = \sqrt{\frac{k}{m}} = \sqrt{\frac{(800)(386.4)}{1200}} = 16.05 \text{ rad/sec}$$

$$C_c = 2m\omega = 2\,\frac{1200}{386.4}\,16.05 = 99.7 \text{ lb-sec/in}$$

$$\zeta = \frac{c}{C_c} = \frac{6}{99.7} = 0.06$$

$$X_0 = \frac{F_0}{k} = \frac{400}{80} = 0.50 \text{ in}$$

Therefore, from Eq. (5.56), the displacement amplitude at resonance is

$$x_{\text{res}} = (\text{DMF})_{\text{res}} X_0 = \frac{0.5}{2(0.06)} = 4.167 \text{ in}$$

and from Eq. (5.55), the maximum displacement is calculated as

$$x_{\max} = (DMF)_{\max}X_0 = \frac{0.5}{2(0.06)\sqrt{1-(0.06)^2}} = 4.174 \text{ in}$$

▲

5.4.1 Sharpness of Resonance

In forced vibration, it is convenient to relate damping to the *sharpness of resonance* or the width of the resonant peak as illustrated in Figure 5.12. As expressed by Eq. (5.56), the dynamic magnification factor at resonance is $(DMF)_{res} = 1/2\zeta$. It is common practice to measure the width of the resonant peak at a point on either side of resonance, at a dynamic magnification factor equal to

$$\frac{1}{\sqrt{2}}(DMF)_{res} = \frac{1}{2\sqrt{2}\zeta} \tag{5.58}$$

These points are referred to as the *half-power points* and are shown in Figure 5.12. The frequency ratios corresponding to the half-power points may be determined from Eq. (5.48) as

$$\frac{1}{2\sqrt{2}\zeta} = \frac{1}{\sqrt{(1-r^2)^2 + (2\zeta r)^2}} \tag{5.59}$$

Solving for r^2 in Eq. (5.59) yields the roots

$$r_{1,2}^2 = 1 - 2\zeta^2 \pm 2\zeta\sqrt{1+\zeta^2} \tag{5.60}$$

Assuming $\zeta \ll 1$ and neglecting higher order terms of ζ, Eq. (5.60) can be written as

$$r_{1,2}^2 = 1 \pm 2\zeta \tag{5.61}$$

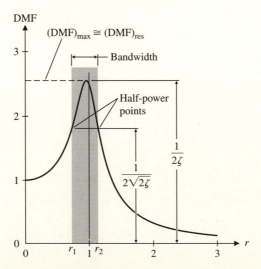

Figure 5.12

Harmonic response curve illustrating bandwidth and half-power points.

or

$$r_{1,2} = 1 \pm \zeta \tag{5.62}$$

The difference between the two frequency ratios defines the *bandwidth* given by

$$r_2 - r_1 = 2\zeta \tag{5.63}$$

The reciprocal of the bandwidth is commonly defined as the *quality factor Q,* given by

$$Q = \frac{1}{r_2 - r_1} = \frac{1}{2\zeta} \tag{5.64}$$

Equation (5.64) provides the basis for an experimental method for determining the equivalent damping in a system. A simple application of the procedure is illustrated in the following example.

EXAMPLE 5.6   ▼

Data collected from a frequency response test of a structure was plotted to construct a response curve similar to that shown in Figure 5.12. From this plot it was determined that the $(DMF)_{max}$ was 1.35 and the DMF at the half-power points was $1.35/\sqrt{2}$, or 0.95. The response ratios r_1 and r_2 corresponding to the half-power points were then determined from the response curve to be 0.91 and 1.05, respectively. Estimate the amount of damping in the system.

Solution

From Eq. (5.64), the quality factor Q is determined as

$$Q = \frac{1}{1.05 - 0.91} = 7.14 \tag{1}$$

Also from Eq. (5.64),

$$Q = \frac{1}{2\zeta} = 7.14 \tag{2}$$

and therefore from Eq. (2) the damping in the system is estimated as

$$\zeta \cong 0.07$$

▲

5.5   HARMONIC EXCITATION CAUSED BY ROTATING UNBALANCE

A very common source of forced harmonic vibration force is imbalance due to the reciprocating mass in rotating machinery such as turbines, centrifugal pumps, and turbogenerators. In this context, imbalance in a rotating machine means the axis of rotation does not coincide with the axis of inertia of the whole unit. In the case of high-speed machines such as turbines, a very small eccentricity can result in large unbalanced forces.

A typical rotating mass–type oscillator is depicted in Figure 5.13a. An unbalanced mass m_r, having eccentricity e from the center of rotation, rotates at angular velocity Ω. If the system is constrained to move in the vertical direction only, and letting m represent the total mass of the machine, the resulting forces are shown in the free-body diagram presented in Figure 5.13b. Applying Newton's second law in the x direction, it follows that

$$\sum F_x = -m\ddot{x} - c\dot{x} - kx + m_r e\Omega^2 \sin \Omega t = 0 \tag{5.65}$$

or

$$m\ddot{x} + c\dot{x} + kx = m_r e\Omega^2 \sin \Omega t \tag{5.66}$$

The term $m_r e\Omega^2$ in Eq. (5.66) represents an applied force comparable to F_0 in Eq. (5.32a).

In standard form, the differential equation of motion is

$$\ddot{x} + \frac{c}{m}\dot{x} + \frac{k}{m}x = \frac{m_r e\Omega^2}{m} \sin \Omega t \tag{5.67}$$

which is completely analogous to Eq. (5.32b). Therefore, the steady-state solution for Eq. (5.67) is completely analogous to Eq. (5.44). Thus

$$x(t) = X_f \sin(\Omega t - \psi) \tag{5.68}$$

where

$$X_f = \frac{X_0}{\sqrt{(1-r^2)^2 + (2\zeta r)^2}} \tag{5.69}$$

For the case of rotating unbalance,

$$X_0 = \frac{F_0}{k} = \frac{m_r e\Omega^2}{k} = \frac{m_r e r^2}{m} \tag{5.70}$$

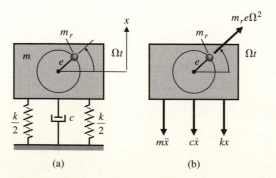

Figure 5.13

Harmonic excitation from rotating unbalance.

(a)   (b)

Figure 5.14

Steady state amplitude response of SDOF system having rotating unbalance.

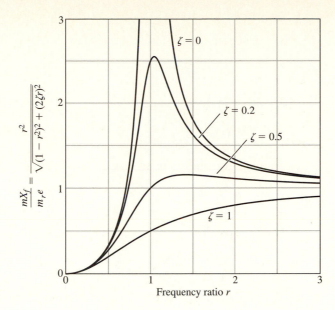

Substituting Eq. (5.70) into Eq. (5.69) gives the amplitude of forced vibration as

$$X_f = \frac{(m_r e r^2)/m}{\sqrt{(1 - r^2)^2 + (2\zeta r)^2}}$$

(5.71)

or in a more convenient nondimensional form as

$$\frac{mX_f}{m_r e} = \frac{r^2}{\sqrt{(1 - r^2)^2 + (2\zeta r)^2}}$$

(5.72)

The ratio $mX_f/m_r e$ plotted as a function of frequency ratio r for several values of damping factor ζ is presented in Figure 5.14. It should be noted that the circular frequency Ω of the forced vibration is the same as the rotating speed. Thus, at low speeds, X_f is small since the exciting inertia force is small, as illustrated in Figure 5.14. At very high speeds, it is noted that the inertia forces predominate and the amplitude X_f approaches $m_r e/m$. The phase angle ψ is identical with the phase angle for a harmonic force $F(t) = F_0 \sin \Omega t$. In this context, ψ refers to the motion of the unbalanced force.

EXAMPLE 5.7   ▼

A rotating machine similar to that depicted in Figure 5.13 has a total weight of 700 lb. At an operating speed of 1800 rpm, the machine has an unbalance of 3.0 lb at an eccentricity of 5 in. Resonance was observed to occur at an operating speed of 900 rpm, and the damping factor ζ was determined to be 0.001. Determine the vibration amplitude for the system when operating at 1800 rpm. If the operating speed is set very high, determine the stationary value of the steady-state vibration amplitude.

Solution

The natural frequency for the system is

$$\omega = 900 \frac{2\pi}{60} = 94.25 \text{ rad/sec}$$

and the operating frequency Ω is

$$\Omega = 1800 \frac{2\pi}{60} = 188.5 \text{ rad/sec}$$

Thus the frequency ratio becomes

$$r = \frac{\Omega}{\omega} = 2.0$$

We also note that

$$\frac{m_r e}{m} = \frac{(3)(5)}{700} = 0.0214 \text{ in}$$

At an operating speed of 1800 rpm, Eq. (5.69) yields the steady-state amplitude as

$$X_f = \frac{m_r e}{m} \frac{r^2}{\sqrt{(1 - r^2)^2 + (2\zeta r)^2}}$$

$$= \frac{4(0.0214)}{\sqrt{(3)^2 + [2(0.001)(2)]^2}} = 0.0285 \text{ in}$$

As the speed is increased, the amplitude of X_f approaches the value $m_r e/m$; that is,

$$X_f \to \frac{m_r e}{m} = 0.0214 \text{ in}$$

▲

5.6 BASE EXCITATION

Another very important class of vibration problems corresponds to the system being excited by the motion of its base or support. The motion of the base results in forces being transmitted to the system, and the resulting vibrations can be severe if the system is not properly designed. This type of induced vibration is generally of concern in the mounting of sensitive equipment in aircraft and other types of moving vehicles. It should be noted that when an excitation motion is applied to the support or base of the system instead of applying to the mass, both the *absolute* motion of the mass and the *relative* motion between the mass and the support are of interest. In this discussion, the absolute motion of the mass is considered.

An illustration of support-induced vibration is depicted in Figure 5.15a. The support exhibits vertical harmonic motion given by $y(t) = Y \sin \Omega t$, in which Y is the displacement amplitude of the support, and the absolute displacement x of the mass m is measured from its at-rest static equilibrium position. The free-body diagram for the sys-

Figure 5.15

SDOF system excited by harmonic base motion.

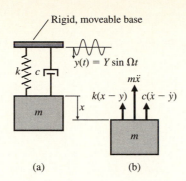

tem is shown in Figure 5.15b, where it is assumed that $x > y$ and $\dot{x} > \dot{y}$. Application of Newton's second law yields the equation of motion

$$m\ddot{x} = -k(x - y) - c(\dot{x} - \dot{y}) \tag{5.73}$$

or

$$m\ddot{x} + c\dot{x} + kx = ky + c\dot{y} \tag{5.74}$$

Substituting $Y \sin \Omega t$ and $\Omega Y \cos \Omega t$ for y and $\dot{y}$, respectively, in Eq. (5.74) yields

$$m\ddot{x} + c\dot{x} + kx = kY \sin \Omega t + c\Omega Y \cos \Omega t \tag{5.75}$$

Taking advantage of the trigonometric identity

$$A \sin bt + B \cos bt = D \sin(bt + \gamma) \tag{5.76}$$

Eq. (5.75) can be expressed as

$$m\ddot{x} + c\dot{x} + kx = Y\sqrt{k^2 + (c\Omega)^2} \sin(\Omega t - \gamma) \tag{5.77}$$

where

$$\gamma = \tan^{-1}\left(-\frac{c\Omega}{k}\right) = \tan^{-1}(-2\zeta r) \tag{5.78}$$

Equation (5.77) is completely analogous to Eq. (5.32a), whereas in this case the amplitude of the forcing function is $Y\sqrt{k^2 + (c\Omega)^2}$. Therefore the solution to Eq. (5.77) will be analogous to that represented by Eqs. (5.44) through (5.47). Thus, the steady-state displacement motion of the mass is given by

$$x(t) = \frac{Y\sqrt{k^2 + (c\Omega)^2}}{\sqrt{(k - m\Omega^2)^2 + (c\Omega)^2}} \sin(\Omega t - \beta) \tag{5.79}$$

or

$$x(t) = \frac{Y\sqrt{1 + (2\zeta r)^2}}{\sqrt{(1 - r^2)^2 + (2\zeta r)^2}} \sin(\Omega t - \beta) \tag{5.80}$$

Representing the amplitude of the steady-state response by X_f such that

$$X_f = Y\frac{\sqrt{1 + (2\zeta r)^2}}{\sqrt{(1 - r^2)^2 + (2\zeta r)^2}} \tag{5.81}$$

Eq. (5.80) becomes

$$x(t) = X_f \sin(\Omega t - \beta) \tag{5.82}$$

where

$$\beta = \gamma + \psi \tag{5.83}$$

and

$$\psi = \tan^{-1}\left(\frac{2\zeta r}{1 - r^2}\right) \tag{5.84}$$

It is convenient to express the vibration amplitude of the mass as a ratio with respect to the amplitude of the base motion. Thus, from Eq. (5.81),

$$\left|\frac{X_f}{Y}\right| = \frac{\sqrt{1 + (2\zeta r)^2}}{\sqrt{(1 - r^2)^2 + (2\zeta r)^2}} \tag{5.85}$$

The ratio X_f/Y is referred to as the *transmission ratio*, or simply the *transmissibility*. It is a measure of the motion that is transmitted to the mass due to the excitation of the base. The concept of transmissibility is examined in detail in the next section.

EXAMPLE 5.8 ▼

An instrument panel weighing 30 lb is installed in a vehicle on a flexible mount that has an equivalent stiffness $k = 400$ lb/in and negligible damping. As a result of engine vibration the vehicle vibrates with a steady-state amplitude of 0.25 in at a frequency of 40 Hz. Determine the steady-state amplitude of vibration of the instrument panel.

Solution

The frequency of the panel is

$$\omega = \sqrt{\frac{k}{m}} = \sqrt{\frac{(400)(386.4)}{30}} = 71.78 \text{ rad/sec}$$

and the frequency ratio is

$$r = \frac{\Omega}{\omega} = \frac{40(2\pi)}{71.78} = 3.5$$

The steady-state displacement amplitude of the mass is determined from Eq. (5.81); noting that $\zeta = 0$, then

$$X_f = \frac{0.25}{1 - (3.5)^2} = -0.022 \text{ in}$$

The minus sign indicates that the motion of the panel is out of phase with that of the vehicle. ▲

5.7 VIBRATION ISOLATION AND TRANSMISSIBILITY

There are two aspects to the principle of vibration isolation: (1) isolation of a force and (2) isolation of a motion. The first category is concerned with the isolation of forces created by rotating and reciprocating machinery, such as fans, compressors, and turbines. The main objective of isolation in this instance is the reduction of the force transmitted to the support from the machinery. The second category of isolation is primarily concerned with the reduction of motions that typically occur in aircraft, ships, and other type vehicles, which are transferred to sensitive equipment mounted within. In either case, the overall objective is essentially the same, that is, to minimize the force or motion being transmitted. This is usually accomplished by mounting the equipment upon resilient supports or *vibration isolators* in such a fashion that the natural frequency of the equipment-isolator system is significantly lower than the disturbing frequency being isolated.

Consider the system excited by a harmonic force depicted in Figure 5.16. In addition to the usual forces acting on the mass, forces acting on the support are also shown in Figure 5.16b. The forces transmitted to the support, F_T, are given by

$$F_T = kx + c\dot{x} \tag{5.86}$$

Substituting the solution for displacement of the mass, x, given by Eq. (5.44), and its first time derivative $\dot{x}$ into Eq. (5.86) yields

$$F_T = kX_f \sin(\Omega t - \psi) + C\Omega X_f \cos(\Omega t - \psi) \tag{5.87}$$

The two harmonic terms on the right-hand side of Eq. (5.87) may be combined to obtain

$$F_T = X_f \sqrt{k^2 + (c\Omega)^2} \sin(\Omega t - \psi - \gamma) \tag{5.88}$$

or

$$F_T = X_f \sqrt{k^2 + (c\Omega)^2} \sin(\Omega t - \beta) \tag{5.89}$$

where

$$\beta = \psi + \gamma \tag{5.90}$$

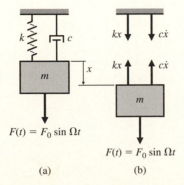

Figure 5.16

Force transmitted to support for a viscously damped SDOF system subject to harmonic exciting force.

$F(t) = F_0 \sin \Omega t$

(a)

(b)

and

$$\gamma = \tan^{-1}\left(-\frac{c\Omega}{k}\right) = \tan^{-1}(-2\zeta r) \tag{5.91}$$

Thus the amplitude, or maximum value, of the transmitted force is determined from Eq. (5.84), neglecting the harmonic term, as

$$(F_T)_{\text{max}} = X_f \sqrt{k^2 + (c\Omega)^2} \tag{5.92}$$

or

$$(F_T)_{\text{max}} = \frac{X_0 \sqrt{k^2 + (c\Omega)^2}}{\sqrt{(1 - r^2)^2 + (2\zeta r)^2}} \tag{5.93}$$

The *transmissibility* TR is defined as the ratio of the maximum transmitted force to the amplitude of the applied force F_0. Thus, noting that $X_0 = F_0/k$,

$$\text{TR} = \frac{(F_T)_{\text{max}}}{F_0} = \frac{\frac{1}{k}\sqrt{k^2 + (c\Omega)^2}}{\sqrt{(1 - r^2)^2 + (2\zeta r)^2}} \tag{5.94}$$

or

$$\text{TR} = \frac{\sqrt{1 + (2\zeta r)^2}}{\sqrt{(1 - r^2)^2 + (2\zeta r)^2}} \tag{5.95}$$

It should be noted that this expression for transmissibility is exactly the same as the ratio $|X_f/Y|$ given by Eq. (5.85). Therefore it is apparent that the transmissibility represented by Eq. (5.95) is applicable to the two aspects of vibration isolation discussed earlier in this section. In the case of vibration being instigated by a machine, transmissibility refers to the *dynamic forces* being transmitted to the support structure. In the case of base excitation discussed in Section 5.5, transmissibility refers to the *dynamic motion* being transmitted from the support to the equipment or machinery (mass).

It is also interesting to note that the transmissibility is the product of the dynamic magnification factor given by Eq. (5.48) and the term $\sqrt{1 + (2\zeta r)^2}$. Therefore Eq. (5.95) may be written alternatively as

$$\text{TR} = \sqrt{1 + (2\zeta r)^2} \ \text{DMF} \tag{5.96}$$

A plot of transmissibility TR versus frequency ratio r for several values of damping ζ is presented in Figure 5.17. Several important observations can be made from this plot. First it is noted that the transmissibility is equal to unity for all values of damping ζ, when $r = \sqrt{2}$, and the individual curves become asymptotic to zero as the frequency ratio increases beyond $\sqrt{2}$. It is also noted that increased damping decreases the transmissibility when $r < \sqrt{2}$; however, for $r > \sqrt{2}$, the transmissibility decreases as damping is decreased. These observations lead to two important conclusions regarding vibration isolation: (1) since TR > 1 for $r < \sqrt{2}$ and *TR* < 1 for $r > \sqrt{2}$, vibration isola-

Figure 5.17

Transmissibility TR as a function of frequency ratio and damping.

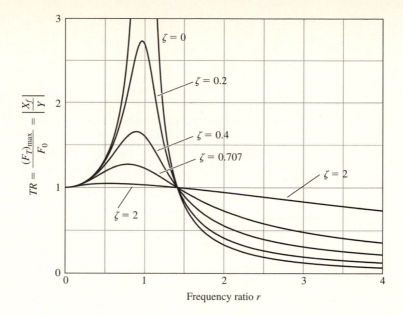

tion can be achieved only in the range of $r > \sqrt{2}$; (2) since damping increases transmissibility in the isolation range (TR $> \sqrt{2}$), the most effective vibration absorbers consist of spring elements with only little or no damping.

EXAMPLE 5.9 ▼

A rotating machine similar to that shown in Figure 5.13 has a total weight of 100 lb and is mounted on a structure having stiffness $k = 1000$ lb/in and damping $\zeta = 0.1$. The machine exhibits a rotating imbalance of 7 lb-in. If the operating speed of the machine is 500 rpm, determine (a) the amplitude of the steady-state vibration, (b) the transmissibility, and (c) the maximum dynamic force transmitted to the base.

Solution

(a) The operating frequency of the machine is

$$\Omega = \frac{500(2\pi)}{60} = 52.36 \text{ rad/sec}$$

and the natural circular frequency of the system is

$$\omega = \sqrt{\frac{(1000)(386.4)}{100}} = 62.16 \text{ rad/sec}$$

Thus the frequency ratio is

$$r = \frac{52.36}{62.16} = 0.842$$

From Eq. (5.71) the steady-state vibration amplitude is given by

$$X_f = \frac{\left(\frac{7}{100}\right)(0.842)^2}{\sqrt{[1 - (0.842)^2]^2 + [2(0.1)(0.842)]^2}} = 0.076 \text{ in}$$

(b) The transmissibility is determined from Eq. (5.95) as

$$TR = \frac{\sqrt{1 + [2(0.1)(0.842)]^2}}{\sqrt{[1 - (0.842)^2]^2 + [2(0.1)(0.842)]^2}} = \frac{1.014}{0.649} = 1.56$$

(c) The force transmitted to the base is given by

$$(F_T)_{max} = TR(F_0)$$

or

$$(F_T)_{max} = 1.56 \frac{(7)(52.36)^2}{386.4} = 77.48 \text{ lb}$$

▲

EXAMPLE 5.10 ▼

An instrument package that weighs 250 lb is to be mounted in a train on a console that experiences a vibration amplitude of 0.4 in at a frequency of 35 Hz while the train is running. Determine (a) the spring constant necessary for a mount in order to achieve 80% isolation and (b) the resulting vibration amplitude of the instrument package. Neglect any damping in the system.

Solution

(a) The excitation frequency is

$$\Omega = 2\pi(35) = 219.9 \text{ rad/sec}$$

and the transmissibility is determined from Eq. (5.95) as

$$TR = \frac{1}{|1 - r^2|} = 0.2 \qquad (1)$$

Solving Eq. (1) for r yields

$$|1 - r^2| = 5 \qquad (2)$$

or

$$r = \sqrt{6} = 2.45 \qquad (3)$$

Noting that

$$r = \frac{\Omega}{\omega} \tag{4}$$

and expressing Eq. (4) as

$$r\omega = 2.45 \sqrt{\frac{k}{m}} \tag{5}$$

we then have from Eqs. (3) and (5),

$$r\omega = 2.45 \sqrt{\frac{k(384.6 \text{ in/sec}^2)}{250 \text{ lb}}} = 219.9 \tag{6}$$

Solving Eq. (6) for k yields

$$k = \frac{(219.9)^2(250)}{(384.6)(2.45)^2} = 5236.59 \text{ lb/in}$$

(b) The steady-state vibration amplitude is determined from Eq. (5.80) as

$$X_f = \frac{0.4}{\left|1 - (2.45)^2\right|} = 0.08 \text{ in}$$

▲

REFERENCES

1 Church, Austin H., *Mechanical Vibrations,* Wiley, New York, 1963.

2 Anderson, Roger A., *Fundamentals of Vibration,* Macmillan, New York, 1967.

3 Dimarogonas, Andrew D., *Vibration Engineering,* West Publishing, St. Paul, MN, 1976.

4 Crede, Charles E., *Vibration and Shock Isolation,* Wiley, New York, 1951.

5 Arya, S., O'Neill, M., and Pincus, G., *Design of Structures and Foundations for Vibrating Machines,* Gulf Publishing, Houston, TX, 1979.

6 Srinivasulu, P. and Vaidyanathan, C.V., *Handbook of Machine Foundations,* McGraw-Hill, New York, 1977.

NOTATION

A, B	constants in free vibration (steady-state) solution	F_o	amplitude of externally applied harmonic force
c	viscous damping coefficient	F_T	transmitted force
C, D	constants in particular solution for a viscously damped system	$F(t)$	externally applied time-varying force
		g	acceleration due to gravity
e	eccentricity of rotating mass	I_0	mass moment of inertia

k	stiffness		X_0	equivalent static deflection
k_T	torsional spring constant		$y(t)$	support displacement motion
m	mass		Y	displacement amplitude of support motion
m_r	rotating mass		DMF	dynamic magnification factor
Q	quality factor		TR	transmissibility (or transmission ratio)
r	frequency ratio		α	constant
t	time		β	constant
T	undamped natural period (sec)		ε	represents a very small quantity
x	displacement		ϕ	phase angle (rad) of free vibration response
$x_h(t)$	homogeneous displacement solution		γ	constant
$x_p(t)$	particular displacement solution		ω	undamped natural circular frequency (rad/sec)
x_0	initial displacement		ω_d	damped natural circular frequency (rad/sec)
$\dot{x}$	velocity		Ω	circular frequency (rad/sec) of externally applied harmonic force; also represents angular velocity of rotating mass
$\dot{x}_0$	initial velocity			
$\ddot{x}$	acceleration			
X	displacement amplitude of free vibration response		ψ	phase angle of steady-state solution
X_f	displacement amplitude of forced vibration (steady-state) response		ζ	damping factor

PROBLEMS

5.1 An undamped spring-mass system is subjected to a harmonic force having an amplitude $F_0 = 80$ N and a frequency of 8 cps. The mass is 6 kg and the spring constant is 3 N/m. Determine the displacement amplitude of the steady-state response, X_f.

5.2 The steady-state vibration displacement amplitude of an undamped spring-mass system was determined to be 0.50 in. If the spring stiffness is 20 lb/in and the applied harmonic force is $50 \sin 3t$ lb, calculate the weight of the mass.

5.3 A system consisting of a spring with a stiffness of 25 lb/in supports a mass weighing 20 lb. The system is subjected to a harmonic force having an amplitude $F_0 = 30$ lb and a frequency of 120 Hz. Determine (a) the amplitude of the steady-state displacement response and (b) the weight of the mass required to reduce the vibration amplitude by 50%.

5.4 An undamped system consisting of a 10-kg mass and a spring of stiffness $k = 4.0$ kN/m is acted upon by a harmonic force of amplitude $F_0 = 0.5$ kN. The displacement amplitude of the steady-state response was observed to be 11 cm. Determine the frequency of the excitation force.

5.5 An undamped system having a mass of 10 kg and a spring constant of 8 N/mm is excited by a harmonic force with an amplitude $F_0 = 200$ N and an operating frequency of 35 rad/sec. If the initial conditions are $x(0) = 21$ mm and $\dot{x}(0) = 175$ mm/sec,

determine the total displacement, velocity, and acceleration of the mass at (a) $t = 2$ sec, (b) $t = 4$ sec, and (c) $t = 6$ sec.

5.6 An undamped system consists of a mass of weight $W = 100$ lb and a spring of stiffness $k = 250$ lb/in. The mass is excited by two harmonic forcing functions: $F_1(t) = 500 \sin 7t$ lb and $F_2(t) = 200 \sin 25t$ lb. Determine the displacement, velocity, and acceleration of the mass at times $t = \pi/3$ sec and $t = \pi/2$ sec.

5.7 A harmonic force having an amplitude $F_0 = 50$ N operating at a frequency of 15 Hz is applied to an undamped spring-mass system producing a steady-state displacement amplitude of 15 mm. Determine the spring constant k of the system.

5.8 An undamped system having a mass of 50 kg is excited by a harmonic force having an amplitude $F_0 = 100$ N and an operating frequency of 10 Hz. The steady-state displacement amplitude of the vibration is observed to be 3.2 mm. Determine the spring constant k for the system.

5.9 A spring-mass system is excited by a harmonic force having an amplitude $F_0 = 40$ N and an operating frequency of 500 cycles/min. The system has a mass of 10 kg and exhibits a steady-state displacement amplitude of vibration of 5 mm. Determine the spring constant k for the system.

5.10 A harmonic torque $T(t) = T_0 \cos \Omega t$ acts on a disk having mass moment of inertia about its center of mass of I_0. The disk is mounted on a shaft between two frictionless bearings such that there is no restoring torque (or spring). Determine an expression for the rotation of the disk as a function of time.

5.11 A uniform slender rod of length L and mass m is acted upon by a harmonic force $F(t) = F_0 \sin \Omega t$ as shown in Figure P5.1. Determine (a) the differential equation of motion for the rod and (b) the angular position of the rod as a function of time when subject to initial conditions $\theta(0) = \theta_0$ and $\dot{\theta}(0) = \dot{\theta}_0$. Assume small oscillations.

5.12 For the uniform slender rod shown in Figure P5.1, let $L = 10$ in, $F_0 = 1$ lb, $\Omega = 8$ rad/sec, and the weight of the rod $= 10$ lb. Determine the steady-state amplitude of angular oscillations. Assume zero initial conditions.

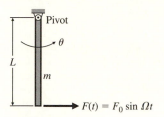

Pivot
θ
L
m
$F(t) = F_0 \sin \Omega t$

Figure P5.1

5.13 A disk of radius R and mass moment of inertia I_0 is mounted on a solid circular shaft of diameter D, length L, and shear modulus G as shown in Figure P5.2. The disk is excited by a harmonic torque $T(t) = T_0 \sin \Omega t$. Derive the differential equation of motion for the disk. For initial conditions $\theta(0) = \theta_0$ and $\dot{\theta}(0) = \dot{\theta}_0$, determine the expression for angular oscillations as a function of time.

5.14 For the torsional system shown in Figure P5.2, the inertial disk has a weight of 200 lb and a radius $R = 4$ in. The shaft has a length $L = 36$ in, diameter $d = 1.5$ in, and a shear modulus $G = 10 \times 10^6$ lb/in^2. The disk is excited by a harmonic torque of amplitude $T_0 = 500$ ft-lb at a frequency of $\Omega = 300$ rad/sec. Determine the amplitude of the steady-state oscillations.

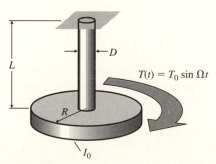

D
L
$T(t) = T_0 \sin \Omega t$
R
I_0

Figure P5.2

5.15 A 10-kg mass is supported by a spring having stiffness $k = 8$ N/m. The mass is driven at resonance by a harmonic force having an amplitude $F_0 = 20$ N. Calculate the amplitude of the steady-state displacement response after (a) 2 1/2 cycles, (b) 5 1/2 cycles, (c) 8 1/2 cycles, and (d) 12 1/2 cycles.

5.16 A spring-mass system is excited by a harmonic force that exhibits a beat period of 0.5 sec. The period of oscillations within the beat is 0.05 sec. Determine the natural frequency of the system and the frequency of the exciting force.

5.17 An undamped system is harmonically forced near resonance, resulting in a beating condition. The natural frequency of the system is 1800 cpm and the frequency of the impressed force is 1785 cpm. Determine the beat period and the number of oscillations within each beat.

5.18 A spring-mass system having a natural frequency of 50 Hz is excited by a harmonic force having a frequency of 49.8 Hz. Calculate the beat period.

5.19 Determine the equation describing the motion of an undamped spring-mass system acted upon by a harmonic force $F(t) = F_0 \cos \Omega t$ for the initial conditions $x(0) = x_0$ and $\dot{x}(0) = \dot{x}_0$ for (a) $r \neq 1$ and (b) $r = 1$.

5.20 A 100-lb weight is suspended by a linear spring of stiffness $k = 15$ lb/in and a viscous dashpot with a damping coefficient $c = 0.10$ lb-sec/in. If the system is excited by a harmonic force of amplitude $F_0 = 50$ lb, determine the steady-state amplitude of motion at resonance.

5.21 The ratio of the steady-state displacement amplitudes corresponding to $r = 0.5$ and $r = 1$ is 1:4 for a certain viscously damped system. Determine the damping factor ζ for the system.

5.22 For the viscously damped system shown in Figure P5.3 the weight of the mass $W = 100$ lb, the spring stiffness $k = 175$ lb/in, damping coefficients $c_1 = 1.5$ lb-sec/in and $c_2 = 2.5$ lb-sec/in, force amplitude $F_0 = 15$ lb, and operating frequency $\Omega = 12$ rad/sec. Determine the amplitude of the steady-state displacement.

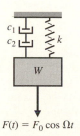

c_1
c_2
k
W
$F(t) = F_0 \cos \Omega t$

Figure P5.3

5.23 Show that the maximum DMF for a viscously damped spring-mass system subject to a harmonic exciting force occurs at a frequency ratio $r = \sqrt{1 - 2\zeta^2}$ for $\zeta \leq 1/\sqrt{2}$.

5.24 A torsional system consists of a disk having a mass moment of inertia $I_0 = 1.5$ lb-in/sec^2 as well as a shaft having a torsional spring constant $k_T = 12,000$ in-lb/rad and a damping coefficient $c_r = 50$ lb-in-sec/rad. If a harmonic torque of amplitude $T_0 = 120$ in-lb is applied to the disk at a frequency of 600 cpm, determine (a) the amplitude of the steady-state displacement and (b) the maximum shear stress in the shaft. Assume the shaft is 12 in long and has a diameter of 0.6 in.

5.25 A mass weighing 100 lb is restrained by a spring having stiffness $k = 15$ lb/in and by a viscous dashpot with a damping coefficient $c = 0.05$ lb-sec/in. If the system is excited by a harmonic force of amplitude $F_0 = 25$ lb, determine the amplitude of the steady-state displacement at resonance.

5.26 A uniform rigid bar of length L and mass m is pinned at one end and restrained by a spring and dashpot as shown in Figure P5.4. One end of the bar is subject to a harmonic force $F(t) = F_0 \cos \Omega t$. Determine the angular amplitude of the steady-state response.

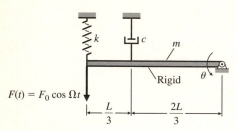

$F(t) = F_0 \cos \Omega t$

Figure P5.4

5.27 For a viscously damped system excited by the harmonic force $F(t) = F_0 \sin \Omega t$, plot the dynamic magnification factor curve (DMF vs. r) for $\zeta = 0.25$. Plot over the domain $0 \le r \le 5$.

5.28 A viscously damped system is excited by a harmonic force $F(t) = F_0 \sin \Omega t$. At resonance the steady-state amplitude was observed to be 0.75 in, and at 75% of the resonant frequency the amplitude was measured to be 0.62 in. Determine the damping factor ζ for the system.

5.29 A damped torsional system is composed of a shaft having a torsional spring constant $k_T = 75,000$ lb-in/rad, as well as a disk having a mass moment of inertia $I_0 = 30$ lb-in/sec^2 and a torsional damping constant $c_r = 900$ lb-in-sec/rad. A harmonic torque with an amplitude $T_0 = 3000$ in-lb acts on the disk producing a steady-state amplitude of angular oscillation of 4.0°. Determine (a) the frequency of the applied torque and (b) the maximum torque transmitted to the support.

5.30 For the system described in Problem 5.22 and illustrated in Figure P5.3, determine the force transmitted to the support.

5.31 A damped system having a mass of 100 kg is supported on springs having a total stiffness of 40,000 N/m and a damping factor $\zeta = 0.15$. The mass is subjected to a harmonic force having amplitude $F_0 = 500$ N and frequency $\Omega = 10$ rad/sec. For the sustained steady-state vibrations, determine (a) the amplitude of the displacement of the mass, (b) its phase with respect to the exciting force, (c) the maximum velocity and acceleration of the mass, and (d) the maximum dynamic force transmitted to the support.

5.32 A machine weighing 3,000 lb is mounted on springs having a total stiffness $k = 11,000$ lb/in. A piston weighing 75 lb moves harmonically in the vertical direction within the machine at a speed of 600 cpm and a stroke of 16 in. The amplitude of the steady-state displacement is 0.507 in. Determine (a) the value of the damping factor ζ and (b) the force transmitted to the support.

5.33 A vibration exciter is attached to a structure as shown in Figure P5.5. The exciter contains two eccentric weights w_0 that rotate with identical speeds but in opposite directions. The unbalance ($w_0 e$) of each eccentric weight is 5 in-lb. The entire arrangement has a weight of 250 lb. When the exciter speed is adjusted to 900 rpm, a resonant condition is observed and the structure exhibits a maximum upward displacement of 1.0 in at the instant the eccentric weights are in the horizontal position. Determine the damping factor ζ for the structure. If the exciter speed is changed to 1300 rpm, determine the amplitude of the steady-state displacement.

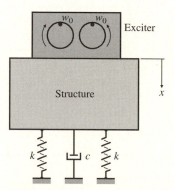

Figure P5.5

5.34 A machine weighing 2000 lb is mounted on isolators and is subjected to a harmonic force during operation. For an operating speed of 1200 rpm, determine the isolator stiffness k required for a transmissibility of 15%. Assume $\zeta = 0$.

5.35 An engine weighing 1000 lb is supported on mounts having a total stiffness $k = 20,000$ lb/in and damping factor $\zeta = 0.1$. The piston weighs 20 lb and has a stroke of 9 in. At an operating frequency of 1500 rpm, determine the force transmitted to the support.

5.36 An instrument package weighing 50 lb is supported on a floor that exhibits harmonic motions with frequencies in the vicinity of 2500 cpm. Determine the required stiffness k for vibration isolators to attain 85% isolation. Assume damping is negligible.

5.37 A package weighing 50 lb is suspended in a box, as shown in Figure P5.6, by springs each having a stiffness $k = 250$ lb/in. The box is placed inside a truck that produces vertical harmonic vibrations during transport of amplitude $y(t) = 1.5 \sin 4t$ in. Determine the maximum displacement, velocity, and acceleration experienced by the package.

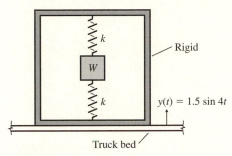

Figure P5.6

5.38 For the system shown in Figure P5.7, determine (a) the amplitude of the steady-state displacement of the mass and (b) the phase relationship of the relative displacement of the mass x and the prescribed motion of the base $y(t)$.

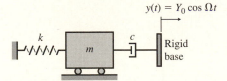

Figure P5.7

5.39 An instrument package weighing 110 N is to be isolated from engine vibrations ranging in frequencies from 1500 cpm to 2000 cpm. Determine the spring stiffness for isolators to achieve 85% isolation.

5.40 A delicate measurement device weighing 15 lb is to be installed on a shop bench. The bench vibrates at 1890 cpm with an amplitude of 0.0023 in. The instrument is mounted on an isolator pad having stiffness $k = 45$ lb/in and a damping factor

$\zeta = 0.12$. Determine the displacement amplitude of the measurement device.

5.41 An electronic instrument having a mass of 125 kg is installed at a location where the acceleration is 20 cm/sec² at a frequency of 25 Hz. The instrument is mounted on isolators having stiffness $k = 3000$ N/cm and damping factor $\zeta = 0.07$. Determine the maximum acceleration experienced by the instrument.

5.42 A simplified representation of a vehicle having mass m, stiffness k, and damping c traveling over a rough road is shown in Figure P5.8. The wheel follows an idealized harmonic surface with amplitude Y and wavelength L. For the vehicle traveling at a velocity v_0, determine (a) the differential equation of relative motion for the vehicle mass m in the vertical direction and (b) the expression for the transmissibility of the vertical motion of the vehicle mass.

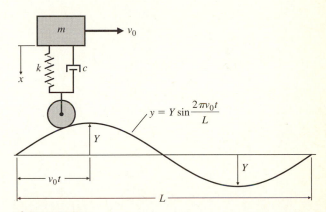

Figure P5.8

5.43 A 4000-lb. automobile is supported on four springs each with stiffness $k = 150$ lb/in. When the vehicle is traveling over a causeway, a resonant vertical motion within it is excited. Each span of the causeway is 80 ft in length and has a 0.75-in sag at midspan. Determine the maximum vertical displacement of the automobile when traveling at 65 mph. Assume no damping in the vehicle.

5.44 The support for a viscously damped system is harmonically oscillated. The body weighs 50 lb and the spring stiffness $k = 340$ lb/in (a) When the support is oscillated at resonance with an amplitude of 1.5 in, the body experiences a steady-state displacement amplitude of 4.65 in. Determine the damping factor ζ for the system. (b) When the support is oscillated at a frequency of 500 cpm, the steady-state displacement amplitude of the body is 1.12 in. Determine the maximum force transmitted to the support.

6 ▲ Response to Periodic and Arbitrary Dynamic Excitation

Harmonic excitation is a very common cause of vibration, especially in industrial and power installations where rotating and reciprocating machinery are commonplace. For such excitation, there is usually a closed-form solution available, as was illustrated in Chapter 5, that is also harmonic. In the vast majority of engineering problems, however, there are exciting forces that vary with time in a nonharmonic fashion. In many such instances, closed-form solutions do not exist; however, approximations can be found to satisfy engineering requirements.

General dynamic excitation can be distinguished in two broad categories: (1) periodic and (2) nonperiodic (arbitrary). Periodic excitation repeats itself at equal time intervals or periods. Harmonic excitation is a special case of periodic excitation, as was discussed in Chapter 1. Nonperiodic excitation does not repeat itself at equal time intervals, and can range from a short-duration blast load lasting only several milliseconds to a long-duration earthquake load lasting several minutes. This chapter addresses the response of SDOF systems to nonharmonic periodic excitation and to nonperiodic or arbitrary excitation.

6.1 RESPONSE TO PERIODIC EXCITATION

Let $F(t)$ represent a periodic excitation force acting on a system. A typical periodic force is depicted in Figure 6.1. By definition, a periodic function must satisfy the equality

$$F(t) = F(t + T_0) \tag{6.1}$$

where T_0 is the *period,* or minimum time required for $F(t)$ to repeat itself. Any periodic function can be expressed by a convergent infinite series of sine and cosine terms known as the *Fourier series*. Thus, a periodic exciting force $F(t)$ having period T_0 can be expressed by the Fourier series given by

$$F(t) = \frac{a_0}{2} + \sum_{n=1}^{\infty} (a_n \cos n\Omega t + b_n \sin n\Omega t) \tag{6.2}$$

Figure 6.1

Illustration of a periodic forcing function with period T_0.

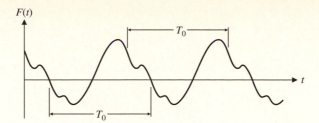

where Ω is the frequency of the forcing function equal to $2\pi/T_0$ and n is the set of positive integers 1, 2, 3, ... The coefficients a_0, a_n, and b_n are given by

$$a_0 = \frac{2}{T_0} \int_0^{T_0} F(t)\, dt \tag{6.3}$$

$$a_n = \frac{2}{T_0} \int_0^{T_0} F(t) \cos(n\Omega t)\, dt \tag{6.4}$$

$$b_n = \frac{2}{T_0} \int_0^{T_0} F(t) \sin(n\Omega t)\, dt \tag{6.5}$$

Therefore, given an excitation force expressed as the Fourier series, the constants a_0, a_n, and b_n must be evaluated by computing the integrals given by Eqs. (6.3), (6.4), and (6.5). Theoretically, convergence of a Fourier series requires an infinite number of terms. However, in practice a relatively small number of terms will render a sufficiently accurate approximation of the forcing function $F(t)$.

Evaluation of the Fourier coefficients may be expedited if the forcing function can be recognized as being odd or even. The forcing function is odd if it is antisymmetric with respect to the time origin, that is, $F(t) = -F(-t)$. For an odd function, the Fourier coefficients $a_0 = a_n = 0$. The function is even if it is symmetric with respect to the time origin, that is, $F(t) = F(-t)$. For an even function, the Fourier coefficients $b_n = 0$. Clearly then, an odd function can be represented by the Fourier sine series, and an even function by the Fourier cosine series. If, however, the function is neither odd nor even, then the full Fourier series must be employed. For practical applications, a computer solution for the Fourier coefficients is recommended.

EXAMPLE 6.1 ▼

Determine the Fourier series expansion for the square wave, periodic forcing function shown in Figure 6.2. Plot the Fourier representation for the first four non-zero harmonic components.

Figure 6.2

Periodic forcing function (square wave) used in Example 6.1.

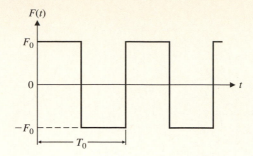

Solution

The Fourier integrals given by Eqs. (6.3), (6.4), and (6.5) can be evaluated over the interval $-T_0/2 \leq t \leq T_0/2$ and written as

$$a_0 = \frac{2}{T_0} \int_{-T_0/2}^{T_0/2} F(t)\, dt \tag{1}$$

$$a_n = \frac{2}{T_0} \int_{-T_0/2}^{T_0/2} F(t) \cos(n\Omega t)\, dt \tag{2}$$

$$b_n = \frac{2}{T_0} \int_{-T_0/2}^{T_0/2} F(t) \sin(n\Omega t)\, dt \tag{3}$$

where $\Omega = 2\pi/T_0$ and

$$F(t) = \begin{cases} -F_0 & \text{for } -\dfrac{T_0}{2} < t < 0 \\[2mm] F_0 & \text{for } 0 < t < \dfrac{T_0}{2} \end{cases} \tag{4}$$

Since $F(t)$ is an *odd* function, the coefficients $a_0 = a_n = 0$. Thus the b_n coefficients are represented by the Fourier sine series, and Eq. (3) becomes

$$b_n = \frac{4F_0}{T_0} \int_0^{T_0/2} \sin(n\Omega t)\, dt \tag{5}$$

or

$$b_n = \frac{4F_0}{T_0} \left(\frac{-1}{n\Omega}\right) \cos(n\Omega t)\, dt \Big|_0^{T_0/2} \tag{6}$$

Noting that $\Omega T_0 = 2\pi$, it follows that

$$b_n = \frac{-2F_0}{n\pi} [\cos(n\pi) - 1] \tag{7}$$

Figure 6.3

Harmonic components of Fourier series representation of square wave. *Mechanical Vibrations, Theory and Applications,* 2/e, by Tse/Morse/Hinkle. Reprinted by permission of authors.

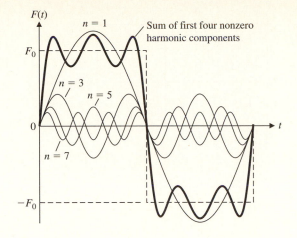

Thus

$$b_n = \begin{cases} \dfrac{4F_0}{n\pi} & \text{for } n = 1, 3, 5, \ldots \\[2mm] 0 & \text{for } n = 2, 4, 6, \ldots \end{cases} \tag{8}$$

The Fourier series representation for the square wave is illustrated in Figure 6.3. Note that consideration of just four nonzero terms yields a fairly accurate representation of $F(t)$, indicating that the series converges rather rapidly. The rapid convergence is attributed to the fact that the b_n coefficients have n in the denominator; thus, higher order terms (i.e., as n increases) contribute at a decreasing rate to the convergence. ▲

Next, consider the response of a viscously damped SDOF system to periodic, non-harmonic excitation of period T_0. The differential equation of motion for the steady-state response is expressed as

$$m\ddot{x} + c\dot{x} + kx = F(t)$$
$$= \frac{a_0}{2} + \sum_{n=1}^{\infty} (a_n \cos n\Omega t + b_n \sin n\Omega t) \tag{6.6}$$

in which the Fourier series expansion was used to represent $F(t)$. The transient response will decay with time and is neglected. To obtain the particular solution for the steady-state response, it is noted that the differential equation is linear, and therefore the principle of superposition is applicable. Hence, the steady-state displacement solution is merely the summation of the individual particular solutions for all harmonic terms representing $F(t)$. Therefore, we obtain for $n = 1, 2, 3,$

$$x_p(t) = \frac{a_0}{2k} + \sum_{n=1}^{\infty} \frac{a_n \cos(n\Omega t - \psi_n) + b_n \sin(n\Omega t - \psi_n)}{k\sqrt{(1 - n^2 r^2)^2 + (2\zeta nr)^2}} \tag{6.7}$$

Figure 6.4

Periodic force of Example 6.2.

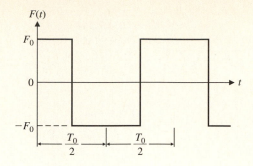

where

$$\psi_n = \tan^{-1}\left(\frac{2\zeta nr}{1 - n^2 r^2}\right) \quad \text{and} \quad r = \frac{\Omega}{\omega} \tag{6.8}$$

in which ψ_n is the phase angle for the steady-state response and r is the ratio of excitation frequency to natural frequency, or simply, the frequency ratio.

EXAMPLE 6.2 ▼

For the nonharmonic periodic forcing function shown in Figure 6.4, determine (a) the Fourier series expansion and (b) the steady-state displacement response for a viscously underdamped SDOF system.

Solution

(a) Noting that the forcing function illustrated in Figure 6.4 is an *even* function, then $b_n = 0$ and only the a_n coefficients must be evaluated. The function $F(t)$ may be decomposed in the time intervals $0 < t < \dfrac{T_0}{4}$ and $\dfrac{T_0}{4} < t < \dfrac{T_0}{2}$ as

$$F(t) = \begin{cases} F_0 & \text{for } 0 < t < \dfrac{T_0}{4} \\[2mm] -F_0 & \text{for } \dfrac{T_0}{4} < t < \dfrac{T_0}{2} \end{cases} \tag{1}$$

Thus the a_0 coefficient is determined as

$$a_0 = \frac{4}{T_0}\int_0^{T_0/4} F_0 \, dt + \frac{4}{T_0}\int_{T_0/4}^{T_0/2} (-F_0) \, dt \tag{2}$$

$$a_0 = \frac{4}{T_0}(F_0 t)\Big|_0^{T_0/4} + \frac{4}{T_0}(-F_0 t)\Big|_{T_0/4}^{T_0/2} \tag{3}$$

or

$$a_0 = F_0 - F_0 = 0 \tag{4}$$

and the a_n coefficients are expressed as

$$a_n = \frac{4}{T_0} \int_0^{T_0/4} F_0 \cos(n\Omega t)\, dt + \frac{4}{T_0} \int_{T_0/4}^{T_0/2} (-F_0)\cos(n\Omega t)\, dt \qquad (5)$$

Since $\Omega = 2\pi/T_0$, Eq. (5) becomes

$$a_n = \frac{2F_0}{n\pi} \frac{\sin 2n\pi t}{T}\Big|_0^{T_0/4} - \frac{2F_0}{n\pi} \frac{\sin 2n\pi t}{T}\Big|_{T_0/4}^{T_0/2} = \frac{4F_0}{n\pi} \sin \frac{n\pi}{2} \qquad (6)$$

or

$$a_n = \frac{4F_0}{n\pi}(-1)^{(n-1)/2}, \qquad n = 1, 3, 5 \qquad (7)$$

Therefore, the Fourier expansion for the function $F(t)$ is

$$F(t) = \frac{4F_0}{\pi} \sum_{n=1}^{\infty} \frac{(-1)^{(n-1)/2}}{n} \cos\left(\frac{2\pi nt}{T_0}\right) \qquad (8)$$

(b) The differential equation of motion is given by Eq. (6.6) and is expressed as

$$m\ddot{x} + c\dot{x} + kx = \frac{4F_0}{\pi} \sum_{n=1}^{\infty} \frac{(-1)^{(n-1)/2}}{n} \cos\left(\frac{2\pi nt}{T_0}\right) \qquad (9)$$

and the steady-state response is given by Eq. (6.7), represented here by

$$x_p(t) = \frac{4F_0}{\pi} + \sum_{n=1}^{\infty} \frac{(-1)^{(n-1)/2} \cos[(2\pi nt/T_0) - \Psi_n]}{nk\sqrt{(1 - n^2 r^2)^2 + (2\zeta nr)^2}} \qquad (10)$$

where $\Psi_n = \tan^{-1}(2\zeta nr/1 - n^2 r^2)$ and $r = \Omega/\omega$. ▲

The following example illustrates the use of the Fourier series to analyze the response of a SDOF system subject to a periodic forcing function.

EXAMPLE 6.3 ▼

The structural building frame shown in Figure 6.5 is constructed of rigid girders and flexible columns that bend about their strong axis in the plane of the paper. The frame supports a uniformly distributed dead load having a total weight of 30 kips and is subject at its girder level to the periodic force described in Figure 6.4 and Example 6.2. (a) Evaluate the Fourier series expansion of the forcing function shown in Figure 6.4 for the first three nonzero terms and plot the result over the time interval $0 < t < 10$ sec. (b) Evaluate the steady-state displacement response of the structure shown in Figure 6.5 and plot displacement versus time over the same time interval. Assume $F_0 = 20$ kips, $T_0 = 2$ sec, $E = 29{,}000$ ksi, and $\zeta = 0.1$.

Figure 6.5

Shear frame structure of
Example 6.3.

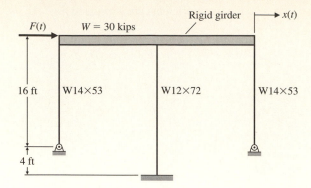

Solution

(a) The BASIC computer program listed in Table 6.1 evaluates Eq. (8) of Example 6.2 and plots the Fourier representation of the loading function for $n = 1$, 3, and 5. The resulting Fourier approximation of the forcing function is shown in Figure 6.6a.

(b) Steady-state response of shear frame:

- The column stiffnesses are calculated as follows:

  For the W14×53 $I_{xx} = 541$ in^4

$$k_1 = \frac{3EI}{L^3} = \frac{3(29,000)(541)}{(16 \times 12)^3} = 6.65 \text{ kips/in}$$

  For the W12×72 $I_{xx} = 597$ in^4

$$k_2 = \frac{12EI}{L^3} = \frac{12(29,000)(597)}{(20 \times 12)^3} = 15.03 \text{ kips/in}$$

TABLE 6.1. **BASIC Computer Program Used for the Fourier Series Representation of the Periodic Force in Figure 6.4**

```
10 READ T1,T2,DT,FO,T,N
20 DATA 0,10,0.05,20,2,5
30 P=3.14159265#
40 A0=0
50 OPEN "O",#1, "2HW5.DAT"
60 PRINT #1,T1,T1
70 PRINT #1,T1,T1
80 FOR I=T1 TO (T2+DT) STEP DT
90 FT=0
100 FOR J=1 TO N STEP 2
110 K=(J-1)/2
120 FT=FT+(4*F0/P)*(1/J)*((-1)^K)*(COS(2*P*J*I/T))
130 NEXT J
140 PRINT #1,I,FT
150 LPRINT USING "#####.###";I,FT
160 NEXT I
170 CLOSE #1
180 END
```

Figure 6.6

(a) Fourier series representation of square wave; shown in Figure 6.4 ($n = 3$); (b) Fourier series representation of displacement response of shear frame structure ($n = 3$).

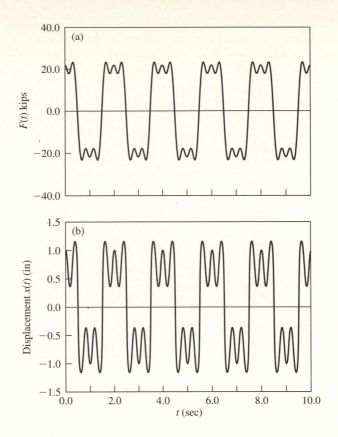

- The total stiffness for the structure is

$$k = 2k_1 + k_2 = 2(6.65) + 15.03 = 28.33 \text{ kips/in}$$

- The mass of the building is

$$m = \frac{W}{g} = \frac{30 \text{ kips}}{386.4 \text{ in/sec}^2} = 0.0776 \text{ kip-sec}^2/\text{in}$$

and the natural frequency of the building is

$$\omega = \sqrt{\frac{k}{m}} = \sqrt{\frac{28.33}{0.0776}} = 19.1 \text{ rad/sec}$$

▲

The BASIC computer program listed in Table 6.2 evaluates the steady-state displacement response of the shear frame subject to the periodic force shown in Figure 6.6a. The expression for the steady-state response is represented by Eq. (10) of Example 6.2. The first three nonzero Fourier terms are evaluated. A plot of the resulting steady-state response is shown in Figure 6.6b. Notice that the inaccuracies of the Fourier series representation of $F(t)$ shown in Figure 6.6a are exaggerated in the steady-state response representation shown in Figure 6.6b. As more terms are considered in the Fourier series, the extent of these inaccuracies will be lessened.

TABLE 6.2. **BASIC Computer Program Used to Evaluate the Response of the Shear Frame Subject to the Periodic Excitation of Figure 6.6a**

```
10  READ T1,T2,DT,FO,T,N
20  READ K,M,Z
30  DATA 0,10,0.01,20,2,5
40  DATA 28.328,0.07764,0.1
50  P=3.14159265#
60  W=SQR(K/M)
70  OPEN "O",#1, "4HW5.DAT"
80  PRINT #1,T1,T1
90  PRINT #1,T1,T1
100 FOR I=T1 TO (T2+DT) STEP DT
110 DP=0
120 FOR J=1 TO N STEP 2
130 RN=2*J*P/(T*W)
140 PHN=ARCTAN(2*Z*RN/(1-RN^2))
150 ANG=(2*P*J*I/T)-PHN
160 AA=SQR((1-RN^2)^2+(2*Z*RN)^2)
170 K1=(J-1)/2
180 DP=DP+(4*FO/P)*(1/J)*((-1)^K1)*(COS(ANG))/(AA*K)
190 NEXT J
200 PRINT #1,I,DP
210 LPRINT USING "#####.###";I,DP
220 NEXT I
230 CLOSE #1
240 END
```

Although Fourier analysis is a convenient method for evaluating the response of systems to periodic excitation, it is generally not conducive for evaluating the response of systems to nonperiodic excitation in the *time domain*. The remainder of this chapter introduces an analytical method for evaluating the response of structures to arbitrary dynamic excitation.

6.2 RESPONSE TO UNIT IMPULSE

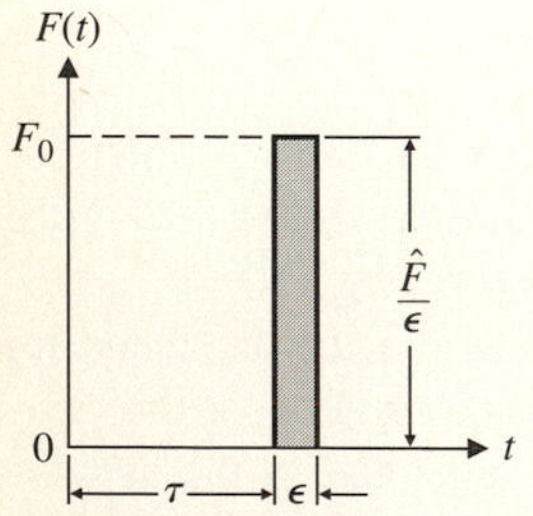

Before proceeding to the development of a general analytical method to evaluate the response of structures to arbitrary dynamic excitation, it is instructive to first consider the response of a SDOF system to a unit impulse. By definition an *impulsive force* is a force of large magnitude that acts over a very short time interval, but having a finite time integral.

The time integral for an impulsive force is called the *linear impulse* $\hat{F}$, expressed as

$$\hat{F} = \int F(t)\, dt \tag{6.9}$$

Figure 6.7
Impulsive force.

An illustration of an impulsive force is presented in Figure 6.7. The magnitude of the force F_0, acting at time $t = \tau$ over a short time interval ε is $\hat{F}/\varepsilon$. As $\varepsilon \to 0$, the impulsive force approaches infinity but $\hat{F}$ becomes equal to unity. This special case is known as the *unit impulse,* symbolized by the *Dirac delta function.* The *Dirac* delta function at time $t = \tau$ is defined by $\delta(t - \tau)$ and has the following properties:

$$\delta(t - \tau) = 0 \qquad \text{for } t \neq \tau \tag{6.10}$$

$$\int_0^\infty \delta(t - \tau)\, dt = 1 \tag{6.11}$$

$$\int_0^\infty \delta(t - \tau)F(t)\, dt = F(\tau) \tag{6.12}$$

in which $0 < \tau < \infty$. Thus, an impulsive force acting at $t = \tau$ can be represented as

$$F(t) = \hat{F}\delta(t - \tau) \tag{6.13}$$

Consider next an underdamped SDOF system subject to a unit impulse defined by Eq. (6.13). The equation of motion is

$$m\ddot{x} + c\dot{x} + kx = F(t) = \hat{F}\delta(t - \tau) \tag{6.14}$$

If the system is initially at rest and $\tau = 0$, then the response of the system immediately after the impulsive force is applied is transient in nature, or free vibration. The initial condition instigating the transient response will be the velocity imparted to the mass by the impulsive force at time $t = \varepsilon$, the duration of the impulse. Since ε is very small, it can be rationalized that the transient response begins at a time slightly greater than 0, or 0^+. To determine the velocity at time $t = 0^+$, $\dot{x}(0^+)$, the impulse-momentum theorem is invoked. Thus from Eqs. (6.12) and (6.13) it follows that

$$\int_0^{0^+} F(t)\, dt = \int_0^{0^+} \hat{F}\delta(t)\, dt = m\dot{x}(0^+) \tag{6.15}$$

Since the integrand in Eq. (6.15) yields a non-zero value for $\hat{F}$, the initial velocity is expressed by

$$\dot{x}(0^+) = \frac{\hat{F}}{m} \tag{6.16}$$

Therefore, the transient or free vibration displacement response of Eq. (6.14) at time $t = 0^+$ (note that for $t > 0^+$ the right-hand side of Eq. (6.14) is zero) for an initial velocity $\dot{x}(0^+) = \hat{F}/m$ and zero initial displacement is

$$x(t) = \frac{\hat{F}}{m\omega_d} e^{-\zeta\omega t} \sin \omega_d t \tag{6.17}$$

in which $\omega_d = \omega\sqrt{1 - \zeta^2}$ is the damped natural circular frequency of the system.

The response to unit impulse is very important in the analysis of transient response and is represented by the *impulsive response function* $h(t)$. The response given by Eq. (6.17) can then be expressed as

$$x(t) = \hat{F}h(t) \tag{6.18}$$

where

$$h(t) = \frac{1}{m\omega_d} e^{-\zeta\omega t} \sin \omega_d t \qquad (6.19)$$

For an undamped system the impulsive response function is given by

$$h(t) = \frac{1}{m\omega} \sin \omega t \qquad (6.20)$$

The impulsive response function will be used extensively throughout the remainder of this chapter for the evaluation of dynamic response to arbitrary forcing functions.

6.3 DUHAMEL INTEGRAL

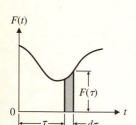

$F(t)$

$F(\tau)$

0

τ

$d\tau$

t

Figure 6.8

Incremental component of an arbitrary force.

The response of a SDOF system to arbitrary forms of excitation can be analyzed with the aid of the impulsive response function $h(t)$ developed in the previous section. Consider the arbitrary nonperiodic forcing function depicted in Figure 6.8. If it is assumed that the arbitrary excitation $F(t)$ consists of a sequence of impulsive forces $F(\tau)$ acting over a very small time interval $d\tau$, then the displacement response to each impulse is valid for all time $t > \tau$. Thus the incremental response dx to each impulse $F(\tau)$ can be expressed as

$$dx = F(\tau)\, d\tau h(t - \tau) \qquad (6.21)$$

To obtain the complete response to $F(t)$, the individual incremental responses dx due to each impulse are summed by superposition. In other words, the complete displacement response $x(t)$ is obtained by integrating the incremental response given by Eq. (6.21) over the entire time interval, resulting in

$$x(t) = \int_0^t F(\tau)h(t - \tau)\, d\tau \qquad (6.22)$$

Equation (6.22) is commonly known as *Duhamel's integral,* or the *convolution integral.* Since the principle of superposition was employed to establish the Duhamel integral, its validity is restricted to the analysis of linear systems.

For a viscously damped system, substitution of the impulsive response function, described by Eq. (6.19), into Eq. (6.22) yields

$$x(t) = \frac{1}{m\omega_d} \int_0^t F(\tau)e^{-\zeta\omega(t-\tau)} \sin \omega_d(t - \tau)\, d\tau \qquad (6.23)$$

which represents the system displacement response to an arbitrary excitation force when the initial conditions are zero. For nonzero initial conditions, the complete solution is the superposition of the particular solution due to the excitation and the homogeneous solution due to the initial conditions. Therefore, substitution of the general initial conditions $x(0) = x_0$ and $\dot{x}(0) = \dot{x}_0$ into Eq. (4.14) gives the homogeneous solution $x_h(t)$ as

$$x_h(t) = e^{-\zeta\omega t}\left(x_0 \cos \omega_d t + \frac{\dot{x}_0 + \zeta\omega x_0}{\omega_d} \sin \omega_d t \right) \qquad (6.24)$$

Superpositioning of Eqs. (6.23) and (6.24) gives the complete solution as

$$x(t) = e^{-\zeta\omega t}\left(x_0 \cos \omega_d t + \frac{\dot{x}_0 + \zeta\omega x_0}{\omega_d} \sin \omega_d t\right)$$

$$+ \frac{1}{m\omega_d}\int_0^t F(\tau)e^{-\zeta\omega(t-\tau)} \sin \omega_d(t - \tau)\, d\tau \qquad (6.25)$$

For undamped systems ($\zeta = 0$), Eqs. (6.24) and (6.25) become, respectively,

$$x(t) = \frac{1}{m\omega}\int_0^t F(\tau) \sin \omega(t - \tau)\, d\tau \qquad (6.26)$$

and

$$x(t) = x_0 \cos \omega t + \frac{\dot{x}_0}{\omega} \sin \omega t + \frac{1}{m\omega}\int_0^t F(\tau)\sin \omega(t - \tau)\, d\tau \qquad (6.27)$$

The Duhamel integral, as represented in Eqs. (6.24) through (6.27), is a powerful tool for the analysis of linear systems. The response of SDOF systems to simply defined loading functions is generally expedient with the Duhamel integral method. However, as the loading function becomes more complicated, an analytical or closed-form solution of the Duhamel integral becomes quite cumbersome. For such instances computer implementation of the procedure is highly recommended.

6.4 RESPONSE TO ARBITRARY DYNAMIC EXCITATION

To illustrate application of the Duhamel integral in evaluating the response of SDOF systems to arbitrary excitation, several classical load functions are considered. Examining the dynamic response of SDOF systems to such load functions will also lend insight to identifying several important factors affecting transient vibrations.

6.4.1 Ideal Step Force

Consider a suddenly applied force of magnitude F_0 that remains constant for all time $t \geq 0$, as illustrated in Figure 6.9a. Such a force is referred to as an *ideal step force*. If an ideal step force is applied to an undamped SDOF system as shown in Figure 6.9b, the displacement response can be determined from the Duhamel integral expression given by either Eq. (6.26) or Eq. (6.27), depending upon whether or not the system is initially at rest. For zero initial conditions, Eq. (6.26) applies, and the response is expressed as

$$x(t) = \frac{F_0}{m\omega}\int_0^t \sin \omega(t - \tau)\, d\tau \qquad (6.28)$$

since $F(\tau) = F_0$ for all $t \geq 0$.

To facilitate the evaluation of Eq. (6.28), the trigonometric identity

$$\sin \omega(t - \tau) = \sin \omega t \cos \omega\tau - \cos \omega t \sin \omega\tau \qquad (6.29)$$

Figure 6.9

(a) Ideal step force;
(b) undamped SDOF system.

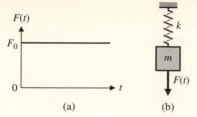

is employed in Eq. (6.28). Therefore, the Duhamel integral expression becomes

$$x(t) = \frac{F_0}{m\omega} \int_0^t (\sin \omega t \cos \omega \tau - \cos \omega t \sin \omega \tau) d\tau \tag{6.30}$$

or

$$x(t) = \frac{F_0}{m\omega} \left(\sin \omega t \int_0^t \cos \omega \tau \, d\tau - \cos \omega t \int_0^t \sin \omega \tau \, d\tau \right) \tag{6.31}$$

Integration of Eq. (6.31) yields

$$x(t) = \frac{F_0}{k} [\sin \omega t (\sin \omega \tau)\big|_0^t - \cos \omega t (\cos \omega \tau)\big|_0^t] \tag{6.32}$$

resulting in

$$x(t) = \frac{F_0}{k} (1 - \cos \omega t) \tag{6.33}$$

For a damped SDOF system, the Duhamel integral expression represented by Eq. (6.23) must be evaluated. The displacement response is therefore represented by

$$x(t) = \frac{F_0}{m\omega_d} \int_0^t e^{-\zeta\omega(t-\tau)} \sin \omega_d(t - \tau) \, d\tau \tag{6.34}$$

Applying a procedure analogous to that employed for the undamped case to Eq. (6.34) yields

$$x(t) = \frac{F_0}{k} \left[1 - e^{-\zeta\omega t} \left(\cos \omega_d t + \frac{\zeta}{\sqrt{1 - \zeta^2}} \sin \omega_d t \right) \right] \tag{6.35}$$

If the responses given by Eqs. (6.33) and (6.35) are divided by the equivalent static deflection, F_0/k, the resulting expressions represent the DMF as a function of time. Such expressions are commonly referred to as the *system response ratio R(t)* for a particular excitation. Thus, the response ratio for the undamped case is

$$R(t) = (1 - \cos \omega t) \tag{6.36}$$

and for the damped case is

$$R(t) = 1 - e^{-\zeta\omega t} \left(\cos \omega_d t + \frac{\zeta}{\sqrt{1 - \zeta^2}} \sin \omega_d t \right) \tag{6.37}$$

Figure 6.10

Response of SDOF systems to ideal step force for several values of damping.

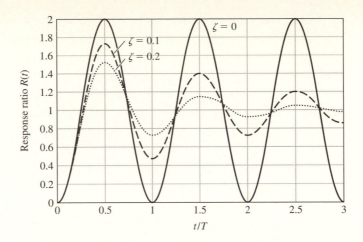

A plot of $R(t)$ vs. t/T, where T is the natural period of the structure, is presented in Figure 6.10 for several levels of damping. $R(t) = 1$ on the response ratio plot corresponds to the static displacement position. Any $R(t) > 1$ represents displacement beyond the static displacement position, or *overshoot*. For an undamped system subject to a step force, the resulting response is oscillatory motion about the static displacement position with a maximum response $R_{max} = 2$. In a damped system, the response ratio gradually approaches the static value of 1 after a number of cycles of damped oscillation. The maximum overshoot in a damped system as well as the rate of decay of the oscillation about the static equilibrium position depends on the damping factor, as illustrated in Figure 6.10.

6.4.2 Rectangular Pulse Force

Next consider the case of a load F_0 applied instantly to a structure, in a similar manner as the ideal step force, except that the force is suddenly removed after a finite time t_d. Such a force is commonly referred to as a *rectangular pulse force,* and t_d is known as the *time of duration* of the force. The ratio t_d/T has a significant effect on the dynamic response of a structure. The rectangular pulse force is a representative example of an impulsive or shock loading of short duration. Consequently, the response is not significantly affected by the presence of damping in the system. The effect of damping is therefore neglected in the presentation.

An illustration of a rectangular pulse force is presented in Figure 6.11. The displacement response of an undamped SDOF system to such a force can be determined by evaluating the Duhamel integral. To capture the complete response, the Duhamel integral must be evaluated over two time intervals: (1) the time over which the force is acting on the system, $0 \leq t \leq t_d$, and (2) the time after the force has been removed, $t > t_d$. The former time interval is generally referred to as the *forced-vibration phase* and the latter as the *transient vibration phase*. The maximum displacement response of the system may occur in either the forced-vibration phase or the transient vibration phase, depending upon the ratio t_d/T.

The forcing function for the rectangular pulse force is defined as

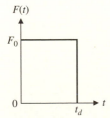

Figure 6.11

Rectangular pulse force.

$$F(t) = \begin{cases} F_0 & \text{for } 0 \leq t \leq t_d \\ 0 & \text{for } (t > t_d) \end{cases} \qquad (6.38)$$

The forced-vibration phase is first considered. Thus, the Duhamel integral expression is given by

$$x(t) = \frac{F_0}{m\omega} \int_0^t \sin \omega(t - \tau) \, d\tau \tag{6.39}$$

It should come as no surprise that the displacement response expressed by Eq. (6.39) is exactly the same as that given by Eq. (6.28), the response to an ideal step force. Hence, the response is

$$x(t) = \frac{F_0}{k}(1 - \cos \omega t) \qquad \text{for } 0 \leq t \leq t_d \tag{6.40}$$

To determine the displacement response in the transient vibration phase, the Duhamel integral must be evaluated over the entire time interval 0 to t, where $t > t_d$. However, since $F(t) = 0$ for all $t > t_d$, the Duhamel integral is evaluated merely in the interval 0 to t_d. Thus

$$x(t) = \frac{F_0}{m\omega} \int_0^{t_d} \sin \omega(t - \tau) \, d\tau + [0] \tag{6.41}$$

or

$$x(t) = \frac{F_0}{m\omega} \int_0^{t_d} (\sin \omega t \cos \omega\tau - \cos \omega t \sin \omega\tau) \, d\tau \tag{6.42}$$

Evaluating the integral yields

$$x(t) = \frac{F_0}{k} \left[\sin \omega t (\sin \omega\tau) \big|_0^{t_d} - \cos \omega t (\cos \omega\tau) \big|_0^{t_d} \right] \tag{6.43}$$

and applying the integration limits gives

$$x(t) = \frac{F_0}{k}(\sin \omega t \sin \omega t_d + \cos \omega t \cos \omega t_d - \cos \omega t) \tag{6.44}$$

Use of the trigonometric identity

$$\cos \omega(t - t_d) = \sin \omega t \sin \omega t_d + \cos \omega t \cos \omega t_d \tag{6.45}$$

in Eq. (6.44) yields

$$x(t) = \frac{F_0}{k} [\cos \omega(t - t_d) - \cos \omega t] \tag{6.46}$$

Therefore, the response ratio for the forced-vibration phase is

$$R_1(t) = (1 - \cos \omega t) \qquad \text{for } 0 \leq t \leq t_d \tag{6.47}$$

and for the transient vibration phase is

$$R_2(t) = \cos \omega(t - t_d) - \cos \omega t \qquad \text{for } t > t_d \tag{6.48}$$

Figure 6.12

Response of undamped SDOF systems to rectangular pulse force.

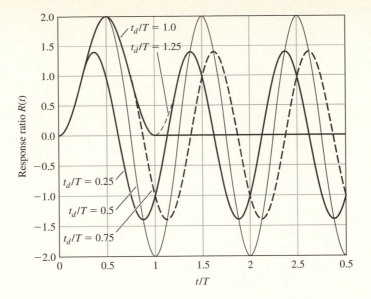

Several plots of $R(t)$ as a function of t/T for several values of t_d/T are presented in Figure 6.12. It is obvious from these plots that during the forced-vibration phase, the system oscillates about the static displacement position at its natural period T. In the transient vibration phase $t > t_d$, the system oscillates about its original equilibrium position (i.e., before application of the force) at its natural period T. It is obvious from these plots that as the t_d/T ratio for the force increases, the response in the forced-vibration phase approaches that for the ideal step force.

Examination of Eq. (6.47) reveals that the maximum displacement response for the forced-vibration phase is $R_{max} = 2$. This can occur only if $\cos \omega t = -1$. Replacing ω with $2\pi/T$ in the cosine term, it is apparent that t must equal $T/2$ to yield $R_{max} = 2$. Thus the maximum displacement response occurs in the forced-vibration phase when $t_d \geq T/2$.

To investigate the maximum displacement response in the transient vibration phase, it is convenient to replace ω in Eq. (6.48) with $2\pi/T$. Thus the resulting expression for the response ratio is

$$R_2(t) = \cos 2\pi\left(\frac{t}{T} - \frac{t_d}{T}\right) - \cos 2\pi\frac{t}{T} \tag{6.49}$$

Taking advantage of several trigonometric identities allows Eq. (6.49) to be expressed as

$$R_2(t) = 2\sin\frac{\pi t_d}{T}\left[2\pi\left(\frac{t}{T} - \frac{t_d}{2T}\right)\right] \tag{6.50}$$

Inspection of Eq. (6.50) indicates that the maximum displacement response in the transient vibration phase is $R_{max} = 2$, which occurs when $2\sin(\pi t_d/T) = 2$. This can occur only if the forcing function ratio $t_d/T \geq 2$.

6.4.3 Step Force with Ramp

In the previous two cases investigated, the constant force F_0 was applied instantaneously at time $t = 0$. Consider the situation in which the force F_0 is applied over a *finite rise time* t_r. Such a force is depicted in Figure 6.13 and is often referred to as a *ramp function*. For a system subject to a ramp function, the dynamic response is significantly affected by the ratio t_r/T.

A typical ramp function, such as that depicted in Figure 6.13, increases linearly from zero to a constant maximum amplitude F_0 over the rise time t_r. Thus the force is described by

$$F(t) = \begin{cases} F_0\left(\dfrac{t}{t_r}\right) & \text{for } 0 \le t \le t_r \\ \\ F_0 & \text{for } t > t_r \end{cases} \tag{6.51}$$

To determine the displacement response of an undamped system to a ramp function by the Duhamel integral method, the integral must be evaluated for the time intervals specified in Eq. (6.51).

The Duhamel integral for the ramp, that is, in the time interval $0 \le t \le t_r$, is

$$x(t) = \frac{F_0}{m\omega} \int_0^t \frac{t}{t_r} \sin \omega(t - \tau)\, d\tau \tag{6.52}$$

or

$$x(t) = \frac{F_0}{m\omega t_r} \left(\sin \omega t \int_0^t \tau \cos \omega\tau\, d\tau - \cos \omega t \int_0^t \tau \sin \omega\tau\, d\tau \right) \tag{6.53}$$

Integrating by parts gives

$$\int_0^t \tau \cos \omega\tau\, d\tau = \frac{\tau}{\omega} \sin \omega\tau \bigg|_0^t - \int_0^t \frac{\sin \omega\tau}{\omega}\, d\tau \tag{6.54}$$

or

$$\int_0^t \tau \cos \omega\tau\, d\tau = \frac{t}{\omega} \sin \omega t + \frac{1}{\omega^2} (\cos \omega t - 1) \tag{6.55}$$

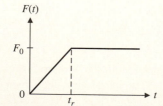

Figure 6.13

Step force with ramp.

Similarly,

$$\int_0^t \tau \sin \omega\tau \, d\tau = \frac{-\tau}{\omega} \cos \omega\tau \bigg|_0^t + \int_0^t \frac{\cos \omega\tau}{\omega} \, d\tau \tag{6.56}$$

or

$$\int_0^t \tau \sin \omega\tau \, d\tau = \frac{-t}{\omega} \cos \omega t + \frac{\sin \omega t}{\omega^2} \tag{6.57}$$

Substituting Eqs. (6.55) and (6.57) into Eq. (6.53) and simplifying yields the displacement response given by

$$x(t) = \frac{F_0}{k}\left[\frac{t}{t_r} - \frac{\sin \omega t}{\omega t_r}\right] \tag{6.58}$$

and the response ratio is expressed as

$$R(t) = \frac{1}{t_r}\left(t - \frac{\sin \omega t}{\omega}\right) \tag{6.59}$$

To facilitate evaluation of the displacement response for the time interval $t > t_r$, the force can be expressed as the sum of the two ramps of equal but opposite magnitude, as shown in Figure 6.14, resulting in a force of constant magnitude F_0. The negative ramp force is given by

$$\bar{F}(t) = \begin{cases} 0 & \text{for } 0 \le t \le t_r \\ -F_0\left(\dfrac{t - t_r}{t_r}\right) & \text{for } t > t_r \end{cases} \tag{6.60}$$

The corresponding displacement response to the force $\bar{F}(t)$ is termed $\bar{x}(t)$; then the actual displacement response for time $t > t_r$ is the sum of the responses $x(t)$, given by Eq. (6.58), and $\bar{x}(t)$.

The Duhamel integral expression for the displacement response $\bar{x}(t)$ to the force $\bar{F}(t)$ is expressed by

$$\bar{x}(t) = \frac{1}{m\omega}\int_0^t \bar{F}(\tau) \sin \omega(t - \tau) \, d\tau \tag{6.61}$$

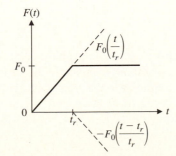

Figure 6.14

Constant force F_0 as the sum of two ramps

Since $F(t) = 0$ for $0 \leq t \leq t_r$, Eq. (6.61) becomes

$$\bar{x}(t) = \frac{F_0}{m\omega} \int_{t_r}^{t} \left(\frac{\tau - t_r}{t_r} \right) \sin \omega(t - \tau) \, d\tau \tag{6.62}$$

or

$$\bar{x}(t) = \frac{F_0}{m\omega}$$

$$\left[\sin \omega t \int_{t_r}^{t} \left(\frac{\tau - t_r}{t_r} \right) \cos \omega\tau \, d\tau - \cos \omega t \int_{t_r}^{t} \left(\frac{\tau - t_r}{t_r} \right) \sin \omega\tau \, d\tau \right]$$

$$\tag{6.63}$$

Evaluating Eq. (6.63) in a manner analogous to that described by Eqs. (6.54) through (6.58) yields

$$\bar{x}(t) = \frac{F_0}{k} \left[\frac{t - t_r}{t_r} - \frac{\sin \omega(t - t_r)}{\omega t_r} \right] \tag{6.64}$$

Thus the complete displacement response for the time interval $t > t_r$ is given by the sum of Eqs. (6.58) and (6.64), resulting in

$$x(t) = \frac{F_0}{k} \left\{ 1 + \left(\frac{1}{\omega t_r} \right) [\sin \omega(t - t_r) - \sin \omega t] \right\} \tag{6.65}$$

The corresponding response ratio is

$$R(t) = 1 + \frac{1}{\omega t_r} [\sin \omega(t - t_r) - \sin \omega t] \tag{6.66}$$

Figure 6.15 depicts the plot of response ratio versus t/T for several values of t_r/T. These plots indicate that as t_r approaches zero (that is, $t_r/T \ll 1$), the response approaches that of an ideal step function having $R_{max} = 2$. For a ratio of $t_r/T \leq 0.25$, the effect is essentially the same as an ideal step function. As illustrated in Figure 6.15, the maximum response will always occur at time $t > t_r$. For increasing values of t_r/T, the overshoot becomes less with smaller oscillations about the static displacement position. Dynamic effects can be ignored for ramp functions with $t_r/T > 3.0$.

6.4.4 Triangular Pulse Force

A load function often employed to simulate a blast is the triangular load pulse shown in Figure 6.16. The load F_0 is instantly applied to the structure and decreases linearly over the time of duration t_d. Similar to the rectangular pulse force, the Duhamel integral must be evaluated over the time intervals $0 \leq t \leq t_d$ and $t > t_d$ to capture the complete response. Depending on the ratio t_d/T, the maximum displacement response may occur in either the force-vibration phase ($0 \leq t \leq t_d$) or the transient vibration phase ($t > t_d$).

Figure 6.15

Response of undamped
SDOF systems to step force
with ramp.

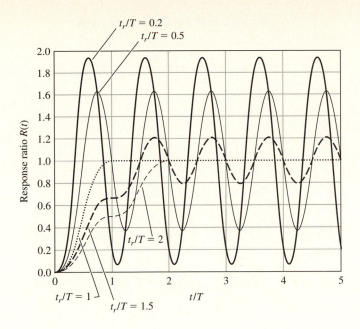

EXAMPLE 6.4 ▼

Determine the displacement response of an undamped SDOF system to the suddenly applied triangular pulse shown in Figure 6.16 by the Duhamel integral method. Assume the system is initially at rest, and consider both the forced-vibration and transient vibration phases of the response.

Solution

The load function $F(t)$ is represented as

Figure 6.16
Triangular pulse force.

$$F(t) = \begin{cases} F_0\left(1 - \dfrac{t}{t_d}\right) & \text{for } 0 \leq t \leq t_d \\[2mm] 0 & \text{for } t > t_d \end{cases}$$

(a) Forced-vibration phase, $0 \leq t \leq t_d$: The Duhamel integral expression is given as

$$x(t) = \frac{F_0}{m\omega}\int_0^t \left(1 - \frac{\tau}{t_d}\right) \sin \omega(t - \tau)\, d\tau \tag{1}$$

or

$$x(t) = \frac{F_0}{m\omega}$$

$$\left[\sin \omega t \int_0^t \left(1 - \frac{\tau}{t_d}\right)\cos \omega\tau\, d\tau - \cos \omega t \int_0^t \left(1 - \frac{\tau}{t_d}\right)\sin \omega\tau\, d\tau \right]$$

$$\tag{2}$$

Integration of Eq. (2) by parts yields

$$\int_0^t \tau(\cos \omega\tau)\, d\tau = \frac{t}{\omega} \sin \omega t - \frac{1}{\omega^2}(\cos \omega t - 1) \tag{3}$$

and

$$\int_0^t \tau(\sin \omega\tau)\, d\tau = -\frac{t}{\omega}\cos \omega t - \frac{\sin \omega t}{\omega^2} \tag{4}$$

Therefore, Eq. (2) can be expressed as

$$x(t) = \frac{F_0}{k}\left[\sin \omega t\left(\sin \omega t - \frac{t \sin \omega t}{t_d} - \frac{\cos \omega t}{\omega t_d} + \frac{1}{\omega t_d} \right) \right.$$

$$\left. - \cos \omega t\left(-\cos \omega t + 1 + \frac{t \cos \omega t}{\omega t_d} - \frac{\sin \omega t}{\omega t_d} \right) \right] \tag{5}$$

Simplification of Eq. (5) results in the displacement response given by

$$x(t) = \frac{F_0}{k}\left(1 - \frac{t}{t_d} - \cos \omega t + \frac{\sin \omega t}{\omega t_d} \right) \tag{6}$$

(b) Transient-vibration phase, $t > t_d$: Since $F(\tau) = 0$ for all $t > t_d$, the Duhamel integral expression becomes

$$x(t) = \frac{F_0}{m\omega} \int_0^{t_d}\left(1 - \frac{\tau}{t_d} \right) \sin \omega(t - \tau)\, d\tau + [0] \tag{7}$$

Note that this integral is the same as the integral expression for the forced-vibration phase, except that it is evaluated at $t = t_d$. Therefore the displacement response is represented by

$$x(t) = \frac{F_0}{k}\left[\sin \omega t\left(-\frac{\cos \omega t_d}{\omega t_d} + \frac{1}{\omega t_d} \right) - \cos \omega t\left(1 - \frac{\sin \omega t_d}{\omega t_d} \right) \right] \tag{8}$$

Simplifying the expression given by Eq. (8) yields

$$x(t) = \frac{F_0}{k\omega t_d}\left[\sin \omega t - \sin \omega(t - t_d) - \omega t_d \cos \omega t \right] \tag{9}$$

▲

Once the Duhamel integral has been evaluated for a specific forcing function, the result may be used to evaluate the response of any SDOF system to that particular type of forcing. Example 6.5 illustrates the application of the previously obtained results for a step force with a finite rise time.

EXAMPLE 6.5 ▼

The shear frame shown in Figure 6.17a is constructed of rigid girders and flexible columns. The frame supports a uniformly distributed dead load having a total weight of 30 kips. The frame is subjected to the step force with a ramp shown in Figure 6.17b at the girder level. Determine the horizontal displacement of the frame at time $t = 0.7$ sec. Assume $E = 30,000$ ksi and $\zeta = 0$.

Solution

(a) Dynamic properties of frame:

$$\text{W}8\times24: \qquad k_1 = \frac{3EI}{L^3} = \frac{3(30,000)(82.8 \text{ in}^4)}{(15 \text{ ft} \times 12 \text{ in})^3} = 1.278 \text{ kips/in}$$

$$\text{W}10\times33: \qquad k_2 = \frac{12EI}{L^3} = \frac{12(30,000)(170 \text{ in}^4)}{(20 \times 12)^3} = 4.427 \text{ kips/in}$$

The total stiffness of the structure is

$$k = 2k_1 + k_2 = 6.983 \text{ kips/in}$$

and the mass of the building is

$$m = \frac{W}{g} = \frac{30 \text{ kips}}{386.4 \text{ in/sec}^2} = 0.0776 \text{ kip-sec}^2/\text{in}$$

The natural frequency is

$$\omega = \sqrt{\frac{k}{m}} = \sqrt{\frac{6.983}{0.0776}} = 9.484 \text{ rad/sec}$$

and the natural period is

$$T = \frac{\omega}{2\pi} = 1.51 \text{ sec}$$

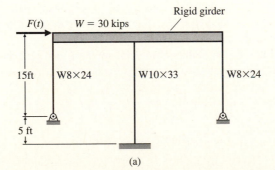

(a)

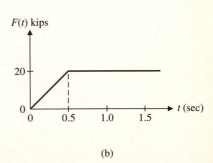

(b)

Figure 6.17
(a) Multibay shear frame;
(b) step force with ramp.

(b) Since the response at $t = 0.7$ is at a time greater than the rise time t_r (0.5 sec), Eq. (6.65) is used to evaluate the response. Thus

$$x(t) = \frac{F_0}{k} \left\{ 1 + \left[\frac{1}{\omega t_r} \right] [\sin \omega(t - t_r) - \sin \omega t] \right\}$$

or

$$x(0.7) = \frac{4}{6.943} \left\{ 1 + \left[\frac{1}{9.484(0.5)} \right] \right.$$

$$\left. [\sin 9.484(0.7 - 0.5) - \sin 9.484(0.7)] \right\}$$

and the displacement at time $t = 0.7$ sec is

$$x(0.7) = 0.645 \text{ in}$$

▲

Although the Duhamel integral method affords the analyst a closed-form procedure for calculating the system response to arbitrary dynamic excitation, evaluation of the integral can prove to be quite cumbersome, as evidenced in this section. Moreover, if the specified input is in digitized form (such as an earthquake ground motion record), this analytical procedure is quite useless. In such cases a numerical procedure for evaluating the dynamic response is required. Numerical techniques for evaluating dynamic response are addressed in Chapters 7, 12, and 13.

6.5   RESPONSE SPECTRUM

Frequently the analyst is concerned with only the maximum dynamic response to a specified input. This is particularly true in the case of *shock,* or *impulsive,* loadings. Such loadings are transient excitations, the duration of which are short in comparison to the natural period of the system. Several such loading conditions were examined in the previous section (i.e., rectangular and triangular pulse forces).

A *response spectrum* (or *shock spectrum*) is a plot of maximum peak response to a specific input versus some system characteristic, such as natural period or natural frequency. The peak response, generally a displacement, velocity, or acceleration, is obtained for a family of SDOF oscillators, each tuned to a different natural period or frequency. The underlying concept of a response spectrum is that the maximum response of a linear SDOF oscillator to any given input depends only on the natural frequency of the system and its damping factor.

With use of the Duhamel integral method, the peak displacement response of an undamped SDOF system to a specified excitation, $F(t)$, is expressed as

$$[x(t)]_{\max} = \left| \frac{1}{m\omega} \int_0^t F(\tau) \sin \omega(t - \tau) d\tau \right|_{\max} \tag{6.67}$$

Then, to construct the response spectrum for the specified excitation, Eq. (6.67) must be evaluated over a wide range of system natural frequencies ω.

To demonstrate the procedure for construction of a response spectrum, consider the rectangular pulse force shown in Figure 6.11. The response ratios for the forced-vibration phase, $t \leq t_d$, and the transient vibration phase, $t > t_d$, are given by Eqs. (6.47) and (6.48), respectively. Whether the maximum or peak response occurs in the forced-vibration phase depends upon the value of time of duration t_d relative to the natural period T.

First assume that the maximum response occurs in the forced-vibration phase, at a time $t < t_d$. The time corresponding to the maximum response, t_m, is determined by setting the first time derivative of the response ratio given by Eq. (6.47) equal to zero. Thus

$$\dot{R}_1(t) = \sin \omega t_m = 0 \tag{6.68}$$

or

$$\omega t_m = \pi \tag{6.69}$$

Equation (6.69) is valid only if

$$t_m = \frac{\pi}{\omega} < t_d \tag{6.70}$$

or

$$\frac{t_d}{T} > \frac{1}{2} \tag{6.71}$$

Then the maximum response ratio $R_{\max}$ is calculated from Eq. (6.47) by setting $t = t_m$, which gives

$$R_{\max} = \text{DMF} = 1 - \cos \omega t_m = 2 \tag{6.72}$$

It should be noted that the maximum response ratio $R_{\max}$ is equivalent to the DMF discussed in Chapter 5.

However, if the condition defined by Eq. (6.70) or (6.71) does not apply, then the maximum response will occur in the transient vibration phase. The time of maximum response is therefore calculated by setting equal to zero the first time derivative of Eq. (6.48), resulting in

$$\dot{R}_2(t) = \omega \sin \omega(t_m - t_d) + \omega \sin \omega t_m = 0 \tag{6.73}$$

which can be simplified to

$$\tan \omega t_m = -\cot \frac{\omega t_d}{2} \tag{6.74}$$

or

$$t_m = \frac{T}{4} + \frac{t_d}{2} \tag{6.75}$$

Substituting t_m given by Eq. (6.75) for t in Eq. (6.48) yields

$$R_{\max} = \text{DMF} = 2 \sin \frac{\pi t}{T} \tag{6.76}$$

Figure 6.18

Response spectrum for rectangular pulse force ($\zeta = 0$).

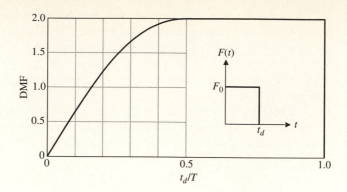

With the use of Eqs. (6.72) and (6.76), a plot of $R_{\max}$ (or DMF) as a function of t_d/T can be generated as shown in Figure 6.18. Figure 6.18 is a *response spectrum* for a rectangular pulse force.

EXAMPLE 6.6 ▼

Construct a response spectrum for the step force with ramp function illustrated in Figure 6.13. Assume a rise time t_r and $\zeta = 0$.

Solution

Expressions for the response ratio $R(t)$ in the time intervals $0 \leq t \leq t_r$ and $t > t_r$ were previously derived in Section 6.4.3 and are represented by Eqs. (6.59) and (6.66), respectively. Therefore

$$
R(t) = \begin{cases} \dfrac{1}{t_r}\left(t - \dfrac{\sin \omega t}{\omega}\right) & 0 \leq t \leq t_r \\[3mm] 1 + \dfrac{1}{\omega t_r}\left[\sin \omega(t - t_r) - \sin \omega t\right] & t > t_r \end{cases} \tag{1}
$$

Equation (1) represents the response of the undamped system to the ramp function shown in Figure 6.13. To determine the maximum response, we set the first time derivative of Eq. (1) for $t > t_r$ equal to zero (since the maximum response will always occur for $t > t_r$) and solve for the time t_m at which the maximum response occurs. The time t_m is then substituted into Eq. (1) for $t > t_r$ to evaluate the maximum response $R_{\max}$. Differentiating Eq. (1) yields

$$
\dot{R}(t_m) = 0 = -\cos \omega t_m + \cos \omega(t_m - t_r) \tag{2}
$$

Applying several basic trigonometric identities to Eq. (1) and solving for ωt_m yields

$$
\tan \omega t_m = \frac{1 - \cos \omega t_r}{\sin \omega t_r} \tag{3}
$$

or

$$\omega t_m = \tan^{-1}\left(\frac{1 - \cos \omega t_r}{\sin \omega t_r}\right) \tag{4}$$

From trigonometric analogy, Eq. (3) corresponds to a right triangle of sides $\sin \omega t_r$ and $(1 - \cos \omega t_r)$. The corresponding expression for the hypotenuse is

$$[\sin^2 \omega t_r + (1 - \cos \omega t_r)^2]^{1/2} = [2(1 - \cos \omega t_r)]^{1/2} \tag{5}$$

Therefore, for $\omega t_m > \pi$, the expression for $\sin \omega t_m$ in Eq. (1) can be represented as

$$\sin \omega t_m = -\sqrt{\frac{1}{2}(1 - \cos \omega t_r)} \tag{6}$$

and

$$\cos \omega t_m = \frac{-\sin \omega t_r}{\sqrt{2(1 - \cos \omega t_r)}} \tag{7}$$

Substitution of Eq. (7) into Eq. (1) evaluated at t_m yields the expression for maximum response (where $R_{max} = R(t_m)$) given by

$$R_{max} = \text{DMF} = 1 + \frac{1}{\omega t_r}\sqrt{2(1 - \cos \omega t_r)} \tag{8}$$

Since

$$\frac{\omega t_r}{2\pi} = \frac{t_r}{T} \tag{9}$$

R_{max} (or DMF) may be plotted versus the dimensionless frequency parameter t_r/T, where T is the natural period of the structure. Figure 6.19 is the response spectrum for the step force with ramp function illustrated in Figure 6.13.

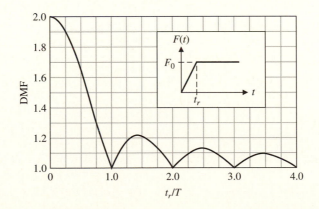

Figure 6.19

Response spectrum for step force with ramp ($\zeta = 0$).

The construction of a response spectrum can be greatly facilitated through numerical evaluation techniques. Construction of a response spectrum by numerical integration is addressed in Chapter 7. Once constructed, the response spectrum offers the design engineer the opportunity to evaluate the maximum response of a wide frequency range of structures to a specific input. Application of the response spectrum technique is illustrated in the following example.

EXAMPLE 6.7 ▼

Determine the maximum horizontal displacement of the structural building frame pictured in Figure 6.17a when subject to the rectangular pulse force illustrated in Figure 6.11 at the girder level. Assume $F_0 = 15$ kips and $t_d = 0.45$ sec.

Solution

From Example 6.5,

$$T = 1.5 \text{ sec} \quad \text{and} \quad k = 6.983 \text{ kips/in}$$

Therefore

$$\frac{t_d}{T} = \frac{0.45}{1.5} = 0.3$$

Then from the response spectrum shown in Figure 6.18,

$$R_{\max} = \text{DMF} \cong 1.7$$

Thus,

$$x_{\max} = R_{\max}\left(\frac{F_0}{k}\right) = 1.7\left(\frac{15 \text{ kips}}{6.983 \text{ kips/in}}\right) = 3.66 \text{ in}$$

▲

REFERENCES

1 Tolstov, Georgi P., *Fourier Series,* Dover Publications, New York, 1962.

2 Humar, Jagmohan L., *Dynamics of Structures,* Prentice Hall, Englewood Cliffs, NJ, 1990.

3 Weaver, W., Timoshenko, S.P., and Young, D.H., *Vibration Problems in Engineering*, 5th ed., Wiley, New York, 1990.

4 *Standard Mathematical Tables,* 23rd ed., CRC Press, Cleveland, OH, 1974.

5   Inman, Daniel J., *Engineering Vibration,* Prentice Hall, Englewood Cliffs, NJ, 1994.

6   Tse, F.S., Morse, I.E., and Hinkle, R.T., *Mechanical Vibrations, Theory and Applications,* 2nd ed. Allyn and Bacon, Boston, 1978.

7   Biggs, John M., *Introduction to Structural Dynamics,* McGraw-Hill, New York, 1964.

NOTATION

a_0, a_n	Fourier coefficients		T	natural period (sec)
b_n	Fourier coefficients		T_0	period (repetition time) of periodic force
c	viscous damping coefficient		x	translational displacement
E	Young's modulus		x_0	initial displacement
F_0	amplitude of forcing function		$x_h(t)$	homogeneous solution for displacement response
$\hat{F}$	linear impulse		$x_p(t)$	particular solution for steady-state of forced displacement response
$F(t)$	time-varying externally applied force			
$F(\tau)$	impulsive force		$\dot{x}$	translational velocity
$h(t)$	impulsive response function		$\dot{x}_0$	initial velocity
I	static moment of inertia		$\ddot{x}$	translational acceleration
k	translational stiffness		DMF	dynamic magnification factor
m	mass		SDOF	single degree of freedom
r	frequency ratio		$\delta(t)$	Dirac delta function
$R(t)$	response ratio		ε	short time interval
$R_{\max}$	maximum response ratio		ω	undamped natural circular frequency (rad/sec)
t	time		ω_d	damped natural circular frequency (rad/sec)
t_d	time of duration of applied force		Ω	frequency of external forcing function
t_m	time of maximum response		ψ_n	phase angle for steady-state response
t_r	rise time of applied force		ζ	damping factor

PROBLEMS

6.1–6.12 Determine the Fourier series representation for the periodic forcing functions shown in Figures P6.1–P6.12.

6.13–6.24 Determine the Fourier series representation of the steady-state displacement response for a viscously damped SDOF system subject to the periodic forcing functions shown in Figures P6.1–P6.12.

6.25 (a) Write a computer program to numerically evaluate the Fourier series expansion of the periodic forcing function shown in Figure P6.9. Consider the first 10 nonzero terms in the series. Plot $F(t)$ versus time in the interval $0 \leq t \leq 10$ sec. Assume $F_0 = 30$ kips and $T_0 = 1.5$ sec.

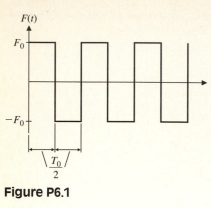

Figure P6.1

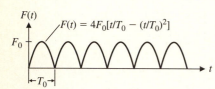

Figure P6.2

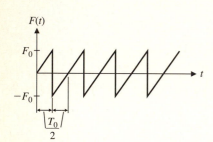

Figure P6.3

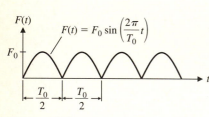

Figure P6.4

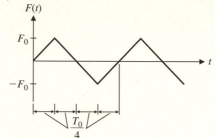

Figure P6.5

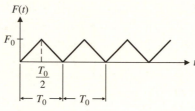

Figure P6.6

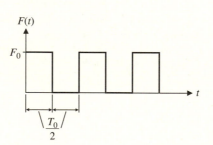

Figure P6.7

Figure P6.8

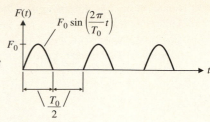

Figure P6.9

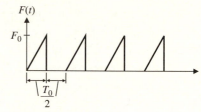

Figure P6.10

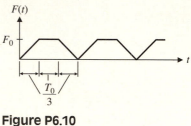

Figure P6.11

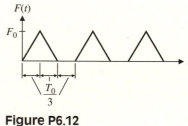

Figure P6.12

(b) The shear frame shown in Figure P6.13 is subject to the periodic forcing function described in part (a) of this problem and shown in Figure P6.9. Write a computer program to numerically evaluate the steady-state displacement response of the frame. Consider the first 10 nonzero terms of the Fourier expansion. Plot displacement versus time in the interval $0 \le t \le 10$ sec. For the frame, assume the columns bend about their major axes, $E = 29,000$ ksi, $\zeta = 0.15$, and $W = 25$ kips.

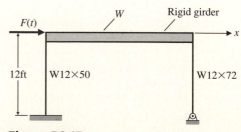

Figure P6.13

6.26 **(a)** Write a computer program to numerically evaluate the Fourier series expansion of the forcing function shown in Figure P6.8. Consider the first 10 nonzero terms in the series. Plot $F(t)$ versus time in the interval $0 \leq t \leq 10$ sec. Assume $F_0 = 10$ kips and $T_0 = 4$ sec.

(b) The shear frame shown in Figure P6.14 is subject to the periodic forcing function described in part (a) of this problem and shown in Figure P6.8. Write a computer program to numerically evaluate the steady-state displacement response of the frame. Consider the first 10 nonzero terms of the expansion. Plot displacement versus time in the interval $0 \leq t \leq 10$ sec. For the frame, assume the columns bend about their major axes, $E = 29,000$ ksi, $\zeta = 0.1$, and $W = 30$ kips.

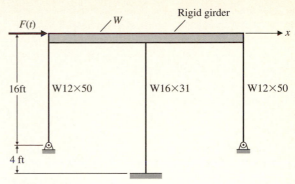

Figure P6.15

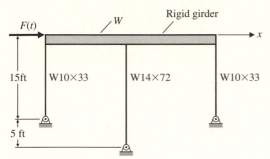

Figure P6.14

6.27 **(a)** Write a computer program to numerically evaluate the Fourier series expansion of the forcing function shown in Figure P6.11. Consider the first 10 nonzero terms in the series. Plot $F(t)$ versus time in the interval $0 \leq t \leq 10$ sec. Assume $F_0 = 15$ kips and $T_0 = 4$ sec.

(b) The shear frame shown in Figure P6.15 is subject to the periodic forcing function described in part (a) of this problem and shown in Figure P6.11. Write a computer program to numerically evaluate the steady-state displacement response of the frame. Consider the first 10 nonzero terms of the expansion. Plot displacement versus time in the interval $0 \leq t \leq 10$ sec. For the frame, assume the columns bend about their major axes, $E = 29,000$ ksi, $\zeta = 0.1$, and $W = 30$ kips.

6.28 **(a)** Write a computer program to numerically evaluate the Fourier series expansion of the forcing function shown in Figure P6.12. Consider the first 10 nonzero terms in the series. Plot $F(t)$ versus time in the interval $0 \leq t \leq 10$ sec. Assume $F_0 = 15$ kips and $T_0 = 1.5$ sec.

(b) The shear frame shown in Figure P6.15 is subject to the periodic forcing function described in part (a) of this problem and shown in Figure P6.12. Write a computer program to numerically evaluate the steady-state displacement response of the frame. Consider the first 10 nonzero terms of the expansion. Plot displacement versus time in the interval $0 \leq t \leq 10$ sec. For the frame, assume the columns bend about their major axes, $E = 29,000$ ksi, $\zeta = 0.1$, and $W = 30$ kips.

6.29 Use the Duhamel integral method to determine expressions for the displacement response of an undamped SDOF system subject to the load function shown in Figure P6.16 over the following time intervals: (a) $0 \leq t \leq t_d$ and (b) $t \geq t_d$. Assume zero initial conditions.

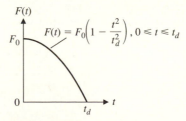

Figure P6.16

6.30 Use the Duhamel integral method to determine expressions for the displacement response of an undamped SDOF system subject to the load function shown in Figure P6.17 over the following time intervals: (a) $0 \le t \le t_1$, (b) $t_1 \le t \le t_2$, and (c) $t \ge t_2$. Assume zero initial conditions.

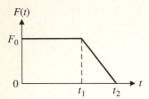

Figure P6.17

6.31 Use the Duhamel integral method to determine expressions for the displacement response of an undamped SDOF system subject to the load function shown in Figure P6.18 over the following time intervals: (a) $0 \le t \le t_d$ and (b) $t \ge t_d$. Assume zero initial conditions.

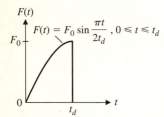

Figure P6.18

6.32 Use the Duhamel integral method to determine expressions for the displacement response of an undamped SDOF system subject to the load function shown in Figure P6.19 over the following time intervals: (a) $0 \le t \le 2\pi/\omega$, (b) $2\pi/\omega \le t \le 4\pi/\omega$, and (c) $t \ge 4\pi/\omega$. Assume zero initial conditions.

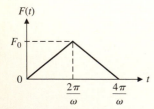

Figure P6.19

6.33 Use the Duhamel integral method to determine expressions for the displacement response of an undamped SDOF system subject to the load function shown in Figure P6.20 over the following time intervals: (a) $0 \le t \le t_1$, (b) $t_1 \le t \le t_2$, and (c) $t \ge t_2$. Assume zero initial conditions.

Figure P6.20

6.34 Use the Duhamel integral method to determine expressions for the displacement response of an undamped SDOF system subject to the load function shown in Figure P6.21 over the following time intervals: (a) $0 \le t \le t_1$, (b) $t_1 \le t \le t_2$, and (c) $t \ge t_2$. Assume zero initial conditions.

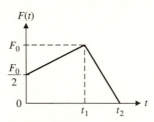

Figure P6.21

6.35 For the shear building shown in Figure P6.22, determine the horizontal deflection due to the load function shown in Figure P6.16 corresponding to the times $t = 0.3$ sec. and $t = 0.7$ sec. Use the solution obtained for Problem 6.29. Assume $F_0 = 20$ kips, $t_d = 0.5$ sec, $E = 29{,}000$ ksi, $W = 15$ kips, and $\zeta = 0$.

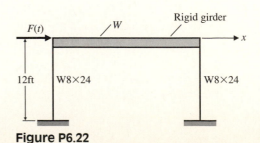

Figure P6.22

6.36 For the shear building shown in Figure P6.14, determine the horizontal deflection due to the load function shown in Figure P6.17 corresponding to the times $t = 0.5$ sec and $t = 0.9$ sec. Use the solution obtained for Problem 6.30. Assume $F_0 = 5$ kips, $t_1 = 0.5$ sec, $t_2 = 1.0$ sec, $E = 29,000$ ksi, $W = 30$ kips, and $\zeta = 0$.

6.37 For the shear building shown in Figure P6.23, determine the horizontal deflection due to the load function shown in Figure P6.18 corresponding to the times $t = 0.5$ sec and $t = 0.8$ sec. Use the solution obtained for Problem 6.31. Assume $F_0 = 30$ kips, $t_d = 0.7$ sec, $E = 29,000$ ksi, $W = 20$ kips, and $\zeta = 0$.

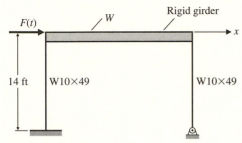

Figure P6.23

6.38 For the shear building shown in Figure P6.24, determine the horizontal deflection due to the load function shown in Figure P6.19 corresponding to the times $t = 0.3$ sec and $t = 0.75$ sec. Use the solution obtained for Problem 6.32. Assume $F_0 = 5$ kips, $2\pi/\omega = 0.25$ sec, $E = 30,000$ ksi, $W = 10$ kips, and $\zeta = 0$.

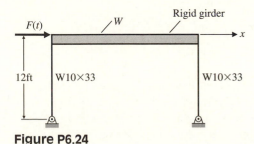

Figure P6.24

6.39 For the shear building shown in Figure P6.24, determine the horizontal deflection due to the load function shown in Figure P6.20 corresponding to the times $t = 0.3$ sec, $t = 0.7$ sec, and $t = 1.2$ sec. Use the solution obtained for Problem 6.33. Assume $F_1 = 7$ kips, $F_2 = 5$ kips, $t_1 = 0.5$ sec, $t_2 = 0.75$ sec, $W = 15$ kips, and $\zeta = 0$.

6.40 Use the Duhamel integral method to construct a response spectrum for a triangular pulse of duration t_d as shown in Figure P6.25. Plot DMF versus t_d/T over the time interval $0 \leq t_d/T \leq 6$.

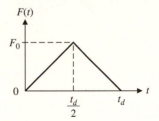

Figure P6.25

6.41 Use the Duhamel integral method to construct a response spectrum for a sine pulse of duration t_d as shown in Figure P6.26. Plot DMF versus t_d/T over the time interval $0 \leq t_d/T \leq 6$.

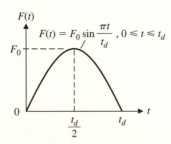

Figure P6.26

6.42 Use the Duhamel integral method to construct a response spectrum for the triangular pulse of duration t_d as shown in Figure P6.27. Plot DMF versus t_d/T over the time interval $0 \leq t_d/T \leq 6$.

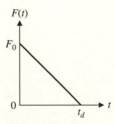

Figure P6.27

7 ⏶ Numerical Evaluation of Dynamic Response

It was clearly demonstrated in Chapter 6 that the analytical, or closed-form, solution of the Duhamel integral can be quite cumbersome, even for relatively simple excitation functions. Furthermore, if the exciting force cannot be expressed as a single mathematical function or exists only in digitized form (such as an earthquake ground record), an analytical solution of the Duhamel integral is precluded. Therefore, for most practical problems some form of numerical evaluation technique must be employed to obtain the dynamic response.

There are two basic approaches to numerically evaluating the dynamic response. The first approach can follow one of two paths: (1) numerical interpolation of the excitation or (2) numerical interpolation of the Duhamel integral. The alternative approach is direct numerical integration of the equation of motion. Both approaches are applicable to linear systems, but only the latter is valid for nonlinear systems. In this chapter both approaches are addressed, but the direct numerical integration procedure is the more commonly utilized in practice.

7.1 INTERPOLATION OF THE EXCITATION

Recall from Chapter 6 that the Duhamel integral expression representing the response of damped and undamped SDOF systems to arbitrary excitation $F(t)$ were given by Eqs. (6.25) and (6.27), respectively. The evaluation of these integrals is usually accomplished through the use of some type of numerical quadrature technique such as Simpson's method or the trapezoidal method. It is generally more convenient, however, to interpolate the excitation function $F(t)$.

Piecewise-linear interpolation of the excitation is an efficient and accurate method for numerically evaluating the response of linear systems, providing the time intervals are short. As shown in Figure 7.1, the excitation is expressed over the time interval $t_i \leq t \leq t_{i+1}$ by piecewise-linear interpolation as

$$F(\tau) = F_i + \left(\frac{\Delta F_i}{\Delta t}\right)\tau \qquad (7.1)$$

Figure 7.1

Piecewise-linear interpolation of excitation function.

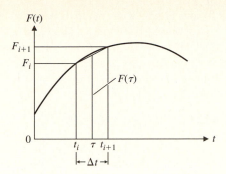

where

$$\Delta F_i = F_{i+1} - F_i \tag{7.2}$$

and

$$\Delta t = t_{i+1} - t_i \tag{7.3}$$

In Eq. (7.1), $F(\tau)$ represents the interpolated force and τ varies from 0 to Δt.

With this interpolated representation of the excitation, the differential equation of motion for an undamped SDOF system becomes

$$\ddot{x} + \omega^2 x = \frac{1}{m}\left[F_i + \left(\frac{\Delta F_i}{\Delta t}\right)\tau\right] \tag{7.4}$$

The solution to Eq. (7.4) is the sum of the homogeneous and particular solutions over the time interval $0 \leq \tau \leq \Delta t$. The homogeneous portion of the solution is evaluated for initial conditions of displacement x_i and velocity $\dot{x}_i$ at $\tau = 0$. The particular solution is composed of two parts: the first is the response to an ideal step force of magnitude F_i, and the second is the response to a ramp function defined by $(\Delta F_i/\Delta t)\tau$. The three components of the solution are represented by Eqs. (3.32), (6.33), and (6.58), respectively, resulting in

$$x_{i+1} = x_i \cos (\omega \, \Delta t) + \left(\frac{x_i}{\omega}\right) \sin (\omega \, \Delta t)$$

$$+ \left(\frac{F_i}{k}\right)[1 - \cos (\omega \, \Delta t)] \tag{7.5}$$

$$+ \left(\frac{\Delta F_i}{k}\right)\left(\frac{1}{\omega \, \Delta t}\right)[\omega \Delta t - \sin(\omega \, \Delta t)]$$

and

$$\frac{\dot{x}_{i+1}}{\omega} = -x_i \sin (\omega \, \Delta t) + \left(\frac{\dot{x}_i}{\omega}\right) \cos (\omega \, \Delta t)$$

$$+ \left(\frac{F_i}{k}\right) \sin (\omega \, \Delta t) \tag{7.6}$$

$$+ \left(\frac{\Delta F_i}{k}\right)\left(\frac{1}{\omega \, \Delta t}\right)[1 - \cos (\omega \, \Delta t)]$$

Equations (7.5) and (7.6) are the *recurrence formulas* for the displacement x_{i+1} and velocity $\dot{x}_{i+1}$ corresponding to time t_{i+1}. To evaluate Eqs. (7.5) and (7.6), the kinematics (x_i and $\dot{x}_i$) must be known for time t_i. Recurrence formulas for a viscously underdamped SDOF system may be derived in a similar manner.

A simpler and more convenient representation of recurrence formulas typified by Eqs. (7.5) and (7.6) is

$$x_{i+1} = AF_1 + BF_{i+1} + Cx_i + D\dot{x}_i \tag{7.7}$$

and

$$\dot{x}_{i+1} = A'F_i + B'F_{i+1} + C'x_i + D'\dot{x}_i \tag{7.8}$$

The recurrence formulas coefficients are presented in Table 7.1 for viscously underdamped SDOF systems. The accuracy of these expressions depends only upon the size of the time step Δt. In practice, Δt should be sufficiently small to closely approximate the excitation force and also to render results at the required time intervals. As a rule of practice, select $\Delta t \leq T/10$, where T is the natural period of the structure, to ensure that important peaks of the structural response are not omitted. Finally, as long as Δt is constant, the recurrence formulas coefficients need be calculated only once.

TABLE 7.1. **Coefficients for Recurrence Formulas for Damped SDOF Systems ($\zeta < 1$)**

$$A = \frac{1}{k\omega_d h}\left\{e^{-\beta h}\left[\left(\frac{\omega_d^2 - \beta^2}{\omega^2} - \beta h\right)\sin\,\omega_d h - \left(\frac{2\omega_d\beta}{\omega^2} + \omega_d h\right)\cos\,\omega_d h\right] + \frac{2\beta\omega_d}{\omega^2}\right\}$$

$$B = \frac{1}{k\omega_d h}\left\{e^{-\beta h}\left[-\left(\frac{\omega_d^2 - \beta^2}{\omega^2}\right)\sin\,\omega_d h + \left(\frac{2\omega_d\beta}{\omega^2}\right)\cos\,\omega_d h\right] + \omega_d h - \frac{2\beta\omega_d}{\omega^2}\right\}$$

$$C = e^{-\beta h}\left[\cos\,\omega_d h + \left(\frac{\beta}{\omega_d}\right)\sin\,\omega_d h\right]$$

$$D = \left(\frac{1}{\omega_d}\right)e^{-\beta h}\sin\,\omega_d h$$

$$A' = \frac{1}{k\omega_d h}\{e^{-\beta h}[(\beta + \omega_d^2 h)\sin\,\omega_d h + \omega_d\cos\,\omega_d h] - \omega_d\}$$

$$B' = \frac{1}{k\omega_d h}\{e^{-\beta h}[\beta\sin\,\omega_d h + \omega_d\cos\,\omega_d h] + \omega_d\}$$

$$C' = -\left(\frac{\omega^2}{\omega_d}\right)e^{-\beta h}\sin\,\omega_d h$$

$$D' = e^{-\beta h}\left[\cos\,\omega_d h - \left(\frac{\beta}{\omega_d}\right)\sin\,\omega_d h\right]$$

where $\beta \equiv \zeta\omega$ and $h \equiv \Delta t$

EXAMPLE 7.1 ▼

The water tower shown in Figure 7.2a is subject to the *blast loading* illustrated in Figure 7.2b. Write a computer program to numerically evaluate the dynamic response of the tower by interpolation of the excitation. Plot the displacement $x(t)$ time and velocity $\dot{x}(t)$ responses in the time interval $0 \leq t \leq 0.5$ sec. Assume $W = 100$ kips, $k = 2800$ kips/ft, $\zeta = 0.05$, $F_0 = 100$ kips, and $t_d = 0.05$ sec. Use the time step $\Delta t = 0.005$ sec.

Solution

(a) Calculate the natural period of the structure

$$k = 2800 \text{ kip/ft} \qquad m = \frac{100}{32.2} = 3.106 \text{ kips-sec}^2/\text{ft}$$

$$\omega = \sqrt{\frac{k}{m}} = \sqrt{\frac{2800}{3.106}} = 30.03 \text{ rad/sec}$$

$$T = \frac{2\pi}{\omega} = 0.209 \text{ sec}$$

$$\Delta t = 0.005 < \frac{T}{10} = 0.02 \qquad \text{OK}$$

(b) Plots for displacement time history $x(t)$ and velocity $\dot{x}(t)$ are presented in Figure 7.3a and b, respectively. The BASIC computer program for the evaluation of the response is listed in Table 7.2.

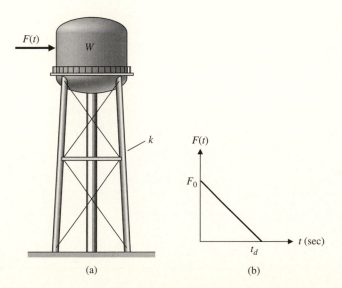

Figure 7.2

(a) Water tower of Example 7.1; (b) approximation for blast loading.

Figure 7.3
Response histories for
Example 7.1: (a) dis-
placement; (b) velocity.

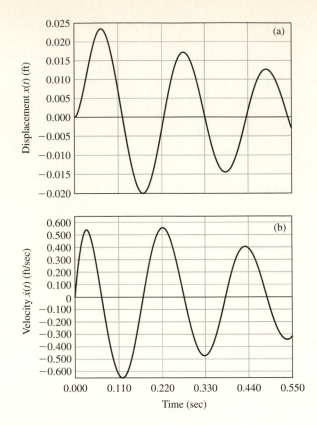

TABLE 7.2. Listing of BASIC Computer Program Used for Example 7.1

```
10 REM THIS IS THE PROGRAM FOR CALCULATING THE DYNAMIC RESPONSE TO THE
20 REM NON-PERIODIC LOADING USING THE PIECEWISE LINEAR INTERPOLATION
30 REM TECHNIQUE.
40 REM
50 OPEN "0",#2,"A:NTX"
60 OPEN "o",#3,"a:ntv"
70 DIM X(500),P(500),V(500)
80 SK=2800000! : W=100000! : ZTA=.05 : F0=100000!
90 REM
100 REM THE TIME INTERVAL IS ASSUMED TO BE CONSTANT
110 WN=SQR(SK*32.2/W) : DT=.005 : WD=WN*SQR(1−ZTA^2) :BT=ZTA*WN
120 F1=WD*DT : F2=BT*DT :F3=SK*WD*DT
130 A=((EXP(−F2)*((((WD^2−BT^2)/WN^2)−F2)*SIN(F1)−((2*WD*BT/WN^2+F1)*COS(F1)))
)+2*BT*WD/WN^2)/F3
140 B=(EXP(−F2)*((WD^2−BT^2)/WN^2*(−SIN(F1))+(2*WD*BT/WN^2)*COS(F1))+F1−2*BT*
WD/WN^2)/F3
150 C=EXP(−F2)*(COS(F1)+(BT/WD)*SIN(F1))
160 D=(1/WD)*(EXP(−F2)*SIN(F1))
170 A1=(EXP(−F2)*((BT+WN^2*DT)*SIN(F1)+WD*COS(F1))−WD)/F3
180 B1=(−EXP(−F2)*(BT*SIN(F1)+WD*COS(F1))+WD)/F3
190 C1=(WN^2/WD)*(−EXP(−F2)*SIN(F1))
200 D1=EXP(−F2)*(COS(F1)−(BT/WD)*SIN(F1))
210 X(0)=0: V(0)=0 :DTI=0 :I=0
220 PRINT #2,"Z"
230 PRINT #3,"z"
```

(continued)

```
240 PRINT #2,DTI,X(0) : PRINT #3,DTI,V(0)
250 FOR DTI=0 TO .5 STEP DT
260 P(I)=F0*(1−(DTI-DT)/.05) : IF ((DTI−DT)>= .05) THEN P(I)=0
270 P(I+1)=F0*(1−DTI/.05) : IF (DTI>=.05) THEN P(I+1)=0
280 X(I+1)=A*P(I)+B*P(I+1)+C*X(I)+D*V(I)
290 V(I+1)=A1*P(I)+B1*P(I+1)+C1*X(I)+D1*V(I)
300 PRINT #2,DTI,X(I)
310 PRINT #3,DTI,V(I)
320 I=I+1
330 NEXT DTI
340 STOP
```

▲

7.2 DIRECT INTEGRATION OF THE EQUATION OF MOTION

In direct integration, the equation of motion is integrated using a step-by-step procedure. There are two fundamental concepts at the essence of direct integration methods: (1) the equation of motion is satisfied at only discrete time intervals Δt, and (2) for any time t, a variation of displacement, velocity, and acceleration within each time interval Δt is assumed.

To illustrate these basic concepts common to all direct integration schemes, consider the equation of motion of a viscously damped SDOF system subject to an arbitrary exciting force $F(t)$ given by

$$m\ddot{x} + c\dot{x} + kx = F(t) \tag{7.9}$$

It is assumed that the displacement, velocity, and acceleration at time $t = 0$ are known and are represented by x_0, $\dot{x}_0$, and $\ddot{x}_0$, respectively. It is also assumed that the solution is sought in the time interval from 0 to t_f. In the solution time interval under consideration, t_f is subdivided into n equal time increments Δt, and the integration scheme approximates a solution at times 0, Δt, $2\Delta t$, . . . , t_f. The integration method calculates the solution at the next required time increment based upon the solution obtained at the previous time increment. Therefore, algorithms can be derived to calculate the solution at some time $t + \Delta t$ based upon the known solution at time t.

In the following sections, several commonly used direct integration methods are presented. A detailed discussion on the accuracy and stability of the direct integration methods is presented in Chapter 13.

7.3 CENTRAL DIFFERENCE METHOD

To derive a direct integration algorithm based upon the central difference method, consider the displacement versus time relationship shown in Figure 7.4. The following approximation can be made regarding the slope (first time derivative) of the displacement curve (velocity):

$$\dot{x}_i = \left(\frac{dx}{dt}\right)_{t_i} = \frac{\Delta x}{\Delta t} = \frac{x_{i+1} - x_{i-1}}{t_{i+1} - t_{i-1}} = \frac{x_{i+1} - x_{i-1}}{2\,\Delta t} \tag{7.10}$$

Figure 7.4

Displacement-time relationship for central difference method.

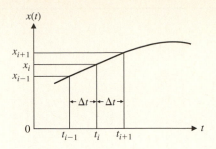

A similar approximation can be made for the second time derivative (acceleration):

$$\ddot{x}_i = \left(\frac{d^2x}{dt^2}\right)_{t_i} = \frac{\left[\left(\frac{dx}{dt}\right)_{t_i+(\Delta t/2)} - \left(\frac{dx}{dt}\right)_{t_i-(\Delta t/2)}\right]}{\Delta t} \tag{7.11}$$

or

$$\ddot{x}_i = \frac{[(x_{i+1} - x_i)/\Delta t - (x_i - x_{i-1})/\Delta t]}{\Delta t} = \frac{x_{i+1} - 2x_i + x_{i-1}}{(\Delta t)^2} \tag{7.12}$$

The displacement solution at time t_{i+1} is obtained by considering Eq. (7.9) at time t_i, that is,

$$m\ddot{x}_i + c\dot{x}_i + kx_i = F_i \tag{7.13}$$

Substituting the relations for $\dot{x}_i$ and $\ddot{x}_i$ given by Eqs. (7.10) and (7.12), respectively, into Eq. (7.13) yields

$$m\left[\frac{x_{i+1} - 2x_i + x_{i-1}}{(\Delta t)^2}\right] + c\left[\frac{x_{i+1} - x_{i-1}}{2\Delta t}\right] + kx_i = F_i \tag{7.14}$$

Equation (7.14) may be rearranged to solve for x_{i+1}, such that

$$x_{i+1} = \left[\frac{1}{m/(\Delta t)^2 + c/2\,\Delta t}\right]\left[\frac{2m}{(\Delta t)^2} - k\right]x_i$$

$$+ \left[\frac{c}{2\,\Delta t} - \frac{m}{(\Delta t)^2}\right]x_{i-1} + F_i \tag{7.15}$$

Equation (7.15) is the recurrence formula for calculating the displacement of the mass at time t_{i+1}, based upon knowledge of the previous displacement history x_{i-1} and x_i and the current magnitude of the external force F_i.

It should be noted that when using the central difference method the calculation of x_{i+1} involves x_i and x_{i-1}. Therefore, to calculate the solution at the first time increment after $t = 0$, a special starting procedure is required to calculate x_{i-1} at time $t = 0$. Since x_0 and $\dot{x}_0$ are known, $\ddot{x}_0$ can be calculated from Eq. (7.13) as

$$\ddot{x}_0 = \frac{1}{m}[F(0) - c\dot{x}_0 - kx_0] \tag{7.16}$$

Then, Eqs. (7.10) and (7.12) evaluated at time $t = 0$ can be used to obtain an expression for x_{i-1} at time $t = 0$ given by

$$x_{i-1} = x_0 - \dot{x}_0 \, \Delta t + \frac{\ddot{x}_0 (\Delta t)^2}{2} \tag{7.17}$$

An important consideration in the case of the central difference method is that the integration method is *conditionally stable* and requires the time step Δt to be smaller than a critical value Δt_{cr}, or

$$\Delta t \leq \Delta t_{cr} \tag{7.18}$$

where

$$\Delta t_{cr} = \frac{T}{\pi} \tag{7.19}$$

and T is the natural period of the system. However, for SDOF systems, the accuracy criteria of $\Delta t < T/10$ will control maximum time step size. The central difference time integration scheme is summarized in Table 7.3.

TABLE 7.3. **Step-by-Step Solution Using the Central Difference Method**

A. Initial calculations:
   1. Calculate k, c, m
   2. Select an appropriate time step Δt
   3. Calculate $\ddot{x}_0$

$$\ddot{x}_0 = \frac{1}{m}[F(0) - c\dot{x}_0 - kx_0]$$

   4. Calculate x_{i-1} at time $t = 0$

$$x_{i-1} = x_0 - \dot{x}_0 \, \Delta t + \frac{\ddot{x}_0 (\Delta t)^2}{2}$$

   5. Calculate effective mass

$$\hat{m} = \frac{m}{(\Delta t)^2} + \frac{c}{2(\Delta t)}$$

B. For each time step:
   1. Calculate effective load at time t_i

$$\hat{F}_i = F_i - \left[k - \frac{2m}{(\Delta t)^2}\right]x_i - \left[\frac{m}{(\Delta t)^2} - \frac{c}{2\Delta t}\right]x_{i-1}$$

   2. Calculate displacement at time t_{i+1}

$$x_{i+1} = \frac{\hat{F}_i}{\hat{m}}$$

   3. Evaluate acceleration and velocity at time t_i

$$\ddot{x}_i = \frac{1}{(\Delta t)^2}[x_{i-1} - 2x_i + x_{i+1}]$$

$$\dot{x}_i = \frac{1}{2(\Delta t)}[-x_{i-1} + x_{i+1}]$$

EXAMPLE 7.2   ▼

The single-story shear frame shown in Figure 7.5a is subject to the arbitrary excitation force specified in Figure 7.5b. The rigid girder supports a uniformly distributed dead load of 1.75 kips/ft. Assume the columns bend about their major axis
and neglect their mass, a damping factor $\zeta = 0.02$, and a modulus of elasticity
$E = 29,000$ ksi. Write a computer program for the central difference method to
evaluate the dynamic response of the frame. Plot the displacement $x(t)$, velocity
$\dot{x}(t)$, and acceleration $\ddot{x}(t)$ responses in the time interval $0 \le t \le 3$ sec.

Solution

(a) The mass of the structure is

$$m = \frac{W}{g} = \frac{1.75 \text{ kips/ft/}(35 \text{ ft})(12 \text{ in})}{386.4 \text{ in/sec}^2} = 0.1585 \text{ kip-sec}^2/\text{in}$$

(b) The stiffness of the structure is

$$k = k_1 + k_2$$

$$k_1 = \frac{12EI}{L^3} = \frac{12(29,000 \text{ ksi})(541 \text{ in}^4)}{(15 \text{ ft} \times 12 \text{ in})^3} = 32.28 \text{ kips/in}$$

$$k_2 = \frac{3EI}{L^3} = \frac{3(29,000 \text{ ksi})(541 \text{ in}^4)}{(15 \text{ ft} \times 12 \text{ in})^3} = 8.07 \text{ kips/in}$$

$$k = 32.28 + 8.07 = 40.35 \text{ kips/in}$$

(c) The dynamic characteristics of the structure are

$$\omega = \sqrt{\frac{k}{m}} = \sqrt{\frac{40.35}{0.1585}} = 15.96 \text{ rad/sec}$$

$$T = \frac{2\pi}{\omega} = \frac{2(3.1416)}{15.96} = 0.394 \text{ sec}$$

(d) The time step is

$$\Delta t < \Delta t_{cr} = \frac{T}{\pi} = 0.125 \text{ sec}$$

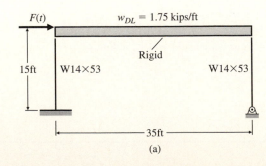

(a)

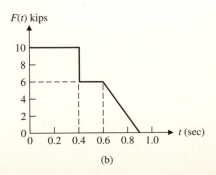

(b)

Figure 7.5
(a) Single-story shear frame
structure of Example 7.2;
(b) excitation force.

or

$$\Delta t \le \frac{T}{10} = 0.0394 \text{ sec}$$

Use a time step $\Delta t = 0.02$ sec.

Time histories of displacement $x(t)$, velocity $\dot{x}(t)$, and acceleration $\ddot{x}(t)$ are presented in Figure 7.6a, b, and c, respectively. A listing of the FORTRAN computer program is presented in Table 7.4.

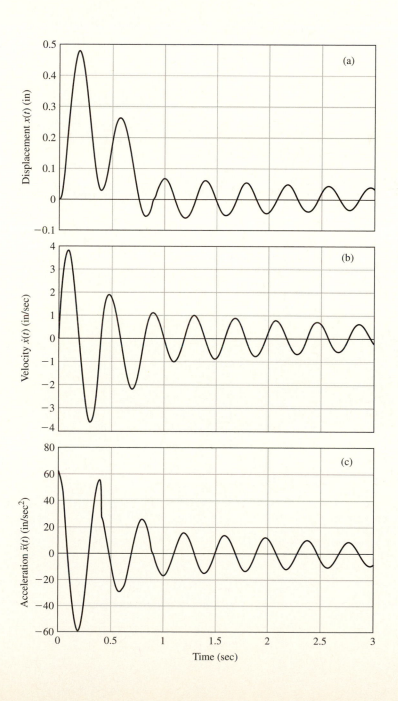

Figure 7.6

Response histories for Example 7.2: (a) displacement; (b) velocity; (c) acceleration.

TABLE 7.4. Listing of FORTRAN Computer Program Used for Example 7.2

```
C
C                      CENTRAL DIFFERENCE METHOD
C
      REAL X(400),XDOT(400),DEL,PERIOD,ZETA,K,XDDOT(400)
      REAL CCRIT,C,M,W,PI,T,F
      INTEGER NSTEPS
      ZETA=0.02
      X(1)=0.0
      XDOT(1)=0.0
      XDDOT(1)=0.0
      T=0
      K=40.35
      M=0.1585145
      W=SQRT(K/M)
      CCRIT=2*M*W
      C=ZETA*CCRIT
      PI=3.1415927
      PERIOD=(2*PI)/W
      DEL=0.02
      NSTEPS=3.0/DEL
      X(2)=X(1)+(DEL*XDOT(1))+(((DEL*DEL)/2)*((10.0/M)−(XDOT(1)*C/M)−
     *  (X(1)*K/M)))
      OPEN(UNIT=6,FILE="x_vs_t.out",STATUS="UNKNOWN")
      OPEN(UNIT=7,FILE="xd_vs_t.out",STATUS="UNKNOWN")
      OPEN(UNIT=8,FILE="xdd_vs_t.out",STATUS="UNKNOWN")
      OPEN(UNIT=9,FILE="centdiff.out",STATUS="UNKNOWN")
      WRITE(9,*)
      WRITE(9,*)' T X(T) V(T) A(T) '
      WRITE(9,*)
      WRITE(9,175)T,X(1),XDOT(1),XDDOT(1)
175   FORMAT(F6.3,2X,F7.4,2X,F7.4,2X,F9.4)
      WRITE(6,120)T,X(1)
      WRITE(7,120)T,XDOT(1)
      WRITE(8,120)T,XDDOT(1)
120   FORMAT(F6.3,1X,F9.4)
      DO 100 I=2,NSTEPS+1
        T=T+DEL
        IF(T.LE.0.4) THEN
          F=10.0
        ELSEIF(T.LE.0.6.AND.T.GT.0.4) THEN
          F=6.0
        ELSEIF(T.LE.0.9.AND.T.GT.0.6) THEN
          F=(−20.0*T)+18.0
        ELSE
          F=0.0
        ENDIF
        XDOT(I)=(((X(I)−X(I−1))/DEL)+((DEL*F)/(2*M))−((DEL*K*X(I))/
     *  (2*M)))/(1+((T*C)/(2*M)))
        XDDOT(I)=((1/M)*(F−(K*X(I))−(C*XDOT(I))))
        X(I+1)=(2*X(I))−X(I−1)+((DEL*DEL)*XDDOT(I))
        WRITE(6,200)T,X(I)
        WRITE(7,200)T,XDOT(I)
        WRITE(8,200)T,XDDOT(I)
200     FORMAT(F6.3,1X,F9.4)
        WRITE(9,300)T,X(I),XDOT(I),XDDOT(I)
300   FORMAT(F6.3,2X,F7.4,2X,F7.4,2X,F9.4)
100   CONTINUE
      STOP
      END
```

▲

7.4   RUNGE-KUTTA METHODS

The Runge-Kutta methods are classified as *single-step*, since they require knowledge of only x_i to determine x_{i+1}. Therefore the methods are *self-starting* and require no special starting procedure as did the central difference method. The most popular of the Runge-Kutta methods is the *fourth-order*, or *classic, form*, which will be discussed in this section.

In application of this Runge-Kutta method, the second-order differential equation of motion is expressed as two first-order equations. Considering the differential equation of motion of a viscously damped SDOF system gives

$$m\ddot{x} + c\dot{x} + kx = F(t) \tag{7.20}$$

which represents a second-order ordinary differential equation in x. However, by setting $\dot{x} = y$, Eq. (7.20) can be expressed as two first-order differential equations. Thus

$$\dot{x} = y \tag{7.21}$$

and

$$\dot{y} = \ddot{x} = \frac{1}{m}[F(t) - c\dot{x} - kx] \tag{7.22}$$

The Runge-Kutta recurrence formulas for x_{i+1} and y_{i+1}, respectively, are given by

$$x_{i+1} = x_i + \frac{\Delta t}{6}(y_1 + 2y_2 + 2y_3 + y_4) \tag{7.23}$$

and

$$y_{i+1} = y_i + \frac{\Delta t}{6}(\dot{y}_1 + 2\dot{y}_2 + 2\dot{y}_3 + \dot{y}_4) \tag{7.24}$$

Equations (7.23) and (7.24) represent an averaging of the velocity and acceleration by Simpson's rule within the time interval Δt. The fourth-order Runge-Kutta time integration scheme is summarized in Table 7.5.

TABLE 7.5.   Step-by-Step Solution Using the Fourth-Order Runge-Kutta Method

A. Initial calculations:
  1. Calculate k, c, m
  2. Initialize variables

$$y_0 = \dot{x}_0, \qquad \dot{y}_0 = \frac{1}{m}[F(0) - cy_0 - kx_0]$$

  3. Select an appropriate time step Δt

(continued)

B. For each time step:
1. Calculations at the beginning of the time interval

$$t = t_i$$

$$x = x_i$$

$$y_1 = y_i$$

$$\dot{y}_1 = \frac{1}{m}[F(t) - cy_1 - kx]$$

2. Calculations at the first midpoint of the time interval

$$t = t_i + \frac{\Delta t}{2}$$

$$x = x_i + \frac{\Delta t}{2}y_1$$

$$y_2 = y_i + \frac{\Delta t}{2}\dot{y}_1$$

$$\dot{y}_2 = \frac{1}{m}[F(t) - cy_2 - kx]$$

3. Calculations at the second midpoint of the time interval

$$t = t_i + \frac{\Delta t}{2}$$

$$x = x_i + \frac{\Delta t}{2}y_2$$

$$y_3 = y_i + \frac{\Delta t}{2}\dot{y}_2$$

$$\dot{y}_3 = \frac{1}{m}[F(t) - cy_3 - kx]$$

4. Calculations at the end of the time interval

$$t = t_i + \Delta t$$

$$x = x_i + y_3 \Delta t$$

$$y_4 = y_i + \Delta t\dot{y}_3$$

$$\dot{y}_4 = \frac{1}{m}[F(t) - cy_4 - kx]$$

(continued)

5. Calculate the displacement and velocity at the end of the time interval

$$x_{i+1} = x_i + \frac{\Delta t}{6}(y_1 + 2y_2 + 2y_3 + y_4)$$

$$\dot{x}_{i+1} = y_i + \frac{\Delta t}{6}(\dot{y}_1 + 2\dot{y}_2 + 2\dot{y}_3 + \dot{y}_4)$$

EXAMPLE 7.3 ▼

Consider the water tower structure shown in Figure 7.2a to be subjected to the loading shown in Figure 7.7. Evaluate the dynamic response of the structure by the fourth-order Runge-Kutta method. Plot the displacement $x(t)$ and velocity $\dot{x}(t)$ responses in the time interval $0 \le t \le 1.5$ sec. Use a time step $\Delta t = 0.005$ sec. Assume $W = 80$ kips, $k = 3500$ kips/ft, and $\zeta = 0.075$.

Solution

(a) Calculate the natural period and time step

$$k = 3500 \text{ kips/ft} \qquad m = \frac{W}{g} = \frac{80}{32.2} = 2.48 \text{ kip-sec}^2/\text{in}$$

$$\omega = \sqrt{\frac{k}{m}} = \sqrt{\frac{3500}{2.48}} = 37.53 \text{ rad/sec}$$

$$T = \frac{2\pi}{\omega} = 0.167 \text{ sec}$$

$$\Delta t \le \frac{T}{10} = 0.0167 \text{ sec} \qquad \text{OK}$$

(b) Time histories for the displacement $x(t)$ and velocity $\dot{x}(t)$ are presented in Figure 7.8a and b, respectively. A listing for the FORTRAN computer program is presented in Table 7.6.

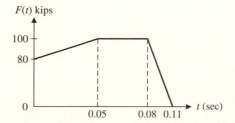

Figure 7.7

Excitation force for Example 7.3.

Figure 7.8

Response histories for
Example 7.3: (a) dis-
placement; (b) velocity.

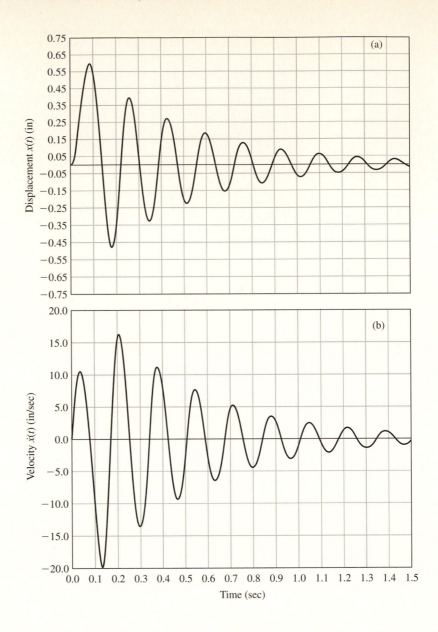

TABLE 7.6.   Listing of FORTRAN Computer Program Used for Example 7.3

```
*
*              :RUNGE KUTTA SOLUTION
*     Program to evaluate the TIME HISTORY response of a structure
*     to a general dynamic load, specified as forcing function
*     values at discrete values of time.
*     As a particular problem, the forcing function given in
*     Example 7.3.
*
      program RK_resp
      implicit real *8 (a-h,o-z)
      real *8 k,mass
```

(continued)

```
       dimension tt(50),ft(50),x(4),y(4),f(4),t(4)
       open(2,file='x.out',status='unknown')
       open(3,file='v.out',status='unknown')

       call input(mass,k,h,c,t1,t2,tt,ft,ift,x0,v0)

       t(1)=t1
       x(1)=x0
       y(1)=v0
       write(2,*)t1,x0
       write(3,*)t1,v0
       ntot=int((t2-t1)/h)+1

       do 10 i=1,ntot
         t(2)=t(1)+h/2
         t(3)=t(2)
         t(4)=t(1)+h
         call inter1(t(1),tt,ft,ift,acc,mass,k,c,x(1),y(1))
         f(1)=acc

         do 11 j=2,4
           x(j)=x(1)+y(j-1)*h/2
           y(j)=y(1)+f(j-1)*h/2
           call inter1(t(j),tt,ft,ift,acc,mass,k,c,x(j),y(j))
           f(j)=acc
11     continue

       x(1)=x(1)+h/6.*(y(1)+2.*y(2)+2.*y(3)+y(4))
       y(1)=y(1)+h/6.*(f(1)+2.*f(2)+2.*f(3)+f(4))
       t(1)=t(1)+h

         write(2,*)t(1),x(1)
         write(3,*)t(1),y(1)
10     continue

       stop
       end
*
*      Subroutine INPUT is used to input the values of the
*      system parameters (mass, damping, and stiffness)
*      as well as the forcing function, whose values are
*      specified at discrete time intervals
*      All input is in free format from the file AS7.IN
*
       subroutine input(mass,k,h,c,t1,t2,tt,ft,i,x0,v0)
       implicit real *8 (a-h,o-z)
       real *8 k,mass
       dimension tt(50),ft(50)

       open(1,file='as7.in',status='old')
       read(1,*)weight,k,zeta,h,x0,v0
       mass=weight/32.2/12.
       k=k/12.
       omega=sqrt(k/mass)
       c=zeta*2.*mass*omega
       read(1,*)t1,t2
       i=1
2      read(1,*,err=3)tt(i),ft(i)
       i=i+1
       goto 2
3      i=i-1

       return
       end
```

(continued)

```
*
*     Subroutine INTER1 is used by the program to generate the
*     acceleration values at each time step:
*     > It first obtains the value of the forcing function
*        at the specified time by suitable interpolation
*        between the discrete forcing function values supplied
*     > It then calculates the acceleration from the system
*        parameters, the forcing function, and the velocity
*        and displacement values
*
      subroutine inter1(t1,t,f,i,a,m,k,c,x,y)
      implicit real *8 (a-h,o-z)
      real *8 k,m
      dimension t(50),f(50)

      if(t1.lt.t(1).or.t1.gt.t(i)) then
        ft=0.0
        goto 75
      endif

      do 78 n=1,(i-1)
      if(t1.ge.t(n).and.t1.le.t(n+1)) then
        ft=f(n)+(f(n+1)-f(n))*(t1-t(n))/(t(n+1)-t(n))
        goto 75
      endif
78    continue

75    a=(ft-k*x-c*y)/m

      return
      end
```

Below is a copy of the INPUT FILE as7.in

All input is in FREE FORMAT

CARD 1: weight, stiffness, damping ratio, time step,
        initial values of displacement, and velocity
CARD 2: t1, t2
        t1 -> initial time value to print response
        t2 -> ending time value to print response
CARD 3: onward : t(i), ft(i)
        t(i) -> time
        ft(i) -> forcing function at time t(i)
        The program interpolates between appropriate values
        of t(i) and corresponding ft(i) to obtain the forcing
        function value at any particular time at which the
        solution is sought. Thus, FORCING FUNCTIONS OF ANY
        GENERAL NATURE CAN BE USED BY THIS PROGRAM AS INPUT
>>>>

80 3500 0.075 0.005 0 0
0 1.5
0 80
0.05 100
0.08 100
0.11 0

>>>>

7.5 AVERAGE ACCELERATION METHOD

In the derivation of recurrence formulas for this method, the assumption is made that within a small increment of time Δt, the acceleration is the average value of the acceleration at the beginning of the interval $\ddot{x}_i$ and the acceleration at the end of the time interval $\ddot{x}_{i+1}$, as illustrated in Figure 7.9. Thus, the acceleration at some time τ between t_i and t_{i+1} is expressed as

$$\ddot{x}(\tau) = \frac{1}{2}(\ddot{x}_i + \ddot{x}_{i+1}) \tag{7.25}$$

Integrating Eq. (7.25) twice yields

$$\dot{x}_{i+1} = \dot{x}_i + \left(\frac{\Delta t}{2}\right)(\ddot{x}_i + \ddot{x}_{i+1}) \tag{7.26}$$

and

$$x_{i+1} = x_i + \Delta t\dot{x}_i + \frac{(\Delta t)^2}{4}(\ddot{x}_i + \ddot{x}_{i+1}) \tag{7.27}$$

To solve for displacement, velocity, and acceleration at time t_{i+1}, the equation of motion, Eq. (7.9), must also be considered at time t_{i+1}; thus

$$m\ddot{x}_{i+1} + c\dot{x}_{i+1} + kx_{i+1} = F_{i+1} \tag{7.28}$$

From Eq. (7.28), $\ddot{x}_{i+1}$ can be solved in terms of x_{i+1}. Then substituting this expression for $\ddot{x}_{i+1}$ into Eq. (7.26), expressions for $\ddot{x}_{i+1}$ and $\dot{x}_{i+1}$, each in terms of the unknown displacement x_{i+1} only, can be determined. These subsequent expressions for $\dot{x}_{i+1}$ and $\ddot{x}_{i+1}$ are then substituted into Eq. (7.28) to solve for x_{i+1}. Thus, the recurrence formula for displacement is given as

$$x_{i+1} = \left[\frac{1}{k + [4m/(\Delta t)^2] + (2c/\Delta t)}\right]$$
$$\left[m\left(\frac{4x_i}{(\Delta t)^2} + \frac{4\dot{x}_i}{\Delta t} + \ddot{x}_i\right) + c\left(\frac{4x_i}{\Delta t} + \dot{x}_i\right) + F_{i+1}\right] \tag{7.29}$$

After x_{i+1} is determined from Eq. (7.29), Eqs. (7.26) and (7.27) may be used to solve for $\ddot{x}_{i+1}$,

$$\ddot{x}_{i+1} = \frac{4}{(\Delta t)^2}(x_{i+1} - x_i) - \frac{4\dot{x}_i}{\Delta t} - \ddot{x}_i \tag{7.30}$$

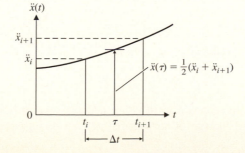

Figure 7.9

Numerical integration using the average acceleration method.

and then $\dot{x}_{i+1}$ can be calculated from Eq. (7.26).

The development of a computational algorithm can be facilitated by employment of incremental quantities for applied load ΔF_i, displacement Δx_i, velocity $\Delta \dot{x}_i$, and acceleration $\Delta \ddot{x}_i$, such that

$$\Delta F_i = F_{i+1} - F_i \tag{7.31a}$$

$$\Delta x_i = x_{i+1} - x_i \tag{7.31b}$$

$$\Delta \dot{x}_i = \dot{x}_{i+1} - \dot{x}_i \tag{7.31c}$$

$$\Delta \ddot{x}_i = \ddot{x}_{i+1} - \ddot{x}_i \tag{7.31d}$$

Equation (7.27) can then be solved for the incremental acceleration given as

$$\Delta \ddot{x}_i = \frac{4}{(\Delta t)^2}(\Delta x_i - \Delta t \dot{x}_i) - 2\ddot{x}_i \tag{7.32}$$

An expression for the incremental velocity $\Delta \dot{x}_i$ can subsequently be established by combining Eqs. (7.26) and (7.27) with Eq. (7.32), resulting in

$$\Delta \dot{x}_i = \left(\frac{2}{\Delta t}\right)\Delta x_i - 2\dot{x}_i \tag{7.33}$$

Since Eq. (7.9) is satisfied at both t_i and t_{i+1}, it can be expressed in incremental form as

$$m\Delta \ddot{x}_i + c\,\Delta \dot{x}_i + k\,\Delta x_i = \Delta F_i \tag{7.34}$$

Equations (7.32) and (7.33) can be substituted into Eq. (7.34) to yield the expression

$$\hat{k}\,\Delta x_i = \Delta \hat{F}_i \tag{7.35}$$

where the *effective stiffness* $\hat{k}$ is given by

$$\hat{k} = k + \left(\frac{2c}{\Delta t}\right) + \frac{4m}{(\Delta t)^2} \tag{7.36}$$

and the *effective incremental force* $\Delta \hat{F}_i$ is expressed as

$$\Delta \hat{F}_i = \Delta F_i + \left(\frac{4m}{\Delta t} + 2c\right)\dot{x}_i + 2m\ddot{x}_i \tag{7.37}$$

Once Δx_i has been determined from Eq. (7.35), $\Delta \dot{x}_i$ can be obtained from Eq. (7.33) and $\Delta \ddot{x}_i$ from Eq. (7.32). With the incremental changes in displacement, velocity, and acceleration between times t_i and t_{i+1} having been established, the updated values of x_i, $\dot{x}_i$, and $\ddot{x}_i$ at times t_{i+1}, can be determined from Eqs. (7.31) as

$$x_{i+1} = x_i + \Delta x_i \tag{7.38a}$$

$$\dot{x}_{i+1} = \dot{x}_i + \Delta \dot{x}_i \tag{7.38b}$$

$$\ddot{x}_{i+1} = \ddot{x}_i + \Delta \ddot{x}_i \tag{7.38c}$$

Although this incremental form is not necessary for analysis of linear systems, it is required for the analysis of nonlinear systems. It is introduced in this section to facili-

TABLE 7.7. Step-by-Step Solution Using the Average Acceleration Method (Incremental Formulation)

A. Initial calculations:
   1. Calculate k, c, m, ω
   2. Calculate $\ddot{x}_0$

$$\ddot{x}_0 = \frac{1}{m}[F(0) - c\dot{x}_0 - kx_0]$$

   3. Select an appropriate time step Δt
   4. Calculate the effective stiffness

$$\hat{k} = k + \frac{4}{(\Delta t)^2}m + \frac{2c}{\Delta t}$$

B. For each time step:
   1. Calculate the effective incremental force

$$\Delta\hat{F}_i = \Delta F_i + \left[\frac{4m}{\Delta t} + 2c\right]\dot{x}_i + 2m\ddot{x}_i$$

   2. Solve for the incremental displacement

$$\Delta x_i = \frac{\Delta\hat{F}_i}{\hat{k}}$$

   3. Calculate the incremental velocity and acceleration

$$\Delta\dot{x}_i = \left(\frac{2}{\Delta t}\right)\Delta x_i - 2\dot{x}_i$$

$$\Delta\ddot{x}_i = \frac{4}{(\Delta t)^2}(\Delta x_i - \Delta t\dot{x}_i) - 2\ddot{x}_i$$

   4. Calculate the displacements, velocities, and accelerations at time t_{i+1}

$$x_{i+1} = x_i + \Delta x_i$$
$$\dot{x}_{i+1} = \dot{x}_i + \Delta\dot{x}_i$$
$$\ddot{x}_{i+1} = \ddot{x}_i + \Delta\ddot{x}_i$$

tate the extension of the algorithm to nonlinear systems discussed in Chapter 15. The algorithm for the incremental formulation of the average acceleration method is summarized in Table 7.7.

The *average acceleration method,* or *constant average acceleration method,* of time integration is equivalent to the trapezoidal rule. It is also a special form of the Newmark method that will be discussed in Chapter 13.

EXAMPLE 7.4 ▼

Solve Example 7.2 by the average acceleration method (incremental formulation). Use a time step $\Delta t = 0.01$ sec and assume a damping factor $\zeta = 0.06$. Compare the results with those obtained in Example 7.2 for $\zeta = 0.02$.

Solution

Time histories for displacement $x(t)$, velocity $\dot{x}(t)$, and acceleration $\ddot{x}(t)$ are presented in Figure 7.10a, b, and c, respectively. A listing of the BASIC computer program is presented in Table 7.8.

It is interesting to note the differences in the time histories presented in Figure 7.10 with those for Example 7.2 presented in Figure 7.6. For the case with light damping, $\zeta = 0.02$, the structure still exhibits appreciable transient vibration (free vibration) after 3 sec, as illustrated in Figure 7.6a. However, for the case with higher damping, $\zeta = 0.06$, the transient vibration is almost completely damped out after only 1.5 sec, as shown in Figure 7.10a.

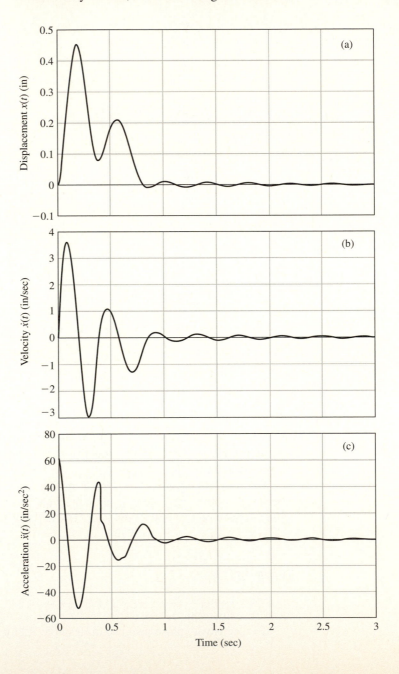

Figure 7.10

Response histories for Example 7.4: (a) displacement; (b) velocity; (c) acceleration.

TABLE 7.8.   Listing of BASIC Computer Program Used for Example 7.4

```
10 REM                          This program performs numerical integration through
20 REM                          the use of the average acceleration method to evaluate
30 REM                          the dynamic response of a shear frame.
40 REM
50 REM                                  DT = TIME STEP
60 REM                                  K = SPRING CONSTANT
70 REM                                  C = DAMPING CONSTANT
80 REM                                  M = MASS
90 REM                                  X(I) = DISPLACEMENT
100 REM                                 V(I) = VELOCITY
110 REM                                 A(I) = ACCELERATION
120 REM                                 DX(I) = INCREMENTAL CHANGE IN DISPLACEMENT
130 REM                                 DV(I) = INCREMENTAL CHANGE IN VELOCITY
140 REM                                 DA(I) = INCREMENTAL CHANGE IN ACCELERATION
150 REM                                 T = TIME
160 REM                                 KI = EFFECTIVE STIFFNESS
170 REM                                 PI = INITIAL LOAD
180 REM                                 P(I) = TIME VARYING LOAD
190 REM                                 XI = INITIAL DISPLACEMENT
200 REM                                 VI = INITIAL VELOCITY
210 REM
220 DIM X(505),V(505),A(505),DX(505),DV(505),DA(505),DP(505)
225 REM
230 OPEN "A:DISP.DAT" FOR OUTPUT AS #1
240 OPEN "A:VEL.DAT" FOR OUTPUT AS #2
250 OPEN "A:ACC.DAT" FOR OUTPUT AS #3
255 REM
260                             K=40.35
270                             C=.3035592
280                             M=.1585
290                             DT=.01
300                             PI=10
310                             X(1)=0
320                             V(1)=0
330 REM --------------------------------
340                                 A(1)=(1/M)*(PI-C*V(1)-K*X(1))
350                                 KI=K+(2*C/DT)+(4*M/DT^2)
360 REM
370 REM
380 PRINT #1,0,X(1):PRINT #2,0,V(1):PRINT #3,0,A(1)
390                                 FOR I = 1 TO 500
400                                     T=(I-1)*DT
410                                     T1=(I)*DT
420                                     IF T1<= .4 THEN P1=10
430                                     IF T<= .4 THEN P=10
440                                     IF T>.4 AND T<= .6 THEN P=6
450                                     IF T1>.4 AND T1<=.6 THEN P1=6
460                                     IF T>.6 AND T<=.9 THEN P=-20*T+18
470                                     IF T1>.6 AND T1<=.9 THEN P1=-20*T1+18
480                                     IF T>.9 THEN P=0
490                                     IF T1>.9 THEN P1=0
500                                     DP(I)=(P1-P)+(4*M/DT+2*C)*V(I)+2*M*A(I)
510                                     DX(I)=DP(I)/KI
520                                     DV(I)=(2/DT)*DX(I)-2*V(I)
530                                     DA(I)=(4/DT^2)*(DX(I)-V(I)*DT)-2*A(I)
540                                     X(I+1)=X(I)+DX(I)
550                                     V(I+1)=V(I)+DV(I)
560                                     A(I+1)=A(I)+DA(I)
570                             PRINT #1,T1,X(I+1):PRINT #2,T1,V(I+1):PRINT #3,T1,A(I+1)
580                     NEXT I
590 END
```

7.6 LINEAR ACCELERATION METHOD

In this temporal integration scheme, a linear variation of acceleration from time t_i to time t_{i+1} is assumed, as illustrated in Figure 7.11. Let τ denote the time within the interval t_i and t_{i+1} such that $0 \le \tau \le \Delta t$; then the acceleration at time τ is expressed as

$$\ddot{x}(\tau) = \ddot{x}_i + \frac{\tau}{\Delta t}(\ddot{x}_{i+1} - \ddot{x}_i) \tag{7.39}$$

Integrating Eq. (7.39) twice yields

$$\dot{x}(\tau) = \dot{x}_i + \tau\ddot{x}_i + \frac{\tau^2}{2\,\Delta t}(\ddot{x}_{i+1} - \ddot{x}_i) \tag{7.40}$$

and

$$x(\tau) = x_i + \tau\dot{x}_i + \frac{1}{2}\tau^2\ddot{x}_i + \frac{\tau^3}{6\,\Delta t}(\ddot{x}_{i+1} + \ddot{x}_i) \tag{7.41}$$

At time t_{i+1}, Eqs. (7.40) and (7.41) become, respectively,

$$\dot{x}_{i+1} = \dot{x}_i + \frac{\Delta t}{2}(\ddot{x}_{i+1} + \ddot{x}_i) \tag{7.42}$$

and

$$x_{i+1} = x_i + \Delta t\dot{x}_i + \frac{(\Delta t)^2}{6}(\ddot{x}_{i+1} + 2\ddot{x}_i) \tag{7.43}$$

from which $\ddot{x}_{i+1}$ and $\dot{x}_{i+1}$ can be solved in terms of x_{i+1}, such that

$$\ddot{x}_{i+1} = \frac{6}{(\Delta t)^2}(x_{i+1} - x_i) - \frac{6}{\Delta t}\dot{x}_i - 2\ddot{x}_i \tag{7.44}$$

and

$$\dot{x}_{i+1} = \frac{3}{\Delta t}(x_{i+1} - x_i) - 2\dot{x}_i - \frac{\Delta t}{2}\ddot{x}_i \tag{7.45}$$

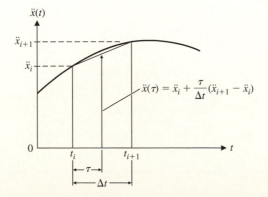

Figure 7.11

Numerical integration using the linear acceleration method.

Substitution of Eqs. (7.44) and (7.45) into the equation of motion, Eq. (7.28), at time t_{i+1} results in an equation from which x_{i+1} can be solved; that is,

$$x_{i+1} = \left\{ \frac{1}{k + [6m/(\Delta t)^2] + (3c/\Delta t)} \right\}$$

$$\left[m\left(\frac{6x_i}{(\Delta t)^2} + \frac{6\dot{x}_i}{\Delta t} + 2\ddot{x}_i \right) + c\left(\frac{3x_i}{\Delta t} + 2\dot{x}_i + \frac{\Delta t \ddot{x}_i}{2} \right) + F_{i+1} \right]$$

$$(7.46)$$

After determination of x_{i+1} from Eq. (7.46), Eqs. (7.44) and (7.45) may be used to calculate the acceleration and velocity, respectively, at time t_{i+1}. The algorithm for the linear acceleration method is summarized in Table 7.9. Note that the computational algorithm presented in Table 7.9 is not an incremental formulation. However, an incremental formulation of the linear acceleration method can be developed in a manner similar to that employed for the average acceleration method in Section 7.5. The linear acceleration method is a specific case of the Wilson-θ method, with $\theta = 1$. The Wilson-θ method is discussed in Chapter 13.

TABLE 7.9. Step-by-Step Solution Using the Linear Acceleration Method

A. Initial calculations:
1. Calculate k, c, m, ω
2. Calculate $\ddot{x}_0$

$$\ddot{x}_0 = \frac{1}{m}[F(0) - c\dot{x}_0 - kx_0]$$

3. Select an appropriate time step Δt
4. Calculate the effective stiffness

$$\hat{k} = k + \frac{6m}{(\Delta t)^2} + \frac{3c}{\Delta t}$$

B. Calculations for each time step:
1. Calculate the effective force at time t_{i+1}

$$\hat{F}_{i+1} = F_i + (F_{i+1} - F_i) + m\left[\frac{6x_i}{(\Delta t)^2} + \frac{6\dot{x}_i}{\Delta t} + 2\ddot{x}_i \right] + c\left(\frac{3x_i}{\Delta t} + 2\dot{x}_i + \frac{\Delta t \ddot{x}_i}{2} \right)$$

2. Solve for the displacements at time t_{i+1}

$$x_{i+1} = \frac{\hat{F}_{i+1}}{\hat{k}}$$

3. Calculate the accelerations and velocities at time t_{i+1}

$$\ddot{x}_{i+1} = \frac{6}{(\Delta t)^2}(x_{i+1} - x_i) - \frac{6\dot{x}_i}{\Delta t} - 2\ddot{x}_i$$

$$\dot{x}_{i+1} = \dot{x}_i + \frac{\Delta t}{2}(\ddot{x}_{i+1} + \ddot{x}_i)$$

EXAMPLE 7.5 ▼

The shear frame shown in Figure 7.12a is subjected to the exponential pulse force shown in Figure 7.12b. Write a computer program for the linear acceleration method (nonincremental formulation) to evaluate the dynamic response of the frame. Plot the time histories for displacement $x(t)$, velocity $\dot{x}(t)$, and acceleration $\ddot{x}(t)$ in the time interval $0 \le t \le 3$ sec. Assume $E = 29,000$ ksi, $W = 25$ kips, $\zeta = 0.07$, $F_0 = 10$ kips, and $t_d = 0.75$ sec. Use a time step Δt of 0.01 sec.

Solution

(a) Calculate the natural period of the frame and verify the time-step size selected.

$$m = \frac{25 \text{ kips}}{386.4 \text{ in/sec}^2} = 0.065 \text{ kip-sec}^2/\text{in}$$

$$k = \frac{3(29,000 \text{ ksi})(248 \text{ in}^4)}{(16 \text{ ft} \times 12 \text{ in})^3}$$

$$+ \frac{12(29,000 \text{ ksi})(248 \text{ in}^4)}{(22 \text{ ft} \times 12 \text{ in})^3} = 7.738 \text{ kips/in}$$

$$\omega = \sqrt{\frac{k}{m}} = \sqrt{\frac{7.738}{0.065}} = 10.9 \text{ rad/sec}$$

$$T = \frac{2\pi}{\omega} = 0.576 \text{ sec}$$

$$\Delta t \le \frac{T}{10} = \frac{0.576}{10} = 0.0576 \text{ sec} \qquad \text{OK}$$

(b) Time history plots of the displacement $x(t)$, velocity $\dot{x}(t)$, and acceleration $\ddot{x}(t)$ are presented in Figure 7.13a, b, and c, respectively. A listing of the BASIC computer program is presented in Table 7.10.

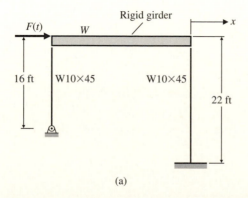

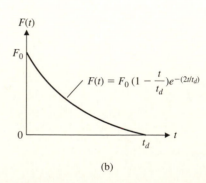

Figure 7.12

(a) Shear frame structure of Example 7.5; (b) excitation force.

Figure 7.13

Response histories for Example 7.5: (a) displacement; (b) velocity; (c) acceleration.

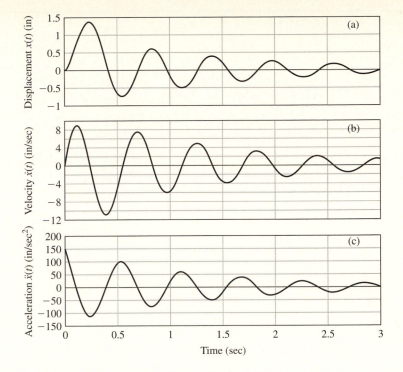

TABLE 7.10. **Listing of BASIC Computer Program Used for Examples 7.5 and 7.6**

```
1 REM DYNAMIC RESPONSE BY THE LINEAR ACCELERATION METHOD
2 CLS
3 DEF FNFORCE(X)=.3*386*(1−(X/TD−1)^2
5 OPEN "try" FOR OUTPUT AS #2: PRINT #2, A: CLOSE #2
6 PI = 3.141592654#
9 INPUT "TYPE OF FORCE 1− FORCE 2−ACCELERATION ",TYP:REM 1=force 2=ground
acceleration
10 INPUT "TYPE OF FUNCTION 1−POINTS 2−EQUATION ",H
11 IF H=2 THEN INPUT "TD ",TD
12 IF TYP=1 THEN A$="APPLIED FORCE" ELSE A$="GROUND ACCELERATION"
14 OPEN "DYN68" FOR OUTPUT AS #1
15 READ M,K,Z,Y0,V0,TMAX,DT,N
16 DATA 64.766,7738,0.07,0,0,3.0,0.01,4
17 IF TYP = 1 THEN KK = 1 ELSE KK = M
18 DIM T(N+1), F(N)
19 IF H=2 THEN 24
20 PRINT #1, "TIME", :IF TYP=1 THEN PRINT #1, "FORCE" ELSE PRINT #1, "ACCELE
RATION"
22 FOR I=1 TO N:READ T(I),F(I): F(I)=KK*F(I): PRINT #1, T(I),F(I): NEXT I
23 DATA 0,0,.25,8000,.75,−8000,1,0
24 PRINT #1, "INPUT DATA"
25 PRINT #1, "MASS=";M, "STIFFNESS=";K, "DAMPING FOEFF.";Z
26 PRINT #1, "Y0= ";Y0 "V0=";V0, "TMAX=";TMAX, "TIMESTEP=:;DT
27 PRINT #1, :PRINT #1, "TYPE OF ANALYSIS =";A$
28 PRINT #1
30 W=SQR(K/M):CR=2*M*W:C=Z*CR:T0=T(1):F0=F(1)
31 IF H=2 THEN T0=0:F0=FNFORCE(T0)*KK
35 T=2*PI/W:IF DT > T/20 THEN PRINT "decrease time step":STOP
```

(continued)

```
45 T1=DT:A1=3/T1:A2=6/T1:A4=6/(T1*T1)
50 Y11 = 1/M*(F0−C*V0−K*Y0)
60 K1=K+A4*M+A1*C
70 Y=Y0:Y1=V0:T(N+1)=TMAX
75 PRINT #1, "OUTPUT DATA"
76 PRINT #1, "*********************************"
80 PRINT #1, " TIME", " DISPL.", "VELOCITY", "ACCELERATION", "FORCE"
81 PRINT #1, T0;",";Y;",";Y1;",";Y11;",";F0
85 N4 = INT((T(N+1))−T(N))/DT)
86 IF H=2 THEN AS1=TMAX/DT−1:N=1
90 FOR J=1 TO N
91 IF H=2 THEN 100
94 IF J=N THEN DF=0:GOTO 96
95 DF=F(J+1)−F(J):DT1=T(J+1)−T(J):N3=DT1/DT
96 IF J=N THEN AS1=N4 ELSE AS1=N3-1
100 FOR G=0 TO AS1
101 IF H=2 THEN GOSUB 400: GOTO 115
102 I=T(J) + G*DT
109 IF J=N AND G <> 0 THEN F=0:F1=0:GOTO 115
110 F=F(J)+DF/DT1*(I−T(J))
111 F1=F(J)+DF/DT1*(I−T(J)+DT)
112 IF J=N AND G=0 THEN F1=0
115 PRINT I,F,F1
120 FI=F+(F1−F)+M*(A4*Y+A2*Y1+2*Y11)+C*(A1*Y+2*Y1+Y11*DT/2)
130 YX=Y:Y=FI/K1
150 Y1X=Y11:Y11=A4*(Y−YX)−A2*Y1−2*Y1X
160 VV=Y1:Y1=Y1+.5*DT*(Y11+Y1X)
170 'Y=YX+DT*VV+DT*DT/6*(Y11+2*Y1X)
220 PRINT #1, I;",";Y;",";Y1;",";Y11;",";F
221 ' PRINT , USING "####.#####   ";I;Y;Y1;Y11
230 IF ABS(Y) > ABS(YM) THEN YM = Y
231 IF ABS(Y1) > ABS (VM) THEN VM = Y1
232 IF ABS(Y11) > ABS (AM) THEN AM=Y11
240 NEXT G
250 NEXT J
251 PRINT #1,
252 PRINT #1, "*********************************"
260 PRINT #1, "         MAXIMUM RESPONSE   "
261 PRINT #1, "*********************************"
265 PRINT #1, :PRINT #1, "MAX. DISPL.", "MAX. VELOC.", "MAX. ACC."
270 PRINT #1, USING "####.#####      ";YM;VM;AM
290 CLOSE #1
300 STOP
400 REM SUBROUTINE FORCING FUNCTION
403 I=G*DT+DT
405 IF I>TD THEN F=0:F1=0:RETURN
410 F=FNFORCE(I)*KK:F1=FNFORCE(I+DT)*KK
430 IF (I+.000001)>TD THEN F1=0
440 RETURN
```

▲

7.7 RESPONSE TO BASE EXCITATION

Evaluating the dynamic response of structures due to arbitrary base or support motion can be greatly facilitated by numerical integration methods. The response of structures to base excitation is a most important consideration in earthquake engineering. A brief discussion of the topic is provided in this section; however, a more detailed analysis of the phenomenon is presented in Chapter 18.

Figure 7.14
Structure subject to base excitation.

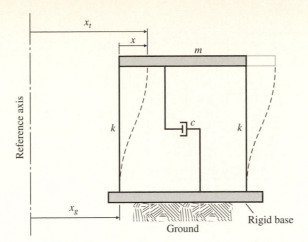

The structure shown in Figure 7.14 experiences an arbitrary ground displacement $x_g(t)$. It is assumed that the structure is attached to a rigid base that moves with the ground. In analyzing the structural response, there are several components of motion that must be considered: specifically, x, the relative displacement of the structure and x_t, the total or absolute displacement of the structure measured from the reference axis.

The equation of motion for this structure is

$$m\ddot{x}_t + c\dot{x} + kx = 0 \tag{7.47}$$

The zero on the right-hand side of Eq. (7.47) would seem to suggest that the structure is not subjected to any external load $F(t)$. This is not entirely true since the ground motion creates inertia forces in the structure. Thus, noting that the total displacement of the mass x_t is given by

$$x_t = x_g + x \tag{7.48}$$

and the total or absolute acceleration of the mass is expressed as

$$\ddot{x}_t = \ddot{x}_g + \ddot{x} \tag{7.49}$$

and substituting Eq. (7.49) into Eq. (7.47), the equation of motion can be expressed as

$$m(\ddot{x} + \ddot{x}_g) + c\dot{x} + kx = 0 \tag{7.50}$$

or

$$m\ddot{x} + c\dot{x} + kx = -m\ddot{x}_g \tag{7.51}$$

The term $-m\ddot{x}_g$ can be thought of as an effective load $F_{\text{eff}}(t)$ applied externally to the mass of the structure as shown in Figure 7.15. Therefore, the equation of motion can be written as

$$m\ddot{x} + c\dot{x} + kx = F_{\text{eff}}(t) \tag{7.52}$$

Note that in Eq. (7.52) x, $\dot{x}$, and $\ddot{x}$ and represent the relative displacement, velocity, and acceleration of the mass, respectively.

Figure 7.15

Effective load for ground acceleration applied to mass of structure.

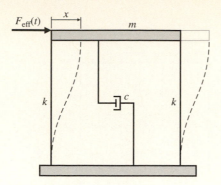

Equation (7.52) can now be integrated by any of the methods discussed in this chapter. The pertinent response quantities for evaluating the dynamic behavior of the structure are the relative displacement $x(t)$ and velocity $\dot{x}(t)$ and the absolute acceleration $\ddot{x}(t)$.

EXAMPLE 7.6   ▼

The base of the elevated water storage tank shown in Figure 7.16a is subject to the horizontal ground acceleration shown in Figure 7.16b. Evaluate the dynamic response of the structure by the linear acceleration method. Plot time histories for relative displacement, relative velocity, and absolute acceleration in the interval $0 \le t \le 3$ sec. Assume $W = 50$ kips, $k = 100$ kips/in, $\zeta = 0.12$, and $\ddot{x}_{g_0} = 0.25g$, where g is the acceleration due to gravity. Use a time step $\Delta t = 0.01$ sec.

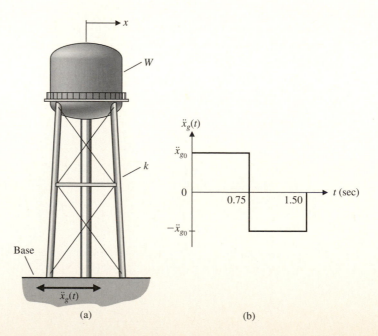

Figure 7.16

(a) Elevated water storage tank of Example 7.6; (b) horizontal ground acceleration.

Solution

(a) Determine the natural period of the structure and verify the time step Δt.

$$m = \frac{50 \text{ kips}}{386.4 \text{ in/sec}^2} = 0.129 \text{ kip-sec}^2/\text{in}$$

$$k = 100 \text{ kips/in}$$

$$\omega = \sqrt{\frac{k}{m}} = \sqrt{\frac{100}{0.129}} = 27.84 \text{ rad/sec}$$

$$T = \frac{2\pi}{\omega} = 0.226 \text{ sec}$$

$$\Delta t \leq \frac{T}{20} = 0.011 \text{ sec} \qquad \text{OK}$$

(b) The BASIC computer program used for Example 7.5 and listed in Table 7.10 is also used in this example. The mass of the structure is subjected to an effective load $F_{\text{eff}}(t) = m\ddot{x}_g(t)$. The effective load applied to the structure is shown in Figure 7.17a and the time histories for the relative displacement $x(t)$, relative velocity $\dot{x}(t)$, and absolute acceleration $\ddot{x}_t(t)$ are shown in Figure 7.17b, c, and d, respectively. Note that the absolute acceleration is the sum of the relative acceleration of the mass and the ground acceleration, as indicated by Eq. (7.49).

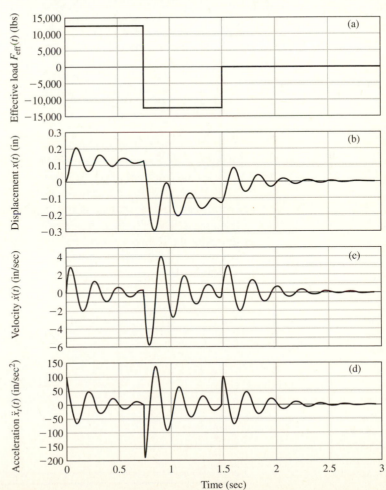

Figure 7.17

Response histories for Example 7.6: (a) effective load for ground acceleration; (b) relative displacement; (c) relative velocity; (d) absolute acceleration.

7.8 RESPONSE SPECTRA BY NUMERICAL INTEGRATION

The concept of a *response spectrum,* or *shock* spectrum, was previously discussed in Chapter 6. The response spectrum is a convenient means of encapsulating the maximum response of SDOF systems to a specified excitation force over a wide range of natural frequencies or periods. Is was clearly demonstrated in Chapter 6, however, that the construction of a response spectrum by analytical evaluation of the Duhamel integral can be quite tedious.

To illustrate a procedure for developing response spectra by numerical integration, consider the family of SDOF oscillators shown in Figure 7.18. Each oscillator has a different natural frequency, or natural period, such that

$$T_1 < T_2 < T_3 < \dots < T_n \tag{7.53}$$

The specified load function $F(t)$, for which the response spectrum is sought, is applied to each oscillator. The response histories of the individual oscillators to the specified load is calculated using one of the numerical integration schemes discussed in this chapter. Thus, for each oscillator we have the equation

$$\ddot{x} + 2\zeta\omega\dot{x} + \omega^2 x = \frac{F(t)}{m} \tag{7.54}$$

for which the maximum response x_{max} is determined from the displacement response history of each oscillator. Then the dynamic magnification factor (DMF) is calculated for each oscillator as

$$\text{DMF} = \frac{x_{max}}{F_0/k} \tag{7.55}$$

where F_0 is the maximum amplitude of the applied force $F(t)$. Finally, the DMF is plotted against some function of natural frequency or natural period, such as t_d/T, to generate the desired response spectrum curve.

A family of response spectrum curves, or *response spectra,* can be produced for a specified load case by evaluating the response maxima of Eq. (7.54) for several values of damping ζ. The construction of response spectrum by numerical integration is demonstrated in the following example. It is emphasized that the underlying concept of a response spectrum is that the maximum response of a linear SDOF system to any specified input depends only on the natural frequency and damping factor of the system.

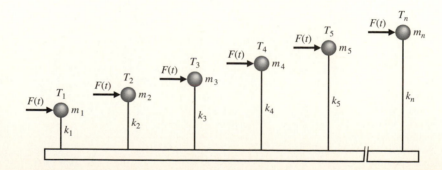

Figure 7.18

Family of SDOF oscillators used for construction of response spectrum.

EXAMPLE 7.7   ▼

Construct a response spectrum for each of the load cases shown in Figure 7.19 by numerical integration. Plot DMF versus t_d/T in the interval $0 \leq t_d/T \leq 10$. Assume $\zeta = 0$.

Solution

The response spectrum curves for the three load cases, plotted on a semi-log graph, are presented in Figure 7.20. These response spectra may be employed to evaluate the dynamic response of SDOF structures subject to the representation load functions in a manner similar to that demonstrated in Example 6.7. The FORTRAN computer program, LRESP, used to generate the response spectra is available on the author's Web site (refer to the Preface).

Figure 7.19

Load cases for Example 7.7: (a) rectangular pulse force; (b) triangular pulse force; (c) symmetric triangular pulse force.

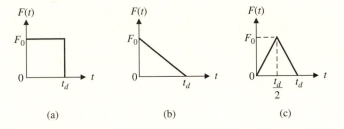

(a)   (b)   (c)

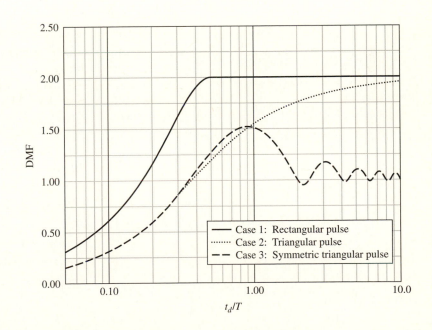

Figure 7.20

Response spectra.

REFERENCES

1 Bathe, Klaus-Jürgen, *Finite Element Procedures in Engineering Analysis,* Prentice Hall, Englewood Cliffs, NJ, 1982.

2 Levy, S. and Wilkinson, J.P.D., *The Component Element Method in Dynamics,* McGraw-Hill, New York, 1976.

3 Ferzinger, Joel H., *Numerical Methods for Engineering Applications,* Wiley, New York, 1981.

4 Potter, C.P. and Goldberg, J., *Mathematical Methods,* 2nd ed., Prentice Hall, New York, 1987.

5 James, M.L., Smith, G.M., and Wolford, J.C., *Applied Numerical Methods for Digital Computation,* 3rd ed., Harper and Row, New York, 1985.

NOTATION

A, A'	
B, B'	constants in recurrence formulas for piecewise-linear interpolation of the excitation
C, C'	
D, D'	
c	viscous damping coefficient
E	Young's modulus
F_i	amplitude of force $F(t)$ at time t_i
F_{i+1}	amplitude of force $F(t)$ at time t_{i+1}
F_0	amplitude of externally applied force
$\hat{F}_{i+1}$	effective force used in the linear acceleration method
$F(t)$	time-varying externally applied force
$F_{\text{eff}}(t)$	effective load for ground acceleration applied to the mass
$F(\tau)$	interpolated force within time interval Δt
g	acceleration due to gravity
k	stiffness
$\hat{k}$	effective stiffness used in the average acceleration and linear acceleration methods
m	mass
$\hat{m}$	effective mass used in the central difference method
t	time
t_d	time of duration of applied force
t_f	final time at which a solution is sought
T	natural period (sec)

x	displacement
x_g	ground displacement
x_o	initial displacement
x_t	total or absolute displacement of the mass
$\dot{x}$	velocity
$\dot{x}_0$	initial velocity
$\ddot{x}$	acceleration
$\ddot{x}_g$	ground acceleration
$\ddot{x}_t$	total or absolute acceleration of the mass
DMF	dynamic magnification factor
SDOF	single degree of freedom
ΔF_i	incremental change in force $F(t)$ within the time interval Δt
$\Delta \hat{F}_i$	effective incremental force used in the average acceleration method
Δt	incremental time interval; also represents time step for numerical integration
Δt_{cr}	critical time step required for stability of the central difference method
Δx	incremental displacement in time interval Δt
$\Delta \dot{x}$	incremental velocity in time interval Δt
$\Delta \ddot{x}$	incremental acceleration in time interval Δt
ω	undamped natural circular frequency (rad/sec)
τ	time variable within incremental time interval Δt
ζ	damping factor

PROBLEMS

7.1 The single-story shear frame shown in Figure P7.1 is subjected to the arbitrary time-varying forcing function specified in Figure P7.2. Calculate the dynamic response of the structure by piecewise-linear interpolation of the excitation. Plot displacement versus time and velocity versus time in the interval $0 \le t \le 3.0$ sec. Use a time step $\Delta t = 0.05$ sec. Assume $m = 0.1$ kip-sec^2/in, $k = 5$ kips/in, and $c = 0.2$ kip-sec/in.

7.8 The shear frame shown in Figure P7.1 is subject to the forcing function shown in Figure P7.3. Calculate the dynamic response using the average acceleration method. Plot time histories for the displacement, velocity, and acceleration in the time interval $0 \le t \le 3.0$ sec. Assume $m = 0.104$ kip-sec^2/in, $k = 10$ kips/in, and $\zeta = 0.1$. Use a time step $\Delta t \le T/10$.

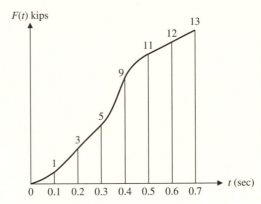

Figure P7.3

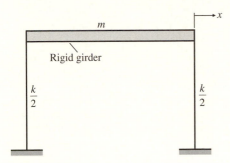

Figure P7.1

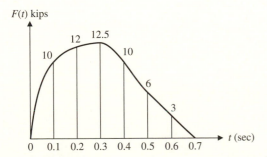

Figure P7.2

7.9 Repeat Problem 7.8 using the linear acceleration method.

7.10 Repeat Problem 7.8 using the central difference method.

7.11 The shear frame shown in Figure 7.12a is subject to the forcing function shown in Figure P7.4. Calculate the dynamic response using the linear acceleration method. Plot time histories for the displacement, velocity, and acceleration in the time interval $0 \le t \le 3.0$ sec. Assume $W = 25$ kips, $E = 29,000$ ksi, and $\zeta = 0.07$. Use a time step $\Delta t \le T/10$.

7.2 Solve Example 7.2 using piecewise-linear interpolation of the excitation.

7.3 Solve Example 7.1 using the central difference method.

7.4 Repeat Problem 7.1 using the central difference method.

7.5 Solve Example 7.2 using the fourth-order Runge-Kutta method.

7.6 Repeat Problem 7.1 using the fourth-order Runge-Kutta method.

7.7 Solve Example 7.3 using the average acceleration method.

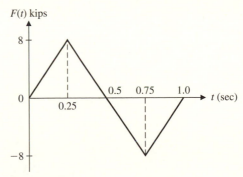

Figure P7.4

7.12 Repeat Problem 7.11 using the average acceleration method.

7.13 The elevated water storage tank shown in Figure 7.16a is subjected to the horizontal ground acceleration $\ddot{x}_g(t)$ shown in Figure P7.5. Calculate the dynamic response of the structure using the linear acceleration method. Plot time histories for the relative displacement, relative velocity, and absolute acceleration in the interval $0 \le t \le 3$ sec. Assume $W = 85$ kips, $k = 15$ kips/in, and $\zeta = 0.1$. Use a time step $\Delta t \le T/10$.

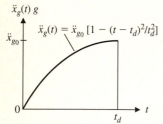

Figure P7.5

7.14 Repeat Problem 7.13 using the Runge-Kutta method.

7.15 Repeat Problem 7.13 for the horizontal ground acceleration shown in Figure P7.6. Assume $W = 50$ kips, $k = 100$ kips/in, $\zeta = 0.12$, $\ddot{x}_{g_0} = 0.3g$, and $t_d = 1.0$ sec. Use a time step $\Delta t \le T/10$.

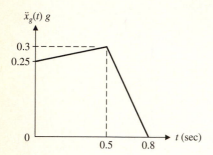

Figure P7.6

7.16 Construct a response spectrum for the load pulse shown in Figure P7.7. Plot DMF versus t_d/T in the interval $0 \le t_d/T \le 5$. Assume $\zeta = 0.05$. Use the numerical method of your choice.

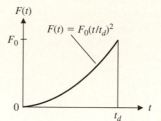

Figure P7.7

7.17 Construct a response spectra for the load pulse shown in Figure P7.8 for $\zeta = 0$ and $\zeta = 0.05$. Plot DMF versus t_d/T in the interval $0 \le t_d/T \le 5$. Use the numerical method of your choice.

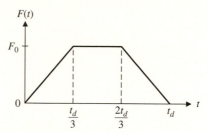

Figure P7.8

7.18 Construct a response spectrum for the ground acceleration shown in Figure P7.9 for $\zeta = 0.07$. Plot DMF versus T in the interval $0 \le T \le 10$ sec. Use the numerical method of your choice.

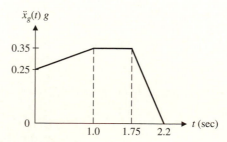

Figure P7.9

8 ▲ Frequency Domain Analysis

The response of SDOF systems to general dynamic excitation was introduced in Chapter 6. This is a very important class of problems since it forms the basis for studying more complex structures. Two general approaches are available for examining the dynamic response of these types of systems: (1) time domain solution and (2) frequency domain solution. To this point, all solutions have been in the time domain; that is, the equation of motion is solved by integrating in time. Except for loading cases represented with simple analytic expressions, these integrals are best evaluated numerically. This topic was addressed in Chapter 7. The second general approach to evaluating dynamic response is to seek solutions in the frequency domain. In this technique, the amplitude coefficients in the Fourier series solution corresponding to each frequency are determined. Again, except for simple analytic loading cases, the Fourier coefficients are numerically evaluated. This chapter presents a discrete form of the Fourier transformation. Next, an algorithm for the rapid numerical computation of Fourier coefficients is developed, the fast Fourier transform, or FFT. Then several potential sources of error and, finally, the Fourier integral that may be employed for nonperiodic forcing are discussed.

8.1 ALTERNATIVE FORMS OF THE FOURIER SERIES

The Fourier series for an arbitrary periodic function of time $p(t)$ was given in Chapter 6 as

$$p(t) = \frac{a_0}{2} + \sum_{n=1}^{\infty} (a_n \cos n\Omega t + b_n \sin n\Omega t) \tag{8.1}$$

in which a_0, a_n, and b_n are real constants, t is time, and Ω is frequency, $= 2\pi/T_0$, where T_0 is the period of the forcing function. In this chapter, $p(t)$ is used to denote force to avoid confusion with f, the frequency. Equation (8.1) is generalized to cases where the period of the forcing is not known or the forcing does not have a period T_0, which simply repeats. For these cases, let $\Omega = 2\pi/T_D$, where T_D is the total time duration of the forcing record or the total duration of time for which the structural response is to be determined. Note that the right-hand side of Eq. (8.1) is periodic in T_D; that is, $f(t) = f(t + T_D)$. For times greater than T_D, the series repeats. If the forcing is periodic with a period T_0, then T_D should be selected to be an integer multiple of T_0.

If the terms $(a_n^2 + b_n^2)^{1/2}$ are factored out, Eq. (8.1) may be written

$$p(t) = \frac{a_0}{2} + \sum_{n=1}^{\infty} (a_n^2 + b_n^2)^{1/2}$$
$$\left[\frac{a_n}{(a_n^2 + b_n^2)^{1/2}} \cos n\Omega t + \frac{b_n}{(a_n^2 + b_n^2)^{1/2}} \sin n\Omega t \right]$$

(8.2)

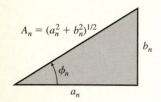

$A_n = (a_n^2 + b_n^2)^{1/2}$

b_n

ϕ_n

a_n

Figure 8.1
Definition of phase angle.

Reference to Figure 8.1 shows that the phase angle ϕ_n can be introduced. This results in the expression

$$p(t) = \frac{a_0}{2} + \sum_{n=1}^{\infty} (a_n^2 + b_n^2)^{1/2}$$
$$(\cos \phi_n \cos n\Omega t + \sin \phi_n \sin n\Omega t)$$

(8.3)

that may be written as

$$p(t) = \frac{a_0}{2} + \sum_{n=1}^{\infty} (a_n^2 + b_n^2)^{1/2} \cos (n\Omega t - \phi_n)$$

(8.4)

Next, the constants A_0 and A_n are defined as

$$A_0 = \frac{a_0}{2} \qquad A_n = (a_n^2 + b_n^2)^{1/2}$$

(8.5)

such that the Fourier series can be written as

$$p(t) = \sum_{n=0}^{\infty} A_n \cos (n\Omega t - \phi_n)$$

(8.6)

in which $\phi_0 = 0$. In Eq. (8.6) the Fourier series is written in terms of a positive real amplitude and a phase shift. This form of the Fourier series replaces a_n and b_n with A_n and ϕ_n, therefore, there are still the same number of coefficients. The plots of A_n and ϕ_n are called the *amplitude spectrum* and *phase spectrum*, respectively. They are discrete and occur at the Fourier frequencies $n\Omega = 2n\pi/T_D$.

EXAMPLE 8.1 ▼

A periodic load function consists of three frequency components given by

$$p(t) = 5 \text{ lb } \cos(\Omega t - 0.5) + 4 \text{ lb } \cos(2\Omega t - 0.2)$$
$$+ 3 \text{ lb } \cos(3\Omega t - 0.1)$$

where $\Omega = 2\pi/T_D$ and $T_D = 5$ sec. Plot the periodic function and the amplitude and phase spectra.

Solution

The equations for the coefficients a_0, a_n, and b_n were given in Chapter 6 [Eqs. (6.3) to (6.5)] as

$$a_0 = \frac{2}{T_D} \int_0^{T_D} p(t)\, dt \tag{1}$$

$$a_n = \frac{2}{T_D} \int_0^{T_D} p(t)\, \cos\left(\frac{2n\pi}{T_D}t\right) dt \tag{2}$$

$$b_n = \frac{2}{T_D} \int_0^{T_D} p(t)\, \sin\left(\frac{2n\pi}{T_D}t\right) dt \tag{3}$$

except that we use T_D as the fundamental period. Evaluating the integrals in Eqs. (1) through (3) yields

n	$n\Omega$	a_n	b_n	$A_n = (a_n^2 + b_n^2)^{1/2}$	$\phi_n = \tan^{-1}(b_n/a_n)$
0	-	0	-	0	-
1	Ω	4.3879	2.3971	5.0	0.5
2	2Ω	3.9203	0.7947	4.0	0.2
3	3Ω	2.9850	0.2995	3.0	0.1
4	4Ω	0	0	0	-
5	5Ω	0	0	0	-
.	.	.	.	.	
.	.	.	.	.	
.	.	.	.	.	

▲

The forcing function $p(t)$ examined in Example 8.1 is a rather simple function. Although, from the plot presented in Figure 8.2, it is not readily apparent that this is in fact a simple function. However, the amplitude and phase spectra plots shown in Figure 8.3 clearly reveal the three cosines and their phases. As functions become more complex with many frequency components, plots of amplitude spectra become even more important in understanding the underlying structure of the functions.

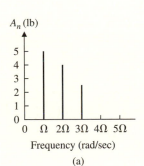

A_n (lb)

(a)

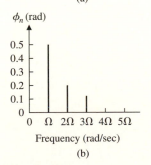

ϕ_n (rad)

(b)

Figure 8.3

Amplitude and phase spectra for Example 8.1.

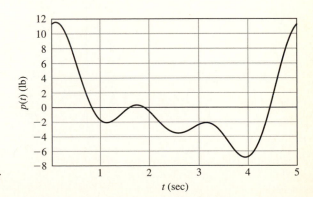

Figure 8.2

Time series for Example 8.1.

EXAMPLE 8.2 ▼

A simple periodic load is characterized by the following deterministic amplitude coefficients:

$$\frac{P_m}{P_p} = \left(\frac{\omega_m}{\omega_p}\right)^4 e^{[1-(\omega_m/\omega_p)^4]} \qquad m = 0, 1, 2, \ldots$$

where $\omega_p =$ peak frequency
$P_p =$ amplitude of the forcing at the peak frequency
$\omega_m =$ Fourier frequencies

Determine the time series for the force for $P_p = 5$ kips, $\omega_p = 2\pi$, and $T_D = 100$ sec. Assume a random phase spectrum.

Solution

The amplitude spectrum is shown in Figure 8.4. The time series is determined using Eq. (8.6),

$$p(t) = \sum_{m=0}^{\infty} P_m \cos(\omega_m t - \phi_m)$$

where $\omega_m = 2m\pi/T_D$ and ϕ_m are random phases between 0 and 2π. It is a simple matter to select random numbers and generate each frequency component in the time series. However, the question remains: At what point to stop the summation? In this example the amplitude coefficients decrease monotonically at high frequencies. Note that the largest amplitude coefficient P_p is at ω_p, which corresponds to $m = 100$. The force coefficient that is 0.001 of this peak value occurs approximately at $m = 179$. Therefore, including 200 terms in the series should be adequate. The first one second of time of the resulting time series is shown in Figure 8.5.

Figure 8.4

Amplitude spectrum for Example 8.2.

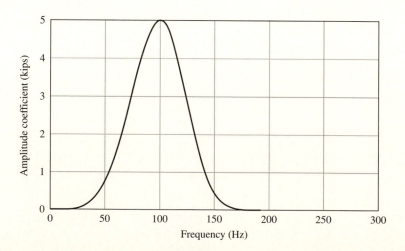

Figure 8.5
Time series for Example 8.2.

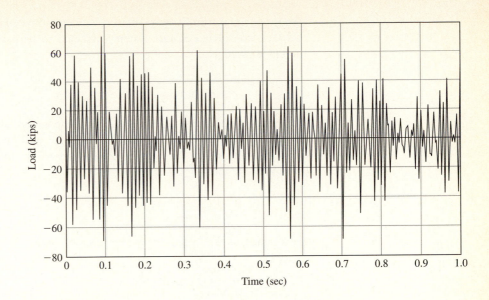

With the use of the Euler formulas, the real Fourier series expressed by Eq. (8.1) may be written using complex variables. Thus Eq. (8.1) becomes

$$p(t) = \frac{a_0}{2} + \sum_{n=1}^{\infty}\left(a_n\frac{e^{in\Omega t} + e^{-in\Omega t}}{2} + b_n\frac{e^{in\Omega t} - e^{-in\Omega t}}{2}\right) \tag{8.7}$$

where $i \equiv \sqrt{-1}$. Collecting similar terms in Eq. (8.7) gives

$$p(t) = \frac{a_0}{2} + \sum_{n=1}^{\infty}\left(\frac{a_n - ib_n}{2}e^{in\Omega t} + \frac{a_n + ib_n}{2}e^{-in\Omega t}\right) \tag{8.8}$$

When new constants P_0, P_n, and P_{-n} are defined as

$$P_o = \frac{a_0}{2} \qquad P_n = \frac{a_n - ib_n}{2} \qquad P_{-n} = \frac{a_n - ib_n}{2} \tag{8.9}$$

the series defined by Eq. (8.8) can be written as

$$p(t) = \sum_{n=-\infty}^{\infty} P_n e^{in\Omega t} \tag{8.10}$$

Equation (8.10) is the exponential form of the Fourier series. Note that the summation on n is now from $-\infty$ to $+\infty$. For cases with $n < 0$, the Fourier frequencies are negative. The coefficients are evaluated as

$$P_n = \frac{1}{T_D}\int_{t}^{(t + T_D)} p(t)e^{-in\Omega t}\, dt \tag{8.11}$$

The exponential form is often more convenient to use than the real form. To illustrate this fact, consider a SDOF system subjected to harmonic forcing. The equation of motion is given as

$$m\ddot{x} + c\dot{x} + kx = f(t) = \sum_{n=-\infty}^{\infty} P_n e^{in\Omega t} \tag{8.12}$$

The steady-state solution to Eq. (8.12) will respond at the same frequencies comprising the forcing function. Since the problem is linear, seek the solution at each distinct frequency and then sum the individual responses to obtain the total response. Assume a solution of the form

$$x(t) = \sum_{n=-\infty}^{\infty} X_n e^{in\Omega t} \tag{8.13}$$

Substitution of Eq. (8.13) into the equation of motion [Eq. (8.12)] results in

$$m \sum_{n=-\infty}^{\infty} X_n (in\Omega)^2 e^{in\Omega t} + c \sum_{n=-\infty}^{\infty} X_n (in\Omega) e^{in\Omega t} + k \sum_{n=-\infty}^{\infty} X_n e^{in\Omega t}$$
$$= \sum_{n=-\infty}^{\infty} P_n e^{in\Omega t} \tag{8.14}$$

Collecting terms at each frequency yields

$$(-n^2\Omega^2 m + icn\Omega + k)X_n = P_n \tag{8.15}$$

The transfer function, relating the response X_n to the forcing P_n, is expressed as

$$H_n = \left(\frac{1}{-n^2\Omega^2 m + icn\Omega + k} \right) = \frac{(-n^2\Omega^2 m + k) - i(cn\Omega)}{[(-n^2\Omega^2 m + k)^2 + (cn\Omega)^2]} \tag{8.16}$$

and the amplitude of the response at frequency $n\Omega$ is given by

$$X_n = H_n P_n \tag{8.17}$$

Written in this way, the response at each frequency is simply a function of the forcing transfer function. For a given structure, H_n depends only on the frequency $n\Omega$. The total solution is the sum or linear superposition over all values of $n\Omega$. This is termed a *frequency domain solution.*

If there is no damping in the system, then $c = 0$, and the transfer function represented by Eq. (8.16) is real. For this case, there is no phase shift between the excitation and the response. If c is nonzero, the transfer function is complex and a phase shift is introduced.

If the frequency ratio is defined as

$$r = \frac{\Omega}{\omega} \tag{8.18}$$

where ω is the undamped natural frequency of the system and using the damping factor defined in Chapter 4

$$\zeta = \frac{c}{2m\omega} \tag{8.19}$$

then the transfer function defined by Eq. (8.16) may be written as

$$H_n = \frac{1}{k[(1 - n^2 r^2) + i(2nr\zeta)]} = \frac{(1 - n^2 r^2) - i(2nr\zeta)}{k[(1 - n^2 r^2)^2 + i(2nr\zeta)^2]} \tag{8.20}$$

The phase is then determined as

$$\phi_n = \tan^{-1}\frac{\text{Im}(H_n)}{\text{Re}(H_n)} = \tan^{-1}\left(\frac{2nr\zeta}{1 - n^2 r^2}\right) \tag{8.21}$$

in which Re( ) denotes the real part and Im( ) denotes the imaginary part of the transfer function. Thus, the system response may be written very succinctly as

$$x(t) = \sum_{n=-\infty}^{\infty} |H_n| P_n\, e^{i(n\Omega t + \phi_n)} \tag{8.22}$$

where the amplitude or *modulus* of the transfer function is represented by

$$|H_n| = [H_n H_n^*]^{1/2} = \frac{1}{k[(1 - n^2 r^2)^2 + (2nr\zeta)^2]^{1/2}} \tag{8.23}$$

where H_n^* is the complex conjugate of P_n and ϕ_n is the phase.

The Fourier coefficients for the forcing P_n can also be complex. Expressing P_n in Eq. (8.22) as a modulus and a phase gives

$$x(t) = \sum_{n=-\infty}^{\infty} |H_n||P_n| e^{i(n\Omega t + \phi_n + \alpha_n)} \tag{8.24}$$

where α_n is the phase corresponding to each P_n. Note that α_n is taken to be the principal value. Equation (8.24) represents a very convenient way to express the response, since the positive, real coefficients $|H_n||P_n|$ plotted against the frequencies $n\Omega$ give the response amplitude spectrum.

Use of the Euler formulas allows Eq. (8.11) to be written as

$$P_n = \frac{1}{T_D} \int_t^{t+T_D} p(t)[\cos(n\Omega t) - i \sin(n\Omega t)]dt \tag{8.25}$$

The time series $p(t)$ is real; therefore, the modulus of P_n is

$$|P_n| = \frac{1}{T_D} \int_t^{t+T_D} \{[p(t)\cos(n\Omega t)]^2 + [p(t)\sin(n\Omega t)]^2\}dt \tag{8.26}$$

Next we consider the frequency $(-n\Omega)$ having the corresponding coefficient P_{-n}. Substituting $(-n\Omega)$ into Eq. (8.26) and comparing with the positive frequencies $(n\Omega)$, it is observed that

$$|P_{-n}| = |P_n| \tag{8.27}$$

Since the spectral coefficients are even, this implies that in a spectrum plot we can consider either all frequencies between $-\infty$ and $+\infty$ or just the frequencies from 0 to ∞ and double the heights. The former case is called a double-sided spectrum and the latter is a single-sided spectrum. Both are completely reasonable and valid spectrum representations, and both are used in practice. Therefore, it is imperative to be aware of which type of spectrum is being used. In a sense, the single-sided alternative is more intuitive because it is conventional to think of frequencies as being positive. An example of single- and double-sided spectra is presented in Example 8.4 for a discrete Fourier transform.

8.2 DISCRETE FOURIER TRANSFORM

Except for very simple cases, Fourier transforms are calculated numerically. This requires that the continuous values of t [for example, in Eq. (8.11)] be expressed discretely and that the discretization be over a finite time interval T_D. This implies that

$$t \rightarrow t_n = n\,\Delta t \tag{8.28}$$

in which Δt is the uniform time increment, such that

$$\Delta t = \frac{T_D}{N} \tag{8.29}$$

where N is the number of points in the time series approximation to $p(t)$.

To this point, the discussion of Fourier series has used the radian frequency $n\Omega = n2\pi/T_D$. It is now somewhat more convenient, and also common practice, to express frequency in hertz. Thus, the Fourier frequencies are written

$$n\Omega = 2n\pi f_D \tag{8.30}$$

where $f_D = 1/T_D$. The discrete Fourier transform is given by

$$p_n = \frac{1}{T_D}\sum_{m=0}^{N-1} P_m\, e^{i2nm\pi/N} \qquad n = 0, 1, \ldots, N-1 \tag{8.31}$$

and the frequency coefficients are expressed as

$$P_m = \frac{T_D}{N}\sum_{n=0}^{N-1} p_n\, e^{-i2nm\pi/N} \qquad m = 0, 1, \ldots, N-1 \tag{8.32}$$

The total number of discrete time values and discrete frequency values are the same.

The N term appearing before the summation in Eq. (8.32) is not unique. In the discrete Fourier transform, it can be used as $1/N$ in Eq. (8.32), N in Eq. (8.31), or $\sqrt{N}$ in both. If a numerical algorithm does not specify which form is being used, it may be determined by examining a test function with known results.

EXAMPLE 8.3   ▼

Determine the discrete Fourier transform for the forcing function defined by

$$p(n\,\Delta t) = 3 + 7\sin\!\left(\frac{n\pi}{2}\right) \qquad n = 0, 1, \ldots, 7$$

Solution

Before developing a solution using the discrete transforms given above, it is useful to first examine the time series. The discrete trigonometric Fourier series is expressed as

$$p_n = a_0 + \sum_{m=1}^{N-1} \left(a_m \cos\frac{2\pi nm}{N} + b_m \sin\frac{2\pi nm}{N} \right) \tag{1}$$

In this example, $N = 8$ and all the coefficients are zero except $a_0 = 3$ and $b_2 = 7$. For this simple problem the solution could be determined by inspection. With use of the Euler formulas, the time series is written as

$$P_n = 3 + 7 \sin\frac{n\pi}{2} = 3 + 7 \frac{e^{in\pi/2} - e^{-in\pi/2}}{2i}$$
$$= 3 - i\frac{7}{2} e^{in\pi/2} + i\frac{7}{2} e^{-in\pi/2} \tag{2}$$

The transform of the series given by Eq. (2) is

$$P_m = \frac{T_D}{N} \sum_{n=0}^{N-1} p_n e^{-i2nm\pi/N} \qquad m = 0, 1, \ldots, N-1$$
$$= \frac{T_D}{N}\left[3\sum_{n=0}^{N-1} e^{-inm\pi/4} + \left(-i\frac{7}{2}\right)\sum_{n=0}^{N-1} e^{in\pi/2}e^{-inm\pi/4} \right.$$
$$\left. + \left(i\frac{7}{2}\right)\sum_{n=0}^{N-1} e^{-in\pi/2}e^{-inm\pi/4} \right] \tag{3}$$

To evaluate the terms in the series defined by Eq. (3), consider the more general expression

$$S_N(k, m) = \sum_{n=0}^{N-1} e^{i2\pi kn/N} e^{-i2\pi nmN} = \sum_{n=0}^{N-1} e^{i2\pi(k-m)n/N}$$
$$= \sum_{n=0}^{N-1} [e^{i2\pi(k-m)/N}]^n \tag{4}$$

in which $S_N(k, m)$ is simply a shorthand notational variable. For $k = m$ in Eq. (4), it is seen that each term is 1; thus the sum from 0 to $N - 1$ is simply N. For $k \neq m$, note that Eq. (4) becomes a geometric series [1]; therefore, Eq. (4) may be written as

$$S_N(k, m) = \frac{1 - e^{i2\pi(k-m)}}{1 - e^{i2\pi(k-m)N}} \tag{5}$$

For any choice of k and m for which $k \neq m$ in Eq. (5), $e^{i2\pi(k-m)} = 1$ and $S_N(k, m) = 0$. Therefore, $S_N(k, m)$ is written as

$$S_N(k, m) = N\delta_{km} \tag{6}$$

where δ_{km} is the Kronecker delta, defined as

$$\delta_{km} = \begin{cases} 1; & k = m \\ 0; & k \neq m \end{cases} \tag{7}$$

Referring to Eq. (3) gives the discrete Fourier transform is given by

$$P_m = \frac{T_D}{N}\left[3(N\delta_{m,0}) - i\frac{7}{2}(N\delta_{m,2}) + i\frac{7}{2}(N\delta_{m,-2})\right] \tag{8}$$

$$m = 0, 1, 2, \ldots, 7$$

or

$$\frac{P_m}{T_D} = 3\delta_{m,0} - i\frac{7}{2}\delta_{m,2} + i\frac{7}{2}\delta_{m,-2} \qquad m = 0, 1, 2, \ldots, 7 \tag{9}$$

The coefficients in Eq. (9) are

$$\begin{array}{ll} \dfrac{P_0}{T_D} = 3 & \dfrac{P_4}{T_D} = 0 \\[2mm] \dfrac{P_1}{T_D} = 0 & \dfrac{P_5}{T_D} = 0 \\[2mm] \dfrac{P_2}{T_D} = -i\dfrac{7}{2} & \dfrac{P_6}{T_D} = i\dfrac{7}{2} \\[2mm] \dfrac{P_3}{T_D} = 0 & \dfrac{P_7}{T_d} = 0 \end{array} \tag{10}$$

These are the same coefficients that were determined by inspection in Eq. (8.2) for this simple example. Recalling that the Fourier transform is periodic in T_D, or for the discrete transform in $N\,\Delta t$, gives $p_n = p_{n+N}$ and $P_m = P_{m+N}$ for real time series. Therefore the term P_{-2} can be written as $P_{-2+N} = P_{-2+8} = P_6$. To test this result, we reconstruct the time series. Thus

$$\begin{aligned} P_n &= \frac{1}{T_D}\sum_{m=0}^{N-1} P_m e^{i2\pi nm/N} \qquad n = 0, 1, \ldots, N-1 \\[2mm] &= \frac{1}{T_D}(3T_D)e^{i2\pi n(0)/8} + \frac{1}{T_D}\left(-i\frac{7}{2}T_D\right)e^{i2\pi n(2)/8} \\[2mm] &\quad + \frac{1}{T_D}\left(i\frac{7}{2}T_D\right)e^{i2\pi n(6)/8} \\[2mm] &= 3 - \frac{7}{2}ie^{in\pi/2} + \frac{7}{2}ie^{-in\pi/2} \\[2mm] &= 3 + 7\frac{e^{in\pi/2} - e^{-in\pi/2}}{2i} \\[2mm] &= 3 + 7\sin\frac{n\pi}{2} \end{aligned} \tag{11}$$

Constructing the transform and then the inverse illustrates why the N is arbitrary. The cycle of transformations defined by Eqs. (9) through (11) multiplies and divides by the coefficient. As long as the usage is consistent between the transform and inverse, the choices of $1/N$, N, and $\sqrt{N}$ are all reasonable. ▲

The forgoing example illustrated the development of the method for calculating the discrete Fourier transform. Although this example exhibited only a mean term and one frequency component, it still required several computation cycles. In general, the number of frequencies equals the number of time series data, such that there are N frequency terms at each of the N time values. Therefore, on the order of N^2 operations are required. For large time series, this approach becomes untenable.

8.3 FAST FOURIER TRANSFORM

Modern frequency domain solutions algorithms are based on the fast Fourier transform (FFT). The FFT is predicated on the observation that if the number of points is a power of 2, (that is, $N = 2^L$, where L is an integer), then the computations can be made using a recursive algorithm. The development of the FFT has had a substantial influence on the solution of engineering problems because it significantly decreases the computation time for evaluating Fourier transforms. The ratio of computation time using an FFT relative to the standard discrete Fourier transform (DFT) is approximately

$$\frac{\text{FFT}}{\text{DFT}} \sim \frac{N \log_2 N}{N^2} \tag{8.33}$$

For example, for a time series with $N = 2^{10} = 1024$ points, the computation factor is $1/102.4$. For a larger time series with $N = 2^{20} = 1,048,576$ points, the computation factor is $1/52,428.8$. If this calculation took 10 minutes to execute using an FFT, it would require 1 year to compute using a DFT. A million points is a very long time series, but it demonstrates that the FFT provides the means to routinely use Fourier analysis by rendering the problem numerically tractable.

In practice, several variations of the FFT are routinely employed. The algorithm developed by Cooley and Tukey [2] is one of the more popular. However, an alternative given by Press et al. [3] is presented since its development is more intuitive. To begin, it is required that the total number of points be a power of 2; that is, $N = 2^L$, where L is an integer. The Fourier coefficients are given by

$$P_m = \sum_{n=0}^{N-1} p_n e^{i2\pi nm/N} \qquad m = 0, 1, \ldots, N-1 \tag{8.34}$$

Dividing the coefficients given in Eq. (8.34) into the even- and odd-numbered terms results in

$$P_m = \sum_{n=0}^{N/2-1} p_{2n} e^{i2\pi m(2n)/N} + \sum_{n=0}^{N/2-1} p_{2n+1} e^{i2\pi m(2n+1)/N} \tag{8.35}$$

$$= \sum_{n=0}^{N/2-1} p_{2n} e^{i2\pi mn/N/2} + e^{i2\pi m} \sum_{n=0}^{N/2-1} p_{2n+1} e^{i2\pi mn/N/2} \tag{8.36}$$

If the term W_m is now defined as

$$W_m = e^{i2\pi m} \tag{8.37}$$

then the Fourier coefficients represented by Eqs. (8.35) and (8.36) can be expressed as

$$P_m = P_m^e + W P_m^o \qquad m = 0, 1, \ldots, N \tag{8.38}$$

where P_m^e and P_m^o are the even and odd series terms, respectively. Note that each is a series over only half of the total data, $N/2$. The even terms can be divided again by taking every other term to again be even and odd. This reduces the data to $N/4$. Since $N = 2^L$, repeated halving will eventually lead to just one coefficient after L divisions.

To illustrate how this algorithm would be implemented, consider the case where $N = 8$ and a specific value of m; thus

$$P = \sum_{n=0}^{N-1} P_n = P_0 + P_1 + P_2 + P_3 + P_4 + P_5 + P_6 + P_7 \tag{8.39}$$

Taking the even and odd terms in Eq. (8.39) gives

$$P^e + P^o = (P_0 + P_2 + P_4 + P_6) + (P_1 + P_3 + P_5 + P_7) \tag{8.40}$$

Again taking the even and odd terms in Eq. (8.40) yields

$$
\begin{aligned}
(P^{ee} + P^{eo}) &+ (P^{oe} + P^{oo}) \\
&= [(P_0 + P_4) + (P_2 + P_6)] + [(P_1 + P_5) + (P_3 + P_7)]
\end{aligned} \tag{8.41}
$$

and finally, taking the even and odd once more in Eq. (8.41) results in

$$
\begin{aligned}
[(P^{eee} &+ P^{eeo}) + (P^{eoe} + P^{eoo})] \\
&+ [(P^{oee} + P^{oeo}) + (P^{ooe} + P^{ooo})] \\
&= (P_0) + (P_4) + (P_2) + (P_6) + (P_1) + (P_5) + (P_3) + (P_7)
\end{aligned} \tag{8.42}
$$

For this example, $N = 2^L$, with $L = 3$, therefore halving the data 3 times reduces each even or odd split down to a single value. These reductions are summarized in Table 8.1.

TABLE 8.1. **Even and Odd Reordering**

	Reversed	*e = 0, o = 1*	*Base 10*
$P^{eee} = P_0$	P^{eee}	P^{000}	P_0
$P^{eeo} = P_4$	P^{oee}	P^{100}	P_4
$P^{eoe} = P_2$	P^{eoe}	P^{010}	P_2
$P^{eoo} = P_6$	P^{ooe}	P^{110}	P_6
$P^{oee} = P_1$	P^{eeo}	P^{001}	P_1
$P^{oeo} = P_5$	P^{oeo}	P^{101}	P_5
$P^{ooe} = P_3$	P^{eoo}	P^{011}	P_3
$P^{ooo} = P_7$	P^{ooo}	P^{111}	P_7

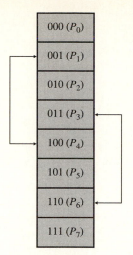

Figure 8.6
Reordering the coefficients.

There is a reordering technique that associates each P^{xyz} with the corresponding P_n. To accomplish this reordering, simply reverse the order of the exponent; that is, $P^{xyz} \rightarrow P^{zyx}$ and replace the e's by 0 and the o's by 1. These two steps are illustrated in Table 8.1. Now, the exponent is the binary value of n. These are given in the last column of Table 8.1 and agree with those determined by splitting the data 3 times. Binary numbers corresponding to the decimal numbers 0 to 7 are given in Table 8.2. The algorithm requires the data are in the order in which it appears in the last column of Table 8.1. The coefficients can be placed in this order by exchanging the positions of the two pairs of coefficients as illustrated in Figure 8.6.

At this point, the FFT is a recursive use of Eq. (8.38). Working back up through the splits for a specific value of m gives

$$P^{eee} = P_0 \tag{8.43}$$

$$P^{ee} = P^{eee} + WP^{eeo} \tag{8.44a}$$

$$P^{eo} = P^{eoe} + WP^{eoo} \tag{8.44b}$$

$$P^{e} = P^{ee} + WP^{eo} = (P^{eee} + WP^{eeo}) + W(P^{eoe} + WP^{eoo}) \tag{8.45a}$$

$$P^{o} = P^{oe} + WP^{oo} = (P^{oee} + WP^{oeo}) + W(P^{ooe} + WP^{ooo}) \tag{8.45b}$$

$$\begin{aligned}P = P^{e} + WP^{o} &= [(P^{eee} + WP^{eeo}) + W(P^{eoe} + WP^{eoo})] \\ &+ W[(P^{oee} + WP^{oeo}) + W(P^{ooe} + WP^{ooo})]\end{aligned} \tag{8.46}$$

The results are developed recursively at each level rather than completely developing a solution at each level. This is the essence of the increased computational efficiency exhibited by the FFT algorithm.

TABLE 8.2.   Binary and Decimal Numbers

Binary	Decimal
000	0
001	1
010	2
011	3
100	4
101	5
110	6
111	7

EXAMPLE 8.4 ▼

A time series with $N = 1024$ points and $\Delta t = 1$ sec is constructed from the sum of 10 cosines of the form

$$p_n = \sum_{m=1}^{10} a_m \cos\left(\frac{N}{2\pi nm} + \phi_m\right)$$

where the amplitudes $a_m = m/10$ kips and the phases ϕ_m are random. Plot the complex amplitude coefficients and the amplitude and phase spectra.

Solution

The FORTRAN program FFT.FOR to determine the Fourier transform and inverse Fourier transform is available from the author's Web site. For this example a short program was written to generate the discrete time series p_n and write the data file required as input by FFT.FOR. A portion of this file is shown in Table 8.3. The file has a descriptive header on the first line. The second line has isign, where isign $= 1$ for a forward transform, isign $= -1$ for an inverse transform. The third line is Δt and NN, where Δt is the time step and NN is the number of data. NN should be a power of 2. Next are the data, either a time series or amplitude coefficients, depending on whether a forward or reverse transform is being made. The first column is the real part and the second column is the imaginary part. If the series is all real, zeros must be entered for the imaginary part.

Figure 8.7 shows the time series. Figure 8.8 shows the real and imaginary parts of the complex coefficients. Note that for the real time series, the real parts of the coefficients are even and the imaginary parts are odd. Figure 8.9 shows the amplitude and phase spectra. These are two-sided spectra since there are positive and negative frequencies. The amplitudes are even and the phases are odd. To create a one-sided spectrum, the amplitudes corresponding to the frequencies above zero are doubled, as shown in Figure 8.10. In this figure, the frequencies are scaled by $f_D = 1/N\,dt$. This illustrates that the amplitude for $m = 1$ is 0.1, $m = 2$ is 0.2, etc.

TABLE 8.3. **Input Data for FFT.FOR**

Time series with 10 frequencies	
1	
1.0000	1024
−0.2833	0.00000
−0.2311	0.00000
−0.1790	0.00000
−0.1269	0.00000
−0.0750	0.00000
.	.
.	.
.	.
.	.

Numerical Recipes in FORTRAN by W.H. Press, S.A. Teukolsky, W.T. Vetterling, and B.P. Flannery, copyright © 1994. Reprinted by permission of Cambridge University Press, Melbourne, Australia.

Figure 8.7

Time series for Example 8.4.

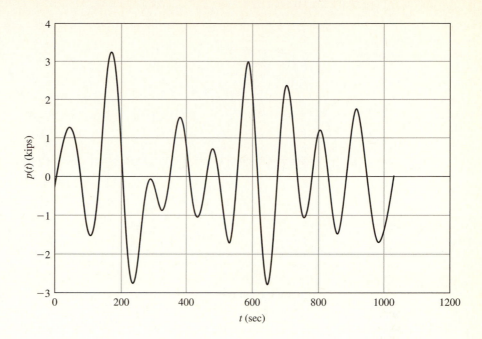

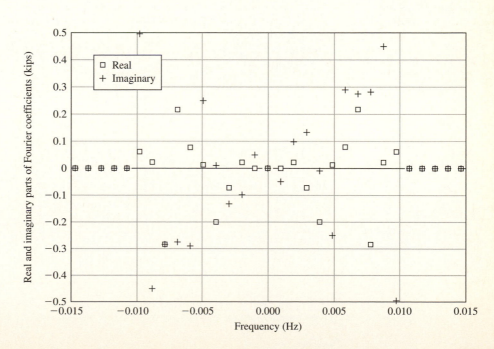

Figure 8.8

Real and imaginary Fourier
coefficients for Example 8.4.

Figure 8.9

Two-sided amplitude and phase spectra for Example 8.4.

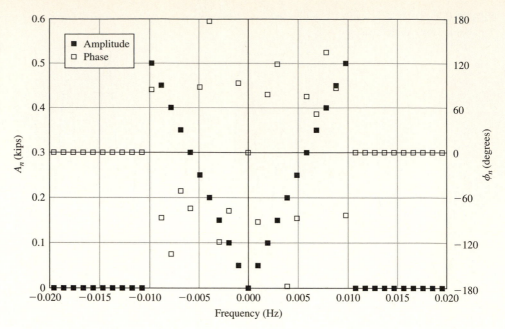

Figure 8.10

One-sided amplitude and phase spectra for Example 8.4.

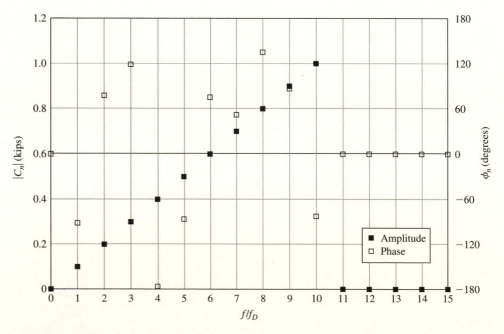

8.4 DISCRETE FOURIER TRANSFORM IMPLEMENTATION CONSIDERATIONS

8.4.1 Energy

The total energy in a time series is the same whether it is calculated in the time domain or the frequency domain. For discrete data, the energy E is given by

$$E = \sum_{n=0}^{N-1} p_n p_n^* = N \sum_{m=-N/2}^{N/2} P_m P_m^* \tag{8.47}$$

where $(\ )^*$ is the complex conjugate. Notice that energy is related to the total variance in the time series. The mean of the time series can also be determined in either the time domain or the frequency domain and is given by

$$\bar{p} = \frac{1}{N} \sum_{n=0}^{N-1} p_n = P_0 \tag{8.48}$$

If these statistics do not vary over time, the process is said to be *stationary*. A specific time series is a function of the phases associated with each amplitude. Different phase spectra will lead to different realizations of the time series. However, if the process is stationary, the variance does not change. This is a very useful result. For many processes, we understand the physics of the amplitude spectrum better than the phase spectrum. In fact, one technique for generating random forcing is to use a deterministic amplitude spectrum and a random, or *white noise*, phase spectrum. This technique was demonstrated in Example 8.2. A white noise spectrum is taken to be a uniform phase distribution in the range $[0, 2\pi)$.

8.4.2 Gibbs' Phenomenon

The Fourier series approximation to a square wave was presented in Chapter 6. The approximation is generally quite good as shown in Figure 8.11 in which $p(t)$ is a square wave scaled by its amplitude, p_o and period, T_o. However, an inaccuracy exists at the corners of the wave. Sines and cosines are smooth, continuous functions and therefore are best suited for approximating other smooth, continuous functions. However, for cases in which jumps or discontinuities exist in the function, the approximation is poor. For the square wave shown in Figure 8.9, a discontinuity occurs at $t/T_0 = 1/2$. At this location, the square wave has two values, $+1$ and -1. When a function has jumps or is double valued, the Fourier series passes through the mean of the two points, which in this case is zero.

Also, at $t/T_0 = 1/2$, the square wave is vertical. The Fourier series tends to overshoot at the corners. This is referred to as *Gibbs' phenomenon*. This overshooting does not disappear, even if a very large number of terms are used in the Fourier series. Figure 8.12 shows the height of the overshoot as a function of the number of terms in the Fourier series. It does not disappear as the number of terms increases. However, the width of the overshoot becomes very narrow for a large number of terms. If a sufficient number of terms are used, Gibbs' phenomenon is very local and the contribution to the total energy is minimal. The approximation is excellent everywhere, except at the jump. Unfortunately, in the case of impact loads, this is precisely a point of interest.

Figure 8.11

Fourier series approximation to a square wave.

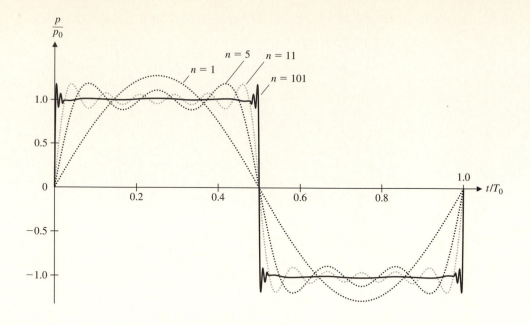

Figure 8.12

Height and location of overshoot for a square wave.

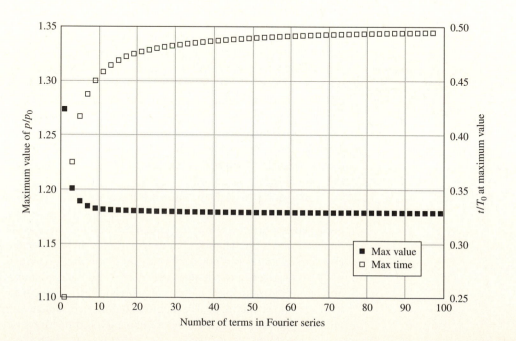

8.4.3 Aliasing

If the data are fit with a series of sines and cosines, the frequencies are thought of as being positive. However, if there are negative frequencies, then

$$a_n \cos(-\omega_n t) + b_n \sin(-\omega_n t) = a_n \cos(\omega_n t) + (-b_n) \sin(\omega_n t) \quad (8.49)$$

In Eq. (8.49), notice that the results at ω_n and $-\omega_n$ cannot be distinguished. These indistinguishable results are termed *aliases*. Next, consider a discrete time series at equal intervals of Δt. As an example, consider a time series that is a cosine of unit amplitude and has no phase shift. At any time $t = n\Delta t$, the value is given by

$$p_n = \cos(\omega n \Delta t) \quad (8.50)$$

If ω is increased, the cosine oscillates more rapidly. If $\omega = \pi/\Delta t$, then

$$p_n = \cos(n\pi) = (-1)^n \quad (8.51)$$

Equation (8.51) represents the most rapid oscillation possible. If the data has oscillations that are faster than this, they will not be properly detected.

The normal range for ω is defined by

$$0 < \omega < \frac{\pi}{\Delta t} \quad (8.52)$$

However, if there are data in the time series with frequencies that exceed this range by an amount ω', then

$$\omega = \frac{\pi}{\Delta t} + \omega' \quad (8.53)$$

The cosine in Eq. (8.50) is expressed by

$$p_n = \cos\left[\left(\frac{\pi}{\Delta t} + \omega'\right) n \, \Delta t\right] = (-1)^n \cos(\omega' n \, \Delta t) \quad (8.54)$$

But if ω' is in the range

$$0 < \omega' < \frac{\pi}{\Delta t} \quad (8.55)$$

then the response at this high frequency will appear in the Fourier series at a lower frequency corresponding to a Fourier frequency. If the high frequency is in the range of $2\pi/\Delta t < \omega < 3\pi/\Delta t$, a ω'' can be defined, and again this will restore energy back into the Fourier frequency range. In fact, all the high-frequency energy will appear in the Fourier frequencies. The frequency $\pi/\Delta t$ is called the *Nyquist frequency,* or folding frequency. Energy at frequencies above the Nyquist frequency is folded back into the interval $[0, \pi/\Delta t]$.

EXAMPLE 8.5   ▼

A force time series is given by

$$p(t) = 1.0 \text{ kips } \sin(2\pi t) + 0.5 \text{ kips } \sin[8(2\pi t)]$$

Determine the amplitude spectrum from samples taken at $\Delta t = 0.01$ sec (100 Hz) and $\Delta t = 0.1$ sec (10 Hz).

Solution

In the time series, there should be spectral amplitude coefficients at 1 and 8 Hz. A sampling rate of 100 Hz will have a Nyquist frequency of

$$\frac{\pi}{\Delta t} = \frac{\pi}{0.01} = 50(2\pi) \quad \text{rad/s} = 50 \text{ Hz}$$

The 10-Hz sampling rate will have a Nyquist frequency of 5 Hz. Therefore, the 100-Hz sampling rate should correctly detect the 8-Hz component in $p(t)$, but it will not be correctly detected by the 10-Hz sampling. Figure 8.13 shows the amplitude coefficients for the two sampling rates. The 100-Hz sample has coefficients at 1 and 8 Hz, as expected. The 10-Hz sample has the proper result for the 1-Hz component, but the 8-Hz component has been folded back into the range of 0 to 5 Hz.

From the 10-Hz time series, it is observed that there are two frequencies in the time series, one at 1 Hz and one at 2 Hz. If the original time series actually did have a component at 2 Hz, then the folded term would have combined with it. This can be taken a step further. Consider the case with two frequencies above the Nyquist frequency. Figure 8.14 shows the amplitude spectra for the above time series with a third term added, $0.5 \sin[16(2\pi t)]$. The spectrum for the 100-Hz time series shows the third amplitude coefficient at 16 Hz. However, the 10-Hz time series folds this back into the range 0 to 5 Hz. If there are many unknown frequencies above the Nyquist frequency, these combine with terms below the Nyquist frequency and it is impossible to separate them. Techniques for addressing this problem are discussed in the following sections.

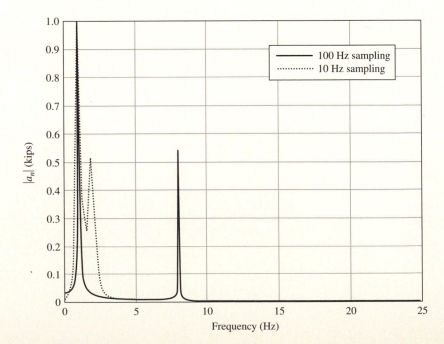

Figure 8.13

Spectra for two frequency-component time series in Example 8.5 showing folding.

Figure 8.14

Spectra for three frequency-component time series in Example 8.5 showing folding.

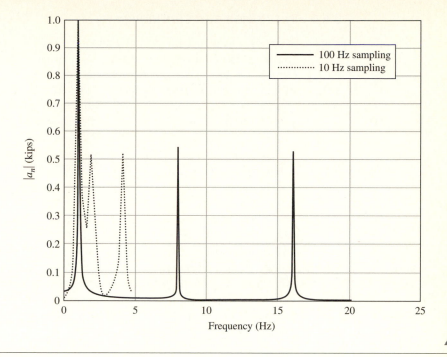

8.4.4 Sampling

The highest frequency that can be sampled in the data is half the Nyquist frequency. If the time series has high frequencies, the errors can be reduced by sampling at a faster rate (i.e., smaller Δt) or by prefiltering the data to remove high frequencies. If the data is sampled at a rate of at least twice the highest frequency in the signal, no high-frequency information will be lost due to sampling. Prefiltering the analog data with an antialiasing filter is a good practice because this will ensure there is no high-frequency information in the sampled time series. Oversampling the data at a high rate while digitizing and then numerically filtering the data does not ensure that aliasing will be eliminated. Consider the simple case illustrated in Figure 8.15. Assume the data is sampled at T_s, such that the highest frequency in the sample is $f_s = 1/2T_s$. If the data are numerically low-pass filtered with a cutoff frequency of $f_c = 1/4f_s$ or $T_c = 4T_s$, then Figure 8.11 represents a case where aliasing occurs that folds energy back into the zero frequency in the filtered results. Once this error has been introduced into the digitized data, there is no straightforward means to remove it.

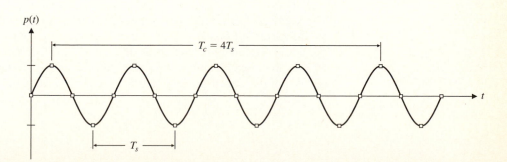

Figure 8.15

Example of aliasing.

8.4.5 Resolution and Confidence

There is a trade-off in spectral estimates between *resolution* and *confidence*. If the spacing of the Fourier frequencies is very close, then the resolution in frequency will be high. Since $\omega_n = 2n\pi/T_D$, this can be accomplished by taking a longer length of data. However, because the frequencies are closely spaced, there may be energy at one frequency, none at the next, etc. This leads to an estimate of the spectrum that appears very rough or jumpy. For this case, the resolution is high but the confidence is low. Confidence is an indicator of how well the spectral estimate at a given frequency agrees with the actual value. Usually the actual value is in a band or confidence interval somewhere around the estimated value. The confidence can be increased by having more widely spaced frequencies so that each Fourier component represents an average over a range. This requires T_D to be smaller.

Thus, resolution and confidence are at odds. One way to improve results is to prefilter the data with a low-pass filter. The filter is chosen to optimize results. This generally requires some type of insight into the physical system to ascertain what ranges of frequencies are meaningful. A common filter is the *box car* or *moving average*. An odd number of consecutive points in the time series are averaged, and this average value is used at the time point corresponding to the center of the box. The width of the box (i.e., 7 points or 21 points) depends on the physics of the problem. Choosing a box width too wide will average out desired frequencies.

Another alternative is to take a long record length, but then analyze the data as a number of shorter segments. In this way, several independent estimates of the spectral coefficients are provided at each Fourier frequency. The mean, variance, and confidence may be determined at each frequency. The determination of the record length and number of segments again depend on the physical system. For a SDOF system with light damping, the length of the data record for estimating a spectrum is approximately

$$T_D \approx \frac{2}{f\zeta\left(\dfrac{\sigma}{\mu}\right)^2} \tag{8.56}$$

where T_D = record length
f = undamped natural frequency
ζ = damping factor, and
(σ/μ) = coefficient of variation (ratio of the standard deviation to the mean)
The required number of segments is approximately

$$n_D \approx \frac{1}{(\sigma/\mu)^2} \tag{8.57}$$

The segments can also overlap with each other. Overlapping the segment means they are no longer independent. However, it does increase the number of estimates at each frequency. Optimum results occur at about a 50% overlap. Figure 8.16 illustrates a 50% overlap with the preceding segments and a 50% overlap with the following segments. For this case the length of each segment T_s is determined by

$$T_s = \frac{2T_D}{n_D + 1} = \frac{4}{f\zeta} \tag{8.58}$$

Figure 8.16

Example of 50% segment overlap.

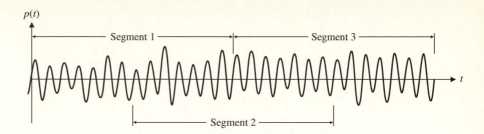

8.5 FOURIER INTEGRAL

Fourier series provide a reasonable approximation for periodic data. For nonperiodic loadings such as single impacts or step loads, this method becomes inadequate. Fourier series are periodic in T_D. Figure 8.17 shows a nonperiodic load function and how it periodically repeats when approximated with a Fourier series. If the response to this load is examined for times greater than T_D, errors will result. To avoid this, T_D must be increased. If the load is zero for all time after the initial application, then $T_D \to \infty$. In Eq. (8.11), the Fourier coefficients are multiplied by $1/T_D$. Since $p(t)$ has finite duration, increasing T_D does not increase the value of the integral, but it does reduce the magnitude of the Fourier coefficients. In the limit as $T_D \to \infty$, the coefficients $P_n \to 0$. For the long-term response to a nonperiodic load, the Fourier series approach breaks down. For these cases, the Fourier integral should be employed. The Fourier frequencies are $\omega_n = 2n\pi/T_D$. As T_D increases, these frequencies become very tightly spaced. In the limit, the summation transforms into an integration. The Fourier integrals are represented as

$$p(t) = \int_{-\infty}^{\infty} P(f)e^{i2\pi ft} \, df \tag{8.59}$$

$$P(f) = \int_{-\infty}^{\infty} p(t)e^{-i2\pi ft} \, dt \tag{8.60}$$

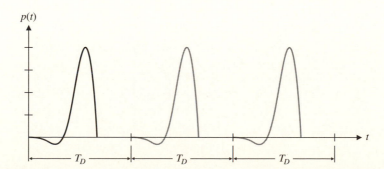

Figure 8.17

Periodicity of the Fourier series approximation.

in which $p(t)$ is the time series and $P(f)$ is the Fourier transform of $p(t)$. $P(f)$ is analogous to the amplitude coefficients for the DFT, but now is a continuous function of frequency. Equations (8.59) and (8.60) are referred to as a transform pair. Given $P(f)$, Eq. (8.59) may be used to determine $p(t)$. Conversely, Eq. (8.60) allows the determination of $P(f)$ given $p(t)$. A commonly employed notation is

$$p(t) \langle = \rangle \ P(f) \tag{8.61}$$

to denote this reciprocity.

If the response of a structure is to be determined using Fourier integrals, then by analogy with the series form given in Eq. (8.22),

$$x(t) = \int_{-\infty}^{\infty} H(f)P(f)e^{i2\pi ft}\,df \tag{8.62}$$

in which $H(f)$ is the transfer function. For a SDOF system, the transfer function is expressed as

$$H(f) = \frac{1}{k[(1 - R^2) + i(2R\zeta)]} = \frac{(1 - R^2) - i(2R\zeta)}{k[(1 - R^2)^2 + i(2R\zeta)^2]^{1/2}} \tag{8.63}$$

where ζ is the damping factor defined in Eq. (8.19) and the frequency ratio R is defined as

$$R = \frac{2\pi f}{\omega} \tag{8.64}$$

in which ω is the undamped natural frequency of the system. This is similar to the frequency ratio given in Eq. (8.18), except that $n\Omega$ in Eq. (8.18) corresponds to discrete frequencies and $2\pi f$ in Eq. (8.64) is continuous.

EXAMPLE 8.6 ▼

A nonperiodic load $p(t)$ having magnitude p_0 and duration s is shown in Figure 8.18. Determine the Fourier transform $P(f)$ and plot the amplitude spectrum for $p_0 = 100$ lb, $s = 1$ sec, and $t_0 = 3$ sec.

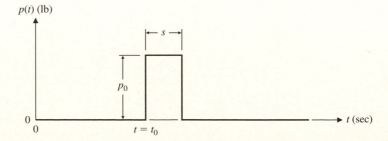

Figure 8.18

Nonperiodic load in Example 8.6.

Solution

$$P(f) = \int_{-\infty}^{\infty} p(t)e^{-i2\pi ft}\, dt$$

$$= \int_{t_0}^{t_0+s} p_0 e^{-i2\pi ft}\, dt \tag{1}$$

$$= p_0 \frac{i}{2\pi f}\left(e^{-i2\pi f(t_0+s)} - e^{-i2\pi ft_0}\right)$$

The amplitude spectrum is the modulus of $P(f)$; that is,

$$|P(f)| = [P(f)P^*(f)]^{1/2} \tag{2}$$

where $P^*(f)$ is the complex conjugate of $P(f)$. The amplitude spectrum is given by

$$|P(f)| = \frac{p_0}{2\pi|f|}[2 - 2\cos(2\pi fs)]^{1/2} \tag{3}$$

An illustration of the amplitude spectrum is shown in Figure 8.19.

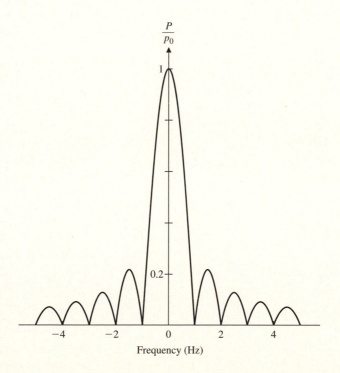

Figure 8.19

Amplitude spectrum in Example 8.6.

Many useful Fourier integrals for analytical load functions are given in tables, such as in [5]. More complex Fourier integrals must be evaluated numerically. There are a number of software packages available to do this, such as MATLAB [6]. To write a computer program for the numerical evaluation of Fourier integrals necessitates a special consideration. Numerical integration requires that the data be represented discretely, then the integral can be written as a summation involving the discrete Fourier transform. However, the DFT has a finite number of frequencies and the Fourier integral has continuous frequencies in the interval $-\infty < f < \infty$, and high-frequency terms are lost. Although discretizing the data at finer intervals increases the range of frequencies, it is not possible to capture the highest frequencies with a finite number of samples and this approach will fail.

An alternative is to use polynomial interpolation formulas to locally fit the data. These interpolation formulas may be integrated analytically over most of the data. The remaining data may be addressed using the DFT or numerically expedient FFT. The implementation of this technique is given in [3] for several interpolation functions.

REFERENCES

1   Ziemer, R.E., Tranter, W.H., and Fannin, D.R., *Signals and Systems: Continuous and Discrete,* MacMillan, New York, 1989.

2   Cooley, J.W. and Tukey, J.W., An Algorithm for the Machine Calculation of Complex Fourier Series, *Mathematics of Computation,* 19:297, 1965.

3   Press, W.H., Teukolsky, S.A., Vetterling, W.T., and Flannery, B.P., *Numerical Recipes in FORTRAN,* Cambridge University Press, Melbourne, Australia, 1994.

4   Bendat, J.S. and Piersol, A.G., *Engineering Applications and Spectral Analysis,* Wiley, New York, 1993.

5   Oberhettinger, F., *Tabellen zur Fourier Transformation,* Springer-Verlag, Berlin, 1957.

6   The MathWorks, Inc., *MATLAB,* Prentice Hall, Upper Saddle River, NJ, 1997.

NOTATION

a_0, a_n, b_n	Fourier series coefficients	n	index
A_0, A_n	Fourier series coefficients	n_D	number of sampling segments
c	damping coefficient	N	number of points in time series
E	energy	r	frequency ratio
$p(t)$	arbitrary function of time	R	frequency ratio
f	natural frequency	s	load duration in Example 8.6
f_c	cutoff frequency	S_N	notational variable
f_D	fundamental Fourier frequency (Hz)	t	time
f_s	sampling frequency	T_c	cut off period
P_0, P_n	complex valued Fourier series coefficients	T_D	duration of time series
H_n	transfer function	T_0	period of forcing function
i	$\sqrt{-1}$	T_s	shortest sampled period
k	stiffness	W_m	notational variable
k	index	$x(t)$	displacement
m	mass	X_n	displacement amplitude
m	index	α_n	phase angle corresponding to F_n

δ_{km}	Kronecker delta		ω_p	peak frequency
Δt	time increment		Ω	fundamental Fourier frequency
ϕ_n	phase angle corresponding to H_n		μ	mean
σ	standard deviation		ζ	damping factor
ω	radian frequency		$\bar{p}$	mean of p
ω'	frequency above Nyquist frequency		$(\ )^*$	complex conjugate
ω_n	Fourier frequencies		T_s	sampling time step

PROBLEMS

8.1 Determine the Fourier series representation for

$$p(t) = \left| \cos\left(\frac{2\pi}{T_0} t\right) \right| \cos\left(\frac{2\pi}{T_0} t\right)$$

using Eq. (8.6). Plot $f(t)$ and the first term in the Fourier series to see how well it approximates the function. *Note:* This is a common technique for linearizing nonlinear drag.

8.2 Determine the Fourier series representation for the forcing function shown in Figure P8.1. The load is given by

$$p(t) = \begin{cases} p_0\left(\dfrac{2t}{T_0}\right)^2 & t < \dfrac{T_0}{2} \\ 0 & t > \dfrac{T_0}{2} \end{cases}$$

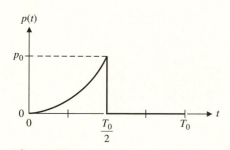

Figure P8.1

8.3 Determine the first five Fourier series terms for the force in Figure P8.2 for $p_0 = 10$ kips and $T_0 = 0.5$ sec.

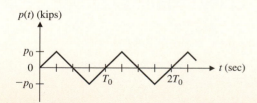

Figure P8.2

8.4 A 20-lb weight is attached to a spring-damper system as shown in Figure P8.3. The weight is acted on by a horizontal force with the following five frequency components:

j	A_j (lb)	f_j (Hz)	ϕ_j
1	1	0.02	219°
2	2	0.04	18°
3	3	0.08	2°
4	2	0.16	246°
5	1	0.32	233°

$$p(t) = \sum_{j=1}^{5} A_j \cos(2\pi f_j t + \phi_j)$$

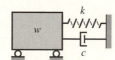

Figure P8.3

Assume stiffness $k = 1$ lb-in and 10% of critical damping $c = 0.1\,(2\sqrt{km})$. Determine the amplitude and phase spectra for the forcing function using the FFT. Plot the amplitudes of the forcing and response spectra.

8.5 The structure in Problem 8.4 is initially at rest when acted upon by the load shown in Figure P8.2. Plot the time history of the response for $p_0 = 3$ lb and $T_0 = 0.5$ sec.

8.6 Determine the horizontal displacement spectrum for the steel frame in Figure P8.4. The periodic horizontal force is given by

$$p(t) = \begin{cases} p_0 \sin\left(2\pi \dfrac{t}{T_0}\right) & 0 < t < \dfrac{T_0}{2} \\ 0 & \dfrac{T_0}{2} < t < T_0 \end{cases}$$

with $p_0 = 1$ kip and $T_0 = 0.5$ sec. The columns are W8×31 and bend about the strong axis.

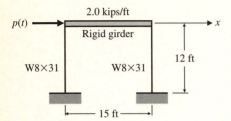

Figure P8.4

8.7 The nonperiodic load in Figure P8.5 is given by

$$p(t) = \frac{p_o}{2}\{1 - \cos\left[\pi(t - t_0 - s)\right]\}$$

Determine the Fourier transform.

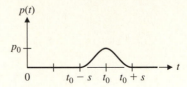

Figure P8.5

PART **II**

Multi-Degree-of-Freedom (MDOF) Systems

9
General Property Matrices for Vibrating Systems

The vibration systems investigated to this point were SDOF systems. That is, only one spatial coordinate was required to define the displacement configuration of the system at all times. Thus, only one differential equation of motion was necessary to characterize the dynamic behavior of the system, and the system had just one natural frequency. In reality, however, most structures exhibit at least several degrees of freedom. The number of degrees of freedom that a structure possesses is equal to the number of independent spatial coordinates required to describe its motion. Moreover, for each degree of freedom, there will be an equal number of equations of motion and natural frequencies.

Therefore, for MDOF systems, matrix formulation is essential to clarify the problem description and to systematize the response calculations. Matrices also provide a convenient format for organizing the computations required to analyze MDOF systems. Furthermore, matrix notation affords the analyst a procedure conducive to programming on digital computers.

There are several general physical properties germane to every MDOF vibrating system: the mass matrix, the stiffness and flexibility matrices, and the damping matrix. Moreover, if the *mode superposition method* is employed for the dynamic analysis, solution of the system *eigenproblem* is necessary. In this chapter the mass and stiffness (and flexibility) properties of MDOF systems are discussed, as is formulation of the general eigenproblem. The concept of reducing the number of DOF in MDOF systems by *static condensation* is also introduced. Damping in MDOF systems will be discussed in Chapters 12 and 13.

9.1 FLEXIBILITY MATRIX

The forces and displacements that exist in a structure can be related to one another by using either *flexibility coefficients* or *stiffness coefficients.* Structural analysts are generally more familiar with procedures for calculating flexibility coefficients than with those for calculating stiffness coefficients. Therefore, construction of the flexibility matrix is discussed first. However, the unique relationship existing between the flexibility matrix and the stiffness matrix will readily become apparent.

Figure 9.1

Simple beam illustrating
flexibility coefficients.

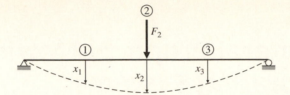

Consider the beam shown in Figure 9.1. We select three stations along the length
of the beam, designated simply as 1, 2, and 3. If a load F_2 is applied at station 2, the
deflections at stations 1, 2, and 3 are x_1, x_2, and x_3, respectively. These deflections can
be expressed as

$$
\begin{aligned}
x_1 &= a_{12}F_2 \\
x_2 &= a_{22}F_2 \\
x_3 &= a_{32}F_2
\end{aligned}
\tag{9.1}
$$

where a_{12}, a_{22}, and a_{32} are termed *flexibility influence coefficients.* The coefficient a_{12} is
defined as the deflection at station 1 due to a unit force at station 2; a_{22} is defined as the
deflection at station 2 due to a unit force at station 2; and a_{32} is defined as the deflection
at station 3 due to a unit force at station 2.

If instead of a single force F_2 at station 2, we apply forces F_1, F_2, and F_3 at stations
1, 2, and 3, respectively, then the expressions for the deflections become

$$
\begin{aligned}
x_1 &= a_{11}F_1 + a_{12}F_2 + a_{13}F_3 \\
x_2 &= a_{21}F_1 + a_{22}F_2 + a_{23}F_3 \\
x_3 &= a_{31}F_1 + a_{32}F_2 + a_{33}F_3
\end{aligned}
\tag{9.2}
$$

These expressions may be written in matrix form as

$$
\begin{Bmatrix} x_1 \\ x_2 \\ x_3 \end{Bmatrix} =
\begin{bmatrix} a_{11} & a_{12} & a_{13} \\ a_{21} & a_{22} & a_{23} \\ a_{31} & a_{32} & a_{33} \end{bmatrix}
\begin{bmatrix} F_1 \\ F_2 \\ F_3 \end{bmatrix}
\tag{9.3}
$$

or

$$
\{x\} = [a]\{F\}
\tag{9.4}
$$

The matrix $[a]$ is the *flexibility matrix* that contains the flexibility influence coefficients
relating the deflections $\{x\}$ to the applied forces $\{F\}$.

For linear elastic systems, the flexibility matrix is symmetric. This is a direct con-
sequence of *Maxwell's law of reciprocal deflections,* which can be stated as follows:
*The deflection of a point i on a structure due to a unit load acting at point j is equal to
the deflection of point j when the unit load is acting at point i.* More succinctly, Maxwell's law may be expressed as

$$
a_{ij} = a_{ji}
\tag{9.5}
$$

Proof of Maxwell's theorem can be found in any textbook on structural analysis.

EXAMPLE 9.1 ▼

Determine the flexibility influence coefficients and construct the flexibility matrix for the uniform cantilever beam modeled as a two-DOF system as shown in Figure 9.2a. The beam is of length L and has flexural rigidity EI.

Solution

The order of the flexibility matrix equals the number of DOF in the system. Thus, in this example the flexibility matrix is 2×2. The first column of the flexibility matrix is obtained by considering the beam in Figure 9.2b. Using the moment area theorem, we obtain

$$a_{11} = \frac{1}{EI}\left[\frac{1}{2}(2L)^2\left(\frac{4L}{3}\right)\right] = \frac{8L^3}{3EI}$$

and

$$a_{21} = \frac{5L^3}{6EI}$$

Similarly, the second column of the flexibility matrix is obtained by considering the beam in Figure 9.2c:

$$a_{12} = \frac{5L^3}{6EI} \qquad \text{(note that } a_{12} = a_{21}\text{)}$$

and

$$a_{22} = \frac{L^3}{3EI}$$

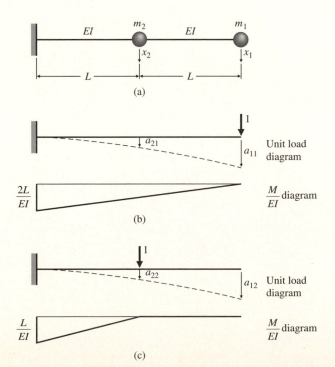

Figure 9.2
(a) Two-DOF representation of cantilever beam;
(b) calculation of first column of flexibility matrix;
(c) calculation of second column of flexibility matrix.

Thus, the flexibility matrix can be written as

$$[a] = \frac{L^3}{EI} \begin{bmatrix} \frac{8}{3} & \frac{5}{6} \\ \frac{5}{6} & \frac{1}{3} \end{bmatrix}$$

Often it is necessary to consider the rotational DOF when conducting a particular analysis. The following example illustrates the construction of a flexibility matrix for a structure exhibiting both translational and rotational DOF. ▲

EXAMPLE 9.2 ▼

Construct the flexibility matrix for the beam shown in Figure 9.3a. The beam exhibits four DOF: two translational degrees and two rotational degrees. The beam is of length L and has flexural rigidity EI.

Solution

To obtain the first column in the matrix, apply a unit vertical force at the free end of the member, as shown in Figure 9.3b. Utilizing the moment area theorem,

$$a_{11} = \frac{8L^3}{3EI} \qquad a_{31} = \frac{5L^3}{6EI}$$

$$a_{21} = \frac{2L^2}{EI} \qquad a_{41} = \frac{3L^2}{2EI}$$

In a similar manner, the second column is obtained by applying a unit moment at the free end of the member, as shown in Figure 9.3c. Thus,

$$a_{12} = \frac{2L^2}{EI} \qquad a_{32} = \frac{L^2}{2EI}$$

$$a_{22} = \frac{2L}{EI} \qquad a_{42} = \frac{L}{EI}$$

The remaining columns of the flexibility matrix, 3 and 4, are obtained by applying a unit force and a unit moment at midspan, as shown in Figure 9.3d and e, respectively, and then calculating the corresponding flexibility coefficients. Assembling all four columns results in the following flexibility matrix:

$$\begin{Bmatrix} x_1 \\ \theta_1 \\ x_2 \\ \theta_2 \end{Bmatrix} = \frac{1}{EI} \begin{bmatrix} \dfrac{8L^3}{3} & 2L^2 & \dfrac{5L^2}{6} & \dfrac{3L^2}{2} \\ 2L^2 & 2L & \dfrac{L^2}{2} & L \\ \dfrac{5L^2}{6} & \dfrac{L^2}{2} & \dfrac{L^3}{3} & \dfrac{L^2}{2} \\ \dfrac{3L^2}{2} & L & \dfrac{L^2}{2} & L \end{bmatrix} \begin{Bmatrix} F_1 \\ M_1 \\ F_2 \\ M_2 \end{Bmatrix}$$

Figure 9.3

(a) Four-DOF representation of cantilever beam; (b) calculation of first column of flexibility matrix; (c) calculation of second column of flexibility matrix; (d) calculation of third column of flexibility matrix; (e) calculation of fourth column of flexibility matrix.

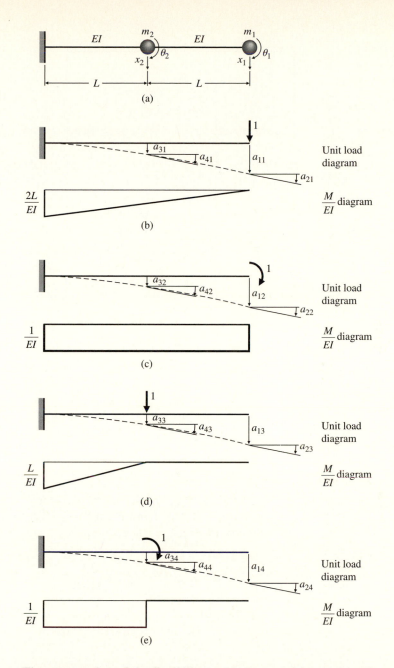

The construction of the flexibility matrix for a multistory shear frame building is quite direct and is illustrated in the following example. ▲

EXAMPLE 9.3 ▼

Construct the flexibility matrix for the three-story shear frame building shown in Figure 9.4a. Assume each story has a height L, the columns have flexural rigidity EI, and the horizontal members are rigid relative to the flexural stiffness of the columns.

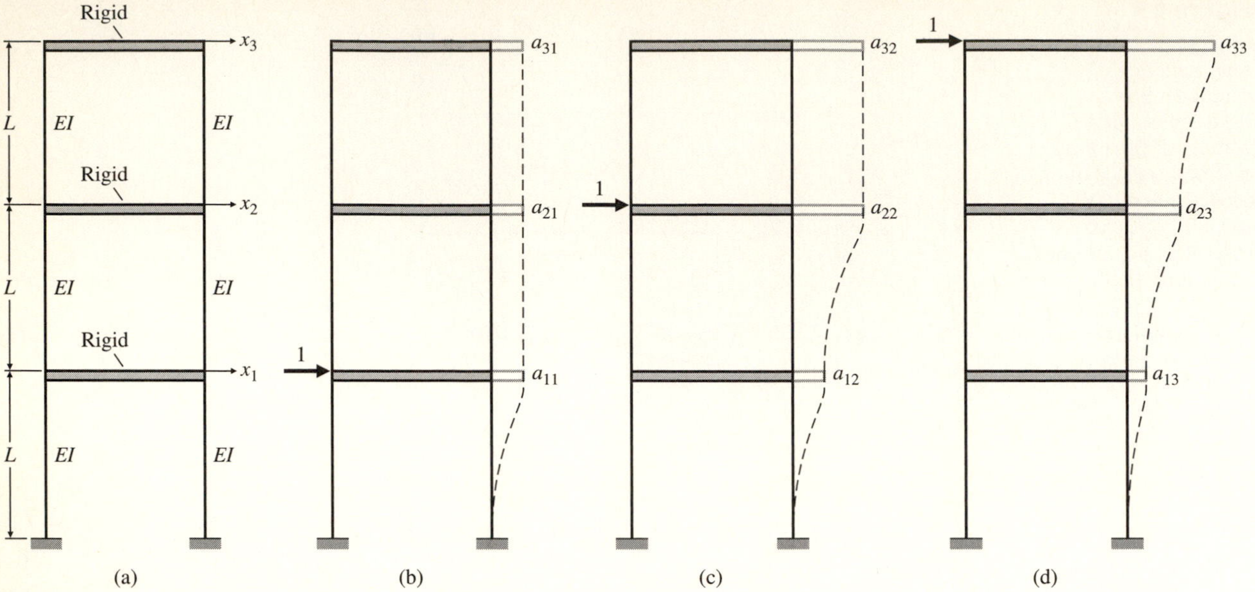

Figure 9.4
(a) Three-DOF shear frame building;
(b) calculation of first column of flexibility matrix;
(c) calculation of second column of flexibility matrix;
(d) calculation of third column of flexibility matrix.

Solution

To obtain the first column in the flexibility matrix, a unit horizontal force is applied at level 1 as shown in Figure 9.4b. The resulting horizontal story translations comprise the influence coefficients for the first column, given as

$$a_{11} = a_{21} = a_{31} = \frac{L^3}{24EI}$$

In a similar manner, the second column is obtained by applying a unit force to level 2 as illustrated in Figure 9.4c. Thus,

$$a_{12} = \frac{L^3}{24EI}$$

$$a_{22} = \frac{L^3}{24EI} + \frac{L^3}{24EI} = \frac{L^3}{12EI}$$

$$a_{32} = \frac{L^3}{12EI}$$

The third column of the flexibility matrix is then obtained by applying a unit force to the top level as shown in Figure 9.4d. Therefore,

$$a_{13} = \frac{L^3}{24EI}$$

$$a_{23} = \frac{L^3}{24EI} + \frac{L^3}{24EI} = \frac{L^3}{12EI}$$

$$a_{33} = \frac{L^3}{24EI} + \frac{L^3}{24EI} + \frac{L^3}{24EI} = \frac{L^3}{8EI}$$

The flexibility matrix may then be expressed as

$$\begin{Bmatrix} x_1 \\ x_2 \\ x_3 \end{Bmatrix} = \frac{1}{EI} \begin{bmatrix} \dfrac{L^3}{24} & \dfrac{L^3}{24} & \dfrac{L^3}{24} \\ \dfrac{L^3}{24} & \dfrac{L^3}{12} & \dfrac{L^3}{12} \\ \dfrac{L^3}{24} & \dfrac{L^3}{12} & \dfrac{L^3}{8} \end{bmatrix} \begin{Bmatrix} F_1 \\ F_2 \\ F_3 \end{Bmatrix}$$

▲

To verify the correctness of the flexibility matrix, symmetry about the main diagonal should be confirmed.

9.2 STIFFNESS MATRIX

Another means of relating the forces and deformations in a structure to one another is through the *stiffness influence coefficients.* Consider the beam in Figure 9.5a. The bending moment M_1 acting at point 1 is related to the rotation θ_1 by the expression

$$M_1 = k_{11}\theta_1 \tag{9.6}$$

The term k_{11} is a stiffness influence coefficient. It is defined as the moment at point 1 due to a unit rotation at point 1.

If the beam is allowed to rotate at point 2 as well as at point 1, as indicated in Figure 9.5b, then the moment at point 1 can be expressed in terms of the rotation at both point 1 and point 2. Therefore,

$$M_1 = k_{11}\theta_1 + k_{12}\theta_2 \tag{9.7}$$

In a similar fashion, the moment M_2 at point 2 can be expressed in terms of the rotations at point 1 and point 2. Thus,

$$M_2 = k_{21}\theta_1 + k_{22}\theta_2 \tag{9.8}$$

Figure 9.5

Beam illustrating
rotational stiffness
influence coefficients.

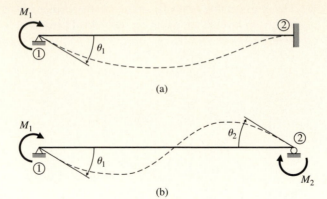

Rewriting Eqs. (9.7) and (9.8) in matrix form yields

$$\begin{Bmatrix} M_1 \\ M_2 \end{Bmatrix} = \begin{bmatrix} k_{11} & k_{12} \\ k_{21} & k_{22} \end{bmatrix} \begin{Bmatrix} \theta_1 \\ \theta_2 \end{Bmatrix} \tag{9.9}$$

or more conventionally,

$$\{F\} = [k]\{x\} \tag{9.10}$$

The matrix $[k]$ contains the stiffness influence coefficients that relate the forces $\{F\}$ to the deformations $\{x\}$ and is referred to as the *stiffness matrix*. The force vector $\{F\}$ generally contains both forces and moments, and the deformation vector $\{x\}$ generally contains both translational deflections and rotations.

If Eq. (9.10) is premultiplied by the inverse of the stiffness matrix $[k]^{-1}$, then

$$[k]^{-1}\{F\} = \{x\} \tag{9.11}$$

or

$$\{x\} = [k]^{-1}\{F\} \tag{9.12}$$

Comparison of Eq. (9.12) with Eq. (9.4) indicates that the inverse of the stiffness matrix is the flexibility matrix, or

$$[k]^{-1} = [a] \tag{9.13}$$

Conversely, the inverse of the flexibility matrix is the stiffness matrix, or

$$[a]^{-1} = [k] \tag{9.14}$$

Similar to the flexibility matrix, the stiffness matrix is also symmetric about its main diagonal, or

$$k_{ij} = k_{ji} \tag{9.15}$$

The stiffness influence coefficient is generally defined as the force (or moment) required to produce a unit displacement (or rotation) at a particular location or joint in a structure if displacement is prevented at all other joints in the structure. The joints or locations of interest are those corresponding to the defined DOF in the structure. This concept is illustrated in its simplest form in the following example.

EXAMPLE 9.4   ▼

Construct the stiffness matrix by columns for the system shown in Figure 9.6a.

Solution

The system has four DOF, one translation associated with each mass. Therefore the resulting stiffness matrix will be 4×4. To obtain the stiffness coefficients for the first column, mass m_1 is given a unit displacement and all other masses are held stationary. The resulting spring forces represent the stiffness coefficients for the first column of the stiffness matrix. Thus, setting $x_1 = 1.0$ and $x_2 = x_3 = x_4 = 0$, the forces in the springs, considered positive if they act to the right, as shown in Figure 9.6b, are

$$k_{11} = k_1 + k_2 \qquad k_{31} = 0$$
$$k_{21} = -k_2 \qquad k_{41} = 0$$

The second column of the stiffness matrix is obtained by setting $x_2 = 1.0$ and $x_1 = x_3 = x_4 = 0$, so that

$$k_{12} = -k_2 \qquad k_{32} = -k_3$$
$$k_{22} = k_2 + k_3 \qquad k_{42} = 0$$

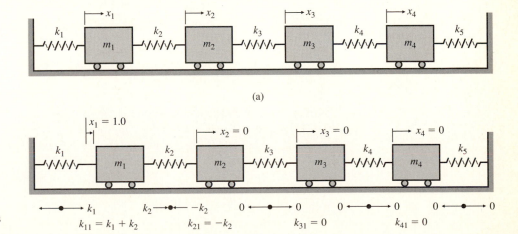

Figure 9.6

(a) Four-DOF system; (b) calculation of stiffness influence coefficients by column.

In a similar manner, the third column of the stiffness matrix is obtained by setting $x_3 = 1.0$ and $x_1 = x_2 = x_4 = 0$. Thus,

$$k_{13} = 0 \qquad k_{33} = k_3 + k_4$$
$$k_{23} = -k_3 \qquad k_{43} = -k_4$$

Finally, the fourth column of the stiffness matrix is obtained by setting $x_4 = 1.0$ and $x_1 = x_2 = x_3 = 0$, which gives

$$k_{14} = 0 \qquad k_{34} = -k_4$$
$$k_{24} = 0 \qquad k_{44} = k_4 + k_5$$

Assembling the columns in order yields the stiffness matrix:

$$[k] = \begin{bmatrix} k_1 + k_2 & -k_2 & 0 & 0 \\ -k_2 & k_2 + k_3 & -k_3 & 0 \\ 0 & -k_3 & k_3 + k_4 & -k_4 \\ 0 & 0 & -k_4 & k_4 + k_5 \end{bmatrix}$$

▲

Many practical structures can be represented by an assemblage of axial and/or flexural members. Thus, if the stiffness coefficients for a typical axial member or flexural member can be determined, the construction of the stiffness matrix for a structure comprised of these members can be greatly facilitated.

Consider the axial force member shown in Figure 9.7a, having cross-sectional area A, modulus of elasticity E, and length L. The ends of the member are the joints or node points where forces F_1 and F_2 are applied and displacements x_1 and x_2 are determined. The directions exhibited by the forces and displacements in Figure 9.7a define the positive sense.

The matrix equation that relates the forces $\{F\}$ to the member displacements $\{x\}$ is of the form

$$\begin{Bmatrix} F_1 \\ F_2 \end{Bmatrix} = \begin{bmatrix} k_{11} & k_{12} \\ k_{21} & k_{22} \end{bmatrix} \begin{Bmatrix} x_1 \\ x_2 \end{Bmatrix} \tag{9.16}$$

To obtain the influence coefficients for the first column of the stiffness matrix, set $x_1 = 1.0$ and $x_2 = 0$. Thus, k_{11} and k_{21} are the axial forces corresponding to a unit contraction at the left end of the member, as shown in Figure 9.7b, given as

$$k_{11} = \frac{EA}{L} \qquad k_{21} = -\frac{EA}{L}$$

Figure 9.7

(a) Two-DOF axial force member; (b) stiffness coefficients for first column; (c) stiffness coefficients for second column.

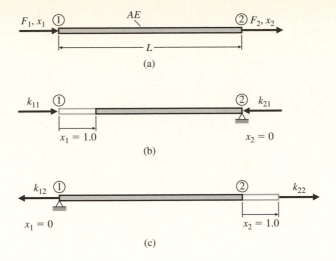

Note that x_1 and x_2 are positive when acting to the right according to the sign convention defined in Figure 9.7a. Therefore, k_{11} is positive and k_{21} is negative.

In a similar manner the second column of the stiffness matrix is obtained by setting $x_2 = 1.0$ and $x_1 = 0$. As illustrated in Figure 9.7c, the forces corresponding to these imposed displacements are

$$k_{12} = -\frac{EA}{L} \qquad k_{22} = \frac{EA}{L}$$

Combining these results with Eq. (9.16) yields

$$\begin{Bmatrix} F_1 \\ F_2 \end{Bmatrix} = \frac{EA}{L} \begin{bmatrix} 1 & -1 \\ -1 & 1 \end{bmatrix} \begin{Bmatrix} x_1 \\ x_2 \end{Bmatrix} \tag{9.17}$$

Equation (9.17) defines the stiffness matrix for an axially loaded member of uniform cross section.

Stiffness coefficients for flexural members can be derived from the slope deflection method or the conjugate beam method. Stiffness coefficients for some commonly encountered flexural conditions are presented in Table 9.1. As shown in the table, the displacements within $\{x\}$ may be either rotations or translations, and the forces within $\{F\}$ may be shear forces or bending moments. The positive directions are indicated by the definition sketch at the top of the table. To establish the appropriate elements of the stiffness matrix, the applied displacements would be set equal to unity.

TABLE 9.1. **Stiffness Coefficients for Flexural Members**

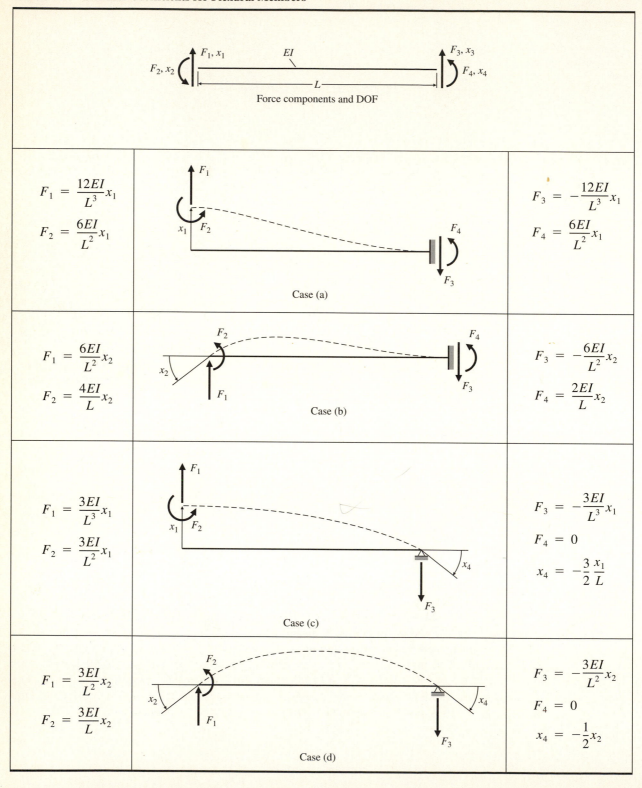

$$F_1 = \frac{12EI}{L^3}x_1$$

$$F_2 = \frac{6EI}{L^2}x_1$$

Case (a)

$$F_3 = -\frac{12EI}{L^3}x_1$$

$$F_4 = \frac{6EI}{L^2}x_1$$

$$F_1 = \frac{6EI}{L^2}x_2$$

$$F_2 = \frac{4EI}{L}x_2$$

Case (b)

$$F_3 = -\frac{6EI}{L^2}x_2$$

$$F_4 = \frac{2EI}{L}x_2$$

$$F_1 = \frac{3EI}{L^3}x_1$$

$$F_2 = \frac{3EI}{L^2}x_1$$

Case (c)

$$F_3 = -\frac{3EI}{L^3}x_1$$

$$F_4 = 0$$

$$x_4 = -\frac{3}{2}\frac{x_1}{L}$$

$$F_1 = \frac{3EI}{L^2}x_2$$

$$F_2 = \frac{3EI}{L}x_2$$

Case (d)

$$F_3 = -\frac{3EI}{L^2}x_2$$

$$F_4 = 0$$

$$x_4 = -\frac{1}{2}x_2$$

EXAMPLE 9.5 ▼

Determine the stiffness matrix for the single-story portal frame shown in Figure 9.8a. Assume the structure to have one translational (x_1) and two rotational (x_2 and x_3) DOF as illustrated in Figure 9.8b. Assume that the directions indicated by the displacements in Figure 9.7b are positive and that each member in the frame has flexural rigidity EI. Neglect the axial stiffness of the members.

Solution

Corresponding to the displacements x_1, x_2, and x_3 will be the resulting forces F_1, F_2, and F_3. Here, F_1 is a translational force and F_2 and F_3 are moments. Since the structure is defined with three DOF, the stiffness matrix will be 3×3. Thus the force-displacement relationship is given by

$$\begin{Bmatrix} F_1 \\ F_2 \\ F_3 \end{Bmatrix} = \begin{bmatrix} k_{11} & k_{12} & k_{13} \\ k_{21} & k_{22} & k_{23} \\ k_{31} & k_{32} & k_{33} \end{bmatrix} \begin{Bmatrix} x_1 \\ x_2 \\ x_3 \end{Bmatrix}$$

The first column of the stiffness matrix is obtained by setting $x_1 = 1.0$ and $x_2 = x_3 = 0$. The resulting deformations and forces are shown in Figure 9.9a, from which

$$k_{11} = \frac{12EI}{L^3} + \frac{3EI}{\left(\frac{3}{4}L\right)^3} = \frac{172EI}{9L^3}$$

$$k_{21} = -\frac{6EI}{L^2}$$

$$k_{31} = -\frac{3EI}{\left(\frac{3}{4}L\right)^2} = -\frac{16EI}{3L^2}$$

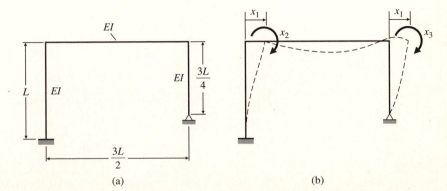

Figure 9.8
Three-DOF portal frame.

(a)   (b)

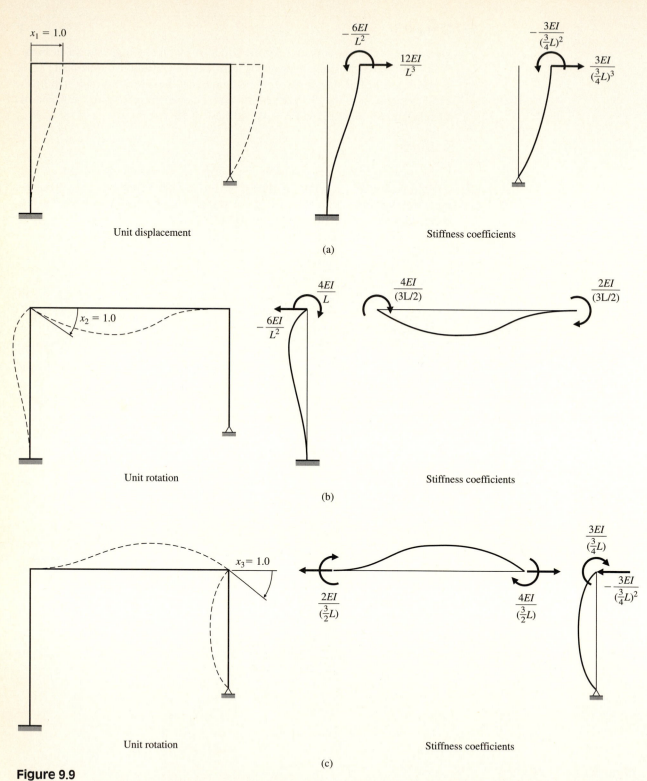

Figure 9.9

(a) Stiffness coefficients for first column of stiffness matrix; (b) stiffness coefficients for second column of stiffness matrix; (c) stiffness coefficients for third column of stiffness matrix.

The second column of the stiffness matrix is obtained by setting $x_2 = 1.0$ and $x_1 = x_3 = 0$. The resulting deformations and forces are shown in Figure 9.9b, from which

$$k_{12} = -\frac{6EI}{L^2}$$

$$k_{22} = \frac{4EI}{L} + \frac{4EI}{\left(\frac{3}{2}L\right)} = \frac{20EI}{3L}$$

$$k_{32} = \frac{2EI}{\left(\frac{3}{2}L\right)} = \frac{4EI}{3L}$$

Finally, the third column of the stiffness matrix is obtained by setting $x_3 = 1.0$ and $x_1 = x_2 = 0$. The resulting deformations and forces are shown in Figure 9.9c, yielding

$$k_{13} = -\frac{3EI}{\left(\frac{3}{4}L\right)^2} = -\frac{16EI}{3L^2}$$

$$k_{23} = \frac{2EI}{\left(\frac{3}{2}L\right)} = \frac{4EI}{3L}$$

$$k_{33} = \frac{4EI}{\left(\frac{3}{2}L\right)} + \frac{3EI}{\left(\frac{3}{4}L\right)} = \frac{20EI}{3L}$$

Thus, the system force-displacement relationship as expressed through the stiffness matrix is

$$\left\{\begin{array}{c} F_1 \\ F_2 \\ F_3 \end{array}\right\} = \frac{EI}{3L} \begin{bmatrix} \dfrac{172}{3L^2} & -\dfrac{18}{L} & -\dfrac{16}{L} \\[2mm] -\dfrac{18}{L} & 20 & 4 \\[2mm] -\dfrac{16}{L} & 4 & 20 \end{bmatrix} \left\{\begin{array}{c} x_1 \\ x_2 \\ x_3 \end{array}\right\}$$

Once again it is noted that the stiffness matrix is symmetric about the main diagonal. ▲

9.3 INERTIA PROPERTIES: MASS MATRIX

There are two basic formulations employed to construct the mass matrix: (1) the lumped mass formulation resulting in a *lumped mass matrix* and (2) the consistent mass formulation resulting in a *consistent mass matrix.* The lumped mass matrix is the simpler to construct and is more frequently employed, and is discussed in this chapter. The consistent mass matrix is constructed by a procedure similar to that used for obtaining the stiffness coefficients for a structure. The consistent mass formulation is generally used for *continuous* or *distributed parameter systems,* rather than for discrete systems. The consistent mass matrix is discussed in Chapter 14.

In the construction of a lumped mass matrix, the distributed mass properties of the structure are lumped (or localized) at the predefined node points, or joints, defining the DOF in the structure. The distribution of the mass properties to the various nodes is determined from statics. Figure 9.10 illustrates the procedure for determining the mass to be lumped at either node of a structural element for uniform, linear, and general mass distributions $\bar{m}$ (x).

The lumped mass matrix will usually be diagonal. It generally incorporates the inertia effect associated with translational DOF only. However, the inertial contributions for rotational DOF may be included if desired. For instance, for the beam segment having a uniform mass distribution $\bar{m}$, as shown in Figure 9.10a, the mass moment of inertia I_0 associated with a rotational DOF (refer to Table 2.1) at points 1 and 2 is

$$I_{0_1} = I_{0_2} = \frac{1}{3} \frac{\bar{m}L}{2} \left(\frac{L}{2}\right)^2 = \frac{\bar{m}L^3}{24} \tag{9.18}$$

Figure 9.10

Lumped mass coefficients for structural elements having distributed mass. *Structural Dynamics: Theory and Computation* by Mario Paz, copyright © 1980. Reprinted by permission of Kluwer Academic Publishers, Norwell, MA.

Mass Distribution	Lumped Mass Coefficients
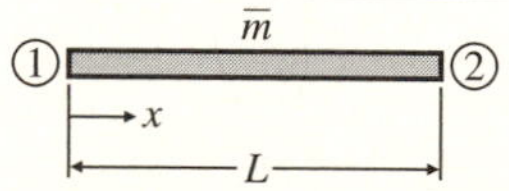 (a) Uniform: $\bar{m}(x) = \bar{m}$	$m_1 = \dfrac{\bar{m}L}{2}$ $m_2 = \dfrac{\bar{m}L}{2}$
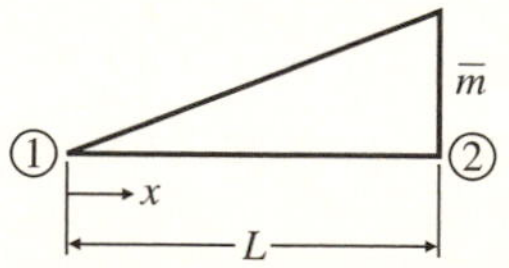 (b) Linear: $\bar{m}(x) = \dfrac{\bar{m}}{L} x$	$m_1 = \dfrac{\bar{m}L}{6}$ $m_2 = \dfrac{\bar{m}L}{3}$
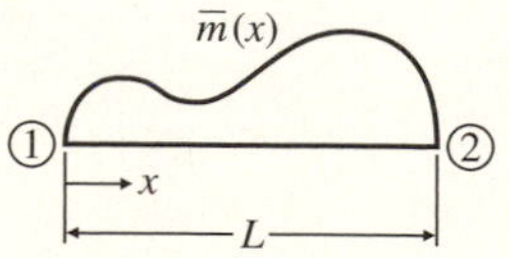 (c) General: $\bar{m}(x)$	$m_1 = \dfrac{\int_0^L \bar{m}(x)(L - x)\, dx}{\int_0^L \bar{m}(x)\, dx}$ $m_2 = \dfrac{\int_0^L \bar{m}(x)x\, dx}{\int_0^L \bar{m}(x)\, dx}$

EXAMPLE 9.6 ▼

The simply supported beam of length L shown in Figure 9.11a has uniform mass per unit length $\overline{m}$. The beam is to be analyzed as a structure having four translational DOF as shown in Figure 9.11b. Construct the lumped mass matrix.

Solution

The beam is subdivided into 5 equal segments of length $L/5$ as shown in Figure 9.11b. The amount of mass lumped (localized) at each node is equal to the uniform mass per unit length $\overline{m}$ times the length between adjacent nodes (on either side of the node). Thus

$$m_1 = m_2 = m_3 = m_4 = 2\,\frac{1}{2}\,\frac{L}{5}\,\overline{m} = \frac{\overline{m}L}{5}$$

Note that the total mass lumped at the four nodes equals $4\,\overline{m}\,L/5$, which is less than the total mass of the beam $\overline{m}\,L$. This is because the lumped mass formulation does not account for that portion of mass between the beam supports and halfway to the exterior masses m_1 and m_4.

The mass matrix relates the inertia forces to the system accelerations. That is,

$$(F_I)_i = m_i\ddot{x}_i$$

The beam is modeled with four DOF; therefore, the mass matrix will be 4×4 and is expressed as

$$
\begin{Bmatrix} (F_I)_1 \\ (F_I)_2 \\ (F_I)_3 \\ (F_I)_4 \end{Bmatrix}
=
\begin{bmatrix}
\dfrac{\overline{m}L}{5} & 0 & 0 & 0 \\
0 & \dfrac{\overline{m}L}{5} & 0 & 0 \\
0 & 0 & \dfrac{\overline{m}L}{5} & 0 \\
0 & 0 & 0 & \dfrac{\overline{m}L}{5}
\end{bmatrix}
\begin{Bmatrix} \ddot{x}_1 \\ \ddot{x}_2 \\ \ddot{x}_3 \\ \ddot{x}_4 \end{Bmatrix}
$$

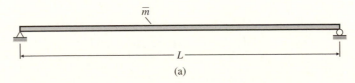

(a)

Figure 9.11

(a) Simply supported beam having uniformly distributed mass;
(b) lumped mass representation for four-DOF system.

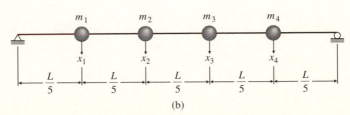

(b)

EXAMPLE 9.7 ▼

Reconstruct the mass matrix for the uniform beam of Example 9.6, but consider the inertia effects associated with the rotational DOF in addition to the translational DOF.

Solution

The beam with the additional DOF is shown in Figure 9.12. The mass moment of inertia I_0 associated with the rotational DOF for each node is calculated in an analagous manner expressed by Eq. (9.18) as

$$I_{0_5} = I_{0_6} = I_{0_7} = I_{0_8} = 2\left[\frac{1}{3}\frac{\overline{m}L}{10}\left(\frac{L}{10}\right)^2\right] = \frac{\overline{m}L^3}{1500}$$

Since the beam in this example has eight DOF, the mass matrix is 8×8. Therefore,

$$
\begin{Bmatrix} (F_I)_1 \\ (F_I)_2 \\ (F_I)_3 \\ (F_I)_4 \\ (F_I)_5 \\ (F_I)_6 \\ (F_I)_7 \\ (F_I)_8 \end{Bmatrix} =
\begin{bmatrix}
\frac{\overline{m}L}{5} & 0 & 0 & 0 & 0 & 0 & 0 & 0 \\
0 & \frac{\overline{m}L}{5} & 0 & 0 & 0 & 0 & 0 & 0 \\
0 & 0 & \frac{\overline{m}L}{5} & 0 & 0 & 0 & 0 & 0 \\
0 & 0 & 0 & \frac{\overline{m}L}{5} & 0 & 0 & 0 & 0 \\
0 & 0 & 0 & 0 & \frac{\overline{m}L^3}{1500} & 0 & 0 & 0 \\
0 & 0 & 0 & 0 & 0 & \frac{\overline{m}L^3}{1500} & 0 & 0 \\
0 & 0 & 0 & 0 & 0 & 0 & \frac{\overline{m}L^3}{1500} & 0 \\
0 & 0 & 0 & 0 & 0 & 0 & 0 & \frac{\overline{m}L^3}{1500}
\end{bmatrix}
\begin{Bmatrix} \ddot{x}_1 \\ \ddot{x}_2 \\ \ddot{x}_3 \\ \ddot{x}_4 \\ \ddot{x}_5 \\ \ddot{x}_6 \\ \ddot{x}_7 \\ \ddot{x}_8 \end{Bmatrix}
$$

It should be noted that the mass matrix must be of the same order as the stiffness matrix, reflecting the same number of DOF.

Figure 9.12

Simply supported beam having four translational DOF and four rotational DOF.

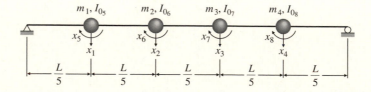

9.4 THE EIGENPROBLEM IN VIBRATION ANALYSIS

Analysis of undamped free vibration in n-DOF MDOF systems requires solution of the *standard eigenproblem.* The standard eigenproblem can be expressed as

$$[D]\{A\} = \lambda\{A\} \tag{9.19}$$

where $[D]$ = a square, symmetric matrix of order n
$\quad\{A\}$ = the solution vector
$\quad\lambda$ = a scalar

There exist n nontrivial solutions to Eq. (9.19). In this context nontrivial means that $\{A\}$ must not be a null vector for which Eq. (9.19) is always satisfied. The rth nontrivial solution is given by the *eigenvalue* λ_r and the corresponding *eigenvector* $\{A\}_r$. Thus

$$[D]\{A\}_r = \lambda_r\{A\}_r \tag{9.20}$$

Therefore, each solution consists of an *eigenpair,* of which there are n solutions given as $(\lambda_1,\{A\}_1), (\lambda_2,\{A\}_2), \ldots, (\lambda_n,\{A\}_n)$, where

$$\lambda_1 \le \lambda_2 \le \ldots \lambda_n \tag{9.21}$$

The proof that there must be n eigenpairs can be obtained by expressing Eq. (9.20) in the form

$$[[D] - \lambda[I]]\{A\} = \{0\} \tag{9.22}$$

where $[I]$ is the *identity matrix* and $[[D] - \lambda[I]]$ is a square matrix of order n and is called the *characteristic matrix* of $[D]$. A necessary and sufficient condition that Eq. (9.22) have a nontrivial solution is that the characteristic matrix be *singular,* or

$$|[D] - \lambda[I]| = 0 \tag{9.23}$$

Equation (9.23), when expanded, is called the *characteristic equation,* the n roots of which are the eigenvalues (λ's) of $[D]$. The corresponding n eigenvectors of $[D]$ are determined from Eq. (9.22).

In the free vibration analysis of structures, $[D]$ is referred to as the *system matrix,* or *dynamic matrix.* It can be formulated from either the stiffness matrix or the flexibility matrix. The n eigenvalues of the system matrix represent the natural frequencies of the structure, and the corresponding eigenvectors represent the normal, or principal, modes of vibration.

An eigenvector can be multiplied by an arbitrary constant. Thus, for $\lambda = \lambda_r$, Eq. (9.20) is also satisfied if $\{A\}_r$ is multiplied by a constant c_r; thus

$$[D]c_r\{A\}_r = \lambda_r c_r\{A\}_r \tag{9.24}$$

Simply stated, each element in $\{A\}_r$ is multiplied by the constant c_r. If $\{\Phi\}_r$ is used to denote $c_r\{A\}_r$, then a more general form of Eq. (9.20) is

$$[D]\{\Phi\}_r = \lambda_r\{\Phi\}_r \tag{9.25}$$

where $\{\Phi\}_r$ is the modal vector corresponding to λ_r. The relative values of the elements in the vector remain unchanged, and the eigenvector $\{A\}_r$ is merely *normalized* by the constant c_r.

Therefore, if all n λ's are considered, then the n number of modal vectors $\{\Phi\}_r$ form the *modal matrix* that describes all the normal modes of vibration for the system. The modal matrix $[\Phi]$ can be expressed in terms of the modal vectors $\{\Phi\}_r$ as

$$[\Phi]_{n \times n} = [\{\Phi\}_1, \{\Phi\}_2, \dots, \{\Phi\}_n] = \begin{bmatrix} \phi_{11} & \phi_{12} & \dots & \phi_{1n} \\ \phi_{21} & \phi_{22} & \dots & \phi_{2n} \\ \dots & \dots & \dots & \dots \\ \phi_{n1} & \phi_{n2} & \dots & \phi_{nn} \end{bmatrix} \quad (9.26)$$

9.4.1 Stiffness Based Formulation of the Eigenproblem

As previously discussed, formulation of the system matrix $[D]$ may be based upon either the stiffness matrix or the flexibility matrix. The stiffness based formulation is more frequently implemented for practical vibration analysis of MDOF systems since structural analysis computer programs are invariably based on the *direct stiffness method*.

For undamped free vibration of a MDOF system, the equations of motion in matrix form are expressed as

$$[m]\{\ddot{x}\} + [k]\{x\} = 0 \quad (9.27)$$

where $[m]$ and $[k]$ are the $n \times n$ mass and stiffness matrices, respectively, and $\{\ddot{x}\}$ and $\{x\}$ are the $n \times 1$ acceleration and displacement vectors, respectively. If Eq. (9.27) is premultiplied by the inverse of the mass matrix $[m]^{-1}$, then

$$[I]\{\ddot{x}\} + [D]\{x\} = 0 \quad (9.28)$$

where $[I]$ is the identity, or unit, matrix formed by the product $[m]^{-1}[m]$. The system, or dynamic, matrix $[D]$ is given by

$$[D] = [m]^{-1}[k] \quad (9.29)$$

Note that if the mass matrix is derived from a lumped mass formulation, it is usually a *diagonal* matrix and therefore its inversion is trivial.

If a harmonic solution is assumed, the displacement vector can be expressed as

$$\{x\} = \{A\}\sin \omega t \quad (9.30)$$

and the acceleration vector becomes

$$\{\ddot{x}\} = -\lambda\{A\}\sin \omega t \quad (9.31)$$

where $\lambda = \omega^2$. Substituting Eqs. (9.30) and (9.31) into Eq. (9.28), and dropping the sine term, yields

$$[[D] - \lambda[I]]\{A\} = \{0\}$$

which we recognize as the standard eigenproblem previously defined by Eq. (9.22).

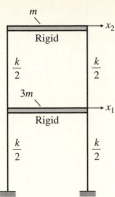

Figure 9.13
Two-story shear frame
building of Example 9.8.

EXAMPLE 9.8 ▼

Determine the natural frequencies and modal matrix for the two-story *shear frame building* (i.e., rigid girders) shown in Figure 9.13.

Solution

The equations of motion in matrix form are

$$
\begin{bmatrix} 3m & 0 \\ 0 & m \end{bmatrix} \begin{Bmatrix} \ddot{x}_1 \\ \ddot{x}_2 \end{Bmatrix} + \begin{bmatrix} 2k & -k \\ -k & k \end{bmatrix} \begin{Bmatrix} x_1 \\ x_2 \end{Bmatrix} = 0 \tag{1}
$$

Premultiplying the equations of motion by the inverse of the mass matrix given by

$$
[m]^{-1} = \begin{bmatrix} \dfrac{1}{3m} & 0 \\ 0 & \dfrac{1}{m} \end{bmatrix} \tag{2}
$$

yields

$$
[I]\{\ddot{x}\} + [D]\{x\} = 0 \tag{3}
$$

where the system matrix $[D]$ is given by Eq. (9.39),

$$
[D] = \begin{bmatrix} \dfrac{2k}{3m} & -\dfrac{k}{3m} \\ -\dfrac{k}{3m} & \dfrac{k}{m} \end{bmatrix} \tag{4}
$$

For $\{x\} = \{A\}$ and $\{\ddot{x}\} = -\lambda\{A\}$, where $\lambda = \omega^2$, Eq. (3) becomes

$$
\begin{bmatrix} \dfrac{2k}{3m} - \lambda & \dfrac{-k}{3m} \\ \dfrac{-k}{3m} & \dfrac{k}{m} - \lambda \end{bmatrix} \begin{Bmatrix} A_1 \\ A_2 \end{Bmatrix} = 0 \tag{5}
$$

Setting the determinant of the coefficient matrix in Eq. (4) equal to zero gives the characteristic equation

$$
\lambda^2 - \dfrac{5k}{3m}\lambda + \dfrac{k^2}{3m^2} = 0 \tag{6}
$$

The roots of Eq. (6) are the system eigenvalues

$$
\lambda_1 = 0.2317\dfrac{k}{m} \qquad \omega_1 = 0.4813\sqrt{\dfrac{k}{m}}
$$

$$
\lambda_2 = 1.345\dfrac{k}{m} \qquad \omega_2 = 1.198\sqrt{\dfrac{k}{m}}
$$

Figure 9.14

Normal modes for two-story shear frame building: (a) mode 1; (b) mode 2.

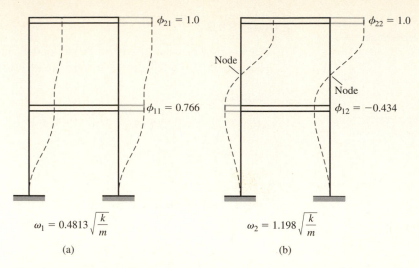

$$\omega_1 = 0.4813\sqrt{\frac{k}{m}}$$

(a)

$$\omega_2 = 1.198\sqrt{\frac{k}{m}}$$

(b)

The eigenvectors can be found from Eq. (5) by substituting the calculated values for λ into either submatrix equation. Thus, for $\lambda = \lambda_1 = 0.2317\ k/m$, from the first submatrix equation of Eq. (5)

$$0.435A_1 = 0.333A_2$$

or

$$\{\Phi\}_1 = \begin{Bmatrix} 0.766 \\ 1.000 \end{Bmatrix}$$

To obtain the second eigenvector, let $\lambda = \lambda_2 = 1.435\ k/m$, and from the first submatrix equation of Eq. (5)

$$-0.768A_1 = 0.333A_2$$

or

$$\{\Phi\}_2 = \begin{Bmatrix} -0.434 \\ 1.000 \end{Bmatrix}$$

The modal matrix is constructed by combining $\{\Phi\}_1$ and $\{\Phi\}_2$; thus

$$[\Phi] = \begin{bmatrix} 0.766 & -0.434 \\ 1.000 & 1.000 \end{bmatrix}$$

The mode shapes are illustrated in Figure 9.14. ▲

9.4.2 Flexibility Based Formulation of the Eigenproblem

As previously mentioned in this section, the system matrix may also be established from the flexibility matrix. Consider once again the matrix form of the equation of motion for a MDOF system undergoing free vibration:

$$[m]\{\ddot{x}\} + [k]\{x\} = \{0\} \tag{9.32}$$

Premultiplying Eq. (9.32) by the structure flexibility matrix $[a]$, which is equal to $[k]^{-1}$, gives

$$[a][m]\{\ddot{x}\} + [I]\{x\} = \{0\} \tag{9.33}$$

or

$$[D]\{\ddot{x}\} + [I]\{x\} = \{0\} \tag{9.34}$$

where $[I]$ represents the unit matrix and $[D]$ is the system, or dynamic, matrix given by

$$[D] = [a][m] \tag{9.35}$$

Assuming a harmonic solution for Eq. (9.34) of $\{x\} = \{A\}\sin \omega t$ and $\{\ddot{x}\} = -\omega^2\{A\}\sin \omega t$,

$$-\omega^2[D]\{A\} + [I]\{A\} = \{0\} \tag{9.36}$$

Equation (9.36) may be rewritten as

$$[[D] - \lambda[I]]\{A\} = \{0\} \tag{9.37}$$

where in this formulation λ is expressed as

$$\lambda = \frac{1}{\omega^2} \tag{9.38}$$

Equation (9.37) represents the standard eigenproblem; however, there is a subtle difference between this case and the eigenproblem based upon the stiffness formulation. For the flexibility based eigenproblem, the smallest eigenvalue does not correspond to the fundamental frequency ω_1. Because of the manner in which the eigenvalue λ is defined in Eq. (9.38), the smallest eigenvalue corresponds to the highest frequency, and vice versa. Therefore in the flexibility based characteristic equation, the n eigenvalues are defined as $\lambda_1 > \lambda_2 > \lambda_3 > \ldots \lambda_n$. In this manner λ_1, the largest eigenvalue, corresponds to ω_1, the lowest frequency, and so forth.

EXAMPLE 9.9 ▼

Determine the natural frequencies and the modal matrix for the beam shown in Figure 9.2a. Use the flexibility matrix obtained in Example 9.1 and assume $m_1 = m$ and $m_2 = 2m$.

Solution

The mass matrix is given by

$$[m] = \begin{bmatrix} m & 0 \\ 0 & 2m \end{bmatrix}$$

and the system matrix $[D]$ is determined from Eq. (9.35) as

$$[D] = \begin{bmatrix} \frac{8}{3} & \frac{5}{6} \\ \frac{5}{6} & \frac{1}{3} \end{bmatrix} \frac{L^3}{EI} \begin{bmatrix} 1 & 0 \\ 0 & 2 \end{bmatrix} m = \begin{bmatrix} \frac{8}{3} & \frac{5}{6} \\ \frac{5}{6} & \frac{1}{3} \end{bmatrix} \frac{mL^3}{EI} \tag{1}$$

The characteristic equation defined by Eq. (9.37) is

$$\frac{mL^3}{EI} \begin{bmatrix} \frac{8}{3} - \lambda\frac{EI}{mL^3} & \frac{5}{3} \\ \frac{5}{6} & \frac{2}{3} - \lambda\frac{EI}{mL^3} \end{bmatrix} \begin{Bmatrix} A_1 \\ A_2 \end{Bmatrix} = 0 \tag{2}$$

When the determinant of the coefficient matrix in Eq. (2) is set equal to zero, the characteristic equation is

$$\lambda^2 - \frac{10}{3}\frac{mL^3}{EI}\lambda + \frac{7}{9}\left(\frac{mL^3}{EI}\right)^2 = 0 \tag{3}$$

The roots of Eq. (3) are the system eigenvalues

$$\lambda_1 = 3.081\frac{mL^3}{EI}$$

$$\lambda_2 = 0.253\frac{mL^3}{EI}$$

Since $\lambda = 1/\omega^2$, the system natural frequencies are

$$\omega_1 = 0.5697\sqrt{\frac{EI}{mL^3}}$$

$$\omega_2 = 1.988\sqrt{\frac{EI}{mL^3}}$$

Substituting $\lambda = \lambda_1$ into Eq. (3) yields

$$0.414A_1 = 1.666A_2$$

or

$$\{\Phi\}_1 = \begin{Bmatrix} 1.000 \\ 0.248 \end{Bmatrix}$$

To obtain the second eigenvector, substitute $\lambda = \lambda_2$ into Eq. (3), which gives

$$2.413A_1 = -1.666A_2$$

or

$$\{\Phi\}_2 = \begin{Bmatrix} 1.000 \\ -1.448 \end{Bmatrix}$$

Figure 9.15

Normal modes for two-DOF cantilever beam: (a) mode 1; (b) mode 2.

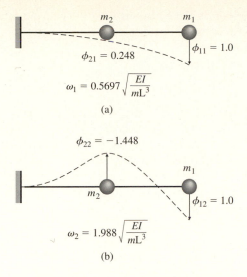

Thus the modal matrix is expressed as

$$[\Phi] = [\{\Phi\}_1 \{\Phi\}_2] = \begin{bmatrix} 1.000 & 1.000 \\ 0.248 & -1.448 \end{bmatrix}$$

The mode shapes are plotted in Figure 9.15. ▲

Solution of the eigenproblem for large MDOF systems can indeed become quite cumbersome. Eigenproblem solution routines generally involve some sort of vector iteration or polynomial iteration technique. Practical solution methods for the eigenproblem in MDOF systems are discussed in Chapter 11.

9.5 STATIC CONDENSATION OF THE STIFFNESS MATRIX

The mass matrix formulation for many structural dynamics problems include inertia terms corresponding to translational DOF only. At the same time, however, a more generalized formulation for the stiffness matrix [i.e., including terms corresponding to rotational degrees of freedom in the matrix, which is often the case in the *finite element method* (FEM) formulation], is often employed. Therefore, it is necessary to eliminate the extraneous degrees of freedom associated with rotation from the stiffness matrix before the equations of motion can be written. The procedure by which these unwanted DOF are eliminated (since they have no corresponding terms in the mass matrix) from the stiffness matrix is called *static condensation*. In effect, static condensation is used to reduce the number of DOF represented by the stiffness matrix to a specified level, without affecting the innate structural response characteristics of the system. Exclusion of rotational DOF in the mass matrix is justified because the kinetic energy component associated with these DOF are generally negligible in comparison to that corresponding to the translational DOF. Thus, if there are no other loads acting in the direction of the rotational DOF, the corresponding terms in the stiffness matrix should also be eliminated.

To establish the equations used in static condensation, assume that the stiffness matrix and corresponding displacement and force vectors are partitioned in the form

$$\begin{bmatrix} \mathbf{K}_{tt} & \mathbf{K}_{t\theta} \\ \mathbf{K}_{\theta t} & \mathbf{K}_{\theta\theta} \end{bmatrix} \begin{bmatrix} \mathbf{X} \\ \boldsymbol{\theta} \end{bmatrix} = \begin{bmatrix} \mathbf{F}_t \\ \mathbf{F}_\theta \end{bmatrix} = \begin{bmatrix} \mathbf{F}_t \\ 0 \end{bmatrix} \tag{9.39}$$

where $\mathbf{X}$ and $\boldsymbol{\theta}$ are the subvectors of the displacements to be retained (translational) and condensed out (rotational), respectively. The submatrices $\mathbf{K}_{tt}$, $\mathbf{K}_{t\theta}$, $\mathbf{K}_{\theta t}$, and $\mathbf{K}_{\theta\theta}$, and the force subvectors $\mathbf{F}_t$ and $\mathbf{F}_\theta$, correspond to the displacement subvectors $\mathbf{X}$ and $\boldsymbol{\theta}$.

If no other force subvectors acting in the structure include any rotational components, then the elastic rotational forces must also vanish, that is, subvector $\mathbf{F}_\theta = 0$. When this static constraint is introduced into Eq. (9.39), it is possible to express the rotational displacements by means of the second submatrix equation of Eq. (9.39) given by

$$\boldsymbol{\theta} = -\mathbf{K}_{\theta\theta}^{-1}\mathbf{K}_{\theta t}\mathbf{X} \tag{9.40}$$

The first submatrix equation of Eq. (9.39) yields

$$\mathbf{K}_{tt}\mathbf{X} + \mathbf{K}_{t\theta}\boldsymbol{\theta} = \mathbf{F}_t \tag{9.41}$$

Substituting the expression for the subvector $\boldsymbol{\theta}$ given by Eq. (9.40) into Eq. (9.41) gives

$$(\mathbf{K}_{tt} - \mathbf{K}_{t\theta}\mathbf{K}_{\theta\theta}^{-1}\mathbf{K}_{\theta t})\mathbf{X} = \mathbf{F}_t \tag{9.42}$$

Equation (9.39) can therefore be represented with the *condensed* stiffness matrix $\mathbf{K}_t$, which includes translational terms only, as follows:

$$\mathbf{K}_t\mathbf{X} = \mathbf{F}_t \tag{9.43}$$

where

$$\mathbf{K}_t = \mathbf{K}_{tt} - \mathbf{K}_{t\theta}\mathbf{K}_{\theta\theta}^{-1}\mathbf{K}_{\theta t} \tag{9.44}$$

This stiffness matrix is now suitable for use with the lumped mass matrix (which includes translational terms only) to write the equations of motion for translational vibration. Note that the relation between the translational and rotational degrees of freedom expressed by Eq. (9.44) is derived by establishing the *static* relation between them, hence the name static condensation method.

To illustrate the concept of static condensation, consider the portal frame shown in Figure 9.16a. The stiffness matrix formulated for this structure was based on the assumption that the frame exhibited three DOF: one translational, x_1, and two rotational, x_2 and x_3, as shown in Figure 9.16b. The force-displacement relationship for this structure, as expressed through the stiffness matrix, can be written as

$$\begin{Bmatrix} F_1 \\ F_2 \\ F_3 \end{Bmatrix} = \frac{EI}{L} \begin{bmatrix} \dfrac{15}{L^2} & \dfrac{-6}{L} & \dfrac{-3}{L} \\ \dfrac{-6}{L} & 8 & 2 \\ \dfrac{-3}{L} & 2 & 7 \end{bmatrix} \begin{Bmatrix} x_1 \\ x_2 \\ x_3 \end{Bmatrix} \tag{9.45}$$

Static condensation can be used on the stiffness matrix (3 × 3), to eliminate the two rotational DOF (x_2 and x_3). The condensed stiffness matrix will in this instance be a

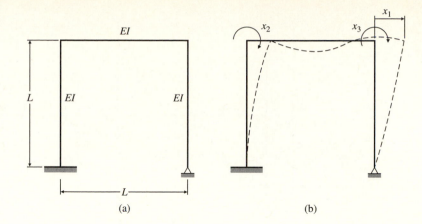

Figure 9.16
(a) Portal frame with flexible girder; (b) illustration of three-DOF.

single stiffness term (one DOF) that can be used to write the SDOF equation of motion for translational vibration of the frame. The expression for the condensed stiffness matrix is thus given by

$$\mathbf{K}_t = \mathbf{K}_{tt} - \mathbf{K}_{t\theta}\mathbf{K}_{\theta\theta}^{-1}\mathbf{K}_{\theta t} \qquad (9.46)$$

where

$$\mathbf{K}_{tt} = \left[\frac{15EI}{L^3}\right] \qquad \mathbf{K}_{\theta\theta} = \frac{EI}{L}\begin{bmatrix} 8 & 2 \\ 2 & 7 \end{bmatrix}$$

$$\mathbf{K}_{t\theta} = \frac{EI}{L}\begin{bmatrix} -\dfrac{6}{L} & -\dfrac{3}{L} \end{bmatrix} \qquad \mathbf{K}_{\theta t} = \frac{EI}{L}\begin{bmatrix} -\dfrac{6}{L} \\[2mm] -\dfrac{3}{L} \end{bmatrix}$$

The equation of motion for free vibration for the SDOF system defined for translational motion only is given as

$$m\ddot{x}_1 + k_t x_1 = 0 \qquad (9.47)$$

It should be noted that the condensed translational stiffness term k_t will be significantly different from the translational stiffness of the structure if calculated on the assumption of a rigid girder (shear frame). This is so because the condensed translational stiffness term k_t incorporates the effect of the flexible girder.

EXAMPLE 9.10 ▼

A two-story building frame is constructed of flexible columns and girders as shown in Figure 9.17a. The top-story girder supports a uniform dead load of 0.75 kip/ft and the bottom-story girder supports a dead load of 1.0 kip/ft. Each level of the building exhibits one translational and two rotational DOF as shown

Figure 9.17

(a) Two-story building frame having flexible girders; (b) illustration of six-DOF.

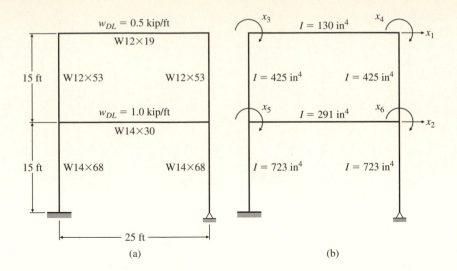

(a)   (b)

in Figure 9.17b. Assume all members bend about their strong axis and modulus of elasticity $E = 29{,}000$ ksi for all members.

(a) Construct the lumped mass matrix (include rotational DOF) [6 × 6].

(b) Construct the stiffness matrix [6 × 6].

(c) Through static condensation, eliminate the rotational terms from the stiffness matrix and construct the resultant [2 × 2] stiffness matrix.

(d) Calculate the two translational natural frequencies of the building frame using the [2 × 2] lump mass matrix and the [2 × 2] condensed stiffness matrix.

(e) Calculate the two (translational) natural frequencies of the building frame based upon the assumption of a shear building (i.e., rigid girders). Compare the results with those obtained in part (d).

Solution

(a) The lumped mass matrix will be a 6 × 6 diagonal matrix. The mass terms m_1 and m_2 are associated with the translational DOF x_1 and x_2, respectively. The remaining mass terms, designated I_{0_3} through I_{0_6} are associated with the rotational DOF x_3 through x_6, respectively. Thus,

$$m_1 = \frac{(0.5 \text{ kip/ft})(25 \text{ ft})}{386.4 \text{ in/sec}^2} = 0.03235 \text{ kip-sec}^2/\text{in}$$

$$m_2 = \frac{(1.0 \text{ kip/ft})(25 \text{ ft})}{386.4 \text{ in/sec}^2} = 0.0647 \text{ kip-sec}^2/\text{in}$$

$$I_{0_3} = I_{0_4} = \frac{1}{3}\left(\frac{0.03235}{2} \frac{\text{kip-sec}^2}{\text{in}}\right)\left[\frac{(25 \text{ ft})(12 \text{ in})}{2}\right]^2$$

$$= 121.31 \text{ kip-sec}^2\text{-in}$$

$$I_{0_5} = I_{0_6} = \frac{1}{3}\left(\frac{0.0647}{2} \frac{\text{kip-sec}^2}{\text{in}}\right)\left[\frac{(25 \text{ ft})(12 \text{ in})}{2}\right]^2 = 242.63 \text{ kip-sec}^2\text{-in}$$

and the mass matrix becomes

$$[m] = \begin{bmatrix} 0.03235 & 0 & 0 & 0 & 0 & 0 \\ 0 & 0.0647 & 0 & 0 & 0 & 0 \\ 0 & 0 & 121.31 & 0 & 0 & 0 \\ 0 & 0 & 0 & 121.31 & 0 & 0 \\ 0 & 0 & 0 & 0 & 242.63 & 0 \\ 0 & 0 & 0 & 0 & 0 & 242.63 \end{bmatrix} \tag{1}$$

(b) The stiffness matrix will be 6 × 6 and have the form

$$\begin{Bmatrix} F_1 \\ F_2 \\ F_3 \\ F_4 \\ F_5 \\ F_6 \end{Bmatrix} = \begin{bmatrix} k_{11} & k_{12} & k_{13} & k_{14} & k_{15} & k_{16} \\ & k_{22} & k_{23} & k_{24} & k_{25} & k_{26} \\ & & k_{33} & k_{34} & k_{35} & k_{36} \\ & & & k_{44} & k_{45} & k_{46} \\ & \text{(symmetric)} & & & k_{55} & k_{56} \\ & & & & & k_{66} \end{bmatrix} \begin{Bmatrix} x_1 \\ x_2 \\ x_3 \\ x_4 \\ x_5 \\ x_6 \end{Bmatrix} \tag{2}$$

In Eq. (2) the forces F_1 and F_2 are translational, and F_3 through F_6 are moments. The displacements x_1 and x_2 are translational, and the displacements x_3 through x_6 are rotations. The six columns of the stiffness matrix are constructed with the aid of Figure 9.18. Refer to Figure 9.18a; the elements of the first column of the stiffness matrix are

$$k_{11} = \frac{12EI}{L^3}(2) = \frac{12(29{,}000 \text{ ksi})(425 \text{ in}^4)}{(180 \text{ in})^3}(2) = 50.72 \text{ kips/in}$$

$$k_{21} = -k_{11} = -50.72 \text{ kips/in}$$

$$k_{31} = \frac{-6EI}{L^2} = \frac{(-6)(29{,}000 \text{ ksi})(425 \text{ in}^4)}{(180 \text{ in})^2} = -2282.41 \text{ kip-in/in}$$

$$k_{41} = k_{51} = k_{61} = k_{31} = -2282.41 \text{ kip-in/in}$$

To obtain the second column of the stiffness matrix, refer to Figure 9.18b; thus,

$$k_{12} = k_{21} = -50.72 \text{ kips/in}$$

$$k_{22} = \frac{12(29{,}000 \text{ ksi})}{(180 \text{ in})^3}[(425 \text{ in}^4)(2) + 723 \text{ in}^4]$$

$$+ \frac{3(29{,}000 \text{ ksi})(723 \text{ in}^4)}{(180 \text{ in})^3}$$

$$= 104.65 \text{ kips/in}$$

$$k_{32} = k_{42} = \frac{6(29{,}000 \text{ ksi})(425 \text{ in}^4)}{(180 \text{ in})^2} = 2282.41 \text{ kip-in/in}$$

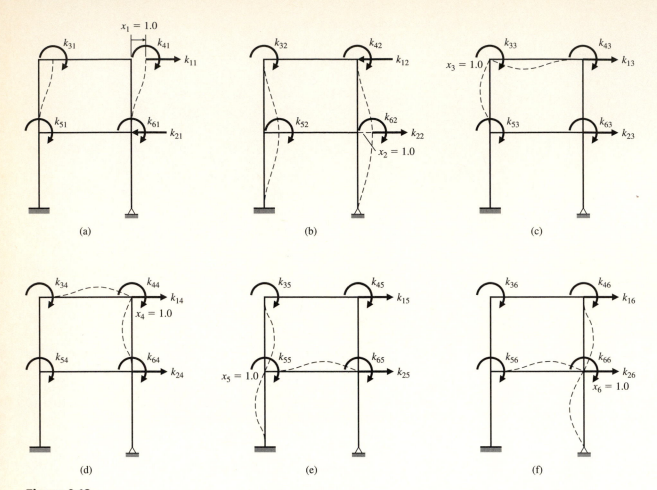

Figure 9.18

Construction of stiffness matrix of two-story building structure by columns: (a) column 1; (b) column 2; (c) column 3; (d) column 4; (e) column 5; (f) column 6.

$$k_{52} = \frac{6(29,000 \text{ ksi})}{(180 \text{ in})^2}(425 \text{ in}^4 - 723 \text{ in}^4) = -1600.37 \text{ kip-in/in}$$

$$k_{62} = \frac{6(29,000 \text{ ksi})(425 \text{ in}^4)}{(180 \text{ in})^2} - \frac{3(29,000 \text{ ksi})(723 \text{ in}^4)}{(180 \text{ in})^2}$$
$$= 341.02 \text{ kip-in/in}$$

Refer to Figure 9.18c; the third column of the stiffness matrix becomes

$$k_{13} = k_{31} = -2282.41 \text{ kip-in/in}$$

$$k_{23} = k_{32} = 2282.41 \text{ kip-in/in}$$

$$k_{33} = \frac{4(29{,}000 \text{ ksi})(130 \text{ in}^4)}{300 \text{ in}} + \frac{4(29{,}000 \text{ ksi})(425 \text{ in}^4)}{180 \text{ in}}$$
$$= 324{,}155 \text{ kip-in}^2/\text{in}$$

$$k_{43} = \frac{2(29{,}000 \text{ ksi})(130 \text{ in}^4)}{300 \text{ in}} = 25{,}133 \text{ kip-in}^2/\text{in}$$

$$k_{53} = \frac{2(29{,}000 \text{ ksi})(425 \text{ in}^4)}{180 \text{ in}} = 136{,}944 \text{ kip-in}^2/\text{in}$$

$$k_{63} = 0$$

The fourth column of the stiffness matrix is established from Figure 9.18d as

$$k_{14} = k_{41} = -2282.41 \text{ kip-in/in}$$

$$k_{24} = k_{42} = 2282.41 \text{ kip-in/in}$$

$$k_{34} = k_{43} = 25{,}133 \text{ kip-in}^2/\text{in}$$

$$k_{44} = k_{33} = 324{,}155 \text{ kip-in}^2/\text{in}$$

$$k_{54} = 0$$

$$k_{64} = k_{53} = 136{,}944 \text{ kip-in}^2/\text{in}$$

With the aid of Figure 9.18e, the fifth column of the stiffness matrix is

$$k_{15} = k_{51} = -2282.41 \text{ kip-in/in}$$

$$k_{25} = k_{52} = -1600.37 \text{ kip-in/in}$$

$$k_{35} = k_{53} = 136{,}944 \text{ kip-in}^2/\text{in}$$

$$k_{45} = k_{54} = 0$$

$$k_{55} = \frac{4(29{,}000 \text{ ksi})}{180 \text{ in}}(425 \text{ in}^4 + 723 \text{ in}^4) + \frac{4(29{,}000 \text{ ksi})(291 \text{ in}^4)}{300 \text{ in}}$$
$$= 852{,}342 \text{ kip-in}^2/\text{in}$$

$$k_{65} = \frac{2(29{,}000 \text{ ksi})(291 \text{ in}^4)}{300 \text{ in}} = 56{,}260 \text{ kip-in}^2/\text{in}$$

Finally, the last column of the stiffness matrix is obtained with the aid of Figure 9.18f,

$$k_{16} = k_{61} = -2282.41 \text{ kip-in/in}$$

$$k_{26} = k_{62} = 341.02 \text{ kip-in/in}$$

$$k_{36} = k_{63} = 0$$

$$k_{46} = k_{64} = 136,944 \text{ kip-in}^2/\text{in}$$

$$k_{56} = k_{65} = 56,260 \text{ kip-in}^2/\text{in}$$

$$k_{66} = \frac{4(29,000 \text{ ksi})(291 \text{ in}^4)}{300 \text{ in}} + \frac{4(29,000 \text{ ksi})(475 \text{ in}^4)}{180 \text{ in}}$$

$$+ \frac{3(29,000 \text{ ksi})(723 \text{ in}^4)}{180 \text{ in}} = 735,858 \text{ kip-in}^2/\text{in}$$

Thus, the complete 6×6 stiffness matrix is given from Eq. (2) as

$$[k] = \begin{bmatrix} 50.72 & -50.72 & -2282.41 & -2292.41 & -2282.41 & -2282.41 \\ & 104.65 & 2283.41 & 2282.41 & -1600.37 & 341.02 \\ & & 324,155 & 25,133 & 136,944 & 0 \\ & & & 324,155 & 0 & 136,944 \\ & \text{(symmetric)} & & & 852,342 & 56,260 \\ & & & & & 735,858 \end{bmatrix}$$

$$(3)$$

(c) The condensed stiffness matrix $\mathbf{K}_t$ is obtained from Eq. (9.44) as

$$\mathbf{K}_t = \mathbf{K}_{tt} - \mathbf{K}_{t\theta}\mathbf{K}_{\theta\theta}^{-1}\mathbf{K}_{\theta t} \tag{4}$$

where

$$\mathbf{K}_{tt} = \begin{bmatrix} 50.72 & -50.72 \\ -50.72 & 104.65 \end{bmatrix} \tag{5}$$

$$\mathbf{K}_{t\theta} = \begin{bmatrix} -2282.41 & -2282.41 & -2282.41 & -2282.41 \\ 2282.42 & 2282.41 & -1600.37 & 341.02 \end{bmatrix} \tag{6}$$

$$\mathbf{K}_{\theta t} = \mathbf{K}_{t\theta}^T \tag{7}$$

$$\mathbf{K}_{\theta\theta} = \begin{bmatrix} 324,155 & 25,133 & 136,944 & 0 \\ 25,133 & 324,155 & 0 & 136,944 \\ 136,944 & 0 & 852,342 & 56,260 \\ 0 & 136,944 & 56,260 & 735,858 \end{bmatrix} \tag{8}$$

$$\mathbf{K}_{\theta\theta}^{-1} = \begin{bmatrix} 0.3337 & -2.999 & -5.426 & 0.9729 \\ -2.999 & 33.77 & 0.9011 & -6.353 \\ -5.426 & 0.9011 & 12.68 & -1.137 \\ 0.9729 & -6.353 & -1.137 & 14.86 \end{bmatrix} \times 10^{-7} \tag{9}$$

Performing the required matrix operations for Eq. (4) results in

$$\mathbf{K}_{t\theta}\mathbf{K}_{\theta\theta}^{-1}\mathbf{K}_{\theta t} = \begin{bmatrix} 34.69 & -24.78 \\ -24.78 & 37.86 \end{bmatrix} \tag{10}$$

Therefore, the condensed stiffness matrix $\mathbf{K}_t$, as determined from Eq. (4), is

$$\mathbf{K}_t = \begin{bmatrix} 50.72 & -50.72 \\ -50.72 & 104.65 \end{bmatrix} - \begin{bmatrix} 34.69 & -24.78 \\ -24.78 & 37.86 \end{bmatrix}$$

$$= \begin{bmatrix} 16.03 & -25.94 \\ -25.94 & 66.79 \end{bmatrix} \text{ kips/in} \tag{11}$$

(d) System natural frequencies are based upon the condensed stiffness matrix (and mass matrix). The condensed mass matrix $[m]_t$ will include only the mass terms associated with the translational DOF. Thus, from Eq. (1) we have

$$[m]_t = \begin{bmatrix} 0.0325 & 0 \\ 0 & 0.0647 \end{bmatrix} \tag{12}$$

The equations of motion are therefore expressed using the condensed mass matrix given by Eq. (12) and the condensed stiffness matrix given by Eq. (11) as

$$\begin{bmatrix} 0.0325 & 0 \\ 0 & 0.0647 \end{bmatrix} \begin{Bmatrix} \ddot{x}_1 \\ \ddot{x}_2 \end{Bmatrix} + \begin{bmatrix} 16.03 & -25.94 \\ -25.94 & 66.79 \end{bmatrix} \begin{Bmatrix} x_1 \\ x_2 \end{Bmatrix} = \{0\} \tag{13}$$

The characteristic determinant as determined from Eq. (13) is

$$\begin{vmatrix} 16.03 - 0.0325\lambda & -25.94 \\ -25.94 & 66.79 - 0.0647\lambda \end{vmatrix} = 0 \tag{14}$$

and the characteristic equation resulting from expansion of Eq. (14) is

$$\lambda^2 - 1527.822\lambda + 190{,}039 = 0 \tag{15}$$

The roots of the characteristic equation are determined from solution of Eq. (15) as

$$\lambda_1 = 136.598 \tag{16a}$$

$$\lambda_2 = 1391.222 \tag{16b}$$

and the two natural frequencies corresponding to translational vibrations of the building frame with flexible girders are

$$\omega_1 = 11.69 \text{ rad/sec} \tag{17a}$$

$$\omega_2 = 37.30 \text{ rad/sec} \tag{17b}$$

(e) For the building considered as a shear frame building, the stiffness matrix for the two translational DOF is exactly equivalent to $\mathbf{K}_{tt}$ given by Eq. (5). Thus

$$[k] = \begin{bmatrix} 50.72 & -50.72 \\ -50.72 & 104.65 \end{bmatrix} \text{ kips/in} \tag{18}$$

and the characteristic determinant for this system is

$$\begin{vmatrix} 50.72 - 0.0325\lambda & -50.72 \\ -50.72 & 104.65 - 0.0647\lambda \end{vmatrix} = 0 \tag{19}$$

The characteristic equation as calculated from expansion of Eq. (19) is

$$\lambda^2 - 3185.316\lambda + 1,306,866 = 0 \tag{20}$$

and the roots of the characteristic equation are

$$\lambda_1 = 483.74 \tag{21a}$$

$$\lambda_2 = 2701.58 \tag{21b}$$

Thus, the corresponding natural frequencies for the *shear frame* building are

$$\omega_1 = 21.99 \text{ rad/sec} \tag{22a}$$

$$\omega_2 = 51.98 \text{ rad/sec} \tag{22b}$$

▲

Comparisons of the natural frequencies for the flexible girder structure and shear frame structure are presented in Table 9.2. It is obvious that the natural frequencies for the shear frame structure, Eqs. (22), are significantly higher than those for the flexible girder structure, Eqs. (17). This is because the rigid girder constraint of no rotation in the shear frame structure greatly increases the overall translational stiffness of the structure, thus yielding the higher frequencies.

TABLE 9.2. Comparison of Natural Frequencies Analysis Assumption

Natural Frequency (rad/sec)	Flexible Girder Structure	Shear Frame Structure	Percent Difference (%)
ω_1	11.69	21.99	+88
ω_2	37.30	51.98	+40

REFERENCES

1 Chajes, Alexander, *Structural Analysis*, 2nd ed., Prentice Hall, Englewood Cliffs, NJ, 1990.

2 McCormac, Jack and Elling, Rudolf E., *Structural Analysis, A Classical and Matrix Approach,* Harper Collins, New York, 1988.

3 Hibbeler, Russel C., *Structural Analysis*, 2nd ed., Macmillan, New York, 1990.

4 McGuire, William and Gallagher, Richard H., *Matrix Structural Analysis,* Wiley, New York, 1979.

5 Paz, Mario, *Structural Dynamics, Theory and Computation,* Van Nostrand Reinhold, New York, 1980.

6 Vanderbilt, M. Daniel, *Matrix Structural Analysis,* Quantum Publishers, New York, 1974.

7 West, Harry H., *Fundamentals of Structural Analysis,* Wiley, New York, 1993.

8 Leet, Kenneth M., *Fundamentals of Structural Analysis,* Macmillan, New York, 1988.

NOTATION

a_{ij}	flexibility coefficients	$[m]$	mass matrix
A	cross-sectional area	$[m]_t$	condensed mass matrix
c_r	scaling constant for rth normal mode	$\{x\}$	displacement vector
E	modulus of elasticity	$\{\ddot{x}\}$	acceleration vector
F_I	inertia force	$\mathbf{F}_t$	force subvector corresponding to $\mathbf{X}$
I	static moment of inertia	$\mathbf{F}_\theta$	force subvector corresponding to $\boldsymbol{\theta}$
I_0	mass moment of inertia	$\mathbf{K}_t$	condensed stiffness matrix
k_{ij}	stiffness coefficients	$\mathbf{K}_{tt}$	submatrix corresponding to $\mathbf{X}$ and $\mathbf{F}_t$
k_t	condensed translational stiffness term	$\mathbf{K}_{t\theta}$	submatrix corresponding to $\boldsymbol{\theta}$ and $\mathbf{F}_t$
L	length	$\mathbf{K}_{\theta t}$	submatrix corresponding to $\mathbf{X}$ and $\mathbf{F}_\theta$
m_{ij}	mass coefficients	$\mathbf{K}_{\theta\theta}$	submatrix corresponding to $\boldsymbol{\theta}$ and $\mathbf{F}_\theta$
$\overline{m}$	mass per unit length	$\mathbf{X}$	translational displacement subvector
M_i	bending moment at station i	$\boldsymbol{\theta}$	rotational displacement subvector
x_i	translational displacement at station i	DOF	degree of freedom
$[a]$	flexibility matrix	MDOF	multidegree of freedom
$\{A\}_r$	eigenvector for rth normal mode	λ_r	eigenvalue for rth normal mode
$[D]$	system matrix	ω	undamped natural circular frequency (rad/sec)
$\{F\}$	applied force vector	θ_i	rotational displacement at station i
$[I]$	identity matrix	$[\Phi]$	modal matrix
$[k]$	stiffness matrix	$\{\Phi\}_r$	modal vector for rth normal mode

PROBLEMS

9.1–9.2 For the structures shown in Figures P9.1 and P9.2, determine (a) the flexibility matrix, (b) the stiffness matrix, and (c) the lumped mass matrix.

9.3–9.10 For the structures shown in Figures P9.3 through P9.10, determine (a) the flexibility matrix and (b) the stiffness matrix. Assume modulus of elasticity $E = 29,000$ ksi and columns bend about their strong axis.

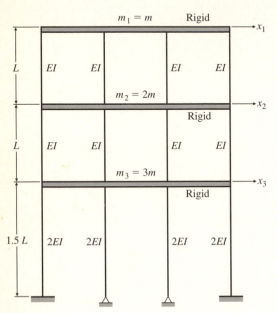

Figure P9.1

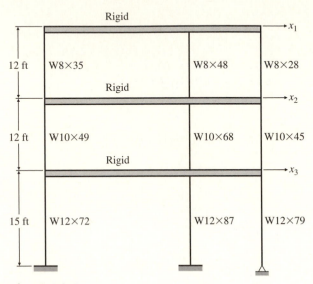

Figure P9.3

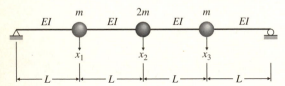

Figure P9.2

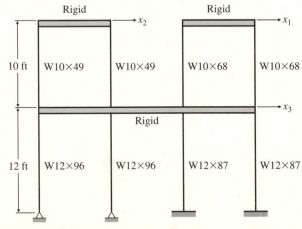

Figure P9.4

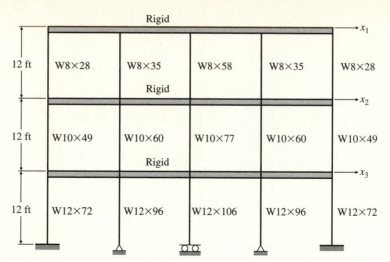

Figure P9.5

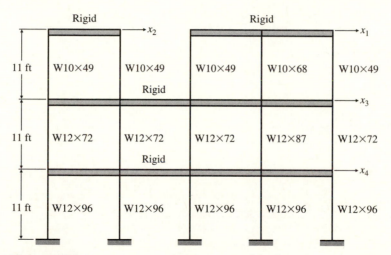

Figure P9.6

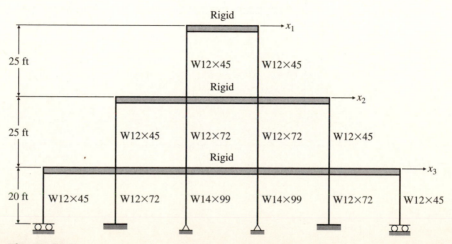

Figure P9.7

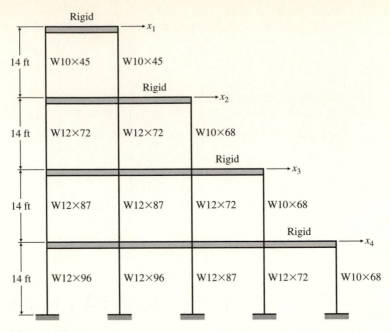

Figure P9.8

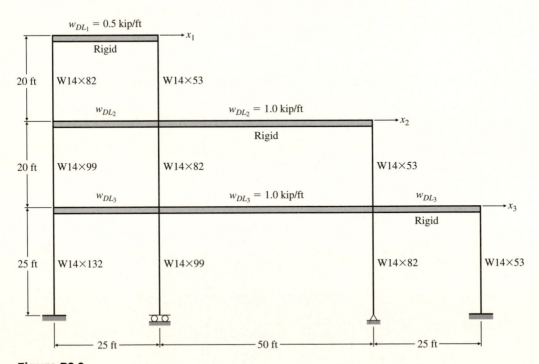

Figure P9.9

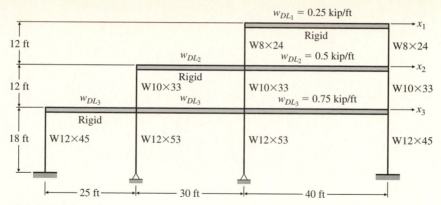

Figure P9.10

9.11 Determine the natural frequencies and mode shapes for the structure shown in Figure P9.9 using the stiffness based formulation.

9.12 Determine the natural frequencies and mode shapes for the structure shown in Figure P9.10 using the stiffness based formulation.

9.13 Repeat Problem 9.11 using the flexibility based formulation.

9.14 Repeat Problem 9.12 using the flexibility based formulation.

9.15–9.22 The building frame structures shown in Figures P9.11 through P9.18 are modeled with flexible girders and columns. Assume all members bend about their major axis, and modulus of elasticity $E = 29,000$ ksi.

(a) Determine the stiffness matrix for each structure. Include both translational and rotational DOF.

(b) Determine the lumped mass matrix for each structure. Include inertia terms for both translational and rotational motion.

(c) Write the equations of motion for each structure, including both translational and rotational motion.

(d) Eliminate the rotational DOF terms from the stiffness matrix by static condensation, and determine the condensed stiffness matrix for translational motion only.

(e) Write the equations of motion for the "condensed" system for translational motion only. Calculate the natural frequencies for translational motion.

(f) Determine the natural frequencies for translational motion if the building frames are modeled as shear frames (all girders rigid). Compare these frequencies with those obtained in part (e).

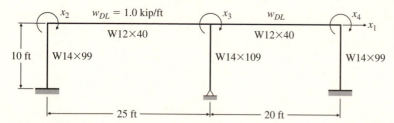

Figure P9.11

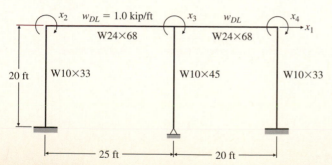

Figure P9.12

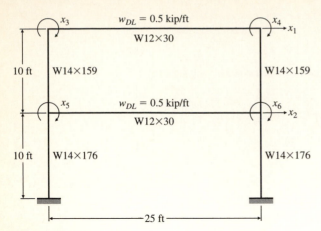

Figure P9.13

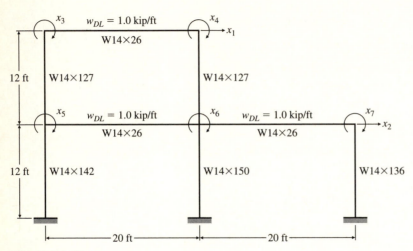

Figure P9.14

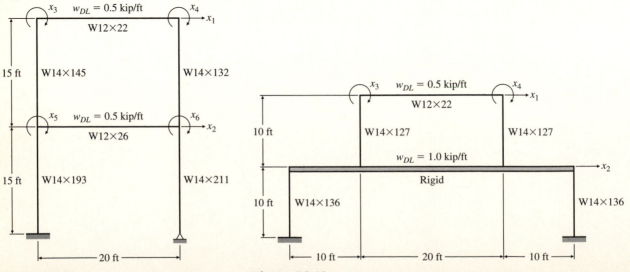

Figure P9.15   **Figure P9.16**

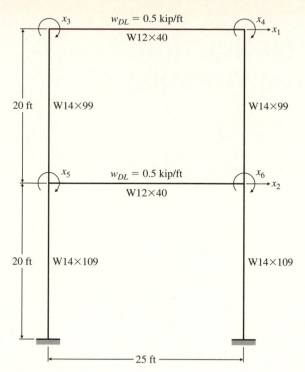

Figure P9.17

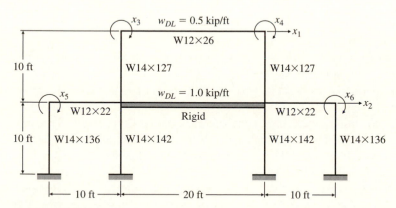

Figure P9.18

10 ▲ Equations of Motion and Undamped Free Vibration

In Chapter 1 it was stated that analytical models for the dynamic analysis of structural systems fell into two basic categories: *continuous models* and *discrete parameter models*. The discrete parameter model (often called a *lumped mass model* because the mass of the system is assumed to be represented by a finite number of point masses, or particles) exhibits a finite number of DOF, since the number of coordinates required to specify the system configuration at any instant of time is finite. It was also stated in Chapter 1 that the governing equations of motion for a discrete parameter model were ordinary differential equations.

A *continuous system* must be treated as being composed of an infinite number of differential elements, each having mass and elastic properties. Thus, a continuous model represents an infinite DOF system. Moreover, the displacement of the differential elements comprising a continuous model must be described by a continuous function of position and time. Therefore, the governing equations of motion for continuous systems are partial differential equations, for which exact solutions can be obtained for relatively few special (simple) cases.

Thus, while continuous models may give valuable insight into the dynamic behavior of a system having simple geometry (e.g., uniform bars and beams) and well defined boundary conditions, the dynamic analysis of most practical structures is based on discrete parameter MDOF models. In this and subsequent chapters (11 through 13), the analysis of discrete parameter MDOF models is presented. The dynamic analysis of continuous systems is addressed in Chapter 14.

This chapter begins with a derivation of Lagrange's equations from Hamilton's principle. Lagrange's equations provide the analyst with a systematic procedure for establishing the equations of motion for MDOF systems. Next is a discussion of systematic methodologies for the determination of the natural frequencies and normal vibration modes of MDOF systems. These procedures are applicable to both positive definite and positive semidefinite systems (i.e., systems admitting rigid body modes) and are feasible for manual calculation for systems exhibiting four DOF or less. The free vibration response of undamped MDOF systems to initial conditions is presented next, and the chapter concludes with a discussion of several approximate methods for estimating the fundamental frequency of MDOF systems.

10.1 HAMILTON'S PRINCIPLE AND THE LAGRANGE EQUATIONS

In many instances, Newton's second law is sufficient to obtain the equations of motion for MDOF systems. However, for complex MDOF systems a more rigorous and systematic method for deriving the equations of motion is often required. The *Lagrange equations* represent an energy method that allows the equations of motion to be written in terms of any set of *generalized coordinates* z_i, $i = 1, 2, \ldots, n$. Generalized coordinates are a set of independent parameters that completely define the system location and are independent of constraints.

The Lagrange equations may be derived from consideration of *Hamilton's principle*. Hamilton's principle is a general variational concept expressed as

$$\int_{t_1}^{t_2} \delta(T - V)\, dt + \int_{t_1}^{t_2} \delta W_{\text{nc}}\, dt = 0 \tag{10.1}$$

where

$$
\begin{aligned}
T &= \text{total kinetic energy}\\
V &= \text{potential energy of the system, including both strain energy and}\\
  &\quad\ \text{potential of any conservative external forces}\\
W_{\text{nc}} &= \text{work done by nonconservative forces acting on the system, including}\\
  &\quad\ \text{damping and any arbitrary external loads}\\
\delta &= \text{variation taken during the indicated time interval}
\end{aligned}
$$

For most structural mechanics problems the potential energy can be expressed in terms of the generalized coordinates, and the kinetic energy can be expressed in terms of the generalized coordinates and their first time derivatives (velocity). Moreover, the virtual work accomplished by the nonconservative forces acting through virtual displacements induced by arbitrary variations in the generalized coordinates can be expressed as a linear function of those variations. Therefore

$$V = V(z_1, z_2, \ldots, z_n) \tag{10.2}$$

$$T = T(z_1, z_2, \ldots, z_n, \dot{z}_1, \dot{z}_2, \ldots, \dot{z}_n) \tag{10.3}$$

$$\delta W_{\text{nc}} = Q_1 \delta z_1 + Q_2 \delta z_2 + \cdots + Q_n \delta z_n \tag{10.4}$$

where $Q_1, Q_2, \ldots, Q_n$ are the *generalized forces* corresponding to the respective generalized coordinate. Substituting Eqs. (10.2), (10.3), and (10.4) into Eq. (10.1) yields the expression

$$
\begin{aligned}
\int_{t_1}^{t_2} \Bigg(&\frac{\partial T}{\partial z_1} \delta z_1 + \frac{\partial T}{\partial z_2} \delta z_2 + \cdots + \frac{\partial T}{\partial z_n} \delta z_n + \frac{\partial T}{\partial \dot{z}_1} \delta \dot{z}_1 \\
&+ \frac{\partial T}{\partial \dot{z}_2} \delta \dot{z}_2 + \cdots + \frac{\partial T}{\partial \dot{z}_n} \delta \dot{z}_n - \frac{\partial V}{\partial z_1} \delta z_1 - \frac{\partial V}{\partial z_2} \delta z_2 \\
&- \cdots - \frac{\partial V}{\partial z_n} \delta z_n + Q_1 \delta z_1 + Q_2 \delta z_2 + \cdots + Q_n \delta z_n \Bigg) dt = 0
\end{aligned}
\tag{10.5}
$$

Integrating the velocity-related terms in Eq. (10.5) by parts results in the general expression

$$\int_{t_1}^{t_2} \frac{\partial T}{\partial \dot{z}_i} \delta \dot{z}_i \, dt = \left[\frac{\partial T}{\partial \dot{z}_i} \delta \dot{z}_i \right]_{t_1}^{t_2} - \int_{t_1}^{t_2} \frac{d}{dt} \left(\frac{\partial T}{\partial \dot{z}_i} \right) \delta z_i \, dt \tag{10.6}$$

To validate Hamilton's principle, the basic condition $\delta z_i\,(t_1) = \delta z_i\,(t_2) = 0$ must be imposed, thus making the first term on the right-hand side of Eq. (10.6) equal to zero. Then, substituting Eq. (10.6) into Eq. (10.1) yields

$$\int_{t_1}^{t_2} \left\{ \sum_{i=1}^{n} \left[-\frac{d}{dt}\left(\frac{\partial T}{\partial \dot{z}_i} \right) + \frac{\partial T}{\partial z_i} - \frac{\partial V}{\partial z_i} + Q_i \right] \delta z_i \right\} dt = 0 \tag{10.7}$$

Since the variations δz_i $(i = 1, 2, \ldots, n)$ are arbitrary and independent, the general satisfaction of Eq. (10.7) can be achieved only when the bracketed expression in Eq. (10.7) vanishes for all i. Thus we have

$$\frac{d}{dt}\left(\frac{\partial T}{\partial \dot{z}_i} \right) - \frac{\partial T}{\partial z_i} - \frac{\partial V}{\partial z_i} = Q_i \tag{10.8}$$

Since Eq. (10.8) was derived under the restrictions that the generalized coordinates z_i be arbitrary and independent and that T, V, and δW_{nc} are defined as indicated in Eqs. (10.2), (10.3), and (10.4), respectively, the Lagrange equations are thus applicable to any system satisfying these conditions. Therefore, the Lagrange equations can be employed for nonlinear as well as linear systems.

Since the potential energy V is not a function of the generalized velocities $\dot{z}_i$, Eq. (10.8) is often written as

$$\frac{d}{dt}\left(\frac{\partial \mathcal{L}}{\partial \dot{z}_i} \right) - \frac{\partial \mathcal{L}}{\partial \dot{z}_i} = Q_i - \frac{\partial v}{\partial z_i} \qquad i = 1, 2, \ldots, n \tag{10.9}$$

where $\mathcal{L} = T - V$ is called the *Lagrangian*. For a conservative system for which there is no damping or externally applied forces, $Q_i = 0$ and Eqs. (10.8) and (10.9), respectively, can be written as

$$\frac{d}{dt}\left(\frac{\partial T}{\partial \dot{z}_i} \right) - \frac{\partial T}{\partial z_i} + \frac{\partial V}{\partial z_i} = 0 \tag{10.10}$$

and

$$\frac{d}{dt}\left(\frac{\partial \mathcal{L}}{\partial \dot{z}_i} \right) - \frac{\partial \mathcal{L}}{\partial z_i} = -\frac{\partial v}{\partial z_i} \tag{10.11}$$

For discrete systems it is convenient to express the potential energy V as

$$V = \frac{1}{2} \sum_{i=1}^{n} \sum_{j=1}^{n} k_{ij} z_i z_j \tag{10.12}$$

where k_{ij} represents the specified element in the stiffness matrix. Similarly, the kinetic energy T can be expressed as

$$T = \frac{1}{2}\sum_{i=1}^{n}\sum_{j=1}^{n}m_{ij}\dot{z}_i\dot{z}_j \tag{10.13}$$

in which m_{ij} is the specified element in the mass matrix.

EXAMPLE 10.1 ▼

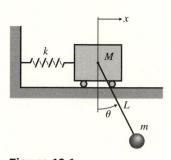

Figure 10.1

Simple pendulum attached to translational mass.

A simple pendulum of length L and end mass m pivots about a body of mass M that vibrates in a horizontal plane as shown in Figure 10.1. Use the Lagrange equations to determine the equations of motion for the system. Assume small oscillations.

Solution

The two generalized coordinates that completely define the motion of the system are $z_1 = x$ and $z_2 = \theta$. Therefore, there will be two Lagrange equations. The free-body diagrams for both the mass M and the pendulum are shown in Figure 10.2a and b, respectively.

(a) The kinetic energy in the system is given by

$$T = \frac{1}{2}M\dot{x}^2 + \frac{1}{2}m\dot{R}^2 \tag{1}$$

Referring to Figure 10.2b and using the law of cosines, the resultant velocity $\dot{R}$ of the mass m is

$$\dot{R}^2 = \dot{x}^2 + L^2\dot{\theta}^2 - 2\dot{x}L\dot{\theta}\cos\beta \tag{2}$$

where

$$\cos\beta = \cos(180° - \theta) = -\cos\theta \tag{3}$$

resulting in

$$\dot{R}^2 = \dot{x}^2 + L^2\dot{\theta}^2 - 2\dot{x}L\dot{\theta}\cos\theta \tag{4}$$

Figure 10.2

Free-body diagrams:
(a) translational mass;
(b) simple pendulum.

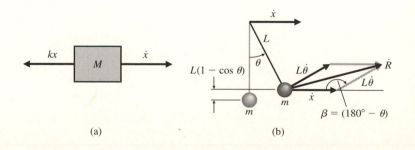

(a)   (b)

Substituting Eq. (4) into Eq. (1) gives

$$T = \frac{1}{2}M\dot{x}^2 + \frac{1}{2}m(\dot{x}^2 + L^2\dot{\theta}^2 - 2\dot{x}L\dot{\theta}\cos\theta) \tag{5}$$

or

$$T = \frac{1}{2}M\dot{z}_1^2 + \frac{1}{2}m(\dot{z}_1^2 + L^2\dot{z}_2^2 + 2\dot{z}_1L\dot{z}_2\cos z_2) \tag{6}$$

(b) The potential energy in the system is given by (referring to Figure 10.2a and b)

$$V = \frac{1}{2}kx^2 + mgL(1 - \cos\theta) \tag{7}$$

or

$$V = \frac{1}{2}kz_1^2 + mgL(1 - \cos z_2) \tag{8}$$

where g is the acceleration due to gravity. Since there are no nonconservative forces acting on the system, both Q_1 and Q_2 are zero.

(c) Substituting Eqs. (6) and (7) into the Lagrange equation (10.3), for $i = 1$ and $i = 2$, individually, yields the two equations of motion.
For $i = 1$, $z_i = z_1$:

$$\frac{\partial T}{\partial \dot{z}_1} = M\dot{z}_1 + m\dot{z}_1 + mL\dot{z}_2\cos z_2 \tag{9}$$

$$\frac{d}{dt}\left(\frac{\partial T}{\partial \dot{z}_1}\right) = M\ddot{z}_1 + m\ddot{z}_1 + mL\ddot{z}_2\cos z_2 - mL\dot{z}_2^2\sin z_2 \tag{10}$$

$$\frac{\partial T}{\partial z_1} = 0 \tag{11}$$

$$\frac{\partial V}{\partial z_1} = kz_1 \tag{12}$$

The equation of motion for the independent variable z_1 is obtained by substituting the expressions given by Eqs. (9) through (12) into the Lagrange equation given by

$$\frac{d}{dt}\left(\frac{\partial T}{\partial \dot{z}_1}\right) - \frac{\partial T}{\partial z_1} + \frac{\partial V}{\partial z_1} = 0$$

resulting in

$$(M + m)\ddot{z}_1 + mL\ddot{z}_2\cos z_2 - mL\dot{z}_2^2\sin z_2 + kz_1 = 0 \tag{13}$$

For small angles of oscillations, $\cos z_2 = 1$, $\sin z_2 = z_2$, and $\dot{z}_2^2 \rightarrow 0$. Therefore Eq. (13) reduces to

$$(M + m)\ddot{z}_1 + mL\ddot{z}_2 + kz_1 = 0 \tag{14}$$

In a similar manner, for $i = 2$ and $z_i = z_2$, the equation of motion for the independent variable z_2 can be determined from

$$\frac{d}{dt}\left(\frac{\partial T}{\partial \dot{z}_2}\right) - \frac{\partial T}{\partial z_2} + \frac{\partial V}{\partial z_2} = 0$$

which results in

$$\frac{\partial T}{\partial \dot{z}_2} = mL^2\dot{z}_2 + m\dot{z}_1 L \cos z_2 = mL^2\dot{z}_2 + m\dot{z}_1 L \tag{15}$$

$$\frac{d}{dt}\left(\frac{\partial T}{\partial \dot{z}_2}\right) = mL^2\ddot{z}_2 + m\ddot{z}_1 L \tag{16}$$

$$\frac{\partial T}{\partial z_2} = 0 \tag{17}$$

$$\frac{\partial V}{\partial z_2} = mgL \sin z_2 = mgLz_2 \tag{18}$$

Substitution of Eqs. (15) through (18) into the Lagrange equation for $z_i = z_2$ yields

$$mL\ddot{z}_2 + m\ddot{z}_1 - mgz_2 = 0 \tag{19}$$

Thus, Eqs. (14) and (19) represent the linearized differential equations of motion in generalized coordinates. Making the substitution $z_1 = x$ and $z_2 = \theta$, Eqs. (14) and (19) become, respectively,

$$(M + m)\ddot{x} + mL\ddot{\theta} + kx = 0 \tag{20}$$

and

$$m\ddot{x} + mL\ddot{\theta} + mg\theta = 0 \tag{21}$$

In matrix form, Eqs. (20) and (21) can be written as

$$\begin{bmatrix} (M + m) & mL \\ m & mL \end{bmatrix} \begin{Bmatrix} \ddot{x} \\ \ddot{\theta} \end{Bmatrix} + \begin{bmatrix} k & 0 \\ 0 & mg \end{bmatrix} \begin{Bmatrix} x \\ \theta \end{Bmatrix} = 0 \tag{22}$$

Note that the mass matrix in Eq. (22) possesses off-diagonal terms, indicating that there is *inertia coupling* of the equations of motion. However, since the stiffness matrix is diagonal, there is no *elastic coupling* of the equations. ▲

EXAMPLE 10.2 ▼

Use the Lagrange equations to determine the general equations of motion for the quadruple pendulum system shown in Figure 10.3. After obtaining the generalized equations of motion, linearize the equations.

Solution

The quadruple pendulum system is shown in an arbitrary displaced configuration in Figure 10.4. The x- and y-coordinate positions for each mass, along with their first time derivatives, can be expressed in terms of the set of generalized coordinates $z_1 = \theta_1$, $z_2 = \theta_2$, $z_3 = \theta_3$, and $z_4 = \theta_4$ as follows:

$$
\begin{aligned}
x_1 &= 1.4L \sin z_1 & \dot{x}_1 &= 1.4\dot{z}_1 \cos z_1 \\
y_1 &= 1.4L \cos z_1 & \dot{y}_1 &= 1.4L\dot{z}_1 \sin z_1 \\
x_2 &= L \sin z_2 & \dot{x}_2 &= L\dot{z}_2 \cos z_2 \\
y_2 &= L \cos z_2 & \dot{y}_2 &= -L\dot{z}_2 \sin z_2 \\
x_3 &= L \sin z_3 & \dot{x}_3 &= L\dot{z}_3 \cos z_3 \\
y_3 &= L \cos z_3 & \dot{y}_3 &= -L\dot{z}_3 \sin z_3 \\
x_4 &= L \sin z_2 + L \sin z_4 & \dot{x}_4 &= L\dot{z}_2 \cos z_2 + L\dot{z}_4 \cos z_4 \\
y_4 &= L \cos z_2 + L \cos z_4 & \dot{y}_4 &= -L\dot{z}_2 \sin z_2 - L\dot{z}_4 \sin z_4
\end{aligned}
\tag{1}
$$

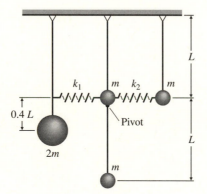

Figure 10.3

Quadruple pendulum system.

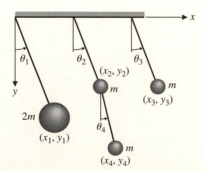

Figure 10.4

Quadruple pendulum system in arbitrary displaced configuration.

(a) The kinetic energy for the system is given by

$$T = \frac{1}{2}[2m(\dot{x}_1^2 + \dot{y}_1^2) + m(\dot{x}_2^2 + \dot{y}_2^2) + m(\dot{x}_3^2 + \dot{y}_3^2) + m(\dot{x}_4^2 + \dot{y}_4^2)] \quad (2)$$

Substituting Eqs. (1) into Eq. (2) yields the following expression for the kinetic energy:

$$T = \frac{mL^2}{2}[3.92\dot{z}_1^2 + 2\dot{z}_2^2 + \dot{z}_3^2 + \dot{z}_4^2 + 2\dot{z}_2\dot{z}_4 \cos(z_4 - z_2)] \quad (3)$$

(b) Reference to Figure 10.4 gives the potential energy in the system by

$$V = (1.4L - 1.4L \cos z_1)2mg + (L - L \cos z_2)mg$$
$$+ (L - L \cos z_3)mg + (L - L \cos z_4)mg$$
$$+ \frac{1}{2}k_1(L \sin z_1 - L \sin z_2)^2 + \frac{1}{2}k_2(L \sin z_2 - L \sin z_3)^2 \quad (4)$$

Expanding and simplifying Eq. (4) results in

$$V = mgL(5.8 - 2.8z_1 - \cos z_2 - \cos z_3 - \cos z_4)$$
$$+ \frac{k_1 L^2}{2}(\sin^2 z_1 - 2 \sin z_1 \sin z_2 + \sin^2 z_2)$$
$$+ \frac{k_2 L^2}{2}(\sin^2 z_2 - 2 \sin z_2 \sin z_3 + \sin^2 z_3) \quad (5)$$

(c) Substitution of Eqs. (3) and (5) into the Lagrange equations (10.3) for $i = 1$, $i = 2$, $i = 3$, and $i = 4$ separately, yields the four equations of motion.
For $i = 1$, $z_i = z_1$:

$$\frac{\partial T}{\partial \dot{z}_1} = 2\left(\frac{mL^2}{2}\right)(3.92\dot{z}_1) = 3.92mL^2\dot{z}_1$$

$$\frac{d}{dt}\left(\frac{\partial T}{\partial \dot{z}_1}\right) = 3.92mL^2\ddot{z}_1 \quad (6)$$

$$\frac{\partial T}{\partial z_1} = 0$$

$$\frac{\partial V}{\partial z_1} = 2.8mgL \sin z_1 + \frac{k_1 L^2}{2}(2 \sin z_1 \cos z_1 - 2 \cos z_1 \sin z_2)$$

Substitution of Eqs. (6) into the Lagrange equation for $z_i = z_1$ given by

$$\frac{d}{dt}\left(\frac{\partial T}{\partial \dot{z}_1}\right) - \frac{\partial T}{\partial z_1} + \frac{\partial V}{\partial z_1} = 0$$

yields the general equation of motion

$$3.92mL^2\ddot{z}_1 + 2.8mgL \sin z_1 + k_1 L^2(\sin z_1 - \sin z_2)\cos z_1 = 0 \quad (Ia)$$

Equation (Ia) can be linearized by noting that

$$\cos z_1 = 1 - \frac{z_1^2}{2} \cong 1 \quad \text{and} \quad \sin z_1 \cong z_1$$

resulting in

$$3.92 mL^2 \ddot{z}_1 + 2.8 mgL z_1 + k_1 L^2 (z_1 - z_2) = 0 \tag{Ib}$$

For $i = 2$, $z_i = z_2$:

$$\frac{\partial T}{\partial \dot{z}_2} = 2\left(\frac{mL^2}{2}\right)(2\dot{z}_2) + mL^2 \dot{z}_4 \cos(z_4 - z_2)$$

$$
\begin{aligned}
\frac{d}{dt}\left(\frac{\partial T}{\partial \dot{z}_2}\right) = {} & 2mL^2 \ddot{z}_2 + mL^2 \dot{z}_4 [-\sin(z_4 - z_2)](\dot{z}_4 - \dot{z}_2) \\
& + mL^2 \ddot{z}_4 \cos(z_4 - z_2)
\end{aligned}
\tag{7}
$$

$$\frac{\partial T}{\partial z_2} = mL^2 \dot{z}_2 \dot{z}_4 [-\sin(z_4 - z_2)](-1) = mL^2 \dot{z}_2 \dot{z}_4 \sin(z_4 - z_2)$$

$$
\begin{aligned}
\frac{\partial V}{\partial z_2} = {} & mgL \sin z_2 + k_1 L^2 (\sin z_2 - \sin z_1)\cos z_2 \\
& + k_2 L^2 (\sin z_2 - \sin z_3)\cos z_2
\end{aligned}
$$

Substitution of Eqs. (7) into the Lagrange equation for $z_i = z_2$ results in the general equation of motion

$$
\begin{aligned}
2mL^2 \ddot{z}_2 - {} & mL^2 \dot{z}_4 [\sin(z_4 - z_2)](\dot{z}_4 - \dot{z}_2) \\
& + mL^2 \ddot{z}_4 \cos(z_4 - z_2) - mL^2 \dot{z}_2 \dot{z}_4 [\sin(z_4 - z_2)] \\
& + mgL \sin z_2 + k_1 L^2 (\sin z_2 - \sin z_1)\cos z_2 \\
& + k_2 L^2 (\sin z_2 - \sin z_3)\cos z_2 = 0
\end{aligned}
\tag{IIa}
$$

Equation (IIa) may be linearized by noting that

$$
\begin{array}{lll}
\cos(z_4 - z_2) = 1 & \cos z_2 = 1 & \sin z_2 = z_2 \\
\sin(z_4 - z_2) = 0 & \sin z_1 = z_1 & \sin z_3 = z_3
\end{array}
$$

which results in

$$
\begin{aligned}
& 2mL^2 \ddot{z}_2 + mL^2 \ddot{z}_4 + mgL z_2 + k_1 L^2 (z_2 - z_1) \\
& + k_2 L^2 (z_2 - z_3) = 0
\end{aligned}
\tag{IIb}
$$

For $i = 3$, $z_i = z_3$:

$$\frac{\partial T}{\partial \dot{z}_3} = mL^2 \dot{z}_3$$

$$\frac{d}{dt}\left(\frac{\partial T}{\partial \dot{z}_3}\right) = mL^2 \ddot{z}_3 \tag{8}$$

$$\frac{\partial T}{\partial z_3} = 0$$

$$\frac{\partial V}{\partial z_3} = mgL \sin z_3 - k_2 L^2 \sin z_2 \cos z_3 - k_2 L^2 \sin z_3 \cos z_3$$

Substituting Eqs. (8) into the Lagrange equation for $z_i = z_3$ gives the general equation of motion as

$$mL^2 \ddot{z}_3 + mgL \sin z_3 + k_2 L^2 (\sin z_3 - \sin z_2)\cos z_3 = 0 \tag{IIIa}$$

Equation (IIIa) can be linearized by noting that $\sin z_2 = z_2$, $\sin z_3 = z_3$, and $\cos z_3 = 1$, resulting in

$$mL^2 \ddot{z}_3 + mgL z_3 + k_2 L^2 (z_3 - z_2) = 0 \tag{IIIb}$$

Finally, for $i = 4$, $z_i = z_4$:

$$\frac{\partial T}{\partial \dot{z}_4} = mL^2 \dot{z}_4 + mL^2 \dot{z}_2 \cos(z_4 - z_2)$$

$$\frac{d}{dt}\left(\frac{\partial T}{\partial \dot{z}_4}\right) = mL^2 \ddot{z}_4 + mL^2 \dot{z}_2 [-\sin(z_4 - z_2)](\dot{z}_4 - \dot{z}_2)$$
$$+ \ mL^2 \ddot{z}_2 \cos(z_4 - z_2) \tag{9}$$

$$\frac{\partial T}{\partial z_4} = -mL^2 \dot{z}_2 \dot{z}_4 \sin(z_4 - z_2)$$

$$\frac{\partial V}{\partial z_4} = mgL \sin z_4$$

Substitution of Eqs. (9) into the Lagrange equation for $z_i = z_4$ gives the general equation of motion

$$mL^2 \ddot{z}_4 - mL^2 \dot{z}_2 \sin(z_4 - z_2)(\dot{z}_4 - \dot{z}_2)$$
$$+ \ mL^2 \ddot{z}_2 \cos(z_4 - z_2) + mL^2 \dot{z}_2 \dot{z}_4 \sin(z_4 - z_2) \tag{IVa}$$
$$+ \ mgL \sin z_4 = 0$$

Note that since $\sin(z_4 - z_2) = 0$, $\cos(z_4 - z_2) = 1$, and $\sin z_4 = z_4$, for small angles of rotation, the linearized equation is

$$mL^2 \ddot{z}_4 + mL^2 \ddot{z}_2 + mgL z_4 = 0 \tag{IVb}$$

Equations (Ib), (IIb), (IIIb), and (IVb) thus represent the linearized equations of motion for the quadruple pendulum system in generalized coordinates. Substituting $z_1 = \theta_1$, $z_2 = \theta_2$, $z_3 = \theta_3$, and $z_4 = \theta_4$ into Eqs. (Ib), (IIb), (IIIb), and

(IVb), respectively, gives the four equations of motion expressed in physical coordinates as

$$3.92mL^2\ddot{\theta}_1 + 2.8mgL\theta_1 + k_1L^2(\theta_1 - \theta_2) = 0 \tag{Va}$$

$$mL^2(2\ddot{\theta}_2 + \ddot{\theta}_4) + mgL\theta_2 + k_1L^2(\theta_2 - \theta_1)$$
$$+ k_2L^2(\theta_2 - \theta_3) = 0 \tag{Vb}$$

$$mL^2\ddot{\theta}_3 + mgL\theta_3 + k_2L^2(\theta_3 - \theta_2) = 0 \tag{Vc}$$

$$mL^2(\ddot{\theta}_2 + \ddot{\theta}_4) + mgL\theta_4 = 0 \tag{Vd}$$

In matrix form the equations can be written as

$$
\begin{bmatrix}
3.92mL^2 & 0 & 0 & 0 \\
0 & 2mL^2 & 0 & mL^2 \\
0 & 0 & mL^2 & 0 \\
0 & mL^2 & 0 & mL^2
\end{bmatrix}
\begin{Bmatrix}
\ddot{\theta}_1 \\
\ddot{\theta}_2 \\
\ddot{\theta}_3 \\
\ddot{\theta}_4
\end{Bmatrix}
$$
$$
+ \begin{bmatrix}
2mgL + k_1L^2 & -k_1L^2 & 0 & 0 \\
-k_1L^2 & (k_1 + k_2)L^2 & -k_2L^2 & 0 \\
0 & -k_2L^2 & k_2L^2 & 0 \\
0 & 0 & 0 & mgL
\end{bmatrix}
\begin{Bmatrix}
\theta_1 \\
\theta_2 \\
\theta_3 \\
\theta_4
\end{Bmatrix} = 0 \tag{VI}
$$

▲

Although the calculations for this example appear to be somewhat tedious, the derivation of the equations of motion by Newton's second law would be much more cumbersome.

10.2 NATURAL VIBRATION FREQUENCIES

The equations of motion for an undamped system having n DOF can be expressed in matrix form as

$$[m]\{\ddot{x}\} + [k]\{x\} = \{0\} \tag{10.14}$$

where $[m]$ and $[k]$ are ($n \times n$) mass and stiffness matrices, respectively, and $\{x(t)\}$ is an ($n \times 1$) vector of physical displacement coordinates.

The problem of vibration analysis consists of determining the conditions under which Eq. (10.14) will permit motions to occur. By analogy with the behavior of SDOF systems, it will be assumed that free vibration motion is simple harmonic, which may be expressed for a MDOF system as

$$\{x(t)\} = \{A\}\sin(\omega t + \phi) \tag{10.15}$$

In Eq. (10.15), $\{A\}$ represents the shape of the system (which does not change with time; only the amplitude varies) and ϕ is the phase angle. When the second time derivative of Eq. (10.15) is taken, the accelerations in free vibration are

$$\{\ddot{x}(t)\} = -\omega^2\{A\}\sin(\omega t + \phi) = -\omega^2\{x(t)\} \tag{10.16}$$

Substituting Eqs. (10.15) and (10.16) into Eq. (10.14) gives

$$-\omega^2[m]\{A\}\sin(\omega t + \phi) + [k]\{A\}\sin(\omega t + \phi) = \{0\} \tag{10.17}$$

which (since the sine term is arbitrary and may be omitted) may be written as

$$[[k] - \omega^2[m]]\{A\} = \{0\} \tag{10.18}$$

Premultiplying Eq. (10.18) by the inverse of the mass matrix $[m]^{-1}$ yields

$$[[D] - \lambda[I]]\{A\} = \{0\} \tag{10.19}$$

which we recognize as the standard eigenproblem discussed in Chapter 9. In Eq. (10.19), $[I]$ is the *unit diagonal matrix* or *identity matrix,* $[D]$ is the *system matrix* or *dynamic matrix* given by

$$[D] = [m]^{-1}[k] \tag{10.20}$$

The matrix $[[D] - \lambda[I]]$ is a square matrix of order n and is referred to as the *characteristic matrix* of $[D]$. A necessary and sufficient condition that Eq. (10.19) have a nontrivial solution is that the characteristic matrix be singular, or

$$\|[D] - \lambda[I]\| = 0 \tag{10.21}$$

Equation (10.21), when expanded, is called the *frequency equation* (or *characteristic equation*) of the system.

When the determinant of Eq. (10.19), given by Eq. (10.21), is expanded, there results a polynomial equation of degree n in λ or $p(\lambda)$ whose n roots (or zeros) are the *eigenvalues,* or squared natural frequencies, such that $\lambda = \omega^2$. (A typical plot of the characteristic polynomial is illustrated in Figure 10.5.) The n roots of the characteristic equation are generally ordered from lowest to highest such that

$$0 \leq \lambda_1 \leq \lambda_2 \leq \dots \lambda_r \leq \dots \lambda_n \tag{10.22}$$

Figure 10.5

Illustration of typical characteristic polynomial for MDOF system.

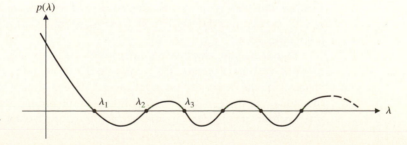

Corresponding to each eigenvalue λ_r is an *eigenvector,* or *natural mode,* representing the displaced configuration of the system for the rth mode. The eigenvector $\{A\}_r$ is expressed as

$$\{A\}_r = \begin{Bmatrix} A_1 \\ A_2 \\ A_3 \\ \cdot \\ \cdot \\ \cdot \\ A_n \end{Bmatrix}_r \qquad r = 1, 2, \ldots, n \tag{10.23}$$

These vectors are unique only in the sense that the ratio between any two elements within the vector, such as A_{ir} and A_{jr}, is constant. However, the value of the elements themselves is arbitrary.

10.3 NATURAL VIBRATION MODES

Since the system eigenvectors, or natural modes, as represented by Eq. (10.23), are determined only to within a constant multiplier, they can be multiplied by any arbitrary constant. Noting that Eq. (10.19) can be written as

$$[D]\{A\} = \lambda\{A\} \tag{10.24}$$

then for the rth mode

$$[D]\{A\}_r = \lambda_r\{A\}_r \tag{10.25}$$

Equation (10.25) is also satisfied if $\{A\}_r$ is multiplied by a constant c_r. Therefore,

$$[D]c_r\{A\}_r = \lambda_r c_r\{A\}_r \tag{10.26}$$

which simply implies that each element in $\{A\}_r$ is multiplied by the constant c_r. If we replace $c_r\{A\}_r$ in Eq. (10.26) with $\{\Phi\}_r$, then a more general form of Eq. (10.26) is

$$[D]\{\Phi\}_r = \lambda_r\{\Phi\}_r \tag{10.27}$$

where $\{\Phi\}_r$ is the *modal vector* corresponding to λ_r. The relative values of the elements within the vector remain unchanged, and the eigenvector $\{A\}_r$ is merely *normalized* by the constant c_r. Thus, although the shape of the natural mode $\{\Phi\}_r$ is unique, the amplitude is not.

One frequently employed normalization scheme consists of setting the largest element of the modal vector $\{\Phi\}_r$ equal to 1, which is convenient for plotting mode shapes. Another often used normalization scheme is merely to set the first element in the modal vector ϕ_{1r} equal to 1. Regardless of the normalization scheme employed, the process bears no physical significance and is merely a convenience.

The *modal matrix* $[\Phi]$ contains the n modal vectors $\{\Phi\}_r$ and describes all the normal modes of vibration for the system. The modal matrix can be expressed in terms of the modal vectors $\{\Phi\}_r$ as

$$[\Phi]_{n \times n} = [\{\Phi\}_1, \{\Phi\}_2, ..., \{\Phi\}_n]$$

$$= \begin{bmatrix} \phi_{11} & \phi_{21} & \cdots & \phi_{1n} \\ \phi_{21} & \phi_{22} & & \phi_{2n} \\ \cdots & \cdots & \cdots & \cdots \\ \phi_{n1} & \phi_{n2} & \cdots & \phi_{nn} \end{bmatrix} \tag{10.28}$$

For very large MDOF systems, solution of the system eigenproblem (that is, the extraction of natural frequencies and mode shapes) given by Eq. (10.19) must be accomplished via a numerical eigensolver. Several commonly employed numerical eigensolvers are discussed in Chapter 11. However, for systems having four DOF or less, manual solution of the eigenproblem is still feasible.

Consider the characteristic matrix of Eq. (10.18). For the rth mode, write

$$[H]_r = [[k] - \omega_r^2 [m]] \tag{10.29}$$

where $[H]_r$ is the characteristic matrix for the rth mode. With the characteristic equations for λ_r or ω_r^2 having been solved, the corresponding modal vector $\{\Phi\}_r$ is sought. To this end Eq. (10.19) is written as

$$[H]_r \{\Phi\}_r = \{0\} \tag{10.30}$$

To facilitate the solution for $\{\Phi\}_r$, partition Eq. (10.30) in the following manner:

$$\begin{bmatrix} \mathbf{H}_{11} & \mathbf{H}_{12} \\ \mathbf{H}_{21} & \mathbf{H}_{22} \end{bmatrix}_r \begin{Bmatrix} 1.0 \\ \{\Phi\}_c \end{Bmatrix}_r = \begin{Bmatrix} 0 \\ \{0\} \end{Bmatrix} \tag{10.31}$$

where the modal vector $\{\Phi\}_r$ has been scaled by setting $\phi_{1r} = 1.0$, $\mathbf{H}_{11}$ through $\mathbf{H}_{22}$ represent the partitioned submatrices of the characteristic matrix $[H]_r$, and $\{\Phi\}_{cr}$ is the *complementary subvector* containing the remaining $n - 1$ unknown terms of $\{\Phi\}_r$. The complementary subvector $\{\Phi\}_{cr}$ may be determined from the second submatrix equation of Eq. (10.31). Therefore, we have for the rth mode

$$\mathbf{H}_{21} + \mathbf{H}_{22} \{\Phi\}_{cr} = \{0\} \tag{10.32}$$

or

$$\{\Phi\}_{cr} = -\mathbf{H}_{22}^{-1} \mathbf{H}_{21} \tag{10.33}$$

This procedure may be used to obtain all or just a few of the modal vectors.

Note that if ω_r is a *distinct frequency,* that is, there are no repeated roots to the characteristic equation such that $\omega_r = \omega_{r+1} = \omega_{r+2} = ...$, then $\mathbf{H}_{11}$ is a (1×1) matrix and $\mathbf{H}_{22}$ is a $(n - 1) \times (n - 1)$ matrix. Moreover, $\mathbf{H}_{22}$ is *positive definite* and its inverse exists.

EXAMPLE 10.3 ▼

Determine the natural frequencies and natural modes of vibration for the three-story shear frame shown in Figure 10.6. Assume modulus of elasticity $E = 29,000$ ksi and moment of inertia $I = 600$ in^4 for all columns in the frame.

Solution

(a) First we establish the mass and stiffness properties of the building frame by considering the simplified mechanical model presented in Figure 10.7.

(i) Mass terms:

$$m_1 = 0.8 \text{ kip/ft} \frac{20 \text{ ft}}{386.4 \text{ in/sec}^2} = 0.04141 \text{ kip-sec}^2/\text{in}$$

$$m_2 = 0.75 \text{ kip/ft} \frac{20 \text{ ft}}{386.4 \text{ in/sec}^2} = 0.03882 \text{ kip-sec}^2/\text{in}$$

$$m_3 = 0.5 \text{ kip/ft} \frac{20 \text{ ft}}{386.4 \text{ in/sec}^2} = 0.02588 \text{ kip-sec}^2/\text{in}$$

(ii) Stiffness terms:

$$k_1 = \frac{12(29,000 \text{ ksi})(2)(600 \text{ in}^4)}{(15 \text{ ft} \times 12 \text{ in})^3} + \frac{3(29,000 \text{ ksi})(2)(600 \text{ in}^4)}{(15 \text{ ft} \times 12 \text{ in})^3}$$
$$= 89.506 \text{ kips/in}$$

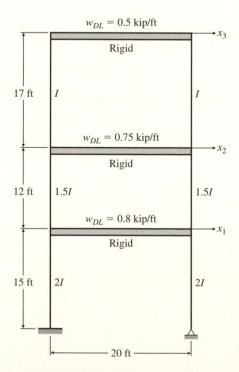

Figure 10.6

Three-story shear frame building of Example 10.3.

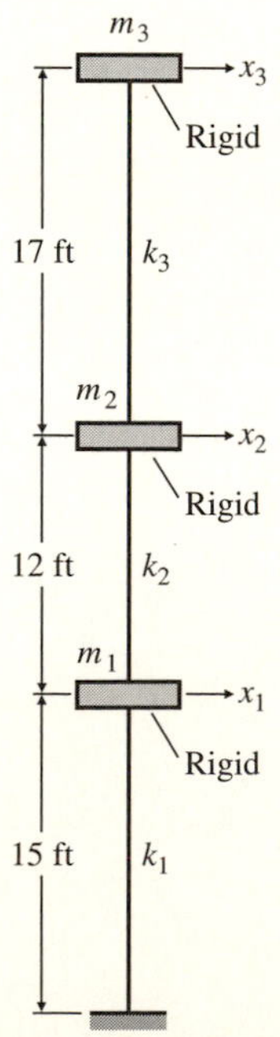

Figure 10.7

Mechanical model for three-story shear frame building of Example 10.3.

$$k_2 = \frac{2(12)(29{,}000 \text{ ksi})(1.5)(600 \text{ in}^4)}{(12 \text{ ft} \times 12 \text{ in})^3} = 209.78 \text{ kips/in}$$

$$k_3 = \frac{2(12)(29{,}000 \text{ ksi})(600 \text{ in}^4)}{(17 \text{ ft} \times 12 \text{ in})^3} = 49.189 \text{ kips/in}$$

The mass matrix is then expressed as

$$[m] = \begin{bmatrix} m_1 & 0 & 0 \\ 0 & m_2 & 0 \\ 0 & 0 & m_3 \end{bmatrix}$$

$$= \begin{bmatrix} 0.04141 & 0 & 0 \\ 0 & 0.03882 & 0 \\ 0 & 0 & 0.02588 \end{bmatrix} \text{kip-sec}^2/\text{in}$$

and the stiffness matrix is given by

$$[k] = \begin{bmatrix} k_1 + k_2 & -k_2 & 0 \\ -k_2 & k_2 + k_3 & -k_3 \\ 0 & -k_3 & k_3 \end{bmatrix}$$

$$= \begin{bmatrix} 299.29 & -209.78 & 0 \\ -209.78 & 258.97 & -49.189 \\ 0 & -49.189 & 49.189 \end{bmatrix} \text{kips/in}$$

(b) To determine the natural frequencies, the determinant of Eq. (10.18) is set equal to zero. It follows that the characteristic determinant is given by

$$\left| [k] - \omega_r^2[m] \right| = 0 \tag{1}$$

or

$$\begin{vmatrix} 299.29 - 0.04141\,\omega^2 & -209.78 & 0 \\ -209.78 & 258.97 - 0.0382\,\omega^2 & -49.189 \\ 0 & -49.189 & 49.189 - 0.02588\,\omega^2 \end{vmatrix}$$

$$= 0 \tag{2}$$

Expanding the characteristic determinant given by Eq. (2) yields the frequency equation in ω^2 of order $n = 3$. Thus

$$(\omega^2)^3 - 15{,}778(\omega^2)^2 + 44{,}778{,}003\,\omega^2 - 2.717 \times 10^{10} = 0 \tag{3}$$

The roots of the characteristic equation given by Eq. (3) are

$$\omega_1^2 = 628.839 \quad \text{or} \quad \omega_1 = 25.08 \text{ rad/sec}$$

$$\omega_2^2 = 2871.18 \quad \text{or} \quad \omega_2 = 53.58 \text{ rad/sec}$$

$$\omega_3^2 = 12{,}277.76 \quad \text{or} \quad \omega_3 = 110.80 \text{ rad/sec}$$

(c) The eigenvectors, or natural modes, corresponding to each frequency are determined from Eqs. (10.31) and (10.33).

(i) For mode 1, $r = 1$ and $\omega_r^2 = \omega_1^2 = 628.839$, the characteristic matrix represented by Eq. (10.31) is given by

$$[\mathbf{H}]_1 = \begin{bmatrix} \mathbf{H}_{11} & \mathbf{H}_{12} \\ \mathbf{H}_{21} & \mathbf{H}_{22} \end{bmatrix}_1 = \left[\begin{array}{c|cc} 273.25 & -209.78 & 0 \\ \hline -209.78 & 234.56 & -49.189 \\ 0 & -49.189 & 32.915 \end{array} \right] \tag{4}$$

and the complementary subvector given by Eq. (10.33) is expressed as

$$\{\Phi\}_{c1} = - \begin{bmatrix} 234.56 & -49.189 \\ -49.189 & 32.915 \end{bmatrix}^{-1} \left\{ \begin{array}{c} -209.78 \\ 0 \end{array} \right\} = \left\{ \begin{array}{c} 1.3026 \\ 1.9466 \end{array} \right\} \tag{5}$$

Therefore, the modal vector for mode 1 is

$$\{\Phi\}_1 = \left\{ \begin{array}{c} 1.0 \\ \{\Phi\}_c \end{array} \right\}_1 = \left\{ \begin{array}{c} 1.0 \\ 1.3026 \\ 1.9466 \end{array} \right\} \tag{6}$$

(ii) For mode 2, $r = 2$ and $\omega_r^2 = \omega_2^2 = 2871.18$, then from Eq. (10.31) the characteristic matrix is

$$[\mathbf{H}]_2 = \left[\begin{array}{c|cc} 180.39 & -209.78 & 0 \\ \hline -209.78 & 147.51 & -49.189 \\ 0 & -49.189 & -25.117 \end{array} \right] \tag{7}$$

and from Eq. (10.33) the complementary subvector is

$$\{\Phi\}_{c2} = - \begin{bmatrix} 147.51 & -49.189 \\ -49.189 & -25.117 \end{bmatrix}^{-1} \left\{ \begin{array}{c} -209.78 \\ 0 \end{array} \right\} = \left\{ \begin{array}{c} 0.8603 \\ -1.685 \end{array} \right\} \tag{8}$$

Thus, the modal vector for mode 2 is

$$\{\Phi\}_2 = \left\{ \begin{array}{c} 1.0 \\ \{\Phi\}_c \end{array} \right\}_2 = \left\{ \begin{array}{c} 1.0 \\ 0.8603 \\ -1.685 \end{array} \right\} \tag{9}$$

(iii) For mode 3, $r = 3$ and $\omega_r^2 = \omega_3^2 = 12,277.76$, then from Eq. (10.31) the characteristic determinant is

$$[\mathbf{H}]_3 = \begin{bmatrix} -209.78 & -209.78 & 0 \\ -209.78 & -217.65 & -49.189 \\ 0 & -49.189 & -268.56 \end{bmatrix} \tag{10}$$

and the complementary subvector from Eq. (10.31) is

$$\{\Phi\}_{c3} = -\begin{bmatrix} -217.65 & -49.189 \\ -49.189 & -268.56 \end{bmatrix}^{-1} \begin{Bmatrix} -209.78 \\ 0 \end{Bmatrix} = \begin{Bmatrix} -1.005 \\ 0.184 \end{Bmatrix} \tag{11}$$

and finally, the modal vector for mode 3 is given by

$$\{\Phi\}_3 = \begin{Bmatrix} 1.0 \\ \{\Phi\}_c \end{Bmatrix}_3 = \begin{Bmatrix} 1.0 \\ -1.005 \\ 0.184 \end{Bmatrix} \tag{12}$$

▲

The three natural mode shapes for the structure are shown in Figure 10.8.

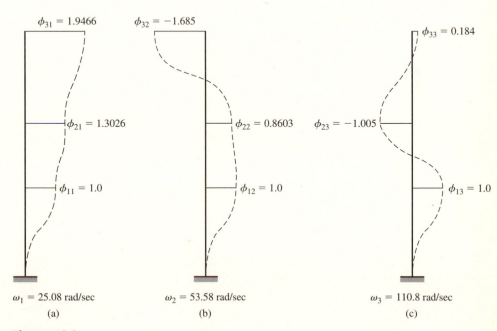

Figure 10.8

Normal modes for three-story shear frame building: (a) mode 1; (b) mode 2; (c) mode 3.

10.4 ORTHOGONALITY OF NATURAL MODES

The natural modes or normal modes of a MDOF vibrating system possess a very important property known as *orthogonality*. The orthogonality of the system eigenvectors is with respect to the mass matrix $[m]$ and the stiffness matrix $[k]$.

To develop these orthogonal relationships, begin by writing Eq. (10.18) for the rth mode as follows:

$$[k]\{A\}_r - \omega_r^2[m]\{A\}_r = \{0\} \tag{10.34}$$

Now, since the modal displacements may be given any amplitude, it is convenient to replace $\{A\}_r$ by $\{\Phi\}_r$ for any particular mode. Thus, if we consider any two distinct solution pairs ω_r^2, $\{\Phi\}_r$ and ω_s^2, $\{\Phi\}_s$ of the eigenproblem expressed by Eq. (10.18), Eq. (10.34) may be rewritten as

$$[k]\{\Phi\}_r - \omega_r^2[m]\{\Phi\}_r = 0 \tag{10.35}$$

and

$$[k]\{\Phi\}_s - \omega_s^2[m]\{\Phi\}_s = 0 \tag{10.36}$$

Premultiplying both sides of Eq. (10.35) by $\{\Phi\}_s^T$ yields

$$\{\Phi\}_s^T[k]\{\Phi\}_r - \omega_r^2\{\Phi\}_s^T[m]\{\Phi\}_r = 0 \tag{10.37}$$

Similarly, premultiplying both sides of Eq. (10.36) by $\{\Phi\}_r^T$ yields

$$\{\Phi\}_r^T[k]\{\Phi\}_s - \omega_s^2\{\Phi\}_r^T[m]\{\Phi\}_s = 0 \tag{10.38}$$

Now, since $[k]$ and $[m]$ are symmetric matrices, then $[k] = [k]^T$ and $[m] = [m]^T$. Therefore, Eq. (10.38) can be *transposed* and rewritten as

$$\{\Phi\}_s^T[k]\{\Phi\}_r - \omega_s^2\{\Phi\}_s^T[m]\{\Phi\}_r = 0 \tag{10.39}$$

Equation (10.37) may be subtracted from Eq. (10.39) to give

$$(\omega_r^2 - \omega_s^2)\{\Phi\}_s^T[m]\{\Phi\}_r = 0 \tag{10.40}$$

For modes with distinct frequencies, that is, $\omega_r \neq \omega_s$, it is necessary that

$$\{\Phi\}_s^T[m]\{\Phi\}_r = 0 \qquad \omega_r \neq \omega_s \tag{10.41}$$

The rth and sth modes are said to be *orthogonal with respect to the mass matrix*. Equation (10.41) can be substituted into Eq. (10.37) to show that the rth mode and the sth modes are also *orthogonal with respect to the stiffness matrix;* that is,

$$\{\Phi\}_s^T[k]\{\Phi\}_r = 0 \tag{10.42}$$

These orthogonality conditions of normal modes are the essence of the *mode superposition method,* or *modal analysis,* discussed in Chapter 12.

EXAMPLE 10.4 ▼

Consider the shear frame building of Example 10.3 and verify that the natural modes are orthogonal with respect to the mass matrix.

Solution

(a) Mode 1 and mode 2:
Let $r = 2$ and $s = 1$; Eq. (10.41) then yields

$$\{\Phi\}_1^T[m]\{\Phi\}_2 = \begin{bmatrix} 1.0 & 1.3026 & 1.9466 \end{bmatrix}$$

$$\begin{bmatrix} 0.04141 & 0 & 0 \\ 0 & 0.03882 & 0 \\ 0 & 0 & 0.02588 \end{bmatrix} \begin{Bmatrix} 1.0 \\ 0.8603 \\ -1.685 \end{Bmatrix} = 0$$

(b) Mode 1 and mode 3:
Let $r = 3$ and $s = 1$; Eq. (10.41) then yields

$$\{\Phi\}_1^T[m]\{\Phi\}_3 = \begin{bmatrix} 1.0 & 1.3026 & 1.9466 \end{bmatrix}$$

$$\begin{bmatrix} 0.04141 & 0 & 0 \\ 0 & 0.03882 & 0 \\ 0 & 0 & 0.02588 \end{bmatrix} \begin{Bmatrix} 1.0 \\ -1.005 \\ 0.184 \end{Bmatrix} = 0$$

(c) Mode 2 and mode 3:
Let $r = 3$ and $s = 2$; Eq. (10.41) then yields

$$\{\Phi\}_2^T[m]\{\Phi\}_3 = \begin{bmatrix} 1.0 & 0.8603 & -1.685 \end{bmatrix}$$

$$\begin{bmatrix} 0.04141 & 0 & 0 \\ 0 & 0.03882 & 0 \\ 0 & 0 & 0.02588 \end{bmatrix} \begin{Bmatrix} 1.0 \\ -1.005 \\ 0.184 \end{Bmatrix} = 0$$

The orthogonality of the modes with respect to the stiffness matrix may be verified by implementing Eq. (10.42). ▲

10.5 SYSTEMS ADMITTING RIGID-BODY MODES

There are a variety of structures, such as aircraft in flight and submarines, that exhibit rigid-body modes of vibration. For such systems, the stiffness matrix $[k]$ is *positive semidefinite,* or the determinant of the stiffness matrix is zero. Physically this implies that when the system undergoes rigid-body motion there are no elastic deformations in the structure.

The frequency corresponding to a rigid-body mode is zero, and some MDOF systems may exhibit as many as six rigid-body modes. Therefore, MDOF systems exhibiting several rigid-body modes will have several frequencies equal to zero. In other words, there will be several *repeated roots* to the characteristic equation, such that $\omega_r = \omega_{r+1} = \ldots \omega_m = 0$, where m is the number of times the frequency is repeated. The eigenvectors corresponding to these repeated eigenvalues (or frequencies) are linearly independent, but generally are not orthogonal. However, it is possible to select the eigenvectors so that they do satisfy the orthogonality conditions of Eqs. (10.41) and (10.42) even though $\omega_r = \omega_s$.

It was discussed in Section 10.3 that for a *positive definite system* (i.e., a system exhibiting no rigid-body modes) the $\mathbf{H}_{11}$ submatrix was (1×1) and the $\mathbf{H}_{22}$ submatrix was $(n - 1) \times (n - 1)$. For a *positive semidefinite system* (i.e., a system admitting rigid-body modes), the $\mathbf{H}_{11}$ submatrix is $(m \times n)$, and therefore, the $\mathbf{H}_{22}$ submatrix is $(n - m) \times (n - m)$. Furthermore, the complementary subvector $\{\Phi\}_c$ will be $(n - m) \times 1$. Thus, for a system admitting m rigid-body modes or having m repeated roots to the characteristic equation, Eq. (10.36) is expressed as

$$\begin{bmatrix} \mathbf{H}_{11} & \mathbf{H}_{12} \\ \mathbf{H}_{21} & \mathbf{H}_{22} \end{bmatrix}_r \begin{Bmatrix} \{\Phi\}_m \\ \{\Phi\}_c \end{Bmatrix}_r = \begin{Bmatrix} \{0\} \\ \{0\} \end{Bmatrix} \tag{10.43}$$

where $\{\Phi\}_{cr}$ is the $(n - m) \times 1$ complementary subvector for the rth mode and $\{\Phi\}_{mr}$ is the $m \times 1$ subvector containing the m assumed elements of the eigenvector $\{\Phi\}_r$. In a manner similar to that employed for positive definite systems [Eq. (10.33)], the complementary subvector $\{\Phi\}_{cr}$ may be determined from the second submatrix equation of Eq. (10.43) as

$$\mathbf{H}_{21}\{\Phi\}_{mr} + \mathbf{H}_{22}\{\Phi\}_{cr} = 0 \tag{10.44}$$

or

$$\{\Phi\}_{cr} = -\mathbf{H}_{22}^{-1}\mathbf{H}_{21}\{\Phi\}_{mr} \tag{10.45}$$

and the eigenvector for the rth mode is then

$$\{\Phi\}_r = \begin{Bmatrix} \{\Phi\}_m \\ \{\Phi\}_c \end{Bmatrix}_r \tag{10.46}$$

Although the eigenvectors determined by this procedure for repeated eigenvalues may not be orthogonal, they may be altered by some linear combination of one another such that they become orthogonal.

EXAMPLE 10.5 ▼

A flying wing having total mass m, length L, and flexural rigidity EI is discretized as shown in Figure 10.9. Determine (a) the natural frequencies and (b) the normal modes of vibration for this system. Consider the translational DOF only. For the rigid-body modes, select the modal vectors such that they are orthogonal.

Figure 10.9

Mechanical model of flying wing of Example 10.5.

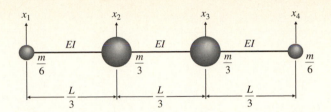

Solution

Determination of the stiffness coefficients k_{ij} for the four translational DOF is illustrated in Figure 10.10 (refer to Section 9.2). The mass and stiffness matrices for the structure are

$$[m] = \frac{m}{6}\begin{bmatrix} 1 & 0 & 0 & 0 \\ 0 & 2 & 0 & 0 \\ 0 & 0 & 2 & 0 \\ 0 & 0 & 0 & 1 \end{bmatrix} \qquad [k] = \frac{54EI}{5L^3}\begin{bmatrix} 4 & -9 & 6 & -1 \\ -9 & 24 & -21 & 6 \\ 6 & -21 & 24 & -9 \\ -1 & 6 & -9 & 4 \end{bmatrix}$$

and the equations of motion are expressed as

$$\frac{m}{6}\begin{bmatrix} 1 & 0 & 0 & 0 \\ 0 & 2 & 0 & 0 \\ 0 & 0 & 2 & 0 \\ 0 & 0 & 0 & 1 \end{bmatrix}\begin{Bmatrix} \ddot{x}_1 \\ \ddot{x}_2 \\ \ddot{x}_3 \\ \ddot{x}_4 \end{Bmatrix}$$

$$+ \frac{54EI}{5L^3}\begin{bmatrix} 4 & -9 & 6 & -1 \\ -9 & 24 & -21 & 6 \\ 6 & -21 & 24 & -9 \\ -1 & 6 & -9 & 4 \end{bmatrix}\begin{Bmatrix} x_1 \\ x_2 \\ x_3 \\ x_4 \end{Bmatrix} = \{0\} \qquad (1)$$

(a) To determine the natural frequencies, let $\gamma = \omega^2(m/6)(5L^3/54EI)$ and set the determinant of Eq. (10.18) equal to zero. Thus, the characteristic determinant is expressed by

$$\left\| [k] - \omega^2[m] \right\| = 0 \qquad (2)$$

or

$$\begin{bmatrix} 4-\gamma & -9 & 6 & -1 \\ -9 & 2(12-\gamma) & -21 & 6 \\ 6 & -21 & 2(12-\gamma) & -9 \\ -1 & 6 & -9 & 4-\gamma \end{bmatrix} = 0 \qquad (3)$$

Expansion of the characteristic determinant given by Eq. (3) yields the frequency equation given by

$$\gamma^2(4\gamma^2 - 128\gamma + 495) = 0 \qquad (4)$$

Figure 10.10
Construction of stiffness matrix by columns for flying wing structure: (a) stiffness coefficients for column 1; (b) stiffness coefficients for column 2; (c) stiffness coefficients for column 3; (d) stiffness coefficients for column 4.

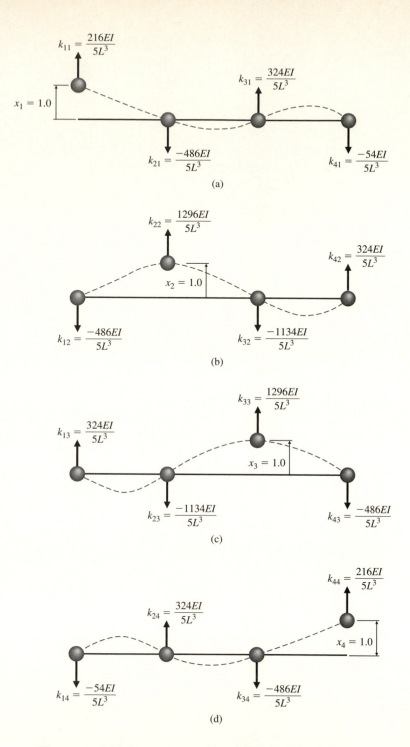

The roots of the frequency equation defined by Eq. (4) are

$$\gamma_1 \; = \; \gamma_2 \; = \; 0, \; \gamma_3 \; = \; 4.5, \quad \text{and} \quad \gamma_4 \; = \; 27.5$$

and the natural frequencies of the system are

$$\omega_1 \; = \; \omega_2 \; = \; 0$$

$$\omega_3 = 17.08 \sqrt{\frac{EI}{mL^3}}$$

$$\omega_4 = 42.21 \sqrt{\frac{EI}{mL^3}}$$

(b) There will be four normal modes of vibration corresponding to the four natural frequencies. Two of the modes are rigid-body modes corresponding to $\omega_1 = \omega_2 = 0$. The remaining two modes are flexural modes of vibration. First begin with the flexural modes corresponding to ω_3 and ω_4.

(i) For mode 3, $r = 3$ and $\omega = \omega_3$. From Eq. (10.31) the characteristic matrix is expressed as

$$[\mathbf{H}]_{3,} = \begin{bmatrix} \mathbf{H}_{11} & \mathbf{H}_{12} \\ \mathbf{H}_{21} & \mathbf{H}_{22} \end{bmatrix}_3 = \left[\begin{array}{c|ccc} -0.5 & -9.0 & 6.0 & -1.0 \\ \hline -9.0 & 15.0 & -21.0 & 6.0 \\ 6.0 & -21.0 & 15.0 & -9.0 \\ -1.0 & 6.0 & -9.0 & -.05 \end{array}\right] \tag{5}$$

and from Eq. (10.33), the complementary subvector for the 3rd mode is

$$\{\Phi\}_{c3} = -\begin{bmatrix} 15.0 & -21.0 & 6.0 \\ -21.0 & 15.0 & -9.0 \\ 6.0 & -9.0 & -.05 \end{bmatrix}^{-1} \begin{Bmatrix} -9.0 \\ 6.0 \\ -1.0 \end{Bmatrix} = \begin{Bmatrix} -0.5 \\ -0.5 \\ 1.0 \end{Bmatrix} \tag{6}$$

and thus the modal vector for the 3rd mode is given as

$$\{\Phi\}_3 = \begin{Bmatrix} 1.0 \\ \{\Phi\}_c \end{Bmatrix}_3 = \begin{Bmatrix} 1.0 \\ -0.5 \\ -0.5 \\ 1.0 \end{Bmatrix} \tag{7}$$

(ii) For mode 4, $r = 4$ and $\omega = \omega_4$; thus, the characteristic matrix from Eq. (10.31) is

$$[\mathbf{H}]_4 = \left[\begin{array}{c|ccc} -23.5 & -9.0 & 6.0 & -1.0 \\ \hline -9.0 & -31.0 & -21.0 & 6.0 \\ 6.0 & -21.0 & -31.0 & -9.0 \\ -1.0 & 6.0 & -9.0 & -23.5 \end{array}\right] \tag{8}$$

and the complementary subvector determined from Eq. (10.33) is

$$\{\Phi\}_{c4} = -\begin{bmatrix} -31.0 & -21.0 & 6.0 \\ -21.0 & -31.0 & -9.0 \\ 6.0 & -9.0 & -23.5 \end{bmatrix}^{-1} \begin{Bmatrix} -9.0 \\ 6.0 \\ -1.0 \end{Bmatrix} = \begin{Bmatrix} -1.5 \\ 1.5 \\ -1.0 \end{Bmatrix} \tag{9}$$

Therefore, the modal vector for the 4th mode is established as

$$\{\Phi\}_4 = \left\{ \begin{matrix} 1.0 \\ \{\Phi\}_c \end{matrix} \right\}_4 = \left\{ \begin{matrix} 1.0 \\ -1.5 \\ 1.5 \\ -1.0 \end{matrix} \right\} \tag{10}$$

For the remaining two rigid-body modes, $\omega = \omega_1 = \omega_2 = 0$. Thus, for $m = 2$, from Eq. (10.43) the characteristic matrices for the rigid-body modes are expressed as

$$[\mathbf{H}]_1 = [\mathbf{H}]_2 = \begin{bmatrix} \mathbf{H}_{11} & \mathbf{H}_{12} \\ \mathbf{H}_{21} & \mathbf{H}_{22} \end{bmatrix}_{1,2}$$

$$= \left[\begin{array}{cc|cc} 4.0 & -9.0 & 6.0 & -1.0 \\ -9.0 & 24.0 & -21.0 & 6.0 \\ \hline 6.0 & -21.0 & 24.0 & -9.0 \\ -1.0 & 6.0 & -9.0 & 4.0 \end{array} \right] \tag{11}$$

From Eq. (10.45) the complementary subvectors are established as

$$\{\Phi\}_{cr} = -\mathbf{H}_{22}^{-1}\mathbf{H}_{21}\{\Phi\}_{mr} \qquad r = 1, 2 \tag{12}$$

or

$$\{\Phi\}_{cr} = -\begin{bmatrix} 24.0 & -9.0 \\ -9.0 & 4.0 \end{bmatrix}^{-1} \begin{bmatrix} 6.0 & -21.0 \\ -1.0 & 6.0 \end{bmatrix} \{\Phi\}_{mr} \qquad r = 1, 2 \tag{13}$$

Therefore, from Eq. (13)

$$\{\Phi\}_{cr} = \begin{bmatrix} -1.0 & 2.0 \\ -2.0 & 3.0 \end{bmatrix} \{\Phi\}_{mr} \qquad \text{for } r = 1, 2 \tag{14}$$

Since the system exhibits two rigid-body modes corresponding to $r = 1$ and $r = 2$, then $m = 2$ and the subvectors $\{\Phi\}_{mr}$ are 2×1. The vectors $\{\Phi\}_{mr}$ must be selected such that they are linearly independent. To aid in the selection process, note that the system is symmetric. Also note that structures of this type experience *plunge* and *roll* rigid-body modes. These modes should also be symmetric.

(iii) For mode 1, the *plunge mode,* it is assumed that

$$\{\Phi\}_{m1} = \left\{ \begin{matrix} 1.0 \\ 1.0 \end{matrix} \right\} \tag{15}$$

Then from Eqs. (14) and (15)

$$\{\Phi\}_{c1} = \begin{bmatrix} -1.0 & 2.0 \\ -2.0 & 3.0 \end{bmatrix} \left\{ \begin{matrix} 1.0 \\ 1.0 \end{matrix} \right\} = \left\{ \begin{matrix} 1.0 \\ 1.0 \end{matrix} \right\} \tag{16}$$

Therefore, from Eq. (10.46) the modal vector is established as

$$\{\Phi\}_1 = \left\{ \begin{matrix} \{\Phi\}_m \\ \{\Phi\}_c \end{matrix} \right\}_1 = \left\{ \begin{matrix} 1.0 \\ 1.0 \\ 1.0 \\ 1.0 \end{matrix} \right\} \tag{17}$$

(iv) For mode 2, the *roll mode*, a linearly independent vector $\{\phi\}_{m1}$ is sought; thus

$$\{\Phi\}_{m2} = \left\{ \begin{matrix} 1.5 \\ 0.5 \end{matrix} \right\} \tag{18}$$

and from Eqs. (14) and (18)

$$\{\Phi\}_{c2} = \begin{bmatrix} -1.0 & 2.0 \\ -2.0 & 3.0 \end{bmatrix} \left\{ \begin{matrix} 1.5 \\ 0.5 \end{matrix} \right\} = \left\{ \begin{matrix} -0.5 \\ -1.5 \end{matrix} \right\} \tag{19}$$

Then from Eq. (10.46) the modal vector is

$$\{\Phi\}_2 = \left\{ \begin{matrix} \{\Phi\}_m \\ \{\Phi\}_c \end{matrix} \right\}_2 = \left\{ \begin{matrix} 1.5 \\ 0.5 \\ -0.5 \\ -1.5 \end{matrix} \right\} \tag{20}$$

Finally, the orthogonality of the rigid-body modes from the expression given by Eq. (10.41) is checked:

$$\{\Phi\}_1^T[m]\{\Phi\}_2 = 0 \quad ? \tag{21}$$

from which it is calculated,

$$[1.0 \quad 1.0 \quad 1.0 \quad 1.0] \begin{bmatrix} \dfrac{m}{6} & 0 & 0 & 0 \\ 0 & \dfrac{m}{3} & 0 & 0 \\ 0 & 0 & \dfrac{m}{3} & 0 \\ 0 & 0 & 0 & \dfrac{m}{6} \end{bmatrix} \left\{ \begin{matrix} 1.5 \\ 0.5 \\ -0.5 \\ -1.5 \end{matrix} \right\} = 0 \tag{22}$$

Therefore, as indicated by Eq. (22), the rigid-body modes are orthogonal. The four normal modes are illustrated in Figure 10.11.

Figure 10.11

Normal modes for flying
wing: (a) mode 1, plunge
mode; (b) mode 2, roll mode;
(c) mode 3, symmetric
flexural mode; (d) mode 4,
antisymmetric flexural mode.

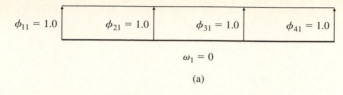

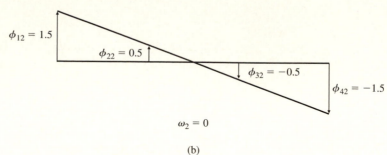

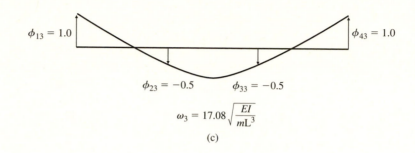

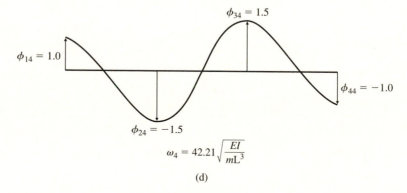

10.6 GENERALIZED MASS AND STIFFNESS MATRICES

At this time reconsider the orthogonality condition for the mass, Eq. (10.41), for which
$r = s$. For this case Eq. (10.41) can be rewritten as

$$\{\Phi\}_r^T[m]\{\Phi\}_r = M_r \tag{10.47}$$

where M_r is referred to as the *generalized mass,* or *modal mass,* for the rth mode.

Similarly, for $r = s$, Eq. (10.42) may be expressed as

$$\{\Phi\}_r^T[k]\{\Phi\}_r = K_r \tag{10.48}$$

where K_r is known as the *generalized stiffness,* or *modal stiffness,* for the rth mode. If next, Eq. (10.18) is written for the rth mode and both sides of the equation are premultiplied by $\{\Phi\}_r^T$, then

$$\{\Phi\}_r^T[k]\{\Phi\}_r = \omega_r^2\{\Phi\}_r^T[m]\{\Phi\}_r \tag{10.49}$$

from which

$$\omega_r^2 = \frac{K_r}{M_r} \tag{10.50}$$

Equation (10.50) represents the natural frequency for the rth mode in terms of the generalized mass and generalized stiffness of the rth mode. It should be noted that K_r and M_r are nonnegative scalar quantities.

If next the relationships of generalized mass and generalized stiffness, as given by Eqs. (10.49) and (10.50), are applied, along with the conditions of mass and stiffness orthogonality, as stated in Eqs. (10.41) and (10.42), then for the entire system

$$[M] = \{\Phi\}^T[m]\{\Phi\} \tag{10.51}$$

and

$$[K] = \{\Phi\}^T[k]\{\Phi\} \tag{10.52}$$

In Eq. (10.51), $[M]$ is known as the *generalized mass matrix,* or *modal mass matrix.* $[M]$ is a diagonal matrix of order n whose elements are the generalized masses for each mode, or

$$[M] = \begin{bmatrix} M_1 & 0 & 0 & \dots & 0 \\ 0 & M_2 & 0 & \dots & 0 \\ 0 & & & & \cdot \\ \dots & \dots & \dots & \dots & \dots \\ 0 & 0 & & \dots & M_n \end{bmatrix}_{n \times n} \tag{10.53}$$

Similarly, in Eq. (10.52), $[K]$ is known as the *generalized stiffness matrix,* or *modal stiffness matrix.* Like $[M]$, $[K]$ is a diagonal matrix of order n and contains the generalized stiffnesses for each mode, or

$$[K] = \begin{bmatrix} K_1 & 0 & 0 & \dots & 0 \\ 0 & K_2 & 0 & \dots & 0 \\ 0 & & & & \cdot \\ \dots & \dots & \dots & \dots & \dots \\ 0 & 0 & & \dots & K_n \end{bmatrix}_{n \times n} \tag{10.54}$$

EXAMPLE 10.6 ▼

Determine the generalized mass matrix $[M]$ and generalized stiffness matrix $[K]$ for the shear frame building of Example 10.3.

Solution

From Example 10.3,

$$[m] = \begin{bmatrix} 0.04141 & 0 & 0 \\ 0 & 0.03882 & 0 \\ 0 & 0 & 0.02588 \end{bmatrix} \text{kip-sec}^2/\text{in}$$

$$[k] = \begin{bmatrix} 299.29 & -209.78 & 0 \\ -209.78 & 258.97 & -49.189 \\ 0 & -49.189 & 49.189 \end{bmatrix} \text{kips/in}$$

$$[\Phi] = \begin{bmatrix} 1.0 & 1.0 & 1.0 \\ 1.3026 & 0.8603 & -1.005 \\ 1.9466 & -1.685 & 0.184 \end{bmatrix}$$

and

$$\{\omega^2\} = \begin{Bmatrix} 628.839 \\ 2871.18 \\ 12{,}277.76 \end{Bmatrix}$$

The generalized mass matrix may be determined directly from Eq. (10.51), or it may be constructed by individual modes from Eq. (10.47). The latter alternative is employed, so that the generalized mass for mode 1 is given by

$$M_1 = \{\Phi\}_1^T[m]\{\Phi\}_1 = \begin{bmatrix} 1.0 & 1.306 & 1.9466 \end{bmatrix}$$

$$\begin{bmatrix} 0.04141 & 0 & 0 \\ 0 & 0.03882 & 0 \\ 0 & 0 & 0.02588 \end{bmatrix} \begin{Bmatrix} 1.0 \\ 1.3026 \\ 1.9466 \end{Bmatrix} \tag{1}$$

or

$$M_1 = 0.04141(1.0)^2 + 0.03882(1.306)^2 + 0.02588(1.9466)^2$$

$$= 0.20534 \text{ kip-sec}^2/\text{in}$$

Similarly, for mode 2

$$M_2 = \{\Phi\}_2^T [m]\{\Phi\}_2 \tag{2}$$

or

$$M_2 = 0.04141(1.0)^2 + 0.03882(0.8603)^2 + 0.02588(-1.685)^2$$
$$= 0.14362 \text{ kip-sec}^2/\text{in}$$

and for mode 3

$$M_3 = \{\Phi\}_3^T [m]\{\Phi\}_3 \tag{3}$$

or

$$M_3 = 0.04141(1.0)^2 + 0.03882(-1.005)^2 + 0.02588(0.184)^2$$
$$= 0.08149 \text{ kip-sec}^2/\text{in}$$

Thus, the generalized mass matrix is then represented by Eq. (10.53) as

$$[M] = \begin{bmatrix} 0.20534 & 0 & 0 \\ 0 & 0.14362 & 0 \\ 0 & 0 & 0.08149 \end{bmatrix}$$

Similarly, the generalized stiffness matrix may be constructed directly from Eq. (10.52), or by individual modes from Eq. (10.48). However, with $[M]$ having already been established, the generalized stiffness terms may be determined from Eq. (10.50), as

$$K_r = \omega_r^2 M_r \tag{4}$$

Therefore, Eq. (4) yields

$$K_1 = \omega_1^2 M_1 = (628.839)(0.20534) = 129.13 \text{ kips/in}$$

$$K_2 = \omega_2^2 M_2 = (2871.18)(0.14362) = 412.36 \text{ kips/in}$$

$$K_3 = \omega_3^2 M_3 = (12{,}277.76)(0.08149) = 1000.51 \text{ kips/in}$$

and the generalized stiffness matrix given by Eq. (10.54) becomes

$$[K] = \begin{bmatrix} 129.13 & 0 & 0 \\ 0 & 412.36 & 0 \\ 0 & 0 & 1000.51 \end{bmatrix} \text{kips/in}$$

▲

10.7 FREE VIBRATION RESPONSE TO INITIAL CONDITIONS

The free vibration response of undamped MDOF systems to initial conditions is greatly facilitated through a systematic procedure of analysis. To this end, the assumed harmonic solution for free vibration motion given by Eq. (10.16) is considered:

$$\{x(t)\} = \{A\}\sin(\omega t + \phi) \tag{10.16}$$

For the rth mode only, Eq. (10.16) can be expressed as

$$\{x(t)\}_r = \{A\}_r\sin(\omega_r t + \phi_r) \tag{10.55}$$

In Eq. (10.26) it was implied that $\{A\}_r$ could be written in the form

$$\{A\}_r = c_r\{\Phi\}_r \tag{10.56}$$

and therefore

$$\{x\} = \sum_{r=1}^{n} c_r\{\Phi\}_r \tag{10.57}$$

where c_r is a scaling constant defined by

$$c_r = \frac{\{\Phi\}_r^T[m]\{x\}}{M_r} \tag{10.58}$$

Thus, Eq. (10.55) can be written as

$$\{x(t)\}_r = c_r\{\Phi\}_r\sin(\omega_r t + \phi_r) \tag{10.59}$$

and the general solution is

$$\{x(t)\} = \sum_{r=1}^{n} c_r\{\Phi\}_r \sin(\omega_r t + \phi_r) \tag{10.60}$$

The unknown coefficients for each mode (c_r, ϕ_r) must be determined from the $2n$ initial conditions of displacement and velocity, that is, $\{x(0)\}$ and $\{\dot{x}(0)\}$.

In a manner analogous to that applied to SDOF systems, the coefficients c_r and ϕ_r can be replaced with two other coefficients, a_r and b_r, respectively. Thus, Eq. (10.60) can be expressed as

$$\{x(t)\} = \sum_{r=1}^{n} \{\Phi\}_r (a_r \sin \omega_r t + b_r \cos \omega_r t) \tag{10.61}$$

and the coefficients a_r and b_r are related to the initial conditions by

$$\{x(0)\} = \sum_{r=1}^{n} b_r\{\Phi\}_r \tag{10.62}$$

and

$$\{\dot{x}(0)\} = \sum_{r=1}^{n} \omega_r a_r\{\Phi\}_r \tag{10.63}$$

Premultiplying Eqs. (10.62) and (10.63) by $\{\Phi\}_s^T[m]$, and considering the *orthogonality condition,* Eq. (10.41), yields

$$a_r = \frac{\{\Phi\}_r^T[m]\{\dot{x}(0)\}}{M_r \omega_r} \qquad (10.64)$$

and

$$b_r = \frac{\{\Phi\}_r^T[m]\{x(0)\}}{M_r} \qquad (10.65)$$

After the $2n$ coefficients (a_r, b_r) from Eqs. (10.64) and (10.65) are determined, Eq. (10.61) may be used to evaluate the free vibration response of the structure due to initial conditions $\{x(0)\}$ and $\{\dot{x}(0)\}$. Eq. (10.61) represents the general solution for free vibration of an undamped MDOF system.

EXAMPLE 10.7 ▼

For the three-story shear frame building shown in Figure 10.12, determine the free vibration response to the following initial conditions:

$$\{x(0)\} = \begin{Bmatrix} 1.0 \\ 2.0 \\ 3.0 \end{Bmatrix} \text{ in } \quad \text{and} \quad \{\dot{x}(0)\} = \begin{Bmatrix} -1.0 \\ 3.0 \\ 4.0 \end{Bmatrix} \text{ in/sec}$$

Assume $k_1 = 25.9$ kips/in, $k_2 = 111.88$ kips/in, $k_3 = 174.82$ kips/in, $m_1 = 0.05823$ kip-sec^2/in, $m_2 = 0.04658$ kip-sec^2/in, and $m_3 = 0.03494$ kip-sec^2/in.

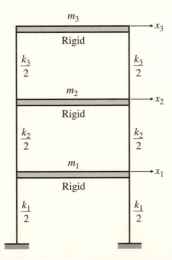

Figure 10.12

Three-story shear frame building of Example 10.7.

Solution

The mass matrix for the structure is given by

$$[m] = \begin{bmatrix} m_1 & 0 & 0 \\ 0 & m_2 & 0 \\ 0 & 0 & m_3 \end{bmatrix}$$

$$= \begin{bmatrix} 0.05823 & 0 & 0 \\ 0 & 0.04658 & 0 \\ 0 & 0 & 0.03494 \end{bmatrix} \text{kip-sec}^2/\text{in}$$

and the stiffness matrix is determined as

$$[k] = \begin{bmatrix} k_1 + k_2 & -k_2 & 0 \\ -k_2 & k_2 + k_3 & -k_3 \\ 0 & -k_3 & k_3 \end{bmatrix}$$

$$= \begin{bmatrix} 137.78 & -111.88 & 0 \\ -111.88 & 286.7 & -174.82 \\ 0 & -174.82 & 174.82 \end{bmatrix} \text{kips/in}$$

The equations of motion are expressed as

$$\begin{bmatrix} 0.05823 & 0 & 0 \\ 0 & 0.04658 & 0 \\ 0 & 0 & 0.03494 \end{bmatrix} \begin{Bmatrix} \ddot{x}_1 \\ \ddot{x}_2 \\ \ddot{x}_3 \end{Bmatrix}$$

$$+ \begin{bmatrix} 137.78 & -111.88 & 0 \\ -111.88 & 286.7 & -174.82 \\ 0 & -174.82 & 174.82 \end{bmatrix} \begin{Bmatrix} x_1 \\ x_2 \\ x_3 \end{Bmatrix} = \{0\}$$

Following the procedure demonstrated in Example 10.3, the natural frequencies are calculated as

$$\omega_1 = 13.02 \text{ rad/sec}$$

$$\omega_2 = 55.36 \text{ rad/sec}$$

$$\omega_3 = 101.44 \text{ rad/sec}$$

and the modal matrix is

$$[\Phi] = \begin{bmatrix} 1.0 & 1.0 & 1.0 \\ 1.143 & -0.364 & -4.118 \\ 1.183 & -0.940 & 3.897 \end{bmatrix}$$

The mode shapes are shown in Figure 10.13. The modal masses M_r determined from Eq. (10.47) are

$$M_1 = \{\Phi\}_1^T [m] \{\Phi\}_1$$

$$M_1 = [1.0 \quad 1.143 \quad 1.183] \begin{bmatrix} 0.0583 & 0 & 0 \\ 0 & 0.04658 & 0 \\ 0 & 0 & 0.0394 \end{bmatrix} \begin{Bmatrix} 1.0 \\ 1.143 \\ 1.183 \end{Bmatrix}$$

$$= 0.1680$$

and similarly,

$$M_2 = \{\Phi\}_2^T [m] \{\Phi\}_2 = 0.0953$$

$$M_3 = \{\Phi\}_3^T [m] \{\Phi\}_3 = 1.3788$$

Then from the initial conditions of displacement

$$[m]\{x(0)\} = \begin{bmatrix} 0.0583 & 0 & 0 \\ 0 & 0.04658 & 0 \\ 0 & 0 & 0.0394 \end{bmatrix} \begin{Bmatrix} 1.0 \\ 2.0 \\ 3.5 \end{Bmatrix} = \begin{Bmatrix} 0.05823 \\ 0.09316 \\ 0.12229 \end{Bmatrix}$$

and from the initial conditions of velocity

$$[m]\{\dot{x}(0)\} = \begin{bmatrix} 0.0583 & 0 & 0 \\ 0 & 0.04658 & 0 \\ 0 & 0 & 0.0394 \end{bmatrix} \begin{Bmatrix} -1.0 \\ 3.0 \\ 4.0 \end{Bmatrix}$$

$$= \begin{Bmatrix} -0.05823 \\ 0.13974 \\ 0.1576 \end{Bmatrix}$$

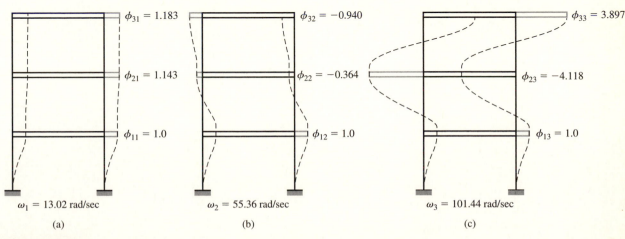

Figure 10.13
Normal modes for three-story shear frame building of Example 10.7: (a) mode 1; (b) mode 2; (c) mode 3.

The coefficients a_r determined from Eq. (10.64) are

$$a_1 = \frac{\{\Phi\}_1^T[m]\{\dot{x}(0)\}}{M_1\omega_1}$$

$$= \frac{[1.0 \quad 1.143 \quad 1.183]\begin{Bmatrix} -0.05823 \\ 0.13974 \\ 0.1576 \end{Bmatrix}}{(0.1680)(13.02)} = 0.121987$$

$$a_2 = \frac{\{\Phi\}_2^T[m]\{\dot{x}(0)\}}{M_2\omega_2}$$

$$= \frac{[1.0 \quad 0.364 \quad 0.940]\begin{Bmatrix} -0.05823 \\ 0.13974 \\ 0.1576 \end{Bmatrix}}{(0.0953)(55.36)} = -0.046374$$

$$a_3 = \frac{\{\Phi\}_3^T[m]\{\dot{x}(0)\}}{M_3\omega_3}$$

$$= \frac{[1.0 \quad -4.118 \quad 3.897]\begin{Bmatrix} -0.05823 \\ 0.13974 \\ 0.1576 \end{Bmatrix}}{(1.3788)(101.44)} = -0.000637$$

and the coefficients b_r determined from Eq. (10.65) are

$$b_1 = \frac{\{\Phi\}_1^T[m]\{x(0)\}}{M_1}$$

$$= \frac{[1.0 \quad 1.143 \quad 1.183]\begin{Bmatrix} 0.05823 \\ 0.09316 \\ 0.12229 \end{Bmatrix}}{(0.1680)} = 1.84155$$

$$b_2 = \frac{\{\Phi\}_2^T[m]\{x(0)\}}{M_2}$$

$$= \frac{[1.0 \quad -0.364 \quad -0.940]\begin{Bmatrix} 0.05823 \\ 0.09316 \\ 0.12229 \end{Bmatrix}}{(0.0953)} = -0.451027$$

$$b_3 = \frac{\{\Phi\}_3^T[m]\{x(0)\}}{M_3}$$

$$= \frac{[1.0 \quad -4.118 \quad 3.897]\begin{Bmatrix} 0.05823 \\ 0.09316 \\ 0.12229 \end{Bmatrix}}{(1.3788)} = 0.19548$$

Finally, from Eq. (10.61), the total response is calculated by summing the responses for all three modes as

$$\{x(t)\} = \sum_{r=1}^{3} \{\Phi\}_r [a_r \sin \omega_r t + b_r \cos \omega_r t]$$

$$= \begin{Bmatrix} 1.0 \\ 1.143 \\ 1.183 \end{Bmatrix}[(0.121987)\sin 13.02t + (1.84155)\cos 13.02t]$$

$$+ \begin{Bmatrix} 1.0 \\ -0.364 \\ -0.940 \end{Bmatrix}[(-0.046374)\sin 55.36t + (-0.951027)\cos 55.36t]$$

$$+ \begin{Bmatrix} 1.0 \\ -4.118 \\ 3.897 \end{Bmatrix}[(-0.000637)\sin 101.44t + (0.109548)\cos 101.44t]$$

or

$$\begin{Bmatrix} x_1 \\ x_2 \\ x_3 \end{Bmatrix} = \begin{Bmatrix} 0.121987 \\ 0.139431 \\ 0.144311 \end{Bmatrix}\sin 13.02t + \begin{Bmatrix} 1.84155 \\ 2.10489 \\ 2.17855 \end{Bmatrix}\cos 13.02t \qquad \text{(mode 1)}$$

$$+ \begin{Bmatrix} -0.046374 \\ 0.016880 \\ 0.043592 \end{Bmatrix}\sin 55.36t + \begin{Bmatrix} -0.951027 \\ 0.346174 \\ 0.893965 \end{Bmatrix}\cos 55.36t \qquad \text{(mode 2)}$$

$$+ \begin{Bmatrix} -0.000637 \\ 0.002623 \\ -0.002482 \end{Bmatrix}\sin 101.44t + \begin{Bmatrix} 0.109548 \\ -0.451119 \\ 0.426909 \end{Bmatrix}\cos 101.44t \qquad \text{(mode 3)}$$

The time-history response for the displacement of the top level of the structure, $x_3(t)$, is presented in Figure 10.14.

Figure 10.14

Free vibration displacement response of top level of three-story shear frame structure.

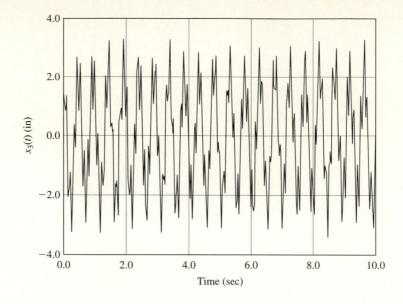

Notice that the general solution for free vibration of an undamped MDOF system given by Eq. (10.61) prescribes the resultant motion of the system as a combination of motions of the system with mode shapes $\{\Phi\}_r$ at corresponding frequencies ω_r. This is clearly illustrated in Figure 10.14 in which the resultant free vibration response is the summation of the responses at the three distinct vibration modes.

10.8 APPROXIMATE METHODS FOR ESTIMATING THE FUNDAMENTAL FREQUENCY

In many practical situations involving MDOF systems, only an accurate estimate of the fundamental frequency is required. In such cases, laborious calculations to extract all the normal vibration modes of the system is not warranted, and approximate analysis methods are desirable. This section discusses two approximate methods for estimating the fundamental frequency of MDOF systems.

The first method, the Rayleigh method, is an *upper bound* method based on energy principles. The second method, Dunkerly's approximation, is based upon the flexibility formulation of the system eigenproblem, and therefore, provides a *lower bound* estimate of the fundamental frequency. Thus the upper bound estimate of the fundamental frequency provided by the Rayleigh method can be complemented by the lower bound estimate afforded by Dunkerly's approximation to envelope the true fundamental natural frequency.

10.8.1 The Rayleigh Method

Use of the Rayleigh method to determine the natural frequency of SDOF systems was discussed in Chapter 4. Application of the Rayleigh method to determine the fundamental frequency of discrete MDOF systems is presented in this section.

To develop the procedure, consider the eigenproblem for a MDOF system represented by Eq. (10.18) and expressed as

$$\lambda[m][\Phi] = [k][\Phi] \tag{10.66}$$

where

$$\lambda = \omega^2$$

$[m]$ and $[k]$ = symmetric matrices

$[\Phi]$ = modal matrix

For the rth mode, Eq. (10.66) may be written as

$$\lambda_r[m][\Phi]_r = [k][\Phi]_r \tag{10.67}$$

where $\{\Phi\}_r$ is the modal vector for the rth mode. Premultiplying both sides of Eq. (10.67) by $\{\Phi\}_r^T$ yields

$$\lambda_r[\Phi]_r^T[m][\Phi]_r = [\Phi]_r^T[k][\Phi]_r \tag{10.68}$$

from which

$$\lambda_r = \frac{[\Phi]_r^T[k][\Phi]_r}{[\Phi]_r^T[m][\Phi]_r} \tag{10.69}$$

In Eq. (10.69), the denominator is related to the kinetic energy for the rth mode, and the numerator is related to the potential energy, or strain energy, for the rth mode.

If the modal vector $\{\Phi\}_r$ is replaced with any arbitrary vector $\{A\}$, Eq. (10.69) may be written as

$$\lambda_R = R(\{A\}) = \frac{\{A\}^T[k]\{A\}}{\{A\}^T[m]\{A\}} \tag{10.70}$$

where $R(\{A\})$ is a scalar quantity referred to as *Rayleigh's quotient*. It is evident from Eq. (10.70) that Rayleigh's quotient is dependent upon the known matrices $[m]$ and $[k]$ and the unknown arbitrary vector $\{A\}$. Obviously, if $\{A\}$ coincides with one of the system's normal modes, then λ_R is the corresponding eigenvalue, or natural frequency, for the system.

An important property of the Rayleigh quotient is that

$$\lambda_1 \le R(\{A\}) \le \lambda_n \tag{10.71}$$

and it also follows that for any vector $\{A\}$, if $[k]$ is positive definite, then $R(\{A\}) > 0$. Also, if $[k]$ is positive semidefinite, then $R(\{A\}) \ge 0$.

Equation (10.71) then indicates that the Rayleigh quotient is never lower than the fundamental eigenvalue. Furthermore, the minimum value the Rayleigh quotient can assume is that of the fundamental eigenvalue itself. Therefore, the Rayleigh quotient is a very good technique to estimate the fundamental frequency of a MDOF system. A reasonable estimate for the vector $\{A\}$ corresponding to the fundamental mode is the vector of static displacements resulting from subjecting the masses in the system to forces proportional to their weights. Many seismic design codes present expressions to estimate the fundamental frequency of high-rise buildings based upon this concept.

The natural frequency obtained by Eq. (10.70) is generally called the Rayleigh frequency ω_R expressed by

$$\lambda_R = \omega_R^2 = R(\{A\}) \tag{10.72}$$

The accuracy of the Rayleigh frequency ω_r depends entirely on the displacement vector $\{A\}$ used to represent the vibration mode shape. In principle, any vector $\{A\}$ may be selected which satisfies the geometric boundary conditions. However, any vector other than the true modal vector requires the action of additional external constraints to maintain equilibrium, which would, in turn, stiffen the structure, resulting in an increased computed frequency. Therefore, the true vibration mode shape will yield the lowest frequency obtainable by the Rayleigh method. Hence, the approximation yielding the lowest frequency for a particular case is the best result.

EXAMPLE 10.8 ▼

For the shear frame building of Example 10.7, determine the fundamental frequency by the Rayleigh method.

Solution

The mass and stiffness matrices for the structure were determined in Example 10.7 as

$$[m] = \begin{bmatrix} 0.05823 & 0 & 0 \\ 0 & 0.04658 & 0 \\ 0 & 0 & 0.03494 \end{bmatrix} \text{ kip-sec}^2/\text{in}$$

$$[k] = \begin{bmatrix} 137.78 & -111.88 & 0 \\ -111.88 & 286.7 & -174.82 \\ 0 & -174.82 & 174.82 \end{bmatrix} \text{ kips/in}$$

As a trial vector $\{A\}$ we use the vector of static displacements due to each mass being subjected to a horizontal force equal to the mass's own weight. Therefore, we seek the vector of displacements of the structure subject to the following force vector $\{F\}$:

$$\{F\} = [m]g = \begin{Bmatrix} 0.05823 \\ 0.04658 \\ 0.03494 \end{Bmatrix} \text{ kip-sec}^2/\text{in} \times (386.4 \text{ in}/\text{sec}^2)$$

$$= \begin{Bmatrix} 22.5 \\ 18.0 \\ 13.5 \end{Bmatrix} \text{ kips}$$

The vector of static displacements $\{A\}$ may be obtained from the expression

$$\{A\} = [a]\{F\} \tag{1}$$

where $[a]$ is the flexibility matrix. Noting that

$$[a] = [k]^{-1} \tag{2}$$

it follows that

$$[a] = \begin{bmatrix} 0.0386 & 0.0386 & 0.0386 \\ 0.0386 & 0.0476 & 0.0476 \\ 0.0386 & 0.0476 & 0.0533 \end{bmatrix} \text{ in/kip}$$

Thus, from Eq. (1)

$$\{A\} = \begin{bmatrix} 0.0386 & 0.0386 & 0.0386 \\ 0.0386 & 0.0476 & 0.0476 \\ 0.0386 & 0.0476 & 0.0533 \end{bmatrix} \begin{Bmatrix} 22.5 \\ 18.0 \\ 13.5 \end{Bmatrix} = \begin{Bmatrix} 2.084 \\ 2.3679 \\ 2.445 \end{Bmatrix} \text{ in}$$

Since vector $\{A\}$ is arbitrary, it is normalized with respect to the largest element in the vector, 2.445 in, so that the resulting $\{A\}$ vector is

$$\{A\} = \begin{Bmatrix} 0.852 \\ 0.968 \\ 1.0 \end{Bmatrix} \tag{3}$$

The denominator of the Rayleigh quotient given by Eq. (10.70) is determined as

$$\{A\}^T[m]\{A\} = [0.852 \quad 0.968 \quad 1.0]$$

$$\begin{bmatrix} 0.05823 & 0 & 0 \\ 0 & 0.04658 & 0 \\ 0 & 0 & 0.03494 \end{bmatrix} \begin{Bmatrix} 0.852 \\ 0.968 \\ 1.0 \end{Bmatrix} \tag{4}$$

$$= 0.1208 \text{ kip-sec}^2/\text{in}$$

and the numerator in the Rayleigh quotient is given by

$$\{A\}^T[k]\{A\} = [0.852 \quad 0.968 \quad 1.0]$$

$$\begin{bmatrix} 137.78 & -111.88 & 0 \\ -111.88 & 286.7 & -174.82 \\ 0 & -174.82 & 174.82 \end{bmatrix} \begin{Bmatrix} 0.852 \\ 0.968 \\ 1.0 \end{Bmatrix} \tag{5}$$

$$= 20.487 \text{ kips/in}$$

Thus the Rayleigh quotient as determined by Eq. (10.70) results in

$$\lambda_R = \omega_R^2 = R(\{A\}) = \frac{\{A\}^T[k]\{A\}}{\{A\}^T[m]\{A\}} = \frac{20.487}{0.1208} = 169.446$$

and the Rayleigh frequency is calculated from Eq. (10.72) as

$$\omega_R = 13.023 \text{ rad/sec}$$

Compared to the exact fundamental frequency, $\omega_1 = 13.02$ rad/sec computed in Example 10.7, the difference is +0.023%. The result is so close because the trial vector $\{A\}$ very closely resembles the actual modal vector $\{\Phi\}_1$. ▲

10.8.2 Dunkerly's Approximation

Dunkerly's equation is another approximate method for estimating the fundamental frequency of MDOF systems. The method yields fairly accurate results for systems in which damping is negligible and the natural frequencies are well separated (i.e., the frequencies of the harmonics are much higher than the fundamental frequency). Dunkerly's equation provides a "lower bound" estimate to the fundamental frequency and is therefore complementary to the Rayleigh method that provides an "upper bound" estimate to the fundamental frequency.

To derive Dunkerly's equation, consider the flexibility formulation of the system eigenproblem discussed in Section 10.4.2. Thus, the system characteristic equation is given by Eq. (9.37) as

$$[[D] - \lambda[I]]\{A\} = 0 \tag{10.73}$$

where $\lambda = 1/\omega^2$ and $[D]$ is the system dynamic matrix given by

$$[D] = [a][m] \tag{10.74}$$

and where $[a]$ is the flexibility matrix and $[m]$ is the mass matrix. The frequency equation is obtained by expanding the determinant of the characteristic matrix in Eq. (10.73).

To clarify the derivation of Dunkerly's equation, we consider a two-DOF system with a lumped mass (diagonal) matrix. Thus, the resulting characteristic determinant from Eq. (10.73) becomes

$$\begin{vmatrix} a_{11}m_1 - \dfrac{1}{\omega^2} & a_{12}m_1 \\[2mm] a_{21}m_1 & a_{22}m_2 - \dfrac{1}{\omega^2} \end{vmatrix} = 0 \tag{10.75}$$

Expanding this determinant results in the system frequency equation, that is, a second-order equation in $\lambda = 1/\omega^2$ given by

$$\left(\frac{1}{\omega^2}\right)^2 - (a_{11}m_1 + a_{12}m_2)\left(\frac{1}{\omega^2}\right) + m_1 m_2 (a_{11}a_{12} - a_{12}a_{21}) = 0 \tag{10.76}$$

If the roots of Eq. (10.76) are assumed to be $1/\omega_1^2$ and $1/\omega_2^2$, the factored form of Eq. (10.76) becomes

$$\left(\frac{1}{\omega^2} - \frac{1}{\omega_1^2}\right)\left(\frac{1}{\omega^2} - \frac{1}{\omega_2^2}\right) = 0 \tag{10.77}$$

or

$$\left(\frac{1}{\omega^2}\right)^2 - \left(\frac{1}{\omega_1^2} + \frac{1}{\omega_2^2}\right)\left(\frac{1}{\omega^2}\right) + \frac{1}{\omega_1^2 \omega_2^2} = 0 \tag{10.78}$$

Equating the coefficients of the $\lambda = 1/\omega^2$ terms in Eqs. (10.76) and (10.78) yields

$$\frac{1}{\omega_1^2} + \frac{1}{\omega_2^2} = a_{11}m_1 + a_{22}m_2 \tag{10.79}$$

The relationship represented by Eq. (10.79) also holds true for systems having n DOF. Thus, Eq. (10.77) may be extended to

$$\frac{1}{\omega_1^2} + \frac{1}{\omega_2^2} + \cdots + \frac{1}{\omega_n^2} = a_{11}m_1 + a_{22}m_2 + \cdots + a_{nn}m_n \tag{10.80}$$

The Dunkerly approximation to the fundamental frequency is made on the assumption that if the fundamental frequency ω_1 is much lower than the higher harmonics (that is, $\omega_2, \omega_3, \ldots, \omega_n$), then the terms $1/\omega_2^2, 1/\omega_3^2, \ldots, 1/\omega_n^2$ on the left-hand side of Eq. (10.80) can be neglected. The elimination of these terms yields an estimate of $1/\omega_1^2$ which is higher than the true value, thereby making the estimate of ω_1, lower than the exact value for the fundamental frequency. Therefore, Dunkerly's *lower bound* estimate of ω_1 is approximated by

$$\frac{1}{\omega_1^2} \cong a_{11}m_1 + a_{22}m_2 + \cdots + a_{nn}m_n = \sum_{i=1}^{n} a_{ii}m_i \tag{10.81}$$

In Eq. (10.81), the terms $a_{ii}m_i$ represent the contribution of each mass to $1/\omega_1^2$ in the absence of all other masses. Thus

$$a_{ii}m_i = \frac{1}{\omega_{ii}^2} \tag{10.82}$$

where ω_{ii}^2 is the natural frequency of an equivalent SDOF system with mass m_i acting alone at station i. Therefore, an alternative form of Dunkerly's equation is given by

$$\frac{1}{\omega_1^2} \cong \frac{1}{\omega_{11}^2} + \frac{1}{\omega_{22}^2} + \cdots + \frac{1}{\omega_{nn}^2} = \sum_{i=1}^{n} \frac{1}{\omega_{ii}^2} \tag{10.83}$$

EXAMPLE 10.9 ▼

Estimate the fundamental frequency of the shear frame building of Examples 10.7 and 10.8 by Dunkerly's method.

Solution

The mass and flexibility matrices presented in Example 10.8 are

$$[m] = \begin{bmatrix} 0.05823 & 0 & 0 \\ 0 & 0.04658 & 0 \\ 0 & 0 & 0.03494 \end{bmatrix} \text{kip-sec}^2/\text{in}$$

$$[a] = \begin{bmatrix} 0.0386 & 0.0386 & 0.0386 \\ 0.0386 & 0.0476 & 0.0476 \\ 0.0386 & 0.0476 & 0.0533 \end{bmatrix} \text{in/kip}$$

The fundamental frequency is estimated from Eq. (10.81) by

$$\frac{1}{\omega_1^2} \cong a_{11}m_1 + a_{22}m_2 + a_{33}m_3$$

$$= (0.0386)(0.05823) + (0.0476)(0.04658) + (0.0533)(0.03494)$$

$$= (0.0063272)$$

or

$$\omega_1 \cong 12.57 \text{ rad/sec}$$

Compared to the exact value of $\omega_1 = 13.02$ rad/sec, the difference is -0.448%. Also notice that the true fundamental frequency falls between the Rayleigh approximation (upper bound) obtained in Example 10.8 and the Dunkerly approximation (lower bound) obtained in this example.   ▲

EXAMPLE 10.10   ▼

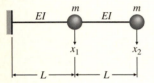

Figure 10.15

Two-mass cantilever beam of Example 10.10.

Estimate the fundamental frequency of the two-mass cantilever beam shown in Figure 10.15 by Dunkerly's method. Assume the beam has flexural rigidity EI.

Solution

The mass coefficients are

$$m_1 = m_2 = m$$

and the pertinent flexibility influence coefficients are

$$a_{11} = \frac{L^3}{3EI} \quad \text{and} \quad a_{22} = \frac{8L^3}{3EI}$$

The natural frequency of the cantilever beam having just the mass m_1 is ω_{11}; thus, from Eq. (10.82)

$$\frac{1}{\omega_{11}^2} = a_{11}m_1 = \frac{mL^3}{3EI} \tag{1}$$

and the natural frequency of the beam having just the mass m_2 is ω_{22}, or

$$\frac{1}{\omega_{22}^2} = a_{22}m = \frac{8mL^3}{3EI} \tag{2}$$

We then express Dunkerly's equation in the form given by Eq. (10.83), or

$$\frac{1}{\omega_1^2} = \frac{1}{\omega_{11}^2} + \frac{1}{\omega_{22}^2} = \frac{mL^3}{3EI} + \frac{8mL^3}{3EI} = \frac{3mL^3}{EI} \tag{3}$$

Thus, solving Eq. (3) yields

$$\omega_1^2 = 0.333 \frac{EI}{mL^3}$$

resulting in a fundamental frequency given by

$$\omega_1 = 0.577 \sqrt{\frac{EI}{mL^3}}$$

The exact value for the fundamental frequency is $\omega_1 = 0.5839\sqrt{EI/mL^3}$, a difference of -1.18% as estimated by Dunkerly's method. ▲

REFERENCES

1 Meirovitch, Leonard, *Analytical Methods in Vibration*, Macmillan, New York, 1967.

2 Craig, Roy R., *Structural Dynamics, An Introduction to Computer Methods*, Wiley, New York, 1981.

3 Dettman, John W., *Introduction to Linear Algebra and Differential Equations*, Dover Publications, New York, 1986.

4 Keller, Herbert B., *Numerical Methods for Two-Point Boundary Value Problems*, Dover Publications, New York, 1992.

5 Hamming, R.W., *Numerical Methods for Scientists and Engineers,* 2nd ed., Dover Publications, New York, 1973.

6 Schneider, H. and Barker, G.P., *Matrices and Linear Algebra*, Dover Publications, New York, 1973.

7 Clough, R.W. and Penzen, J., *Dynamics of Structures,* 2nd ed., McGraw Hill, New York, 1993.

8 Anderson, Roger A., *Fundamentals of Vibrations*, Macmillan, New York, 1967.

NOTATION

a_{ij}	terms in flexibility matrix	M_r	generalized (modal) mass for rth normal mode
a_r	scalar coefficient for rth normal mode	Q_i	generalized forces
b_r	scalar coefficient for rth normal mode	$R(\{A\})$	Rayleigh's quotient
c_r	scaling constant for rth normal mode	T	kinetic energy
E	modulus of elasticity	V	potential energy
g	acceleration due to gravity	W_{nc}	work done by nonconservative forces
I	static moment of inertia	x	translational displacement
I_0	mass moment of inertia	$\dot{x}$	translational velocity
k_{ij}	stiffness coefficients	$\ddot{x}$	translational acceleration
k_R	rotational stiffness	z_i	generalized coordinates
K_r	generalized (modal) stiffness for rth normal mode	DOF	degree of freedom
$\mathscr{L}$	the Lagrangian	MDOF	multi-degree-of-freedom
L	length	$[a]$	flexibility matrix
m_{ij}	mass coefficient	$\{A\}_r$	eigenvector for rth normal mode

$[D]$	system matrix		δ_{z_i}	virtual displacements
$\{F\}$	force vector		δW_{nc}	virtual work done by nonconservative forces
$[H]_r$	characteristic matrix for rth normal mode		ϕ_r	phase angle for rth normal mode
$[I]$	identity matrix		ϕ_{ir}	ith element in modal vector $\{\Phi\}_r$
$[k]$	stiffness matrix		λ_r	eigenvalue for rth normal mode
$[K]$	generalized (modal) stiffness matrix		λ_R	Rayleigh's quotient
$[m]$	mass matrix		ω_r	undamped natural circular frequency for rth normal mode
$[M]$	generalized (modal) mass matrix			
$\{x\}$	displacement vector		ω_R	Rayleigh's frequency
$\{x(t)\}_r$	displacement response for rth normal mode		θ	rotational displacement
$\{x(0)\}$	initial displacement vector		$\dot{\theta}$	rotational velocity
$\{\dot{x}(0)\}$	initial velocity vector		$\ddot{\theta}$	rotational acceleration
$\{\ddot{x}\}$	acceleration vector		$[\Phi]$	modal matrix
			$\{\Phi\}_r$	modal vector for rth normal mode
$\left.\begin{array}{l}\mathbf{H}_{11}, \mathbf{H}_{12},\\ \mathbf{H}_{21}, \mathbf{H}_{22}\end{array}\right\}$ partitioned submatrices for $[H]_r$			$\{\Phi\}_{cr}$	complementary modal subvector of $\{\Phi\}_r$
			$\{\Phi\}_{mr}$	modal subvector of $\{\Phi\}_r$ containing rigid-body terms

PROBLEMS

10.1–10.12 Use the Lagrange equations to derive the equations of motion for the systems shown in Figures P10.1–P10.12.

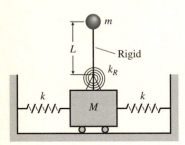

Figure P10.1

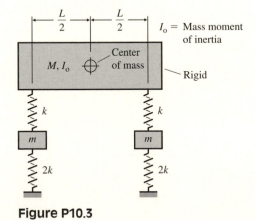

Figure P10.3

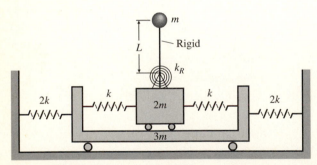

Figure P10.2

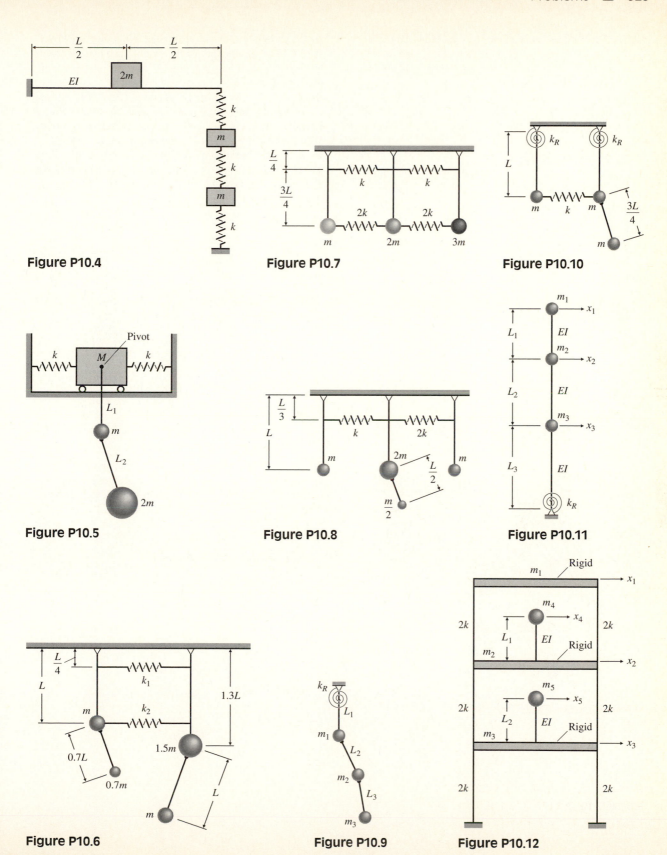

Figure P10.4

Figure P10.7

Figure P10.10

Figure P10.5

Figure P10.8

Figure P10.11

Figure P10.6

Figure P10.9

Figure P10.12

10.13 For the three-story shear frame building shown in Figure P10.13, determine the three natural frequencies and corresponding mode shapes. Check for orthogonality of modes and sketch the three mode shapes.

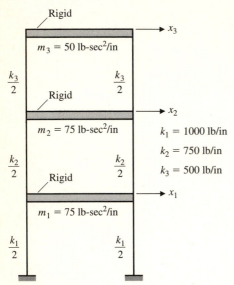

$k_1 = 1000$ lb/in
$k_2 = 750$ lb/in
$k_3 = 500$ lb/in

Figure P10.13

10.14 A two-story shear frame building has an interior substructure supported on its first level as shown in Figure P10.14. Determine the three natural frequencies and corresponding mode shapes for this structure. Check for modal orthogonality.

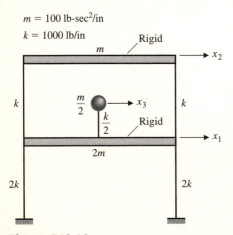

Figure P10.14

10.15 A W14×48 beam spanning 25 ft is supported by a W10×33 beam of length 20 ft at point A and by a W12×36 beam also of 20 ft length at point B, as shown in the framing plan in Figure P10.15. The W10×33 is fixed at both ends, while the W12×36 is simply supported. The W14×48 is pin connected at points A and B and carries a concentrated dead load W as shown in the figure. Assume the self weight of the W10×33 and W12×36 to be lumped at points A and B, respectively. Determine the natural frequencies and mode shapes for this system. Check for modal orthogonality. Assume all beams bend about their major axis and $E = 29,000$ ksi. Assume 3 DOF for the system.

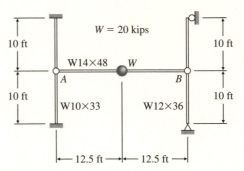

Figure P10.15

10.16 Determine the three natural frequencies and corresponding mode shapes for the beam shown in Figure P10.16. Check for modal orthogonality and sketch the mode shapes.

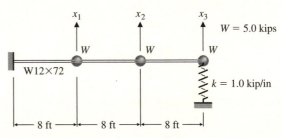

Figure P10.16

10.17 Determine the three natural frequencies and corresponding mode shapes for the system shown in Figure P10.17. Check for orthogonality of modes.

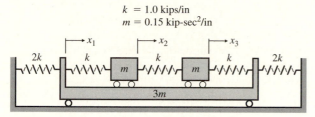

$k = 1.0$ kips/in
$m = 0.15$ kip-sec^2/in

Figure P10.17

10.18 A rigid rectangular slab weighing 200 kips is supported by four columns of 12-ft height and flexural rigidity EI as shown in Figure P10.18. Determine the three natural frequencies and corresponding mode shapes of this structure. Check for orthogonality of modes. Assume $a = \dfrac{b}{2} = 5.0$ ft.

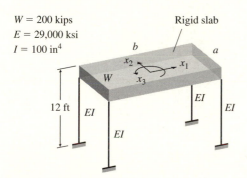

$W = 200$ kips
$E = 29,000$ ksi
$I = 100$ in^4

Figure P10.18

10.19 A rigid rectangular slab weighing 150 kips is supported on three columns of 10-ft height and flexural rigidity EI as shown in Figure P10.19. Determine the three natural frequencies and mode shapes for this structure. Check for modal orthogonality. Assume $a = \dfrac{b}{2} = 6.0$ ft.

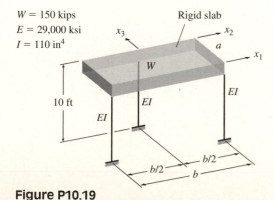

$W = 150$ kips
$E = 29,000$ ksi
$I = 110$ in^4

Figure P10.19

10.20 Determine the natural frequencies and mode shapes for the semidefinite system shown in Figure P10.20. Check for modal orthogonality.

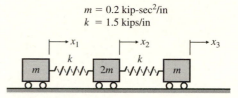

$m = 0.2$ kip-sec^2/in
$k = 1.5$ kips/in

Figure P10.20

10.21 An airplane is modeled using three lumped masses as shown in Figure P10.21. Determine the natural frequencies and vibration modes. Verify that the rigid-body modes are orthogonal.

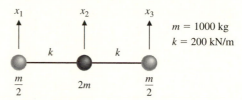

$m = 1000$ kg
$k = 200$ kN/m

Figure P10.21

10.22 A flying wing is idealized as an unconstrained four-mass beam as depicted in Figure P10.22. Assuming that each mass moves in the transverse direction only, determine (1) the four natural frequencies for this system and (2) determine and sketch the corresponding mode shapes. Select the two rigid-body modes such that they are orthogonal with one another. Assume $E = 29,000$ ksi.

$I = 350$ in^4
$m = 26.8$ lb-sec^2/in
$L = 10$ ft

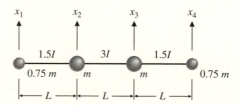

Figure P10.22

10.23 Determine the four natural frequencies and mode shapes for the semidefinite system shown in Figure P10.23. Verify that the modes are orthogonal.

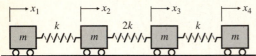

$$m = 0.5 \text{ kip-sec}^2/\text{in}$$
$$k = 250 \text{ kips/in}$$

Figure P10.23

10.24 Repeat Problem 10.22 for length $L = 12$ ft, $m = 40$ lb-sec²/in, and moment of inertia $I = 500$ in⁴.

10.25–10.30 Determine the free vibration displacement response $\{x(t)\}$ for the structures shown in Figures P10.13 through P10.18 subject to the following initial conditions:

10.31–10.37 Estimate the fundamental frequency of the structures shown in Figures P10.13 through P10.19 by the Rayleigh method (or Rayleigh's quotient). For the trial vector (assumed mode shape), use the static displacement vector resulting from subjecting each mass to a force equal to its own weight. Compare the results with those obtained in Problems 10.13 through 10.19.

10.38–10.44 Estimate the fundamental frequency of the structures shown in Figures P10.13 through P10.19 by Dunkerly's method.

Problem No.	Figure No.	Initial Conditions $\{x(0)\}$	Initial Conditions $\{\dot{x}(0)\}$	Problem No.	Figure No.	Initial Conditions $\{x(0)\}$	Initial Conditions $\{\dot{x}(0)\}$
10.25	P10.13	$\begin{Bmatrix} 0 \\ -1.0 \\ -2.0 \end{Bmatrix}$ in	$\{0\}$	10.28	P10.16	$\begin{Bmatrix} 1.0 \\ 0 \\ 0 \end{Bmatrix}$ in	$\begin{Bmatrix} 0 \\ 0 \\ -4.0 \end{Bmatrix}$ in/sec
10.26	P10.14	$\begin{Bmatrix} 1.0 \\ 0 \\ -1.0 \end{Bmatrix}$ in	$\{0\}$	10.29	P10.17	$\begin{Bmatrix} 1.0 \\ 0 \\ 0 \end{Bmatrix}$ in	$\begin{Bmatrix} 0 \\ 0 \\ 4.0 \end{Bmatrix}$ in/sec
10.27	P10.15	$\{0\}$	$\begin{Bmatrix} 0 \\ 3.0 \\ -3.0 \end{Bmatrix}$ in/sec	10.30	P10.18	$\begin{Bmatrix} 2.0 \\ 0 \\ 0 \end{Bmatrix}$ in	$\begin{Bmatrix} 0 \\ 5.0 \\ 0 \end{Bmatrix}$ in/sec

11 ▲ Numerical Solution Methods for Natural Frequencies and Mode Shapes

A critical step in the dynamic analysis of MDOF systems is quite often the solution of the eigenproblem, or the determination of the system natural frequencies and corresponding normal vibration modes. This is especially true if a mode superposition analysis is to be conducted (dynamic analysis by the mode superposition method is discussed in Chapter 12). Several procedures for solving the eigenproblem were discussed in Chapters 9 and 10 that were predicated upon finding the roots of the characteristic (polynomial) equation. These techniques are satisfactory for systems having only several degrees of freedom. For large MDOF systems (i.e., systems having many DOF), extracting the roots of the characteristic polynomial manually requires tedious computational effort and is quite often an indeterminable task.

Over the years, especially since the evolvement of the digital computer, various methods and techniques for solving the eigenproblem have been developed. These techniques are all iterative in nature and therefore can represent a very significant computational endeavor for systems with many DOF. This chapter discusses the basic solution schemes for the MDOF system eigenproblem and describes several commonly employed eigensolvers.

11.1 GENERAL SOLUTION METHODS FOR EIGENPROBLEMS

In structural dynamics the basic eigenproblem for a MDOF system having n degrees of freedom is expressed as

$$[k]\{\Phi\} = \lambda[m]\{\Phi\} \tag{11.1}$$

where $[k]$ is the stiffness matrix of order n and $[m]$ is the mass matrix, also of order n. For most structural systems, $[k]$ is a *narrowly banded* matrix and $[m]$ is a diagonal matrix for a lumped mass formulation (without inertia coupling), or a narrowly banded matrix for a consistent mass formulation. There are n eigenvalues and corresponding eigenvectors satisfying Eq. (11.1). The rth *eigenpair* is denoted as $(\lambda_r, \{\Phi\}_r)$, where the eigenvalues λ_r are the free vibration frequencies squared ω_r^2 and $\{\Phi\}_r$ are the corresponding mode shape vectors. The eigenvalues are ordered according to their magnitudes such that

$$0 < \lambda_1 < \lambda_2 < \dots < \lambda_n \tag{11.2}$$

The dynamic response of MDOF systems having a large number of DOF is generally confined to a relatively small subset of the lowest vibration modes of the system. Therefore, for such systems only p eigenpairs need to be solved for, where $p << n$. The solution for p eigenvalues and corresponding eigenvectors of Eq. (11.1) can be written as

$$[k]\{\Phi\} = [m]\{\Phi\}[\Lambda] \tag{11.3}$$

where $[\Phi]$ is the $n \times p$ modal matrix whose columns are the p eigenvectors and $[\Lambda]$ is a $p \times p$ diagonal matrix containing the corresponding eigenvalues. However, for the purpose of general discussion of eigenproblem solution techniques, it will be assumed that $p = n$.

The majority of eigenproblem solution techniques can be classified into three basic categories [1]: (1) *vector iteration methods,* (2) *transformation methods,* and (3) *polynomial iteration methods.* Clearly, all methods must be iterative in nature, because solution of the eigenproblem as defined by Eq. (11.1) is tantamount to solving the characteristic polynomial of order n. Since explicit formulas for the determination of roots to the characteristic polynomial having an order higher than 4 do not exist, an iterative solution is mandatory.

The essence of each method is very distinctive. The vector iteration methods are based on the property that

$$[k]\{\Phi\}_r = \lambda_r[m]\{\Phi\}_r \tag{11.4}$$

The transformation methods are characterized by the general relationships

$$[\Phi]^T[k][\Phi] = [K] \tag{11.5}$$

and

$$[\Phi]^T[m][\Phi] = [M] \tag{11.6}$$

where $[K]$ and $[M]$ are the modal stiffness and modal mass matrices, respectively. The polynomial iteration techniques are formulated on the property that

$$p(\lambda_r) = 0 \tag{11.7}$$

where

$$p(\lambda) = \det([k] - \lambda[m]) \tag{11.8}$$

is the characteristic polynomial of order n.

In this chapter, several vector iteration techniques (i.e., the inverse iteration method and the forward iteration method) are presented along with one transformation method (i.e., the generalized Jacobi method). A general discussion of solution techniques for large eigenproblems is also presented. These large eigensolver techniques are formulated by a combination of two or more basic methods, and the polynomial iteration techniques are discussed within that context.

11.2   INVERSE VECTOR ITERATION

Inverse iteration is a very popular vector iteration technique used to solve eigenproblems in structural mechanics (i.e., dynamics and stability) because it converges to the eigenvector corresponding to the lowest eigenvalue. All vector iteration techniques begin with consideration of the basic relation given by Eq. (11.1). The intent in inverse iteration is to satisfy Eq. (11.1) by initially assuming a trial eigenvector and eigenvalue and iterating to convergence.

To illustrate the basic concept in inverse iteration, we assume a trial vector $\{x\}_1$ for $\{\Phi\}$ and assume a value for λ. If it is also assumed that $\lambda = 1$, then the right-hand side of Eq. (11.1) can be expressed as

$$\{R\}_1 = [m]\{x\}_1 \tag{11.9}$$

Since $\{x\}_1$ is an arbitrary vector, then it is not generally true that $\{R\}_1 = [m]\{x\}_1$ (if this relation is satisfied then $\{x\}_1$ is an eigenvector). Therefore we select another vector $\{x\}_2$, which is the static displacement vector corresponding to the forces $\{R\}_1$, such that $\{x\}_2 \neq \{x\}_1$ and arrive at the equilibrium condition

$$[k]\{x\}_2 = \{R\}_1 \tag{11.10}$$

Thus $\{x\}_2$ represents the approximation of the eigenvector $\{\Phi\}$ after one cycle of iteration. Indeed, $\{x\}_2$ is a better approximation to the eigenvector than $\{x\}_1$. Repeating the cycle yields an increasingly better approximation to the eigenvector, and convergence is eventually attained.

The algorithm describing the solution for the fundamental eigenpair $(\lambda_1, \{\Phi\}_1)$ is summarized in Table 11.1. In this formulation, $[k]$ is assumed to be positive definite, whereas $[m]$ can be either positive definite or positive semidefinite.

In Eq. (11.16), TOL is a previously set tolerance. TOL should be 10^{-2s} or smaller when the eigenvalue is required to $2s$-digit accuracy. The eigenvector will be accurate to s digits.

Equation (11.13) is a Rayleigh quotient estimate of the eigenvalue. Eqs. (11.14) and (11.15) represent normalization of the vectors $\{\bar{y}\}_{k+1}$ and $\{\bar{x}\}_{k+1}$, respectively, so that the normalized vectors $\{y\}_{k+1}$ and $\{x\}_{k+1}$ have a *Euclidian vector norm* of unity.

11.2.1   Higher Modes by Inverse Vector Iteration

The basic inverse iteration technique just discussed converges to the numerically smallest eigenvalue λ_1 and corresponding eigenvector $\{\Phi\}_1$. Because of this condition of convergence, λ_1 is said to be the *least dominant eigenvalue* and $\{\Phi\}_1$ is called the *least dominant eigenvector*. However, the method may be modified to calculate eigenvalues and the corresponding eigenvectors for higher modes by either *matrix deflation* or deflation of the iteration vectors.

For the inverse iteration method, it is convenient to deflate the iteration vector in order to converge to an eigenpair other than $(\lambda_1, \{\Phi\}_1)$. The basic premise for *vector deflation* is that, for an iteration vector to converge to a required eigenvector, the iteration vector must not be orthogonal to the eigenvector. Therefore, for the case at hand, this can be interpreted as meaning that if the iteration vector is orthogonalized to the eigenvectors already calculated (for example, $\{\Phi\}_1$), the convergence to these eigenvectors is

TABLE 11.1. **Algorithm for Inverse Vector Iteration**

1. Assume an arbitrary trial vector $\{x\}_1$
2. Define $\{y\}_1 = [m]\{x\}_1$
3. Evaluate the following vectors for $k = 1, 2, 3, \ldots$ iterations:

$$\{\bar{x}\}_{k+1} = [k]^{-1}\{y\}_k \tag{11.11}$$

$$\{\bar{y}\}_{k+1} = [m]\{\bar{x}\}_{k+1} \tag{11.12}$$

$$\rho(\{\bar{x}\}_{k+1}) = \frac{\{\bar{x}\}_{k+1}^T\{y\}_k}{\{\bar{x}\}_{k+1}^T\{\bar{y}\}_{k+1}} \tag{11.13}$$

$$\{y\}_{k+1} = \frac{\{\bar{y}\}_{k+1}}{[\{\bar{x}\}_{k+1}^T\{\bar{y}\}_{k+1}]^{1/2}} \tag{11.14}$$

$$\{x\}_{k+1} = \frac{\{\bar{x}\}_{k+1}}{[\{\bar{x}\}_{k+1}^T\{\bar{y}\}_{k+1}]^{1/2}} \tag{11.15}$$

Then as $k \to \infty$,

$$\rho(\{\bar{x}\}_{k+1}) \to \lambda_1 \quad \text{and} \quad \{x\}_{k+1} \to \{\Phi\}_1$$

Convergence is achieved when

$$\frac{|\lambda_{k+1} - \lambda_k|}{\lambda_{k+1}} \leq \text{TOL} \tag{11.16}$$

precluded and occurs instead to another (higher) eigenvector. More succinctly, an eigenpair other than $(\lambda_1, \{\Phi\}_1)$ becomes the *least dominant eigenpair.*

A commonly employed *vector orthogonalization* procedure is the Gram-Schmidt method. To illustrate general deployment of Gram-Schmidt orthogonalization to inverse vector iteration, it is assumed that the eigenvectors $\{\Phi\}_1, \{\Phi\}_2, \ldots, \{\Phi\}_m$ have been calculated. It is desired to $[m]$-orthogonalize $\{x\}_1$ to these eigenvectors. Thus, a vector $\{z\}_1$ that is $[m]$-orthogonal to the eigenvectors $\{\Phi\}_i, i = 1, \ldots, m$, is calculated as

$$\{z\}_1 = [S]_m\{x\}_1 \tag{11.17}$$

where the vector $\{z\}_1$, in *Gram-Schmidt orthogonalization,* is defined as

$$\{z\}_1 = \{x\}_1 - \sum_{i=1}^{m} \alpha_i\{\Phi\}_i \tag{11.18}$$

The coefficients α_i in Eq. (11.18) are obtained using the conditions that

$$\{\Phi\}_i^T[m]\{z\}_1 = 0 \qquad i = 1, 2, \ldots, m \tag{11.19}$$

and

$$\{\Phi\}_i^T[m]\{\Phi\}_j = \delta_{ij} \tag{11.20}$$

where δ_{ij} is the Kronecker delta. Premultiplying both sides of Eq. (11.18) by $\{\Phi\}_i^T[m]$ and solving for α_i yields

$$\alpha_i = \frac{\{\Phi\}_i^T[m]\{x\}_1}{\{\Phi\}_i^T[m]\{\Phi\}_i} \qquad i = 1, 2, \ldots, m \tag{11.21}$$

Then substituting Eq. (11.21) into Eq. (11.18) results in

$$\{z\}_1 = \{x\}_1 - \sum_{i=1}^{m} \frac{\{\Phi\}_i\{\Phi\}_i^T[m]\{x\}_1}{\{\Phi\}_i^T[m]\{\Phi\}_i} \tag{11.22}$$

or

$$\{z\}_1 = \left([I] - \sum_{i=1}^{m} \frac{\{\Phi\}_i\{\Phi\}_i^T[m]}{\{\Phi\}_i^T[m]\{\Phi\}_i}\right)\{x\}_1 = [S]_m\{x\}_1 \tag{11.23}$$

in which $[I]$ is the identity matrix. Solving Eq. (11.23) for $[S]_m$ yields

$$[S]_m = [I] - \sum_{i=1}^{m} \frac{\{\Phi\}_i\{\Phi\}_i^T[m]}{\{\Phi\}_i^T[m]\{\Phi\}_i} \tag{11.24}$$

where $[S]_m$ is called the *sweeping matrix* for the mth mode. The purpose of $[S]_m$ is to eliminate, or "sweep out," the effect of modes 1 through m and allow the mode $m + 1$ to become least dominant. This vector deflation technique can be incorporated into the inverse vector iteration algorithm defined by Eqs. (11.11) through (11.15) by substituting $\{z\}_1$, given by Eq. (11.17), for $\{x\}_1$ in Eq. (11.11) for the initial approximation of the trial vector $\{y\}_1$.

To summarize the inverse vector iteration method, recurrence formulas for calculating the fundamental (least dominant) eigenpair $(\lambda_1, \{\Phi\}_1)$ are given by Eqs. (11.11) through (11.15). Vector deflation of modes 1 through m by the Gram-Schmidt orthogonalization method is given by Eqs. (11.17) and (11.24), which may be implemented into the iteration scheme to solve for additional eigenpairs in ascending order.

EXAMPLE 11.1 ▼

(a) Write a computer program to extract any user specified number of natural frequencies and corresponding mode shapes for an arbitrary MDOF system having n degrees of freedom by the *inverse vector iteration* technique. Specify a convergence tolerance TOL $= 10^{-6}$.

(b) Determine the three natural frequencies and corresponding modal vectors for the three-story shear frame building shown in Figure 11.1. Assume modulus of elasticity $E = 29,000$ ksi and moment of inertia $I = 750$ in^4 for all columns.

Figure 11.1

Three-story shear frame building of Example 11.1.

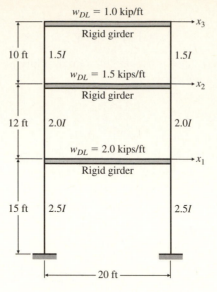

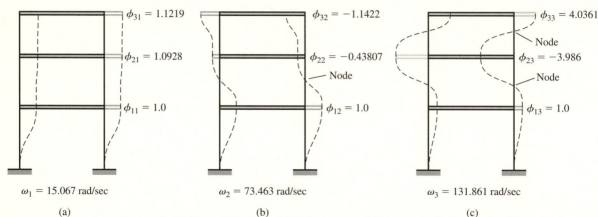

$\omega_1 = 15.067$ rad/sec $\omega_2 = 73.463$ rad/sec $\omega_3 = 131.861$ rad/sec

(a)   (b)   (c)

Figure 11.2

Normal modes for three-story shear frame building of Example 11.1: (a) mode 1; (b) mode 2; (c) mode 3.

Solution

(a) The input parameters for the structure shown in Figure 11.1 are

$$[m] = \begin{bmatrix} 0.1035 & 0 & 0 \\ 0 & 0.0776 & 0 \\ 0 & 0 & 0.0518 \end{bmatrix} \text{kip-sec}^2/\text{in}$$

and

$$[k] = \begin{bmatrix} 405.57 & -349.630 & 0 \\ -349.63 & 802.76 & -453.13 \\ 0 & -453.13 & 453.13 \end{bmatrix} \text{kip/in}$$

(b) The results of the frequency analysis are summarized in Table 11.2, the mode shapes are shown in Figure 11.2, and a listing of the FORTRAN computer program is provided in Table 11.3.

TABLE 11.2.　**Eigenvalues and Eigenvectors for Example 11.1**

Degrees of freedom	= 3
Number of eigenvalues extracted	= 3
Tolerance used for convergence	= 0.10E−05

(a) Eigenvalues

No.	Eigenvalue	Omega (ω) (rad/sec)	Iterations
1	0.227015E+03	0.150670E+02	4
2	0.539681E+04	0.734630E+02	5
3	0.173873E+05	0.131861E+03	4

(b) Eigenvectors

Mode 1	Mode 2	Mode 3
0.10000E+01	0.10000E+01	0.10000E+01
0.10928E+01	−0.43807E+00	−0.39864E+01
0.11219E+01	−0.11422E+01	0.40361E+01

TABLE 11.3.　**Listing of FORTRAN Computer Program Used For Example 11.1**

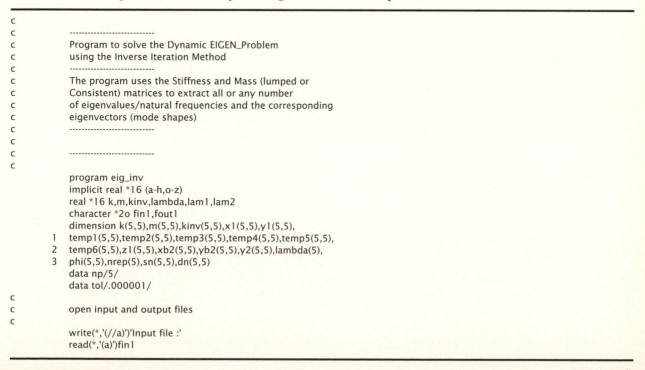

```
c
c            ----------------------------
c            Program to solve the Dynamic EIGEN_Problem
c            using the Inverse Iteration Method
c            ----------------------------
c            The program uses the Stiffness and Mass (lumped or
c            Consistent) matrices to extract all or any number
c            of eigenvalues/natural frequencies and the corresponding
c            eigenvectors (mode shapes)
c            ----------------------------
c
c            ----------------------------
c
            program eig_inv
            implicit real *16 (a-h,o-z)
            real *16 k,m,kinv,lambda,lam1,lam2
            character *2o fin1,fout1
            dimension k(5,5),m(5,5),kinv(5,5),x1(5,5),y1(5,5),
      1     temp1(5,5),temp2(5,5),temp3(5,5),temp4(5,5),temp5(5,5),
      2     temp6(5,5),z1(5,5),xb2(5,5),yb2(5,5),y2(5,5),lambda(5),
      3     phi(5,5),nrep(5),sn(5,5),dn(5,5)
            data np/5/
            data tol/.000001/
c
c            open input and output files
c
            write(*,'(//a)')'Input file :'
            read(*,'(a)')fin1
```

(continued)

```
               write(*,'(/a)')' Output file :'
               read(*,'(a)')fout1
               open(1,file=fin1,status='old')
               open(2,file=fout1,status='unknown')
c
c              read input data
c
               call input1(nd,k,m,neig,np)
c
c              invert k matrix
c
               call matinv(k,nd,np,kinv)
c
c              setup trial vector X1
c
               do 1001 i=1,nd
                     x1(i,1)=1.0
1001           continue
               call matmul(m,x1,nd,nd,1,temp,np,1,0.)
c
c              neig=number of eigenvalues to be extracted
c
               do 1005 nloop=1,neig
c              fundamental frequency
                     if (nloop.eq.1) then
                             call matmul(m,x1,nd,nd,1,y1,yp,1,0.)
                             icount=1
                             goto 1101
                     else
c              selection of trial vector for higher modes
                     do 1006 ii=1,nd
                             z1(ii,1)=x1(ii,1)
1006                 continue
c              Setup Sn = I (identity matrix)
c              Sn will be used as the sweeping matrix to orthogonalize
c              the trial vector with respect to the previous modes
                     do 1019 ii=1,nd
                             do 1020 jj=1,nd
                                  sn(ii,jj)=0.0
1020                         continue
                             sn(ii,ii)=1.0
1019                 continue
                     do 1010 i=1,nloop-1
                             do 1011 ii=1,nd
                                  temp2(ii,1)=phi(ii,i)
1011                         continue
                             call trans(temp2,nd,1,temp3,np)
                             call matmul(temp3,temp1,1,nd,1,temp4,np,1,0.)
                             call matmul(temp2,temp3,nd,1,1,temp5,np,2,temp4(1,1))
                             call matsub(z1,temp5,nd,1,np,temp6,1)
                             do 1012 ii=1,nd
                                  z1(ii,1)=temp6(ii,i)
1012                         continue
c              Setup Sweeping Matrix Sn
                             call matmul(temp3,m,1,nd,nd,temp5,np,1,0.)
                             call matmul(temp2,temp5,nd,1,nd,temp6,np,1,0.)
                             call matmul(temp5,temp2,1,nd,1,temp4,np,1,0.)
                             temp4(1,1)=1.0/temp4(1,1)
                             call matmul(temp6,temp3,nd,nd,1,temp5,np,2,temp4(1,1))
                             call matsub(sn,temp5,nd,nd,np,temp6,1)
                             do 1018 ii=1,nd
                                  do 1018 jj=1,nd
                                       sn(ii,jj)=temp6(ii,jj)
```

(continued)

```
1018              continue
1010           continue
           endif
c
c          additional purification of the iteration vector
c          at each iteration using the sweeping matrix Sn
c          Y1 = ([K]−1).[M].{Sn}
               call matmul(kinv,m,nd,nd,nd,temp2,np,1,0.)
                   call matmul(temp2,sn,nd,nd,nd,dn,np,1,0.)
                   do 1021 ii=1,nd
                       y1(ii,1)=z1(ii,1)/z1(1,1)
1021               continue
           icount=1
c
c          iterative loop to extract eigenvalue
c
1101       continue
               if(nloop.eq.1) then
                       call matmul(kinv,y1,nd,nd,1,xb2,np,1,0.)
                   else
                       call matmul(dn,y1,nd,nd,1,xb2,np,1,0.)
               endif
                   call matmul(m,xb2,nd,nd,1,yb2,np,1,0.)
                   call trans(xb2,nd,1,temp3,np)
                   call matmul(temp3,y1,1,nd,1,temp4,np,1,0.)
                   call matmul(temp3,yb2,1,nd,1,temp5,np,1,0.)
                   lam2=temp4(1,1)/temp5(1,1)
                   call matmul(temp3,yb2,1,nd,1,temp4,np,1,0.)
                   temp4(1,1)=1.0/sqrt(temp4(1,1))
                   call matmul (xb2,temp2,nd,1,1,temp5,np,2,temp4(1,1))
                   call matmul(yb2,temp2,nd,1,1,y2,np,2,temp4(1,1))
                   if(icount.gt.1) then
c
c          check for tolerance
c          store eigenvalue/eigenvector if satisfied
c
                       t1=abs((lam2−lam1)/lam2)
                           if(t1.le.tol) then
                                   lambda(nloop)=lam2
                                   do 1103 ii=1,nd
                                       phi(ii,nloop)=temp5(ii,1)
1103                               continue
                           nrep(nloop)=icount
                                   goto 1005
                               endif
                       endif
c
c          setup for next iteration if t1>tol
c
                   do 1105 ii=1,nd
                       y1(ii,1)=y2(ii,1)
1105               continue
                   lam1=lam2
                   icount=icount+1
                   goto 1101
1005       continue
c
c          print output
c
                   write(2,1114)nd,neig,tol
1114               format(' Degrees of Freedom          =',i5,/,
          1                ' Number of Eigenvalues Extracted =',i5,/,
          1                ' Tolerance used for Convergence ='e12.2,//,
```

(continued)

```
      2        '    No.     Eigenvalue              omega              nrep',/,
      3        '                                  (rad/sec)',/,
      4        ' -----------------------'
      5        ,)
               do 1111 i=1,neig
                   omega=sqrt(lambda(i))
                        write(2,1112)i,lambda(i),omega,nrep(i)
1112                    format(i5,2e15.6,i11)
1111           continue
c
c         scale eigenvectors with respect to 1st DOF
c
               do 1120 j=1,neig
                   scale=1.0/phi(1,j)
                        do 1120 i=1,nd
                            phi(i,j)=phi(i,j)*scale
1120           continue
                   write(2,1115)
1115      format(//,' EIGENVECTORS')
               do 1113 i=1,nd
                   write(2,1116)(phi(i,j),j=1,neig)
1116               format(5e15.5)
1113           continue
      stop
           end
c
c         subroutine INPUT1 is used to read in input data
c         nd = degrees of freedom
c         neig = no of eigenvalues to be extracted
c         K = stiffness matrix -- must be positive definite, symmetric
c         M = mass matrix -- may be lumped or consistent
c
               subroutine input1(nd,k,m,neig,np)
               implicit real *16 (a-h,o-z)
               real *16 k,m
               dimension k(np,np),m(np,np)
               read(1,*)nd,neig
               do 9101 i=1,nd
                   read(1,*)(k(i,j),j=1,nd)
9101           continue
           do 9102 i=i,nd
                   read(1,*) (m(i,j),j=1,nd)
9102      continue
      return
           end
c
c         subroutine MATMUL is used to multiply matrices
c         a and b for isol = 1
c         for isol = 2 multiplies a by a scalar alpha
c
               subroutine matmul(a,b,m,l,n,c,np,isol,alpha)
               implicit real *16 (a-h,o-z)
               dimension a(np,np),b(np,np),c(np,np)

               if(isol.eq.1) then
                   do 9030 i=1,m
                       do 9030 j=1,n
                           c(i,j)=0.0
                               do 9030 ii=1,l
                                   c(i,j)=c(i,j)+a(i,ii)*b(ii,j)
9030               continue
           elseif(isol.eq.2)then
               do 9031 i=1,m
```

(continued)

```
                        do 9031 j=1,l
                            c(i,j)=alpha*a(i,j)
9031             continue
           endif
           return
           end
c
c        subroutine MATSUB is used to subtract/add matrices
c        isol = 1 >> A−B
c        isol = 2 >> A+B
           subroutine matsub(a,b,m,n,np,c,isol)
                   implicit real *16 (a-h,o-z)
                   dimension a(np,np),n(np,np),c(np,np)

                   do 9055 i=1,m
                       do 9055 j=1,n
                           c(i,j)=a(i,j)+((−1.)**isol)*b(i,j)
9055             continue
           return
           end
c
c        subroutine TRANS is used to transpose the matrix X
c
           subroutine trans(x,m,n,y,np)
           implict real *16 (a-h,o-z)
           dimension x(np,np),y(np,np)
               do 9050 i=1,m
                   do 9050 j=1,n
                       y(j,i)=x(i,j)
9050               continue
           return
           end
c
c        subroutine MATINV is used to inverse the K matrix
c        this routine uses the CHOLESKI DECOMPOSITION technique
c        which is efficient for positive definite and symmetric
c        matrices (e.g. the stiffness matrix)
c
                   subroutine matinv(k,n,np,y)
                       implicit real *16 (a-h,o-z)
                       real *16 k
                       dimension a(5,5),k(np,np),y(np,np),p(5),b(5),x(5)
c
c        Set up Identity Matrix
c
                   do 9012 i=1,n
                           do 9011 j=1,n
                               y(i,j)=0.0
9011                   continue
                       y(i,i)=1.0
9012           continue
c
c        assign K matrix to A
c
               do 9013 i=1,n
                   do 9013 j=1,n
                       a(i,j)=k(i,j)
9013       continue
c
c        Construct Choleski decomposition of A
c
           call choldc(a,n,np,p)
c
```

(continued)

```fortran
c         define columns of inverse matrix y
c
          do 9014 j=1,n
              do 9020 i=1,n
                      b(i)=y(i,j)
9020          continue
              call cholsl(a,n,np,p,b,x)
              do 9021 i=1,n
                      y(i,j)=x(i)
9021          continue
9014      continue
      return
          end
c
c         Constructs the Cholesky decomposition of the matrix A
c         A = L.L'
c
          subroutine choldc(a,n,np,p)
              implicit real *16 (a-h,o-z)
              dimension a(np,np),p(n)
              do 9015 i=1,n
                  do 9016 j=i,n
                      sum=a(i,j)
                      do 9017 k=i-1,1,-1
                          sum=sum-a(i,k)*a(j,k)
9017                  continue
                      if(i.eq.j)then
                          if(sum.le.0.0)then
                                  write(*,*)' Choleski Failed'
                                  stop
                          endif
                          p(i)=sqrt(sum)
                      else
                          a(j,i)=sum/p(i)
                      endif
9016              continue
9015          continue
      return
          end
c
c         Solves n linear equations A.x = B
c         x returns the corresponding column of the inverse of A
c
                  subroutine cholsl(a,n,np,p,b,x)
                  implicit real *16 (a-h,o-z)
                  dimension a(np,np),p(n),b(n),x(n)
                      do 9022 i=1,n
                          sum=b(i)
                          do 9023 k=i-1,1,-1
                              sum=sum-a(i,k)*x(k)
9023                      continue
                          x(i)=sum/p(i)
9022                  continue
      do 9024 i=n,1,-1
                          sum=x(i)
                          do 9025 k=i+1,n
                              sum=sum-a(k,i)*x(k)
9025                      continue
                          x(i)=sum/p(i)
9024              continue
      return
          end
```

▲

11.3 FORWARD VECTOR ITERATION

The forward vector iteration method, also known as the *direct iteration method* or the *power iteration method,* is the complement to the inverse iteration method in that the dominant eigenpair is the numerically largest. Therefore, for a MDOF system having n degrees of freedom the method will converge to the eigenpair $(\lambda_n, \{\Phi\}_n)$. To take advantage of this method in dynamic analysis, the flexibility-based formulation of the eigenproblem must be used.

To develop the iteration procedure, begin by expressing the eigenproblem as

$$[[a][m] - \lambda[I]][\Phi] = [0] \tag{11.25}$$

where $[a]$ is the flexibility matrix, $[m]$ is the mass matrix assumed to be positive definite, and the eigenvalue $\lambda = 1/\omega^2$. Expressing Eq. (11.25) in standard form, analogous to Eq. (11.1),

$$[D][\Phi] = \lambda\{\Phi\} \tag{11.26}$$

where $[D]$ is the dynamic matrix given by

$$[D] = [a][m] \tag{11.27}$$

Then operate on Eq. (11.26) as described in Table 11.4.

Then as $k \to \infty$,

$$\rho(\{\bar{x}\}_{k+1}) \to \lambda_1 \quad \text{and} \quad \{x\}_{k+1} \to \{\Phi\}_1$$

The method converges to the highest (*dominant*) eigenvalue λ_1 ($\lambda_1 > \lambda_2 > \lambda_3 > \ldots > \lambda_n$). Since $\lambda = 1/\omega^2$, the method converges to the eigenvector corresponding to the *lowest* natural frequency. Similar to the inverse iteration procedure, Eq. (11.29) is a Rayleigh quotient estimate of the eigenvalue, and Eq. (11.30) provides the normalized vector $\{x\}_{k+1}$ with a Euclidian vector norm of unity. Convergence is verified by Eq. (11.16).

TABLE 11.4. **Algorithm for Forward Vector Iteration**

1. Assume an arbitrary trial vector $\{x\}_1$
2. Evaluate the following vectors for $k = 1, 2, 3, \ldots$ iterations:

$$\{\bar{x}\}_{k+1} = [D]\{x\}_k \tag{11.28}$$

$$\rho(\{\bar{x}\}_{k+1}) = \frac{\{\bar{x}\}_{k+1}^T \{\bar{x}\}_{k+1}}{\{\bar{x}\}_{k+1}^T \{x\}_k} \tag{11.29}$$

$$\{x\}_{k+1} = \frac{\{\bar{x}\}_{k+1}}{[\{\bar{x}\}_{k+1}^T \{\bar{x}\}_{k+1}]^{1/2}} \tag{11.30}$$

11.3.1   Higher Modes by Forward Vector Iteration

Approximations of successively less *dominant eigenpairs* can be achieved by matrix deflation of the $[D]$ matrix. The Gram-Schmidt orthogonalization procedure may be used for this purpose, in a manner analogous to that employed in the inverse iteration technique.

With λ_1 and $\{\Phi\}_1$ having been determined, a new matrix $[D]_1$ is defined such that

$$[D]_1 = [D][S]_1 \tag{11.31}$$

and from Eq. (11.24) for iteration $m = 1$,

$$[S]_1 = [I] - \frac{\{\Phi\}_1\{\Phi\}_1^T[m]}{\{\Phi\}_1^T[m]\{\Phi\}_1} \tag{11.32}$$

where $[S]_1$ is the sweeping matrix that eliminates $\{\Phi\}_1$ as an eigenvector of $[D]_1$. Thus the new matrix $[D]_1$ has eigenvalues $\lambda_2, \lambda_3, \ldots, \lambda_n$ and associated eigenvectors $\{\Phi\}_2$, $\{\Phi\}_3, \ldots, \{\Phi\}_n$. Therefore, iteration of the equilibrium equation given by

$$[D]_1\{\Phi\} = \lambda\{\Phi\} \tag{11.33}$$

converges to $\{\Phi\}_2$ and its corresponding eigenvalue λ_2.

Similarly, deflation of the matrix $[D]_1$ is given by

$$[D]_2 = [D]_1[S]_2 \tag{11.34}$$

where

$$[S]_2 = [I] - \frac{\{\Phi\}_2\{\Phi\}_2^T[m]}{\{\Phi\}_1^T[m]\{\Phi\}_1} \tag{11.35}$$

which sweeps out $\{\Phi\}_2$ as an eigenvector of $[D]_2$, leaving $(\lambda_3, \{\Phi\}_3)$ as the dominant eigenpair. Therefore, iteration of the equilibrium equation

$$[D]_2\{\Phi\} = \lambda\{\Phi\} \tag{11.36}$$

converges to λ_3 and its associated eigenvector $\{\Phi\}_3$. This deflation procedure may be extended to extract any desired number of successive dominant eigenpairs. Thus, having calculated the eigenvectors $\{\Phi\}_1, \{\Phi\}_2, \ldots, \{\Phi\}_m$, the next dominant eigenpair $(\lambda_{m+1}, \{\Phi\}_{m+1})$ can be determined by iteration of the equations

$$[D]_m\{\Phi\} = \lambda\{\Phi\} \tag{11.37}$$

where

$$[D]_m = [D][S]_m \tag{11.38}$$

and $[S]_m$ is the sweeping matrix for the mth mode given by Eq. (11.24).

In summary, recurrence formulas for determining the dominant eigenpair are stated as Eqs. (11.28), (11.29), and (11.30). The introduction of modal constraints and the use of sweeping matrices to eliminate the first mode and second mode are given by Eqs. (11.31) and (11.32), and Eqs. (11.34) and (11.35), respectively. General expressions for calculating the successively dominant eigenpairs $(\lambda_{m+1}, \{\Phi\}_{m+1})$ are represented by Eqs. (11.37), (11.38), and (11.24).

Although the forward vector iteration technique is somewhat less computationally intense than the inverse vector iteration technique, it is not commonly employed in practical vibration analysis. For the forward iteration method to be useful, it must operate on the flexibility based eigenproblem, whereas all major structural analysis computer programs are based upon a stiffness formulation.

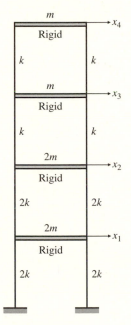

Figure 11.3

Four-story shear frame building of Example 11.2.

EXAMPLE 11.2 ▼

(a) Write a computer program to extract any user defined number of natural frequencies and corresponding mode shapes for an arbitrary MDOF system having up to four degrees of freedom by the *forward vector iteration* technique. Specify a convergence tolerance of TOL $= 10^{-6}$.

(b) Determine the four natural frequencies and associated modal vectors for the four-story shear frame building shown in Figure 11.3. Assume $m = 1.0$ kip-sec^2/in and $k = 125$ kips/in.

Solution

(a) The input parameters for the structure shown in Figure 11.3 are

$$[m] = \begin{bmatrix} 2.0 & 0 & 0 & 0 \\ 0 & 2.0 & 0 & 0 \\ 0 & 0 & 1.0 & 0 \\ 0 & 0 & 0 & 1.0 \end{bmatrix} \text{ kip-sec}^2/\text{in}$$

and the flexibility matrix is

$$[a] = \begin{bmatrix} 0.002 & 0.002 & 0.002 & 0.002 \\ 0.002 & 0.004 & 0.004 & 0.004 \\ 0.002 & 0.004 & 0.008 & 0.008 \\ 0.002 & 0.004 & 0.008 & 0.012 \end{bmatrix} \text{ in/kip}$$

(b) The results of the frequency analysis are presented in Table 11.5, the mode shapes are shown in Figure 11.4, and a listing of the BASIC computer program is given in Table 11.6.

Figure 11.4

Normal modes for four-story
shear frame building of
Example 11.1: (a) mode 1;
(b) mode 2; (c) mode 3;
(d) mode 4.

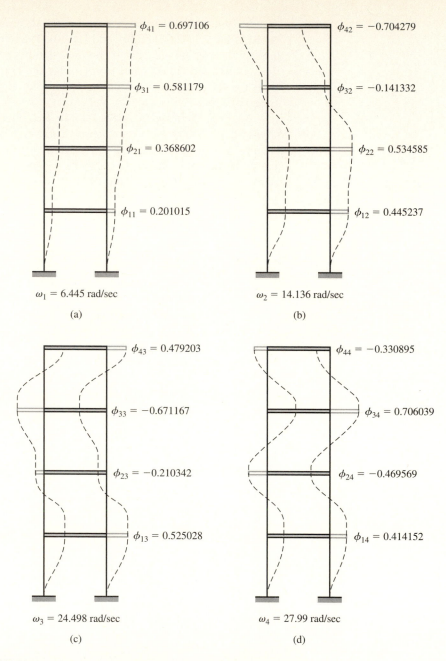

$\omega_1 = 6.445$ rad/sec

(a)

$\omega_2 = 14.136$ rad/sec

(b)

$\omega_3 = 24.498$ rad/sec

(c)

$\omega_4 = 27.99$ rad/sec

(d)

TABLE 11.5. Eigenvalues and Eigenvectors for Example 11.2

(a) Eigenvalues

Mode 1	Mode 2	Mode 3	Mode 4
41.57	199.83	600.15	783.44

(b) Natural frequencies (rad/sec)

Mode 1	Mode 2	Mode 3	Mode 4
6.445	14.136	24.498	27.990

(continued)

(c) Mode shapes

Mode 1	Mode 2	Mode 3	Mode 4
0.201015	0.445237	0.525028	0.414152
0.368602	0.534585	−0.210342	−0.469569
0.581179	−0.141332	−0.671167	0.706039
0.697106	−0.704279	0.479203	−0.330895

(d) Number of iterations for each mode

Mode	Iterations
1	10
2	12
3	37
4	4

TABLE 11.6. **Listing of BASIC Computer Program Used for Example 11.2**

```
1      REM Eigensolution by the Forward Iteration Method
2      OPEN "try" FOR INPUT AS #2
3      INPUT #2, A:CLOSE #2: IF A = 5 * INT(A/5) THEN SAVE "iter":PRINT "saving please wait"
4      A=A+1
5      OPEN "try" FOR OUTPUT AS #2:PRINT #2,A:CLOSE #2
10     OPEN "freq" FOR OUTPUT AS #1
15     CLS
20     TOL = .000001:READ N
25     DIM F(N,N),M(N,N),X(N),A(N,N),X1(N),X2(N),X3(N),S(N,N),B(N,N),I(N,N),
W(N,N),C(N,N),N(N)
26     FOR I = 1 TO N:I(I,I)=1:NEXT I
27     PRINT #1,"**** INPUT DATA ***** ":PRINT #1
28     PRINT #1, "Flexibility Matrix":PRINT#1
30     FOR I=1 TO N: FOR J=1 TO N:READ F(I,J):PRINT #1,USING "#.###### ";F(I,J);:NEXT J:PRINT #1,:NEXT I
39     PRINT #1:PRINT #1,"Mass Matrix":PRINT #1
40     FOR I=1 TO N:FOR J=1 TO N:READ M(I,J):PRINT #1,USING "###.### ";M(I,J);:NEXT J:PRINT #1,:NEXT I
50     FOR I=1 TO N:FOR J=1 TO N:FOR K=1 TO N:A(I,J)=A(I,J)+F(I,K)*M(K,J):NEXT K:NEXT J,I
70     FOR I=1 TO N:X(I)=1:NEXT I
80     Q=1:P=6.283185
100    FOR G=1 TO 1000
1001   R1 = R
105    FOR I=1 TO N:X1(I)=0:NEXT I
110    FOR I=1 TO N:FOR J=1 TO N:X1(I)=X1(I)+A(I,J)*X(J):NEXT J:NEXT I
115    S1=0:S2=0
120    FOR I=1 TO N:S1=S1+X1(I)*X1(I):S2=S2+X1(I)*X(I):NEXT I
125    S3=SQR(S1)
130    R=S1/S2
140    FOR I=1 TO N:X(I)=X1(I)/S3:NEXT I
150    IF ABS(R−R1)/ABS(R)<TOL THEN 160 ELSE GOTO 200
160    W(Q,1)=1/R:W(Q,2)=SQR(1/R)/P
170    FOR I=1 TO N:C(Q,I)=X(I):NEXT I
180    N(Q)=G
185    PRINT G,1/R
```

(continued)

```
190    IF Y=0 THEN GOTO 290 ELSE RETURN
200    NEXT G
290    Y=1
300    FOR Q=2 TO N
305    FOR I=1 TO N:X1(I)=0:X2(I)=0:NEXT I
310    FOR I=1 TO N:FOR J=1 TO N:X2(I)=X2(I)+X(J)*M(J,I):NEXT J:NEXT I
320    FOR I=1 TO N:FOR J=1 TO N:X3(I)=X3(I)+M(I,J)*X(J):NEXT J:NEXT I
330    S4=0:FOR I=1 TO N:S4=S4+X(I)*X3(I):NEXT I
340    FOR I=1 TO N:FOR J=1 TO N:S(I,J)=X(I)*X2(J):NEXT J:NEXT I
345    FOR I=1 TO N:FOR J=1 TO N:S(I,J)=S(I,J)/S4:NEXT J:NEXT I
350    FOR I=1 TO N:FOR J=1 TO N:S(I,J)=I(I,J)-S(I,J):NEXT J:NEXT I
360    FOR I=1 TO N:FOR J=1 TO N:B(I,J)=A(I,J):NEXT J:NEXT I
370    FOR I=1 TO N:FOR J=1 TO N:A(I,J)=0:FOR K=1 TO N:A(I,J)=A(I,J)+B(I,K)*S(K,J):NEXT K,J,I
375    FOR I=1 TO N:X(I)=1:NEXT I
380    GOSUB 100
400    NEXT Q
500    REM output phase
504    PRINT #1,:PRINT #1,"****************"
505    PRINT #1," OUTPUT DATA"
506    PRINT #1,"****************"
510    PRINT #1,
520    PRINT #1,"Eigenvalues ":PRINT #1,
530    FOR I=1 TO N:PRINT #1, USING "######.## ";W(I,1);:NEXT I
540    PRINT #1, :PRINT #1, :PRINT #1, "Natural Frequencies (Hz)":PRINT #1,
550    FOR I=1 TO N: PRINT #1, USING "##.##### ";W(I,2);:NEXT I
560    PRINT #1, :PRINT #1, :PRINT #1, "Mode Shapes": PRINT #1,
570    FOR I=1 TO N
580    FOR J=1 TO N
590    PRINT #1, USING "##.###### ";C(J,I);
600    NEXT J
610    PRINT #1,
615    NEXT I
620    PRINT #1,:PRINT #1, "No. of iterations for each mode": PRINT #1,
630    FOR I=1 TO N: PRINT #1,I,N(I):NEXT I
900    CLOSE #1
999    DATA 4: REM No. of degrees of freedom
```

```
**** INPUT DATA *****
Flexibility Matrix
0.002000 0.002000 0.002000 0.002000
0.002000 0.004000 0.004000 0.004000
0.002000 0.004000 0.008000 0.008000
0.002000 0.004000 0.008000 0.012000
Mass Matrix
2.0000 0.0000 0.0000 0.0000
0.0000 2.0000 0.0000 0.0000
0.0000 0.0000 1.0000 0.0000
0.0000 0.0000 0.0000 1.0000
```

▲

11.4   GENERALIZED JACOBI METHOD

The *generalized Jacobi method* is a popular *transformation method* for solution of the eigenproblem. The transformation methods utilize the basic properties of eigenvectors in the modal matrix [Φ], specifically:

$$[\Phi]^T[k][\Phi] = [K]$$

(11.39)

and

$$[\Phi]^T[m][\Phi] \; = \; [M] \tag{11.40}$$

where $[K]$ is the diagonalized *modal stiffness matrix* and $[M]$ is the diagonalized *modal mass matrix*. In the solution process, $[k]$ may be either positive definite or positive semidefinite, and $[m]$ may be either diagonal, with or without zero diagonal elements, or banded.

The basic scheme involves the reduction of $[k]$ and $[m]$ into diagonal form (that is, $[K]$ and $[M]$, respectively) by successive premultiplication and postmultiplication by transformational matrices $[T]_k^T$ and $[T]_k$, respectively, where $k = 1, 2, \ldots$, iterations. Specifically, if $[k]_1 = [k]$ and $[m]_1 = [m]$,

$$[k]_2 \; = \; [T]_1^T[k]_1[T]_1$$

$$[k]_3 \; = \; [T]_2^T[k]_2[T]_2$$

$$[k]_4 \; = \; [T]_3^T[k]_3[T]_3$$

$$\cdot$$
$$\cdot \tag{11.41}$$
$$\cdot$$

$$[k]_{k+1} \; = \; [T]_k^T[k][T]_k$$

$$\cdot$$
$$\cdot$$
$$\cdot$$

Similarly,

$$[m]_2 \; = \; [T]_1^T[m]_1[T]_1$$

$$[m]_3 \; = \; [T]_2^T[m]_2[T]_2$$

$$[m]_4 \; = \; [T]_3^T[m]_3[T]_3$$

$$\cdot$$
$$\cdot \tag{11.42}$$
$$\cdot$$

$$[m]_{k+1} \; = \; [T]_k^T[m][T]_k$$

$$\cdot$$
$$\cdot$$
$$\cdot$$

The selection of the $[T]_k$ matrices is made in such a manner as to bring $[k]_k$ and $[m]_k$ closer and closer to diagonal form with each iteration. That is, as $k \to \infty$,

$$[k]_{k+1} \to [K] \quad \text{and} \quad [m]_{k+1} \to [M]$$

After a sufficient number of iterations m, such that $[m]$ and $[k]$ are diagonalized, then concurrently solve for the eigenvalues and eigenvectors by

$$[\Lambda] \; = \; \text{diag}\left(\frac{K_r^{(m+1)}}{M_r^{(m+1)}}\right) \; = \; \text{diag}(\lambda) \; = \; \text{diag}(\omega^2) \tag{11.43}$$

and

$$[\Phi] = [T]_1[T]_2...[T]_m \text{diag}\left(\frac{1}{\sqrt{M_r^{(m+1)}}}\right) \tag{11.44}$$

where K_r and M_r represent the diagonal elements of the modal stiffness matrix $[K]$ and modal mass matrix $[M]$, respectively. This procedure generally does not yield the eigenvalues (or eigenvectors) in ascending order (that is, $\lambda_1, \lambda_2, ..., \lambda_n$).

In the generalized Jacobi method, the $[T]_k$ transformation matrices are constructed in the following manner for all i and j, where $i < j$:

$$[T]_k = \begin{matrix} & & & i & j & & \\ \begin{bmatrix} 1 & 0 & & \cdots & & & 0 \\ 0 & 1 & \cdots & & & & \\ \cdots & \cdots & \cdots & \cdots & \cdots & \cdots & \cdots \\ 0 & 0 & \cdots & 1 & \alpha & \cdots & 0 \\ 0 & 0 & \cdots & \gamma & 1 & \cdots & 0 \\ \cdots & \cdots & \cdots & \cdots & \cdots & \cdots & \cdots \\ \cdots & & & & & 1 & 0 \\ 0 & & \cdots & & & 0 & 1 \end{bmatrix} & \begin{matrix} \\ \\ \\ i \\ j \\ \\ \\ \end{matrix} \end{matrix} \tag{11.45}$$

For instance, for 4×4 $[k]$ and $[m]$ matrices, the total number of iterations m required to complete one *sweep* is 6, determined as follows:

$[T]_1$	for $i = 1, j = 2$	$[T]_4$	for $i = 2, j = 3$
$[T]_2$	for $i = 1, j = 3$	$[T]_5$	for $i = 2, j = 4$
$[T]_3$	for $i = 1, j = 4$	$[T]_6$	for $i = 3, j = 4$

The constants α and γ in the $[T]_k$ matrices are formulated in such a manner as to simultaneously reduce to zero all off-diagonal elements (i, j, where $i \neq j$) in the matrices $[k]_k$ and $[m]_k$. The general solution scheme for α and γ is defined by the following recurrence equations:

$$\bar{k}_{ii}^{(k)} = k_{ii}^{(k)}m_{ij}^{(k)} - m_{ii}^{(k)}k_{ij}^{(k)} \tag{11.46}$$

$$\bar{k}_{jj}^{(k)} = k_{ii}^{(k)}m_{ij}^{(k)} - m_{jj}^{(k)}k_{ij}^{(k)} \tag{11.47}$$

$$\bar{k}^{(k)} = k_{ii}^{(k)}m_{jj}^{(k)} - k_{jj}^{(k)}m_{ii} \tag{11.48}$$

and then,

$$\alpha = \frac{\bar{k}_{jj}^{(k)}}{x} \tag{11.49}$$

and

$$\gamma = -\frac{\bar{k}_{ii}^{(k)}}{x} \tag{11.50}$$

where

$$x = \frac{\bar{k}^{(k)}}{2} + [\operatorname{sign}(\bar{k}^{(k)})]\sqrt{\left(\frac{\bar{k}^{(k)}}{2}\right)^2 + \bar{k}_{ii}^{(k)}\bar{k}_{jj}^{(k)}} \tag{11.51}$$

where $k_{ij}^{(k)}$ and $m_{ij}^{(k)}$ are the elements of the $[k]_k$ and $[m]_k$ matrices, respectively. Note that if $[m]$ is a diagonal matrix and all $m_{ii} > 0$, then

$$\bar{k}_{ii}^{(k)} = -m_{ii}^{(k)}k_{ij}^{(k)} \tag{11.52}$$

and

$$\bar{k}_{jj}^{(k)} = -m_{jj}^{(k)}k_{ij}^{(k)} \tag{11.53}$$

Equations (11.46) through (11.51) represent the calculations required for one sweep. Convergence is determined by comparing successive eigenvalue approximations and also by checking if the off-diagonal terms in the $[k]_k$ and $[m]_k$ matrices are sufficiently close to zero. Thus, for the last two successive iterations within a sweep, convergence has been achieved if the eigenvalues satisfy

$$\frac{\left|\lambda_i^{(k+1)} - \lambda_i^{(k)}\right|}{\lambda_i^{(k+1)}} \le \text{TOL} \qquad i = 1, 2, \ldots, n \tag{11.54}$$

where

$$\lambda_i^{(k)} = \frac{k_{ii}^{(k)}}{m_{ii}^{(k)}} \tag{11.55}$$

and

$$\lambda_i^{(k+1)} = \frac{k_{ii}^{(k+1)}}{m_{ii}^{(k+1)}} \tag{11.56}$$

and the off-diagonal terms in $[k]_k$ and $[m]_k$ satisfy

$$\left[\frac{(k_{ij}^{(k+1)})^2}{k_{ii}^{(k+1)}k_{jj}^{(k+1)}}\right]^{1/2} \le \text{TOL} \tag{11.57}$$

and

$$\left[\frac{(m_{ij}^{(k+1)})^2}{m_{ii}^{(k+1)}m_{jj}^{(k+1)}}\right]^{1/2} \le \text{TOL} \tag{11.58}$$

TOL should not be less that 10^{-s}, where s is the required digit accuracy.

Transformation methods such as the generalized Jacobi method yield the *simultaneous solution of all eigenvalues* for the system. Therefore, these methods are most useful when the matrices on which they operate (that is, $[m]$ and $[k]$ matrices) are of relatively small order and have large bandwidths. These methods are not particularly efficient for sparsely populated matrices, since the transformations destroy matrix sparsity.

EXAMPLE 11.3 ▼

(a) Write a computer program to extract all the eigenvalues and corresponding eigenvectors for an arbitrary MDOF system having n degrees of freedom by the generalized Jacobi method. Specify a convergence tolerance TOL = 10^{-6}.

(b) Determine the three natural frequencies and corresponding modal vectors for the three-story shear frame building shown in Figure 11.5. Assume all columns bend about their major axes, and modulus of elasticity $E = 29,000$ ksi.

Solution

(a) The input parameters for the structure shown in Figure 11.5 are

$$[m] = \begin{bmatrix} 0.1035 & 0 & 0 \\ 0 & 0.2070 & 0 \\ 0 & 0 & 0.2070 \end{bmatrix} \text{ kip-sec}^2/\text{in}$$

$$[k] = \begin{bmatrix} 79.017 & -79.017 & 0 \\ -79.017 & 274.93 & -195.91 \\ 0 & -195.91 & 383.61 \end{bmatrix} \text{ kips/in}$$

(b) The results of the frequency analysis are presented in Table 11.7, the mode shapes are shown in Figure 11.6, and a listing of the FORTRAN computer program is provided in Table 11.8.

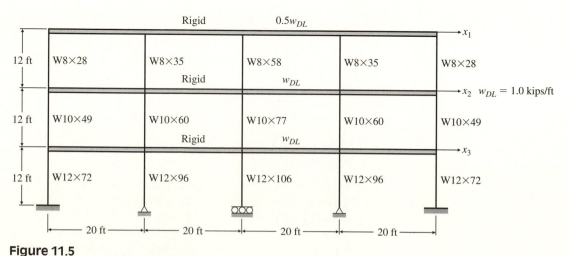

Figure 11.5
Three-story shear frame building of Example 11.3.

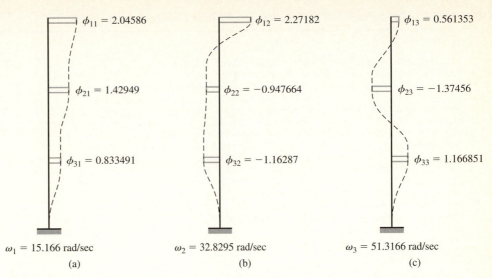

Figure 11.6
Normal modes for three-story shear frame building of Example 11.1: (a) mode 1; (b) mode 2; (c) mode 3.

TABLE 11.7. **Eigenvalues and Eigenvectors for Example 11.3**

* * *7 Sweeps Completed * * *

(a) Frequencies (rad/sec)

Mode 2	Mode 1	Mode 3
0.328295E+02	0.151662E+02	0.513116E+02

(b) Eigenvectors

Mode 2	Mode 1	Mode 3
0.227182E+01	0.204586E+01	0.561353E+00
−0.947664E+00	0.142949E+01	−0.137456E+01
−0.116287E+01	0.833491E+00	0.166851E+01

TABLE 11.8. Listing of FORTRAN Computer Program Used for Example 11.3

```
        program jacobi
*
*         This program uses the generalized Jacobi method to calculate
* the frequencies (eigenvalues) and mode shapes (eigenvectors) of
* a multi-degree-of-freedom structural system. The program requires
* an input file labeled 'jacobi.in'. This input file contains the
* number of degrees of freedom and the stiffness and mass matrices
* in single entry form. The program will create an output file
* labeled 'jacobi.out'. This output file will contain the number of
* sweeps performed, the natural frequencies of the system (in no
* particular order), and the corresponding mode shape. By changing
* the parameter Size and TOL, this program can be adapted to any
* MDOF system with any desired accuracy. To incorporate this program
* into a larger program (such as a finite element code), change the
* command 'program jacobi' to 'subroutine jacobi(n,k,m,lambda,phi)'
* and add the command 'return' before the last line ('end').

        integer i,ii,j,jj,n,nsweep,size
        double precision tol
        parameter (size=100,tol=1.0d-6)
        double precision alpha,gamma,kbar,kbarii,kbarjj
1       k(size,size),lambda(size),m(size,size),oldlambda,
2       p(size,size),phi(size,size),tmp(size,size),x1,x,
3       kcheck,ktol,lcheck,ltol,mcheck,mtol
        logical converged
        open(unit=10,file='jacobi.in',status='old')
        open(unit=11,file='jacobi.out',status='new')
        read(10,*)n
        read(10,*)

        do 10 i=1,n
            read (10,*)
            do 9 j=1,n
              read(10,*)k(i,j)
9       continue
10      continue

        read(10,*)

        do 20 i=1,n
            read(10,*)
            do 19 j=1,n
            read(10,*)m(i,j)
19      continue
20      continue

        do 80 i=1,n
            lambda(i)=0.0d+00
              do 79 j=1,n
            if(i.eq.j) then
                    phi (i,j)=1.0d+00
                    p(i,j)=1.0d+00
                else
                  phi(i,j)=0.0d+00
                    p(i,j)=0.0d+00
                endif
79      continue
80      continue

        i=1
1       converged=.true.
        nsweep=nsweep+1
        j=j+1
        if(j.gt.n)then
```

(continued)

```
        i=i+1
            if(i.ge.n) i=1
            j=i+1
     endif

     kbar=k(i,i)*m(j,j)−k(j,j)*m(i,i)
     kbarii=k(i,i)*m(i,j)−m(i,i)*k(i,j)
     kbarjj=k(j,j)*m(i,j)−m(j,j)*k(i,j)
     x1=(dsqrt((kbar*kbar/4.0d+00+kbarii*kbarjj)))
     x=kbar/2.0d+00+dsign(x1,kbar)
     if (kbar.eq.0.0d+00.and.kbarii.eq.0.0d+00) then
            gamma=0.0d+00
                goto 2
     endif
     gamma=−kbarii/x

2    if (kbar.eq.0.0d+00.and.kbarjj.eq.0.0d+00) then
          alpha=0.0d+00
                goto 3
     endif
     alpha=kbarjj/x

3    p(i,j)=alpha
     p(j,i)=gamma
     call matprob(k,p,tmp,n)
     p(i,j)=gamma
     p(j,i)=alpha
     call matprod(p,tmp,k,n)
     p(i,j)=alpha
     p(j,i)=gamma
     call matprod(m,p,tmp,n)
     p(i,j)=gamma
     p(j,i)=alpha
     call matprod(p,tmp,m,n)
     p(i,j)=alpha
     p(j,i)=gamma
     call matprod(phi,p,tmp,n)
     do 95 ii=1,n
         do 94 jj=1,n
             phi(ii,jj)=tmp(ii,jj)
94       continue
95    continue
     p(i,j)=0.0d+00
     p(j,i)=0.0d+00
     do 101 ii=1,n
         oldlambda=lambdal(ii)
             lambda(ii)=k(ii,ii)/m(ii,ii)
             if (lambda(ii).ne.0.0d+00)then
                 lcheck=(lambda(ii)−oldlambda)**2
                 ltol=(lambda(ii)*tol)**2
                 if (lcheck.gt.ltol) converged=.false.
             endif
101   continue

     if(.not.converged) goto 1
     kcheck=k(i,j)*k(i,j)
     ktol=k(i,i)*k(j,j)*tol*tol
     mcheck=m(i,j)*m(i,j)
     mtol=m(i,i)*m(j,j)*tol*tol
     if (kcheck.gt.ktol.or.mcheck.gt.mtol) goto 1

     do 105 i=1,n
         do 104 j=1,n
             if (i.ne.j) m(i,j)=0.0d+00
104   continue
```

(continued)

```
             m(i,i)=dsqrt(1/m(i,i))
105   continue

          call matprod(phi,m,tmp,n)
          do 107 ii=1,n
              do 106 jj=1,n
                  phi(ii,jj)=tmp(ii,jj)
106       continue
107   continue
          write(11,108)nsweep

108       format('*',/,i3,' Sweeps Completed')
          write(11,*)'*'
          write(11,*) 'Frequencies'
          write(11,*)'*'
          do 110 i=1,n
              write(11,109)i,dsqrt(lambda(i))
109           format (2x,i3,3x,d13.6)
110   continue

          write(11,*)'*'
          write(11,*)'Modal Matrix'
          do 120 j=1,n
              write(11,*)'*'
              do 119 i=1,n
                  write(11,118)i,j,phi(i,j)
118       format (3x,i3,',',i3,3x,d13.6)
119       continue
120   continue
          end
*
*         subroutine matprod
*
          subroutine matprod(a,b,c,n)
          integer i,j,k,n,size
          parameter (size=100)
          double precision a(size,size),b(size,size),c(size,size)
          do 3 i=1,n
              do 2 j=1,n
                  c(i,j)=0.0d+00
                  do 1 k=1,n
                      c(i,j)=c(i,j)+a(i,k)*b(k,j)
1             continue
2             continue
3         continue

          return
          end
```

▲

11.5 SOLUTION METHODS FOR LARGE EIGENPROBLEMS

The three basic groups of eigenproblem solution techniques were discussed in Section 11.1. Specific routines for two of these groups (i.e., matrix iteration methods and transformation methods) were presented in Sections 11.2 through 11.4. As described in this chapter, these methods are computationally efficient for "small eigenproblems," and are generally not suitable for the solution of "large eigenproblems." The difference between large and small eigenproblems is indeed relative. However, the distinguishing factor is that for a large eigenproblem, a substantially less computational effort is required to solve for only a small percentage of the system eigenvalues and eigenvectors than to simply calculate all the eigenpairs for the system. This is especially true in the majority of finite element method (FEM) analyses.

In recent years, more efficient solution algorithms for large eigenproblems have been developed. These algorithms are generally composed of some combination of several of the basic group methods. Two such combination eigensolvers are the *determinant search method* and the *subspace iteration method,* which have been incorporated into several commercially available FEM computer codes, such as ADINA [2].

The determinant search method [3, 4, 5] is a combination of the characteristic polynomial iteration technique, with the use of the *Sturm sequence* property of the characteristic polynomial, and inverse vector iteration. The method is most effective when applied to well-banded systems for which a relatively few number of eigenpairs are sought.

The subspace iteration method [6] is an efficient eigensolver for large systems when only the lower vibration modes are of interest. In essence, it is similar to the inverse vector iteration method, except that simultaneous iteration is performed on multiple trial vectors, and an eigensolution in a reduced subspace is generated at the end of each iteration. Because the iterations are performed within a subspace, convergence of individual iteration vectors to an eigenvector is not required, merely convergence of the subspace. Moreover, if the iteration vectors are linear combinations of the required eigenvectors, the solution converges in a single step.

REFERENCES

1 Bathe, K.J, *Finite Element Procedures,* Prentice Hall, Englewood Cliffs, NJ, 1996.

2 ADINA, A Finite Element Computer Program for Automatic Dynamic Incremental Nonlinear Analysis Report ARD 92-4, ADINA R & D, Watertown, MA, 1994.

3 Bathe, K.J. and Wilson, E.L., "Large Eigenvalue Problems in Dynamic Analysis," ASCE, *Journal of Engineering Mechanics Division,* Vol. 98, 1972, pp. 1471–1485.

4 Bathe, K.J. and Wilson, E.L., "Eigensolution of Large Structural Systems With Small Bandwidth," ASCE, *Journal of Engineering Mechanics Division,* Vol. 99, 1973, pp. 467–479.

5 Bathe, K.J. and Wilson, E.L., "Solution Methods for Eigenvalue Problems in Structural Mechanics," *International Journal for Numerical Methods in Engineering,* Vol. 6, 1973, pp. 213–236.

6 Bathe, K.J., "Solution Methods of Large Generalized Eigenproblems in Structural Engineering," Report UC SESM 71-20, Civil Engineering Department, University of California, Berkeley, 1971.

7 Humar, J.L., *Dynamics of Structures,* Prentice Hall, Englewood Cliffs, NJ, 1990.

8 Burden, R.L. and Faires, R.L., *Numerical Analysis,* 5th ed., PWS-Kent Publishing, Boston, 1993.

9 Hammerlin, G. and Hoffmann, K.H., *Numerical Mathematics,* Springer-Verlag, New York, 1989.

10 Bathe, K.J. and Wilson, E.L., *Numerical Methods in Finite Element Analysis,* Prentice Hall, Englewood Cliffs, NJ, 1976.

NOTATION

k_{ij}	terms in stiffness matrix	MDOF	multi-degree-of-freedom
K_r	generalized (modal) stiffness for rth normal mode	TOL	prescribed iteration tolerance
m_{ij}	terms in mass matrix	$[a]$	flexibility matrix
M_r	generalized (modal) mass for rth normal mode	$[D]$	dynamic (system) matrix
$p(\lambda)$	characteristic polynomial	$[I]$	identity matrix
DOF	degree of freedom	$[k]$	stiffness matrix
FEM	finite element method	$[K]$	generalized (modal) stiffness matrix

$[m]$	mass matrix		δ_{ij}	Kronecker delta
$[M]$	generalized (modal) mass matrix		γ	coefficient in generalized Jacobi method
$\{R\}_1$	equilibrium vector representing the right-hand side of Eq. (11.1)		λ_r	eigenvalue for rth normal mode
$[S]_m$	sweeping matrix for mth mode		ω_r	natural circular frequency for rth normal mode
$[T]_k$	transformation matrix for kth iteration		$\rho(\{\bar{x}\}_{k+1})$	Rayleigh quotient estimate of eigenvalue for iteration $k + 1$
$\{x\}_i$	trial eigenvector for ith iteration		$[\Phi]$	modal matrix
$\{y\}_i$	trial vector for ith iteration		$\{\Phi\}_r$	modal vector for rth normal mode
α	coefficient in generalized Jacobi method		$[\Lambda]$	system eigenvalue matrix
α_i	coefficient in Gram-Schmidt orthogonalization			

PROBLEMS

11.1–11.4 Use the inverse vector iteration method to determine the fundamental frequency and corresponding modal vector for the structures shown in Figures P11.1 through P11.4. Deflate the modal vector of the first mode characteristics by the Gram-Schmidt orthogonalization method and determine the second eigenpair $(\lambda_2, \{\Phi\}_2)$ by inverse vector iteration. (These problems are suitable for manual calculations.)

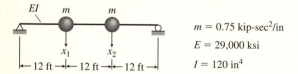

$m = 0.75$ kip-sec^2/in

$E = 29,000$ ksi

$I = 120$ in^4

Figure P11.1

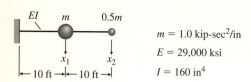

$m = 1.0$ kip-sec^2/in

$E = 29,000$ ksi

$I = 160$ in^4

Figure P11.2

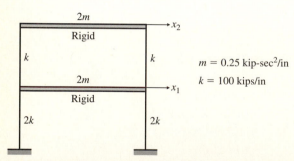

$m = 0.25$ kip-sec^2/in

$k = 100$ kips/in

Figure P11.3

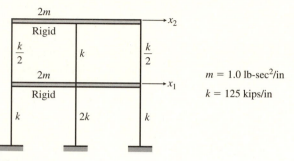

$m = 1.0$ lb-sec^2/in

$k = 125$ kips/in

Figure P11.4

11.5–11.8 Repeat Problems 11.1 through 11.4 using the forward matrix iteration method. The iterations should operate on the flexibility formulation of the eigenproblem. (These problems are suitable for manual calculations.)

11.9–11.12 Determine all the eigenpairs for the structures shown in Figures P11.5 through P11.8 using the generalized Jacobi Transformation method. (These problems are suitable for manual calculations.)

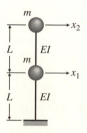

Figure P11.5

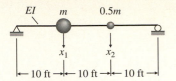

$m = 0.5$ kip-sec²/in
$E = 29{,}000$ ksi
$I = 85$ in⁴

Figure P11.6

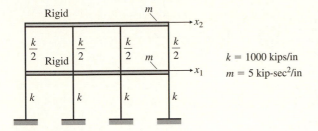

$k = 1000$ kips/in
$m = 5$ kip-sec²/in

Figure P11.7

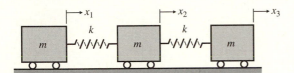

Figure P11.8

11.13–11.16 Write a computer program to extract any user specified number of eigenpairs for an arbitrary MDOF system having n DOF by the inverse vector iteration method. Determine all the eigenpairs for the systems shown in Figures P11.9 through P11.12. Specify a convergence tolerance of TOL = 10^{-6}.

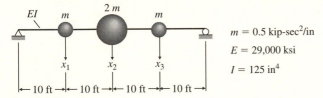

$m = 0.5$ kip-sec²/in
$E = 29{,}000$ ksi
$I = 125$ in⁴

Figure P11.9

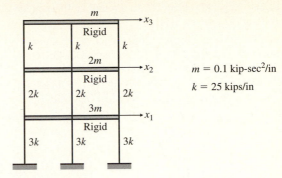

$m = 0.1$ kip-sec²/in
$k = 25$ kips/in

Figure P11.10

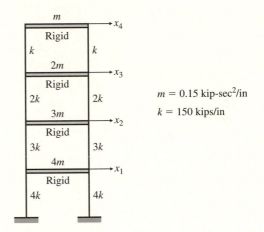

$m = 0.15$ kip-sec²/in
$k = 150$ kips/in

Figure P11.11

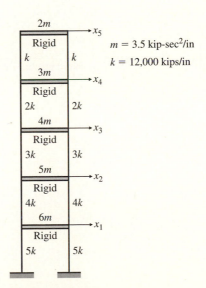

$m = 3.5$ kip-sec²/in
$k = 12{,}000$ kips/in

Figure P11.12

11.17–11.20 Repeat Problems 11.13 through 11.16 using the forward vector iteration method.

11.21–11.24 Write a computer program to extract all the eigenpairs for an arbitrary MDOF system having n DOF by the generalized Jacobi transformation method. Determine all the eigenpairs for the systems shown in Figures P11.13 through P11.16. Specify a convergence tolerance of TOL $= 10^{-8}$.

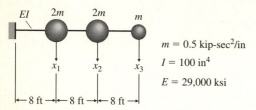

$m = 0.5$ kip-sec^2/in
$I = 100$ in^4
$E = 29,000$ ksi

Figure P11.13

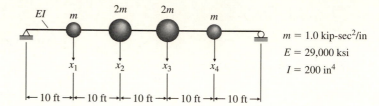

$m = 1.0$ kip-sec^2/in
$E = 29,000$ ksi
$I = 200$ in^4

Figure P11.14

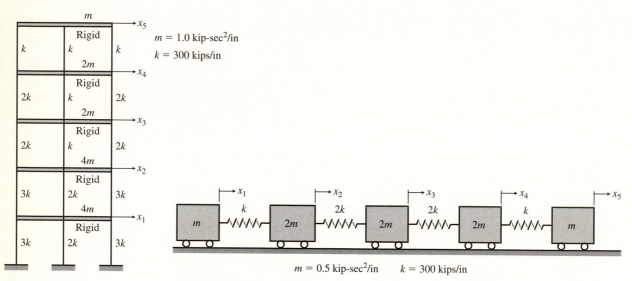

$m = 1.0$ kip-sec^2/in
$k = 300$ kips/in

Figure P11.15

$m = 0.5$ kip-sec^2/in $k = 300$ kips/in

Figure P11.16

12

Analysis of Dynamic Response by Mode Superposition

If the *principal modes* (normal modes) of vibration for a MDOF system are used as generalized coordinates for defining the system response, the *n* equations of motion become *uncoupled*. In these coordinates each uncoupled equation may be solved independently as if each equation pertained to an independent system having only one degree of freedom. Therefore, the individual responses of the *n* uncoupled equations of the MDOF systems (one for each mode of vibration), to any form of excitation may be determined by applying the techniques of analysis for SDOF systems discussed in Chapters 2 through 8. The response of the MDOF system by mode superposition is then defined as the summation of the responses of the individual modes. This procedure of dynamic analysis is referred to as the *mode superposition method, normal mode method,* or simply *modal analysis.*

The mode superposition method is a very popular vibration analysis technique, but it has several important limitations. The method is valid for linear systems only, and damping in the system must be *proportional* (or *classical*). Moreover, there are two forms of the mode superposition method: (1) The *mode displacement method* and (2) the *mode acceleration method.* The mode displacement method is predicated upon a *stiffness formulation* of the structural response, whereas the mode acceleration method is based upon a *flexibility formulation* and therefore is not widely used (although it exhibits improved convergence properties). For this reason the mode displacement method is emphasized in this chapter.

Finally, the mode superposition procedure is most useful when the system response can be accurately evaluated by considering only a relatively small subset of all the vibration modes for the system. It should be noted, however, that for most types of dynamic loadings on structural and mechanical systems, the contributions of the various modes to the dynamic response generally are greatest for the lowest frequencies and tend to decrease for higher frequencies. Consequently, it is usually not necessary to include any of the higher modes of vibration in the superposition process.

12.1 MODE DISPLACEMENT METHOD FOR UNDAMPED SYSTEMS

The general matrix formulation of the equations of motion for an undamped *n*-DOF MDOF system subject to external excitation is given by

$$[m]\{\ddot{x}\} + [k]\{x\} = \{F(t)\} \tag{12.1}$$

where $[k]$ and $[m]$ are the symmetric $n \times n$ stiffness and mass matrices, respectively, $\{x\}$ and $\{\ddot{x}\}$ are the $n \times 1$ displacement and acceleration vectors (in physical coordinates), respectively, and $\{F(t)\}$ is the $n \times 1$ external force vector. Equations (12.1) are *coupled equations;* that is, they cannot be solved independently, only simultaneously. The coupling is generally manifested through the stiffness matrix (*elastic coupling*), but may also be exhibited through the mass matrix (*inertia coupling*), if the mass matrix contains off-diagonal terms (e.g., consistent mass matrix). The mode superposition method cannot be applied to the coupled system of equations, such as Eqs. (12.1).

Finding a coordinate system that exhibits neither form of coupling is the essence of the mode superposition procedure. Once Eqs. (12.1) are uncoupled, they may be solved independently of one another. Coordinates that enable the decoupling of the equations of motion are called *principal coordinates,* or *normal coordinates.*

To uncouple the differential equations, we introduce an alternative set of coordinates

$$\{q\} = \{q(x_1, x_2, \ldots, x_n)\} \tag{12.2}$$

such that Eqs. (12.1) can be transformed into a set of n independent equations whose solutions can be determined independently.

The normal (or principal) coordinates $\{q\}$ are defined by the matrix transformation

$$\{x\} = [\Phi]\{q\} \tag{12.3}$$

where $[\Phi]$ is the $n \times n$ modal matrix determined from the solution of the system eigenproblem. Therefore, the transformation of the equations of motion from physical to normal coordinates is perpetrated by the following matrix premultiplication of Eq. (12.1):

$$[\Phi]^T([m]\{\ddot{x}(t)\} + [k]\{x(t)\} = \{F(t)\}) \tag{12.4}$$

where $[\Phi]^T$ is the transpose of $[\Phi]$,

$$\{\ddot{x}\} = [\Phi]\{\ddot{q}\} \tag{12.5}$$

and $\{x\}$ is defined by Eq. (12.3). Thus, the uncoupled equations of motion in normal coordinates are expressed as

$$[\Phi]^T[m][\Phi]\{\ddot{q}(t)\} + [\Phi]^T[k][\Phi]\{q(t)\} = [\Phi]^T\{F(t)\} \tag{12.6}$$

Equation (12.6) can be written more succinctly as

$$[M]\{\ddot{q}(t)\} + [K]\{q(t)\} = \{P(t)\} \tag{12.7}$$

where $[M]$ and $[K]$ are the $n \times n$ *modal mass matrix* and *modal stiffness matrix,* respectively (which have been diagonalized by the modal transformation), and $\{P(t)\}$ is called the *modal force vector.* Thus

$$[M] = [\Phi]^T[m][\Phi] \tag{12.8}$$

$$[K] = [\Phi]^T[k][\Phi] \tag{12.9}$$

$$\{P(t)\} = [\Phi]^T\{F(t)\} \tag{12.10}$$

Since the modal mass and modal stiffness matrices are diagonal, Eqs. (12.7) can be written as n uncoupled equations

$$M_r\ddot{q}_r + K_r q_r = P_r(t) \qquad r = 1, 2, \dots, n \tag{12.11}$$

where

$$M_r = \{\Phi\}_r^T[m][\Phi]_r \tag{12.12}$$

$$K_r = \{\Phi\}_r^T[k][\Phi]_r \tag{12.13}$$

and

$$P_r(t) = \{\Phi\}_r^T\{F(t)\} \tag{12.14}$$

The complete response for the rth mode can be expressed as the sum of the response due to modal initial conditions and the modal response due to $P_r(t)$.

The initial conditions in physical coordinates can be expressed as

$$\{x(0)\} = [\Phi]\{q(0)\}$$
$$\{\dot{x}(0)\} = [\Phi]\{\dot{q}(0)\} \tag{12.15}$$

Multiplying Eqs. (12.15) by $[\Phi]^T[m]$ yields

$$[\Phi]^T[m]\{x(0)\} = [M]\{q(0)\}$$
$$[\Phi]^T[m]\{\dot{x}(0)\} = [M]\{\dot{q}(0)\} \tag{12.16}$$

Since $[M]$ is a diagonal matrix, Eqs. (12.16) can be solved for the modal initial conditions, resulting in

$$\left.\begin{array}{l} q_r(0) = \left(\dfrac{1}{M_r}\right)\{\Phi\}_r^T[m]\{x(0)\} \\[4mm] \dot{q}_r(0) = \left(\dfrac{1}{M_r}\right)\{\Phi\}_r^T[m]\{\dot{x}(0)\} \end{array}\right\} \qquad r = 1, 2, \dots, n \tag{12.17}$$

Thus the complete response for the rth mode in normal coordinates can be represented by the Duhamel integral expression, similar to that given by Eq. (6.27), as

$$q_r(t) = q_r(0)\cos(\omega_r t) + \left(\frac{1}{\omega_r}\right)\dot{q}_r(0)\sin(\omega_r t)$$

$$+ \left(\frac{1}{M_r\omega_r}\right)\int_0^t P_r(\tau)\sin\omega_r(t - \tau)\,d\tau \tag{12.18}$$

The procedures discussed in Chapter 6 for evaluating the Duhamel integral are also applicable to Eq. (12.18). However, for arbitrary dynamic excitation, it is generally necessary to employ one of the numerical evaluation techniques discussed in Chapter 7 to obtain the time-history response $q_r(t)$.

The exact response in physical coordinates is obtained by summing all the individual modal responses in normal coordinates given by Eq. (12.18) for $r = 1, 2, 3, \ldots, n$, while simultaneously employing the transformation defined by Eq. (12.3). Thus

$$\{x(t)\} = \sum_{r=1}^{n} \{\Phi\}_r \{q_r(t)\} \tag{12.19}$$

As mentioned at the beginning of this chapter, the mode superposition method is most effective when only a small subset of the total number of modes (eigenpairs) of the system are required to render an accurate solution. For example, consider a 50-story building having 50 degrees of freedom (one at each story). The entire structure therefore possesses 50 eigenpairs (i.e., 50 eigenvalues and 50 eigenvectors) that describe its normal vibration modes. If it is known that the frequency content of the excitation force is in the vicinity of the lowest few frequencies of the building, the higher modes will not be excited and the forced response can be determined as the *superposition* of only these few lower frequency modes. Therefore the displacement $\{x(t)\}$, given by Eq. (12.19), can be approximated by summing a number of modes $p \ll n$. Thus a truncated superposition solution is attained given by

$$\{\hat{x}(t)\} = \{\hat{\Phi}\}\{q(t)\} = \sum_{r=1}^{p} \{\Phi\}_r \{q_r(t)\} \tag{12.20}$$

where $[\Phi]$ is the $n \times p$ truncated modal matrix so that

$$\{\hat{\Phi}\} = [\{\Phi\}_1 \{\Phi\}_2 \{\Phi\}_3 \ldots \{\Phi\}_p] \tag{12.21}$$

Equation (12.20) represents the coordinate transformation for mode superposition with truncation of modes from $p + 1$ to n (for example, $n = 50$, $p = 5$).

To illustrate implementation of the mode superposition method, consider the steady-state response of an undamped MDOF system subject to harmonic excitation given by

$$\{F(t)\} = \{F_0\} \sin \Omega t \tag{12.22}$$

From Eqs. (12.10) and (12.22)

$$\{P(t)\} = \{\Phi\}^T \{F_0\} \sin \Omega t \tag{12.23}$$

or for each mode r from Eq. (12.14)

$$P_r(t) = P_r \sin \Omega t \tag{12.24}$$

where

$$P_r = \{\Phi\}_r^T \{F_0\}$$

The steady-state modal response for the rth mode in normal coordinates is analogous to that for a SDOF system as expressed by Eq. (5.12), resulting in

$$q_r(t) = \left(\frac{P_r}{K_r}\right)\left[\frac{1}{1 - (\Omega / \omega_r)^2}\right] \sin \Omega t \tag{12.25}$$

Combining Eq. (12.25) with Eq. (12.19), the exact steady-state response in physical coordinates is given by

$$\{x(t)\} = \sum_{r=1}^{n} \{\Phi\}_r^T \left(\frac{P_r}{K_r}\right) \left[\frac{1}{1 - (\Omega/\omega_r)^2}\right] \sin \Omega t \tag{12.26}$$

An approximated (truncated) solution ($p \ll n$) is given by

$$\{\hat{x}(t)\} = \sum_{r=1}^{p} \{\Phi\}_r^T \left(\frac{P_r}{K_r}\right) \left[\frac{1}{1 - (\Omega/\omega_r)^2}\right] \sin \Omega t \tag{12.27}$$

EXAMPLE 12.1   ▼

The three-story shear frame building shown in Figure 12.1 is subject to the harmonic forces given by

$$\{F(t)\} = \begin{Bmatrix} F_1 \cos \Omega_1 t \\ F_2 \cos \Omega_2 t \\ F_3 \cos \Omega_3 t \end{Bmatrix}$$

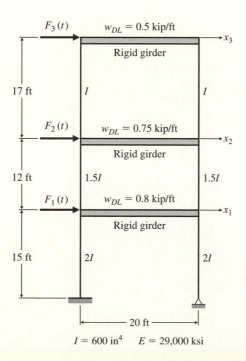

Figure 12.1

Three-story shear frame building of Example 12.1.

where $F_1 = 8$ kips
$F_2 = 10$ kips
$F_3 = 18$ kips
$\Omega_1 = 10$ rad/sec
$\Omega_2 = 20$ rad/sec
$\Omega_3 = 50$ rad/sec

Determine the steady-state response $\{x(t)\}$. Provide time-history plots for the truncated solution $\hat{x}_3 t$ for $p = 1$ and $p = 2$. Compare the truncated solutions with the exact solution $x_3(t)$, for $p = n = 3$.

Solution

The mass and stiffness matrices for the structure are

$$[m] = \begin{bmatrix} 0.04140 & 0 & 0 \\ 0 & 0.0388 & 0 \\ 0 & 0 & 0.025880 \end{bmatrix} \text{ kip-sec}^2/\text{in}$$

$$[k] = \begin{bmatrix} 299.286 & -209.780 & 0 \\ -209.780 & 258.969 & -49.189 \\ 0 & -49.184 & 49.189 \end{bmatrix} \text{ kips/in}$$

A frequency analysis of the structure yielded the following frequencies and mode shapes:

$$\{\omega\} = \begin{Bmatrix} 25.076 \\ 53.578 \\ 110.907 \end{Bmatrix} \text{ rad/sec} \qquad [\Phi] = \begin{bmatrix} 1.0 & 1.0 & 1.0 \\ 1.303 & 0.860 & -1.000 \\ 1.947 & -1.685 & 0.183 \end{bmatrix}$$

The mode shapes for the structure are shown in Figure 12.2.

(a) The modal mass matrix is determined from Eq. (12.8) as

$$[M] = [\Phi]^T[m][\Phi] = \begin{bmatrix} 0.2054 & 0 & 0 \\ 0 & 0.1436 & 0 \\ 0 & 0 & 0.08117 \end{bmatrix}$$

(b) The modal stiffness matrix is determined from Eq. (12.9) as

$$[K] = [\Phi]^T[k][\Phi] = \begin{bmatrix} 129.166 & 0 & 0 \\ 0 & 412.216 & 0 \\ 0 & 0 & 998.42 \end{bmatrix}$$

(c) The modal force vector is determined from Eq. (12.10) as

$$\{P(t)\} = [\Phi]^T\{F(t)\} \tag{1}$$

Figure 12.2

Normal modes for three-story shear frame building of Example 12.1. (a) mode 1; (b) mode 2; (c) mode 3.

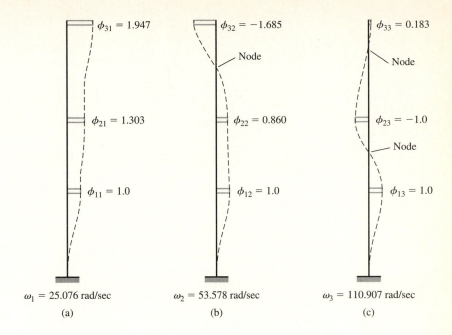

$\phi_{31} = 1.947$ $\phi_{32} = -1.685$ $\phi_{33} = 0.183$

Node   Node

$\phi_{21} = 1.303$ $\phi_{22} = 0.860$ $\phi_{23} = -1.0$

Node

$\phi_{11} = 1.0$ $\phi_{12} = 1.0$ $\phi_{13} = 1.0$

$\omega_1 = 25.076$ rad/sec $\omega_2 = 53.578$ rad/sec $\omega_3 = 110.907$ rad/sec

(a)   (b)   (c)

or

$$\{P(t)\} = \begin{Bmatrix} 8 \cos 10t + 13.03 \cos 20t + 35.046 \cos 50t \\ 8 \cos 10t + 8.6 \cos 20t - 30.33 \cos 50t \\ 8 \cos 10t - 10.01 \cos 20t + 3.294 \cos 50t \end{Bmatrix} \text{ kips}$$

(d) The expressions for steady-state response, $q_r(t)$, in modal (normal) coordinates are determined from Eq. (12.25) as

$$q_r(t) = \left(\frac{P_r}{K_r}\right)\left[\frac{1}{1 - (\Omega/\omega_r)^2}\right]\sin \Omega t \tag{2}$$

Thus for $r = 1$, Eq. (2) yields

$$q_1(t) = 0.00808 \cos 10t - 0.01036 \cos 20t + 0.00414 \cos 50t$$

for $r = 2$,

$$q_2(t) = 0.02108 \cos 10t + 0.02424 \cos 20t - 0.569918 \cos 50t$$

and for $r = 3$,

$$q_3(t) = 0.008078 \cos 10t - 0.0103678 \cos 20t + 0.00414 \cos 50t$$

(e) The exact response $\{x(t)\}$ in physical coordinates is determined from Eq. (12.26) as

$$\{x(t)\} = \sum_{r=1}^{3} \{\Phi\}_r \{q_r(t)\} \tag{3}$$

Thus,

$$\{x(t)\}$$

$$= \left\{ \begin{array}{l} 0.1018 \cos 10t + 0.2911 \cos 20t - 0.6570 \cos 50t \\ 0.1052 \cos 10t + 0.3925 \cos 20t - 0.3755 \cos 50t \\ 0.1110 \cos 10t + 0.4970 \cos 20t + 0.7835 \cos 50t \end{array} \right\} \text{ in}$$

(f) The truncated response $\hat{x}_3(t)$ in physical coordinates is determined from Eq. (12.27) as

$$\hat{x}_3(t) = \sum_{r=1}^{p} \{\phi_3\}_r \{q_r(t)\} \tag{4}$$

Thus, for $p = 1$,

$$\hat{x}_3(t) = 0.14339 \cos 10t + 0.53977 \cos 20t - 0.1775 \cos 50t \quad \text{in}$$

and for $p = 2$,

$$\hat{x}_3(t) = 0.1095 \cos 10t + 0.4989 \cos 20t + 0.78279 \cos 50t \quad \text{in}$$

for $p = 3 = n$ (exact solution),

$$\hat{x}_3(t) = x_3(t) = 0.11098 \cos 10t + 0.49703 \cos 20t$$
$$+ \ 0.78354 \cos 50t \quad \text{in}$$

Plots of the truncated displacement response histories $\hat{x}_3(t)$ are presented in Figures 12.3 and 12.4, respectively, for $p = 1$ and $p = 2$. The exact displacement time history $x_3(t)$ considering all three modes is presented in Figure 12.5. Comparison of these plots indicates that a single-mode solution (Figure 12.3) is not very accurate. This is because the forcing frequency Ω_3 (50 rad/sec) is very close to ω_2 (53.578 rad/sec); therefore, exclusion of the mode-2 response eliminates an important contribution of the total response. There appears to be no significant difference between the truncated response for $p = 2$ (Figure 12.4) and the exact response given by $x_3(t)$ in Figure 12.5. This is because none of the forcing frequencies (Ω_1, Ω_2, or Ω_3) are in close proximity to ω_3 (110.907 rad/sec), and the contribution of the third mode is therefore relatively insignificant. Thus, a 2-mode solution for this particular problem would be sufficient. However, in general, for MDOF systems having just a few DOF, it is standard practice to consider all the modes in a mode superposition analysis.

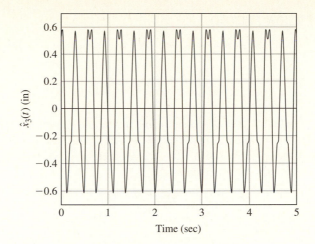

Figure 12.3

Truncated response $\hat{x}_3(t)$, mode 1 only ($p = 1$).

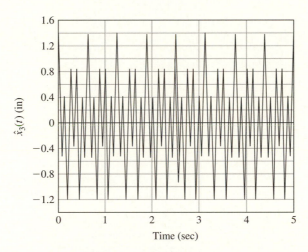

Figure 12.4

Truncated response $\hat{x}_3(t)$, modes 1 and 2 ($p = 2$).

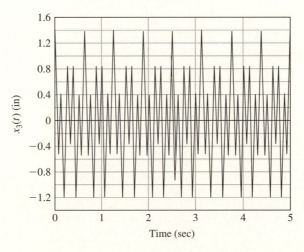

Figure 12.5

Exact response $x_3(t)$, all three modes considered ($p = 3$).

12.2 MODAL PARTICIPATION FACTOR

The forced response for a MDOF system given by Eq. (12.19) can also be expressed as

$$\{x(t)\} = \sum_{r=1}^{n} \{\Phi\}_r \Gamma_r R_r(t) \tag{12.28}$$

where Γ_r is referred to as the *modal participation factor* for the rth mode and $R_r(t)$ is the response ratio corresponding to the rth mode (refer to Chapter 6 for a discussion of the response ratio). Thus the dynamic response for the rth mode expressed in principal coordinates is given as

$$q_r(t) = \Gamma_r R_r(t) \tag{12.29}$$

where

$$\Gamma_r = \frac{P_r}{K_r} = \frac{\{\Phi\}_r^T\{F\}}{\{\Phi\}_r^T[k]\{\Phi\}_r} = \frac{\{\Phi\}_r^T\{F\}}{(\omega_r)^2\{\Phi\}_r^T[m]\{\Phi\}_r} \tag{12.30}$$

The modal participation factor is particularly useful when used in conjunction with the response spectrum analysis of MDOF systems, which is discussed in Section 12.7.

EXAMPLE 12.2 ▼

The second level of the three-story shear frame building shown in Figure 12.6 is subjected to the triangular load pulse shown in Figure 12.7. Determine the dynamic response of the top level of the building, $x_3(t)$, in the time intervals $0 \leq t \leq t_d$ and $t > t_d$. Assume $t_d = 0.5$ sec and $F_0 = 10$ kips.

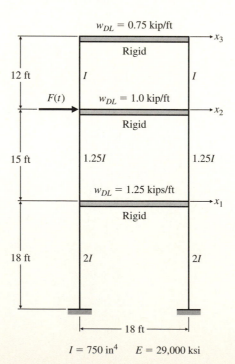

$I = 750 \text{ in}^4 \qquad E = 29{,}000 \text{ ksi}$

Figure 12.6

Three-story shear frame building of Example 12.2.

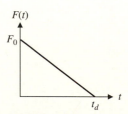

Figure 12.7

Triangular load pulse used in Example 12.2.

Solution

The mass and stiffness matrices for the structure are

$$[m] = \begin{bmatrix} 0.05823 & 0 & 0 \\ 0 & 0.04658 & 0 \\ 0 & 0 & 0.03494 \end{bmatrix} \text{ kip-sec}^2/\text{in}$$

$$[k] = \begin{bmatrix} 137.782 & -111.883 & 0 \\ -111.883 & 286.70 & -174.817 \\ 0 & -174.817 & 174.817 \end{bmatrix} \text{ kips/in}$$

and the load vector is

$$\{F\} = \begin{Bmatrix} 0 \\ 10 \\ 0 \end{Bmatrix} \text{ kips}$$

The eigensolution for the system yielded the following frequencies and mode shapes:

$$\{\omega\} = \begin{Bmatrix} 13.02 \\ 55.36 \\ 101.44 \end{Bmatrix} \text{ rad/sec} \qquad [\Phi] = \begin{bmatrix} 1.0 & 1.0 & 1.0 \\ 1.143 & -0.364 & -4.12 \\ 1.183 & -0.939 & 3.90 \end{bmatrix}$$

The normal modes for the frame are shown in Figure 12.8. The mode superposition displacement solution in physical coordinates for the top story, considering all three modes, is given by Eq. (12.28) as

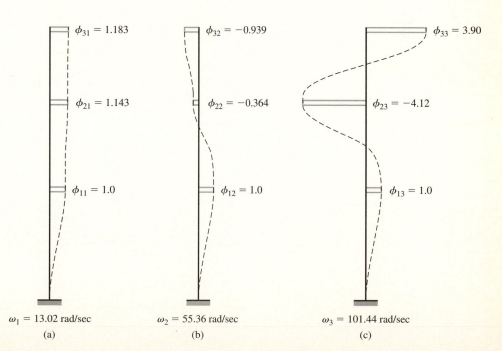

Figure 12.8

Normal modes for three-story shear frame building of Example 12.2: (a) mode 1; (b) mode 2; (c) mode 3.

$\phi_{31} = 1.183$ $\phi_{32} = -0.939$ $\phi_{33} = 3.90$

$\phi_{21} = 1.143$ $\phi_{22} = -0.364$ $\phi_{23} = -4.12$

$\phi_{11} = 1.0$ $\phi_{12} = 1.0$ $\phi_{13} = 1.0$

$\omega_1 = 13.02$ rad/sec $\omega_2 = 55.36$ rad/sec $\omega_3 = 101.44$ rad/sec

(a)   (b)   (c)

(a) for $0 \le t \le t_d$,

$$x_3(t) = \sum_{r=1}^{3} (\phi_3)_r \Gamma_r [R_1(t)]_r \tag{1}$$
$$= \phi_{31} \Gamma_1 [R_1(t)]_1 + \phi_{32} \Gamma_2 [R_1(t)]_2 + \phi_{33} \Gamma_3 [R_1(t)]_3$$

(b) for $t > t_d$,

$$x_3(t) = \sum_{r=1}^{3} (\phi_3)_r \Gamma_r [R_2(t)]_r \tag{2}$$
$$= \phi_{31} \Gamma_1 [R_2(t)]_1 + \phi_{32} \Gamma_2 [R_2(t)]_2 + \phi_{33} \Gamma_3 [R_2(t)]_3$$

where

$$(\phi_3)_1 = 1.183 \qquad (\phi_3)_2 = -0.939 \qquad (\phi_3)_3 = 3.90$$

$R_1(t)$ and $R_2(t)$ are the response ratios for the triangular load pulse for the intervals $0 \le t \le t_d$ and $t > t_d$, respectively. $R_1(t)$ and $R_2(t)$ were previously determined in Example 6.4. The modal participation factors are determined from Eq. (12.30) as

$$\Gamma_1 = \frac{\{\Phi\}_1^T \{F\}}{(\omega_1)^2 \{\Phi\}_1^T [m] \{\Phi\}_1} = 0.401 \text{ in}$$

$$\Gamma_2 = \frac{\{\Phi\}_2^T \{F\}}{(\omega_2)^2 \{\Phi\}_2^T [m] \{\Phi\}_2} = -0.0125 \text{ in}$$

$$\Gamma_3 = \frac{\{\Phi\}_3^T \{F\}}{(\omega_3)^2 \{\Phi\}_3^T [m] \{\Phi\}_3} = -0.002906 \text{ in}$$

Therefore the third-story response is given as
(a) For $0 \le t \le t_d$, from Eq. (1)

$$x_3(t) = 0.474 [R_1(t)]_1 + 0.0117 [R_1(t)]_2 - 0.0113 [R_1(t)]_3 \tag{3}$$

where (from Example 6.4)

$$[R_1(t)]_1 = 1 - \left(\frac{t}{t_d}\right) - \cos \omega_1 t + \left(\frac{1}{\omega_1 t_d}\right) \sin \omega_1 t_d \tag{4}$$

$$[R_1(t)]_2 = 1 - \left(\frac{t}{t_d}\right) - \cos \omega_2 t + \left(\frac{1}{\omega_2 t_d}\right) \sin \omega_2 t_d \tag{5}$$

$$[R_1(t)]_3 = 1 - \left(\frac{t}{t_d}\right) - \cos \omega_3 t + \left(\frac{1}{\omega_3 t_d}\right) \sin \omega_3 t_d \tag{6}$$

(b) For $t > t_d$, from Eq. (2)

$$x_3(t) = 0.474 [R_2(t)]_1 + 0.0117 [R_2(t)]_2 - 0.0113 [R_2(t)]_3 \tag{7}$$

where

$$[R_2(t)]_1$$

$$= \left(\frac{t}{\omega_1 t_d}\right)[\sin \omega_1 t(1 - \cos \omega_1 t_d) - \cos \omega_1 t(\omega_1 t_d - \sin \omega_1 t_d)] \tag{8}$$

$$[R_2(t)]_2$$

$$= \left(\frac{t}{\omega_2 t_d}\right)[\sin \omega_2 t(1 - \cos \omega_2 t_d) - \cos \omega_2 t(\omega_2 t_d - \sin \omega_2 t_d)] \tag{9}$$

$$[R_2(t)]_3$$

$$= \left(\frac{1}{\omega_3 t_d}\right)[\sin \omega_3 t(1 - \cos \omega_3 t_d) - \cos \omega_3 t(\omega_3 t_d - \sin \omega_3 t_d)] \tag{10}$$

An interesting feature of the modal participation factor is that it gives an indication of the relative contribution of each mode to the total dynamic response. Comparing the modal participation factors for the previous example problem, it is quite apparent that the major portion of the response is associated with the first mode, and the relative contributions of modes 2 and 3 are significantly less. ▲

12.3 MODE SUPERPOSITION SOLUTION FOR SYSTEMS WITH CLASSICAL DAMPING

The equations of motion for a *general viscously damped* MDOF system are expressed as

$$[m]\{\ddot{x}\} + [c]\{\dot{x}\} + [k]\{x\} = \{F(t)\} \tag{12.31}$$

where $[c]$ is the $n \times n$ viscous damping matrix. For the mode superposition method to be valid for viscously damped MDOF systems, Eqs. (12.31) must be expressed in uncoupled form (in principal coordinates); that is,

$$[M]\{\ddot{q}\} + [C]\{\dot{q}\} + [K]\{q\} = \{P(t)\} \tag{12.32}$$

where $[M]$, $[K]$, and $\{P(t)\}$ are the previously defined modal mass matrix, modal stiffness matrix, and modal force vector, respectively, and $[C]$ is the *modal damping matrix* defined by

$$[C] = [\Phi]^T[c][\Phi] \tag{12.33}$$

$[C]$ is an $n \times n$ symmetric matrix that is diagonal for only a special case of $[c]$. This special case of damping is referred to as *classical damping* or *proportional damping*, for which $[c]$ is proportional to $[m]$ and/or $[k]$. Thus, if it is assumed that the damping matrix $[c]$ is so constructed that

$$\{\Phi\}_r^T[c]\{\Phi\}_s = 0, \qquad r \neq s \tag{12.34}$$

then $[C]$ is diagonal and Eqs. (12.32) are uncoupled and may be written in the form

$$\ddot{q}_r + 2\zeta_r\omega_r + \omega_r^2 q_r = \left(\frac{1}{M_r}\right)P_r(t) \qquad r = 1, 2, \ldots \tag{12.35}$$

where ζ_r is the modal damping factor for the rth mode defined by

$$\zeta_r = \frac{C_r}{2M_r\omega_r} = \left(\frac{1}{2M_r\omega_r}\right)\{\Phi\}_r^T[c]\{\Phi\}_r \tag{12.36}$$

and C_r is the *modal damping coefficient* for the rth mode.

The solution of Eq. (12.35) for the rth mode can be expressed in the same form as the SDOF solution given by Eq. (6.25); that is,

$$\begin{aligned}
q_r(t) = {}& \left(\frac{1}{M_r\omega_{dr}}\right)\!\int_0^t P_r(\tau)e^{-\zeta_r\omega_r(t-\tau)}\sin\omega_{dr}(t-\tau)\,d\tau \\
& + q_r(0)e^{-\zeta_r\omega_r t}\cos\omega_{dr}t \\
& + \left(\frac{1}{\omega_{dr}}\right)[\dot{q}_r(0) + \zeta_r\omega_r q_r(0)]e^{-\zeta_r\omega_r t}\sin\omega_{dr}t
\end{aligned} \tag{12.37}$$

where ω_{dr} is the damped natural frequency for the rth mode given by

$$\omega_{dr} = \omega_r\sqrt{1 - \zeta_r^2} \tag{12.38}$$

Frequently, in the absence of more definitive information about damping, modal damping as in Eq. (12.35) is simply assumed to be valid, and "reasonable" values of ζ_r are assumed.

If the solutions to the n modal equations given by Eq. (12.35) are substituted into Eq. (12.19), then the exact system response $\{x(t)\}$ in physical coordinates for all n modes is obtained. However, if only p modes ($p \ll n$) are retained in the solution and Eq. (12.20) rather than Eq. (12.19) is used to define the truncated response, the resulting mode superposition solution omits the contribution of the modes ($p + 1$) to n.

In the principal coordinate equations of motion given by Eq. (12.32), it was assumed that the principal coordinate transformation defined by Eq. (12.33) uncoupled the damping forces in the same manner as the inertia and elastic forces are uncoupled. Therefore, the vibration mode shapes for the damped system are the same as the mode shapes for the undamped system. However, as previously stated, the principal coordinate transformation is valid only for classical or proportional damping. A very popular form of proportional damping is known as *Rayleigh damping*.

Rayleigh proposed that a damping matrix having the form

$$[c] = \alpha[m] + \beta[k] \tag{12.39}$$

where α and β are arbitrary proportionality factors, will satisfy the orthogonality condition given by Eq. (12.34). This can be easily verified by applying the orthogonality operation to both sides of Eq. (12.39). Therefore, it is apparent that a damping matrix proportional to the mass and/or stiffness matrices will permit uncoupling of the equations of motion. Rayleigh damping is just one example of *proportional damping*. As stated previously, proportional damping is also referred to as *classical damping, orthogonal damping,* and *modal damping*.

To illustrate implementation of Rayleigh damping, consider the rth mode eigenpair ($\omega_r, \{\Phi\}_r$) corresponding to the eigensystem defined by

$$[[k] - \omega_r^2[m]]\{\Phi\}_r = \{0\} \tag{12.40}$$

The orthogonality conditions specify

$$\{\Phi\}_r^T[m]\{\Phi\}_s = M_r \qquad \text{for } r = s \tag{12.41}$$

and

$$\{\Phi\}_r^T[k]\{\Phi\}_s = \omega_r^2 M_r \qquad \text{for } r = s \tag{12.42}$$

Then for Rayleigh damping defined by Eq. (12.39),

$$\{\Phi\}_r^T[c]\{\Phi\}_s = (\alpha + \beta\omega_r^2)M_r \qquad \text{for } r = s \tag{12.43}$$

Note that the modal damping coefficients given by Eq. (12.36) are defined as

$$C_r = \{\Phi\}_r^T[c]\{\Phi\}_s = 2M_r\omega_r\zeta_r \tag{12.44}$$

then the modal damping factors are given by

$$\zeta_r = \frac{1}{2}\left(\frac{\alpha}{\omega_r} + \beta\omega_r\right) \tag{12.45}$$

and the modal damping matrix $[C]$ is determined as

$$\begin{aligned}
[C] &= [\Phi]^T[c][\Phi] \\
&= 2\begin{bmatrix}
\zeta_1\omega_1 M_1 & 0 & 0 & \cdots \\
0 & \zeta_2\omega_2 M_2 & 0 & \cdots \\
0 & 0 & \zeta_3\omega_3 M_3 & \cdots \\
\cdots & \cdots & \cdots & \cdots & \cdots & \cdots & \cdots
\end{bmatrix} \\
&= \text{diag}(2\zeta_r\omega_r M_r)
\end{aligned} \tag{12.46}$$

Rayleigh damping is therefore defined for a MDOF system by specifying ζ_r for two different and unequal frequencies of vibration, and then solving for α and β by simultaneous solution of Eq. (12.45). With α and β having been calculated, the damping in the remaining modes is subsequently determined from Eq. (12.45). Note that the $\alpha[m]$ (mass proportional) contribution to damping in Eq. (12.39) renders a contribution to ζ_r in Eq. (12.45) that is inversely proportional to ω_r. However, the $\beta[k]$ (stiffness proportional) term yields a contribution to ζ_r that increases linearly with ω_r. Therefore, for large values of ω_r, the stiffness proportional term dominates the system damping. This trend generally leads to unrealistically high damping ratios in the higher vibration modes for large MDOF systems.

Fortunately, in most superposition analyses of large MDOF systems, since the number of vibration modes actually considered in the analysis, p, is significantly less than the total n modes of vibration the system possesses, damping in the higher modes is generally not a critical issue. Moreover, to incorporate damping in a mode superposition analysis, it is not necessary to formulate the damping matrix $[c]$ explicitly, since the system response can be determined from the solution of the set of independent equations represented by Eqs. (12.37) in which only the values of ζ_r are required. However, explicit formulation of the system damping matrix is required to evaluate the dynamic response of a MDOF system by *direct numerical integration*. Direct numerical integration procedures for MDOF systems and alternative methods for formulating the system damping matrix are discussed in Chapter 13.

EXAMPLE 12.3 ▼

For the three-story shear building of Example 12.2, the modal damping factors for the first two modes of vibration have been determined to be $\zeta_1 = 0.05$ and $\zeta_2 = 0.06$, respectively. Calculate ζ_3 on the assumption of Rayleigh damping.

Solution

The vibration frequencies and corresponding vibration mode shapes are

$$\{\omega\} = \begin{Bmatrix} 13.02 \\ 55.36 \\ 101.44 \end{Bmatrix} \text{ rad/sec} \qquad [\Phi] = \begin{bmatrix} 1.0 & 1.0 & 1.0 \\ 1.143 & -0.364 & -4.12 \\ 1.183 & -0.939 & 3.90 \end{bmatrix}$$

The Rayleigh damping factor for the rth mode is given by Eq. (12.45) as

$$\zeta_r = \frac{1}{2}\left(\frac{\alpha}{\omega_r} + \beta\omega_r\right) \tag{1}$$

for $\omega_r = \omega_1 = 13.02$ rad/sec, Eq. (1) yields

$$\zeta_1 = \frac{1}{2}\left(\frac{\alpha}{13.02} + \beta 13.02\right) = 0.05 \tag{2}$$

and for $\omega_r = \omega_2 = 55.36$ rad/sec, Eq. (1) yields

$$\zeta_2 = \frac{1}{2}\left(\frac{\alpha}{55.36} + \beta 55.36\right) = 0.06 \tag{3}$$

Expanding Eqs. (2) and (3) yields

$$\alpha + \beta(13.02)^2 = 1.302 \tag{4}$$

and

$$\alpha + \beta(55.36)^2 = 6.644 \tag{5}$$

Simultaneous solution of Eqs. (4) and (5) for α and β gives

$$\alpha = 0.9893$$

and

$$\beta = 0.001845$$

Therefore ζ_3 is determined from Eq. (12.45) as

$$\zeta_3 = \frac{1}{2}\left(\frac{\alpha}{101.44} + \beta 101.44\right) = 0.0984$$

and the modal damping matrix can be determined from Eq. (12.46) as

$$[C] = \begin{bmatrix} \zeta_1\omega_1 M_1 & 0 & 0 \\ 0 & \zeta_2\omega_2 M_2 & 0 \\ 0 & 0 & \zeta_3\omega_3 M_3 \end{bmatrix} \tag{6}$$

or

$$[C] = \begin{bmatrix} 0.2187 & 0 & 0 \\ 0 & 0.6325 & 0 \\ 0 & 0 & 27.531 \end{bmatrix}$$

▲

12.4 NUMERICAL EVALUATION OF MODAL RESPONSE

The numerical integration procedures for evaluating the dynamic response of SDOF systems discussed in Chapter 7 can be readily applied to the analysis of MDOF systems when used in conjunction with the mode superposition method. The MDOF system is treated essentially as n independent SDOF systems in normal coordinates. The independent (uncoupled) differential equations are then evaluated individually at discrete time intervals Δt by any numerical procedure discussed in Chapter 7. The transformation back to physical coordinates is effected after each time step.

Numerical evaluation of the dynamic response of MDOF systems is essential when the forcing function is arbitrary or cannot be expressed by means of a simple analytical function. A general procedure for the implementation of any of the numerical integration schemes (developed for SDOF systems) for evaluating the dynamic response of MDOF systems by mode superposition is illustrated in Table 12.1. To ensure accuracy in the analysis, a time step $\Delta t \leq T_n/10$ should be selected, where T_n is the natural period corresponding to the highest vibration mode associated with the system. For very large MDOF systems, when only p modes ($p << n$) are considered in the superposition analysis, select $\Delta t \leq T_p/10$, where T_p is the natural period corresponding to the pth mode.

TABLE 12.1. **Modal Response of MDOF Systems by Numerical Integration**

A. Initial Calculations:
 1. Establish $[m]$, $[k]$, $\{F(t)\}$
 2. Calculate the natural frequencies and mode shapes. For large MDOF systems, consider only p modes ($p << n$)
 3. Calculate the proportional damping factors in the system if required
 4. Calculate the modal mass, stiffness, and damping matrices, and the modal force vector

$$[M] = [\Phi]^T[m][\Phi]$$

$$[K] = [\Phi]^T[k][\Phi]$$

$$[C] = \text{diag}(2\zeta_r\omega_r M_r)$$

$$\{P(t)\} = [\Phi]^T\{F(t)\}$$

 5. Express the independent (uncoupled) equations of motion in normal coordinates

$$\ddot{q}_r + 2\zeta_r\omega_r + \omega_r^2 q_r = \left(\frac{1}{M_r}\right)P_r(t) \qquad r = 1, 2, \dots, n$$

 6. Select an appropriate time step Δt

$$\Delta t \le \frac{T_n}{10} \quad \text{or} \quad \Delta t \le \frac{T_p}{10} \qquad \text{for large systems}$$

 7. Make the remaining initial calculations required for the numerical procedure selected
B. Calculations for each time step:
 1. Perform the required calculations on each independent equation, corresponding to the numerical method selected
 2. Establish the displacements, velocities, and accelerations in normal coordinates

$$\{q\}_{i+1}, \{\dot{q}\}_{i+1}, \{\ddot{q}\}_{i+1}$$

 3. Transform the displacements, velocities, and accelerations back to the physical coordinates

$$\{x\}_{i+1} = [\Phi]\{q\}_{i+1}$$

$$\{\dot{x}\}_{i+1} = [\Phi]\{\dot{q}\}_{i+1}$$

$$\{\ddot{x}\}_{i+1} = [\Phi]\{\ddot{q}\}_{i+1}$$

EXAMPLE 12.4 ▼

The top level of the three-story shear building shown in Figure 12.9 is subjected to the arbitrary time-varying force specified in Figure 12.10. Determine the dynamic response by the mode superposition method. Write a computer program to numerically evaluate the dynamic response by the average acceleration method. Plot the displacement, velocity, and acceleration time histories for the top level of the building, x_3. Let $\zeta_1 = 0.03$ and $\zeta_2 = 0.04$, and calculate ζ_3 based upon the assumption of Rayleigh damping. Assume moment of inertia $I = 1500$ in^4 and modulus of elasticity $E = 29,000$ ksi for all columns.

Solution

The mass and stiffness matrices for the building are

$$[m] = \begin{bmatrix} 0.1035 & 0 & 0 \\ 0 & 0.0776 & 0 \\ 0 & 0 & 0.0518 \end{bmatrix} \text{ kip-sec}^2/\text{in}$$

$$[k] = \begin{bmatrix} 405.57 & -349.63 & 0 \\ -349.63 & 802.76 & -453.13 \\ 0 & -453.13 & 453.13 \end{bmatrix} \text{ kips/in}$$

The system natural frequencies and corresponding vibration modes are

$$\{\omega\} = \begin{Bmatrix} 15.067 \\ 73.464 \\ 131.86 \end{Bmatrix} \text{ rad/sec} \qquad [\Phi] = \begin{bmatrix} 1.0 & 1.0 & 1.0 \\ 1.093 & -0.438 & -3.986 \\ 1.122 & -1.142 & 4.036 \end{bmatrix}$$

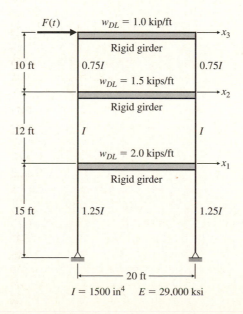

$F(t)$ $w_{DL} = 1.0$ kip/ft

10 ft   0.75I Rigid girder   0.75I →x_3

$w_{DL} = 1.5$ kips/ft

12 ft I Rigid girder I →x_2

$w_{DL} = 2.0$ kips/ft

15 ft   1.25I Rigid girder   1.25I →x_1

|← 20 ft →|

$I = 1500$ in^4 $E = 29,000$ ksi

Figure 12.9

Three-story shear frame building of Example 12.4.

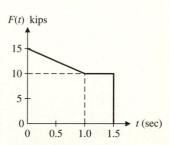

$F(t)$ kips

Figure 12.10

Time-varying force applied to top level of shear frame building in Example 12.4.

The mode shapes are shown in Figure 12.11. The natural period corresponding to the highest frequency ω_3 is $T_3 = 0.048$ sec. The integration time step is selected such that $\Delta t < T_3/10$, or 0.0048 sec. In this example, a time step $\Delta t = 0.002$ sec is employed to ensure accuracy of the calculations.

The modal mass and stiffness matrices are determined from Eqs. (12.8) and (12.9), respectively, as

$$[M] = [\Phi]^T[m][\Phi] = \begin{bmatrix} 0.2614 & 0 & 0 \\ 0 & 0.1859 & 0 \\ 0 & 0 & 2.1802 \end{bmatrix}$$

and

$$[K] = [\Phi]^T[k][\Phi] = \begin{bmatrix} 59.345 & 0 & 0 \\ 0 & 1003.499 & 0 \\ 0 & 0 & 37907.85 \end{bmatrix}$$

The modal damping matrix is determined from Eq. (12.46) as

$$[C] = \text{diag}(2\zeta_r\omega_r M_r) \tag{1}$$

and from Eq. (12.45) the modal damping factors are expressed as

$$\alpha + \beta\omega_r^2 = 2\omega_r\zeta_r \tag{2}$$

Then for $\omega_r = \omega_1$ and $\zeta_r = \zeta_1$, Eq. (2) yields

$$\alpha + 227\beta = 0.90402 \tag{3}$$

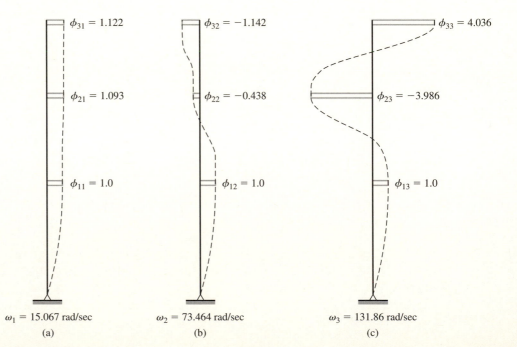

Figure 12.11

Normal modes for three-story shear frame building of Example 12.4: (a) mode 1; (b) mode 2; (c) mode 3.

$\phi_{31} = 1.122$ $\phi_{32} = -1.142$ $\phi_{33} = 4.036$

$\phi_{21} = 1.093$ $\phi_{22} = -0.438$ $\phi_{23} = -3.986$

$\phi_{11} = 1.0$ $\phi_{12} = 1.0$ $\phi_{13} = 1.0$

$\omega_1 = 15.067$ rad/sec $\omega_2 = 73.464$ rad/sec $\omega_3 = 131.86$ rad/sec

(a)   (b)   (c)

and for $\omega_r = \omega_2$ and $\zeta_r = \zeta_2$, Eq. (2) gives

$$\alpha + 5397\beta = 5.87712 \tag{4}$$

Simultaneous solution of Eqs. (3) and (4) for α and β gives

$$\alpha = 0.68567$$

and

$$\beta = 0.0009619$$

Substituting these values for α and β into Eq. (2) for $\omega_r = \omega_3$ and solving for ζ_3 results in

$$\zeta_3 = \frac{1}{2}\left(\frac{\alpha}{\omega_3} + \beta\omega_3\right) = 0.066$$

Then from Eq. (1), the modal damping matrix is given by

$$[C] = \begin{bmatrix} 0.2363 & 0 & 0 \\ 0 & 1.0926 & 0 \\ 0 & 0 & 37.9578 \end{bmatrix}$$

Plots of the time histories for $x_3(t)$, $\dot{x}_3(t)$, and $\ddot{x}_3(t)$ are presented in Figures 12.12, 12.13, and 12.14, respectively. The FORTRAN computer program, MODES, used to evaluate the dynamic response, is available on the authors' web site (refer to the Preface).

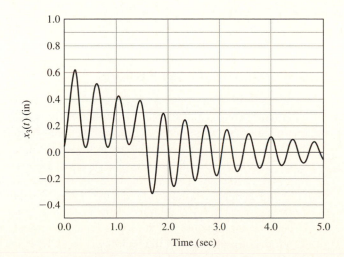

Figure 12.12

Displacement response history $x_3(t)$.

Figure 12.13

Velocity response history $\dot{x}_3(t)$.

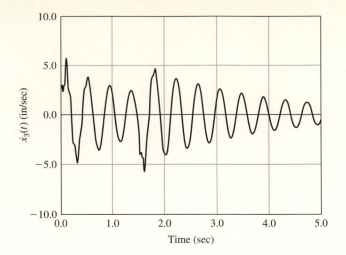

Figure 12.14

Acceleration response history $\ddot{x}_3(t)$.

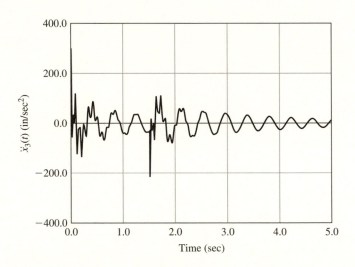

12.5 NORMAL MODE RESPONSE TO SUPPORT MOTIONS

In many instances, interest is focused on the response of MDOF systems caused by support motions rather than by applied actions, such as the earthquake response of structures. A typical multi-degree-of-freedom system often used to study vibrations of tall buildings or similar structures is shown in Figure 12.15. In this particular model, the masses m_i are lumped (localized) at the floor levels and are interconnected with massless columns that have the equivalent spring constants k_i. Dashpots characterized by the damping constants c_i model the energy dissipation in the system. The shear force at each floor level is designated by V_i, and V_n corresponds to the base shear acting between the structure and the soil; $\ddot{x}_g(t)$ is the ground acceleration, and x_i represents the *relative displacement* of the building at story level i. The other pertinent parameters in

Figure 12.15

Mechanical model representing MDOF building structure subject to ground acceleration.

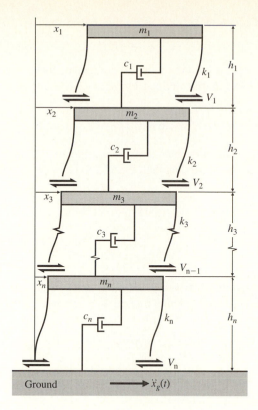

assessing the dynamic response of the structure subject to base excitation are the relative velocities $\dot{x}_i$ and the total (absolute) accelerations x_{ti}, at each story.

The system equilibrium equations in physical coordinates, one of which corresponds to each mass m_i, are written as

$$[m]\{\ddot{x}\} + [m]\{I\}\ddot{x}_g(t) + [c]\{\dot{x}\} + [k]\{x\} = 0 \tag{12.47}$$

where $\ddot{x}_g(t)$ is the ground acceleration and $\{I\}$ is the *unit vector* of dimension n. To expedite formulation and solution of the problem, it is convenient to consider *effective loads* acting at the various floor levels of the structure equal to the product of the lumped mass at the floor level and the ground acceleration. That is, at any floor i the effective load acting on the mass may be written as

$$F_{\text{eff}}(t)_i = m_i \ddot{x}_g(t) \tag{12.48}$$

Thus, the complete effective load vector is given by the product of the mass matrix and the ground acceleration $\ddot{x}_g(t)$, which may be expressed as

$$\{F_{\text{eff}}(t)\} = [m]\{I\}\ddot{x}_g(t) \tag{12.49}$$

Equation (12.47) can then be rewritten as

$$[m]\{\ddot{x}\} + [c]\{\dot{x}\} + [k]\{x\} = -[m]\{I\}\ddot{x}_g(t) = \{F_{\text{eff}}(t)\} \tag{12.50}$$

The modal equations of motion are thus expressed as

$$[M]\{\ddot{q}\} + [C]\{\dot{q}\} + [K]\{q\} = \{P_{\text{eff}}(t)\} \qquad (12.51)$$

where $[M]$ and $[K]$ are the modal mass and stiffness matrices, respectively, $[C]$ is the modal damping matrix defined by Eq. (12.46), and the effective modal force vector $\{P_{\text{eff}}(t)\}$ is given by

$$\{P_{\text{eff}}(t)\} = [\Phi]^T[m]\{I\}\ddot{x}_g(t) = [\Phi]^T\{F_{\text{eff}}(t)\} \qquad (12.52)$$

Since the modal mass, modal stiffness, and modal damping (assuming proportional damping) matrices are diagonal, Eq. (12.51) can be written in normal coordinates as n uncoupled equations:

$$M_r\ddot{q}_r + C_r\dot{q}_r + K_rq_r = P_r(t) \qquad r = 1, 2, \ldots, n \qquad (12.53)$$

where

$$P_r(t) = \{\Phi\}_r^T[m]\{I\}\ddot{x}_g(t) \qquad (12.54)$$

The response of the rth mode in normal coordinates at any time t may be obtained by evaluation of the Duhamel integral expression for the given ground motion

$$q_r(t) = \left(\frac{1}{M_r\omega_r}\right)\int_0^t P_r(\tau)e^{-\zeta_r\omega_r(t-\tau)} \sin \omega_r(t-\tau)\, d\tau \qquad (12.55)$$

For an undamped system, Eq. (12.55) reduces to

$$q_r(t) = \left(\frac{1}{M_r\omega_r}\right)\int_0^t P_r(\tau) \sin \omega_r(t-\tau)\, d\tau \qquad (12.56)$$

The system response in physical coordinates, for relative displacement, is then given as

$$\{x(t)\} = \sum_{r=1}^n \{\Phi\}_r q_r(t) \qquad (12.57)$$

For most practical applications, numerical evaluation of the system response is essential. Once the effective load vector $\{F(t)\}_{\text{eff}}$ has been established, the numerical evaluation procedure for modal response described in Section 12.4 is applicable. Generally, the n-modal equations represented by Eq. (12.53) are numerically evaluated individually, and the response in physical coordinates after each discrete time step Δt is obtained from the transformation represented by Eq. (12.57).

EXAMPLE 12.5 ▼

The base of the four-story shear frame building shown in Figure 12.16 is excited by the horizontal ground acceleration $\ddot{x}_g(t)$ shown in Figure 12.17. Determine the dynamic response of the structure by the mode superposition method. Plot the time histories for relative displacement and relative velocity, and the total (absolute) acceleration for the top level of the building in the time interval $0 \leq t \leq 5$ sec. Let $\zeta_1 = 0.02$ and $\zeta_2 = 0.03$, and calculate ζ_3 and ζ_4 based upon the assumption of Rayleigh damping. Evaluate the dynamic response numerically. Assume $m = 0.5$ kip-sec^2/in and $k = 150$ kips/in.

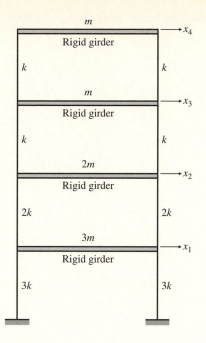

Figure 12.16
Four-story shear frame
building of Example 12.5.

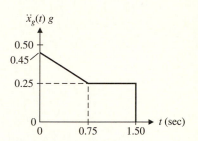

Figure 12.17
Ground acceleration used in
Example 12.5.

Solution

The mass and stiffness matrices for the structure are

$$
[m] = \begin{bmatrix} 1.5 & 0 & 0 & 0 \\ 0 & 1.0 & 0 & 0 \\ 0 & 0 & 0.5 & 0 \\ 0 & 0 & 0 & 0.5 \end{bmatrix} \text{ kip-sec}^2/\text{in}
$$

$$
[k] = \begin{bmatrix} 1500 & -600 & 0 & 0 \\ -600 & 900 & -300 & 0 \\ 0 & -300 & 600 & -300 \\ 0 & 0 & -300 & 300 \end{bmatrix} \text{ kips/in}
$$

The system natural frequencies and vibration modes are

$$
\omega = \begin{Bmatrix} 10.62 \\ 22.18 \\ 35.94 \\ 42.41 \end{Bmatrix} \text{ rad/sec}
$$

$$
[\Phi] = \begin{bmatrix} 0.200 & 0.475 & 0.579 & 0.254 \\ 0.445 & 0.602 & -0.425 & -0.508 \\ 0.767 & -0.135 & -0.600 & 1.016 \\ 0.945 & -0.761 & 0.519 & -0.508 \end{bmatrix}
$$

The vibration modes are illustrated in Figure 12.18. The modal mass and stiffness matrices are calculated from Eq. (12.8) and (12.9), respectively, as

$$[M] = [\Phi]^T[m][\Phi] = \begin{bmatrix} 1.0 & 0 & 0 & 0 \\ 0 & 1.0 & 0 & 0 \\ 0 & 0 & 1.0 & 0 \\ 0 & 0 & 0 & 1.0 \end{bmatrix}$$

$$[K] = [\Phi]^T[k][\Phi] = \begin{bmatrix} 112.8 & 0 & 0 & 0 \\ 0 & 493.8 & 0 & 0 \\ 0 & 0 & 1293 & 0 \\ 0 & 0 & 0 & 1800 \end{bmatrix}$$

The modal damping matrix is determined from Eq. (12.46) as

$$[C] = \text{diag}(2\zeta_r\omega_r M_r) \qquad (1)$$

From Eq. (12.45) the modal damping factors are expressed as

$$\alpha + \beta\omega_r^2 = 2\omega_r\zeta_r \qquad (2)$$

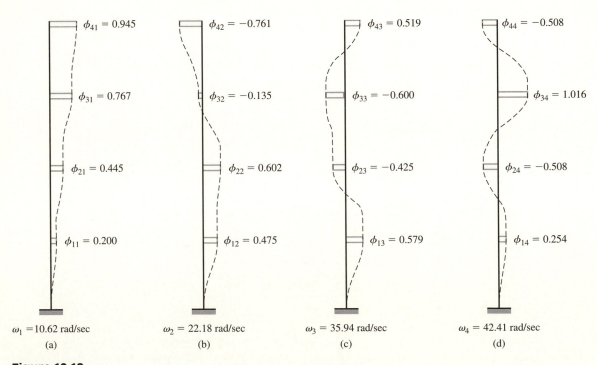

$\phi_{41} = 0.945$ $\phi_{42} = -0.761$ $\phi_{43} = 0.519$ $\phi_{44} = -0.508$

$\phi_{31} = 0.767$ $\phi_{32} = -0.135$ $\phi_{33} = -0.600$ $\phi_{34} = 1.016$

$\phi_{21} = 0.445$ $\phi_{22} = 0.602$ $\phi_{23} = -0.425$ $\phi_{24} = -0.508$

$\phi_{11} = 0.200$ $\phi_{12} = 0.475$ $\phi_{13} = 0.579$ $\phi_{14} = 0.254$

$\omega_1 = 10.62$ rad/sec $\omega_2 = 22.18$ rad/sec $\omega_3 = 35.94$ rad/sec $\omega_4 = 42.41$ rad/sec

(a)   (b)   (c)   (d)

Figure 12.18
Normal modes for four-story shear frame building of Example 12.5: (a) mode 1; (b) mode 2; (c) mode 3; (d) mode 4.

Then for $\omega_r = \omega_1$ and $\zeta_r = \zeta_1$, Eq. (2) yields

$$\alpha + 112.8\beta = 0.425 \tag{3}$$

and for $\omega_r = \omega_2$ and $\zeta_r = \zeta_2$, Eq. (2) yields

$$\alpha + 493.8\beta = 1.333 \tag{4}$$

Simultaneous solution of Eqs. (3) and (4) for α and β gives

$$\alpha = 0.157$$

and

$$\beta = 0.00237$$

Substituting these values for α and β into Eq. (2) for $\omega_r = \omega_3$, and for $\omega_r = \omega_4$, respectively, results in the modal damping factors ζ_3 and ζ_4 as follows

$$\zeta_3 = \frac{1}{2}\left(\frac{\alpha}{\omega_3} + \beta\omega_3\right) = 0.044$$

and

$$\zeta_4 = \frac{1}{2}\left(\frac{\alpha}{\omega_4} + \beta\omega_4\right) = 0.052$$

Then from Eq. (1), the modal damping matrix is established as

$$[C] = \text{diag}(2\zeta_r\omega_r M_r) = \begin{bmatrix} 0.425 & 0 & 0 & 0 \\ 0 & 1.33 & 0 & 0 \\ 0 & 0 & 3.16 & 0 \\ 0 & 0 & 0 & 4.41 \end{bmatrix}$$

The dynamic response was evaluated using computer program MODES, which is available on the authors' web site (refer to the Preface). Time-history plots for relative displacement $x_4(t)$, relative velocity $\dot{x}_4(t)$, and total (absolute) acceleration $\ddot{x}_{t4}(t)$ are presented in Figures 12.19, 12.20, and 12.21, respectively. The acceleration $\ddot{x}_{t4}(t)$ represents the total or absolute acceleration of the mass m_4. It is calculated as the sum of the relative acceleration $\ddot{x}_4(t)$ and the ground acceleration $\ddot{x}_g(t)$.

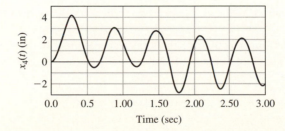

Figure 12.19

Relative displacement response history $x_4(t)$.

Figure 12.20

Relative velocity response history $\dot{x}_4(t)$.

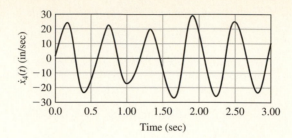

Figure 12.21

Total (absolute) acceleration response history $\ddot{x}_{t4}(t)$.

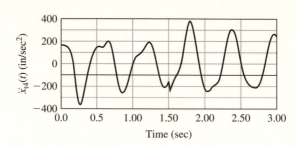

12.6 RESPONSE SPECTRUM ANALYSIS

The construction of response spectra or shock spectra for a specified dynamic distur-bance was discussed in Chapters 6 and 7. Its application to SDOF systems was also illustrated. The technique is also applicable to MDOF systems when used in conjunc-tion with the mode superposition procedure.

In many practical situations, the design engineer is primarily interested in the maximum response of the system to a specified input. In such instances a response spectrum analysis is quite useful. The maximum *modal response* for the rth mode of a MDOF system to a specified input, expressed in physical coordinates, is given by

$$(\{x\}_r)_{\max} = \left|\{\Phi\}_r \Gamma_r (\text{DMF})_r\right| \tag{12.58}$$

where

$(\{x\}_r)_{\max}$ = maximum response for the rth mode

$\{\Phi\}_r$ = modal vector for the rth mode

Γ_r = modal participation factor (refer to Section 12.2 for a discussion) for the rth mode

$(\text{DMF})_r$ = dynamic magnification factor for the rth mode determined from the appropriate response spectrum

An upper bound for the total response may be obtained by summing numerically the maximum response of each mode. Therefore, the maximum response can be expressed as

$$\{x\}_{\max} = \sum_{r=1}^{n} \left|\{\Phi\}_r \Gamma_r (\text{DMF})_r\right| \tag{12.59}$$

The numerical sum of the modal responses given by Eq. (12.59) is a very conservative estimate of the actual response, since it is implicit in Eq. (12.59) that the maximum responses for all modes occur simultaneously. This is neither reasonable nor correct. Many statistical methods for combining the maximum modal responses have been developed, and several of these modal combination techniques are discussed in Chapter 18. A very popular method for combining modal maxima, known as the square root of the sum of the squares (SRSS), is given by

$$\{x\}_{max} = \left[\sum_{r=1}^{n} (\{\Phi\}_r \Gamma_r (DMF)_r)^2 \right]^{1/2} \tag{12.60}$$

The following example illustrates application of the response spectrum method of analysis to MDOF systems.

EXAMPLE 12.6 ▼

The top level of the three-story shear frame building of Example 12.2 and shown in Figure 12.6 is subjected to a triangular pulse force similar to that shown in Figure 12.7. Determine the maximum displacement of the top level $(x_3)_{max}$ by the response spectrum technique. Combine the individual modal maxima by the SRSS method. Assume $F_0 = 8$ kips and $t_d = 0.5$ sec.

Solution

The natural frequencies and mode shapes for this structure are given in Example 12.2. The natural periods of vibration are

$$T_1 = \frac{2\pi}{\omega_1} = 0.4826 \text{ sec}$$

$$T_2 = \frac{2\pi}{\omega_2} = 0.1135 \text{ sec}$$

$$T_3 = \frac{2\pi}{\omega_3} = 0.0620 \text{ sec}$$

and the force vector is

$$\{F\} = \begin{Bmatrix} 0 \\ 0 \\ 8 \end{Bmatrix} \text{ kips}$$

The modal participation factors are determined from Eq. (12.30) as

$$\Gamma_1 = \frac{\{\Phi\}_1^T \{F\}}{(\omega_1)^2 \{\Phi\}_1^T [m] \{\Phi\}_1} = 0.3223 \text{ in}$$

$$\Gamma_2 = \frac{\{\Phi\}_2^T\{F\}}{(\omega_2)^2\{\Phi\}_2^T[m]\{\Phi\}_2} = -0.0257 \text{ in}$$

$$\Gamma_3 = \frac{\{\Phi\}_3^T\{F\}}{(\omega_3)^2\{\Phi\}_3^T[m]\{\Phi\}_3} = 0.0022 \text{ in}$$

The response spectrum for the triangular load pulse is given in Figure 7.20. Therefore, the DMF for each mode is (referring to Figure 7.20):

Mode 1: $\quad \dfrac{t_d}{T_1} = 1.0361 \quad$ $(DMF)_1 = 1.55$

Mode 2: $\quad \dfrac{t_d}{T_2} = 4.4053 \quad$ $(DMF)_2 = 1.87$

Mode 3: $\quad \dfrac{t_d}{T_3} = 8.0645 \quad$ $(DMF)_3 = 1.92$

The individual modal maxima expressed in physical coordinates are determined from Eq. (12.58) as

$$[(x_3)_{max}]_1 = \phi_{31}\Gamma_1(DMF)_1 = (1.183)(0.3323)(1.55) = 0.609 \text{ in}$$

$$[(x_3)_{max}]_2 = \phi_{32}\Gamma_2(DMF)_2$$
$$= (-0.939)(-0.0257)(1.87) = 0.0451 \text{ in}$$

$$[(x_3)_{max}]_3 = \phi_{33}\Gamma_3(DMF)_3 = (3.9)(0.0022)(1.92) = 0.0165 \text{ in}$$

Combining the modal maxima by the SRSS method given by Eq. (12.60) yields

$$(x_3)_{max} = [(0.609)^2 + (0.0451)^2 + (0.0165)^2]^{1/2} = 0.6109 \text{ in}$$

Construction of response spectra was previously discussed in Chapters 6 and 7. However, it should be noted that the computer program LRESP, available on the authors' web site, can be used to develop response spectra for any arbitrary disturbances. Moreover, computer program MODES, which is also available on the authors' web site, has the option to conduct a response spectrum analysis. ▲

12.7 MODE ACCELERATION METHOD

An alternative formulation for mode superposition analysis is the *mode acceleration method*. As previously mentioned, the mode acceleration method is not as frequently employed in practical structural dynamics problems because it is flexibility based rather than stiffness based. However, it does exhibit improved convergence properties over the mode displacement method. That is, usually a lesser number of vibration modes is required from the eigenproblem to yield an accurate superposition solution.

Consider the equations of motion for an undamped MDOF represented by Eq. (12.1). The truncated mode displacement solution $\{\hat{x}(t)\}$ is given by Eq. (12.20), in which the contribution from modes $(p + 1)$ to n are not included. The corresponding mode acceleration solution is based on the following expression of Eq. (12.1):

$$\{x(t)\} = [k]^{-1}[\{F(t)\} - [m]\{\ddot{x}\}] \tag{12.61}$$

where $[k]^{-1} \equiv [a]$. If the acceleration vector $[\ddot{x}]$ in Eq. (12.61) is approximated by $\{\hat{x}\}$, then the truncated mode acceleration solution $\{\hat{x}(t)\}$ is given by

$$\{\hat{x}(t)\} = [k]^{-1}[\{F(t)\} - [m]\{\hat{\ddot{x}}\}] \tag{12.62}$$

From Eq. (12.20), $\{\hat{\ddot{x}}(t)\}$ can be written as

$$\{\ddot{x}(t)\} = \sum_{r=1}^{p} \{\Phi\}_r \ddot{q}_r(t) \tag{12.63}$$

Combining Eqs. (12.62) and (12.63) results in

$$\{\hat{x}(t)\} = [k]^{-1}\{F(t)\} - [k]^{-1}\sum_{r=1}^{p} [m]\{\Phi\}_r \ddot{q}_r(t) \tag{12.64}$$

Recalling the frequency equation

$$[[k] - \omega_r^2[m]]\{\Phi\}_r = \{0\} \tag{12.65}$$

and noting that

$$[k]\{\Phi\}_r = \omega_r^2[m]\{\Phi\}_r \tag{12.66}$$

it follows that

$$[m] = \frac{1}{\omega_r^2}[k] \tag{12.67}$$

Combining Eqs. (12.64) and (12.66) yields the truncated displacement solution as defined by the mode acceleration method expressed as

$$\{\hat{x}(t)\} = [k]^{-1}\{F(t)\} - \sum_{r=1}^{p} \left(\frac{1}{\omega_r^2}\right)\{\Phi\}_r \ddot{q}_r(t) \tag{12.68}$$

The improved convergence of the mode acceleration method relative to the mode displacement method is illustrated by Eq. (12.68). The first term on the right hand side of Eq. (12.68) represents the *pseudostatic response* and the second term is the *mode acceleration*, which gives the method its name. The presence of ω_r^2 in the denominator of the second term provides quadratic convergence. However, despite the improved convergence exhibited by the mode acceleration method, it is still very important to consider all vibration modes whose natural frequencies are in close proximity to any excitation frequency.

EXAMPLE 12.7 ▼

The top level of an undamped three-story shear frame building is subjected to a harmonic force $F(t) = F_0 \sin \Omega t$ as shown in Figure 12.22. The mass and stiffness matrices for the structure are

$$[m] = \begin{bmatrix} 0.05823 & 0 & 0 \\ 0 & 0.04658 & 0 \\ 0 & 0 & 0.03494 \end{bmatrix} \text{kip-sec}^2/\text{in}$$

$$[k] = \begin{bmatrix} 137.78 & -111.88 & 0 \\ -111.88 & 286.70 & -174.82 \\ 0 & -174.82 & 174.82 \end{bmatrix} \text{kips/in}$$

Determine the displacement response of the top level $x_3(t)$ by the mode acceleration method. Compare the responses of the truncated solutions corresponding to $p = 1$ and $p = 2$ to the exact solution given by $p = 3$. Consider excitation frequencies of $\Omega = 0.75\omega_1$, $\Omega = 2.5\omega_1$, $\Omega = 5.0\omega_1$, and $\Omega = 1.2\omega_3$.

Solution

The natural frequencies and modal matrix for the structure are

$$\{\omega\} = \begin{Bmatrix} 13.02 \\ 55.36 \\ 101.44 \end{Bmatrix} \text{rad/sec} \qquad [\Phi] = \begin{bmatrix} 1.0 & 1.0 & 1.0 \\ 1.143 & -0.364 & -4.118 \\ 1.183 & -0.940 & 3.897 \end{bmatrix}$$

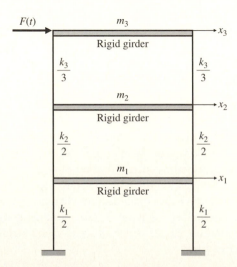

Figure 12.22

Three-story shear frame building of Example 12.7.

The mode shapes are shown in Figure 12.23. The flexibility matrix is calculated as

$$[a] = [k]^{-1} = \begin{bmatrix} 0.0386 & 0.0386 & 0.0386 \\ 0.0386 & 0.0476 & 0.0476 \\ 0.0386 & 0.0476 & 0.0533 \end{bmatrix} \text{ in/kip}$$

The truncated solution for the top level of the structure is expressed in physical coordinates by Eq. (12.68) as

$$\hat{x}_3(t) = a_{33}F_0 \sin \Omega t - \sum_{r=1}^{p} \left(\frac{1}{\omega_r^2}\right)(\phi_3)_r \ddot{q}_r(t) \tag{1}$$

Since $\ddot{q}_r = -\Omega^2 q_r$, Eq. (1) yields

$$\hat{x}_3(t) = a_{33}F_0 \sin \Omega t + \sum_{r=1}^{p} \left(\frac{\Omega^2}{\omega_r^2}\right)(\phi_3)_r q_r \tag{2}$$

Then, combining Eq. (2) with Eq. (12.25) results in the truncated solution $\hat{x}_3(t)$ as

$$\hat{x}_3(t) = a_{33}F_0 \sin \Omega t$$
$$+ \sum_{r=1}^{p} \left(\frac{\Omega^2}{\omega_r^2}\right)(\phi_3)_r \left(\frac{P_r}{K_r}\right)\left[\frac{1}{1 - (\Omega / \omega_r)^2}\right] \sin \Omega t \tag{3}$$

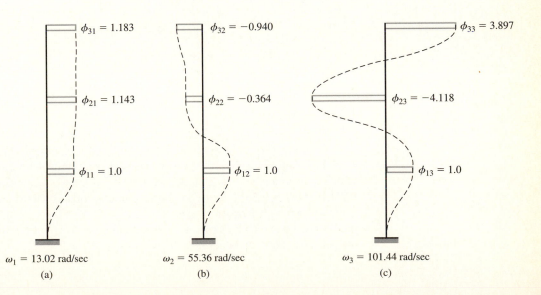

Figure 12.23

Normal modes for three-story shear frame building of Example 12.7: (a) mode 1; (b) mode 2; (c) mode 3.

$\phi_{31} = 1.183$ $\phi_{32} = -0.940$ $\phi_{33} = 3.897$

$\phi_{21} = 1.143$ $\phi_{22} = -0.364$ $\phi_{23} = -4.118$

$\phi_{11} = 1.0$ $\phi_{12} = 1.0$ $\phi_{13} = 1.0$

$\omega_1 = 13.02$ rad/sec $\omega_2 = 55.36$ rad/sec $\omega_3 = 101.44$ rad/sec

(a)   (b)   (c)

The modal stiffness matrix is established from Eq. (12.8) as

$$[K] = [\Phi]^T[k][\Phi] = \begin{bmatrix} 28.479 & 0 & 0 \\ 0 & 292.069 & 0 \\ 0 & 0 & 14187.953 \end{bmatrix}$$

and the modal force vector from Eq. (12.10) as

$$\{P(t)\} = \{\Phi\}^T\{F(t)\} = \begin{Bmatrix} 1.183F_0 \\ -0.940F_0 \\ 3.897F_0 \end{Bmatrix} \sin \Omega t$$

Thus, the truncated response given by Eq. (3) is expressed as

$$\hat{x}_3(t) = 0.0533F_0 \sin \Omega t$$

$$\left. \begin{array}{l} \left. \begin{array}{l} \left. + \dfrac{\left(\dfrac{\Omega^2}{169.52}\right)(1.183)^2 F_0 \sin \Omega t}{(28.79)\left[1 - \left(\dfrac{\Omega^2}{169.52}\right)\right]} \right\} p = 1 \\[2em] + \dfrac{\left(\dfrac{\Omega^2}{3064.73}\right)(-0.940)^2 F_0 \sin \Omega t}{(292.069)\left[1 - \left(\dfrac{\Omega^2}{3064.73}\right)\right]} \end{array} \right\} p = 2 \\[4em] + \dfrac{\left(\dfrac{\Omega^2}{10290.07}\right)(3.897)^2 F_0 \sin \Omega t}{(14187.953)\left[1 - \left(\dfrac{\Omega^2}{10290.07}\right)\right]} \end{array} \right\} p = 3$$

▲

A summary of the responses $\hat{x}_3(t)$ for the four values of Ω is presented in Table 12.2. The results in Table 12.2 indicate that the exact static solution at $\Omega = 0$ is attained without any contributions from normal modes. At the low excitation frequencies, $\Omega = 0.75\omega_1$ and $\Omega = 2.5\omega_1$, even a one-term solution produces fairly accurate results. However, for the high-frequency excitation cases, $\Omega = 5.0\omega_1$ and $\Omega = 1.2\omega_3$, all three modes must be considered for an accurate solution. This is because the excitation frequencies for these loadings are in the vicinity of ω_3.

TABLE 12.2. **Summary of Results for Example 12.7**

Excitation Frequency	Response $\hat{x}_3(t) \times F_0 \sin \Omega t$ (in)		
Ω	$p = 1$	$p = 2$	$p = 3$ (exact)
0	0.0533	0.0533	0.0533
$0.75\omega_1$	0.11576	0.11585	0.11586
$2.5\omega_1$	0.02496	0.02656	0.02668
$5.0\omega_1$	0.002665	−0.008263	−0.007513
$1.2\omega_3$	0.004122	0.000312	−0.003189

A general expression for $\ddot{q}_r(t)$, for any arbitrary excitation, to substitute into the mode acceleration equation, Eq. (12.68), can be determined from Eq. (12.18). Therefore

$$q_r(t) = -\omega_r^2 q_r(0)\cos \omega_r t - \omega_r^2 \dot{q}_r(0)\sin \omega_r t$$
$$+ \frac{P_r(t)}{M_r} - \frac{\omega_r}{M_r}\int_0^t P_r(\tau)\sin \omega_r(t - \tau)\, d\tau \tag{12.69}$$

Integration by parts may be employed to convert Eq. (12.27) to the alternative form given by

$$\ddot{q}_r(t) = -\omega_r^2 q_r(0)\cos \omega_r t - \omega_r^2 \dot{q}_r(0)\sin \omega_r t$$
$$+ \frac{1}{M_r}\{P_r(0)\}\cos \omega_r t - \int_0^t \frac{d}{d\tau}[P_r(\tau)]\cos \omega_r(t - \tau)\, d\tau \tag{12.70}$$

The numerical integration procedures discussed in Chapter 7 may also be used to evaluate Eqs. (12.69) and (12.70) to determine modal acceleration time histories.

REFERENCES

1 Timoshenko, S., Young, D.H., and Weaver, W., *Vibration Problems in Engineering,* Wiley, New York, 1974.

2 Humar, J.L., *Dynamics of Structures,* Prentice Hall, Englewood Cliffs, NJ, 1990.

3 Craig, Roy R., *Structural Dynamics: An Introduction to Computer Methods,* Wiley, New York, 1981.

4 Meirovitch, Leonard, *Computational Methods in Structural Dynamics,* Sijthoff and Noordhoff, Rockville, MD, 1980.

5 Biggs, John M., *Introduction to Structural Dynamics,* McGraw-Hill, New York, 1964.

6 Clough, R.W. and Penzien, J., *Dynamics of Structures,* 2nd ed., McGraw-Hill, New York, 1993.

7 Fertis, Demeter G., *Dynamics and Vibration of Structures,* Robert E. Krieger Publishing, Malabar, FL, 1984.

NOTATION

C_r	modal damping coefficient for rth normal mode
$F_{\text{eff}}(t)_i$	effective load for ground acceleration applied to ith mass m_i
K_r	generalized (modal) stiffness for rth normal mode
M_r	generalized (modal) mass for rth normal mode
p	number of modes considered in the mode superposition
$P_r(t)$	modal force for rth normal mode
$R_r(t)$	response ratio for rth normal mode
t	time
t_d	time of duration of externally applied force
t_r	rise time of externally applied force
T_n	natural period of nth normal mode
T_p	natural period of pth normal mode
V_i	story shear in ith story of building
$\ddot{x}_g(t)$	ground acceleration
$[c]$	viscous damping matrix
$[C]$	modal damping matrix
$[D]$	dynamic (system) matrix
$\{F_0\}$	amplitude vector for external harmonic forces
$\{F(t)\}$	external force vector
$\{F_{\text{eff}}(t)\}$	effective load vector for ground acceleration
$\{I\}$	unit vector
$[k]$	stiffness matrix
$[K]$	generalized (modal) stiffness matrix
$[m]$	mass matrix
$[M]$	generalized (modal) mass matrix
$\{P(t)\}$	generalized (modal) force vector
$\{P_{\text{eff}}(t)\}$	effective modal force vector for ground acceleration
$\{q\}$	principal (normal) coordinates vector

$\{q\}_r$	normal coordinate displacement vector for rth normal mode
$\{x\}$	displacement vector
$\{x(0)\}$	initial displacement vector
$\{\dot{x}\}$	velocity vector
$\{\dot{x}(0)\}$	initial velocity vector
$\{\ddot{x}\}$	acceleration vector
$\{\ddot{x}_t\}$	total (absolute) acceleration vector
$\{\hat{x}(t)\}$	displacement vector for truncated solution
$\{\hat{\ddot{x}}(t)\}$	acceleration vector for truncated solution
DOF	degree of freedom
$(\text{DMF})_r$	dynamic magnification factor for rth normal mode
MDOF	multi-degree-of-freedom
RAM	random access memory
SDOF	single degree of freedom
α	mass proportionality constant for Rayleigh damping
β	stiffness proportionality constant for Rayleigh damping
Δt	numerical integration time step
Γ_r	modal participation factor for rth normal mode
ω_{dr}	damped natural circular frequency for rth normal mode
ω_r	undamped natural circular frequency for rth normal mode
Ω	frequency of external harmonic force
ζ_r	modal damping factor for rth normal mode
$[\Phi]$	modal matrix
$[\hat{\Phi}]$	truncated modal matrix
$\{\Phi\}_r$	modal vector for rth normal mode

PROBLEMS

12.1 Determine expressions for the steady-state displacement response for the three-story shear frame building shown in Figure P12.1 by the mode superposition method. The structure is subjected to harmonic forces given by

$$\{F(t)\} = \begin{Bmatrix} F_1 \cos \Omega_1 t \\ F_2 \cos \Omega_2 t \\ F_3 \cos \Omega_3 t \end{Bmatrix}$$

where $F_1 = 8$ kips, $F_2 = 10$ kips, $F_3 = 18$ kips, $\Omega_1 = 10$ rad/sec, $\Omega_2 = 20$ rad/sec, and $\Omega_3 = 50$ rad/sec.

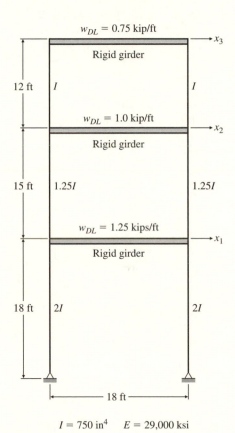

$I = 750$ in^4 $E = 29,000$ ksi

Figure P12.1

12.2 Repeat Problem 12.1 for the structure shown in Figure P12.2.

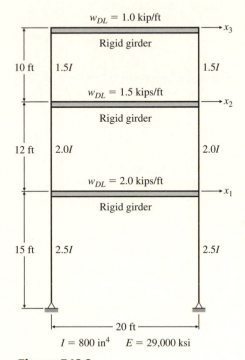

$I = 800$ in^4 $E = 29,000$ ksi

Figure P12.2

Let $F_1 = 10$ kips, $F_2 = 15$ kips, $F_3 = 20$ kips, $\Omega_1 = 15$ rad/sec, $\Omega_2 = 25$ rad/sec, and $\Omega_3 = 75$ rad/sec.

12.3 Repeat Problem 12.1 for the structure shown in Figure P12.3. Let $F_1 = 5$ kips, $F_2 = 7$ kips, $F_3 = 15$ kips, $\Omega_1 = 5$ rad/sec, $\Omega_2 = 15$ rad/sec, and $\Omega_3 = 25$ rad/sec.

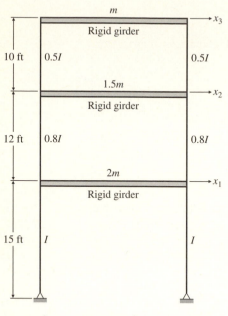

$m = 10$ kip-sec^2/in $I = 250$ in^4 $E = 29,000$ ksi

Figure P12.3

12.4 Repeat Problem 12.1 for the structure shown in Figure P12.4. Let $F_1 = 66.7$ kN, $F_2 = 0$, $F_3 = 111.2$ kN, $\Omega_1 = 5$ rad/sec, and $\Omega_3 = 25$ rad/sec.

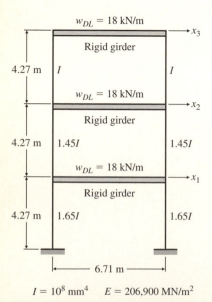

$I = 10^8$ mm^4 $E = 206,900$ MN/m^2

Figure P12.4

12.5 Determine the displacement response of the structure shown in Figure P12.5 subject to the rectangular pulse force shown in Figure P12.6 by the mode superposition method. Determine expressions for $\{x(t)\}$ in the intervals (a) $0 \le t \le t_d$ and (b) $t > t_d$. Assume $F_0 = 75$ kips and $t_d = 0.5$ sec.

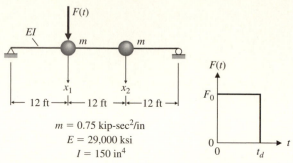

$m = 0.75$ kip-sec^2/in
$E = 29,000$ ksi
$I = 150$ in^4

Figure P12.5   **Figure P12.6**

12.6 The structure shown in Figure P12.7 is subjected to the triangular pulse force shown in Figure P12.8. By the mode superposition method, determine expressions for the displacement response $\{x(t)\}$ in the time intervals (a) $0 \le t \le t_d$ and (b) $t > t_d$. Assume $F_0 = 80$ kips and $t_d = 0.75$ sec.

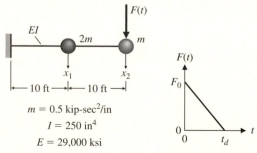

$m = 0.5$ kip-sec^2/in
$I = 250$ in^4
$E = 29,000$ ksi

Figure P12.7   **Figure P12.8**

12.7 The top level of the three-story shear frame building shown in Figure P12.2 is subjected to the ramp force shown in Figure P12.9. By the mode superposition method, determine expressions for the displacement response $\{x(t)\}$ in the time intervals (a) $0 \le t \le t_r$ and (b) $t > t_r$. Assume $F_0 = 25$ kips and $t_r = 0.5$ sec.

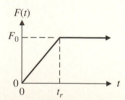

Figure P12.9

12.8 The top level of the shear frame building shown in Figure P12.2 is subject to the symmetric triangular load pulse shown in Figure P12.10. By the mode superposition method, determine expressions for the displacement response $\{x(t)\}$ in the time intervals (a) $0 \le t \le t_d/2$ and (b) $t_d/2 < t \le t_d$. Assume $F_0 = 40$ kips and $t_d = 1.0$ sec.

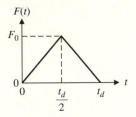

Figure P12.10

12.9 The top level of the shear frame building shown in Figure P12.11 is subjected to the triangular load pulse shown in Figure P12.8. By the mode superposition method, determine expressions for the displacement response $\{x(t)\}$ in the time intervals (a) $0 \le t \le t_d$ and (b) $t > t_d$. Assume $F_0 = 150$ kips and $t_d = 1.0$ sec.

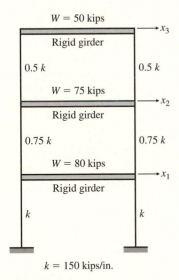

$k = 150$ kips/in.

Figure P12.11

12.10 The top level of the shear frame building shown in Figure P12.12 is subjected to the arbitrary time-varying force specified in Figure P12.13. By the mode superposition method, calculate the dynamic response of the top story, x_3. Evaluate the response by using one of the numerical methods discussed in Chapter 7. Plot time histories for $x_3(t)$, $\dot{x}_3(t)$, and $\ddot{x}_3(t)$ in the time interval $0 \le t \le 5$ sec. Let $\zeta_1 = 0.03$ and $\zeta_2 = 0.05$, and calculate ζ_3 based on the assumption of Rayleigh damping.

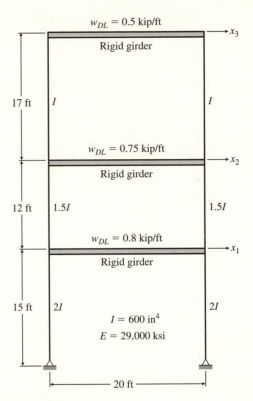

Figure P12.12

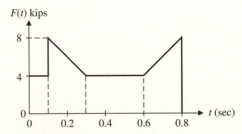

Figure P12.13

12.11 The top level of the shear frame building shown in Figure P12.1 is subjected to the arbitrary time-varying force specified in Figure P12.14. By the mode superposition method, calculate the dynamic response of the top level, x_3. Evaluate the response by using one of the numerical methods discussed in Chapter 7. Plot time histories for $x_3(t)$, $\dot{x}_3(t)$, and $\ddot{x}_3(t)$ in the time interval $0 \leq t \leq 5$ sec. Let $\zeta_1 = 0.05$ and $\zeta_2 = 0.06$, and calculate ζ_3 based on the assumption of Rayleigh damping.

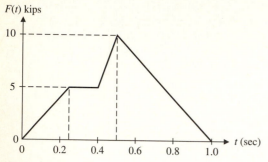

Figure P12.14

12.13 The second level of the shear frame building shown in Figure P12.4 is subject to the arbitrary time-varying force shown in Figure P12.16. By the mode superposition method, calculate the dynamic response of the top story, x_3. Evaluate the response by using one of the numerical methods discussed in Chapter 7. Plot $x_3(t)$, $\dot{x}_3(t)$, and $\ddot{x}_3(t)$ in the interval $0 \leq t \leq 5$ sec. Let $\zeta_1 = 0.02$ and $\zeta_2 = 0.03$, and calculate ζ_3 based on the assumption of Rayleigh damping.

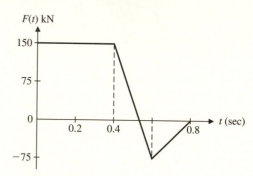

Figure P12.16

12.12 The top level of the shear frame building shown in Figure P12.3 is subjected to the time varying force specified in Figure P12.15. By the mode superposition method, calculate the dynamic response of the top level, x_3. Evaluate the response by using one of the numerical methods discussed in Chapter 7. Plot time histories for $x_3(t)$, $\dot{x}_3(t)$, and $\ddot{x}_3(t)$ in the time interval $0 \leq t \leq 5$ sec. Let $\zeta_1 = 0.05$ and $\zeta_2 = 0.07$, and calculate ζ_3 based on the assumption of Rayleigh damping.

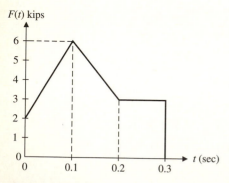

Figure P12.15

12.14 The base of the shear frame structure shown in Figure P12.1 is subject to the horizontal ground acceleration $\ddot{x}_g(t)$ specified in Figure P12.17. By the mode superposition method, calculate the dynamic response of the top story, x_3. Evaluate the response by using one of the numerical methods discussed in Chapter 7. Plot relative displacement $x_3(t)$, relative velocity $\dot{x}_3(t)$, and absolute acceleration $\ddot{x}_{t3}(t)$ in the interval $0 \leq t \leq 5$ sec. Let $\zeta_1 = 0.15$ and $\zeta_2 = 0.17$, and calculate ζ_3 based on the assumption of Rayleigh damping.

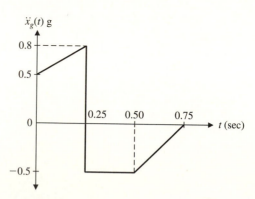

Figure P12.17

12.15 Repeat Problem 12.14 for the four-story shear frame structure shown in Figure P12.18 subject to the horizontal ground acceleration $\ddot{x}_g(t)$ specified in Figure P12.19. Let $\zeta_1 = 0.02$ and $\zeta_2 = 0.03$, and calculate ζ_3 and ζ_4 based on the assumption of Rayleigh damping. Plot relative displacement $x_3(t)$, relative velocity $\dot{x}_3(t)$, and absolute acceleration $[\ddot{x}_{t3}(t)]_t$ in the interval $0 \le t \le 5$ sec.

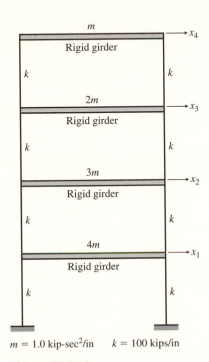

$m = 1.0$ kip-sec²/in $k = 100$ kips/in

Figure P12.18

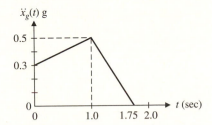

Figure P12.19

12.16 Repeat Problem 12.14 for the shear frame structure shown in Figure P12.2 subject to the horizontal ground acceleration $\ddot{x}_g(t)$ specified in Figure P12.20. Let $\zeta_1 = 0.08$ and $\zeta_2 = 0.10$, and calculate ζ_3 based on the assumption of Rayleigh damping.

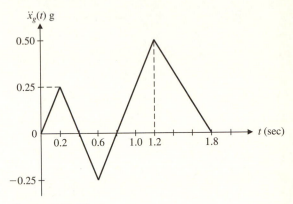

Figure P12.20

12.17 The top level of the three-story shear frame building shown in Figure P12.1 is subjected to the rectangular pulse force shown in Figure P12.6. Determine the maximum displacement of the top level, $(x_3)_{max}$, by the response spectrum technique. Use the appropriate spectrum curve in Figure 7.20. Assume $F_0 = 18$ kips and $t_d = 1.5$ sec. Combine the individual modal responses by the SRSS method.

12.18 The second level of the three-story shear frame shown in Figure P12.2 is subject to the triangular load pulse shown in Figure P12.8. Determine the maximum displacement of the top level, $(x_3)_{max}$, by the response spectrum technique. Use the appropriate spectrum curve in Figure 7.20. Assume $F_0 = 20$ kips and $t_d = 0.75$ sec. Combine the individual modal responses by the SRSS method.

12.19 The top level of the three-story shear frame building shown in Figure P12.3 is subjected to the symmetric triangular pulse force shown in Figure P12.10. Determine the maximum displacement of the top level, $(x_3)_{max}$, by the response spectrum technique. Use the appropriate spectrum curve in Figure 7.20. Combine the individual modal responses by the SRSS method. Assume $F_0 = 15$ kips and $t_d = 1.0$ sec.

12.20 The third level of the four-story shear frame building shown in Figure P12.18 is subjected to the triangular load pulse shown in Figure P12.8. Determine the maximum displacement of the top level, $(x_4)_{max}$, by the response spectrum technique. Use the appropriate spectrum curve in Figure 7.20. Combine the individual modal responses by the SRSS method. Assume $F_0 = 25$ kips and $t_d = 0.85$ sec.

12.21–12.24 Repeat Problems 12.1 through 12.4 using the mode acceleration method.

13 ▲ Analysis of Dynamic Response by Direct Integration

In Chapter 12, the mode superposition method was employed to obtain the dynamic response of linear MDOF systems by transforming the equations of motion to principal (normal) coordinates and solving the resulting set of *uncoupled* equations of motion. For simple excitation functions, the uncoupled modal equations could be solved in closed form. However, for more complex or arbitrary excitations it was necessary to implement one of the numerical methods discussed in Chapter 7 to evaluate the response.

The mode superposition method is not valid for MDOF systems with nonlinearities or with *nonclassical* damping (i.e., damping that is not orthogonal or that is coupled). Such systems require *direct integration* of a set of *coupled differential equations* to evaluate the dynamic response. In this chapter, we first consider direct integration methods for general linear MDOF systems whose equations are coupled. Then the accuracy and stability properties of the numerical integration algorithms are examined. Finally, the relative advantages of both the direct numerical integration method and the mode superposition method are discussed.

13.1 BASIC CONCEPTS OF DIRECT INTEGRATION METHODS

The equations of motion for a general MDOF system are expressed as

$$[m]\{\ddot{x}\} + [c]\{\dot{x}\} + [k]\{x\} = \{F(t)\} \tag{13.1}$$

where $[m]$, $[c]$, and $[k]$ = mass, damping, and stiffness matrices, respectively

$\{F(t)\}$ = external force vector

$\{\ddot{x}\}$, $\{\dot{x}\}$, and $\{x\}$ = acceleration, velocity, and displacement vectors, respectively.

In *direct integration* the differential equations represented by Eq. (13.1) are integrated using a numerical step-by-step procedure. "Direct" in this context means that there is no transformation of the equations to an alternative coordinate system (such as the transformation to normal coordinates performed in a mode superposition analysis) to facilitate the numerical integration. Quite conversely, the system differential equations are integrated directly in coupled form as they exist in physical coordinates.

There are two basic concepts that define the crux of direct numerical integration [1]. The first underlying concept is that a continuous solution to Eq. (13.1) is not sought, but rather the intention is to satisfy equilibrium at discrete time intervals Δt apart. In essence, this is tantamount to establishing an equivalent "static equilibrium," which includes the effects of inertia forces and energy dissipation (damping forces), at these discrete time intervals within the solution time domain. The second fundamental concept underscoring direct integration is that some variation of the system kinematics (i.e., displacements, velocities, and accelerations) is assumed within each discrete time interval Δt. The exact nature of this assumption defines the accuracy, stability, and computational effectiveness of the particular integration scheme.

As previously mentioned, the solution process for direct numerical integration methods is step by step in nature. This implies that over the time domain for which we seek the solutions to Eq. (13.1) at discrete time intervals Δt, the solution procedure must begin at a point in time at which the solution is already known. This generally occurs at time $t = 0$, for which the initial conditions for displacement, velocity, and acceleration (that is, $\{x\}_0$, $\{\dot{x}\}_0$, and $\{\ddot{x}\}_0$) are indeed known, or at least can be specified. Since the solution time domain is divided into discrete time intervals Δt apart, direct integration algorithms can be developed that yield a solution at some time $t + \Delta t$ based on the "known" solution at time t, beginning at time $t = 0$.

The following sections of this chapter present several commonly employed direct integration schemes, discuss *stability* and *accuracy* considerations for each method, and address relative advantages of one method over another.

13.2 THE CENTRAL DIFFERENCE METHOD

The discrete MDOF system equilibrium equations represented by Eq. (13.1) are a set of simultaneous ordinary differential equations with constant coefficients. Therefore, any convenient finite difference expressions that approximate the acceleration and velocities in terms of the displacements may be used. However, only a small number of finite difference expressions would render an effective solution scheme. One particular finite difference algorithm that has proved to be quite effective in these types of applications is the *central difference method*. In the central difference method, the acceleration is expressed in terms of the displacement as

$$\{\ddot{x}\}_t = \frac{1}{\Delta t^2}[\{x\}_{t-\Delta t} - 2\{x\}_t + \{x\}_{t+\Delta t}] \tag{13.2}$$

The error in the expansion of Eq. (13.2) is of the order $(\Delta t)^2$. To maintain the same order of error in the velocity expansion, the velocity is approximated as

$$\{\dot{x}\}_t = \frac{1}{2\,\Delta t}[-\{x\}_{t-\Delta t} + \{x\}_{t+\Delta t}] \tag{13.3}$$

The displacement solution at time $t + \Delta t$ is obtained by consideration of Eq. (13.1) at time t. Thus

$$[m]\{\ddot{x}\}_t + [c]\{\dot{x}\}_t + [k]\{x\}_t = \{F(t)\}_t \tag{13.4}$$

Substituting the expressions for $\{\ddot{x}\}_t$ and $\{\dot{x}\}_t$ given by Eqs. (13.2) and (13.3), respectively, into Eq. (13.4) results in

$$\left[\frac{1}{(\Delta t)^2}[m] + \frac{1}{2\,\Delta t}[c]\right]\{x\}_{t+\Delta t} = \{F\}_t - \left[[k] + \frac{2}{(\Delta t)^2}[m]\right]\{x\}_t$$

$$\qquad\qquad (13.5)$$

$$- \left[\frac{1}{(\Delta t)^2}[m] - \frac{1}{2\,\Delta t}[c]\right]\{x\}_{t-\Delta t}$$

from which the displacements $\{x\}_{t+\Delta t}$ can be calculated. An important feature of this solution technique is that the displacement solution $\{x\}_{t+\Delta t}$ is determined from equilibrium equations established at time t. That is, the solution at time $t + \Delta t$, denoted by $\{x\}_{t+\Delta t}$, is approximated from the equilibrium conditions represented by Eq. (13.4), which are formulated at time t. This type of integration procedure is referred to as an *explicit integration method*. In contrast, integration methods that yield a solution at time $t + \Delta t$ based upon equilibrium established at time $t + \Delta t$ are called *implicit integration methods*. Explicit integration methods do not require a factorization of the (effective) stiffness matrix, whereas the implicit methods do. This implies that explicit methods can be computationally more efficient than implicit methods in certain applications, some of which are discussed in Section 14.7.

Another noteworthy feature of the central difference method is that the procedure is not self-starting. Note from Eq. (13.5) that the calculation of $\{x\}_{t+\Delta t}$ requires knowledge of the displacements at times t and $t - \Delta t$, that is, $\{x\}_t$ and $\{x\}_{t-\Delta t}$. Therefore, to initiate the solution method at time $t = 0$, a special starting procedure must be implemented. Since the initial conditions of displacement, $\{x\}_0$, and velocity, $\{\dot{x}\}_0$, are known, the initial acceleration can be calculated from Eq. (13.1) evaluated at time $t = 0$ as

$$\{\ddot{x}\}_0 = [m]^{-1}(\{F(0)\} - [c]\{\dot{x}\}_0 + [k]\{x\}_0) \qquad\qquad (13.6)$$

The expressions given by Eqs. (13.2) and (13.3) can be used to obtain $\{x\}_{-\Delta t}$ as

$$\{x\}_{-\Delta t} = \{x\}_0 - \Delta t\{\dot{x}\}_0 + \frac{(\Delta t)^2}{2}\{\ddot{x}\}_0 \qquad\qquad (13.7)$$

The central difference algorithm, defined for implementation on a digital computer, is summarized in Table 13.1.

At this point in the discussion of the central difference scheme, it is noteworthy that the integration method requires a time step Δt that is smaller than a critical value Δt_{cr}. That is, a stable solution can be obtained only by selecting a time step $\Delta t \leq \Delta t_{cr}$, given by

$$\Delta t_{cr} = \frac{T_n}{\pi} \qquad\qquad (13.8)$$

where T_n is the smallest natural period (corresponding to nth or highest natural frequency) of the MDOF system.

Integration schemes such as the central difference method that require a time step Δt smaller than a critical time step Δt_{cr} are considered to be *conditionally stable*. This is because if a time step larger than Δt_{cr} is used in the analysis, the integration becomes unstable, resulting in a solution that is divergent (i.e., round-off errors grow at an increasing rate after each time step). The concept of integration stability is indeed a very important one and is discussed further in Section 13.6.

TABLE 13.1. Step-by-Step Solution Using the Central Difference Method

A. *Initial calculations*:
1. Input the stiffness matrix $[k]$, mass matrix $[m]$, and damping matrix $[c]$
2. Input initial conditions $\{x\}_0$ and $\{\dot{x}\}_0$
3. Calculate $\{\ddot{x}\}_0$ from Eq. (13.1)

$$\{\ddot{x}\}_0 = [m]^{-1}(\{F(0)\} - [c]\{\dot{x}\}_0 + [k]\{x\}_0)$$

4. Select time step Δt, $\Delta t < \Delta t_{cr}$, and calculate integration constants

$$a_0 = \frac{1}{(\Delta t)^2} \qquad a_1 = \frac{1}{2\,\Delta t} \qquad a_2 = 2a_0 \qquad a_3 = \frac{1}{a_2}$$

5. Calculate $\{x\}_{-\Delta t}$

$$\{x\}_{-\Delta t} = \{x\}_0 - \Delta t\{\dot{x}\}_0 + a_3\{\ddot{x}\}_0$$

6. Form effective mass matrix

$$[\hat{m}] = a_0[m] + a_1[c]$$

B. *For each time step*:
1. Calculate the effective force vector at time t

$$\{\hat{F}\}_t = \{F\}_t - [[k] - a_2[m]]\{x\}_t - [a_0[m] - a_1[c]]\{x\}_{t-\Delta t}$$

2. Solve for displacements at time $t + \Delta t$

$$[\hat{m}]\{x\}_{t+\Delta t} = \{\hat{F}\}_t \rightarrow \{x\}_{t+\Delta t} = [\hat{m}]^{-1}\{\hat{F}\}_t$$

3. Evaluate accelerations and velocities at time t

$$\{\ddot{x}\}_t = a_0(\{x\}_{t-\Delta t} - 2\{x\}_t + \{x\}_{t+\Delta t})$$

$$\{\dot{x}\}_t = a_1(-\{x\}_{t-\Delta t} + \{x\}_{t+\Delta t})$$

EXAMPLE 13.1 ▼

(a) Write a computer program for the central difference method to evaluate the dynamic response of a nonclassically damped *n*-DOF MDOF system subject to arbitrary dynamic forces.

(b) The four-story shear frame building with general damping shown in Figure 13.1 is subject to the forcing function shown in Figure 13.2. Determine the dynamic response by the central difference method. Plot time histories for $\{x(t)\}$ in the time interval $0 \leq t \leq 3.0$ sec. Assume $k = 60$ kips/in, $W = 300$ kips, and $c = 0.5$ kip-sec/in. Use a time step $\Delta t = 0.001$ sec.

Figure 13.1

Four-story shear frame building of Example 13.1.

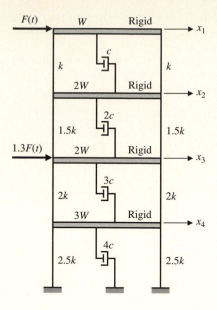

Figure 13.2

Time-varying force used in Example 13.1.

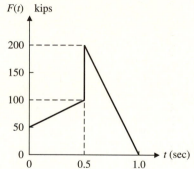

Solution

(a) The input parameters for the shear frame building are

$$
[m] = \begin{bmatrix} 0.7764 & 0 & 0 & 0 \\ 0 & 1.5528 & 0 & 0 \\ 0 & 0 & 1.5528 & 0 \\ 0 & 0 & 0 & 2.3292 \end{bmatrix} \text{ kip-sec}^2\text{/in}
$$

$$
[k] = \begin{bmatrix} 120 & -120 & 0 & 0 \\ -120 & 300 & -180 & 0 \\ 0 & -180 & 420 & -240 \\ 0 & 0 & -240 & 540 \end{bmatrix} \text{ kips/in}
$$

$$
[c] = \begin{bmatrix} 0.5 & -0.5 & 0 & 0 \\ -0.5 & 1.5 & -1.0 & 0 \\ 0 & -1.0 & 2.5 & -1.5 \\ 0 & 0 & -1.5 & 3.5 \end{bmatrix} \text{ kip-sec/in}
$$

(b) The integration constants are (refer to A.4 in Table 13.1)

$$a_0 = \frac{1}{(\Delta t)^2} = 10^6 \qquad a_1 = \frac{1}{2\,\Delta t} = 500$$

$$a_2 = 2a_0 = 2 \times 10^6 \qquad a_3 = \frac{1}{a_2} = 0.5 \times 10^{-6}$$

(c) Time histories for the floor displacements are presented in Figure 13.3. The listing for the BASIC computer program is presented in Table 13.2.

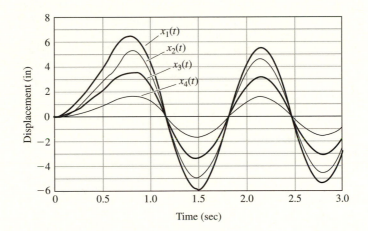

Figure 13.3

Time histories for floor displacements for Example 13.1.

TABLE 13.2. **Listing of BASIC Computer Program Used for Example 13.1**

```
10    REM
20    REM
30    REM                 THIS PROGRAM WILL EVALUATE THE DYNAMIC RESPONSE OF AN
40    REM                 N-DOF SYSTEM BY DIRECT NUMERICAL INTEGRATION. THE TIME
50    REM                 INTEGRATION SCHEME USED IS THE CENTRAL DIFFERENCE METHOD.
60    REM
70    REM                 DATA STATEMENTS (TO BE INCLUDED AT THE END OF THE PROGRAM)
80    REM                 SHOULD BE INPUT IN THE FOLLOWING ORDER:
90    REM                 ENTER NUMBER OF DEGREES OF FREEDOM (NDOF)
100   REM                 ENTER <K> ROW BY ROW
110   REM                 ENTER DIAGONAL TERMS OF <M>
120   REM                 ENTER <C> ROW BY ROW
130   REM                 ENTER INITIAL TIME (TI), FINAL TIME (TF),TIME INCREMENT (DT)
140   REM                 ENTER INITIAL DISPLACEMENTS
150   REM                 ENTER INITIAL VELOCITIES
160   REM                 ENTER LOAD VECTOR FOR TIME = 0
170   REM
180   REM
```

(continued)

```
190    REM
200    REM              -------------------------READ NDOF--------------------------
210    READ NDOF
220    REM              -------------------------READ STIFFNESS MATRIX--------------------
230    FOR I = 1 TO NDOF
240            FOR J = 1 TO NDOF
250            READ K(I,J)
260            NEXT J
270    NEXT I
280    REM              -------------------------READ DIAGONAL TERMS OF MASS MATRIX--------
290    FOR I = 1 TO NDOF
300            READ M(I)
310    NEXT I
320    REM            ---------READ DAMPING MATRIX---------
330    FOR I = 1 TO NDOF
340            FOR J = 1 TO NDOF
350            READ C(I,J)
360            NEXT J
370    NEXT I
380    REM
390    READ TI,TF,DT
400    REM          --------READ INITIAL DISPLACEMENTS------
410    FOR I = 1 TO NDOF
420            READ X(2,I)
430    NEXT I
440    REM          --------READ INITIAL VELOCITIES-------
450    FOR I = 1 TO NDOF
460            READ V(1,I)
470    NEXT I
480    REM          --------READ LOAD VECTOR AT TIME = 0-----
490    FOR I = 1 TO NDOF
500            READ PO(I)
510    NEXT I
520    REM
530    OPEN "C:DISP.DAT" FOR OUTPUT AS #1
540    OPEN "C:VEL.DAT" FOR OUTPUT AS #2
550    OPEN "C:ACC.DAT" FOR OUTPUT AS #3
560    PRINT #2,O,V(1,1),V(1,2),V(1,3),V(1,4)
570    REM
580    REM                                    CALCULATE INITIAL ACCELERATIONS
590    FOR I = 1 TO NDOF
600            FOR J = 1 TO NDOF
610            CXX(I)=CXX(I)+C(I,J)*V(1,I)
620            KXX(I)=KXX(I)+K(I,J)*X(2,I)
630            NEXT J
640    NEXT I
650    FOR I =1 TO NDOF
660    A(1,I)=(PO(I)−CXX(I)−KXX(I))/M(I)
670    NEXT I
680    PRINT #3,O,A(1,1),A(1,2),A(1,3),A(1,4)
690    REM
700    REM                                    CALCULATE INTEGRATION CONSTANTS
710    A0=1/(DT^2)
720    A1=1/(2*DT)
730    A2=2*A0
740    A3=1/A2
750    REM
760    REM                                    CALCULATE INITIAL X VALUES (THIS IS A SPECIAL STARTING
770    REM                                    PROCEDURE NEEDED IN THE CENTRAL DIFFERENCE METHOD)
780    FOR I = 1 TO NDOF
790            X(1,I)=X(2,I)−DT*V(1,I)+A3*A(1,I)
800    NEXT I
810    REM
820    PRINT #1,O,X(1,1),X(1,2),X(1,3),X(1,4)
830    PRINT #1,O,X(2,1),X(2,2),X(2,3),X(2,4)
```

(continued)

```
840    REM
850    REM                                    CALCULATE EFFECTIVE MASS MATRIX (EMM)
860    REM        EMM = A0 * <M> + A1 + <C>
870    FOR I = 1 TO NDOF
880            FOR J = 1 TO NDOF
890    AOMM(I,J) = 0
900    NEXT J
910    NEXT I
920    FOR I = 1 TO NDOF
930    AOMM(I,I) = A0*M(I)
940    NEXT I
950    FOR I = 1 TO NDOF
960            FOR J = 1 TO NDOF
970            A1CC(I,J)=A1*C(I,J)
980            NEXT J
990    NEXT I
1000   FOR I = 1 TO NDOF
1010     FOR J =  1 TO NDOF
1020     EMM(I,J) =  AOMM(I,J) +  A1CC(I,J)
1030     NEXT J
1040   NEXT I
1050   REM
1060   REM        USE GAUSS ELIMINATION TO INVERT EFFECTIVE MASS MATRIX (EMM)
1070   REM        THE INVERTED MATRIX WILL BE LABELED (IEMM)
1080   REM
1090   FOR I = 1 TO NDOF
1100     FOR J = 1 TO NDOF
1110     IEMM(I,J) = 0
1120     NEXT J
1130   NEXT I
1140   FOR I = 1 TO NDOF
1150     IEMM(I,I) = 1
1160   NEXT I
1170   REM        FOR BELOW THE DIAGONAL
1180   FOR I = 1 TO NDOF−1
1190     FOR J = I + 1 TO NDOF
1200     MJI=EMM(J,I)/EMM(I,I)
1210            FOR K = 1 TO NDOF
1220            EMM(J,K)=EMM(J,K)−MJI*EMM(I,K)
1230            IEMM(J,K)=IEMM(J,K)−MJI*IEMM(I,K)
1240            NEXT K
1250     NEXT J
1260   NEXT I
1270   REM        FOR ABOVE THE DIAGONAL
1280   FOR I = NDOF TO 2 STEP −1
1290     FOR J = I−1 TO 1 STEP −1
1300     MJI = EMM(J,I)/EMM(I,I)
1310            FOR K = 1 TO NDOF
1320            EMM(J,K)=EMM(J,K)−MJI*EMM(I,K)
1330            IEMM(J,K)=IEMM(J,K)−MJI*IEMM(I,K)
1340            NEXT K
1350     NEXT J
1360   NEXT I
1370   FOR I = 1 TO NDOF
1380   XX = EMM(I,I)
1390     FOR J = 1 TO NDOF
1400     EMM(I,J)=EMM(I,J)/XX
1410     IEMM(I,J)=IEMM(I,J)/XX
1420     NEXT J
1430   NEXT I
1440   REM
1450   REM        BEGIN TIME STEP LOOP
1460   FOR T = TI TO TF STEP DT
1470   REM DEFINE TIME VARYING FORCE F(T)
1480     IF T <= 0.5 THEN P=(100*T)+50
```

(continued)

```
1490    IF T > 0.5 AND T <= 1 THEN P=400−(400*T)
1500    IF T > 1 THEN P=0
1510  P(1)=2*P
1520  P(2)=0
1530  P(3)=1.5*P
1540  P(4)=0
1550  REM
1560  REM       CALCULATE EFFECTIVE LOADS (EP) AT TIME T
1570  REM       LET <K>−A2<M> = Z1
1580  REM       LET Z1*(X) = Z2
1590  REM       LET (A0*<M>−A1*<C>) = Z3
1600  REM       LET Z3 * X(T−DT) = Z4
1610  REM       THEREFORE, (EP) = P(T) − Z2 − Z4
1620  FOR I = 1 TO NDOF
1630    FOR J = 1 TO NDOF
1640    Z1(I,J)=K(I,J)
1650    NEXT J
1660  NEXT I
1670  FOR I = 1 TO NDOF
1680  Z1 (I, I) = K (I,I) − A2*M(I)
1690  NEXT I
1700  FOR I = 1 TO NDOF
1710  Z2(I)=0
1720  Z4(I)=0
1730  NEXT I
1740  FOR I = 1 TO NDOF
1750    FOR J = 1 TO NDOF
1760    Z2(I)=Z2(I)+Z1(I,J)*X(2,J)
1770    NEXT J
1780  NEXT I
1790  FOR I = 1 TO NDOF
1800    FOR J = 1 TO NDOF
1810    Z3(I,J)=0
1820    NEXT J
1830  NEXT I
1840  FOR I = 1 TO NDOF
1850    FOR J = 1 TO NDOF
1860    Z3(I,J)=A0MM(I,J)−A1CC(I,J)
1870    NEXT J
1880  NEXT I
1890  FOR I = 1 TO NDOF
1900    FOR J = 1 TO NDOF
1910    Z4(I)=Z4(I)+Z3(I,J)*X(1,J)
1920    NEXT J
1930  NEXT I
1940  FOR I = 1 TO NDOF
1950  EP(I)=P(I)-Z2(I)−Z4(I)
1960  NEXT I
1970  REM
1980  REM       SOLVE FOR DISPLACEMENTS AT TIME (T+DT)
1990  FOR I = 1 TO NDOF
2000    FOR J = 1 TO NDOF
2010    X(3,I)=X(3,I)+IEMM(I,J)*EP(J)
2020    NEXT J
2030  NEXT I
2040  PRINT #1,T,X(3,1),X(3,2),X(3,3),X(3,4)
2050  REM
2060  REM       EVALUATE ACCELERATIONS AND VELOCITIES AT TIME (T)
2070  FOR I = 1 TO NDOF
2080  AA(I)=A0*(X(1,I)−2*X(2,I)+X(3,I))
2090  VV(I)=A1*(−1*X(1,I)+X(3,I))
2100  NEXT I
2110    PRINT #2,T,VV(1),VV(2),VV(3),VV(4)
2120    PRINT #3,T,AA(1),AA(2),AA(3),AA(4)
```

(continued)

```
2130  FOR I = 1 TO NDOF
2140    X(1,I)=X(2,I)
2150    X(2,I)=X(3,I)
2160    X(3,I)=0
2170  NEXT I
2180  NEXT T
2190  REM
2200  REM
2210  REM ----------------INPUT DATA FOR EXAMPLE 13.1----------------
2220  DATA 4
2230  REM
2240  DATA 120,−120,0,0
2250  DATA −120,300,−180,0
2260  DATA 0,−180,420,−240
2270  DATA 0,0,−240,540
2280  REM
2290  DATA .7764,1.5528,1.5528,2.3292
2300  REM
2310  DATA .5,-.5,0,0
2320  DATA −.5,1.5,−1,0
2330  DATA 0,−1,2.5,−1.5
2340  DATA 0,0,−1.5,3.5
2350  REM
2360  DATA 0,3,.001
2370  REM
2380  DATA 0,0,0,0
2390  DATA 0,0,0,0
2400  REM
2410  DATA 50,0,65,0
2420  CLOSE
2430  END
```

▲

13.3   THE WILSON-θ METHOD

The *Wilson-θ method* is a special case of the linear acceleration method discussed in Section 7.6. Similar to the linear acceleration method, the basic kinematic assumption characterizing the temporal integration in the Wilson-θ method is linear variation of the acceleration over the time interval Δt. In the Wilson-θ method the acceleration is assumed to vary linearly from time t to time $t + \theta \Delta t$ as illustrated in Figure 13.4. The parameter θ, which is greater than or equal to 1.0, influences the stability and accuracy of the solution scheme. When $\theta = 1.0$, the method is exactly equivalent to the linear acceleration method. However, for *unconditional stability* during the integration, it is necessary that $\theta \geq 1.37$. Stability and accuracy of direct integration schemes are examined in Section 13.6.

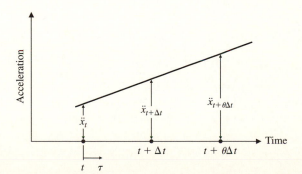

Figure 13.4

Linear variation of acceleration assumed in Wilson-θ method.

From reference to Figure 13.4, a time interval between time t and time $t + \theta \Delta t$ is defined over which it is assumed that the acceleration varies linearly. For any time τ within that interval, where $0 < \tau \leq \theta \Delta t$, the acceleration is approximated by

$$\{\ddot{x}\}_{t+\tau} = \{\ddot{x}\}_t + \frac{\tau}{\theta \Delta t}(\{\ddot{x}\}_{t+\theta \Delta t} - \{\ddot{x}\}_t) \tag{13.9}$$

Integrating Eq. (13.9) twice yields

$$\{\dot{x}\}_{t+\tau} = \{\dot{x}\}_t + \{\ddot{x}\}_t \tau + \frac{\tau^2}{2\theta \Delta t}(\{\ddot{x}\}_{t+\theta \Delta t} - \{\ddot{x}\}_t) \tag{13.10}$$

and

$$\{x\}_{t+\tau} = \{x\}_t + \{\dot{x}\}_t \tau + \frac{1}{2}\{\ddot{x}\}_t \tau^2 + \frac{\tau^3}{6\theta \Delta t}(\{\ddot{x}\}_{t+\theta \Delta t} - \{\ddot{x}\}_t) \tag{13.11}$$

Evaluating Eqs. (13.10) and (13.11) at time $t + \theta \Delta t$ results in

$$\{\dot{x}\}_{t+\theta \Delta t} = \{\dot{x}\}_t + \frac{\theta \Delta t}{2}(\{\ddot{x}\}_{t+\theta \Delta t} + \{\ddot{x}\}_t) \tag{13.12}$$

and

$$\{x\}_{t+\theta \Delta t} = \{x\}_t + \theta \Delta t \{\dot{x}\}_t + \frac{\theta^2 (\Delta t)^2}{6}(\{\ddot{x}\}_{t+\theta \Delta t} + 2\{\ddot{x}\}_t) \tag{13.13}$$

from which $\{\ddot{x}\}_{t+\theta \Delta t}$ and $\{\dot{x}\}_{t+\theta \Delta t}$ can be solved for in terms of $\{x\}_{t+\theta \Delta t}$ as

$$\{\ddot{x}\}_{t+\theta \Delta t} = \frac{6}{\theta^2 (\Delta t)^2}(\{x\}_{t+\theta \Delta t} - \{x\}_t) - \frac{6}{\theta \Delta t}\{\dot{x}\}_t - 2\{\ddot{x}\}_t \tag{13.14}$$

and

$$\{\dot{x}\}_{t+\theta \Delta t} = \frac{3}{\theta \Delta t}(\{x\}_{t+\theta \Delta t} - \{x\}_t) - 2\{\dot{x}\}_t - \frac{\theta \Delta t}{2}\{\ddot{x}\}_t \tag{13.15}$$

To approximate the solution at time $t + \Delta t$, the system equilibrium equations represented by Eq. (13.1) are satisfied at time $t + \theta \Delta t$. However, since a linear variation of the acceleration is assumed, then a linearly projected force vector $\{\hat{F}\}_{t+\theta \Delta t}$ is employed in conjunction with Eq. (13.1), resulting in the equilibrium condition given by

$$[m]\{\ddot{x}\}_{t+\theta \Delta t} + [c]\{\dot{x}\}_{t+\theta \Delta t} + [k]\{x\}_{t+\theta \Delta t} = \{\hat{F}(t)\}_{t+\theta \Delta t} \tag{13.16}$$

where

$$\{\hat{F}\}_{t+\theta \Delta t} = \{F\}_t + \theta(\{F\}_{t+\theta \Delta t} - \{F\}_t) \tag{13.17}$$

Substituting Eqs. (13.14) and (13.15) into Eq. (13.16) results in an equation from which $\{x\}_{t+\theta \Delta t}$ can be solved. By substituting this result for $\{x\}_{t+\theta \Delta t}$ into Eq. (13.14), the acceleration $\{\ddot{x}\}_{t+\theta \Delta t}$ is obtained. This solution for $\{\ddot{x}\}_{t+\theta \Delta t}$ is then used in Eqs. (13.9), (13.10), and (13.11), all evaluated at $\tau = \Delta t$, to calculate $\{\ddot{x}\}_{t+\Delta t}$, $\{\dot{x}\}_{t+\Delta t}$, and $\{x\}_{t+\Delta t}$, respectively. The algorithm for the Wilson-θ method is summarized in Table 13.3.

TABLE 13.3. Step-by-Step Solution Using Wilson-θ Integration Method

A. Initial calculations:
  1. Input stiffness matrix $[k]$, mass matrix $[m]$, and damping matrix $[c]$
  2. Input initial conditions $\{x\}_0$ and $\{\dot{x}\}_0$.
  3. Calculate $\{\ddot{x}\}_0$ from Eq. (13.1)

$$\{\ddot{x}\}_0 = [m]^{-1}(\{F(0)\} - [c]\{\dot{x}\}_0 + [k]\{x\}_0)$$

  4. Select a time step Δt and calculate integration constants ($\theta = 1.4$ usually, for unconditional stability)

$$a_0 = \frac{6}{(\theta\,\Delta t)^2} \qquad a_1 = \frac{3}{\theta\,\Delta t} \qquad a_2 = 2a_1 \qquad a_3 = \frac{\theta\,\Delta t}{2} \qquad a_4 = \frac{a_0}{\theta}$$

$$a_5 = \frac{-a_2}{\theta} \qquad a_6 = 1 - \frac{3}{\theta} \qquad a_7 = \frac{\Delta t}{2} \qquad a_8 = \frac{(\Delta t)^2}{2}$$

  5. Form the effective stiffness matrix $[\hat{k}]$

$$[\hat{k}] = [k] + a_0[m] + a_1[c]$$

B. For each time step:
  1. Calculate the effective force vector at time $t + \theta\,\Delta t$

$$\{\hat{F}\}_{t + \theta\Delta t} = \{F\}_t + \theta(\{F\}_{t + \Delta t} - \{F\}_t) + [m](a_0\{x\}_t + a_2\{\dot{x}\}_t + 2\{\ddot{x}\}_t)$$
$$+ [c](a_1\{x\}_t + 2\{\dot{x}\}_t + a_3\{\ddot{x}\}_t)$$

  2. Solve for the displacements at time $t + \theta\,\Delta t$

$$[\hat{k}]\{x\}_{t + \theta\Delta t} = \{\hat{F}\}_{t + \theta\Delta t} \rightarrow \{x\}_{t + \theta\Delta t} = [\hat{k}]^{-1}\{\hat{F}\}_{t + \theta\Delta t}$$

  3. Calculate the displacements, velocities, and accelerations at time $t + \Delta t$

$$\{\ddot{x}\}_{t + \Delta t} = a_4(\{x\}_{t + \theta\Delta t} - \{x\}_t) + a_5\{\dot{x}\}_t + a_6\{\ddot{x}\}_t$$

$$\{\dot{x}\}_{t + \Delta t} = \{\dot{x}\}_t + a_7(\{\ddot{x}\}_{t + \Delta t} + \{\ddot{x}\}_t)$$

$$\{x\}_{t + \Delta t} = \{x\}_t + \Delta t\{\dot{x}\}_t + a_8(\{\ddot{x}\}_{t + \Delta t} + \{\ddot{x}\}_t)$$

As previously mentioned, the Wilson-θ method is an unconditionally stable integration scheme when $\theta \geq 1.37$. Unconditionally stable implies that accuracy can be obtained in the integration without imposing a requirement on the time step Δt such as given in Eq. (13.8). Methods that pose no restrictions on time-step size to ensure stability are *implicit* integration schemes.

EXAMPLE 13.2 ▼

(a) Write a computer program for the Wilson-θ method to evaluate the dynamic response of a nonclassically damped n-DOF MDOF system subject to arbitrary dynamic forces.

(b) The four-story shear building with general damping shown in Figure 13.5 is subject to dynamic forces described by Figure 13.2. Determine the dynamic response by the Wilson-θ method. Plot time histories for $\{x(t)\}$ in the time interval $0 \leq t \leq 5.0$ sec. Select $\theta = 1.4$ and a time step $\Delta t = 0.001$ sec. Assume $k = 60$ kips/in, $c = 0.5$ kip-sec/in, and $W = 300$ kips.

Solution

(a) The input parameters for the shear frame building are

$$
[m] = \begin{bmatrix} 1.1641 & 0 & 0 & 0 \\ 0 & 1.5528 & 0 & 0 \\ 0 & 0 & 1.5528 & 0 \\ 0 & 0 & 0 & 1.5528 \end{bmatrix} \text{ kip-sec}^2\text{/in}
$$

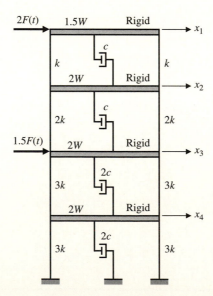

Figure 13.5

Four-story shear frame building with general damping of Example 13.2.

$$[k] = \begin{bmatrix} 120 & -120 & 0 & 0 \\ -120 & 360 & -240 & 0 \\ 0 & -240 & 600 & -360 \\ 0 & 0 & -360 & 720 \end{bmatrix} \text{kips/in}$$

$$[c] = \begin{bmatrix} 0.5 & -0.5 & 0 & 0 \\ -0.5 & 1.0 & -0.5 & 0 \\ 0 & -0.5 & 1.5 & -1.0 \\ 0 & 0 & -1.0 & 2.0 \end{bmatrix} \text{kip-sec/in}$$

(b) The integration constants are (refer to A.4 in Table 13.3)

$$a_0 = \frac{6}{(\theta \, \Delta t)^2} = 3,061,224.5 \qquad a_1 = \frac{3}{\theta \, \Delta t} = 2142.86$$

$$a_2 = 2a_1 = 4285.72 \qquad a_3 = \frac{\theta \, \Delta t}{2} = 0.0007$$

$$a_4 = \frac{a_0}{\theta} = 2,186,588.9 \qquad a_5 = -\frac{a_2}{\theta} = -3061.23$$

$$a_6 = 1 - \frac{3}{\theta} = -1.1429 \qquad a_7 = \frac{\Delta t}{2} = 0.0005$$

$$a_8 = \frac{(\Delta t)^2}{6} = 0.167 \times 10^{-6}$$

(c) Time histories for the floor displacements are presented in Figure 13.6. The listing for the FORTRAN computer program is presented in Table 13.4.

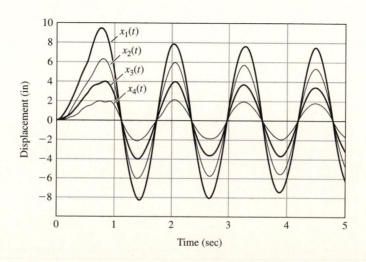

Figure 13.6

Time histories for floor displacements for Example 13.2.

TABLE 13.4. **Listing of FORTRAN Computer Program Used for Example 13.2**

```
c           ***********************************************
c
c           Step-by-step solution using Wilson-Theta
c                         wilson.f
c           ***********************************************
c
c           Step 1: Input [k], [m], [c]
c
            integer n,node
            real k(50,50),kinv(50,50),khat(50,50),m(50,50)
            real minv(50,50),c(50,50),sum
            real cx(4),kx(4),store(4),a0,a1,a2,a3,a4,a5,a6,a7,a8
            real x(4,500),xd(4,500),xdd(4,500),p(4,500),px(500)
            real ptt(4,500),xtt(4,500),term1(4,500),term2(4,500)
            real prod1(4,500),prod2(4,500),time,theta,khatinv(50,50)

            open(unit=5,file="wilson.dat",status="old")
            open(unit=6,file="wilson4.dis",status="unknown")
            open(unit=7,file="wilson4.vel",status="unknown")
            open(unit=8,file="wilson4.acc",status="unknown")
            open(unit=9,file="wilson.out",status="unknown")

            n=4
            node=4
            theta=1.4
            time=0.00
            dt=0.001

            write(9,*)'                    [k] Matrix'
            write(9,*)' *********************************************'
            do 25 i=1,n
            read(5,*)(k(i,j),j=1,n)
            write(9,10)(k(i,j), j=1,n)
25          continue
            write(9,*)

            call invert(k,kinv,n)
            write(9,*)'                    [kinv] Matrix'
            write(9,*)' *********************************************'
            do 30 i=1,n
            write(9,11)(kinv(i,j),j=1,n)
30          continue
            write(9,*)

            write(9,*)'                    [m] Matrix'
            write(9,*)' *********************************************'
            do 35 i=1,n
            read(5,*) (m(i,j),j=1,n)
            write(9,11)(m(i,j),j=1,n)
35          continue
            write(9,*)

            call invert (m,minv,n)

            write(9,*)'                    [minv] Matrix'
            write(9,*)' *********************************************'
            do 40 i=1,n
            write(9,11) (minv(i,j),j=1,n)
40          continue
            write(9,*)

            write(9,*)'                    [c] Matrix'
```

(continued)

```
                   write(9,*)' ***********************************************'
                   do 45 i=1,n
                   read(5,*) (c(i,j),j=1,n)
                   write(9,10) (c(i,j),j=1,n)
45                 continue
10                 format(2x,4(f10.3,2x))
11                 format(2x,4(f10.5,2x))

c                  ************************************************************
c
c                  Step 2: Input initial conditions / calculate p
c
                   write(9,*)
                   write(9,*)'          Initial Conditions'
                   write(9,*)'          ******************'
                   do 55 i=1,n
                   read(5,*)x(i,1)
                   write(9,12)i,x(i,1)
12                 format(3x,'x(',i1,') = ',f5.2)
55                 continue
                   write(9,*)

                   do 57 i=1,n
                   read(5,*)xd(i,1)
                   write(9,13)i,xd(i,1)
13                 format(3x,'xd(',i1,') =',f5.2)
57                 continue
                   write(9,*)

                   do 62 i=1,500

                   if(time.le.0.5)then
                   px(i)=50.00+time*50.00/0.5
                   elseif(time.le.1.0)then
                   px(i)=200.0-(time-0.5)*200.00/0.5
                   else
                   px(i)=0.00
                   endif
                   p(1,i)=2*px(i)
                   p(2,i)=0.00
                   p(3,i)=1.5*px(i)
                   p(4,i)=0.00

                   time=time+dt

62                 continue

c                  ************************************************************
c
c                  Step 3: Calculate xdd(1) from Eq. (13.1)
c                   . . . . also (Reset 'time' variable)
c
                   time=0.00

                   do 75 r=1,n
                   cx(r)=0.00
                   kx(r)=0.00
75                 continue

                   do 70 i=1,n
                   do 80 j=1,n
                            sum=c(i,j)*xd(j,i)
                            cx(i)=cx(i)+sum
                            sum=k(i,j)*x(i,j)
```

(continued)

```
                                  kx(i)=kx(i)+sum
80                 continue
70                 continue

                   do 90 i=1,n
                   store(i)=p(i,1)−cx(i)−kx(i)
90                 continue

                   do 93 i=1,n
                   xdd(i,1)=0.00
93                 continue

                   do 100 i=1,n
                   do 95 j=1,n
                          sum=minv(i,j)*store(j)
                          xdd(i,1)=xdd(i,1)+sum
95                 continue
                          write(9,14)i,xdd(i,1)
14                 format(3x,'xdd(',i1,') =',f6.2)
100                continue
                   write(9,*)
c                  ************************************************************
c
c                  Step 4: Calculate integration constants
c
                   a0=6.00/((theta*dt)**2)
                   a1=3.00/(theta*dt)
                   a2=2.00a1
                   a3=theta*dt/2.00
                   a4=a0/theta
                   a5=-a2/theta
                   a6=1.00−3.00/theta
                   a7=dt/2.00
                   a8=(dt**2)/6.00
c                  ************************************************************
c
c                  Step 5: Form effective stiffness matrix [khat]
c
                   write(9,*)' [khat] Matrix'
                   write(9,*)' *******************************************'
                   do 110 i=1,n
                   d0 105 j=1,n
                   khat(i,j)=k(i,j)+a0*m(i,j)+a1*c(i,j)
105                continue
                   write(9,10) (khat(i,hh),hh=1,n)
110                continue
                   write(9,*)

                   call invert(khat,khatinv,n)

                   write(9,*)' [khatinv] Matrix'
                   write(9,*)' *******************************************'
                   do 112 i=1,n
                   write(9,22) (khatinv(i,hh),hh=1,n)
112                continue
22                 format(3x,4(f10.8,2x))
                   write(9,*)

c                  ************************************************************
c
c                  Step 6: Calculate effective loads at time= t+(theta*dt)
c
                   do 200 i=1,500

                   do 150 s=1,n
```

```
                  term1(s,i)=a0x(s,i)+a2*xd(s,i)+2*xdd(s,i)
                  term2(s,i)=a1x(s,i)+2*xd(s,i)+a3*xdd(s,i)
150               continue

                  do 153 s=1,n
                  prod1(s,i)=0.00
                  prod2(s,i)=0.00
153               continue

                  do 160 s=1,n
                  do 155 t=1,n
                  sum=m(s,t)*term1(t,i)
                  prod1(s,i)=prod1(s,i)+sum
                  sum=c(s,t)*term2(t,i)
                  prod2(s,i)=prod2(s,i)+sum
155               continue
160               continue

                  do 170 s=1,n
                  ptt(s,i)=p(s,i)+theta*(p(s,i+1)-p(s,i))
     *                      + prod1(s,i)+prod2(s,i)
170               continue

c                 ************************************************************
c
c
c                 Step 7: Solve for displacements at time=t+(theta*dt)
c
                  do 172 s=1,n
                  xtt(s,i)=0.00
172               continue

                  do 173 s=1,n
                  do 174 t=1,n
                  sum=khatinv(s,t)*ptt(t,i)
                  xtt(s,i)=xtt(s,i)+sum
174               continue
173               continue

c                 ************************************************************
c
c
c                 Step 8: Calculate displacements, velocities and
c                  accelerations at time (t + dt)
c
                  do 180 s=1,n
                  xdd(s,(i+1))=a4*(xtt(s,i)-x(s,i))+a5*xd(s,i)+a6*xdd(s,i)
                  xd(s,(i+1)) =xd(s,i)+a7*(xdd(s,(i+1))+xdd(s,i))
                  x(s,(i+1)) =x(s,i)+dt*xd(s,i)+a8*(xdd(s,(i+1))+2*xdd(s,i))
180               continue

                  write(6,*)time,x(node,i)
                  write(7,*)time,xd(node,i)
                  write(8,*)time,xdd(node,i)

                  time=time+dt
200               continue
                  end

c                 ************************************************************
c
c                           This subroutine inverts a matrix (n x n)
c
c
c                           Notation
c                                   mat1 . . . matrix to be inverted
c                                   mat2 . . . inverted matrix
c
```

(continued)

```
c                              Note: range of line numbers (5000−5500)
c
c              **********************************************************
               subroutine invert(mat1,mat2,n)
               integer n
               real mat1(50,50),mat2(50,50),dummy, hold(50,50)

c              **************** copy [mat1] into [hold] *****************
               do 5004 i=1,n
               do 5005 j=1,n
               hold(i,j)=mat1(i,j)
5005  continue
5004  continue
c              **********************************************************
c                       Set up the identity matrix [mat2]
c              **********************************************************
               do 5010 i=1,n
               do 5000 j=1,n
               mat2(i,j)=0.000
5000  continue
5010  continue
c              **********************************************************
c                       Begin the loop turning [mat1] into
c                       upper triangular matrix
c              **********************************************************
c
c              **************** divide terms by diagonal term ***********
               do 5200 i=1,n
               dummy=mat1(i,i)
               do 5400 j=1,n
               mat1(i,j)=mat1(i,j)/dummy
               mat2(i,j)=mat2(i,j)/dummy
5400  continue
c              **************** remove non-diagonal terms ***************
               do 5050 k=(i+1),n
               dummy=mat1(k,i)
               do 5045 j=1,n
               mat1(k,j)=mat1(k,j)−mat1(i,j)*dummy
               mat2(k,j)=mat2(k,j)−mat2(i,j)*dummy
5045  continue
5050  continue
5200  continue
c              **********************************************************
c                       Begin loop that reduces upper triangular
c                       matrix [mat1] into identity matrix
c              **********************************************************
               do5475 k=n,2,−1
               do 5450 i=(n−1),1,−1
               dummy=mat1(i−(n−k),k)
               do 5425 j=n,1,−1
               mat1(i−(n−k),j)=mat1(i−(n−k),j)−mat1(k,j)*dummy
               mat2(i−(n−k),j)=mat2(i−(n−k),j)−mat2(k,j)*dummy
5425  continue
5450  continue
5475  continue
c              **************** copy [hold] into [mat1] ****************
               do 5500 i=1,n
               do 5490 j=1,n
               mat1(i,j)=hold(i,j)
5490  continue
5500  continue

               return
               end
```

13.4 THE NEWMARK METHOD

The *Newmark method* of direct integration can be regarded as a special case of the linear acceleration method discussed in Section 7.6 (or the Wilson-θ method discussed in the previous section). In the Newmark method, the following assumptions for velocity and displacement in the time interval from t to $t + \Delta t$ are employed:

$$\{\dot{x}\}_{t+\Delta t} = \{\dot{x}\}_t + [(1 - \delta)\{\ddot{x}\}_t + \delta\{\ddot{x}\}_{t+\Delta t}]\,\Delta t \tag{13.18}$$

$$\{x\}_{t+\Delta t} = \{x\}_t + \{\dot{x}\}_t\,\Delta t + \left[\left(\frac{1}{2} - \alpha\right)\{\ddot{x}\}_t + \alpha\{\ddot{x}\}_{t+\Delta t}\right](\Delta t)^2 \tag{13.19}$$

where α and δ are integration parameters that determine the stability and accuracy of the method. For the condition where $\alpha = 1/6$ and $\delta = 1/2$, Eqs. (13.18) and (13.19) correspond to the linear acceleration method [Eqs. (7.34) and (7.35)] or the Wilson-θ method with $\theta = 1.0$ [Eqs. (13.12) and (13.13)].

For an unconditionally stable integration scheme, Newmark originally proposed that $\alpha = 1/4$ and $\delta = 1/2$, in which case Eqs. (13.18) and (13.19) correspond to the *constant average acceleration method* previously defined by Eqs. (7.29) and (7.30). The basic kinematic assumption defining the essence of the constant average acceleration method is illustrated in Figure 13.7, which describes the acceleration within the time interval Δt to be the average of the accelerations at the beginning of the interval $\ddot{x}_t$ and at the end of the interval $\ddot{x}_{t+\Delta t}$.

Like the Wilson-θ method, the Newmark method is an implicit integration scheme. Therefore, the system equilibrium equations given by Eq. (13.1) are considered at time $t + \Delta t$, resulting in

$$[m]\{\ddot{x}\}_{t+\Delta t} + [c]\{\dot{x}\}_{t+\Delta t} + [k]\{x\}_{t+\Delta t} = \{F\}_{t+\Delta t} \tag{13.20}$$

From Eq. (13.19), $\{\ddot{x}\}_{t+\Delta t}$ may be solved for in terms of $\{x\}_{t+\Delta t}$. Then by substitution of this expression for $\{\ddot{x}\}_{t+\Delta t}$ into Eq. (13.18), expressions are obtained for both $\{\ddot{x}\}_{t+\Delta t}$ and $\{\dot{x}\}_{t+\Delta t}$ in terms of the unknown displacements $\{x\}_{t+\Delta t}$ only. These two resulting expressions for $\{\dot{x}\}_{t+\Delta t}$ and $\{\ddot{x}\}_{t+\Delta t}$ can then be substituted into Eq. (13.20) to solve for $\{x\}_{t+\Delta t}$. Finally, after determining $\{x\}_{t+\Delta t}$, $\{\dot{x}\}_{t+\Delta t}$ and $\{\ddot{x}\}_{t+\Delta t}$ can be calculated from Eqs. (13.18) and (13.19). The algorithm for the Newmark integration method is presented in Table 13.5. A discussion of stability and accuracy considerations for the Newmark method is presented in Section 13.6.

Figure 13.7

Constant average acceleration assumption originally proposed for Newmark method ($\alpha = 1/4$ and $\delta = 1/2$).

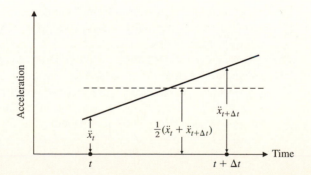

TABLE 13.5. **Step-by-Step Solution Using Newmark Integration Method**

A. Initial calculations:
1. Input stiffness matrix $[k]$, mass matrix $[m]$, and damping matrix $[c]$
2. Input initial conditions $\{x\}_0$ and $\{\dot{x}\}_0$
3. Calculate $\{\ddot{x}\}_0$ from Eq. (13.1)

$$\{\ddot{x}\}_0 = [m]^{-1}(\{F(0)\} - [c]\{\dot{x}\}_0 + [k]\{x\}_0)$$

4. Select a time step Δt, parameters α and δ, and calculate integration constants

$$\delta \geq 0.50 \qquad \alpha \geq 0.25(0.5 + \delta)^2$$

$$a_0 = \frac{1}{\alpha(\Delta t)^2} \qquad a_1 = \frac{\delta}{\alpha \Delta t} \qquad a_2 = \frac{1}{\alpha \Delta t} \qquad a_3 = \frac{1}{2\alpha} - 1$$

$$a_4 = \frac{\delta}{\alpha} - 1 \qquad a_5 = \frac{\Delta t}{2}\left(\frac{\delta}{\alpha} - 2\right) \qquad a_6 = \Delta t(1 - \delta) \qquad a_7 = \delta \Delta t$$

5. Form the effective stiffness matrix $[\hat{k}]$

$$[\hat{k}] = [k] + a_0[m] + a_1[c]$$

B. For each time step:
1. Calculate the effective force vector at time $t + \Delta t$

$$\{\hat{F}\}_{t+\Delta t} = \{F\}_{t+\Delta t} + [m](a_0\{x\}_t + a_2\{\dot{x}\}_t + a_3\{\ddot{x}\}_t)$$
$$+ [c](a_1\{x\}_t + a_4\{\dot{x}\}_t + a_5\{\ddot{x}\}_t)$$

2. Solve for the displacements at time $t + \Delta t$

$$[\hat{k}]\{x\}_{t+\Delta t} = \{\hat{F}\}_{t+\Delta t} \rightarrow \{x\}_{t+\Delta t} = [\hat{k}]^{-1}\{\hat{F}\}_{t+\Delta t}$$

3. Calculate the accelerations and velocities at time $t + \Delta t$

$$\{\ddot{x}\}_{t+\Delta t} = a_0(\{x\}_{t+\Delta t} - \{x\}_t) - a_2\{\dot{x}\}_t - a_3\{\ddot{x}\}_t$$

$$\{\dot{x}\}_{t+\Delta t} = \{\dot{x}\}_t + a_6\{\ddot{x}\}_t + a_7\{\ddot{x}\}_{t+\Delta t}$$

EXAMPLE 13.3 ▼

(a) Write a computer program for the Newmark method to evaluate the dynamic response of a nonclassically damped n-DOF MDOF system subject to arbitrary dynamic forces.

(b) The four-story shear building with general damping shown in Figure 13.8 is subject to dynamic forces described by Figure 13.9. Determine the dynamic response by the Newmark method. Plot time histories for $\{x(t)\}$ in the time interval $0 \leq t \leq 5$ sec. Select integration parameters $\alpha = 1/4$ and $\delta = 1/2$ and a time step $\Delta t = 0.002$ sec. Assume $k = 50$ kips/in, $c = 0.5$ kip-sec/in, and $W = 200$ kips.

Solution

(a) The input parameters for the shear frame building are

$$[m] = \begin{bmatrix} 1.0352 & 0 & 0 & 0 \\ 0 & 1.0352 & 0 & 0 \\ 0 & 0 & 1.0352 & 0 \\ 0 & 0 & 0 & 0.5176 \end{bmatrix} \text{kip-sec}^2/\text{in}$$

$$[k] = \begin{bmatrix} 400 & -200 & 0 & 0 \\ -200 & 300 & -100 & 0 \\ 0 & -100 & 200 & -100 \\ 0 & 0 & -100 & 100 \end{bmatrix} \text{kips/in}$$

$$[c] = \begin{bmatrix} 3.5 & -1.5 & 0 & 0 \\ -1.5 & 2.5 & -1.0 & 0 \\ 0 & -1.0 & 1.5 & -0.5 \\ 0 & 0 & -0.5 & 0.5 \end{bmatrix} \text{kip-sec/in}$$

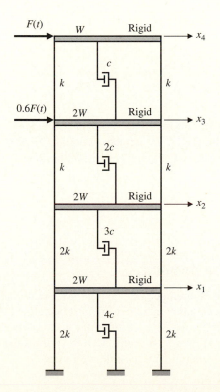

Figure 13.8

Four-story shear frame building with general damping of Example 13.3.

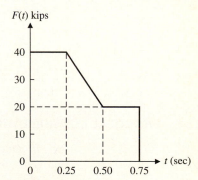

Figure 13.9

Time-varying force used in Example 13.3.

Figure 13.10

Time histories for floor
displacements for Example 13.3.

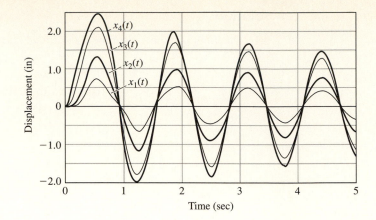

(b) The integration constants are (refer to A.4 of Table 13.5)

$$a_0 = \frac{1}{\alpha(\Delta t)^2} = 4 \times 10^6 \quad a_1 = \frac{\delta}{\alpha \, \Delta t} = 2000 \quad a_3 = \frac{1}{2\alpha} - 1 = 1.0$$

$$a_4 = \frac{\delta}{\alpha} - 1 = 1.0 \qquad a_5 = \frac{\Delta t}{2}\left(\frac{\delta}{\alpha} - 2\right) = 0$$

$$a_6 = \Delta t(1 - \delta) = 0.0005 \qquad a_7 = \delta\Delta t = 0.0005$$

(c) Time histories for the floor displacements are presented in Figure 13.10.
The FORTRAN computer program, NEWB, used to evaluate the dynamic
response, is available from the author's web site (refer to Preface).

13.5 PRACTICAL CONSIDERATIONS FOR DAMPING

If damping is to be included in the direct numerical integration analysis of MDOF systems, the physical damping matrix $[c]$ must be defined explicitly. In the previous examples, discrete dashpots were specified for each DOF, from which the damping matrix
was calculated. However, for most MDOF systems the specification of discrete damping elements is generally not feasible nor desirable.

In Chapter 12, a procedure for constructing a system damping matrix for proportional damping (specifically, Rayleigh damping) was defined by

$$[c] = \alpha[m] + \beta[k] \tag{12.39}$$

where α and β are the proportionality parameters determined from the simultaneous solution of the expression

$$\zeta_r = \left(\frac{\alpha}{\omega_r} + \beta\omega_r\right) \tag{13.21}$$

for two specified values of ζ_r, the damping factor corresponding to the rth normal mode. The disadvantage of Rayleigh damping is that the higher modes of large MDOF systems are considerably more damped than the lower modes for which the parameters α and β have been selected.

An alternative procedure for formulating the physical damping matrix can be derived from consideration of the modal damping matrix relationship described in Chapter 12, given by

$$[C] = [\Phi]^T[c][\Phi] = \text{diag}(2\zeta_r\omega_r M_r) \tag{13.22}$$

where $[C]$ = modal damping matrix
$[\Phi]$ = modal matrix
ω_r = natural circular frequency for the rth normal mode
M_r = modal mass for the rth normal mode

The physical damping matrix $[c]$ can then be determined from Eq. (13.22) by

$$[c] = ([\Phi]^T)^{-1}[C][\Phi]^{-1} \tag{13.23}$$

An expedient procedure for expressing the inverse of the modal matrix can be evolved from consideration of the mass orthogonality relationship given by

$$[M] = [\Phi]^T[m][\Phi] \tag{13.24}$$

where $[M]$ is the modal mass matrix. Noting that the identity matrix $[I]$ may be expressed as

$$[I] = [M]^{-1}[M] = [M]^{-1}[\Phi]^T[m][\Phi] = [\Phi]^{-1}[\Phi] \tag{13.25}$$

it is evident that the modal matrix inverse is

$$[\Phi]^{-1} = [M]^{-1}[\Phi]^T[m] \tag{13.26}$$

The physical damping matrix is now given by substituting Eq. (13.26) into Eq. (13.23),

$$[c] = [[m][\Phi][M]^{-1}][C][[M]^{-1}[\Phi]^T[m]] \tag{13.27}$$

Because $[M]$ and $[C]$ are diagonal, Eqs. (13.23) and (13.27) can be combined to give

$$[c] = \sum_{r=1}^{n}\left(\frac{2\zeta_r\omega_r}{M_r}\right)[m]\{\Phi\}_r[[m]\{\Phi\}_r]^T \tag{13.28}$$

where $\{\Phi\}_r$ is the modal vector for the rth normal mode. From the orthogonality of modes condition, Eq. (13.28) yields terms in the damping matrix corresponding only to those modes for which ζ_r is specified, and all other modes will be undamped.

Equation (13.28) may be modified to include damping in higher modes. Let ω_p be the natural frequency corresponding to the highest mode in an n-DOF system ($p < n$) for which damping is specified. Then the coefficient corresponding to stiffness proportional modal damping in the pth mode is given by

$$a_p = \frac{2\zeta_p}{\omega_p} \tag{13.29}$$

and the stiffness proportional damping at all other modes under consideration is given by

$$\overline{\zeta}_r = \frac{a_p \omega_r}{2} = \zeta_p\left(\frac{\omega_r}{\omega_p}\right) \tag{13.30}$$

Then the damping (mass proportional) required to supplement the stiffness proportional damping is given by

$$\hat{\zeta}_r = \zeta_r - \overline{\zeta}_r = \zeta_r - \zeta_p\left(\frac{\omega_r}{\omega_p}\right) \tag{13.31}$$

Finally, Eq. (13.28) can be modified to yield a damping matrix expressed by

$$[c] = a_p[k] + \sum_{r=1}^{p-1}\left(\frac{2\hat{\zeta}_r \omega_r}{M_r}\right)[m]\{\Phi\}_r[[m]\{\Phi\}_r]^T \tag{13.32}$$

which provides the specified modal damping for frequencies less than and equal to ω_p, and which specifies linearly increasing damping for frequencies higher than ω_p.

The proportional damping defined by Eqs. (13.28) and (13.32) is generally more realistic (for higher modes) than Rayleigh damping [Eq. (12.39)]. Proportional-type damping may be adequate for many analyses and can be implemented in either a mode superposition analysis or a direct integration analysis. However, in the analysis of structures involving substantially different material properties, nonproportional damping may need to be implemented. A frequently encountered condition of nonproportional damping occurs in the analysis of soil-structure interaction problems, where significantly more damping is exhibited by the soil than by the structure supported by it. In such instances, in the construction of the damping matrix, it is common to assign different proportionality parameters to the soil and structure resulting in an overall system damping matrix that is not proportional (even though it may be segmentally proportional). In such instances a direct integration analysis is mandatory.

Regardless of the type of damping employed in the dynamic analysis, it is indeed a difficult task to quantify damping in physical structures. The level of damping a structure exhibits is dependent on several parameters, such as intensity and type of the excitation, stress level in the material, the type of material, and the type and physical condition of the structure. Damping in existing structures may be ascertained experimentally by vibration testing, but this is often time consuming and expensive for large structures, and is generally applicable only to a narrow band of frequency ratios or low level of excitation. For structures in the planning stages, damping values cannot be measured, but are generally estimated from measured data obtained for similar existing structures.

TABLE 13.6. **Typical Damping Values**

Stress Level	Type and Condition of Structure	Percentage of Critical Damping
1. Low, well below proportional limit, stresses below 1/4 yield point	Steel, reinforced or prestressed concrete, wood; no cracking; no joint slip	0.5–1.0
2. Working stress, no more than about 1/2 yield point	Welded steel, prestressed concrete, well-reinforced concrete (only slight cracking)	2
	Reinforced concrete with considerable cracking	3–5
	Bolted and/or riveted steel, wood structures with bolted joints	5–7
3. At or just below yield point	Welded steel, prestressed concrete (without complete loss of prestress)	5
	Reinforced concrete and prestressed concrete with no prestress left	7–10
	Bolted and/or riveted steel, wood structures with bolted joints	10–15
	Wood structures with nailed joints	15–20
4. Beyond yield point, with permanent strain greater than yield point limit strain	Welded steel	7–10
	Reinforced concrete and prestressed concrete	10–15
	Bolted and/or riveted steel, and wood	10–15

Adapted from *Earthquake Spectra and Design* by Newmark and Hall, copyright © 1982. Reprinted by permission of Earthquake Engineering Research Institute, Oakland, CA.

Recommended damping levels for various types of structures and stress levels are presented in Table 13.6. These damping values should be evaluated in conjunction with existing data for similar structures before being implemented into an analysis.

EXAMPLE 13.4 ▼

The base of the four-story shear frame building shown in Figure 13.11 is excited by the horizontal ground acceleration $\ddot{x}_g(t)$ described by Figure 13.12. Evaluate the dynamic response of the structure by the Wilson-θ method of numerical integration. Plot time histories for relative displacement $\{x(t)\}$ and absolute (total) acceleration $\{\ddot{x}_t(t)\}$ in the time interval $0 \leq t \leq 6.0$ sec. Assume modal damping factors of $\zeta_1 = 0.03$ and $\zeta_2 = 0.05$, and determine the physical damping matrix from Eq. (13.32). Assume $k = 150$ kips/in and $W = 150$ kips. Use a time step $\Delta t = 0.001$ sec.

Figure 13.11

Four-story shear frame building of Example 13.4.

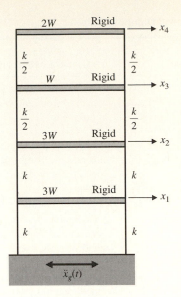

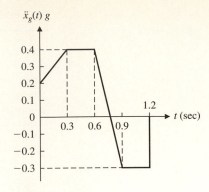

Figure 13.12

Horizontal ground acceleration used for Example 13.4.

Solution

The mass and stiffness properties for the structure are

$$[m] = \begin{bmatrix} 1.16 & 0 & 0 & 0 \\ 0 & 1.16 & 0 & 0 \\ 0 & 0 & 0.338 & 0 \\ 0 & 0 & 0 & 0.778 \end{bmatrix} \text{ kip-sec}^2/\text{in}$$

$$[k] = \begin{bmatrix} 600 & -300 & 0 & 0 \\ -300 & 450 & -150 & 0 \\ 0 & -150 & 300 & -150 \\ 0 & 0 & -150 & 150 \end{bmatrix} \text{ kips/in}$$

The natural frequencies and first normal mode for the structure are

$$\{\omega\} = \begin{Bmatrix} 6.38 \\ 13.70 \\ 25.74 \\ 31.30 \end{Bmatrix} \text{ rad/sec} \qquad \{\Phi\}_1 = \begin{Bmatrix} 0.277 \\ 0.419 \\ 0.669 \\ 0.850 \end{Bmatrix}$$

The effective load vector for ground acceleration is determined from Eq. (12.49) as (refer to Section 12.5)

$$\{F_{\text{eff}}(t)\} = [m]\{I\}\ddot{x}_g(t) \tag{1}$$

The damping matrix is defined by Eq. (13.32); thus, from Eq. (13.29) the coefficient relating stiffness proportional damping in the mode $p = 2$ is first established as

$$a_p = \frac{2\zeta_p}{\omega_p} = \frac{2\zeta_2}{\omega_2} = \frac{2(0.05)}{(13.7)} = 7.3 \times 10^{-3} \tag{2}$$

Then from Eq. (13.31), $\hat{\zeta}_1$ is determined as

$$\hat{\zeta}_1 = 0.03 - 0.05\left(\frac{6.38}{13.7}\right) = 0.00671 \tag{3}$$

and from Eq. (13.30) the stiffness proportional modal damping factors that will result for modes 3 and 4 are established, given by

$$\overline{\zeta}_3 = \zeta_2\left(\frac{\omega_3}{\omega_2}\right) = 0.05\left(\frac{25.74}{13.7}\right) = 0.0094 \tag{4}$$

$$\overline{\zeta}_4 = \zeta_2\left(\frac{\omega_4}{\omega_2}\right) = 0.05\left(\frac{31.3}{13.7}\right) = 0.114 \tag{5}$$

Then from Eq. (13.32), the physical damping matrix is constructed as

$$[c] = a_p[k] + \left(\frac{2\zeta_1\omega_1}{M_1}\right)[m]\{\Phi\}_1[[m]\{\Phi\}_1]^T \tag{6}$$

Since

$$[m]\{\Phi\}_1 = \begin{Bmatrix} 0.321 \\ 0.486 \\ 0.259 \\ 0.661 \end{Bmatrix} \tag{7}$$

and

$$M_1 = \{\Phi\}_1^T[m]\{\Phi\}_1 = 1.0 \tag{8}$$

Eq. (6) yields

$$[c] = \begin{bmatrix} 4.38 & -2.176 & 0.0071 & 0.018 \\ -2.176 & 3.31 & -1.08 & 0.03 \\ 0.0071 & -1.08 & 2.20 & -1.08 \\ 0.018 & 0.03 & -1.08 & 1.13 \end{bmatrix} \text{kip-sec/in}$$

The integration constants for the Wilson-θ method are (refer to A.4 in Table 13.3)

$$a_0 = \frac{6}{(\theta \, \Delta t)^2} = 3,061,224.5 \qquad a_1 = \frac{3}{\theta \, \Delta t} = 2142.86$$

$$a_2 = 2a_1 = 4285.72 \qquad a_3 = \frac{\theta \, \Delta t}{2} = 0.0007$$

$$a_4 = \frac{a_0}{\theta} = 2,186,588.9 \qquad a_5 = \frac{a_2}{\theta} = -3061.23$$

$$a_6 = 1 - \frac{3}{\theta} = -1.1429 \qquad a_7 = \frac{\Delta t}{2} = 0.0005$$

$$a_8 = \frac{(\Delta t)^2}{6} = 0.167 \times 10^{-6}$$

The computer program listed for the Wilson-θ method in Table 13.4 was used to evaluate the response. Time histories for relative displacement $\{x(t)\}$ and absolute acceleration $\{\ddot{x}_t(t)\}$ are presented in Figures 13.13 and 13.14, respectively. Note that the absolute or total acceleration is the sum of the ground acceleration $\ddot{x}_g(t)\{I\}$ and the relative acceleration vector $\{\ddot{x}(t)\}$.

Figure 13.13

Time history for relative displacement for Example 13.4.

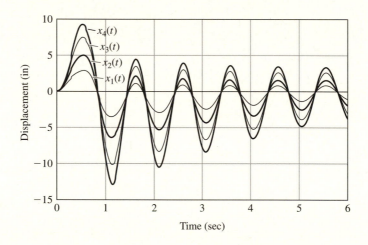

Figure 13.14

Time history for absolute acceleration for Example 13.4.

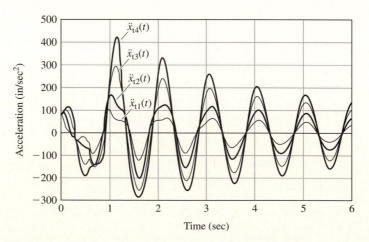

13.6 STABILITY AND ACCURACY OF DIRECT INTEGRATION METHODS

The selection of an appropriate time-step size Δt is one of the most critical aspects in the dynamic analysis of large MDOF systems. The time-step size affects not only the stability and accuracy of the solution, but the computational effort expended as well. A time step that is too large may render highly inaccurate results, or cause the solution scheme to become unstable, or "blow up." Conversely, a time step that is too small often results in unnecessary overexpenditure of computational resources. Indeed, time-step size is integrally related to the accuracy and stability of the integration method and dramatically influences the computational efficiency of the solution.

To lend insight to this intricate interrelationship of time-step size with stability and accuracy of the integration method, we first examine the fundamental nature of integration methods. Direct integration methods are categorized as either *explicit* or *implicit*. In explicit (or *open*, or *predictor*) methods, such as the central difference method, dynamic equilibrium is considered at time t to evaluate the solution at time $t + \Delta t$. In *implicit* (or *closed*, or *corrector*) methods, such as the Wilson-θ and Newmark methods, the solution $\{x\}_{t+\Delta t}$ is found from expressions that consider equilibrium at time $t + \Delta t$. Usually, explicit methods mandate a small Δt but result in equations requiring relatively little computational effort to solve. In contrast, implicit methods permit a relatively large Δt but result in equations requiring greater computational effort to solve. That is, in an implicit integration scheme, inversion of an effective stiffness matrix (which is not diagonal) is required, whereas, in an explicit scheme, it is necessary only to invert an effective mass matrix, which is quite often diagonal. The increased computational effort required in implicit integration is even more pronounced in a nonlinear analysis.

Explicit methods are *conditionally stable*. If Δt exceeds a certain fraction of the smallest vibration period of the structure, computed displacements and velocities grow without bound. The central difference method is an explicit method that is conditionally stable. The limiting time step for stability is $\Delta t_{\mathrm{cr}} \leq T_n/\pi$, where T_n is the natural period of the structure corresponding to the highest mode of vibration (Δt_{cr} is often referred to as the stability limit). Invariably, the maximum time-step size for an explicit conditionally stable integration scheme is determined by the stability limit.

Unconditionally stable integration methods are those for which the size of Δt is governed by considerations of accuracy rather than stability. Unconditionally stable methods are implicit. The Newmark method becomes unconditionally stable, provided that the integration parameters $\delta \geq 0.5$ and $\alpha \geq 0.25(\delta + 0.5)^2$. A common choice of integration parameters is $\delta = 0.5$ and $\alpha = 0.25$, for which the method is also known as the *constant-average-acceleration method*, or the *trapezoidal method*. The Wilson-θ method becomes unconditionally stable when $\theta \geq 1.37$. This information must be supplemented by accuracy considerations to arrive at an optimum value of θ, but $\theta = 1.4$ is generally employed.

The choice of integration scheme to be used in a particular analysis is, to a large extent, controlled by the computational resources required for the solution, which in turn is directly proportional to the number of time steps required in the integration. If a conditionally stable algorithm is employed, such as the central difference method, the time-step requirement is governed by the critical time step Δt_{cr}, with no other choice available. However, if an unconditionally stable integration method is used, only accuracy considerations are pertinent and a time step substantially larger than Δt_{cr} is generally permitted.

Figure 13.15

(a) Percent period elongation (PE); (b) Percent amplitude decay (AD). Adapted from *Finite Element Procedures* by Klaus-Jürgen Bathe, copyright © 1996 by permission of Prentice Hall, Inc., Upper Saddle River, NJ.

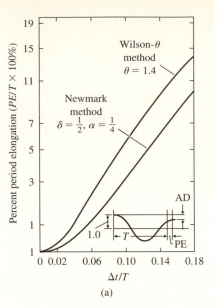

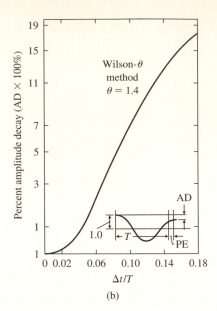

(a)   (b)

The accuracy of numerical integration algorithms can be characterized by two attributes: *period elongation* (PE) and *amplitude decay* (AD). It has been determined [1] that sufficient accuracy can be achieved (for a SDOF system) when $\Delta t/T$ is smaller than about 0.1. Such a time step used in conjunction with $\theta = 1.4$ in the Wilson-θ method has been shown to exhibit very little period elongation and amplitude decay. The same time step used with the Newmark *constant-average-acceleration* method (that is, $\delta = 0.5$ and $\alpha = 0.25$) introduces only slight period elongation and no amplitude decay. The percentage of period elongations for the Wilson and Newmark methods as a function of the time step size are shown in Figure 13.15a, and the percentage of amplitude decay versus time step size for the Wilson method is illustrated in Figure 13.15b. These figures confirm that high accuracy can be achieved (for a SDOF system) from numerical integration when $\Delta t/T$ is less than 0.1. To achieve integration accuracy for a MDOF system, the most important concern is the choice of an appropriate time-step-to-period ratio $\Delta t/T_n$. By use of the central difference method, the time step has to be chosen such that Δt is smaller than Δt_{cr}, which means that Δt will, in general, be small enough to obtain accuracy in the integration of all n equations of dynamic equilibrium. However, with use of one of the unconditionally stable schemes, the time step Δt can be much larger and should be only small enough that the response in all modes that *significantly contribute* to the total structural response are calculated accurately. The other modal response components are not evaluated accurately, but these errors are unimportant because the actual contribution from those modes to the total response measured in those components is negligible. Stated in another way, the time-step-to-period ratio of 0.1 can be defined by $\Delta t/T_p$, where T_p is the natural period of the pth mode of vibration, where $p \ll n$ or $T_p \gg T_n$. Thus, the period T_p corresponds to the highest vibration mode for which there is a significant contribution to the dynamic response.

13.7 DIRECT INTEGRATION VERSUS MODE SUPERPOSITION

In this chapter our discussions were directed toward the dynamic analysis of MDOF systems by direct numerical integration. In Chapter 12 the response of MDOF systems by the mode superposition method was examined. In this section the relative advantages of each method of analysis are investigated.

Obviously, if the system under consideration exhibits any nonlinearities or is characterized by nonclassical damping, then a mode superposition analysis is precluded. Mode superposition analysis is most effective for large systems when the dynamic response can be accurately evaluated by consideration of only a few of the lowest modes of vibration. This is because the majority of the computational effort expended for a mode superposition analysis is associated with the eigenproblem solution (that is, the extraction of the system natural frequencies and vibration mode shapes). For example, if it is necessary to calculate a large number of vibration modes to accurately represent the dynamic response, to a specific excitation, of a system having several thousand degrees of freedom, then solution of the eigenproblem may be computationally prohibitive.

In short, a mode superposition analysis is ideally suited for situations where the dynamic disturbance is confined to the lowest few modes of vibration of the system, and the duration of the disturbance is relatively long. The response of structures to earthquake excitation is generally quite conducive to a mode superposition analysis. The computational effort for the eigensolution is minimized, and only a small number of uncoupled equations are required to be integrated over the solution time domain. The selection of a particular integration operator and time-step size for a mode superposition analysis should also be made in light of the considerations discussed in Section 13.6.

A direct numerical integration analysis should be conducted in situations when a large number of vibration modes must be included in the response calculations. This is generally the scenario for structures subjected to high-intensity, short-duration impulse-type loading, such as a shock or blast load. In such instances, as many as two-thirds of the total number of vibration modes for the system may contribute to the response [1]. Also, direct integration analysis is more suitable for *wave propagation* problems. In wave propagation problems a large number of vibration modes contribute significantly to the system response, and therefore, mode superposition analyses do not yield computationally efficient results. Finally, for nonclassically damped systems or systems exhibiting any nonlinear characteristics, a direct integration analysis is required.

REFERENCES

1 Bathe, Klaus-Jürgen, *Finite Element Procedures,* Prentice Hall, Englewood Cliffs, NJ, 1996.

2 Bathe, K.J. and Wilson, E.L., *Numerical Methods in Finite Element Analysis,* Prentice Hall, Englewood Cliffs, NJ, 1976.

3 Bathe, K.J. and Wilson, E.L., "Stability and Accuracy of Direct Integration Methods," *International Journal of Earthquake Engineering and Structural Dynamics,* Vol. 1, 1973, pp. 238–291.

4 Newmark, Nathan M., "*A Method of Computation for Structural Dynamics,*" ASCE, *Journal of Engineering Mechanics Division,* Vol. 85, 1959, pp. 67–94.

5 Clough, R.W. and Penzien, J., *Dynamics of Structures,* 2nd ed., McGraw-Hill, New York, 1993.

6 Biggs, John M., *Introduction to Structural Dynamics,* McGraw-Hill, New York, 1964.

7 Craig, Roy R., *Structural Dynamics: An Introduction to Computer Methods,* Wiley, New York, 1981.

8 Chopra, Anil K., *Dynamics of Structures: Theory and Applications to Earthquake Engineering,* Prentice Hall, Englewood Cliffs, NJ, 1995.

9 Newmark, Nathan M., "Earthquake Response Analysis of Reactor Structures," *Nuclear Engineering and Design,* Vol. 20, No. 2, July 1972, pp. 303–322.

10 Newmark, N.M. and Hall, W.J., *Earthquake Spectra and Design,* Earthquake Engineering Research Institute, Berkeley, CA, 1982.

NOTATION

a_p	coefficient relating stiffness proportional damping in pth normal mode
$a_0, a_1, \ldots, a_8$	integration constants
E	modulus of elasticity
g	acceleration due to gravity
I	moment of inertia
K_r	generalized (modal) stiffness for rth normal mode
M_r	generalized (modal) mass for rth normal mode
t	time
$\ddot{x}_g(t)$	ground acceleration
AD	amplitude decay
DOF	degree of freedom
MDOF	multi-degree-of-freedom
PE	period elongation
SDOF	single-degree-of-freedom
$[c]$	viscous damping matrix
$[C]$	modal damping matrix
$\{\hat{F}\}$	effective force vector used in numerical integration schemes
$\{F(t)\}$	external force vector
$\{F_{\text{eff}}(t)\}$	effective load vector for ground acceleration
$[I]$	identity matrix
$\{I\}$	identity vector
$[k]$	stiffness matrix
$[\hat{k}]$	effective stiffness matrix
$[K]$	generalized (modal) stiffness matrix
$[m]$	mass matrix
$[\hat{m}]$	effective mass matrix
$[M]$	generalized (modal) mass matrix

$\{x\}$	displacement vector
$\{x\}_0$	initial displacement vector
$\{\dot{x}\}$	velocity vector
$\{\dot{x}\}_0$	initial velocity vector
$\{\ddot{x}\}$	acceleration vector
$\{\ddot{x}\}_0$	initial acceleration vector
$\{\ddot{x}_t\}$	total (absolute) acceleration vector
α	integration parameter used in Newmark method; also, mass proportionality constant for Rayleigh damping
β	stiffness proportionality constant for Rayleigh damping
δ	integration parameter used in Newmark method
Δt	numerical integration time step
Δt_{cr}	critical time step
τ	time parameter
θ	integration parameter used in Wilson-θ method
ω_p	undamped natural circular frequency for pth normal mode
ω_r	undamped natural circular frequency for rth normal mode
ζ_p	modal damping factor for pth normal mode
ζ_r	modal damping factor for rth normal mode
$\overline{\zeta}_r$	stiffness proportional damping in modes for which damping is not specified
$\hat{\zeta}_r$	damping complementary to stiffness proportional damping in modes for which damping is not specified
$[\Phi]$	modal matrix
$\{\Phi\}_r$	modal vector for rth normal mode

PROBLEMS

13.1 Evaluate by the central difference method the dynamic response of the four-story shear frame building shown in Figure P13.1 subjected to the time-varying force shown in Figure P13.2. Plot the time-history responses $\{x(t)\}$, $\{\dot{x}(t)\}$, and $\{\ddot{x}(t)\}$ in the time interval $0 \le t \le 5.0$ sec. Assume $k = 50$ kips/in, $W = 200$ kips, and $c = 0.3$ kip-sec/in. Use a time step $\Delta t = 0.001$ sec.

13.2 Repeat Problem 13.1 for the four-story shear frame building with suspended mass at level 3 shown in Figure P13.3 subjected to the time-varying force shown in Figure P13.4. Assume $k = 600$ kips/in, $m = 1.2$ kip-sec^2/in, and $c = 0.5$ kip-sec/in.

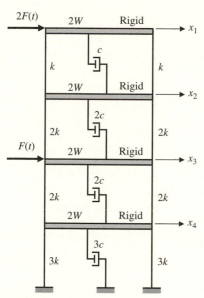

Figure P13.1

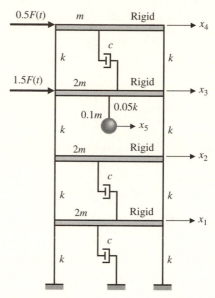

Figure P13.3

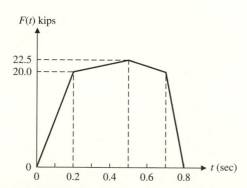

Figure P13.2

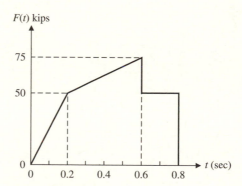

Figure P13.4

13.3 Repeat Problem 13.1 for the three-story shear frame building shown in Figure P13.5 subjected to the time-varying force shown in Figure P13.6. Assume $k = 10$ kips/in, $m = 0.1$ kip-sec²/in, and $c = 0.1$ kip-sec/in.

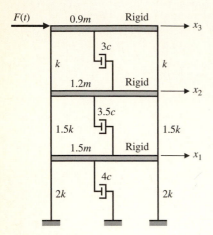

Figure P13.5

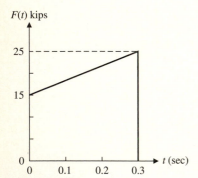

Figure P13.6

13.4 Repeat Problem 13.1 for the four-story shear frame structure of Example 13.1 subjected to the time-varying force specified in Figure P13.4.

13.5 Evaluate by the Newmark method of time integration the dynamic response of the three-story shear frame building shown in Figure P13.5 subjected to the time-varying force specified in Figure P13.7. Assume $k = 350$ kips/in, $m = 6.0$ kip-sec²/in, and $c = 5$ kip-sec/in. Plot response histories for $\{x(t)\}$, $\{\dot{x}(t)\}$, and $\{\ddot{x}(t)\}$ in the time interval $0 \le t \le 5.0$ sec.

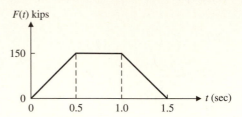

Figure P13.7

13.6 Evaluate by the Newmark method of time integration the dynamic response of the five-story shear frame building shown in Figure P13.8 subjected to the time-varying force specified in Figure P13.9. Plot response histories for $\{x(t)\}$, $\{\dot{x}(t)\}$, and $\{\ddot{x}(t)\}$ in the time interval $0 \le t \le 5$ sec. Assume $\zeta_1 = 0.02$, $\zeta_2 = 0.07$, and calculate the $[c]$ matrix from a Rayleigh damping formulation. Let $k = 500$ kips/in and $m = 0.6$ kip-sec²/in. Select an appropriate time step.

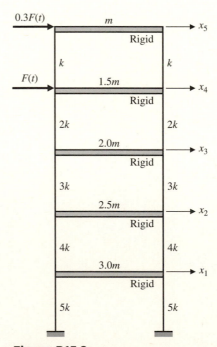

Figure P13.8

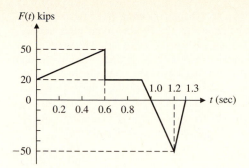

Figure P13.9

13.7 Repeat Problem 13.6 for the condition where the base of the structure is subjected to the horizontal ground acceleration shown in Figure P13.10.

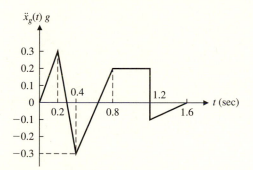

Figure P13.10

13.8 Repeat Problem 13.5 for the horizontal ground acceleration specified in Figure 13.18.

13.9 The base of the shear frame structure shown in Figure P13.11 is subjected to the horizontal ground acceleration shown in Figure P13.12. Evaluate the dynamic response by the Wilson-θ method of time integration. Plot response histories for $\{x(t)\}$, $\{\dot{x}(t)\}$, and $\{\ddot{x}_t(t)\}$ in the time interval $0 \leq t \leq 6$ sec. Assume $W = 100$ kips and $k = 80$ kips/in. Let $\zeta_1 = 0.015$, $\zeta_2 = 0.017$, and calculate the $[c]$ matrix by the procedure discussed in Section 13.5. Select an appropriate time step.

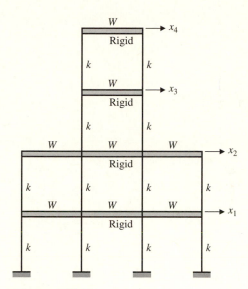

Figure P13.11

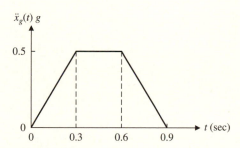

Figure P13.12

13.10 Repeat Problem 13.9 for the case where the second level (x_2) of the structure is subjected to the time-varying force shown in Figure P13.9 in lieu of the base excitation.

13.11 Evaluate the dynamic response of the shear frame structure shown in Figure P13.13 subjected to the time-varying force specified in Figure P13.14. Use any direct numerical integration scheme discussed in Chapter 13. Assume $W = 50$ kips and $k = 60$ kips/in. Let $\zeta_1 = 0.05$, $\zeta_2 = 0.07$, and calculate the $[c]$ matrix by the method discussed in Section 13.5. Plot response histories for $\{x(t)\}$, $\{\dot{x}(t)\}$, and $\{\ddot{x}(t)\}$ in the time interval $0 \leq t \leq 5$ sec. Select an appropriate time step.

13.13 Evaluate the dynamic response of the fixed-ended beam shown in Figure P13.15 subjected to the time-varying force shown in Figure P13.16. Use any direct integration scheme discussed in Chapter 13. Assume $W = 5$ kips, modulus of elasticity $E = 29{,}000$ ksi, and moment of inertia $I = 5760$ in^4 for the beam. Let $\zeta_1 = 0.1$, $\zeta_2 = 0.15$, and calculate the damping matrix $[c]$ by the method discussed in Section 13.5. Plot response histories for $\{x(t)\}$, $\{\dot{x}(t)\}$, and $\{\ddot{x}(t)\}$ in the time interval $0 \leq t \leq 6$ sec. Select an appropriate time step.

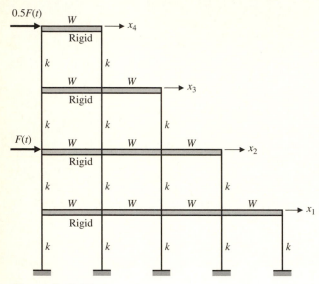

Figure P13.13

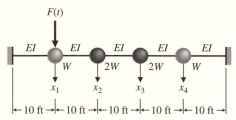

Figure P13.15

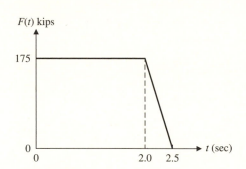

Figure P13.16

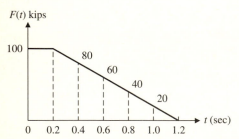

Figure P13.14

13.14 Repeat Problem 13.13 for the condition where the supports of the beams are subjected to the vertical acceleration as specified in Figure 13.10 in lieu of the time varying force.

13.12 Repeat Problem 13.11 for the case where the base of the structure is subjected to the horizontal ground acceleration specified in Figure P13.10 in lieu of the time varying force.

Part III

Continuous Systems

14 ⏶ Vibrations of Continuous Systems

Up to this point in the text, attention has been focused on the response of *discrete systems.* In discrete systems, both stiffness and mass, as well as damping, are modeled as discrete properties. The mathematical models for discrete systems are ordinary differential equations, which thereby render themselves quite conducive to numerical solution techniques. An alternative method of modeling physical systems is based on the principle of distributed mass and stiffness characteristics. Systems for which stiffness and mass are considered to be distributed properties (rather than discrete) are referred to as *distributed systems,* or *continuous systems.*

Unlike discrete systems that possess a finite number of degrees of freedom, continuous systems, which are considered to be composed of an infinite number of infinitesimal mass particles, theoretically possess an infinite number of degrees of freedom. Moreover, the governing equations or mathematical models for continuous systems are partial differential equations. Several common examples of continuous systems include the longitudinal vibration of bars, the transverse vibration of cables and beams, and the vibration of membranes and shells.

In consideration of the vibrations of continuous systems, three very restrictive but necessary assumptions must be made concerning the systems' material properties. The first is that the material is homogeneous, the second is that the material is elastic and obeys Hooke's law, and the third is that the material is isotropic. These assumptions are necessary so that the dynamic response may be calculated as the sum of the normal mode contributions. Consequently, these assumptions preclude the analysis of systems exhibiting nonlinear characteristics or possessing nonclassical damping.

Vibrations of continuous systems is indeed a very broad subject area, and many textbooks have been written on specific topics in this category. This chapter is intended as only an introduction to the subject area; it discusses closed-form solutions for several classical problems. It will readily become apparent, however, that analytical or closed-form solutions can be obtained only for relatively simple continuous systems with well-defined boundary conditions. Therefore, this chapter also discusses several approximate methods of analysis.

14.1 LONGITUDINAL VIBRATION OF A UNIFORM ROD

Consider the longitudinal vibration of a uniform elastic rod of cross-sectional area A, modulus of elasticity E, and material mass density ρ as shown in Figure 14.1a. The free-body diagram of a differential element of the rod dx in its deformed position is shown in Figure 14.1b. The undeformed position of the element is denoted by x and the deformed position is u. The differential element has also been elastically strained, such that its deformed length is $(1 + \partial u/\partial x)dx$. Summing forces in the x direction yields from Figure 14.1b

$$\sum F_x = m\ddot{x} \tag{14.1}$$

or

$$F + \frac{\partial F}{\partial x}dx - F = \rho A \, dx \frac{\partial^2 u}{\partial t^2} \tag{14.2}$$

where $\partial^2 u/\partial t^2$ is the acceleration of the differential element. Noting that the axial force F can be expressed as

$$F = EA\frac{\partial u}{\partial x} \tag{14.3}$$

where $\partial u/\partial x$ is the axial strain, it follows that

$$\frac{\partial F}{\partial x} = EA\frac{\partial^2 u}{\partial x^2} \tag{14.4}$$

Substituting Eq. (14.4) into Eq. (14.2) and rearranging terms yields

$$\frac{\partial^2 u}{\partial t^2} = \frac{E}{\rho} \frac{\partial^2 u}{\partial x^2} \tag{14.5}$$

or

$$\frac{\partial^2 u}{\partial x^2} = \frac{1}{c^2} \frac{\partial^2 u}{\partial t^2} \tag{14.6}$$

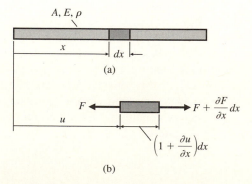

Figure 14.1

Longitudinal vibrations in a uniform rod.

Equation (14.6) is referred to as the one-dimensional *wave equation,* and $c = \sqrt{E/\rho}$ is the velocity of propagation of the stress wave, or the *wave speed.*

The wave equation can be solved by the method of separation of variables by assuming a solution $u(x, t)$ as the product of two separate functions, one of x and one of t. Therefore

$$u(x, t) = f(x)g(t) \tag{14.7}$$

Then from Eq. (14.7)

$$\frac{\partial^2 u}{\partial t^2} = \frac{\partial^2 g(t)}{\partial t^2} f(x) \tag{14.8}$$

and

$$\frac{\partial^2 u}{\partial x^2} = \frac{\partial^2 f(x)}{\partial x^2} g(t) \tag{14.9}$$

Substituting Eqs. (14.8) and (14.9) into the wave equation, Eq. (14.6), yields

$$\frac{\partial^2 f(x)}{\partial x^2} g(t) = \frac{1}{c^2} \frac{\partial^2 g(t)}{\partial t^2} f(x) \tag{14.10}$$

The functions $f(x)$ and $g(t)$ can be separated in Eq. (14.10) such that

$$\frac{1}{f(x)} \frac{\partial^2 f(x)}{\partial x^2} = \frac{1}{c^2} \frac{1}{g(t)} \frac{\partial^2 g(t)}{\partial t^2} \tag{14.11}$$

Since the functions are separated, partial derivatives are no longer needed. Moreover, the only solution permitting two functions of different independent variables x and t to be equal to one another is that both be equal to a constant. Since the constant is arbitrary, for reasons of convenience we let that constant be equal to $-(\omega/c)^2$, where ω is the undamped natural circular frequency. Thus

$$\frac{1}{f(x)} \frac{d^2 f(x)}{dx^2} = -\left(\frac{\omega}{c}\right)^2 \tag{14.12}$$

and

$$\frac{1}{c^2} \frac{1}{g(t)} \frac{d^2 g(t)}{dt^2} = -\left(\frac{\omega}{c}\right)^2 \tag{14.13}$$

This leads to two separate ordinary differential equations of motion,

$$\frac{d^2 f(x)}{dx^2} + \left(\frac{\omega}{c}\right)^2 f(x) = 0 \tag{14.14}$$

and

$$\frac{d^2 g(t)}{dt^2} + \omega^2 g(t) = 0 \tag{14.15}$$

The solution to Eqs. (14.14) and (14.15), respectively, are

$$f(x) = C_1 \sin \frac{\omega}{c}x + C_2 \cos \frac{\omega}{c}x \tag{14.16}$$

and

$$g(t) = A \sin \omega t + B \cos \omega t \tag{14.17}$$

The complete solution for the wave equation is obtained by substituting Eqs. (14.16) and (14.17) into Eq. (14.7), resulting in

$$u(x, t) = \left(C_1 \sin \frac{\omega}{c}x + C_2 \cos \frac{\omega}{c}x \right)(A \sin \omega t + B \cos \omega t) \tag{14.18}$$

The four arbitrary constants in Eq. (14.18) are determined from the initial conditions and the boundary conditions.

The topic of wave propagation in elastic media is discussed in detail in Chapter 16.

EXAMPLE 14.1 ▼

Determine expressions for the natural frequencies and displacement response for longitudinal vibrations of a uniform rod with one end fixed and the other end free, as shown in Figure 14.2.

Solution

The equation of motion for the longitudinal displacement $u(x, t)$ of a uniform rod is given by the wave equation

$$\frac{\partial^2 u}{\partial x^2} = \frac{1}{c^2} \frac{\partial^2 u}{\partial t^2} \tag{1}$$

The general solution for the displacement u is given by Eq. (14.18),

$$u(x, t) = \left(C_1 \sin \frac{\omega}{c}x + C_2 \cos \frac{\omega}{c}x \right)(A \sin \omega t + B \cos \omega t) \tag{2}$$

The boundary condition at the fixed end at $x = 0$ is

$$u(0, t) = 0 \tag{3}$$

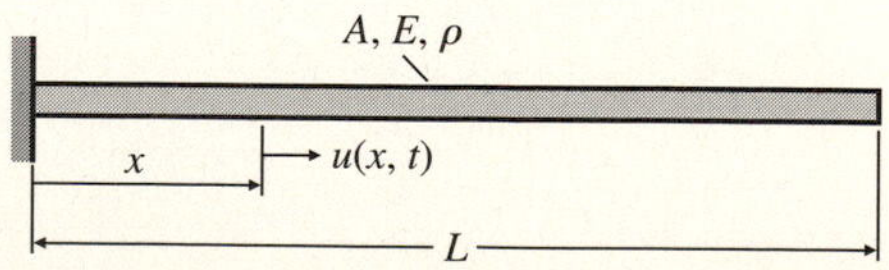

Figure 14.2
Uniform rod of Example 14.1.

and evaluating Eq. (2) for this boundary condition gives

$$0 = C_2(A \sin \omega t + B \cos \omega t) \tag{4}$$

for which $C_2 = 0$. The free end of the rod at $x = L$ must be free of stress. Using this as the second boundary condition yields

$$EA\left(\frac{\partial u}{\partial x}\right) = 0 \tag{5}$$

or

$$\frac{\partial u}{\partial x} = 0 = (A \sin \omega t + B \cos \omega t)C_1 \frac{\omega}{c} \cos \frac{\omega}{c}L = 0 \tag{6}$$

Neglecting the trivial solution of $C_1 = 0$, Eq. (6) can be satisfied only if

$$\cos \frac{\omega}{c}L = 0 \tag{7}$$

Equation (7) is the frequency equation that leads to the natural frequencies

$$\omega_n = \frac{\pi c}{2L}, \frac{3\pi c}{2L}, \ldots, \frac{(2n-1)\pi c}{2L} \tag{8}$$

or

$$\omega_n = \frac{(2n-1)\pi}{2L}\sqrt{\frac{E}{\rho}} \qquad n = 1, 2, \ldots, \infty \tag{9}$$

From Eq. (2), the displacement u_n for a particular principal mode is

$$u_n = (A_n \sin \omega_n t + B_n \cos \omega_n t)\sin \frac{\omega_n}{c}x \tag{10}$$

The total displacement response $u(x, t)$ is the superposition of the individual responses of all the principal modes, resulting in

$$u(x, t) = \sum_{n=1}^{\infty} (A_n \sin \omega_n t + B_n \cos \omega_n t)\sin \frac{\omega_n}{c}x \tag{11}$$

where $A_n = AC_1$ and $B_n = BC_1$ for each mode n.

The arbitrary constants A_n and B_n must be determined from the initial conditions, and ω_n is given by Eq. (9). ▲

EXAMPLE 14.2 ▼

A uniform rod of length L, mass density ρ, cross-sectional area A, and elastic modulus E is fixed at one end and has a concentrated mass m attached to its free end as shown in Figure 14.3. Determine the expression for the frequency equation of longitudinal vibrations.

Figure 14.3

Uniform rod with concentrated mass at free end.

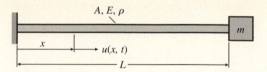

Solution

The general solution for longitudinal vibrations of uniform bars is given by Eq. (14.18) and is expressed using principal modes as

$$u(x, t) = \sum_{n=1}^{\infty} \left(C_n \sin \frac{\omega_n x}{c} + D_n \cos \frac{\omega_n x}{c} \right)(A_n \sin \omega_n t + B_n \cos \omega_n t) \tag{1}$$

There is no displacement at the fixed end ($x = 0$), and a dynamic force in the bar at the free end is equal to the inertia force of the concentrated mass. Therefore, the boundary conditions are

$$u(0, t) = 0 \tag{2}$$

and

$$AE \frac{\partial u(L, t)}{\partial x} = -m \frac{\partial^2 u(L, t)}{\partial t^2} \tag{3}$$

Applying the first boundary condition, Eq. (2), to the general solution, Eq. (1), results in

$$u(0, t) = \sum_{n=1}^{\infty} D_n(A_n \sin \omega_n t + B_n \cos \omega_n t) = 0 \tag{4}$$

from which $D_n = 0$. Applying the second boundary condition, Eq. (3), to Eq. (1) yields the transcendental equation

$$m \omega_n^2 \sin \frac{\omega_n L}{c} = AE \frac{\omega_n}{c} \cos \frac{\omega_n L}{c} \tag{5}$$

or

$$\frac{AE}{m \omega_n c} = \tan \frac{\omega_n c}{L} \tag{6}$$

in which ω_n is the only unknown. To clarify the solution, we rewrite the frequency equation as

$$\frac{AEL}{mc^2} = \frac{\omega_n L}{c} \tan \frac{\omega_n L}{c} \tag{7}$$

Since $c^2 = E/\rho$, Eq. (7) becomes

$$\frac{\rho AL}{m} = \frac{\omega_n L}{c} \tan \frac{\omega_n L}{c} \tag{8}$$

Equation (8) is then the frequency equation for the system and illustrates that the natural frequencies depend upon the magnitude of mass of the shaft (ρAL) relative to the concentrated mass m. When $\rho AL/m \to \infty$, that is, when the concentrated mass is very small compared to the mass of the bar, the frequency equation becomes

$$\tan \frac{\omega_n L}{c} = \infty \tag{9}$$

which yields the same value for natural frequencies ω_n given by Eqs. (8) and (9) of Example 14.1. Thus, the system becomes that of a uniform bar fixed at one end and free at the other.

When m is very large compared to the total mass of the bar ρAL, the system behavior corresponds to that of a simple SDOF spring-mass system and the fundamental frequency is closely approximated by

$$\omega_1 = \sqrt{\frac{EA}{mL}} \tag{10}$$

▲

14.2 TRANSVERSE VIBRATION OF A PRETENSIONED CABLE

The transverse vibrations of a taut cable or string have an equation of motion similar to that governing the longitudinal vibrations in a uniform rod. Consider a uniform elastic cable, having mass per unit length γ, to be stretched under a tension T between two fixed points as shown in Figure 14.4a. The free-body diagram of a differential element dx of the cable is shown in Figure 14.4b. Assuming small deflections and slopes, the equation of motion for transverse (y direction) vibration is given by

$$T\left(\theta + \frac{\partial \theta}{\partial x}dx\right) - T\theta = \gamma \, dx \frac{\partial^2 y}{\partial t^2} \tag{14.19}$$

or

$$\frac{\partial \theta}{\partial x} = \frac{\gamma}{T}\frac{\partial^2 y}{\partial t^2} \tag{14.20}$$

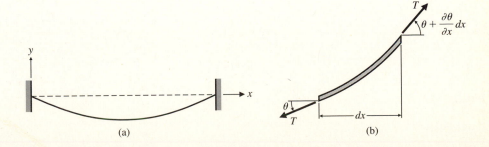

Figure 14.4

Transverse vibration of a pretensioned cable.

(a)   (b)

Since the slope of the cable $\theta = \partial y/\partial x$, Eq. (14.20) reduces to

$$\frac{\partial^2 y}{\partial x^2} = \frac{1}{c^2}\frac{\partial^2 y}{\partial t^2} \tag{14.21}$$

where $c = \sqrt{T/\gamma}$ is the velocity of the wave propagation along the cable. We immediately notice that Eq. (14.21) is identical to the one-dimensional wave equation previously given by Eq. (14.6). Therefore, we obtain the solution to Eq. (14.21) by the same procedure discussed in Section 14.1.

EXAMPLE 14.3 ▼

A uniform cable of length L and mass per unit length γ is fixed at its ends and stretched under a tension T. If the cable is displaced into an initial shape $y(x, 0) = Y_0 \sin(\pi/L)x$, as shown in Figure 14.5, and suddenly released, determine expressions for its natural frequencies and free vibration displacement response.

Solution

The general free vibration solution is given by Eq. (14.18) for $y(x, t)$ as

$$y(x, t) = \left(C_1 \sin\frac{\omega}{c}x + C_2 \cos\frac{\omega}{c}x\right)(A \sin \omega t + B \cos \omega t) \tag{1}$$

The boundary conditions are $y(0, t) = y(L, t) = 0$. The boundary condition that $y(0, t) = 0$ will require that $C_2 = 0$; therefore, Eq. (1) becomes

$$y(x, t) = (A \sin \omega t + B \cos \omega t)C_1 \sin\frac{\omega}{c}x \tag{2}$$

Applying the boundary condition $y(L, t) = 0$ will satisfy Eq. (2) if $C_1 = 0$, which is the trivial solution. However, Eq. (2) will also be satisfied by

$$\sin\frac{\omega L}{c} = 0 \tag{3}$$

from which

$$\frac{\omega L}{c} = n\pi \qquad n = 1, 2, 3, \ldots, \infty \tag{4}$$

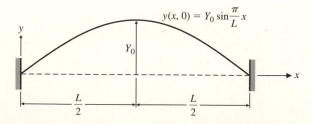

$$y(x, 0) = Y_0 \sin\frac{\pi}{L}x$$

Figure 14.5

Pretensioned cable with initial displacement.

Equation (4) yields an infinite number of natural frequencies

$$\omega_n = \frac{n\pi c}{L} = \frac{n\pi}{L}\sqrt{\frac{T}{\gamma}} \qquad n = 1, 2, 3, \ldots, \infty \tag{5}$$

Each ω_n corresponds to a normal mode vibration having a mode shape y_n with a sinusoidal distribution given by

$$y_n = \sin\frac{n\pi x}{L} \tag{6}$$

The general expression for the free vibration transverse displacements can be expressed from Eq. (2) as

$$y(x, t) = \sum_{n=1}^{\infty}(A_n \sin \omega_n t + B_n \cos \omega_n t)\sin\frac{\omega_n}{c}x \tag{7}$$

where $A_n = AC_1$ and $B_n = BC_1$. Applying the initial conditions of displacement

$$y(x, 0) = Y_0 \sin\frac{\pi}{L}x \tag{8}$$

and velocity

$$\frac{\partial y(x, 0)}{\partial t} = 0 \tag{9}$$

to Eq. (7) yields

$$A_n = 0 \qquad \text{for all } n$$

$$B_1 = Y_0$$

$$B_n = 0 \qquad n \neq 1$$

Thus, the expression for the free vibration displacement response of the cable considering all n modes becomes

$$y(x, t) = \sum_{n=1}^{\infty} Y_0 \cos \omega_n t \sin\frac{n\pi}{L}x \tag{10}$$

▲

14.3 FREE TRANSVERSE VIBRATION OF UNIFORM BEAMS

The transverse vibration of beams is another vibration problem that typifies distributed mass and elasticity properties. Consider the uniform beam shown in Figure 14.6a having cross-sectional area A, flexural rigidity EI, and mass density ρ. The deflection $y(x, t)$ of the beam is considered to be a result of bending moment effects only. Such a model is often referred to as the *Bernoulli-Euler beam*.

Figure 14.6

Transverse vibration of a uniform beam.

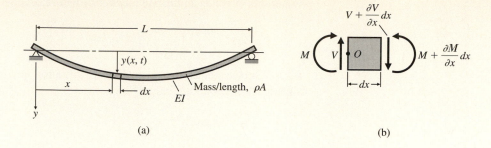

(a)   (b)

A differential element dx of the beam is isolated in Figure 14.6b, which illustrates the bending moments M and shear forces V acting upon it. As the beam vibrates, the differential element dx moves vertically up and down in the y direction and also rotates very slightly. Ignoring the inertial properties of the element in rotation and summing moments about point O yields

$$\sum M_0 = 0 \tag{14.22}$$

$$V \, dx = \frac{\partial M}{\partial x} dx \tag{14.23}$$

or

$$V = \frac{\partial M}{\partial x} \tag{14.24}$$

Next, forces in the y direction are summed, and from Newton's second law

$$\sum F = \rho A \, dx \frac{\partial^2 y}{\partial t^2} \tag{14.25}$$

or

$$\frac{\partial V}{\partial x} dx = \rho A \, dx \frac{\partial^2 y}{\partial t^2} \tag{14.26}$$

Combining Eqs. (14.24) and (14.26) results in

$$\frac{\partial^2 M}{\partial x^2} dx = \rho A \, dx \frac{\partial^2 y}{\partial t^2} \tag{14.27}$$

From strength of materials, it is known that bending moment is a function of the change in slope of the beam's elastic curve, or

$$M = -EI \frac{\partial^2 y}{\partial x^2} \tag{14.28}$$

Substituting Eq. (14.28) into Eq. (14.27) yields

$$\frac{\partial^2}{\partial x^2}\left(-EI\frac{\partial^2 y}{\partial x^2}\right) = \rho A\frac{\partial^2 y}{\partial t^2} \tag{14.29a}$$

or

$$EI\frac{\partial^4 y}{\partial x^4} + \rho A\frac{\partial^2 y}{\partial t^2} = 0 \tag{14.29b}$$

Although Eq. (14.29b) is a fourth-order partial differential equation, its form resembles the one-dimensional wave equation, and Eq. (14.29b) can be rewritten as

$$-\frac{\partial^4 y}{\partial x^4} = \frac{1}{c^2}\frac{\partial^2 y}{\partial t^2} \tag{14.30}$$

where $c = \sqrt{EI/\rho A}$. Equation (14.30) can therefore be solved by the method of separation of variables, and a separable solution of the following form can be sought:

$$y(x, t) = f(x)g(t) \tag{14.31}$$

where

$$\frac{\partial^4 y}{\partial x^4} = \frac{d^4 f(x)}{dx^4}g(t) \tag{14.32}$$

and

$$\frac{\partial^2 y}{\partial t^2} = \frac{d^2 g(t)}{dt^2}f(x) \tag{14.33}$$

Substituting Eqs. (14.32) and (14.33) into Eq. (14.30) and separating the variables yields

$$\frac{-c^2}{f(x)}\frac{d^4 f(x)}{dx^4} = \frac{1}{g(t)}\frac{d^2 g(t)}{dt^2} \tag{14.34}$$

For these two independent functions to be equal, they must be equal to a constant. For convenience, this constant is selected to be $-\omega_n^2$. Therefore, Eq. (14.34) may be expressed as two independent ordinary differential equations:

$$\frac{d^2 g(t)}{dt^2} + \omega_n^2 g(t) = 0 \tag{14.35}$$

and

$$\frac{d^4 f(x)}{dx^4} - \frac{\omega_n^2}{c^2}f(x) = 0 \tag{14.36}$$

The solution to Eq. (14.35) is simply

$$g(t) = A \sin \omega t + B \cos \omega t \qquad (14.37)$$

The solution to Eq. (14.36) is not as obvious. For this equation, assume a linear exponential solution of the form

$$f(x) = Ae^{sx} \qquad (14.38)$$

where A and s are constants. Substituting Eq. (14.38) into Eq. (14.36) gives the auxiliary equation

$$s^4 - \frac{\omega_n^2}{c^2} = 0 \qquad (14.39)$$

Let $\beta^4 = \omega_n^2/c^2$ and Eq. (14.39) may be written as

$$s^4 - \beta^4 = 0 \qquad (14.40)$$

which has the four roots

$$s_1 = \beta \qquad s_2 = -\beta$$
$$s_3 = i\beta \qquad s_4 = -i\beta \qquad (14.41)$$

where $i = \sqrt{-1}$. The general solution then becomes

$$f(x) = A_1 e^{\beta x} + A_2 e^{-\beta x} + A_3 e^{i\beta x} + A_4 e^{-i\beta x} \qquad (14.42)$$

Equation (14.42) may be transformed by making use of the following trigonometric relationships:

$$e^{\beta x} = \cosh \beta x + \sinh \beta x$$
$$e^{-\beta x} = \cosh \beta x - \sinh \beta x$$
$$e^{i\beta x} = \cosh \beta x + i \sinh \beta x$$
$$e^{-i\beta x} = \cosh \beta x - i \sinh \beta x \qquad (14.43)$$

Substituting these relations into Eq. (14.42) results in

$$f(x) = C_1 \sinh \beta x + C_2 \cosh \beta x + C_3 \sin \beta x + C_4 \cos \beta x \qquad (14.44)$$

Finally, the solution to Eq. (14.30), as represented by Eq. (14.31), is

$$y(x, t) = (A \sin \omega t + B \cos \omega t)$$
$$(C_1 \sinh \beta x + C_2 \cosh \beta x + C_3 \sin \beta x + C_4 \cos \beta x) \qquad (14.45)$$

The natural frequencies of the system and the constants C_1 through C_4 appearing in Eq. (14.45) are determined by applying the appropriate boundary conditions for the beam and the initial conditions of displacement and velocity.

EXAMPLE 14.4 ▼

Determine expressions for the natural frequencies and normal vibration modes for a simply supported beam of length L and uniform cross section as shown in Figure 14.7. Assume flexural rigidity EI, cross-sectional area A, and mass density ρ.

Solution

The general solution for displacement is given by Eq. (14.44) as

$$y(x) = C_1 \sinh \beta x + C_2 \cosh \beta x + C_3 \sin \beta x + C_4 \cos \beta x \qquad (1)$$

and the appropriate boundary conditions for a simply supported beam are

$$y(0) = y(L) = 0 \qquad (2)$$

$$\left.\frac{\partial^2 y}{\partial x^2}\right|_{x=0} = \left.\frac{\partial^2 y}{\partial x^2}\right|_{x=L} = M = 0 \qquad (3)$$

Applying these four boundary conditions to Eq. (1) results in

$$C_2 + C_4 = 0 \qquad (4)$$

$$C_1 \sinh \beta L + C_2 \cosh \beta L + C_3 \sin \beta L + C_4 \cos \beta L = 0 \qquad (5)$$

$$\beta^2 (C_2 - C_4) = 0 \qquad (6)$$

$$\beta^2 (C_1 \sinh \beta L + C_2 \cosh \beta L + C_3 \sin \beta L + C_4 \cos \beta L) = 0 \qquad (7)$$

From Eqs. (4) and (6), $C_2 = C_4 = 0$, which leaves

$$C_1 \sinh \beta L + C_3 \sin \beta L = 0 \qquad (8)$$

$$\beta^2 (C_1 \sinh \beta L - C_3 \sin \beta L) = 0 \qquad (9)$$

Equations (8) and (9) represent a pair of linear homogeneous algebraic equations. A nontrivial solution exists only if the determinant of the coefficient matrix vanishes. Thus

$$\begin{vmatrix} \sinh \beta L & \sin \beta L \\ \beta^2 \sinh \beta L & -\beta^2 \sin \beta L \end{vmatrix} = 0 \qquad (10)$$

or

$$-2\beta^2 \sinh \beta L \sin \beta L = 0 \qquad (11)$$

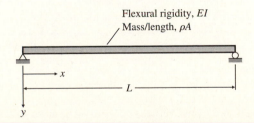

Flexural rigidity, EI
Mass/length, ρA

x

L

y

Figure 14.7
Uniform beam of Example 14.4.

Since $\beta \neq 0$, it follows that

$$\sinh \beta L \sin \beta L = 0 \tag{12}$$

In Eq. (12), $\sinh \beta L = 0$ only if $\beta L = 0$; therefore, the only nontrivial solution to Eq. (12) exists for

$$\sin \beta L = 0 \tag{13}$$

Equation (13) is the frequency equation, and if Eq. (13) is substituted back into Eq. (8), then $C_1 = 0$. To satisfy Eq. (13), it is necessary that

$$\beta L = n\pi \qquad n = 1, 2, 3, \ldots \tag{14}$$

Since $\beta_n^2 = \omega_n/c$, the natural frequencies of the system are

$$\omega_n = \beta_n^2 \sqrt{\frac{EI}{\rho A}} = (\beta_n L)^2 \sqrt{\frac{EI}{\rho A L^4}} \tag{15}$$

in which β_n is dependent upon the beams boundary conditions. Values for $\beta_n L$ for some typical boundary conditions are presented in Table 14.1. The free vibration response for the simply supported beam is obtained by superposition of the normal modes and is expressed as

$$y(x, t) = \sum_{n=1}^{\infty} (A_n \sin \omega_n t + B_n \cos \omega_n t) \sin \frac{n\pi x}{L} \tag{16}$$

where $A_n = C_2 A$ and $B_n = C_2 B$. Then the constants A_n and B_n are evaluated by applying the initial conditions. The individual displacement responses for each of the first three modes are given, respectively, by

$$[y(x, t)]_1 = (A_1 \sin \omega_1 t + B_1 \cos \omega_1 t) \sin \frac{\pi x}{L} \tag{17}$$

$$[y(x, t)]_2 = (A_2 \sin \omega_2 t + B_2 \cos \omega_2 t) \sin \frac{2\pi x}{L} \tag{18}$$

$$[y(x, t)]_3 = (A_3 \sin \omega_3 t + B_3 \cos \omega_3 t) \sin \frac{3\pi x}{6} \tag{19}$$

The corresponding mode shapes are shown in Figure 14.8.

TABLE 14.1. **Natural Frequencies for Single-Span Beams**

Boundary Conditions	Frequency Equation	$\beta_1 L$	$\beta_2 L$	$\beta_3 L$
Pinned-pinned	$\sin \beta L = 0$	3.141	6.282	9.423
Fixed-free	$\cos \beta L \cosh \beta L + 1 = 0$	1.875	4.694	7.855
Fixed-pinned (and pinned-free)	$\tan \beta L = \tanh \beta L$	3.927	7.069	10.210
Fixed-fixed (and free-free)	$\cos \beta L \cosh \beta L = 1$	4.730	7.853	10.996
Fixed-sliding (and free-sliding)	$\tan \beta L + \tanh \beta L = 0$	2.365	5.498	8.639

Figure 14.8

Normal modes for uniform beam
of Example 14.4: (a) mode 1;
(b) mode 2; (c) mode 3.

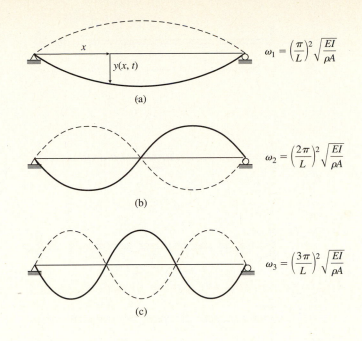

$$\omega_1 = \left(\frac{\pi}{L}\right)^2 \sqrt{\frac{EI}{\rho A}}$$

(a)

$$\omega_2 = \left(\frac{2\pi}{L}\right)^2 \sqrt{\frac{EI}{\rho A}}$$

(b)

$$\omega_3 = \left(\frac{3\pi}{L}\right)^2 \sqrt{\frac{EI}{\rho A}}$$

(c)

EXAMPLE 14.5 ▼

Determine the frequency equation for the two-span beam shown in Figure 14.9a. Assume each span has flexural rigidity *EI*, cross-sectional area *A*, and mass density ρ.

Solution

For each span, select the exterior support for the origin as shown in Figure 14.9b. The general displacement solution is given by Eq. (14.44) for each span. Thus, for span 1

$$y_1(x_1)$$
$$= (A_1 \sin \beta_1 x_1 + B_1 \cos \beta_1 x_1 + C_1 \sinh \beta_1 x_1 + D_1 \cosh \beta_1 x_1) \quad (1)$$

and for span 2

$$y_2(x_2)$$
$$= (A_2 \sin \beta_2 x_2 + B_2 \cos \beta_2 x_2 + C_2 \sinh \beta_2 x_2 + D_2 \cosh \beta_2 x_2) \quad (2)$$

Since the spans are symmetric, $\beta_1 = \beta_2 = \beta$. The boundary conditions are applied next. For span 1

(a) $y_1 = 0$ at $x_1 = 0$:

$$y_1(0) = B_1 + D_1 = 0 \tag{1}$$

or

$$D_1 = -B_1 \tag{2}$$

Figure 14.9

Two-span uniform beam of Example 14.5.

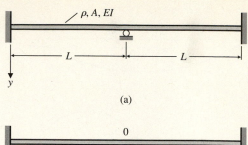

(a)

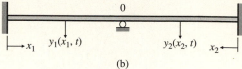

(b)

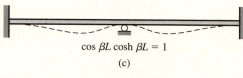

(c)

$$\cos \beta L \cosh \beta L = 1$$

(d)

$$\tan \beta L = \tanh \beta L$$

(b) $dy_1/dx_1 = 0$ at $x_1 = 0$:

$$\frac{dy_1(0)}{dx_1} = A_1\beta + C_1\beta = 0 \qquad (3)$$

or

$$C_1 = -A_1 \qquad (4)$$

(c) $y_1 = 0$ at $x_1 = L$:

$$y_1(L) = A_1(\sin \beta L - \sinh \beta x) + B_1(\cos \beta L - \cosh \beta L) = 0 \qquad (5)$$

therefore

$$B_1 = -A_1 \frac{(\sin \beta L - \sinh \beta L)}{\cos \beta L - \cosh \beta L} \qquad (6)$$

Similarly, applying the boundary conditions to span 2

(a) $y_2 = 0$ at $x_2 = 0$:

$$y_2(0) = B_2 + D_2 = 0 \qquad (7)$$

or

$$D_2 = -B_2 \qquad (8)$$

(b) $dy_2/dx_2 = 0$ at $x_2 = 0$:

$$\frac{dy_2(0)}{dx_2} = A_2\beta + C_2\beta = 0 \qquad (9)$$

or

$$C_2 = -A_2 \tag{10}$$

(c) $y_2 = 0$ at $x_2 = L$:

$$y_2(L) = A_2(\sin \beta L - \sinh \beta x) + B_2(\cos \beta L - \cosh \beta L) = 0 \tag{11}$$

or

$$B_2 = -A_2 \frac{(\sin \beta L - \sinh \beta L)}{\cos \beta L - \cosh \beta L} \tag{12}$$

At the interior support, point O, $x_1 = x_2 = L$, the condition for slope of the elastic curve is

$$\frac{dy_1(L)}{dx_1} = -\frac{dy_2(L)}{dx_2} \tag{13}$$

Expansion of Eq. (13) gives

$$A_1(\cos \beta L - \cosh \beta L) + B_1(-\sin \beta L - \sinh \beta x)$$
$$= A_2(\cos \beta L - \cosh \beta L) - B_2(-\sin \beta L - \sinh \beta x) \tag{14}$$

Also at point O, the bending moments in each span must be equal. Thus,

$$\frac{d^2 y_1(L)}{dx_1^2} = -\frac{d^2 y_2(L)}{dx_2^2} \tag{15}$$

which results in

$$A_1(-\sin \beta L - \sinh \beta x) + B_1(-\cos \beta L - \cosh \beta L)$$
$$= -A_2(-\sin \beta L - \sinh \beta x) + B_2(-\cos \beta L - \cosh \beta L) \tag{16}$$

Substituting the expressions for B_1 and B_2 given by Eqs. (6) and (12), respectively, into Eqs. (14) and (16) yields

$$(2A_1 + 2A_2)(1 - \cos \beta L \cosh \beta L) = 0 \tag{17}$$

$$(2A_1 - 2A_2)(-\sin \beta L \cosh \beta L + \cos \beta L \sin \beta x) = 0 \tag{18}$$

From Eqs. (17) and (18), the frequency equations are found as

$$\cos \beta L \cosh \beta L = 1 \tag{19}$$

and

$$\tan \beta L = \tanh \beta L \tag{20}$$

The displacement configuration corresponding to Eqs. (19) and (20) are illustrated in Figure 14.9c and d, respectively. Also note that these frequency equations correspond to the single-span case of fixed-fixed [Eq. (19)] and fixed-pinned [Eq. (20)], as illustrated in Table 14.1. ▲

14.3.1 Rotary Inertia and Shear Effects

In the foregoing discussion of transverse beam vibrations, the inertial properties of the beam section in rotation and beam deformation due to shear were ignored. When the effects of shear deformation and rotary inertia are considered, the beam is often referred to as a *Timoshenko beam*. These additional considerations may be significant in the analysis of high-frequency vibration of beams and in the analysis of deep beams (i.e., beams exhibiting large depth to length ratios).

As illustrated in Figure 14.10, as a beam undergoes transverse vibration, a typical differential beam element will rotate in addition to its translational motion. The angle of rotation is equal to the slope of the elastic curve and is expressed by $\partial y/\partial x$. The corresponding angular acceleration is then given by

$$\frac{\partial^3 y}{\partial x\, \partial t^2}$$

Therefore, the inertial moment of the differential element about an axis passing through its mass center and perpendicular to the *x-y* plane is

$$-\rho I\, \frac{\partial^3 y}{\partial x\, \partial t^2}\, dx$$

where I is the static moment of inertia of the cross section. Equation (14.29b) can thus be modified to account for rotary inertia by including the above term, so that the beam equation becomes

$$EI\, \frac{\partial^4 y}{\partial x^4} + \rho A\, \frac{\partial^2 y}{\partial t^2} - \rho I\, \frac{\partial^4 y}{\partial x^2\, \partial t^2} = 0 \tag{14.46}$$

When deformation due to shear effects is accounted for, an even more accurate differential equation of motion can be formulated. The slope of the deflection curve will therefore depend upon shearing deformations as well as on rotation of the cross section. Let θ represent the slope of the deflection curve where shear deformation is neglected and ψ is the angle of shear at the neutral axis. The slope of the deflection curve due to both shear force V and bending moment M is given by

$$\frac{dy}{dx} = \theta + \psi \tag{14.47}$$

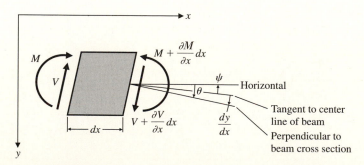

Figure 14.10

Transverse vibration of a beam considering rotary inertia and shear deformation.

in which

$$\psi = \frac{V}{kAG} \tag{14.48}$$

where A = cross-sectional area
G = shear modulus
k = a numerical factor related to the shape of the cross section [10]
The differential equation for rotation of an element will thus become

$$V\,dx - \frac{\partial M}{\partial x}\,dx - \rho I\,\frac{\partial^2 \theta}{\partial t^2}\,dx = 0 \tag{14.49}$$

where

$$M = -EI\,\frac{d\theta}{dx} \tag{14.50}$$

and

$$V = k\psi AG = k\left(\frac{dy}{dx} - \theta\right)AG \tag{14.51}$$

Substituting Eqs. (14.50) and (14.51) into Eq. (14.49) yields

$$EI\,\frac{\partial^2 \theta}{\partial x^2} + k\left(\frac{\partial y}{\partial x} - \theta\right)AG - \rho I\,\frac{\partial^2 \theta}{\partial t^2} = 0 \tag{14.52}$$

For translational motion in the vertical direction, the differential equation is still given by

$$\frac{\partial V}{\partial x} - \rho A\,\frac{\partial^2 \theta}{\partial t^2} = 0 \tag{14.53}$$

Substitution of Eq. (14.51) into Eq. (14.53) gives

$$k\left(\frac{\partial^2 y}{\partial x^2} - \frac{\partial \theta}{\partial x}\right)G - \rho\,\frac{\partial^2 y}{\partial t^2} = 0 \tag{14.54}$$

Eliminating θ from Eqs. (14.52) and (14.54) and combining with Eq. (14.29b) results in the following more comprehensive differential equation for the transverse vibration of beams:

$$EI\,\frac{\partial^4 y}{\partial x^4} + \rho A\,\frac{\partial^2 y}{\partial t^2} - \rho I\left(1 + \frac{E}{kG}\right)\frac{\partial^4 y}{\partial x^2\,\partial t^2} + \frac{\rho^2 I}{kG}\,\frac{\partial^4 y}{\partial t^4} = 0 \tag{14.55}$$

For most practical situations, the increased accuracy obtained by including shear and rotation effects is much less than the modeling errors. The correction for shear stress is generally less than the correction for rotation, but both effects can generally be ignored for shallow (thin) beams.

14.3.2 The Effect of Axial Loading

In practice, situations may arise in which a beam undergoing transverse vibrations is also subjected to an axial load. If the magnitude of the axial load is substantial, it will have a significant effect on the natural frequencies of the beam.

Consider the free-body diagram of the differential element dx of a long beam experiencing transverse vibration as shown in Figure 14.11. The tension T is assumed constant for small deflections, and the effects of shear deformation and rotary inertia are ignored. The forces involved are (1) that of the vibrating string in Eq. (14.19) and (2) that of the vibrating beam in Eq. (14.26). Assume the beam is of uniform cross-sectional area A, flexural rigidity EI, and mass density ρ.

From Newton's second law, the equation of motion for transverse vibration is given as

$$\frac{\partial V}{\partial x}dx + T\frac{\partial^2 y}{\partial x^2}dx = \rho A\,dx\frac{\partial^2 y}{\partial t^2} \tag{14.56}$$

Applying the previously defined relations for deflection, shear, and bending moment to Eq. (14.56) results in

$$\rho A\frac{\partial^2 y}{\partial t^2} + EI\,\frac{\partial^4 y}{\partial x^4} - T\frac{\partial^2 y}{\partial x^2} = 0 \tag{14.57}$$

The solution of Eq. (14.57) may be obtained by the method of separation of variables, represented as

$$y(x, t) = f(x)g(t) \tag{14.58}$$

and from which

$$\frac{\partial^2 y}{\partial t^2} = \frac{\partial^2 g(t)}{\partial t^2}f(x) \tag{14.59}$$

$$\frac{\partial^2 y}{\partial x^2} = \frac{\partial^2 f(x)}{\partial x^2}g(t) \tag{14.60}$$

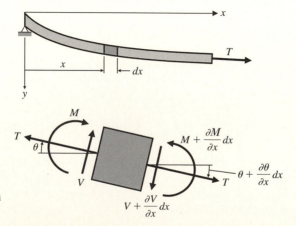

Figure 14.11

Transverse vibration of a beam
with axial tension.

and

$$\frac{\partial^4 y}{\partial x^4} = \frac{\partial^4 f(x)}{\partial x^4} g(t)$$

(14.61)

The separate ordinary differential equations are

$$\frac{\partial^2 g(t)}{\partial t^2} + \omega_n^2 g(t) = 0$$

(14.62)

and

$$EI \frac{\partial^4 f(x)}{\partial x^4} - T \frac{\partial^2 f(x)}{\partial x^2} - \rho A \omega_n^2 f(x) = 0$$

(14.63)

As previously demonstrated, the time-dependent solution is given by

$$g(t) = A \sin \omega t + B \cos \omega t$$

(14.64)

For the time-independent solution, assume

$$f(x) = Ce^{sx}$$

(14.65)

which results in the characteristic equation

$$EIs^4 - Ts^2 - \rho A \omega_n^2 = 0$$

(14.66)

The quadratic roots of Eq. (14.66) are

$$s_{1,2}^2 = \frac{T}{2EI} \left[1 \pm \left(1 + 4\rho A \frac{\omega_n^2 EI}{T^2} \right)^{1/2} \right]$$

(14.67)

Let $s_1^2 = \lambda^2$ and $s_2^2 = -\beta^2$ (that is, $s_2 = \pm i\beta$) and the time-independent solution is given by

$$y(x) = C_1 \sinh \lambda_1 x + C_2 \cosh \lambda_1 x + C_3 \sin \beta x + C_4 \cos \beta x$$

(14.68)

where the constants C_1, C_2, C_3, and C_4 are evaluated by satisfying the appropriate boundary conditions.

For a simply supported beam of length L, the boundary conditions are

$$y(x)|_{x=0} = y(x)|_{x=L} = 0$$

(14.69)

and

$$\frac{d^2 y}{dx^2}\bigg|_{x=0} = \frac{d^2 y}{dx^2}\bigg|_{x=0} = 0$$

(14.70)

Applying these boundary conditions to Eq. (14.68) and its derivatives results in $C_1 = C_2 = C_4 = 0$, and the frequency equation becomes

$$C_3 \sin \beta L = 0$$

(14.71)

or

$$\beta L = n\pi \qquad n = 1, 2, 3, \ldots \tag{14.72}$$

Thus the natural frequency is obtained by equating Eqs. (14.72) and (14.67),

$$\beta_n^2 = \left(\frac{n\pi}{L}\right)^2 = -\frac{T}{2EI}\left[1 - \left(1 + 4\rho A\frac{\omega_n^2 EI}{T^2}\right)^{1/2}\right] \tag{14.73}$$

such that

$$\omega_n = \left(\frac{n\pi}{L}\right)^2 \sqrt{\frac{EI}{\rho A}} + \frac{n\pi}{L}\sqrt{\frac{T}{\rho A}} \tag{14.74}$$

or

$$\omega_n = (\beta_n L)^2 \sqrt{\frac{EI}{\rho A L^4}} + \frac{n\pi}{L}\sqrt{\frac{T}{\rho A}} \tag{14.75}$$

The natural frequency expressed in the form given by Eq. (14.75) extends its application to single-span beams with different boundary conditions by selecting the appropriate value of $\beta_n L$ from Table 14.1. From Eq. (14.75), note that if $T = 0$, the natural frequency is that of a single-span beam as given by Eq. (15) of Example 14.4. If $EI = 0$, the problem degenerates to that of a flexible taut cable discussed in Section 14.2. The tensile axial load "stiffens" the beam and thereby increases its natural frequencies. Conversely, if the applied axial load is compressive, the net effect is that of "stiffness degradation," which will decrease the natural frequencies of the beam. Therefore, in the event that the applied axial load is compressive, the second term on the right-hand side of either Eq. (14.74) or Eq. (14.75) should be subtracted (made negative).

14.4 ORTHOGONALITY OF NORMAL MODES

Like discrete systems, systems having distributed mass and elasticity exhibit the same orthogonality properties of normal vibration modes. To illustrate the concept, consider the normal mode vibration of a single-span beam of length L with arbitrary boundary conditions. The system natural frequencies and normal vibration modes are governed by Eq. (14.36), which may be expressed as

$$\frac{d}{dx^2}\left[EI\frac{d^2 f(x)}{dx^2}\right] - \omega^2 \rho A f(x) = 0 \tag{14.76}$$

or

$$[EIf''(x)]'' - \omega^2 \rho A f(x) = 0 \tag{14.77}$$

Equation (14.77) represents the frequency equation for a continuous system. Solution of the equation yields the natural frequencies and normal vibration modes for the system. Thus, for the rth mode, Eq. (14.77) may be written as

$$[EI\phi_r''(x)]'' = \omega_r^2 \rho A \phi_r(x) \tag{14.78}$$

where $\phi_r(x)$ represents the normal vibration mode shape for the rth mode.

To derive the orthogonality property, multiply both sides of Eq. (14.78) by $\phi_s(x)$, where $\phi_s(x) \neq \phi_r(x)$, and integrate from 0 to L, resulting in

$$\int_0^L [EI\phi_r''(x)]'' \phi_s(x) \, dx = \omega_r^2 \int_0^L \rho A \phi_r(x)\phi_s(x) \, dx \tag{14.79}$$

Integrating the left-hand side of Eq. (14.79) by parts and applying the procedure a second time yields

$$\int_0^L [EI\phi_r''(x)]'' \phi_s(x) \, dx = \{[EI\phi_r''(x)]'\phi_s(x)\}_0^L - \{[EI\phi_r''(x)]\phi_s'(x)\}_0^L$$

$$+ \int_0^L EI\phi_r''(x)\phi_s''(x) \, dx \tag{14.80}$$

If only those beams having some combination of fixed, hinged, or free-boundary conditions are considered, then the first two terms on the right-hand side of Eq. (14.80) vanish. Then, substituting Eq. (14.80) into (14.79) gives

$$\int_0^L EI\phi_r''(x)\phi_s''(x) \, dx = \omega_r^2 \int_0^L \rho A \phi_r(x)\phi_s(x) \, dx \tag{14.81}$$

Beginning again with Eq. (14.79) written for the sth mode, multiplying by $\phi_r(x)$, and integrating from 0 to L, the result is

$$\int_0^L EI\phi_r''(x)\phi_s''(x) \, dx = \omega_s^2 \int_0^L \rho A \phi_r(x)\phi_s(x) \, dx \tag{14.82}$$

Subtracting Eq. (14.82) from Eq. (14.81) yields

$$(\omega_r^2 - \omega_s^2)\int_0^L \rho A \phi_r(x)\phi_s(x) \, dx = 0 \tag{14.83}$$

Therefore, for two vibration modes r and s where $\omega_r \neq \omega_s$, the orthogonality property with respect to the mass distribution is

$$\int_0^L \rho A \phi_r(x)\phi_s(x) \, dx = 0 \tag{14.84}$$

Substitution of Eq. (14.84) into Eq. (14.79) gives the orthogonality property with respect to the stiffness distribution as

$$\int_0^L [EI\phi_r''(x)]'' \phi_s(x) \, dx = 0 \tag{14.85}$$

14.5 UNDAMPED FORCED VIBRATION OF BEAMS BY MODE SUPERPOSITION

In Chapter 12 the equations of motion for discrete MDOF systems were decoupled through the modal transformation to obtain the forced vibration solution in terms of the principal or normal coordinates of the system. A similar technique may be applied to continuous systems by expanding the displacement function $y(x, t)$ in terms of the normal modes of the system.

Consider the general motion of a non-prismatic beam subjected to an arbitrary distributed force $F(x, t)$ as shown in Figure 14.12. The equation of motion for the system is given by

$$[EI(x)y''(x, t)]'' + \rho A(x)\ddot{y}(x, t) = F(x, t) \tag{14.86}$$

The normal vibration modes for the beam $\phi_r(x)$ must satisfy Eq. (14.77), represented as

$$[EI(x)\phi_r''(x)]'' - \omega_r^2 \rho A(x)\phi_r(x) = 0 \tag{14.87}$$

and the boundary conditions. The normal modes $\phi_r(x)$ are orthogonal functions that must satisfy the mass orthogonality relationship

$$\int_0^L \rho A(x)\phi_r(x)\phi_s(x)\, dx = \begin{cases} 0 & \text{for } r \neq s \\ M_r & \text{for } r = s \end{cases} \tag{14.88}$$

where M_r is the generalized (modal) mass for the continuous system for the rth mode. Then by representing the general solution in terms of the normal modes $\phi_r(x)$,

$$y(x, t) = \sum_{r=1}^{\infty} \phi_r(x)q_r(t) \tag{14.89}$$

where the normal coordinates $q_r(t)$ can be determined from Lagrange's equations, discussed in Section 12.1.

Establishing the kinetic energy, in recognition of the orthogonality relation given by Eq. (14.88), gives

$$T = \frac{1}{2}\int_0^L \dot{y}^2(x, t)\rho A(x)\, dx = \frac{1}{2}\sum_{r=1}^{\infty}\sum_{s=1}^{\infty} \dot{q}_r \dot{q}_s \int_0^L \rho A(x)\phi_r(x)\phi_s(x)\, dx$$

$$= \frac{1}{2}\sum_{r=1}^{\infty} M_r \dot{q}_r^2 \tag{14.90}$$

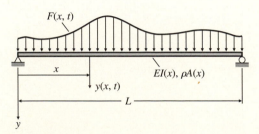

Figure 14.12

Uniform beam subject to arbitrary load $F(x, t)$.

where the generalized (modal) mass is given by

$$M_r = \int_0^L \rho A(x) \phi_r^2(x) \, dx \tag{14.91}$$

In a similar manner, the potential energy is determined by

$$V = \int_0^L EI(x)[y''(x, t)]^2 \, dx = \frac{1}{2} \sum_{r=1}^{\infty} \sum_{s=1}^{\infty} q_r q_s \int_0^L EI(x) \phi_r''(x) \phi_s''(x) \, dx \tag{14.92}$$

$$= \frac{1}{2} \sum_{r=1}^{\infty} K_r q_r^2 = \frac{1}{2} \sum_{r=1}^{\infty} \omega_r^2 M_r q_r^2$$

where K_r is the generalized (modal) stiffness for the rth mode expressed as

$$K_r = \int_0^L EI(x)[\phi_r''(x)]^2 \, dx \tag{14.93}$$

Finally, the generalized (modal) force of the rth mode P_r must be determined from the work done by the applied force $F(x, t)$ acting through the virtual displacement δq_r. Therefore

$$\delta W = \int_0^L F(x, t) \left[\sum_{r=1}^{\infty} \phi_r(x) \delta q_r \right] dx = \sum_{r=1}^{\infty} \delta q_r \int_0^L F(x, t) \phi_r(x) \, dx \tag{14.94}$$

and

$$P_r = \int_0^L F(x, t) \phi_r(x) \, dx \tag{14.95}$$

Substituting the above expressions for T, V, and P_r into Lagrange's equations

$$\frac{d}{dt} \left(\frac{\partial T}{\partial \dot{q}_r} \right) - \frac{\partial T}{\partial q_r} + \frac{\partial V}{\partial q_r} = P_r \tag{14.96}$$

the differential equation of motion in normal coordinates is

$$\ddot{q}_r + \omega_r^2 q_r = \frac{1}{M} \int_0^L F(x, t) \phi_r(x) \, dx \tag{14.97}$$

Equation (14.97) represents the uncoupled equations of motion for $r = 1, 2, 3, \ldots, \infty$.

As a convenience, the loading $F(x, t)$ can be considered separable in the form

$$F(x, t) = \frac{F_0}{L} f(x) g(t) \tag{14.98}$$

where $f(x)$ is the position function, F_0 is the maximum force amplitude and $g(t)$ is the time function. Equation (14.97) may then be expressed as

$$\ddot{q}_r + \omega_r^2 q_r = \frac{F_0}{M_r} \Gamma_r g(t) \tag{14.99}$$

where Γ_r is the *modal participation factor* for the rth mode defined as

$$\Gamma_r = \frac{1}{L}\int_0^L f(x)\phi_r(x)\,dx \tag{14.100}$$

The complete solution to Eq. (14.99) may be expressed for the rth mode by the Duhamel integral expression

$$q_r(t) = q_r(0)\cos\omega_r t + \frac{1}{\omega_r}\dot{q}_r(0)\sin\omega_r t$$

$$+ \left(\frac{F_0\Gamma_r}{M_r\omega_r^2}\right)\omega_r\int_0^L g(\tau)\sin\omega_r(t-\tau)\,d\tau \tag{14.101}$$

The term $F_0\Gamma_r/M_r\omega_r^2$ in Eq. (14.101) represents the *equivalent static deflection* for the rth mode. Therefore the response ratio $R_r(t)$ (refer to Chapter 6 for a discussion) for the rth mode is given by

$$R_r(t) = \omega_r\int_0^L g(\tau)\sin\omega_r(t-\tau)\,d\tau \tag{14.102}$$

The solution in physical coordinates $y(x, t)$ is then obtained from Eq. (14.89).

EXAMPLE 14.6 ▼

The simply supported uniform beam shown in Figure 14.13a is subjected to a suddenly applied distributed force $F(x, t)$ whose time function is described in Figure 14.13b. Determine the expression for the response of the beam. Assume the beam has cross sectional area A, flexural rigidity EI, mass density ρ, and zero initial conditions.

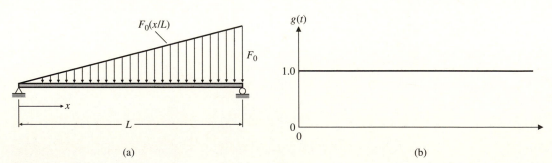

(a) (b)

Figure 14.13
Uniform beam subject to suddenly applied distributed loads: (a) load distribution; (b) time function for loading.

Solution

The eigenfunction $\phi_r(x)$ for a simply supported beam is given by (see Example 14.4)

$$\phi_r(x) = C_r \sin \frac{r\pi x}{L} \qquad r = 1, 2, 3, \ldots \tag{1}$$

where C_r is an arbitrary constant for the rth mode. The natural frequencies for the system are given by Eq. (15) of Example 14.4 and Table 14.1 as

$$\omega_r^2 = \frac{(r\pi)^4 EI}{\rho A L^4} \qquad r = 1, 2, 3, \ldots \tag{2}$$

The generalized (modal) mass for the rth mode determined from Eq. (14.91) is

$$M_r = \rho A \int_0^L \phi_r^2(x)\, dx = \rho A C_r^2 \frac{L}{2} \tag{3}$$

Since the constant C_r is arbitrary, for convenience we let $C_r = \sqrt{2}$. Thus the normal mode eigenfunction and the modal mass, respectively, become

$$\phi_r(x) = \sqrt{2} \sin \frac{r\pi x}{L} \tag{4}$$

$$M_r = \rho A L \tag{5}$$

The distributed force $F(x, t)$ is given by Eq. (14.98) as

$$F(x, t) = \frac{F_0}{L} f(x) g(t) = \frac{F_0 x}{L} g(t) \tag{6}$$

where $g(t)$ is the load time history. Substitution of Eq. (6) into Eq. (14.95) gives the generalized force

$$
\begin{aligned}
P_r &= F_0 g(t) \int_0^L \frac{x}{L} \sqrt{2} \sin \frac{r\pi x}{L}\, dx \\
&= g(t) \frac{F_0 \sqrt{2}}{L} \left[\frac{\sin(r\pi x / L)}{(r\pi x / L)^2} - \frac{x \cos(r\pi x / L)}{(r\pi x / L)} \right]_0^L \\
&= -g(t) \frac{F_0 \sqrt{2}\, L}{r\pi} \cos r\pi = (-1)^{r+1} g(t) \frac{F_0 \sqrt{2}\, L}{r\pi}
\end{aligned}
\tag{7}
$$

The uncoupled equations of motion in normal coordinates are given by Eq. (14.99) as

$$\ddot{q}_r + \omega_r^2 q_r = (-1)^{r+1} g(t) \frac{F_0 \sqrt{2}\, L}{r\pi \rho A L} \tag{8}$$

The solution to Eq. (8) is given by Eq. (14.101). For zero initial conditions the solution becomes

$$q_r(t) = \frac{(-1)^{r+1}\sqrt{2}}{\rho A r \pi \omega_r} \int_0^L g(\tau) \sin \omega_r(t - \tau) \, d\tau \tag{9}$$

The Duhamel integral solution of Eq. (9) is given by Eq. (6-33),

$$q_r(t) = (-1)^{r+1} \frac{F_0 \sqrt{2}}{\rho A r \pi \omega_r^2}(1 - \cos \omega_r t) \qquad r = 1, 2, 3, \ldots \tag{10}$$

and the response of the beam in physical coordinates is given by Eq. (14.89), resulting in

$$y(x, t) = \sum_{r=1}^{\infty} \phi_r(x) q_r(t) = \sum_{r=1}^{\infty} \sqrt{2} \sin \frac{r\pi x}{L} q_r(t) \tag{11}$$

▲

EXAMPLE 14.7 ▼

The simply supported uniform beam shown in Figure 14.14 is subjected to a concentrated harmonic force $F_0 \sin \Omega t$. Determine the expression for the response of the beam. Assume zero initial conditions.

Solution

For this case, the simple beam eigenfunction and natural frequency, respectively, for the rth mode are

$$\phi_r(x) = \sin \frac{r\pi x}{L} \tag{1}$$

$$\omega_r^2 = \frac{(r\pi)^4 EI}{\rho A L^4} \tag{2}$$

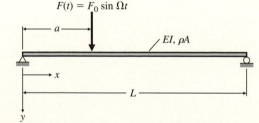

Figure 14.14

Simply supported beam subject to concentrated harmonic forces.

The equation of motion in normal coordinates for the rth mode given by Eq. (14.97) can be expressed as

$$M_r \ddot{q}_r + K_r q_r = P_r(t) \tag{3}$$

where M_r is determined from Eq. (14.91),

$$M_r = \int_0^L \rho A \, \phi_r^2(x) \, dx = \int_0^L \rho A \left(\sin \frac{r \pi x}{L} \right)^2 dx = \frac{\rho A L}{2} \tag{4}$$

and K_r is determined from Eq. (14.93),

$$K_r = \int_0^L EI(x)[\phi_r''(x)]^2 \, dx = EI \left(\frac{r\pi}{L} \right)^4 \int_0^L \left(-\sin \frac{r \pi x}{L} \right)^2 dx$$

$$= \frac{EI(r\pi)^4}{2L^3} \tag{5}$$

The concentrated force acting at $x = a$ can be expressed as $F(t)\delta(x - a)$, where $\delta(x - a)$ is the Dirac delta function (see Section 6.2) at $x = a$. Thus, the generalized (modal) force for the rth mode determined from Eq. (14.95) is

$$P_r = \int_0^L (F_0 \sin \Omega t)\delta(x - a)\phi_r(x) \, dx = \phi_r(a)F_0 \sin \Omega t$$

$$= \left(\sin \frac{r \pi a}{L} \right) F_0 \sin \Omega t \tag{6}$$

Then the (modal) equation of motion for the rth mode, Eq. (3), becomes

$$\frac{\rho A L}{2} \ddot{q}_r + \frac{EI(r\pi)^4}{2L^3} q_r = \left(\sin \frac{r \pi a}{L} \right) F_0 \sin \Omega t \tag{7}$$

or

$$\ddot{q}_r + \omega_r^2 q_r = \frac{2}{\rho A L} \left(\sin \frac{r \pi a}{L} \right) F_0 \sin \Omega t \tag{8}$$

where ω_r^2 is given by Eq. (1). The solution to Eq. (8) for zero initial conditions is given by

$$q_r(t) = \frac{F_0 \sin (r\pi a / L)}{(\rho A L / 2)\omega_r} \int_0^t \sin \Omega \tau \sin \omega_r(t - \tau) \, d\tau \tag{9}$$

or

$$q_r(t) = \frac{2F_0 \sin (r\pi a / L)}{\rho A L \omega_r} \left(\frac{\omega_r \sin \Omega t - \Omega \sin \omega_r t}{\omega_r^2 - \Omega^2} \right) \tag{10}$$

and the steady-state solution in physical coordinates, $y(x, t)$, is determined from Eq. (14.89) as

$$y(x, t) = \frac{2F_0L^3}{\pi^4 EI} \sum_{r=1}^{\infty} \left[\frac{\sin{(r\pi a/L)}}{r^4 - (\Omega/\omega_r)^2} \right]$$

$$\sin\frac{r\pi x}{L}\left[\sin{\Omega t} - \left(\frac{\Omega}{r^2\omega_1}\right)\sin{r^2\omega_1 t} \right] \qquad (11)$$

where $\omega_1 = \pi^4 EI/\rho AL^4$. ▲

EXAMPLE 14.8 ▼

A simply supported uniform bridge girder is subjected to a moving load having constant velocity v as shown in Figure 14.15. Determine the expression for the deflection of the bridge girder $y(x, t)$. Assume zero initial conditions.

Solution

The concentrated force F_0 is applied at $x = a = vt$ in the interval $0 \leq vt \leq L$. Thus the eigenfunction for the rth mode is expressed as

$$\phi_r(x) = \phi_r(vt) = \sin\left(\frac{r\pi vt}{L}\right) \qquad (1)$$

and the natural frequency for the rth mode is given by

$$\omega_r^2 = \frac{(r\pi)^4 EI}{\rho AL^4} \qquad (2)$$

From Example 14.7, the generalized (modal) mass for the rth mode is determined as $M_r = \rho AL/2$. The generalized (modal) force determined from Eq. (14.95) is given by

$$P_r(t) = \int_0^L [F_0\delta(x - vt)]\phi_r(vt)\,dx$$

$$= \begin{cases} F_0\phi_r(vt) & \text{for } 0 \leq t \leq L/v \\ 0 & \text{for } t > L/v \end{cases} \qquad (3)$$

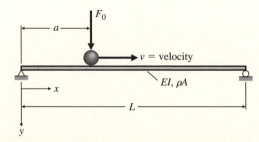

Figure 14.15

Simply supported bridge girder subject to moving concentrated load.

where $\delta(x - vt)$ is the Dirac delta function (refer to Section 6.2). The equation of motion in normal coordinates for the rth mode is then expressed as

$$\ddot{q}_r + \omega_r^2 q_r = \frac{P_r(t)}{M_r} \tag{4}$$

The Duhamel integral solution to Eq. (4) for zero initial conditions is given by

$$P_r(t) = \frac{1}{M_r \omega_r} \int_0^L \left(F_0 \sin \frac{r\pi v \tau}{L} \right) \sin \omega_r(t - \tau) \, d\tau$$

$$= \frac{F_0}{M_r \omega_r} \left[\frac{1}{(r\pi v/L)^2 - \omega_r^2} \right] \left(\frac{r\pi v}{L} \sin \omega_r t - \omega_r \sin \frac{r\pi v}{L} \right) \tag{5}$$

Finally, the solution $y(x, t)$ in physical coordinates given by Eq. (14.89) is

$$y(x, t) = \sum_{r=1}^{\infty} \frac{2F_0}{\rho A L \omega_r} \left[\frac{1}{(r\pi v/L)^2 - \omega_r^2} \right] \left(\frac{r\pi v}{L} \sin \omega_r t - \omega_r \sin \frac{r\pi v}{L} \right) \tag{6}$$

Equation (6) is valid for the time interval $0 \le t \le L/v$. For $t > L/v$, the girder is in free vibration with initial conditions $q_r(L/v)$ and $\dot{q}_r(L/v)$, and the free vibration or transient response for the rth mode in normal coordinates can be evaluated from the expression

$$q_r(t) = q_r\left(\frac{L}{v}\right) \cos \omega_r t + \frac{1}{\omega_r} \dot{q}_r\left(\frac{L}{v}\right) \sin \omega_r t \tag{7}$$

The transient response in physical coordinates is again obtained from Eq. (14.89). ▲

14.6 APPROXIMATE METHODS

The governing equations and boundary conditions for the continuous systems discussed to this point have been simple enough that an exact analytical (closed form) solution could be obtained. Although the solutions appear in the form of infinite series of normal modes, each such normal mode is an exact solution. Unfortunately, in most practical applications, exact solutions for the vibration of continuous systems cannot be attained, and approximate methods must be employed. Generally, an approximate analysis must be conducted for continuous systems that support discrete masses at nonboundary positions and for many forced vibration problems. This section discusses several commonly employed approximate analysis techniques.

14.6.1 Rayleigh Method

The Rayleigh method has been previously discussed in Section 3.4 for its application to SDOF systems and in Section 10.8 for its application to discrete MDOF systems. The Rayleigh method is the basis for the majority of approximate techniques used in vibration analysis. The Rayleigh method may also be employed to approximate the fundamental frequency of continuous systems.

First the application of the Rayleigh method to continuous beams is investigated. To generalize the formulation, consider the transverse vibrations of a single-span beam of nonuniform cross section as shown in Figure 14.16. For this system, both the cross-sectional area $A(x)$ and the flexural rigidity $EI(x)$ are functions of the position coordinate x. If ρ is assumed to be the material mass density, the mass per unit length $\rho A(x)$ is also a function of the position coordinate x.

Let the transverse deflection of the beam $y(x, t)$ be approximated by

$$y(x, t) = \psi(x)z(t) \tag{14.103}$$

where $\psi(x)$ is an assumed deflected shape and $z(t)$ is the unknown generalized coordinate selected at a convenient location on the structure. The strain energy in the beam due to flexural deformations is given by

$$V = \frac{1}{2}z^2 \int_0^L EI(x)[\psi''(x)]^2 \, dx \tag{14.104}$$

and the kinetic energy in the system is determined by

$$T = \frac{1}{2}\dot{z}^2 \int_0^L \rho A(x)[\psi(x)]^2 \, dx \tag{14.105}$$

Assuming that the free vibrations are harmonic in nature, the displacement $z(t)$ can be expressed by

$$z(t) = C \sin \omega t \tag{14.106}$$

where C is the amplitude of vibration at the generalized coordinate location. Then, substituting Eq. (14.106) into Eqs. (14.104) and (14.105) gives expressions for the maximum potential energy (represented entirely by the strain energy of the beam) and kinetic energy, respectively:

$$V_{\text{max}} = \frac{1}{2}C^2 \int_0^L EI(x)[\psi''(x)]^2 \, dx \tag{14.107}$$

and

$$T_{\text{max}} = \frac{1}{2}C^2\omega^2 \int_0^L \rho A(x)[\psi(x)]^2 \, dx \tag{14.108}$$

From the conservation of energy,

$$T_{\text{max}} = V_{\text{max}} \tag{14.109}$$

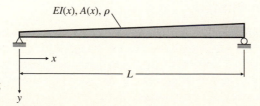

EI(x), A(x), ρ

Figure 14.16
Simply supported beam having nonuniform cross section.

x

L

y

from which the Rayleigh quotient λ_R is obtained, given by

$$\lambda_R = \omega_R^2 = \frac{\displaystyle\int_0^L EI(x)[\psi''(x)]^2 \, dx}{\displaystyle\int_0^L \rho A(x)[\psi(x)]^2 \, dx} \tag{14.110}$$

where ω_R is the Rayleigh frequency.

Obviously, the more closely the assumed shape function $\psi(x)$ approximates the exact vibration mode, the more accurate will be the estimate of the fundamental frequency. As a minimum requirement, the assumed shape function should satisfy the prescribed geometric boundary conditions for the problem and also be differentiable at least to the order appearing in the strain energy expression.

EXAMPLE 14.9   ▼

Determine the fundamental frequency of the uniform beam shown in Figure 14.17 by the Rayleigh method. Assume a shape function consistent with the deflected shape due to a uniformly distributed static load. Also assume flexural rigidity EI, cross-sectional area A, and mass density ρ.

Solution

The shape function $\psi(x)$ is given by

$$\psi(x) = \frac{1}{48EI}(3Lx^3 - 2x^4 - L^3x) \tag{1}$$

and

$$\psi''(x) = \frac{1}{8EI}(3Lx - 4x^2) \tag{2}$$

Substituting Eqs. (1) and (2) into Eq. (14.110) yields

$$\omega_R^2 = \frac{(1/8EI)^2 \displaystyle\int_0^L (3Lx - 4x^2)^2 \, dx}{(1/48EI)^2 \displaystyle\int_0^L (3Lx^3 - 2x^4 - L^3x)^2 \, dx} \tag{3}$$

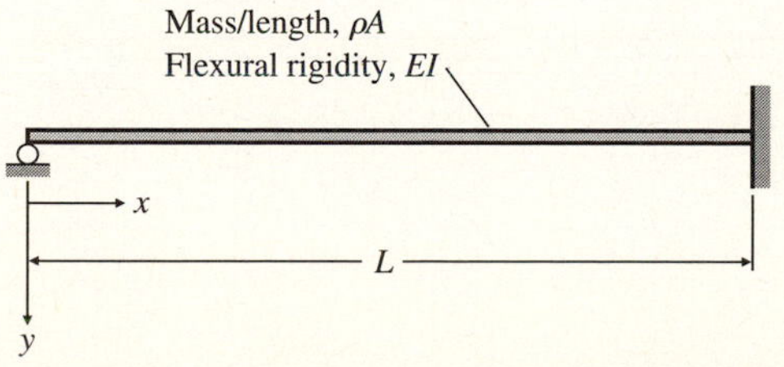

Mass/length, ρA
Flexural rigidity, EI

x

L

y

Figure 14.17

Uniform cantilever beam of Example 14.9.

Integrating Eq. (3) and simplifying results in a Rayleigh frequency given by

$$\omega_R = 15.45 \sqrt{\frac{EI}{\rho A L^4}} \tag{4}$$

The exact frequency is given by Eq. (15) in Example 14.4 as

$$\omega_1 = (\beta_1 L)^2 \sqrt{\frac{EI}{\rho A L^4}} \tag{5}$$

From Table 14.1, $\beta_1 L = 3.927$, thus yielding

$$\omega_1 = 15.42 \sqrt{\frac{EI}{\rho A L^4}} \tag{6}$$

The relative error between the Rayleigh approximation and the exact value of the fundamental frequency is only $+0.19\%$. ▲

EXAMPLE 14.10 ▼

A uniform cantilever beam of length L and having flexural rigidity EI, cross-sectional area A, and mass density ρ has a concentrated mass m at its free end that is pinned to a spring of stiffness k as shown in Figure 14.18. Determine the fundamental frequency by the Rayleigh method. Assume a shape function $\psi(x) = (x/L)^2$.

Solution

Noting that

$$\psi(x) = \left(\frac{x}{L}\right)^2 \tag{1}$$

then

$$\psi(L) = 1.0 \tag{2}$$

and

$$\psi''(x) = \frac{2}{L^2} \tag{3}$$

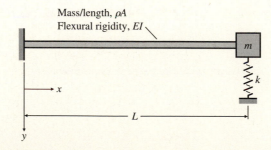

Mass/length, ρA
Flexural rigidity, EI

m

k

x

L

y

Figure 14.18
Uniform cantilever beam having concentrated tip mass and elastic support.

Substituting Eqs. (1) through (3) into the Rayleigh quotient given by Eq. (14.110) yields

$$\omega_R^2 = \frac{EI \int_0^L (2/L^3)^2 \, dx + \frac{1}{2} k (1.0)^2}{\rho A \int_0^L \left[(x/L)^2 \right] dx + m (1.0)^2} \tag{4}$$

Integrating Eq. (4) and simplifying results in a Rayleigh frequency of

$$\omega_R = \sqrt{\frac{(4EI/L^3) + (k/2)}{(\rho AL/5) + m}} \tag{5}$$

Note in Eq. (5) that as k and m approach zero, $\omega_R = 4.472 \sqrt{EI/\rho AL^4}$. For the same condition Table 14.1 yields a value for ω_1 of $3.516 \sqrt{EI/\rho AL^4}$, resulting in a relative error of $+27.2\%$. This indicates that the assumed shape function does not provide an accurate representation of the actual vibration mode. ▲

14.6.2 Rayleigh-Ritz Method

The Rayleigh-Ritz method is an extension of the Rayleigh method with several important improvements. In addition to increased accuracy for estimating the fundamental frequency, the Rayleigh-Ritz method furnishes estimates to higher frequencies and to the associated mode shapes as well.

Like the Rayleigh method, the Rayleigh-Ritz method is formulated on the premise that the deflection of a continuous system can be approximated by

$$y(x, t) = \psi(x) z(t) \tag{14.111}$$

However, in the Rayleigh-Ritz method the assumed deflected shape is specified as a finite series given by

$$\psi(x) = \sum_{i=1}^{n} C_i \gamma_i(x) \tag{14.112}$$

where the C_i's are constants that together with $\gamma_i(x)$ form a set of linearly independent functions satisfying all the boundary conditions. Since the set of functions is limited to a finite number, the analysis can be interpreted as approximating a continuous system as a discrete n-DOF MDOF system. Therefore, the analysis techniques used for MDOF systems can also be applied to continuous systems expressed in this manner.

In the Rayleigh-Ritz method, the Rayleigh quotient is expressed as

$$\lambda_R = \omega_R^2 = \frac{V_{max}}{T_{max}^*} \tag{14.113}$$

where T_{max}^* is called the *reference kinetic energy* and is related to the kinetic energy by

$$T_{max} = \omega^2 T_{max}^* \tag{14.114}$$

and where

$$V_{\max} = \frac{1}{2}\sum_{i=1}^{n}\sum_{j=1}^{n}C_i C_j k_{ij} \tag{14.115}$$

$$T_{\max}^* = \frac{1}{2}\sum_{i=1}^{n}\sum_{j=1}^{n}C_i C_j m_{ij} \tag{14.116}$$

The mass coefficients m_{ij} are defined by

$$m_{ij} = \int m(x)\gamma_i(x)\gamma_j(x)\,dx \tag{14.117}$$

and the stiffness coefficients k_{ij} depend upon the particular type of vibration. For example, for flexural vibrations of beams

$$k_{ij} = \int EI(x)\gamma_i''(x)\gamma_j''(x)\,dx \tag{14.118}$$

and for axial vibrations

$$k_{ij} = \int EA(x)\gamma_i'(x)\gamma_j'(x)\,dx \tag{14.119}$$

The underlying concept of the method is to minimize λ (or ω^2) as defined in Eq. (14.113) with respect to the C_i's by

$$\frac{\partial \lambda_R}{\partial C_i} = 0 \tag{14.120}$$

Thus from Eq. (14.113)

$$\begin{aligned}\frac{\partial \lambda_R}{\partial C_i} &= \frac{\partial}{\partial C_i}\left(\frac{V_{\max}}{T_{\max}^*}\right)\\[2mm] &= \frac{T_{\max}^*(\partial V_{\max}/\partial C_i) - V_{\max}(\partial T_{\max}^*/\partial C_i)}{(T_{\max}^*)^2} = 0\end{aligned} \tag{14.121}$$

To facilitate the differentiation, substitute Eqs. (14.115) and (14.116) into Eq. (14.113) and express the Rayleigh quotient as

$$\lambda_R = \frac{V_{\max}}{T_{\max}^*} = \frac{\displaystyle\sum_{i=1}^{n}\sum_{j=1}^{n}C_i C_j k_{ij}}{\displaystyle\sum_{i=1}^{n}\sum_{j=1}^{n}C_i C_j m_{ij}} \tag{14.122}$$

or in matrix notation,

$$\lambda_R = \frac{\{C\}^T[k_{ij}]\{C\}}{\{C\}^T[m_{ij}]\{C\}} \tag{14.123}$$

Taking the partial derivatives of V_{max} and T^*_{max} with respect to C_i and noting that $k_{ij} = k_{ji}$ and $m_{ij} = m_{ji}$ gives

$$\frac{\partial V_{max}}{\partial C_i} = \frac{\partial}{\partial C_i}(\{C\}^T[k_{ij}]\{C\}) = 2\sum_{j=1}^{n} C_j k_{ij} \tag{14.124}$$

and

$$\frac{\partial T^*_{max}}{\partial C_i} = \frac{\partial}{\partial C_i}(\{C\}^T[m_{ij}]\{C\}) = 2\sum_{j=1}^{n} C_j m_{ij} \tag{14.125}$$

Substituting Eqs. (14.113), (14.124), and (14.125) into Eq. (14.121) and equating the numerator to zero for the minimum results in

$$\sum_{j=1}^{n}(k_{ij} - \omega^2 m_{ij})C_j = 0 \tag{14.126}$$

or

$$[[k] - \omega^2[m]]\{C\} = 0 \tag{14.127}$$

Equation (14.127) represents a set of n homogeneous algebraic equations with the C_j's as the unknowns. For a nontrivial solution the determinant of the coefficient matrix must vanish, or

$$\left|[k] - \omega^2[m]\right| = 0 \tag{14.128}$$

Expansion of Eq. (14.128) results in the frequency equation, which resembles that of a discrete system.

It should be noted that formulation of the mass coefficients m_{ij} prescribed by Eq. (14.117) is similar to or *consistent* with the formulation of the stiffness coefficients prescribed by Eqs. (14.118) and (14.119). A mass matrix formulated in this manner is referred to as a *consistent mass matrix*. The consistent mass matrix is generally not diagonal.

EXAMPLE 14.11 ▼

Determine the first two natural frequencies and corresponding mode shapes for the trapezoidal plate shown in Figure 14.19 by the Rayleigh-Ritz method. The mass and cross-sectional area of the plate per unit length are defined by, respectively,

$$m(x) = m_0\left(1 - \frac{x}{2L}\right)$$

$$A(x) = A_0\left(1 - \frac{x}{2L}\right)$$

Assume a Ritz deflection function, $y(x) = C_1 x^2 + C_2 x^3$. Also assume $m_0 = 0.09$ lb-sec^2/in^2, $A_0 = 120$ in^2, modulus of elasticity $E = 29,000$ ksi, length $L = 10$ ft, and the plate thickness $t = 1$ in.

Figure 14.19

Trapezoidal plate: (a) plan view; (b) side view.

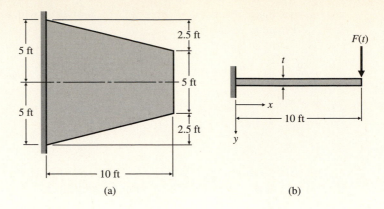

(a)   (b)

Solution

For the specified Ritz deflection function

$$\gamma_1(x) = x^2 \qquad \gamma_2(x) = x^3$$

$$\gamma_1'(x) = 2x \qquad \gamma_2'(x) = 3x^3$$

$$\gamma_1''(x) = 2 \qquad \gamma_2''(x) = 6x$$

The coefficients m_{ij} in the consistent mass matrix are evaluated by Eq. (14.117) as

$$m_{ij} = \int_0^L m(x)\gamma_i(x)\gamma_j(x)\, dx \tag{1}$$

Therefore

$$m_{11} = \int_0^L m_0\left(1 - \frac{x}{2L}\right)(x^2)(x^2)\, dx = m_0\left[\frac{x^5}{5} - \frac{x^6}{12L}\right]_0^L = \frac{7m_0 L^5}{60} \tag{2}$$

$$m_{21} = m_{12} = \int_0^L m_0\left(1 - \frac{x}{2L}\right)(x^2)(x^3)\, dx$$
$$= m_0\left[\frac{x^6}{6} - \frac{x^7}{14L}\right]_0^L = \frac{2m_0 L^6}{21} \tag{3}$$

$$m_{22} = \int_0^L m_0\left(1 - \frac{x}{2L}\right)(x^3)(x^3)\, dx = m_0\left[\frac{x^7}{7} - \frac{x^8}{16L}\right]_0^L = \frac{9m_0 L^7}{112} \tag{4}$$

The coefficients k_{ij} in the stiffness matrix are evaluated by Eq. (14.118) as

$$k_{ij} = \int EI(x)\gamma_i''(x)\gamma_j''(x)\, dx \tag{5}$$

Thus

$$k_{11} = \int_0^L \frac{EA_0 t^2}{12}\left(1 - \frac{x}{2L}\right)(2)(2) \, dx = \frac{EA_0 t^2}{12}\left[4x - \frac{x^2}{L}\right]_0^L$$

$$= \frac{3EA_0 t^2 L}{12} \tag{6}$$

$$k_{21} = k_{12} = \int_0^L \frac{EA_0 t^2}{12}\left(1 - \frac{x}{2L}\right)(2)(6x) \, dx$$

$$= \frac{EA_0 t^2}{12}\left[6x^2 - \frac{2x^3}{L}\right]_0^L = \frac{4EA_0 t^2 L^2}{12} \tag{7}$$

$$k_{22} = \int_0^L \frac{EA_0 t^2}{12}\left(1 - \frac{x}{2L}\right)(6x)(6x) \, dx = \frac{EA_0 t^2}{12}\left[12x^3 - \frac{9x^4}{2L}\right]_0^L$$

$$= \frac{15EA_0 t^2 L^3}{24} \tag{8}$$

Then the mass and stiffness matrices are given by

$$[m] = \begin{bmatrix} \dfrac{7}{60} & \dfrac{2L}{21} \\[2mm] \dfrac{2L}{21} & \dfrac{9L^2}{112} \end{bmatrix} m_0 L^5 \qquad [k] = \begin{bmatrix} \dfrac{3}{12} & \dfrac{4L}{12} \\[2mm] \dfrac{4L}{12} & \dfrac{15L^2}{24} \end{bmatrix} EA_0 t^2 L$$

and the characteristic matrix is determined from Eq. (14.127) as

$$[[k] - \omega^2[m]] = \begin{bmatrix} \dfrac{3}{12}\alpha - \dfrac{7\omega^2}{20} & \dfrac{4L}{12}\alpha - \dfrac{2L\omega^2}{21} \\[3mm] \dfrac{4L}{12}\alpha - \dfrac{2L\omega^2}{21} & \dfrac{15L^2}{24}\alpha - \dfrac{7L^2\omega^2}{112} \end{bmatrix} m_0 L^5 \tag{9}$$

where

$$\alpha = \frac{EA_0 t^2}{m_0 L^4} \tag{10}$$

Expanding the determinant of the characteristic matrix in Eq. (9) and equating to zero yields the frequency equation

$$0.0003047\omega^4 - 0.0295139\alpha\omega^2 + 0.0451387\alpha^2 = 0 \tag{11}$$

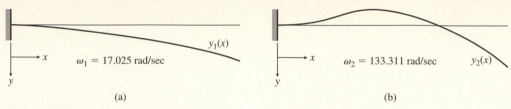

Figure 14.20

Normal modes for trapezoidal plate: (a) mode 1; (b) mode 2.

The positive real roots of Eq. (11) are

$$\omega_1^2 = 1.55436\alpha \quad \text{and} \quad \omega_2^2 = 95.30622\alpha$$

and the two natural circular frequencies are

$$\omega_1 = 17.025 \text{ rad/sec} \quad \text{and} \quad \omega_2 = 133.311 \text{ rad/sec}$$

The modal matrix $[\Phi]$, which contains the C coefficients for each mode, is determined by any method applicable to a discrete system. Thus

$$[\Phi] = \begin{bmatrix} C_{11} & C_{12} \\ C_{21} & C_{22} \end{bmatrix} = \begin{bmatrix} 1.0 & 1.0 \\ -0.003088 & -0.010357 \end{bmatrix} \tag{12}$$

Therefore the mode shapes are expressed by the Ritz deflection function as

$$y_1(x) = C_1 x^2 + C_2 x^3 = x^2 - 0.003088 x^3 \tag{13}$$

$$y_2(x) = C_1 x^2 + C_2 x^3 = x^2 - 0.010357 x^3 \tag{14}$$

The mode shapes are shown in Figure 14.20.

▲

It should be noted, when using the Rayleigh-Ritz procedure, that as the number of terms in the Ritz series defined by Eq. (14.112) is increased, the approximations for the system natural frequencies are improved. For instance, if the number of terms in the series used in this problem were increased from $n = 2$ to $n = 3$, then a better estimate for ω_1 and ω_2 would be achieved while providing a first estimate for ω_3.

Example 14.11 illustrates how the Rayleigh-Ritz method allows a continuous system to be analyzed as an analogous discrete MDOF system. This analysis paradox may also be employed to evaluate the dynamic response of continuous systems subject to arbitrary loading. The procedure is illustrated in the following example.

EXAMPLE 14.12 ▼

The free end of the cantilever plate of Example 14.11 and shown in Figure 14.19b is subjected to the dynamic force illustrated in Figure 14.21. Determine the dynamic response of the plate by the mode superposition method. Plot the transverse deflection of the cantilever tip, $y(L, t)$, in the time interval $0 \leq T \leq 5$ sec. Assume modal damping factors of $\zeta_1 = 0.04$ and $\zeta_2 = 0.05$.

Solution

In Example 14.11, the cantilever plate has essentially been represented as a discrete two-DOF system. Thus we can apply the modal analysis techniques discussed in Chapter 13 to obtain the solution. The uncoupled equations of motion may be expressed in normal coordinates as

$$\ddot{q}_r + 2\omega_r\zeta_r\dot{q}_r + \omega_r^2 q_r = \frac{P_r(t)}{M_r} \qquad r = 1, 2 \tag{1}$$

where ω_r, M_r, and $P_r(t)$ are the natural frequency, modal mass, and modal force for the rth mode. The modal masses are determined from the expression

$$M_r = \{\Phi\}_r^T[m]\{\Phi\}_r \tag{2}$$

and therefore

$$M_1 = [1.0 \quad -0.003068] \begin{bmatrix} \dfrac{7}{60} & \dfrac{2L}{21} \\ \dfrac{2L}{21} & \dfrac{4L^2}{112} \end{bmatrix} m_0 L^5 \begin{Bmatrix} 1.0 \\ -0.003068 \end{Bmatrix}$$

$$= 127.926 \times 10^6 \text{ lb-sec}^2/\text{in}$$

$$M_2 = [1.0 \quad -0.010357] \begin{bmatrix} \dfrac{7}{60} & \dfrac{2L}{21} \\ \dfrac{2L}{21} & \dfrac{4L^2}{112} \end{bmatrix} m_0 L^5 \begin{Bmatrix} 1.0 \\ -0.010357 \end{Bmatrix}$$

$$= 9.095 \times 10^6 \text{ lb-sec}^2/\text{in}$$

Similarly, the modal forces are determined from the expression

$$P_r = \{\Phi\}_r^T\{F(t)\} \tag{3}$$

Figure 14.21

Time-varying concentrated force applied to free end of cantilever plate in Figure 14.19.

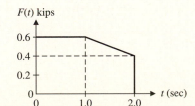

Thus

$$P_1 = [1.0 \quad -0.003068] \begin{Bmatrix} F(t)\gamma_1(L) \\ F(t)\gamma_2(L) \end{Bmatrix} = 9064.45 F(t)$$

$$P_2 = [1.0 \quad -0.010357] \begin{Bmatrix} F(t)\gamma_1(L) \\ F(t)\gamma_2(L) \end{Bmatrix} = -8165.95 F(t)$$

Substituting the above values of M_r and P_r into Eq. (1) results in

$$\ddot{q}_1 + 1.362\dot{q}_1 + 289.8938 q_1 = 7.086 \times 10^{-5} F(t) \tag{4}$$

$$\ddot{q}_2 + 13.331\dot{q}_2 + 17771.876 q_2 = -8.978 \times 10^{-4} F(t) \tag{5}$$

Equations (4) and (5) are evaluated numerically by the Newmark method using a time step $\Delta t = 0.002$ sec. The transformation back to physical coordinates after each time step is effected by

$$\{C\} = [\Phi]\{q\} \tag{6}$$

and the deflection of the cantilever tip in physical coordinates at each discrete time step is expressed as

$$y(x = L, \Delta t) = \{C\}^T \{\gamma(L)\} = C_1\gamma_1(L) + C_2\gamma_2(L)$$
$$= C_1 L^2 + C_2 L^3 \tag{7}$$

A time history of the deflection of the cantilever tip, $y(L, t)$, is presented in Figure 14.22. The listing for the FORTRAN computer program used to evaluate the dynamic response is presented in Table 14.2.

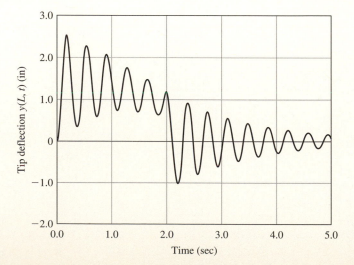

Figure 14.22

Displacement response history of cantilever plate tip $y(L, t)$.

TABLE 14.2. Listing of FORTRAN Computer Program Used for Example 14.12

```
*
        Program Newmark
*       Integrates Eqs.(4) and (5) in Example 14.12 by the Newmark Method
*
        integer i,j,n
        parameter (n=2)
        double precision k(n),m(n),c(n),phi(n,n),a0,a1,a2,a3,
   1          a4,a5,a6,a7,x,xdot,x2dot,p,q(n),qdot(n),q2dot(n),
   2          qdt,q2dotdt,alpha,delta,ti,tf,dt,t
*
        open(unit=11,file='x1',status="new")
        open(unit=12,file='x2',status="new")
*
        open(unit=13,file='x1dot',status="new")
        open(unit=14,file='x2dot',status="new")
*
        open(unit=15,file='x1dot2',status="new")
        open(unit=16,file='x2dot2',status="new")
*
        dt=0.002
        ti=0.00
        tf=5.00
        alpha=0.25
        delta=0.50
*
        a0=1.00/(alpha*dt*dt)
        a1=delta/(alpha*dt)
        a2=1.00/(alpha*dt)
        a3=1.00/(2.0*alpha)-1.00
        a4=delta/alpha-1.00
        a5=(dt/2.0)*((delta/alpha)-2.0)
        a6=dt*(1.00-delta)
        a7=delta*dt
*
        m(1)=1.0
        m(2)=1.0
*
        c(1)=1.3620
        c(2)=13.331
*
        k(1)=289.8938
        k(2)=17771.8760
*
        q(1)=0.00
        q(2)=0.00
*
        qdot(1)=0.00
        qdot(2)=0.00
*
        phi(1,1)=1.000
        phi(1,2)=1.000
        phi(2,1)=-0.003088
        phi(2,2)=-0.010357
*
        do 1 i=1,n
              call load(i,ti,phi,p)
              q2dot(i)=(p-c(i)*qdot(i)-k(i)*q(i))/m(i)
              k(i)=k(i)+a0*m(i)+a1*c(i)
   1    continue
*
        do 4 i=1,n
              x=0.0
              xdot=0.0
              x2dot=0.0
```

(continued)

```
              do 2 j=1,n
                     x=x+phi(i,j)*q(j)
                     xdot=xdot+phi(i,j)*qdot(j)
                     x2dot=x2dot+phi(i,j)*q2dot(j)
2             continue
              call load(i,ti,phi,p)
              write(10+i,3) ti,x
              write(12+i,3) ti,xdot
              write(14+i,3) ti,x2dot
3             format(1x,e13.6,3x,e13.6,3x,e13.6)
4     continue
*
      do 8 t=ti+dt,tf,dt
              do 5 i=1,n
                     call load(i,t,phi,p)
                     qdt=(p+m(i)*(a0*q(i)+a2*qdot(i)+a3*q2dot(i))+
    +                        c(i)*(a1*q(i)+a4*qdot(i)+a5*q2dot(i)))/k(i)
                     q2dotdt=a0*(qdt-q(i))-a2*qdot(i)-a3*q2dot(i)
                     qdot(i)=qdot(i)+a6*q2dot(i)+a7*q2dotdt
                     q(i)=qdt
                     q2dot(i)=q2dotdt
5             continue
              do 7 i=1,n
                     x=0.0
                     xdot=0.0
                     x2dot=0.0
                     do 6 j=1,n
                            x=x+phi(i,j)*q(j)
                            xdot=xdot+phi(i,j)*qdot(j)
                            x2dot=x2dot+phi(i,j)*q2dot(j)
6                    continue
                     call load(i,t,phi,p)
                     write(10+i,3) t,x
                     write(12+i,3) t,xdot
                     write(14+i,3) t,x2dot
7             continue
8     continue
*
      end
*
*
*
      subroutine load(i,t,phi,p)
*
      integer i,n
      parameter (n=2)
      double precision f,p,phi(m,n),t
*
      if (t.le.1.0) then
              f=6.0e2
              goto 1
      endif
      if (t.le.2.0) then
              f=8.0e2-2.0e2*t
              goto 1
              endif
      f=0.0
1     if (i.eq.1) then
              p=7.0857e-5*f
      endif
      if (i.eq.2) then
              p=-8.9784e-4*f
      endif
      return
      end
```

(continued)

```
*
*
      program mode
*
*
      integer i,n
      double precision f1,f2,x,x1,x2,ti,tf,dt,t
*
      f1(x)=x*x
      f2(x)=x*x*x
*
      open (unit=11,file='x1',status="old")
      open (unit=12,file='x2',status="old")
      open (unit=13,file='tip',status="new")
*
      dt=0.002
      ti=0.00
      tf=5.00
      n=1+(tf-ti)/dt
      x=120
*
      do 2 i=1,n
            read(11,1)t,x1
            read(12,1)t,x2
            write(13,1)t,f1(x)*x1+f2(x)*x2
    1       format(1x,e13.6,3x,e13.6)
    2 continue
    4 continue
      end
*
```

▲

14.6.3 Assumed Modes Method

In a manner similar to the Rayleigh-Ritz method, the assumed modes method permits the continuous system to be analyzed as an analogous discrete system to obtain estimates of the system natural frequencies, and therefore, the dynamic response may be evaluated by procedures applicable to discrete systems. The solution of the continuous system is assumed to be described by the expression

$$y(x, t) = \sum_{i=1}^{n} \psi_i(x)z_i(t) \qquad i = 1, 2, \ldots, n \tag{14.129}$$

where $z_i(t)$ are the n generalized coordinates and $\psi_i(x)$ are the n shape functions that satisfy the boundary conditions for the system. Effectively, this methodology is tantamount to representing the continuous system as a discrete system having n degrees of freedom.

The kinetic energy for the system may be expressed as

$$T = \frac{1}{2} \sum_{i=1}^{n} \sum_{j=1}^{n} m_{ij} \dot{z}_i(t) \dot{z}_j(t) \tag{14.130}$$

where the m_{ij} are determined by the mass distribution of the system and by the function $\psi_i(x)$ (note that the m_{ij} coefficients are symmetric). That is,

$$m_{ij} = \int \rho A(x) \psi_i(x) \psi_j(x) \, dx \tag{14.131}$$

where ρ is the material mass density and $A(x)$ is the cross-sectional area. The mass matrix of a system determined by Eq. (14.132) is called a *consistent mass matrix* (see Section 14.6.2). For a localized mass m, the corresponding mass coefficient is determined by

$$m_{ij} = m\psi_i(x)\psi_j(x) \tag{14.132}$$

where the shape functions are evaluated at the x location of m.

Similarly, the potential energy is expressed as

$$V = \frac{1}{2}\sum_{i=1}^{n}\sum_{j=1}^{n}k_{ij}z_i(t)z_j(t) \tag{14.133}$$

where k_{ij} are also symmetric and depend on the stiffness distribution as well as on $\psi_i(x)$. That is,

$$k_{ij} = \int E(x)A(x)\psi_i'(x)\psi_j' \, dx \tag{14.134}$$

are the stiffness coefficients for axial stiffness. The coefficients for flexural stiffness are given by

$$k_{ij} = \int E(x)I(x)\psi_i''(x)\psi_j''(x) \, dx \tag{14.135}$$

The stiffness coefficients corresponding to a localized translational spring of stiffness k are given by

$$k_{ij} = k\psi_i(x)\psi_j(x) \tag{14.136}$$

and the stiffness coefficients corresponding to a localized rotational spring having rotational stiffness k_R are given by

$$k_{ij} = k_R\psi_i'(x)\psi_j'(x) \tag{14.137}$$

In Eqs. (14.136) and (14.137), the shape functions are evaluated at the x location of the spring elements. The generalized forces acting on the system are obtained from the expression

$$F_i(t) = \int F(x, t)\psi_i(x) \, dx + \sum_{l=1}^{n}F_l(t)\psi_i(x_l) \tag{14.138}$$

where $F(x, t)$ represents distributed forces and $F_l(t)$ represents concentrated forces.

Since normal mode vibration is associated with conservative systems, Lagrange's equations may be employed to establish the equations of motion. Thus

$$\frac{d}{dt}\left(\frac{\partial T}{\partial \dot{z}_i}\right) - \frac{\partial T}{\partial \dot{z}_i} + \frac{\partial V}{\partial z_i} = 0 \qquad i = 1, 2, \ldots, n \tag{14.139}$$

Noting that T does not depend on the coordinates $z_i(t)$ and V does not depend on the velocities $\dot{z}_i(t)$, and substituting Eqs. (14.130) and (14.133) into Eq. (14.139), the equations of motion in generalized coordinates are obtained as

$$\sum_{j=1}^{n}m_{ij}\ddot{z}_j(t) + \sum_{j=1}^{n}k_{ij}z_j(t) = 0 \qquad j = 1, 2, \ldots, n \tag{14.140}$$

which can be expressed in the familiar matrix form

$$[m]\{\ddot{z}\} + [k]\{z\} = 0 \qquad (14.141)$$

Once again the equations of motion for the continuous system have been expressed in the standard form of a discrete MDOF system. Therefore, analysis techniques for evaluating the dynamic response of discrete systems become available to continuous systems approximated in this manner.

EXAMPLE 14.13 ▼

Determine the first two natural frequencies of the uniform cantilever beam of length L shown in Figure 14.23. Assume $L = 10$ ft, modulus of elasticity $E = 29,000$ ksi, moment of inertia $I = 597$ in^4, and the mass per unit length $\rho A = 0.12336 \times 10^{-3}$ kip-sec^2/in^2. Use the transverse deflection of the cantilever tip as the generalized coordinate and assume the following shape functions:

$$\psi_1(x) = 3\left(\frac{x}{L}\right)^2 - 2\left(\frac{x}{L}\right)^3$$

$$\psi_2(x) = \left[\left(\frac{x}{L}\right)^3 - \left(\frac{x}{L}\right)^2\right]L$$

Solution

The shape functions and the appropriate derivatives are

$$\psi_1(x) = 3\left(\frac{x}{L}\right)^2 - 2\left(\frac{x}{L}\right)^3 \qquad \psi_2(x) = \left[\left(\frac{x}{L}\right)^3 - \left(\frac{x}{L}\right)^2\right]L$$

$$\psi_1'(x) = \frac{6x}{L^2} - \frac{6x^2}{L^3} \qquad \psi_2'(x) = \frac{3x^2}{L^2} - \frac{2x}{L}$$

$$\psi_1''(x) = \frac{6}{L^2} - \frac{12x}{L^3} \qquad \psi_2''(x) = \frac{6x}{L^2} - \frac{2}{L}$$

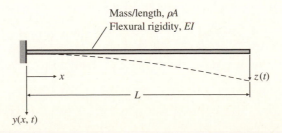

Mass/length, ρA
Flexural rigidity, EI

x

$z(t)$

L

$y(x, t)$

Figure 14.23

Uniform cantilever beam of Example 14.13.

The mass coefficients are determined from Eq. (14.131) as

$$m_{ij} = \rho A \int_0^L \psi_i(x) \psi_j(x) \, dx \tag{1}$$

Thus

$$m_{11} = \rho A \int_0^{120} \left(\frac{3x^2}{L^2} - \frac{2x^3}{L^3} \right) dx = \rho A(44.5714)$$

$$m_{21} = m_{12} = \rho A \int_0^{120} \left(\frac{3x^2}{L^2} - \frac{2x^3}{L^3} \right) \left(\frac{x^3}{L^2} - \frac{x^2}{L} \right) dx = \rho A(-754.286)$$

$$m_{22} = \rho A \int_0^{120} \left(\frac{x^3}{L^2} - \frac{x^2}{L} \right) dx = \rho A(16457.143)$$

and the consistent mass matrix is expressed as

$$[m] = \rho A \begin{bmatrix} 44.5714 & -754.286 \\ -754.286 & 16457.143 \end{bmatrix}$$

$$= \begin{bmatrix} 5.498 \times 10^{-3} & -0.09305 \\ -0.09305 & 2.03015 \end{bmatrix} \text{ kip-sec}^2/\text{in}$$

The stiffness coefficients are evaluated from Eq. (14.135) as

$$k_{ij} = EI \int \psi_i''(x) \psi_j''(x) \, dx \tag{2}$$

Therefore

$$k_{11} = EI \int_0^{120} \left(\frac{6}{L^2} - \frac{12x}{L^3} \right)^2 dx = EI(6.944 \times 10^{-6})$$

$$k_{21} = k_{12} = EI \int_0^{120} \left(\frac{6}{L^2} - \frac{12x}{L^3} \right) \left(\frac{6x}{L^2} - \frac{2}{L} \right) dx = EI(-4.1667 \times 10^{-4})$$

$$k_{22} = EI \int_0^{120} \left(\frac{6x}{L^2} - \frac{2}{L} \right)^2 dx = EI(0.03333)$$

and the stiffness matrix is given by

$$[k] = \begin{bmatrix} 6.944 \times 10^{-6} & -4.1667 \times 10^{-4} \\ -4.1667 \times 10^{-4} & 0.03333 \end{bmatrix} EI$$

$$= \begin{bmatrix} 120.229 & -7213.75 \\ -7213.75 & 577100.0 \end{bmatrix} \text{kips/in}$$

The system characteristic determinant is expressed by Eq. (14.128) as

$$\left| [k] - \omega^2 [m] \right| = 0 \tag{3}$$

The system eigenproblem defined by Eq. (3) may be solved by any discrete MDOF system eigensolver discussed in Chapters 10 and 11. Thus the system natural frequencies and normal vibration modal matrix are

$$\{\omega\} = \begin{Bmatrix} 91.911 \\ 905.652 \end{Bmatrix} \text{rad/sec} \qquad [\Phi] = \begin{bmatrix} 1.0 & 1.0 \\ 0.0115 & 0.0635 \end{bmatrix}$$

The "exact" frequencies are determined from Eq. (15) of Example 14.4 as

$$\omega_n = (\beta_n L)^2 \sqrt{\frac{EI}{\rho A L^4}}$$

where the terms $(\beta_n L)$ are given in Table 14.1. A comparison of the approximate frequencies with the exact frequencies is presented in Table 14.3.

It is noted that the two-mode approximation of the continuous system provides an excellent estimate for ω_1 but a rather inaccurate "first approximation" for ω_2. If the system was approximated with three assumed modes, then an improved estimate to ω_1 and ω_2 would be achieved while providing a first estimate to ω_3.

In a manner similar to that illustrated with the Rayleigh-Ritz method in Example 14.12, the dynamic response of continuous systems modeled by the assumed modes method may be analyzed by procedures developed for discrete MDOF systems.

TABLE 14.3. Comparison of Natural Frequencies

	Natural Frequency (rad/sec)		
	Approximate Value	Exact Value	% difference
ω_1	91.99	91.46	+ 0.579
ω_2	905.652	573.21	+ 57.99
ω_3	—	1605.15	—

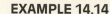

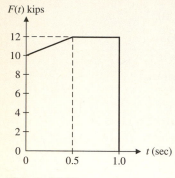

Figure 14.24

Time-varying concentrated force applied to free end of cantilever beam in Figure 14.23.

EXAMPLE 14.14 ▼

The free end of the cantilever beam of Example 14.13 is subject to the vertical concentrated force $F(t)$ shown in Figure 14.24. Evaluate the dynamic response of the beam by the mode superposition method. Plot a time history of the transverse deflection of the cantilever tip, $y(L, t)$, in the interval $0 \leq t \leq 3$ sec. Also determine the rotation of the elastic curve at the cantilever tip, $y'(L, t)$, in the same time interval. Assume modal damping factors of $\zeta_1 = 0.03$ and $\zeta_2 = 0.04$.

Solution

The uncoupled equations of motion may be expressed in normal coordinates as

$$\ddot{q}_r + 2\omega_r\zeta_r\dot{q}_r + \omega_r^2 q_r = \frac{P_r(t)}{M_r} \qquad r = 1, 2 \tag{1}$$

With the results of Example 14.13, the modal masses are determined from the expression

$$M_r = \{\Phi\}_r^T [m]\{\Phi\}_r \tag{2}$$

and therefore

$$M_1 = M_2 = 1.0$$

Similarly, the modal forces are determined from the expression

$$P_r = \{\Phi\}_r^T \{F(t)\} \tag{3}$$

and therefore

$$P_1(t) = 0.5257F(t)$$

$$P_2(t) = 0.7562F(t)$$

Substituting the above values for M_r and P_r into Eq. (1) yields the uncoupled equations of motion in normal coordinates:

$$\ddot{q}_1 + 5.5248\dot{q}_1 + 8476.9q_1 = 0.5257F(t) \tag{4}$$

$$\ddot{q}_2 + 73.9672\dot{q}_2 + 854802q_2 = 0.7562F(t) \tag{5}$$

Equations (4) and (5) were evaluated by the Newmark method of time integration in a manner similar to that performed in Example 14.12. Since the assumed modes $\psi_1(x)$ and $\psi_2(x)$ are independent of each other, then the normal coordinates $q_r(t)$ are the same as the generalized coordinates $z_i(t)$, and the transverse beam deflection in physical coordinates as given by Eq. (14.129) is expressed as

$$y(x, t) = \psi_1(x)q_1(t) + \psi_2(x)q_2(t) \tag{6}$$

The deflection of the cantilever tip at $x = L$ is thus given by

$$y(L, t) = \psi_1(L)q_1(t) + \psi_2(L)q_2(t) \tag{7}$$

Figure 14.25

Displacement response history for free end of cantilever beam $y(L, t)$.

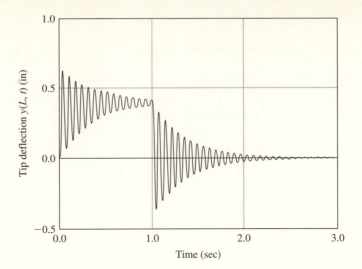

Since $\psi_1(L) = 1.0$ and $\psi_2(L) = 0$, then Eq. (7), which represents the time history of the transverse deflection of the cantilever tip, becomes

$$y(L, t) = q_1(t) \tag{8}$$

The time-history displacement response in physical coordinates is presented in Figure 14.25. The rotation or slope of the cantilever tip is expressed as

$$y'(L, t) = \psi_1'(L)q_1(t) + \psi_2'(L)q_2(t) \tag{9}$$

and since $\psi_1'(L) = 0$ and $\psi_2'(L) = 1$, then Eq. (9) becomes

$$y'(L, t) = q_2(t) \tag{10}$$

The time history for the rotation of the cantilever tip in physical coordinates is presented in Figure 14.26.

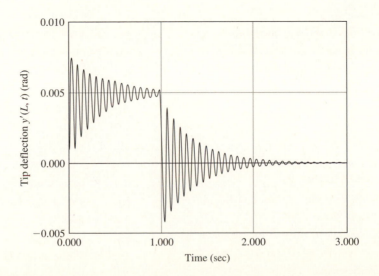

Figure 14.26

Rotation response history of free end of cantilever beam $y'(L, t)$.

REFERENCES

1 Church, Austin H., *Mechanical Vibrations,* Wiley, New York, 1963.

2 Steidel, Robert F., *An Introduction to Mechanical Vibrations,* Wiley, Inc., New York, 1971.

3 Jacobsen, L.S. and Ayre, R.S., *Engineering Vibrations,* McGraw-Hill, New York, 1958.

4 Hartog, J.P. Den, *Mechanical Vibrations,* 4th ed., McGraw-Hill, New York, 1956.

5 Blevins, Robert D., *Formulas for Natural Frequency and Mode Shape,* Van Nostrand Rheinhold, New York, 1979.

6 Timoshenko, S., Young, D.H., and Weaver, W., *Vibration Problems in Engineering,* 4th ed., Wiley, New York, 1974.

7 Thompson, William T., *Theory of Vibration with Applications,* 2nd ed., Prentice Hall, Englewood Cliffs, NJ, 1981.

8 Tse, F.S., Morse, I.E., and Hinkle, R.T., *Mechanical Vibrations, Theory and Applications,* 2nd ed., Allyn and Bacon, Boston, 1963.

9 Craig, Roy R. *Structural Dynamics: An Introduction to Computer Methods,* Wiley, New York, 1981.

10 Timoshenko, S., Young, D.H., *Elements of Strength of Materials,* 5th ed., Van Nostrand Rheinhold, New York, 1968.

11 Hutton, David V., *Applied Mechanical Vibrations,* McGraw-Hill, New York, 1981.

NOTATION

A	cross-sectional area	M	bending moment
A_n	constants	M_r	generalized (modal) mass for rth normal mode
B_n	constants	P_r	generalized (modal) force for rth normal mode
c	wave speed	$q_r(t)$	normal coordinates for rth normal mode
C_n	constants	$R_r(t)$	response ratio for rth normal mode
D_n	constants	s	constant in linear exponential function
e^{sx}	linear exponential function	s_i	roots of the auxiliary equation
E	modulus of elasticity	t	time
EI	flexural rigidity	T	axial tension force; also represents kinetic energy
$f(x)$	position function	T^*_{max}	reference kinetic energy
F	internal axial force	u	axial displacement
F_0	amplitude of externally applied force	$u(x, t)$	axial displacement solution
$F(x, t)$	externally applied distributed force	V	shear force; also represents potential energy
$F(t)$	externally applied concentrated force	x	position coordinate
$g(t)$	time function	$y(x, t)$	transverse displacement solution
G	shear modulus	$z(t)$	generalized coordinate
i	imaginary number ($\sqrt{-1}$)	DOF	degrees of freedom
I	static moment of inertia of cross section	MDOF	multi-degree-of-freedom
k	cross-section shape factor; also represents translational stiffness	$\{C\}$	vector of Rayleigh-Ritz method constants
k_{ij}	stiffness coefficients	β_n	beam frequency constants ($\sqrt{\omega_n / c}$)
k_R	rotational stiffness	δq_r	virtual displacement of normal coordinate
K_r	generalized (modal) stiffness for rth normal mode	$\delta(x-vt)$	Dirac delta function
L	length	δW	virtual work
m_{ij}	mass coefficients	$\delta z(t)$	virtual displacement of generalized coordinate
		Δt	time step for numerical integration

$\phi_r(x)$	mode shape (eigenfunction) for rth normal mode	ω_R	Rayleigh frequency
γ	mass per unit length	θ	slope of elastic deflection curve
$\gamma_i(x)$	displacement functions used in Rayleigh-Ritz method	ρ	mass density
Γ_r	modal participation factor for rth normal mode	ψ	shear angle
λ_R	Rayleigh quotient	$\psi(x)$	shape function
ω_r	undamped natural circular frequency for rth normal mode (rad/sec)	ζ_r	damping factor for rth normal mode
		$[\Phi]$	modal matrix

PROBLEMS

14.1–14.4 Determine expressions for the natural frequencies and normal vibration modes for longitudinal vibrations for the uniform rods shown in Figures P14.1 through P14.4. For all rods assume cross-sectional area A, modulus of elasticity E, mass density ρ, and length L.

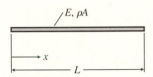

Figure P14.1

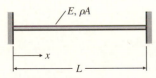

Figure P14.2

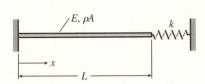

Figure P14.3

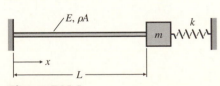

Figure P14.4

14.5 Determine the fundamental frequency for the longitudinal vibration of a uniform slender steel rod having a material specific weight of 0.283 lb/in³ and elastic modulus $E = 29,000$ ksi. The rod has a length of 5 ft and a cross-sectional area of 0.5 in.² The rod is free at both ends as shown in Figure P14.1.

14.6 Determine the velocity of propagation of the longitudinal waves (wave speed) along the rod of Problem 14.5.

14.7 Determine the fundamental frequency for the longitudinal vibration of a uniform copper rod having a length of 6 ft and a cross-sectional area of 0.75 in.² Consider one end of the rod to be fixed and the other end to be free. The material has a specific weight of 0.322 lb/in³ and an elastic modulus $E = 17,000$ ksi.

14.8 Determine the velocity of propagation of the longitudinal waves along the bar of Problem 14.7.

14.9 Determine the fundamental frequency of a 0.2-in-diameter steel wire stretched between two supports. The tension in the wire is 100 lb, the specific weight of the wire material is 0.293 lb/in³, and the length of the wire is 20 ft.

14.10 Determine the wave speed in a taut cable having a mass per unit length $\gamma = 0.5$ kg/m and tension $T = 650\ N$.

14.11 A uniform cable of length L and mass per unit length γ is fixed at its ends and stretched into an initial shape shown in Figure P14.5 and suddenly released. Determine expressions for its natural frequencies and free vibration response.

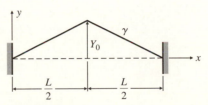

Figure P14.5

14.12 Repeat Problem 14.11 for the initial displaced configuration shown in Figure P14.6.

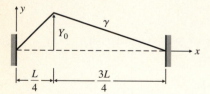

Figure P14.6

14.13 A cable of length L and mass per unit length γ is stretched with tension T. The left end is fixed and the right end is attached to a spring of stiffness k as shown in Figure P14.7. Determine the expression for the natural frequencies.

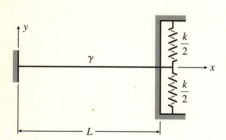

Figure P14.7

14.14 A cable of length L and mass per unit length γ is stretched with tension T. The left end is fixed and the right end is attached to a spring-mass system as shown in Figure P14.8. Determine the expression for the natural frequencies.

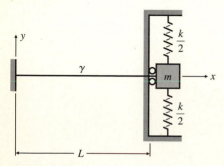

Figure P14.8

14.15–14.18 Determine the frequency equation for transverse vibration for the uniform beams shown in Figures P14.9 through P14.12. For all beams assume cross-sectional area A, flexural rigidity EI, and mass density ρ.

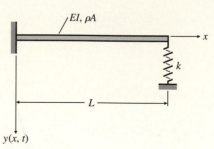

Figure P14.9

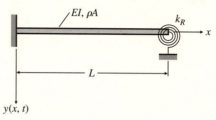

Figure P14.10

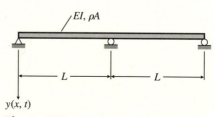

Figure P14.11

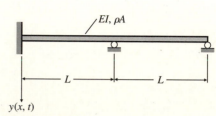

Figure P14.12

14.19 Determine the fundamental frequency for the transverse vibration of a W12×65 beam having a length of 25 ft. The beam is simply supported at both ends and has a specific weight (including superimposed dead load) of 6.0 lb/in³. Assume modulus of elasticity $E = 29,000$ ksi.

14.20 Determine the fundamental frequency for the transverse vibration of a W14×61 beam having a length of 30 ft. The beam is fixed at both ends and has a specific weight of 10.5 lb/in³. Assume modulus of elasticity $E = 29,000$ ksi.

14.21 Repeat Problem 14.19 if (a) the beam is subject to an axial tensile force of 50 kips and (b) the beam is subject to an axial compressive force of 50 kips.

14.22 Repeat Problem 14.20 if (a) the beam is subject to an axial tensile force of 75 kips and (b) the beam is subject to an axial compressive force of 75 kips.

14.23 Determine the expression for the displacement response $y(x, t)$ of the uniform simply supported beam subjected to the uniformly distributed harmonic force $F(x, t) = F_0 \sin \Omega t$ as shown in Figure P14.13. Assume zero initial conditions.

14.24 Repeat Problem 14.23 if the harmonic uniformly distributed force is replaced with the suddenly applied uniformly distributed force given by $F(x, t) = F_0 g(t)$, where the load time-history function $g(t)$ is described by Figure 14.13.

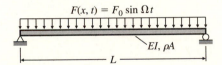

$$F(x, t) = F_0 \sin \Omega t$$

$EI, \rho A$

L

Figure P14.13

14.25 Determine the expression for the displacement response $y(x, t)$ of a uniform simply supported beam subjected to a suddenly applied distributed force of amplitude F_0 as shown in Figure P14.14, where the force time history is an ideal step input as shown in Figure 14.13. Assume zero initial conditions.

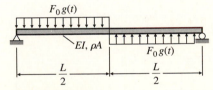

$F_0 g(t)$

$EI, \rho A$

$F_0 g(t)$

$\dfrac{L}{2}$ $\dfrac{L}{2}$

Figure P14.14

14.26 Repeat Problem 14.25 if the time variation of the distributed force is harmonic so that $F(x, t) = F_0 \sin \Omega t$.

14.27 A simply supported beam is subjected to the concentrated force $F(t) = F_0 e^{-\Omega t}$ as shown in Figure P14.15. Determine the expression for the forced displacement response $y(x, t)$.

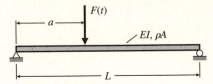

$F(t)$

a

$EI, \rho A$

L

Figure P14.15

14.28 Repeat Problem 14.27 if the force $F(t)$ is defined by Figure P14.16.

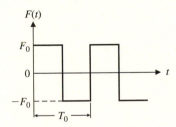

$F(t)$

F_0

0 t

$-F_0$

T_0

Figure P14.16

14.29 By the Rayleigh method, estimate the fundamental frequency for transverse vibration of a uniform cable of length L, mass per unit length γ, and under tension T as shown in Figure 14.4. Assume a shape function $\psi(x) = C(L - x)x$.

14.30 By the Rayleigh method estimate the fundamental frequency for longitudinal vibration of a uniform rod of length L, cross-sectional area A, elastic modulus E, and mass density ρ. Assume the bar is fixed at one end and free at the other. Use the shape function $\psi(x) = C(x/L)$.

14.31 By the Rayleigh method estimate the fundamental frequency for transverse vibration for the nonprismatic beam shown in Figure P14.17. Use the shape function $\psi(x) = C(4x/L)(1 - x/L)$ and assume the beam material has mass density ρ.

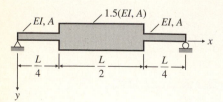

Figure P14.17

14.32 By the Rayleigh method estimate the fundamental frequency for transverse vibration for the cantilever beam with tip mass m shown in Figure P14.18. The beam has cross-sectional area A, flexural rigidity EI, and material mass density ρ. Use the shape function $\psi(x) = C(x/L)^2$.

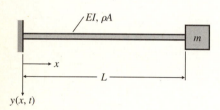

Figure P14.18

14.33 A uniform beam of length L and flexural rigidity EI is fixed at both ends. Determine the fundamental frequency for transverse vibration by the Rayleigh method. Assume cross sectional area A and mass density ρ of the beam material. Use the shape function $\psi(x) = Cx(L - x)^2$.

14.34 Estimate the first two natural frequencies and mode shapes for transverse vibration for the cable of Problem 14.29 by the Rayleigh-Ritz method. Assume a Ritz deflection function $y(x) = C_1(L - x) + C_2(L - x)^2x^2$.

14.35 Estimate the first two natural frequencies and mode shapes for longitudinal vibration for the uniform bar of Problem 14.30 by the Rayleigh-Ritz method. Assume a Ritz deflection function $y(x) = C_1(x/L) + C_2(x/L)^2$.

14.36 By the Rayleigh-Ritz method estimate the first two natural frequencies and mode shapes for transverse vibration for the uniform beam shown in Figure P14.19. Assume the beam has flexural rigidity EI, cross-sectional area A, and mass density ρ. Use a Ritz deflection function
$$y(x) = C_1(x/L) + C_2 \sin(\pi x/L).$$

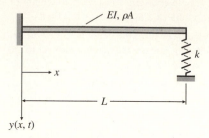

$y(x, t)$

Figure P14.19

14.37 A cantilever beam has a constant width b, but has a linearly varying depth $d(x)$ as shown in Figure P14.20. Estimate the first two natural frequencies and mode shapes for transverse vibration by the Rayleigh-Ritz method. Assume the beam has material modulus of elasticity E and mass density ρ. Use the Ritz deflection function $y(x) = C_1x^2 + C_2x^3$.

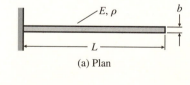

(a) Plan

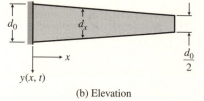

(b) Elevation

Figure P14.20

14.38 Determine the first two natural frequencies and mode shapes for longitudinal vibration for the nonprismatic rod shown in Figure P14.21 by the Rayleigh-Ritz method. Assume the beam material has modulus of elasticity E and mass density ρ. Use a Ritz deflection function $u(x) = C_1 \sin(\pi x/2L) + C_2 \sin(3\pi x/2L)$.

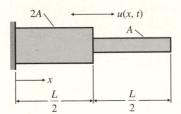

Figure P14.21

14.39 The free end of the tapered cantilever beam of Problem 14.37 is subjected to the transverse time varying force shown in Figure P14.22. By the mode superposition method determine the dynamic response of the beam. Assume modal damping factors of $\zeta_1 = 0.02$ and $\zeta_2 = 0.05$. Evaluate the dynamic response by one of the numerical integration methods discussed in Chapter 7. Plot the time history for the transverse deflection of the cantilever tip in the time interval $0 \le t \le 4.0$ sec. Assume the specific weight = 0.283 lb/in³, modulus of elasticity $E = 29,000$ ksi, $b = 2$ in, $d_0 = 24$ in, and $L = 12$ ft.

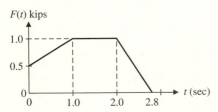

Figure P14.22

14.40 The free end of the nonprismatic rod of Problem 14.38 is subjected to an axial time-varying force shown in Figure P14.23. Determine the dynamic response of the bar by the mode superposition method. Evaluate the dynamic response by one of the numerical integration methods discussed in Chapter 7. Plot the time history for the axial displacement of the midpoint of the bar (at $x = L/2$) in the time interval $0 \le t \le 3$ sec. Assume modal damping factors $\zeta_1 = \zeta_2 = 0.05$, specific weight = 0.344 lb/in³, modulus of elasticity $E = 17,000$ ksi, cross-sectional area $A = 3$ in², and length $L = 8$ ft.

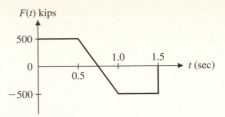

Figure P14.23

14.41 Estimate the first two natural frequencies for the axial vibration of a uniform bar fixed at one end, and pinned to an axial spring of stiffness k at the other end, as shown in Figure P14.3, by the method of assumed modes. Assume the bar has elastic modulus E, cross-sectional area A, and mass density ρ. Use the shape functions $\psi_1(x) = \sin(\pi x/2L)$ and $\psi_2(x) = \sin(3\pi x/2L)$.

14.42 Repeat Problem 14.41 for the uniform rod with end mass m as shown in Figure P14.4. Use the shape functions $\psi_1(x) = (x/L)$ and $\psi_2(x) = (x/L)^2$.

14.43 Estimate the first two natural frequencies for the transverse vibration for the tapered cantilever beam shown in Figure P14.24 by the method of assumed modes. The beam has constant depth d but variable width $b(x)$. Assume the beam material has elastic modulus E, and mass density ρ. Use the shape functions $\psi_1(x) = (x/L)^2$ and $\psi_2(x) = (x/L)^3$.

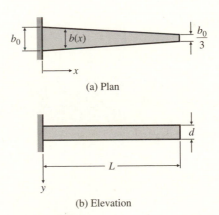

(a) Plan

(b) Elevation

Figure P14.24

14.44 Estimate the first two natural frequencies for the transverse vibration for a simply supported uniform beam of length L by the method of assumed modes. Assume the beam has cross-sectional area A, elastic modulus E, and mass density ρ. Use the shape functions and $\psi_1(x) = \sin(\pi x/L)$ and $\psi_2(x) = \sin(2\pi x/L)$.

14.45 The free end of the tapered cantilever beam of Problem 14.43 is subjected to a transverse time-varying force shown in Figure P14.25. By the mode superposition method determine the dynamic response of the beam. Evaluate the dynamic response by one of the numerical procedures discussed in Chapter 7. Plot the time history for the transverse deflection of the cantilever tip in the time interval $0 \leq t \leq 4$ sec. Assume modal damping factors $\zeta_1 = 0.03$ and $\zeta_2 = 0.06$, modulus of elasticity $E = 29{,}000$ ksi, specific weight $= 0.283$ lb/in^3, length $L = 15$ ft, depth $d = 30$ in, and $b_0 = 6$ in.

14.46 The simply supported uniform beam of Problem 14.44 is subjected to the transverse time-varying force shown in Figure P14.26 at its midspan. Determine the dynamic response of the beam by the mode superposition method. Evaluate the dynamic response by one of the numerical integration methods discussed in Chapter 7. Plot the time history for the transverse deflection of midspan in the time interval $0 \leq t \leq 5$ sec. Assume modal damping factors $\zeta_1 = 0.02$ and $\zeta_2 = 0.07$, modulus of elasticity $E = 29{,}000$ ksi, specific weight $= 0.395$ lb/in^3, length $L = 20$ ft, cross-sectional area $A = 15$ in^2, and moment of inertia $I = 400$ in.^{4}

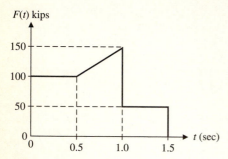

Figure P14.25

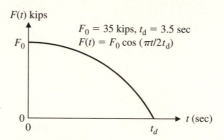

Figure P14.26

PART **IV**

Nonlinear Dynamic Response

15 ⏶ Analysis of Nonlinear Response

When discussing the vibrational characteristics of linear systems, it had always been assumed that the force in the spring was proportional to its deformation. It was also recognized that viscous damping, in which the damping force is proportional to the velocity, is much easier to describe analytically than other forms of energy dissipation. Therefore, to avoid mathematical difficulties, the concept of equivalent viscous damping was introduced (Section 4.3) to characterize other forms of damping less conduitive to an analytical formulation. Furthermore, the mass of the vibrating system was always assumed to be constant with time. Consequently, the equations of motion for such discrete systems are linear, second-order ordinary differential equations with constant coefficients. For a SDOF system the equation of motion is recognized as

$$m\ddot{x} + c\dot{x} + kx = F(t) \tag{15.1}$$

This equation adequately represents many practical problems and forms the essence of linear vibration theory. However, there are also many physical systems for which linear differential equations with constant coefficients are inadequate to describe the motion, and the analysis of such systems requires the solution of nonlinear differential equations.

Elimination of the possibility of variable mass allows the general form of the differential equation describing a nonlinear SDOF vibrating system to be expressed as

$$\ddot{x} + f(x, \dot{x}, t) = F(t) \tag{15.2}$$

Such equations are distinguished from linear equations in that the principle of superposition does not apply for their solution. For example, if the magnitude of a forcing function is doubled, the response of a nonlinear system is not necessarily doubled. In general, nonlinear vibrations are not harmonic, and their frequencies vary with amplitude. Also, a linear system has only one position of equilibrium, whereas a nonlinear system could exhibit multiple equilibrium positions, depending on the operating conditions.

In recent years an increasing demand for nonlinear dynamic analysis for a variety of engineering problems has emerged. Consequently, a relatively large amount of research effort has been dedicated to the development of efficient solution procedures for nonlinear vibration problems. At the present time there is a relatively small subset of nonlinear dynamic problems that can be solved accurately and efficiently. However, most nonlinear problems are either very difficult and highly computationally intensive to solve, or they cannot be solved at all.

An accurate and efficient methodology for nonlinear dynamic analysis must be based on the use of appropriate kinematic formulations, constitutive models, time integration schemes, and an appropriate analytical model of the system under consideration. The complexity of practical nonlinear dynamic analyses lie in the interdependency between these important considerations.

15.1 CLASSIFICATION OF NONLINEAR ANALYSES

In all previous formulations of the equations of motion, it was assumed that the displacements of the system were infinitessimally small and that the material behavior was linear elastic. It was further assumed that the system boundary conditions remained unchanged during application of the loads. With these assumptions, the resulting equations corresponded to a *linear* analysis because the displacement response was a linear function of the applied loads. This basic premise for a linear analysis defines the meaning of a nonlinear analysis, and also suggests how to categorize different types of nonlinear analyses.

Indeed, nonlinear phenomena are manifested in a multiplicity of forms in physical systems. However, for convenience, nonlinear analyses can be categorized into three basic classifications: (1) materially nonlinear only (MNO), (2) large displacements (large rotations) and small strains, and (3) large displacements (large rotations) and large strains [1]. The nature of the nonlinearity significantly influences the formulation of the problem and the type of analysis to be conducted. Table 15.1 summarizes a classification of nonlinear analyses that separately considers material nonlinear effects and kinematic (geometric) nonlinear effects.

TABLE 15.1.   **Classification of Nonlinear Analyses**

Type of Analysis	Description	Typical Formulation Used
Materially nonlinear only	Infinitesimal displacements and strains; the stress-strain relation is nonlinear	Materially nonlinear only (MNO)
Large displacements, large rotations, but small strains	Displacements and rotations of fibers are large, but fiber extensions and angle changes between fibers are small; the stress-strain relation may be linear or nonlinear	Total Lagrangian (TL) / Updated Lagrangian (UL)
Large displacements, large rotations, and large strains	Fiber extensions and angle changes between fibers are large, fiber displacements and rotations may also be large; the stress-strain relation may be linear or nonlinear	Total Lagrangian (TL) / Updated Lagrangian (UL)

Adapted from *Finite Element Procedures* by Klaus-Jürgen Bathe, copyright © 1996 by permission of Prentice Hall, Inc., Upper Saddle River, NJ.

15.1.1 Material Nonlinearity

The most frequently conducted type of nonlinear dynamic analysis is the MNO. In a MNO analysis, geometric nonlinearity is excluded and the nonlinear effect resides solely in the nonlinear stress-strain relation. Since the displacements and strains are infinitesimally small, the standard engineering stress and strain measures can be employed in the response description. If the stress-strain relation is nonlinear, but elastic, there is a unique relation between stress and strain. However, if there are plastic strains, the stress-strain relation is path dependent and not unique.

Consider the uniform rod of cross-sectional area A and length L subjected to a time-varying force $F(t)$ as shown in Figure 15.1a. As the bar elongates, a differential element in the bar experiences a deformation du as shown in Figure 15.1b. The stress in the element at any instant of time is simply

$$\sigma = \frac{F}{A} \tag{15.3}$$

and the strain ε in the element is prescribed by the nonlinear stress-strain relation shown in Figure 15.1c. If the stress is less than the yield stress σ_y, then the response is linear elastic and the strain is determined by Hooke's law. However, if $\sigma > \sigma_y$, then the response is nonlinear and the strain is determined by

$$\varepsilon = \frac{du}{dx} = \frac{\sigma_y}{E} + \frac{\sigma - \sigma_y}{E_T} \tag{15.4}$$

where E is the elastic modulus, E_T is the *tangent modulus,* and $\varepsilon < 0.04$ for small strains. A summary of some typical material models that may be employed in a MNO analysis is presented in Table 15.2.

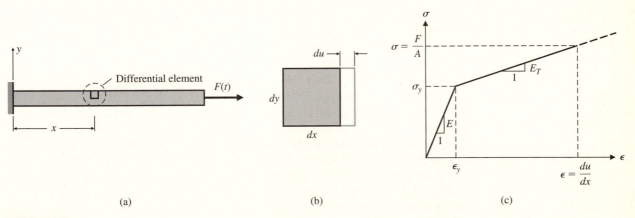

(a) (b) (c)

Figure 15.1

Materially nonlinear only (MNO); infinitesimal displacements, but nonlinear stress-strain relation.

TABLE 15.2. **Typical Material Models**

Material Model	Characteristics	Examples
Elastic, linear or nonlinear	Stress is a function of strain only; same stress path on unloading as on loading Linear elastic: $\sigma = E\varepsilon$ E is constant Nonlinear elastic: E varies as a function of strain	Almost all materials, provided the stresses are small enough: steel, cast iron, glass, rock, wood, etc., before yielding or fracture
Hyperelastic	Stress is calculated from a strain energy functional	Rubberlike materials, e.g., Mooney-Rivlin and Ogden models
Hypoelastic	Stress increments are calculated from strain increments $$d\sigma = Ed\varepsilon$$ The material modulus E is defined as a function of stress, strain, fracture criteria, loading and unloading parameters, maximum strains reached, etc.	Concrete, rock, and masonry models
Elastoplastic	Linear elastic behavior until yield, use of yield condition, flow rule, and hardening rule to calculate stress and plastic strain increments; plastic strain increments are instantaneous	Metals, soils, rocks when subjected to high stresses

Adapted from *Finite Element Procedures* by Klaus-Jürgen Bathe, copyright © 1996 by permission of Prentice Hall, Inc., Upper Saddle River, NJ.

15.1.2 Geometric Nonlinearity

The essential feature of geometric nonlinearity is that equilibrium equations must be written with respect to the deformed geometry, which is not known in advance. In MDOF systems (especially large finite element models), large displacement problems can be analyzed in either *Lagrangian coordinates* or *Eulerian coordinates*. The Lagrangian formulation is more suitable to the solutions of structural mechanics problems than the Eulerian formulation, which is typically used in the analysis of fluid mechanics problems. However, Eulerian formulations have been successfully employed in *penetration mechanics* applications, involving high velocity projectiles penetrating a solid material. In these formulations a "hydrodynamics" theory of medium penetration is implemented which permits flow of the target material around the penetrator. Finite element method (FEM) computer programs based on this theory are referred to as *hydrodynamics codes* or *hydrocodes*.

There are two basic Lagrangian formulations, the *total Lagrangian* (TL) and the *updated Lagrangian* (UL). In the TL kinematic formulation, the original reference frame remains stationary and all displacements, differentiations, and integrations are with respect to the original reference frame. In the UL formulation, the reference frame moves as the body deforms. Differentiations and integrations are done with

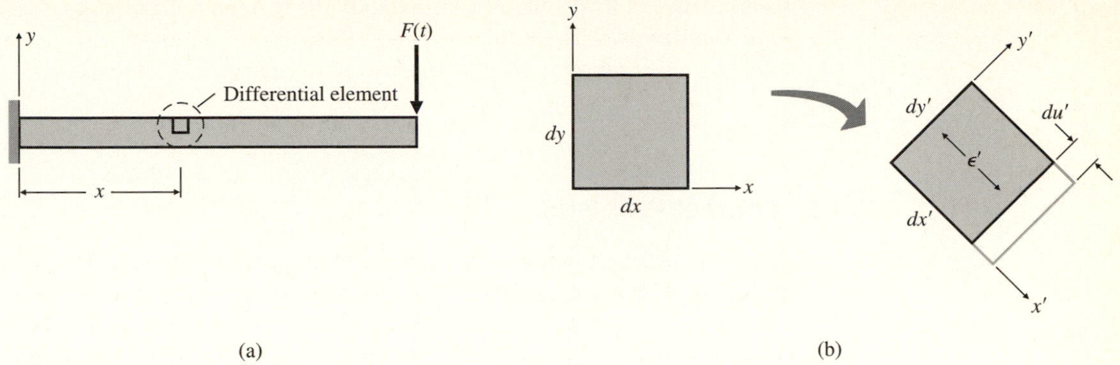

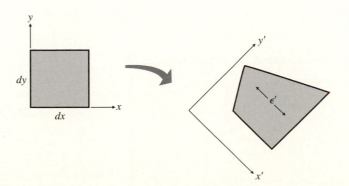

Figure 15.2

Large displacements and large rotations, but small strains; linear or nonlinear material behavior.

respect to the current deformed state. Thus the only theoretical difference between the two formulations is the choice of a reference configuration for the kinematic and static variables. In practice, the choice of using either the TL or the UL formulation depends on the type of structure being analyzed (i.e., beam, plate, cable, etc.) and the constitutive law used.

To illustrate the concepts of geometric nonlinearity, consider the uniform cantilever beam in Figure 15.2a. For the case of large displacements but small strains, the differential element shown in Figure 15.2b is subjected to infinitesimally small strains measured in the deformed configuration (x', y') while it undergoes large rigid-body displacements and rotations. The stress and strain relation of the material can be either linear or nonlinear, but the strains in the deformed state remain small, that is [1],

$$\varepsilon' = \frac{du'}{dx} \leq 0.04 \tag{15.5}$$

The condition of large displacements and large strains for the case shown in Figure 15.2a is illustrated in Figure 15.3. This is the most general case of geometric

Figure 15.3

Large displacements, large rotations, and large strains; linear or nonlinear material behavior.

nonlinearity. In its deformed configuration the differential element is significantly distorted and $\varepsilon' \geq 0.04$. In this case the stress-strain relation is also usually nonlinear. This type of behavior is generally restricted to rubberlike materials.

15.1.3 Contact Problems

A final category of nonlinear analyses is often given the general description of contact problems. These are problems in which the boundary conditions change during the motion of the system under consideration. Loading may cause parts of a structure to come in contact (or separate). For example, if during loading a degree of freedom in the system, which was initially free, it becomes restrained at a certain load level, the response is only linear prior to the changes in boundary condition. Also, contact areas may change in size as the load changes. These types of problems may include both geometric and material nonlinearity. Thus the kinematic formulation can be any combination of the previously discussed types of nonlinear analyses.

A simple example of a contact problem is illustrated in Figure 15.4a. A beam having stiffness k_b is separated from a "stopper" having stiffness k_S by a gap x_{gap}. As the load is applied to the beam, it eventually comes in contact with the stopper. The force-displacement relation for this system is described in Figure 15.4b. If the midspan displacement x of the beam is less than x_{gap}, then the total stiffness of the system $k_{\text{tot}} = k_b$ (other types of nonlinearities notwithstanding). However, once the beam comes in contact with the stopper (that is, $x > x_{\text{gap}}$), then the system resistance R or stiffness increases such that $k_{\text{tot}} = k_b + k_s$.

It should be noted at this point that the dynamic analysis of large MDOF systems exhibiting one or more types of nonlinearity generally represents a significant computational endeavor. There are several commercially available FEM computer programs, such as ADINA [2], that can accommodate a variety of nonlinear analyses for large systems. In the next section several examples of simple systems exhibiting nonlinear characteristics are discussed; however, the remainder of the chapter addresses MNO analyses of *elastoplastic* systems.

Figure 15.4

Change in boundary conditions at displacement x_{gap}:
(a) beam separated from elastic stopper;
(b) system resistance function.

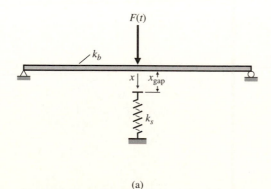

(a)

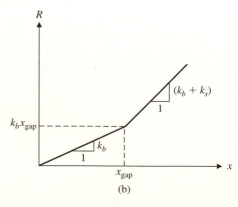

(b)

15.2 SYSTEMS WITH NONLINEAR CHARACTERISTICS

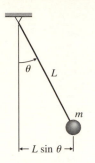

Figure 15.5
Simple pendulum.

To illustrate how nonlinearities may develop in the equations of motion, several simple examples of systems possessing nonlinear characteristics are investigated. As a first example consider the simple pendulum shown in Figure 15.5. The well-known *linearized* (i.e., small angles of oscillation θ) equation of motion for this system is

$$\ddot{\theta} + \frac{g}{L}\theta = 0 \tag{15.6}$$

and the motion is simple harmonic. However, for oscillations that cannot be considered small, then $\sin \theta \neq \theta$ and Eq. (15.6) becomes

$$\ddot{\theta} + \frac{g}{L} \sin \theta = 0 \tag{15.7}$$

Noting that the series expansion for $\sin \theta$ is given by

$$\sin \theta = \theta - \frac{\theta^3}{3!} + \frac{\theta^5}{5!} - \frac{\theta^7}{7!} + \cdots \tag{15.8}$$

then, neglecting the terms higher than third order in the series, Eq. (15.7) may be expressed as

$$\ddot{\theta} + \frac{g}{L}\left(\theta - \frac{\theta^3}{6}\right) = 0 \tag{15.9}$$

This is a simple example of geometric nonlinearity.

Several interesting observations can be made by comparing Eqs. (15.7) and (15.9). Other than the simple fact that Eq. (15.9) possesses a nonlinear term, notice that the restoring force in the linearized system, Eq. (15.6), is linearly proportional to the amplitude of oscillation θ, and that the natural period (frequency) of the system is constant for any θ. Such a system is referred to as *isochronous*. However, in the nonlinear system, Eq. (15.9), the restoring force is proportional to $(\theta - \theta^3/6)$ and therefore decreases as the amplitude increases, exhibiting the characteristic known as *softening* as illustrated in Figure 15.6. Moreover, as θ increases, the natural period of the non-linear system increases (or frequency decreases). Systems in which the period (frequency) is dependent on the amplitude are *nonisochronous*.

As another example of a system exhibiting geometric nonlinearity, consider the large amplitude vibration of a pretensioned cable with a concentrated mass m at midlength and having length $2L$ as shown in Figure 15.7a. From the free-body diagram of Figure 15.7b, the nonlinear equation of motion may be expressed as (neglecting the mass of the cable)

$$m\ddot{x} + 2\left(T_i + \frac{AE\delta}{L}\right)\sin \theta = 0 \tag{15.10}$$

where T_i = initial tension in the cable
A and E = cross-sectional area and modulus of elasticity of the cable, respectively
δ = axial elongation of half the cable length

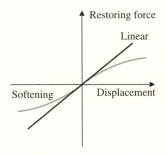

Figure 15.6
Softening phenomenon.

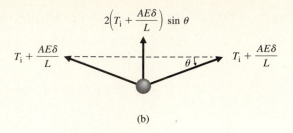

(a)   (b)

Figure 15.7
Nonlinear vibration of pretensioned cable.

Note that the length of the elongated cable half is

$$L + \delta = \sqrt{L^2 + x^2}$$ (15.11)

or

$$\frac{\delta}{L} = \sqrt{1 + \left(\frac{x}{L}\right)^2} - 1 \cong \frac{x^2}{2L^2}$$ (15.12)

and that

$$\sin \theta = \frac{x}{\sqrt{L^2 + x^2}} \cong \frac{x}{L}$$ (15.13)

Then substitution of Eqs. (15.12) and (15.13) into Eq. (15.10) yields the nonlinear equation of motion

$$m\ddot{x} + \frac{2}{L}\left(T_i + \frac{AEx^2}{2L^2}\right)x = 0$$ (15.14)

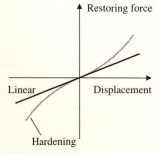

Figure 15.8
Hardening phenomenon.

Despite the use of the two approximations described in Eqs. (15.12) and (15.13), the equation of motion is still nonlinear. Similar to the nonlinear simple pendulum system, this system is also nonisochronous. However, from Eq. (15.14) note that the restoring force increases as the amplitude of oscillation x increases. This characteristic is referred to as *hardening* and is illustrated in Figure 15.8. For systems exhibiting hardening, the natural period decreases (or frequency increases) as the vibration amplitude increases. For small oscillations, the x^3 term in Eq. (15.14) can be neglected, and the linearized equation of motion for the corresponding isochronous system is

$$m\ddot{x} + \frac{2}{L}T_i x = 0$$ (15.15)

For simple systems with either nonlinear hardening or softening restoring forces, the equation of motion may be expressed in the general form

$$m\ddot{x} + k(x \pm \mu x^3) = 0 \tag{15.16}$$

where μ is a constant and the plus sign refers to the hardening characteristic and the minus sign to the softening characteristic. Analytical solutions to Eq. (15.16) for simple SDOF systems can generally be achieved by *perturbation methods,* where μ is the "small" perturbation parameter associated with the nonlinearity. However, for non-linear MDOF systems the perturbation method of analysis is not feasible, and an incremental formulation of the equations of motion is necessary.

As a final example of geometric nonlinearity, consider the system shown in Figure 15.9a. The system is representative of the contact problem discussed in the previous section. If the springs in the system are otherwise linear, the restoring force is described by Figure 15.9b. Although nonlinear, the system exhibits *piecewise-linear* characteristics. That is, the system response can be viewed as linear in different stages. For instance, if the displacement $|x| \leq \pm |x_1|$, the motion is simple harmonic and the equation of motion for this stage is

$$m\ddot{x} + k_1 x = 0 \tag{15.17}$$

When the displacement x is numerically larger than x_1, then the equation of motion is

$$m\ddot{x} + k_1 x + k_2(x - x_1) = 0 \tag{15.18}$$

and the motion for this stage is still simple harmonic, albeit at a higher frequency.

The response of the system may therefore be determined by evaluating the linear response for each stage individually (piecewise) and combining the results. A similar solution technique was employed for the systems characterized with Coulomb (or frictional) damping discussed in Section 4.4. Another common example of a piecewise-linear system is one exhibiting multilinear inelastic material behavior such as *elastoplasticity.*

Figure 15.9
Piecewise-linear

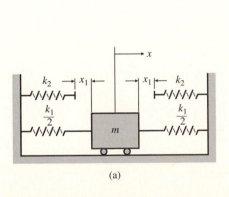

(a)

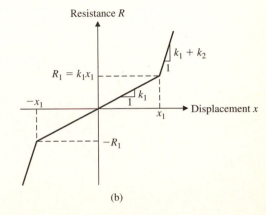

(b)

15.3 FORMULATION OF INCREMENTAL EQUATIONS OF EQUILIBRIUM

The concept of incremental equations of equilibrium has been previously discussed in Chapter 7 for linear SDOF systems and in Chapter 13 for linear MDOF systems. This section extends the concept to nonlinear systems. However, the discussion will be restricted to material nonlinearities only or MNO analyses.

To clarify the presentation, a SDOF system as represented in Figure 15.10a is considered. From the free-body diagram in Figure 15.10b, the equation of motion can be expressed as

$$F_I(t) + F_D(t) + F_S(t) = F(t) \tag{15.19}$$

where $F_I(t)$ = inertia force
$F_D(t)$ = damping force
$F_S(t)$ = spring force
$F(t)$ = externally applied force

The (material) nonlinearity is assumed to be associated entirely with the spring force (or restoring force) as illustrated in Figure 15.11.

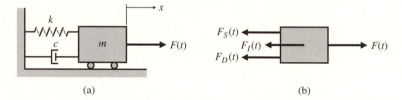

Figure 15.10

Response of SDOF system.

(a)   (b)

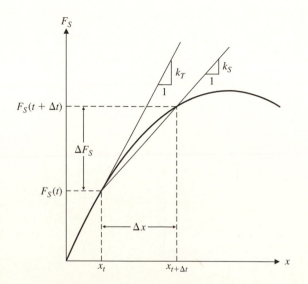

Figure 15.11

Variation of spring force for material nonlinearity.

The equilibrium at some small increment of time Δt after the current time t (i.e., at time $t + \Delta t$) can be expressed as

$$F_I(t + \Delta t) + F_D(t + \Delta t) + F_S(t + \Delta t) = F(t + \Delta t) \tag{15.20}$$

Subtracting Eq. (15.19) from Eq. (15.20) yields the incremental equations of motion given by

$$\Delta F_I + \Delta F_D + \Delta F_S = \Delta F \tag{15.21}$$

where ΔF_I, ΔF_D, ΔF_S, and ΔF represent the incremental forces of inertia, damping, spring force, and external force, respectively, given by

$$
\begin{aligned}
\Delta F_I &= F_I(t + \Delta t) - F_I(t) \\
\Delta F_D &= F_D(t + \Delta t) - F_D(t) \\
\Delta F_S &= F_S(t + \Delta t) - F_S(t) \\
\Delta F &= F(t + \Delta t) - F(t)
\end{aligned}
\tag{15.22}
$$

Since the (material) nonlinearity is associated only with the incremental force ΔF_S, then

$$\Delta F_I = m\,\Delta \ddot{x} \tag{15.23}$$

and

$$\Delta F_D = c\,\Delta \dot{x} \tag{15.24}$$

where $\Delta \ddot{x}$ and $\Delta \dot{x}$ represent incremental acceleration and velocity, respectively. Then the incremental equation of motion given by Eq. (15.21) may be rewritten as

$$m\,\Delta \ddot{x} + c\,\Delta \dot{x} + F_S(t + \Delta t) - F_S(t) = \Delta F \tag{15.25}$$

If it is assumed that the stiffness is constant over the small time interval Δt and equal to the stiffness at the beginning of the time interval, then

$$F_S(t + \Delta t) - F_S(t) = k_T\,\Delta x = \Delta F \tag{15.26}$$

where k_T is the *tangent stiffness* as shown in Figure 15.11 and Δx is the displacement increment over the time interval Δt. Therefore the incremental equation of motion can be written as

$$m\,\Delta \ddot{x} + c\,\Delta \dot{x} + k_T\,\Delta x = \Delta F \tag{15.27}$$

The equivalence represented by Eq. (15.26) is not exact as illustrated in Figure 15.11. The exact equivalence is represented by

$$F_S(t + \Delta t) - F_S(t) = k_S\,\Delta x \tag{15.28}$$

where k_S is the *secant stiffness*. However, the secant stiffness cannot be determined because the displacement at time $t + \Delta t$ is not known and therefore k_T must be used. The error introduced into the approximation given by Eq. (15.26) can be minimized if a sufficiently small integration time step is employed in the analysis. Several solution methods for the nonlinear equations of motion are discussed in Section 15.4.

15.4 NUMERICAL SOLUTION OF NONLINEAR EQUILIBRIUM EQUATIONS

The numerical solution of the nonlinear dynamic response of vibrating systems can be summarized in three steps: (1) the incremental formulation of the equations of motion as discussed in Section 15.3, (2) application of an appropriate time integration algorithm as discussed in Chapters 7 and 13, and (3) implementation of an iterative solution procedure (for implicit temporal integration schemes only). The specific details of the solution methodology are highly dependent on the nature of the time integration method selected, that is, *implicit* or *explicit.*

15.4.1 Implicit Integration Without Equilibrium Iterations

The implicit time integration schemes discussed in Chapters 7 and 13 can also be employed in nonlinear dynamic response calculations with slight modifications. The algorithms for the Newmark method and the Wilson method presented in Chapter 13 can be altered to accommodate nonlinear response calculations by substituting the *initial tangent stiffness* (i.e., the tangent stiffness calculated at the beginning of the time step Δt as was illustrated in Figure 15.11) for the linear stiffness term.

For the unconditionally stable Newmark method with integration constants $\alpha = 1/4$ and $\delta = 1/2$, the step-by-step solution algorithm (without iterations) for nonlinear SDOF systems is presented in Table 15.3. The step-by-step solution algorithm without iterations for the Wilson method with $\theta = 1.4$ is presented in Table 15.4.

15.4.1.1 Newmark Method ($\alpha = 1/4$ and $\delta = 1/2$)

For the unconditionally stable Newmark method (constant average acceleration method with integration parameters $\alpha = 1/4$ and $\delta = 1/2$), the acceleration at time $t + \Delta t$ is given by [see Eq. (7.30)]

$$\ddot{x}_{t+\Delta t} = \frac{4}{(\Delta t)^2}(x_{t+\Delta t} - x_t - \Delta t \dot{x}_t) - \ddot{x}_t \tag{15.29}$$

To express Eq. (15.29) in incremental form, assume $x_{t+\Delta t} = x_t + \Delta x$ and $\ddot{x}_{t+\Delta t} = \ddot{x}_t + \Delta \ddot{x}$. Then from Eq. (15.29)

$$\Delta \ddot{x} = \frac{4}{(\Delta t)^2}\left[\Delta x - \Delta t \dot{x}_t - \frac{(\Delta t)^2}{2}\ddot{x}_t\right] \tag{15.30}$$

Similarly, the velocity at time $t + \Delta t$ [see Eq. (7.26)] is expressed as

$$\dot{x}_{t+\Delta t} = \dot{x}_t + \frac{\Delta t}{2}(\ddot{x}_t + \ddot{x}_{t+\Delta t}) \tag{15.31}$$

With $\dot{x}_{t+\Delta t} = \dot{x}_t + \Delta \dot{x}$ and Eq. (15.30) substituted into Eq. (15.31), the incremental expression for $\Delta \dot{x}$ becomes

$$\Delta \dot{x} = \frac{2\,\Delta x}{\Delta t} - 2\dot{x}_t \tag{15.32}$$

TABLE 15.3. **Newmark's Method for Nonlinear SDOF Systems: Without Equilibrium Iterations ($\alpha = 1/4$, $\delta = 1/2$)**

A. Initial calculations:
   1. Input m, c, k_{T_0} (initial tangent stiffness)
   2. Calculate $\ddot{x}_0$ from initial conditions

$$\ddot{x}_0 = \frac{1}{m}[F(0) - c\dot{x}_0 - k_{T_0}x_0]$$

   3. Select time-step size Δt
B. For each time step:
   1. Calculate the incremental force ΔF

$$\Delta F = F_{t + \Delta t} - F_t$$

   2. Calculate the effective incremental force $\Delta \hat{F}$

$$\Delta \hat{F} = \Delta F + m\left(\frac{4\dot{x}_t}{\Delta t} + 2\ddot{x}_t\right) + 2c\dot{x}_t$$

   3. Determine the tangent stiffness k_T at time t
   4. Calculate the effective tangent stiffness $\hat{k}_T$

$$\hat{k}_T = \left[\frac{4m}{(\Delta t)^2} + \frac{2c}{\Delta t} + k_T\right]$$

   5. Solve for incremental displacements Δx from

$$\Delta x = \frac{\Delta \hat{F}}{\hat{k}_T}$$

   6. Calculate $\Delta \dot{x}$ and $\Delta \ddot{x}$

$$\Delta \dot{x} = \frac{2\,\Delta x}{\Delta t} - 2\dot{x}_t$$

$$\Delta \ddot{x} = \frac{4}{(\Delta t)^2}\left[\Delta x - \Delta t\dot{x}_t - \frac{(\Delta t)^2}{2}\ddot{x}_t\right]$$

   7. Calculate displacements, velocities, and accelerations at time $t + \Delta t$

$$x_{t + \Delta t} = x_t + \Delta x$$
$$\dot{x}_{t + \Delta t} = \dot{x}_t + \Delta \dot{x}$$
$$\ddot{x}_{t + \Delta t} = \ddot{x}_t + \Delta \ddot{x}$$

TABLE 15.4. **Wilson-θ Method for Nonlinear SDOF Systems: Without Equilibrium Iterations**

A. Initial calculations:
  1. Input m, c, k_{T_0} (initial tangent stiffness)
  2. Calculate $\ddot{x}_0$ from initial conditions

$$\ddot{x}_0 = \frac{1}{m}[F(0) - c\dot{x}_0 - k_{T_0}x_0]$$

  3. Select time-step size Δt and integration parameter θ
B. For each time step:
  1. Calculate the incremental force $\theta\,\Delta F$

$$\theta\,\Delta F = F_{t+\theta\Delta t} - F_t$$

  2. Calculate the effective incremental force $\theta\,\Delta\hat{F}$

$$\theta\,\Delta\hat{F} = \theta\,\Delta F + m\left(\frac{6\dot{x}_t}{\theta\,\Delta t} + 3\ddot{x}_t\right) + c\left(3\dot{x}_t + \frac{\theta\,\Delta t}{2}\ddot{x}_t\right)$$

  3. Determine the tangent stiffness k_T at time t
  4. Calculate the effective tangent stiffness $\hat{k}_T$

$$\hat{k}_T = \frac{6m}{(\theta\,\Delta t)^2} + \frac{3}{\theta\,\Delta t} + k_T$$

  5. Solve for incremental displacements $\theta\,\Delta x$ from

$$\hat{k}_T\,\theta\,\Delta x = \theta\,\Delta\hat{F} \rightarrow \theta\,\Delta x = \frac{\theta\,\Delta F}{\hat{k}_T}$$

  6. Calculate $\theta\,\Delta\ddot{x}$ and $\theta\,\Delta\dot{x}$

$$\theta\,\Delta\ddot{x} = \frac{6}{(\theta\,\Delta t)^2}\left[\theta\,\Delta x - \theta\,\Delta t\dot{x}_t - \frac{(\theta\,\Delta t)^2}{2}\ddot{x}_t\right]$$

$$\theta\,\Delta\dot{x} = \theta\,\Delta t\ddot{x}_t + \frac{\theta\,\Delta t}{2}(\theta\,\Delta\ddot{x})$$

  7. Calculate displacements, velocities, and accelerations at time $t + \Delta t$

$$x_{t+\Delta t} = x_t + \theta\frac{\Delta x}{\theta}$$

$$\dot{x}_{t+\Delta t} = \dot{x}_t + \theta\frac{\Delta\dot{x}}{\theta}$$

$$\ddot{x}_{t+\Delta t} = \ddot{x}_t + \theta\frac{\Delta\ddot{x}}{\theta}$$

Substitution of Eqs. (15.30) and (15.32) into Eq. (15.25) results in

$$\left[\frac{4m}{(\Delta t)^2} + \frac{2c}{\Delta t} + k_T\right]\Delta x = \Delta F + m\left(\frac{4\dot{x}_t}{\Delta t} + 2\ddot{x}_t\right) + 2c\dot{x}_t \qquad (15.33)$$

or

$$\hat{k}_T\,\Delta x = \Delta\hat{F} \qquad (15.34)$$

where $\hat{k}_T$ is the *effective tangent stiffness* given by

$$\hat{k}_T = \frac{4m}{(\Delta t)^2} + \frac{2c}{\Delta t} + k_T \qquad (15.35)$$

and $\Delta\hat{F}$ is the *effective incremental force* expressed as

$$\Delta\hat{F} = \Delta F + m\left(\frac{4\dot{x}_t}{\Delta t} + 2\ddot{x}_t\right) + 2c\dot{x}_t \qquad (15.36)$$

The incremental displacement is obtained from Eq. (15.34). Substitution of this value into Eq. (15.32) yields the incremental velocity. The incremental acceleration can then by determined from Eq. (15.30). The system kinematics at time $t + \Delta t$ are then obtained by adding Δx, $\Delta\dot{x}$, and $\Delta\ddot{x}$ to x_t, $\dot{x}_t$, and $\ddot{x}_t$, respectively. The incremental solution algorithm for the Newmark method ($\alpha = 1/4$, $\delta = 1/2$) is presented in Table 15.3.

15.4.1.2 Wilson-θ Method

As a second implicit integration scheme, consider the Wilson-θ method, where θ is the integration parameter for stability and accuracy. From Eq. (13.14), the displacement at time $t + \theta\,\Delta t$ is given by

$$x_{t+\theta\Delta t} = x_t + \theta\,\Delta t\dot{x}_t + \frac{(\theta\,\Delta t)^2}{3}\ddot{x}_t + \frac{(\theta\,\Delta t)^2}{6}\ddot{x}_{t+\theta\Delta t} \qquad (15.37)$$

To express Eq. (15.37) in incremental form, assume $x_{t+\theta\,\Delta t} = x_t + \theta\,\Delta x$ and $\ddot{x}_{t+\theta\Delta t} = \ddot{x}_t + \theta\,\Delta\ddot{x}$. Equation (15.37) may then be expressed as

$$\theta\,\Delta x = \theta\,\Delta t\dot{x}_t + \frac{(\theta\,\Delta t)^2}{2}\ddot{x}_t + \frac{(\theta\,\Delta t)^2}{6}\theta\,\Delta\ddot{x} \qquad (15.38)$$

With $\theta\,\Delta\ddot{x}$ then expressed in terms of $\theta\,\Delta x$, Eq. (15.38) can be written as

$$\theta\,\Delta\ddot{x} = \frac{6}{(\theta\,\Delta t)^2}\left[\theta\,\Delta x - \theta\,\Delta t\dot{x}_t - \frac{(\theta\,\Delta t)^2}{2}\ddot{x}_t\right] \qquad (15.39)$$

Next, considering the velocity at time $t + \theta\,\Delta t$, from Eq. (13.12)

$$\dot{x}_{t+\theta\Delta t} = \dot{x}_t + \frac{\theta\,\Delta t}{2}(\ddot{x}_t + \ddot{x}_{t+\theta\Delta t}) \qquad (15.40)$$

Noting that $\dot{x}_{t+\theta\Delta t} = \dot{x}_t + \theta\,\Delta\dot{x}$, $\ddot{x}_{t+\theta\Delta t} = \ddot{x}_t + \theta\,\Delta\ddot{x}$ and using Eq. (15.40) results in

$$\theta\,\Delta\dot{x} = \theta\,\Delta t\ddot{x}_t + \frac{\theta\,\Delta t}{2}(\theta\,\Delta\ddot{x}) \tag{15.41}$$

Then substituting Eq. (15.39) into Eq. (15.41) yields

$$\theta\,\Delta\dot{x} = \frac{3\theta\,\Delta x}{\theta\,\Delta t} - 3\dot{x}_t - \frac{\theta\,\Delta t}{2}\ddot{x}_t \tag{15.42}$$

Substitution of Eqs. (15.39) and (15.42) into Eq. (15.27) gives

$$\left[\frac{6m}{(\theta\,\Delta t)^2} + \frac{3c}{\theta\,\Delta t} + k_T\right]\theta\,\Delta x = \theta\,\Delta F + m\left(\frac{6\dot{x}_t}{\theta\,\Delta t} + 3\ddot{x}_t\right)$$
$$+\ c\left(3\dot{x}_t + \frac{\theta\,\Delta t}{2}\ddot{x}_t\right) \tag{15.43}$$

or

$$\hat{k}_T\theta\,\Delta x = \theta\,\Delta\hat{F} \tag{15.44}$$

where

$$\hat{k}_T = \frac{6m}{(\theta\,\Delta t)^2} + \frac{3c}{\theta\,\Delta t} + k_T \tag{15.45}$$

and the effective incremental force is

$$\theta\,\Delta\hat{F} = \theta\,\Delta F + m\left(\frac{6\dot{x}_t}{\theta\,\Delta t} + 3\ddot{x}_t\right) + c\left(3\dot{x}_t + \frac{\theta\,\Delta t}{2}\ddot{x}_t\right) \tag{15.46}$$

in which $\theta\,\Delta F = F_{t+\theta\,\Delta t} - F_t$.

The incremental displacement $\theta\,\Delta x$ is obtained from Eq. (15.44). Substitution of this value into Eq. (15.42) provides the incremental velocity $\theta\,\Delta\dot{x}$. The incremental acceleration $\theta\,\Delta\ddot{x}$ can then be determined from Eq. (15.39).

$$\ddot{x}_{t+\Delta t} = \ddot{x}_t + \theta\frac{\Delta\ddot{x}}{\theta}$$

Once the incremental kinematics $\theta\,\Delta x$, $\theta\,\Delta\dot{x}$, and $\theta\,\Delta\ddot{x}$ have been determined, dividing them by θ and adding them to x_t, $\dot{x}_t$, and $\ddot{x}_t$, respectively, yields the solution at time $t + \Delta t$.

The solution algorithm for the Wilson-θ method is presented in Table 15.4. For unconditional stability, θ must be greater than 1.37. A value of 1.4 for θ is generally chosen. Note that for $\theta = 1$, the method becomes the linear acceleration method.

15.4.2 Implicit Integration with Equilibrium Iterations

The numerical solution of nonlinear dynamic response can be obtained by using implicit time integration algorithms without equilibrium iterations for a few special cases (one of these cases will be discussed in the next section). However, in general, if equilibrium iterations are not performed in an implicit incremental analysis, errors will be admitted into the incremental solution. These errors are attributed to two sources: (1) use of the tangent stiffness k_T in place of the secant stiffness k_S (refer to Figure 15.11) and (2) delay in the detection of transitions in the force-displacement relationship. Moreover, since any errors admitted in the incremental solution at a particular time affects in a path-dependent manner the solution at any subsequent time, omission of equilibrium iterations from the incremental solution scheme could render the analysis highly inaccurate.

The most commonly employed equilibrium iteration schemes for the solution of nonlinear equations in structural mechanics are the Newton-Raphson iteration and closely related techniques. We consider first the *modified Newton-Raphson* technique illustrated in Figure 15.12. For both the Newmark and Wilson methods the iterations are performed on the incremental equilibrium condition given by

$$\hat{k}_T \, \Delta x \; = \; \Delta \hat{F} \tag{15.47}$$

where $\hat{k}_T$ and $\Delta\hat{F}$ are defined for the Newmark and Wilson methods, respectively, by Eqs. (15.35) and (15.45), and Eqs. (15.36) and (15.46). In the modified Newton-Raphson method the tangent stiffness k_T is determined at the beginning of the time step (i.e., at time t) and remains unchanged through the iterative procedure, and is thus referred to in this context as the *initial tangent stiffness*.

With reference to Figure 15.12, the iterative solution of Eq. (15.47) is initiated as

$$\hat{k}_T \, \Delta x^{(1)} \; = \; \Delta \hat{F} \tag{15.48}$$

to determine $\Delta x^{(1)}$ as the first approximation of Δx. Associated with $\Delta x^{(1)}$ is the actual incremental force $\Delta P^{(1)}$ that is less than $\Delta\hat{F}$, resulting in an incremental residual force

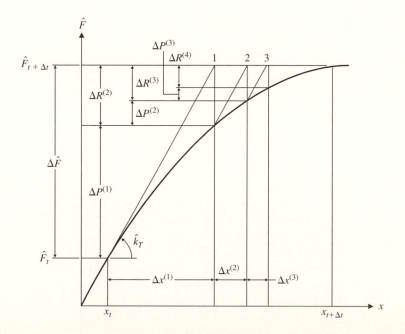

Figure 15.12

Modified Newton-Raphson iteration.

$\Delta R^{(2)} = \Delta \hat{F} - \Delta \mathrm{P}^{(1)}$. The increase in incremental displacement corresponding to the incremental residual force $\Delta R^{(2)}$ is given by

$$\hat{k}_T \, \Delta x^{(2)} = \Delta \hat{F} - \Delta P^{(1)} = \Delta R^{(2)} \tag{15.49}$$

This additional incremental displacement $\Delta x^{(2)}$ is then used to determine a new residual incremental force, and the procedure is continued until convergence is achieved. Within a time step, the iteration procedure is described as follows:

$$\Delta x^{(k)} = \frac{R^{(k)}}{\hat{k}_T} \tag{15.50}$$

$$x_{t+\Delta t}^{(k)} = x_{t+\Delta t}^{(k-1)} + \Delta x^{(k)} \tag{15.51}$$

$$\Delta P^{(k)} = F_S^{(k)} - F_S^{(k-1)} + (\hat{k}_T - k_T) \, \Delta x^{(k)} \tag{15.52}$$

$$\Delta R^{(k+1)} = \Delta R^{(k)} - \Delta P^{(k)} \tag{15.53}$$

where F_S represents the restoring force in the system and the superscript k represents the iteration counter. The procedure is initialized at $k = 1$ with the initial conditions $x_{t+\Delta t}^{(0)} = x_t$, $\Delta R^{(1)} = \Delta \hat{F}$, $F_S^{(0)} = (F_S)_t$. The iterations are continued until convergence, that is, when the incremental displacements $\Delta x^{(k)}$ have become sufficiently small in comparison to the current estimate of Δx. Thus since the total incremental displacement after m iterations is given by

$$\Delta x = \sum_{k=1}^{m} \Delta x^{(k)} \tag{15.54}$$

iterations may be terminated at iteration m when

$$\frac{\Delta x^{(m)}}{\Delta x} \le \mathrm{TOL} \tag{15.55}$$

where TOL is the specified convergence tolerance.

The incremental solution algorithms for the Newmark and Wilson methods described in Tables 15.3 and 15.4, respectively, may be modified to include equilibrium iterations by implementing the iteration procedure defined by Eqs. (15.50) to (15.53) in step B.5 of either table to solve for the incremental displacement Δx.

The modified Newton-Raphson procedure discussed above is adequate for systems exhibiting mild to moderate nonlinearities (i.e., the nonlinearities are localized, such as to the spring element). For systems manifesting general nonlinear behavior and/or exhibiting several different types of nonlinearity the modified Newton-Raphson method exhibits a slow convergence rate or may not converge at all. An improvement to the modified Newton-Raphson method is the full Newton-Raphson method or simply the Newton-Raphson method illustrated in Figure 15.13. In this method the *current tangent stiffness* $k_T^{(k)}$ (and the corresponding value of the effective stiffness $\hat{k}_T^{(k)}$) is used in place of the initial tangent stiffness k_T in Eqs. (15.50) through (15.53). The current tangent stiffness is evaluated at each iteration, resulting in a smaller incremental residual force $\Delta R^{(k)}$ that provides an improved rate of convergence. However, evaluation of the tangent stiffness at each iteration requires significant additional computational effort, particularly for large MDOF systems.

As a final note on the convergence of nonlinear solution schemes, dynamic problems generally exhibit better convergence characteristics than the corresponding static problem. This is because the inertia of the system renders its dynamic response "more smooth" than its static response, and the convergence behavior of dynamic systems can be improved by decreasing Δt. Consider the expression for $\hat{k}_T$ given by Eq. (15.35) or (15.45). It is recognized that $\hat{k}_T$ is the slope of the nonlinear force-displacement relationships illustrated in Figures 15.12 and 15.13, which is dependent on displacement and time. In a dynamic analysis the presence of the inertia term in the expression for $\hat{k}_T$ decreases the severity of its nonlinearity and ultimately becomes dominant as the time step decreases. In static response, $\hat{k}_T$ is equal to k_T and the nonlinearity is subsequently greater.

15.4.3 Explicit Integration

The central difference method is probably the most frequently used explicit time integration operator in nonlinear dynamic analysis. Similar to linear analysis, when using the control difference method in nonlinear dynamic analysis, equilibrium of the system is considered at time t to calculate the solution at time $t + \Delta t$. This is in direct contrast to the implicit methods that must satisfy equilibrium at time $t + \Delta t$. Therefore, equilibrium iterations are not performed, and we operate on the equation

$$m\ddot{x}_t + c\dot{x}_t = F_t - (F_S)_t \tag{15.56}$$

where $(F_S)_t$ is the restoring force in the system at time t that replaces kx_t. From Eqs. (7.10) and (7.11), respectively, the central difference approximations for velocity and acceleration are given by

$$\dot{x}_t = \frac{x_{t+\Delta t} - x_{t-\Delta t}}{2\,\Delta t} \tag{15.57}$$

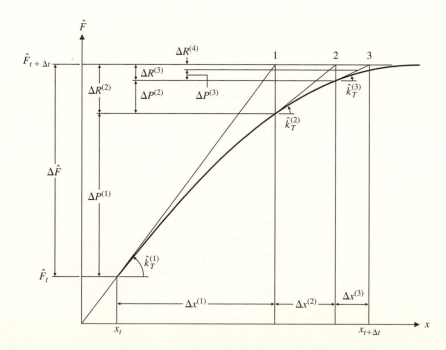

Figure 15.13
Full Newton-Raphson iteration.

and

$$\ddot{x}_t = \frac{x_{t+\Delta t} - 2x_t + x_{t-\Delta t}}{(\Delta t)^2} \tag{15.58}$$

Substituting Eqs. (15.57) and (15.58) into Eq. (15.56) results in

$$m\left(\frac{x_{t+\Delta t} - 2x_t + x_{t-\Delta t}}{(\Delta t)^2}\right) + c\left(\frac{x_{t+\Delta t} - x_{t-\Delta t}}{2\Delta t}\right) = F_t - (F_S)_t \tag{15.59}$$

The only unknown quantity in Eq. (15.59) is the displacement $x_{t+\Delta t}$. Isolating this unknown quantity on the left-hand side results in

$$\left[\frac{m}{(\Delta t)^2} + \frac{c}{2\Delta t}\right]x_{t+\Delta t} = F_t - (F_S)_t$$

$$-\left[\frac{m}{(\Delta t)^2} - \frac{c}{2\Delta t}\right]x_{t-\Delta t} + \frac{2m}{(\Delta t)^2}x_t \tag{15.60a}$$

or

$$\hat{m}x_{t+\Delta t} = \hat{F}_t \tag{15.60b}$$

where $\hat{m}$ is called the effective mass expressed by

$$\hat{m} = \frac{m}{(\Delta t)^2} + \frac{c}{2\Delta t} \tag{15.61}$$

and $\hat{F}_t$ is the effective force given by

$$\hat{F}_t = F_t - (F_S)_t + \frac{2m}{(\Delta t)^2}x_t - \left[\frac{m}{(\Delta t)^2} - \frac{c}{2\Delta t}\right]x_{t-\Delta t} \tag{15.62}$$

The displacement at time $t + \Delta t$ is then determined from Eq. (15.60b), and the velocity and acceleration at time t are determined from Eqs. (15.57) and (15.58), respectively. The solution therefore simply corresponds to a forward marching in time. The step-by-step solution algorithm for the central difference method is presented in Table 15.5.

The central difference method is very effective for nonlinear dynamic analysis of SDOF systems. The major shortcoming with the use of the central difference method lies in the severe time-step restriction for large MDOF systems. For stability of the method, the time-step size Δt must be smaller than a critical time step Δt_{cr}, which is equal to T_n/π, where T_n is the smallest vibration period exhibited by the system. In linear analyses T_n does not change; however, when a nonlinear analysis is conducted, the stiffness properties change during the response calculations, and therefore the value of T_n is not constant throughout the solution time domain. If the system "stiffens" (that is, T_n decreases), Δt must be decreased in a conservative manner so that the condition $\Delta t \leq T_n/\pi$ is satisfied at all times and stability is maintained.

TABLE 15.5. Central Difference Method for Nonlinear SDOF Systems

A. Initial calculations:
    1. Input m, c, k_{T_0} (initial tangent stiffness at time $t = 0$)
    2. Calculate $\ddot{x}_0$ from initial conditions

$$\ddot{x}_0 = \frac{1}{m}[F(0) - c\dot{x}_0 - k_{T_0}x_0]$$

    3. Select time step size Δt
    4. Calculate $x_{t-\Delta t}$ at time $t = 0$

$$x_{t-\Delta t} = x_0 - \dot{x}_0\,\Delta t + \ddot{x}_0\frac{(\Delta t)^2}{2}$$

    5. Calculate effective mass $\hat{m}$

$$\hat{m} = \frac{m}{(\Delta t)^2} + \frac{c}{2\,\Delta t}$$

B. For each time step:
    1. Determine the restoring force $(F_S)_t$ at time t
    2. Calculate the effective force at time t

$$\hat{F}_t = F_t - (F_S)_t + \frac{2m}{(\Delta t)^2}x_t - \left[\frac{m}{(\Delta t)^2} - \frac{c}{2\,\Delta t}\right]x_{t-\Delta t}$$

    3. Calculate displacements at time $t + \Delta t$

$$x_{t+\Delta t} = \frac{\hat{F}_t}{\hat{m}}$$

    4. Evaluate accelerations and velocities at time t

$$\ddot{x}_t = \frac{1}{(\Delta t)^2}(x_{t-\Delta t} - 2x_t + x_{t+\Delta t})$$

$$\dot{x}_t = \frac{1}{2\,\Delta t}(-x_{t-\Delta t} + x_{t+\Delta t})$$

15.5 RESPONSE OF ELASTOPLASTIC SDOF SYSTEMS

The inelastic response of many structural steel and reinforced concrete structures subject to extreme loadings can be characterized by elastoplastic behavior. Although excursions beyond the elastic range are usually not permitted under normal conditions of operation, the extent of permanent damage a structure may sustain when subjected to extreme conditions, such as severe blast or earthquake loading, is frequently of interest to the design engineer.

To clarify the discussion, consider the one-story structural steel shear frame building subjected to a horizontal static force F as shown in Figure 15.14. The girder in the structure is considered to be infinitely rigid relative to the columns. Therefore, as the load is monotonically increased, plastic hinges will eventually form at the ends of the columns. The plot of the resistance versus displacement relationship presented in Figure 15.15 is linear up to point a corresponding to a resistance R_y, where *first yielding* in the cross section occurs. As the load is increased, the resistance curve becomes nonlinear as the column cross section plasticizes and the system softens. Plastic hinges form when full plastification of the cross section is attained at point b corresponding to maximum resistance R_m. Upon unloading, the system rebounds elastically along the line bc, parallel to the initial linear portion of the curve Oa. Upon complete unloading, the resistance-displacement curve terminates at point c. If the loading is then reversed, the resistance-displacement curve remains linear until first yielding is again attained at point d, corresponding to a resistance $-R_y$. As the load is increased further, plastic hinges reform at $-R_m$, corresponding to point e. Unloading will be linear elastic along a curve parallel to line cd. If the maximum positive and negative restoring forces R_m and $-R_m$ are numerically equal, the *hysteresis loop* formed by the cyclic loading will be symmetric with respect to the origin. Energy is dissipated during each load cycle by an amount that is proportional to the area within the hysteresis loop.

The behavior illustrated in Figure 15.15 is often simplified by assuming linear behavior to the point of full plastification (i.e., when the resistance is R_m or $-R_m$) beyond which additional displacement occurs at a constant value for the restoring force without any further increase in load. This type of behavior is referred to as *elastoplastic*. The corresponding resistance-displacement curve is illustrated in Figure 15.16. The slope of the elastic loading and unloading curves is proportional to the elastic stiffness of the structure.

For a SDOF system exhibiting elastoplastic behavior, expressions for the restoring force are easily written and incorporated into the time integration algorithms discussed in Section 15.4. These expressions depend on the magnitude of the restoring force as well as on whether the displacement in the system is increasing ($\dot{x} > 0$) or decreasing ($\dot{x} < 0$). Consider the general elastoplastic cycle for a SDOF system shown in Figure 15.17. The displacement response history of an elastoplastic system can be described in five stages to complete one hysteresis loop (with reference to Figure 15.17):

Stage 1. Elastic loading

In this stage, defined by the segment Oa on the resistance-displacement curve, $0 \leq x \leq x_{el}$ and $\dot{x} > 0$, where x_{el} is the elastic limit displacement at which the system plasticizes and is equal to R_m/k. The restoring force for this stage is given by

$$F_S = kx \tag{15.63}$$

Unloading in this stage occurs when $\dot{x} < 0$.

Stage 2. Plastic loading

This stage is represented by the segment ab on the resistance-displacement curve and corresponds to the conditions $x_{el} < x < x_{max}$ and $\dot{x} > 0$, where x_{max} is the maximum displacement value within the hysteresis loop. The restoring force in this stage is given by

$$F_S = R_m \tag{15.64}$$

Stage 3. Elastic rebound

This stage is defined by the segment bc on the resistance-displacement curve and corresponds to the conditions $(x_{max} - 2x_{el}) < x < x_{max}$ and $\dot{x} < 0$. The system resistance is defined by

$$F_S = R_m - k(x_{max} - x) \tag{15.65}$$

Note that load reversal in this stage occurs when $x < (x_{max} - 2x_{el})$ and $\dot{x} < 0$.

Stage 4. Plastic loading

The system response in this stage is represented by segment cd on the resistance-displacement curve and corresponds to the conditions $x_{min} < x < (x_{max} - 2x_{el})$ and $\dot{x} < 0$, where x_{min} is the minimum displacement in this stage. The system resistance in this stage is given by

$$F_S = -R_m \tag{15.66}$$

Stage 5. Elastic rebound

One cycle of hysteresis is completed and the system unloads elastically along segment de of the resistance-displacement curve. This stage corresponds to the condition $x_{min} < x < (x_{min} + 2x_{el})$ and $\dot{x} > 0$. The expression for the restoring force in this stage is given by

$$F_S = k(x - x_{min}) - R_m \tag{15.67}$$

As a final note on elastoplastic response, the system stiffness is equal to the elastic stiffness k for all stages of elastic loading and unloading (i.e., stages 1, 3, and 5) and is equal to zero in the plastic stages (i.e., stages 2 and 4). Also, the tangent stiffness k_T is equal to k in the elastic stages and equal to zero in the plastic stages. This information, when used in conjunction with the simplified restoring force expression given by Eqs. (15.63) through (15.67), may be implemented into both the implicit and explicit temporal integration algorithms discussed in Section 15.4 to evaluate the dynamic response of elastoplastic SDOF systems. Moreover, if a sufficiently small integration time step is used, accurate results can be obtained from the implicit algorithms without performing equilibrium iterations. Numerical solutions for the response of elastoplastic SDOF systems are illustrated in the following examples.

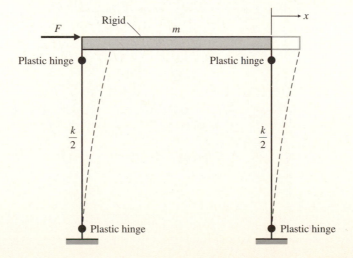

Figure 15.14

Elastoplastic shear frame structure.

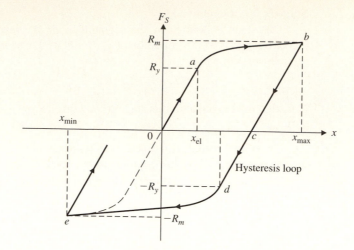

Figure 15.15
General plastic behavior.

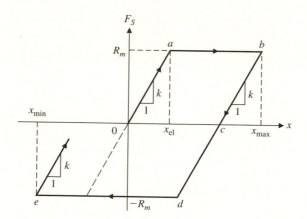

Figure 15.16
Idealized elastoplastic
behavior.

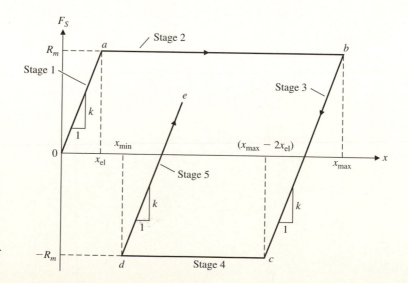

Figure 15.17
Elastoplastic resistance-
displacement relationship for
SDOF systems.

EXAMPLE 15.1 ▼

The shear frame structure shown in Figure 15.18 is subjected to the time-varying force shown in Figure 15.19. The system exhibits an elastoplastic response function similar to that shown in Figure 15.17. Evaluate the elastoplastic response of the structure by the Newmark method without equilibrium iterations. Compare the elastoplastic response with the pure elastic response. Assume $k = 12.35$ kips/in, $c = 0.27$ kip-sec/in, $m = 0.20$ kip-sec^2/in, and $R_m = -R_m = 15$ kips. Use a time step $\Delta t = 0.001$ sec.

Solution

The yield or elastic limit displacement x_{el} is given as

$$x_{el} = \frac{R_m}{k} = 1.215 \text{ in}$$

The natural period for the elastic structure is $T = 0.8$ sec. The algorithm for the Newmark method summarized in Table 15.3 was employed to evaluate both the elastic and elastoplastic responses. The FORTRAN computer program, PLAS,

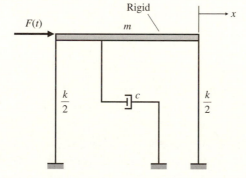

Figure 15.18

Elastoplastic shear frame structure of Example 15.1.

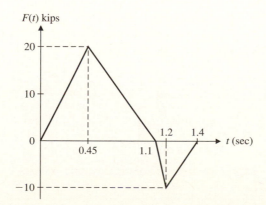

Figure 15.19

Time-varying force.

Figure 15.20

Displacement response histories $x(t)$ for Example 15.1.

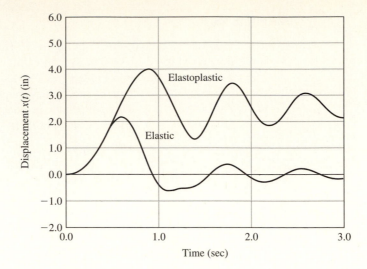

used to evaluate the dynamic response is available on the authors' Web site (refer to the Preface). The time-history responses $x(t)$ for the elastic and elastoplastic systems are presented in Figure 15.20. It is clearly illustrated that the elastoplastic structure exhibits substantial permanent deformation (approximately 2.5 in). It is also evident from Figure 15.20 that the natural period of the elastoplastic structure, T_{ep}, has increased ($T_{ep} \cong 0.9$ sec) approximately 12.5% compared to the natural period of purely elastic structure. This increase in period is typical of "softening" structures.

▲

EXAMPLE 15.2 ▼

A simply supported W18×50 steel beam spanning 20 ft supports a dead load of 10 kips at its midspan as shown in Figure 15.21. The midspan of the beam is subjected to an ideal step force of 30 kips as shown in Figure 15.22. Evaluate both the elastic and elastoplastic response of the beam by the Newmark method (without equilibrium iterations) for damping factors $\zeta = 0$ and $\zeta = 0.04$. Assume modulus of elasticity $E = 29,000$ ksi and yields stress $F_y = 36$ ksi for the steel beam. Use a time step $\Delta t = 0.001$ sec.

Figure 15.21

Elastoplastic beam of Example 15.2.

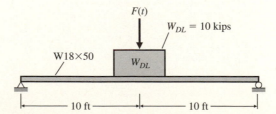

Figure 15.22
Ideal step force.

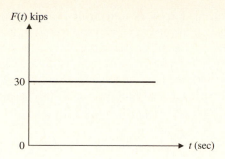

$F(t)$ kips

30

0   t (sec)

Solution

The elastic properties of the system are

$$k = \frac{48EI}{L^3} = \frac{48(29{,}000 \text{ ksi})(802 \text{ in}^4)}{(240 \text{ in})^3} = 80.757 \text{ kips/in}$$

$$T = 2\pi\sqrt{\frac{m}{k}} = 2\pi\sqrt{\frac{0.02588 \text{ kip-sec}^2/\text{in}}{80.757 \text{ kips/in}}} = 0.112 \text{ sec}$$

$$c = 0 \qquad \text{for } \zeta = 0, \text{ and for } \zeta = 0.04$$

$$c = \zeta C_c = \zeta 2\sqrt{mk} = 0.04(2)\sqrt{(0.02588)(80.757)}$$
$$= 0.11565 \text{ kip-sec/in}$$

The *total plastic resistance* R_{mT} is determined as

$$R_{mT} = \frac{4M_P}{L} \tag{1}$$

where L is the span length and M_P is the plastic bending moment given by

$$M_P = ZF_y \tag{2}$$

in which Z is the plastic section modulus. Therefore from Eq. (1)

$$R_{mT} = \frac{4(101 \text{ in}^3)(36 \text{ ksi})}{(240 \text{ in})} = 60.6 \text{ kips}$$

Since the beam supports a dead load of 10 kips, the maximum force available to resist the dynamic load is

$$R_m = R_{mT} - 10 \text{ kips} = 50.6 \text{ kips}$$

And therefore the elastic limit displacement is

$$x_{\text{el}} = \frac{R_m}{k} = 0.626$$

The dynamic responses were again evaluated using the computer program PLAS available on the authors' Web site (refer to the Preface). Displacement response histories $x(t)$ for the elastic and elastoplastic solutions are presented for damping factors of $\zeta = 0$ and $\zeta = 0.04$ in Figures 15.23 and 15.24, respectively. These time histories indicate only slight additional deformation in the elastoplastic beam and that the period increase for the elastoplastic system is negligible.

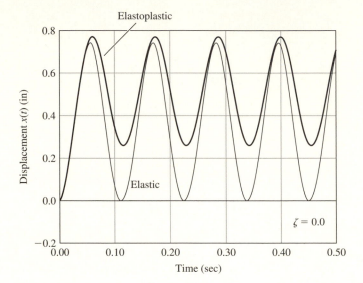

Figure 15.23

Displacement response histories $x(t)$ for Example 15.2 ($\zeta = 0.0$).

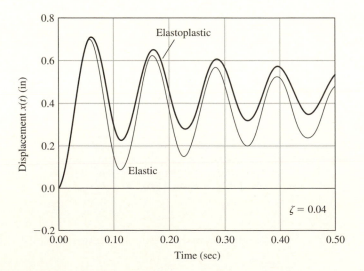

Figure 15.24

Displacement response histories $x(t)$ for Example 15.2 ($\zeta = 0.4$).

EXAMPLE 15.3   ▼

The shear frame structure shown in Figure 15.25 is subjected to the time-varying force shown in Figure 15.26. Evaluate the elastic and elastoplastic response of the structure by the Wilson-θ method without equilibrium iterations for damping factors of $\zeta = 0$ and $\zeta = 0.06$. Assume $m = 0.25$ kip-sec²/in, modulus of elasticity $E = 29,000$ ksi, moment of inertia $I = 110$ in⁴, and maximum resistance $R_m = -R_m = 15$ kips. Use a time step $\Delta t = 0.001$ sec.

Solution

The elastic properties of the system are

$$k = 2\left(\frac{12EI}{L^3}\right) = \frac{2(12)(29,000 \text{ ksi})(110 \text{ in}^4)}{(192 \text{ in})^3} = 10.817 \text{ kips/in}$$

$$T = 2\pi\sqrt{\frac{m}{k}} = 0.955 \text{ sec}$$

and $c = 0$ for $\zeta = 0$; for $\zeta = 0.06$

$$c = \zeta C_c = \zeta 2\sqrt{mk} = 0.06(2)\sqrt{(0.25)(10.817)}$$
$$= 0.1973 \text{ kip-sec/in}$$

and the elastic limit displacement is

$$x_{\text{el}} = \frac{R_m}{k} = 1.3867 \text{ in}$$

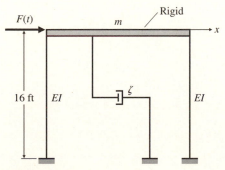

Figure 15.25
Elastoplastic shear frame
structure of Example 15.3.

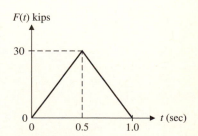

Figure 15.26
Time-varying force.

The algorithm for the Wilson-θ method summarized in Table 15.4 (with $\theta = 1.4$) was employed to evaluate both the elastic and elastoplastic responses. The listing for the FORTRAN computer program is presented in Table 15.6. The displacement response histories $x(t)$ for the elastic and elastoplastic responses are presented for $\zeta = 0$ and $\zeta = 0.06$ in Figures 15.27 and 15.28, respectively. These figures clearly demonstrate the significant permanent deformation sustained by the elastoplastic system.

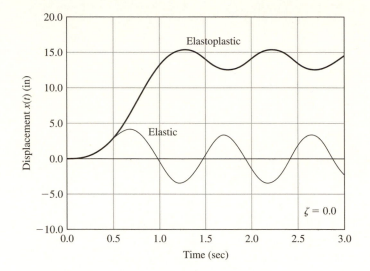

Figure 15.27

Displacement response histories $x(t)$ for Example 15.3 ($\zeta = 0.0$).

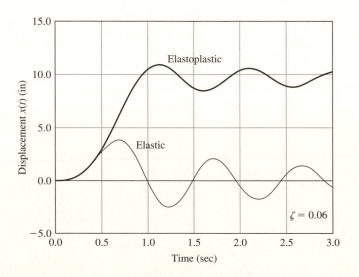

Figure 15.28

Displacement response histories $x(t)$ for Example 15.3 ($\zeta = 0.06$).

TABLE 15.6. Listing of FORTRAN Computer Program Used for Example 15.3

```
*
            program wilson
*
*           Elastoplastic response of SDOF systems
            integer iflag
            double precision k,m,c,a0,a1,a2,a3,a4,a5,a6,
      1         a7,a8,xel,xt,xdt,xdth,xtdot,xdtdot,xt2dot,xdt2dot,
      2         khat,phatdth,pt,pdt,rm,theta,ti,tf,dt,t
*
            open(unit=10,file='x',status="new")
            open(unit=11,file='xdot',status="new")
            open(unit=12,file='xdot2',status="new")
*
            dt=0.001
            ti=0.00
            tf=3.0
            theta=1.4
            xm=1.0e4
*
            a0=6.00/(theta*theta*dt*dt)
            a1=3.00/(theta*dt)
            a2=2.00*a1
            a3=theta*dt/2.00
            a4=a0/theta
            a5=-a2/theta
            a6=1.00-(3.00/theta)
            a7=dt/2.00
            a8=dt*dt/6.00
*
            m=0.25
            c=0.0
            k=10.817
            rm=15.0
            xel=rm/k
*
            xt=0.00
            xtdot=0.00
*
            call load(ti,pt)
            xt2dot=(pt-c*xdot-k*xt)/m
            write(10,3)ti,xt
            write(11,3)ti,xtdot
            write(12,3)ti,xt2dot
      3     format(1x,e13.6,3x,e13.6,3x,e13.6)
*
            do 8 t=ti+dt, tf, dt
              if (xt.lt.xel.and.iflag.eq.0)then
                khat=k+a0*m+a1*c
                call load(t,pdt)
                phatdth=pt+theta*(pdt-pt)+m*(a0*xt+a2*xtdot+2*xt2dot)+
      +           c*(a1*xt+2*xtdot+a3*xt2dot)
                xdth=phatdth/khat
                xdt2dot=a4*(xdth-xt)+a5*xtdot+a6*xt2dot
                xdtdot=xtdot+a7*(xdt2dot+xt2dot)
                xdt=xt+dt*xtdot+a8*(xdt2dot+2*xt2dot)
              endif
*
              if(xt.ge.xel.and.xt.lt.xm.and.iflag.eq.0)then
                khat=a0*m+a1*c
```

(continued)

```
            call load(t,pdt)
            phatdth=pt+theta*(pdt-pt)+m*(a0*xt+a2*xtdot+2*xt2dot)+
     +        c*(a1*xt+2*xtdot+a3*xt2dot)
            xdth=(phatdth-rm)/khat
            xdt2dot=a4*(xdth-xt)+a5*xtdot+a6*xt2dot
            xdtdot=xtdot+a7*(xdt2dot+xt2dot)
            xdt=xt+dt*xtdot+a8*(xdt2dot+2*xt2dot)
            if (xtdot*xdtdot.le.0.00.and.iflag.eq.0)then
              xm=xdt
              iflag=1
            endif
          endif
*
*           if(xt.ge.(xm-2*xel).and.xt.le.xm.and.iflag.eq.1)then
          if (iflag.eq.1)then
            khat=k+a0*m+a1*c
            call load(t,pdt)
            phatdth=pt+theta*(pdt-pt)+m*(a0*xt+a2*xtdot+2*xt2dot)+
     +        c*(a1*xt+2*xtdot+a3*xt2dot)
            xdth=(phatdth-(rm-k*xm))/khat
            xdt2dot=a4*(xdth-xt)+a5*xtdot+a6*xt2dot
            xdtdot=xtdot+a7*(xdt2dot+xt2dot)
            xdt=xt+dt*xtdot+a8*(xdt2dot+2*xt2dot)
            if (xtdot*xdtdot.le.0.00.and.iflag.eq.0)then
              xm=xdt
              iflag=1
            endif
          endif
*
          xt=xdt
          xtdot=xdtdot
          xt2dot=xdt2dot
          pt=pdt
          write(10,3)t,xt
          write(11,3)t,xtdot
          write(12,3)t,xt2dot
8         continue
*
          end
*
*
*
          subroutine load(t,p)
*
          double precision p,t
*
          if(t.le.0.5)then
            p=60.0*t
            goto 1
          endif
          if(t.le.1.0)then
            p=60.0-60.0*t
            goto 1
          endif
          p=0.0
1         return
          end
```

EXAMPLE 15.4 ▼

The shear frame structure shown in Figure 15.29 has its base excited by the horizontal component of ground acceleration $\ddot{x}_g(t)$ specified in Figure 15.30. Evaluate the elastoplastic response of the system by the central difference method. To ensure stability of the explicit central difference algorithm, evaluate the elastic response by the Wilson-θ method for comparison. Assume $m = 0.6$ kip-sec^2/in, damping factor $\zeta = 0.15$, height $h = 12$ ft, $k = 30$ kips/in, and plastic moment $M_P = 105$ ft-kip. Use a time step $\Delta t = 0.001$ sec.

Solution

The elastic properties of the system are

$$T = 2\pi\sqrt{\frac{m}{k}} = 2\pi\sqrt{\frac{(0.6 \text{ kip-sec}^2/\text{in})}{30 \text{ kips/in}}} = 0.888 \text{ sec}$$

$$c = \zeta C_c = \zeta 2\sqrt{mk} = 1.273 \text{ kip-sec/in}$$

The system becomes fully plastic when a plastic hinge has formed at the ends of both columns as illustrated in Figure 15.14. Thus the plastic resistance is given by

$$R_m = \frac{4M_p}{h} = \frac{4(105 \text{ ft-kip})}{12 \text{ ft}} = 35 \text{ kips}$$

and elastic limit displacement is

$$x_{el} = \frac{R_m}{k} = \frac{35 \text{ kips}}{30 \text{kips/in}} = 1.167 \text{ in}$$

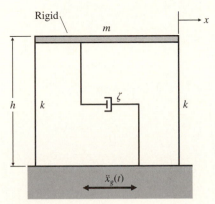

Figure 15.29

Elastoplastic shear frame structure of Example 15.4.

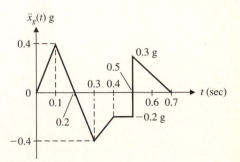

Figure 15.30

Horizontal ground acceleration.

The effective load for ground acceleration is determined by

$$F_{\text{eff}}(t) = m\ddot{x}_g(t)$$

The algorithm for the central difference method summarized in Table 15.5 was employed to evaluate the elastoplastic response and the BASIC computer program used for this example is listed in Table 15.7. The elastic response was evaluated by the Wilson-θ method (with $\theta = 1.4$). The response histories for relative displacement $x(t)$, relative velocity $\dot{x}(t)$, and total (absolute) acceleration $\ddot{x}_t(t)$ are presented in Figures 15.31, 15.32, and 15.33, respectively. The stability and accuracy of the central difference algorithm is corroborated by the elastic response obtained by the implicit Wilson-θ method. [Note that the total acceleration $\ddot{x}_t(t)$ is the sum of the ground acceleration $\ddot{x}_g(t)$ and the relative acceleration $\ddot{x}(t)$.]

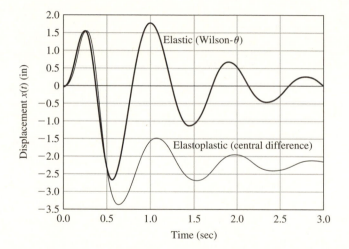

Figure 15.31

Relative displacement response histories $x(t)$ for Example 15.4.

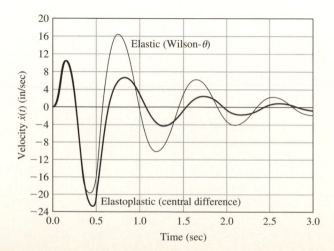

Figure 15.32

Relative velocity response histories $\dot{x}(t)$ for Example 15.4.

Figure 15.33

Absolute (total) acceleration response histories $\ddot{x}(t)$ for Example 15.4.

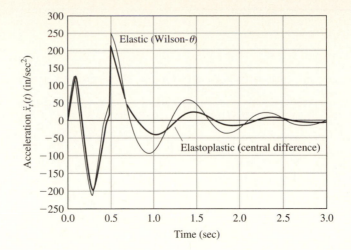

TABLE 15.7. Listing of BASIC Computer Program Used for Example 15.4

```
1       CLS
2       OPEN "XXX" FOR INPUT AS#2
3       INPUT #2,A:CLOSE#2:IF A=5*INT(A/5) THEN SAVE "ELST":PRINT"saving please wait"
4       A=A+1
5       OPEN "XXX" FOR OUTPUT AS#2:PRINT #2,A:CLOSE#2
6       OPEN "P2" FOR OUTPUT AS#1
10      REM Elastoplastic response of SDOF systems by central difference method
20      REAM M,K,Z,DT,N,TMAX
30      DATA 0.60,30.0,0.15,0.001,7,3
31      PRINT "M";M,"K";K,"Z";Z
32      READ X1,X,H:PRINT "X1";X1,"X";X
33      DATA 0,0,2
35      READ RM,RL
36      DATA 35,−35
37      PRINT "RM";RM,"RL";RL
38      ML=1
40      W=SQR(K/M)
45      XEL=RM/K
50      C=Z*2*W*M:DT2=DT*DT:PRINT "DAMPING";C
51      IF H=1 THEN KK=1 ELSE KK=M*386
55      PRINT:PRINT "FORCE"
60      FOR I=1 TO N
70      READ T(I),F(I):F(I)=F(I)*KK:PRINT T(I),F(I)
80      NEXT I
84      T(N+1)=TMAX
85      N4=INT((T(N+1)−T(N))/DT)
90      DATA 0,0,0.1,0.4,0.3,−0.4,0.4,−0.2,0.5,−0.2,0.5,0.3,0.7,0.0
100     X11=F(1)−C*X1−K*X)/M:PRINT "X11O ",X11
110     XP=X−X1*DT+X11*DT2/2
120     M1=M/DT2+.5*C/DT
130     PRINT:PRINT "M1";M1,"X";X,"X11";X11
199     PRINT
210     FOR J=1 TO N
230     IF J=N THEN DF=0:GOTO 250
240     DF=F(J+1)−F(J):DT1=T(J+1)−T(J):N3=DT1/DT
250     IF J=N THEN AS1=N4 ELSE AS1=N3−1
260     FOR G=0 TO AS1
280     I=T(J)+G*DT
290     IF J=N AND G <>0 THEN F=0:GOTO 310
300     F=F(J)+DF/DT1*(I−T(J))
```

(continued)

```
310     'PRINT I,F
316     IF XMAX>X THEN GOTO 320 ELSE XMAX=X
320     IF X < XEL AND X1>=0 AND S=0 THEN FS=K*X:MN=1:GOTO 450
330     IF X>XEL AND X1>0 AND S=0 THEN FS=RM:MN=2:XX=X:GOTO 450
335     S=1
340     IF X>(XX−2*XEL) AND X1<0 AND ML=1 THEN FS=RM−K*(XX−X):MN=3:ML=1:GOTO 450
350     IF ML=1 OR ML=2 AND X1<0 THEN FS=RL:MN=4:XX=X:ML=2:GOTO 450
360     IF X<(XX+2*XEL) AND X1>0 AND ML>=2 THEN FS=RL+K*(X−XX):MN=5:ML=3:GOTO450
370     IF ML<>2 AND X1<0 THEN FS=RM:MN=6:XX=X:ML=1:GOTO 450
380     PRINT "ERROR":FS=RM
450     F1=F−FS+2*M*X/DT2−(M/DT2-C/(2*DT))*XP
460     XN=F1/M1
465     PRINT USING "###.### ";I,X,X1,FS,MN,ML
470     X11=(XP−2*X+XN)/DT2
480     X1=(−XP+XN)/(2*DT)
485     PRINT #1,I;",";X",";X1",";X11
490     XP=X:X=XN
500     NEXT G
510     NEXT J
800     CLOSE #1
900     STOP
```

▲

15.6 RESPONSE OF ELASTOPLASTIC MDOF SYSTEMS

The concepts for the analysis of elastoplastic SDOF systems discussed in Section 15.5 can be modified and extended to MDOF systems. The equations of motion for an elastoplastic MDOF system may be expressed as

$$[m]\{\ddot{x}\} + [c]\{\dot{x}\} + \{F_s(t)\} = \{F(t)\} \tag{15.68}$$

Since the nonlinearity is confined to the restoring forces in the system, the only difference between Eq. (15.68) and the corresponding equations for a linear system, Eq. (12.31), is the last term, $\{F_S(t)\}$, on the left-hand side of the equations. Thus the expressions for the elastoplastic restoring force given by Eqs. (15.63) through (15.67) for SDOF systems are also applicable to MDOF systems.

To clarify the discussion, consider the undamped three-story shear frame structure shown in Figure 15.34. The elastoplastic restoring force-displacement relationship for each story i ($i = 1, 2, 3$) is described in Figure 15.35, where R_{mi} and k_i are the maximum story restoring force and elastic story stiffness, respectively. Also, from this relationship

$$R_{mi} = \frac{4M_{pi}}{h_i} \tag{15.69}$$

and the corresponding elastic limit displacements are given by

$$(x_{\text{el}})_i = \frac{R_{mi}}{k_i} \tag{15.70}$$

Figure 15.34

Three-story shear frame structure.

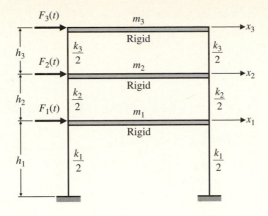

Figure 15.35

Elastoplastic resistance-displacement relationship for MDOF systems.

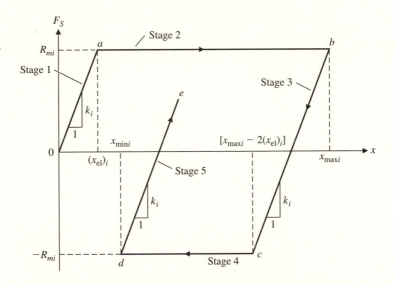

where M_{Pi} is the plastic moment capacity for the columns in story i and h_i is the height of story i. Similar to the hysteresis response of SDOF systems discussed in Section 15.5, the response of elastoplastic MDOF systems for one complete hysteresis cycle can be described in five stages, as illustrated in Figure 15.35. The displacement bounds, equations of motion, and restoring force vector for each stage are described as follows (with reference to Figure 15.35):

Stage 1. Elastic loading (segment Oa)
  (a) Displacement bounds at time t

$$0 < x_1 < (x_{el})_1 \tag{15.71a}$$

$$0 < (x_2 - x_1) < (x_{el})_2 \tag{15.71b}$$

$$0 < (x_3 - x_2) < (x_{el})_3 \tag{15.71c}$$

(b) Equations of motion at time t

$$m_1\ddot{x}_1 + k_1x_1 - k_2(x_2 - x_1) = F_1(t) \tag{15.72a}$$

$$m_2\ddot{x}_2 + k_2(x_2 - x_1) - k_3(x_3 - x_2) = F_2(t) \tag{15.72b}$$

$$m_3\ddot{x}_3 + k_3(x_3 - x_2) = F_3(t) \tag{15.72c}$$

(c) Restoring force vector at time t

$$\{F_S\}_t = [k]\{x\}_t \tag{15.73}$$

Stage 2. Plastic loading (segment ab)
(a) Displacement bounds at time t

$$(x_{\text{el}})_1 < x_1 < x_{\text{max}1} \tag{15.74a}$$

$$(x_{\text{el}})_2 < (x_2 - x_1) < x_{\text{max}2} \tag{15.74b}$$

$$(x_{\text{el}})_3 < (x_3 - x_2) < x_{\text{max}3} \tag{15.74c}$$

(b) Equations of motion at time t

$$m_1\ddot{x}_1 + R_{m1} - R_{m2} = F_1(t) \tag{15.75a}$$

$$m_2\ddot{x}_2 + R_{m2} - R_{m3} = F_2(t) \tag{15.75b}$$

$$m_3\ddot{x}_3 + R_{m3} = F_3(t) \tag{15.75c}$$

(c) Restoring force vector at time t

$$\{F_S\}_t = \begin{Bmatrix} R_{m1} - R_{m2} \\ R_{m2} - R_{m3} \\ R_{m3} \end{Bmatrix} \tag{15.76}$$

Stage 3. Elastic rebound (segment bc)
(a) Displacement bounds at time t

$$[x_{\text{max}1} - 2(x_{\text{el}})_1] < x_1 < x_{\text{max}1} \tag{15.77a}$$

$$[x_{\text{max}2} - 2(x_{\text{el}})_2] < (x_2 - x_1) < x_{\text{max}2} \tag{15.77b}$$

$$[x_{\text{max}3} - 2(x_{\text{el}})_3] < (x_3 - x_2) < x_{\text{max}3} \tag{15.77c}$$

(b) Equations of motion at time t

$$m_1\ddot{x}_1 + [R_{m1} - k_1(x_{\text{max}1} - x_1)] \\ - \{(R_{m2} - k_2[(x_{\text{max}2} - x_2) - (x_{\text{max}1} - x_1)])\} = F_1(t) \tag{15.78a}$$

$$m_2\ddot{x}_2 + [R_{m2} - k_2(x_{max2} - x_2)]$$
$$- \{R_{m3} - k_3[(x_{max3} - x_3) - (x_{max2} - x_2)]\} = F_2(t) \qquad (15.78b)$$

$$m_3\ddot{x}_3 + R_{m3} - k_3[(x_{max3} - x_3) - (x_{max2} - x_2)] = F_3(t) \qquad (15.78c)$$

(c) Restoring force vector at time t

$$\{F_S\}_t = \begin{Bmatrix} R_{m1} - R_{m2} \\ R_{m2} - R_{m3} \\ R_{m3} \end{Bmatrix} - [k]\{x_{max} - x\}_t \qquad (15.79)$$

Stage 4. Plastic loading (segment cd)
   (a) Displacement bounds at time t

$$x_{min1} < x_1 < [x_{max1} - 2(x_{el})_1] \qquad (15.80a)$$

$$x_{min2} < (x_2 - x_1) < [x_{max2} - 2(x_{el})_2] \qquad (15.80b)$$

$$x_{min3} < (x_3 - x_2) < [x_{max3} - 2(x_{el})_3] \qquad (15.80c)$$

   (b) Equations of motion at time t

$$m\ddot{x}_1 - R_{m1} + R_{m2} = F_1(t) \qquad (15.81a)$$

$$m\ddot{x}_2 - R_{m2} + R_{m3} = F_2(t) \qquad (15.81b)$$

$$m\ddot{x}_3 - R_{m3} = F_3(t) \qquad (15.81c)$$

   (c) Restoring force vector at time t

$$\{F_S\}_t = \begin{Bmatrix} -R_{m1} + R_{m2} \\ -R_{m2} + R_{m3} \\ -R_{m3} \end{Bmatrix} \qquad (15.82)$$

Stage 5. Elastic rebound (segment de)
   (a) displacement bounds at time t

$$x_{min1} < x_1 < [x_{min1} + 2(x_{el})_1] \qquad (15.83a)$$

$$x_{min2} < (x_2 - x_1) < [x_{min2} + 2(x_{el})_2] \qquad (15.83b)$$

$$x_{min3} < (x_3 - x_2) < [x_{min3} + 2(x_{el})_3] \qquad (15.83c)$$

   (b) Equations of motion at time t

$$m_1\ddot{x}_1 - R_{m1} + k_1(x_1 - x_{min1}) + R_{m2}$$
$$- k_2[(x_2 - x_{min2}) - (x_1 - x_{min1})] = F_1(t) \qquad (15.84a)$$

$$m_2\ddot{x}_2 - R_{m2} + k_2(x_2 - x_{\min2}) + R_{m3}$$
$$-k_3[(x_3 - x_{\min3}) - (x_2 - x_{\min2})] = F_2(t) \qquad (15.84b)$$

$$m_3\ddot{x}_3 - R_{m3} + k_3[(x_3 - x_{\min3}) - (x_2 - x_{\min2})] = F_3(t) \qquad (15.84c)$$

(c) Restoring force vector at time t

$$\{F_S\}_t = \begin{Bmatrix} -R_{m1} + R_{m2} \\ -R_{m2} + R_{m3} \\ -R_{m3} \end{Bmatrix} + [k]\{x - x_{\min}\}_t \qquad (15.85)$$

Although the five stages exist for each story, they are dependent on the load intensity and motion of the system, and they need not exist uniquely. That is, it is not likely that each story will always be in the same stage at the same time.

15.6.1 Explicit Integration

Numerical evaluation of the elastoplastic response of MDOF systems may be accomplished by either the implicit or explicit algorithms discussed in Section 15.5 for SDOF systems. However, if an implicit integration operator is employed in the response calculations for an elastoplastic MDOF system, it is highly recommended that equilibrium iterations be performed. For large MDOF systems, implicit integration with iterations invariably requires significant computational effort and can be computationally prohibitive at times.

For many types of nonlinear dynamic analyses, explicit integration is an efficient alternative to implicit integration with equilibrium iterations. Although in most cases a smaller time step is required for stability of the explicit integration operator, the solution of the explicit equations requires significantly less computation than implicit equations with equilibrium iterations. The central difference algorithm, presented in Table 13.1 for the analyses of elastic MDOF systems, can easily be altered to accommodate elastoplastic response calculations of MDOF systems, simply by modifying the expression for the effective load vector $\{\hat{F}\}_t$.

For an elastoplastic system, the modified effective load vector is determined by replacing the term $[k]\{x\}_t$ in the expression for the elastic effective load vector by $\{F_S\}_t$, as determined from some combination of Eqs. (15.73), (15.76), (15.79), (15.82), and (15.85), depending on the resistance stage. Therefore the resulting expression for the effective load vector for an elastoplastic MDOF system to be used with the central difference method is similar to that found in Eq. (15.62) and is determined by

$$\{\hat{F}\}_t = \{F\}_t - \{F_S\}_t + \frac{2[m]}{(\Delta t)^2}\{x\}_t - \frac{1}{\Delta t}\left(\frac{[m]}{\Delta t} - \frac{[c]}{2}\right)\{x\}_{t-\Delta t} \qquad (15.86)$$

The step-by-step central difference algorithm for the response of elastoplastic MDOF systems is summarized in Table 15.8.

TABLE 15.8. **Central Difference Method for Elastoplastic MDOF Systems**

A. Initial calculations:
1. Input the stiffness matrix $[k]$, mass matrix $[m]$, and damping matrix $[c]$
2. Input initial conditions $\{x\}_0$ and $\{\dot{x}_0\}$
3. Calculate $\{\ddot{x}_0\}$ from initial conditions

$$\ddot{x}_0 = [m]^{-1}[\{F(0)\} - [c]\{\dot{x}_0\} - [k]\{x_0\}]$$

4. Select time step Δt, $\Delta t < \Delta t_{cr}$, and calculate integration constants

$$a_0 = \frac{1}{(\Delta t)^2} \qquad a_1 = \frac{1}{2\Delta t} \qquad a_2 = 2a_0 \qquad a_3 = \frac{1}{a_2}$$

5. Calculate $\{x\}_{-\Delta t}$

$$\{x\}_{-\Delta t} = \{x\}_0 - \Delta t\{\dot{x}\}_0 + a_3\{\ddot{x}\}_0$$

6. Form effective mass matrix $[\hat{m}]$

$$[\hat{m}] = a_0[m] + a_1[c]$$

B. For each time step:
1. Determine restoring force vector $\{F_S\}_t$
2. Calculate the effective force vector at time t

$$\{\hat{F}\}_t = \{F\}_t - \{F_S\}_t + a_2[m]\{x\}_t - (a_0[m] - a_1[c])\{x\}_{t-\Delta t}$$

3. Solve for displacements at time $t + \Delta t$

$$[\hat{m}]\{x\}_{t+\Delta t} = \{\hat{F}\}_t \rightarrow \{x\}_{t+\Delta t} = [\hat{m}]^{-1}\{\hat{F}\}_t$$

4. Evaluate accelerations and velocities at time t

$$\{\ddot{x}\}_t = a_0[\{x\}_{t-\Delta t} - 2\{x\}_t + \{x\}_{t+\Delta t}]$$

$$\{\dot{x}\}_t = a_1[-\{x\}_{t-\Delta t} + \{x\}_{t+\Delta t}]$$

EXAMPLE 15.5 ▼

Evaluate the elastoplastic response of the three-story shear frame structure shown in Figure 15.36 subject to the loads shown in Figure 15.37 by the central difference method. Compare the elastoplastic solution with the elastic solution. Use a time step $\Delta t = 0.001$ sec. Assume $m_1 = 0.141$ kip-sec^2/in, $m_2 = 0.132$ kip-sec^2/in, $m_3 = 0.066$ kip-sec^2/in, $k_1 = 30.5$ kips/in, $k_2 = k_3 = 44.5$ kips/in, $R_{m1} = 40$ kips, $R_{m2} = R_{m3} = 26$ kips, and modal damping factors of $\zeta_1 = \zeta_2 = \zeta_3 = 0.07$.

Solution

(a) The dynamic properties for the elastic system are

$$
\{\omega\} = \begin{Bmatrix} 8.325 \\ 24.062 \\ 35.086 \end{Bmatrix} \text{ rad/sec} \qquad [\Phi] = \begin{bmatrix} 1.000 & 1.000 & 1.000 \\ 1.471 & -0.146 & -2.220 \\ 1.639 & -1.041 & 2.680 \end{bmatrix}
$$

(Note that the critical time step Δt_{cr} for integration stability is equal to T_3/π or $0.179/\pi = 0.056$ sec. Therefore a time step $\Delta t = 0.001$ sec ensures stability and accuracy of the computations.)

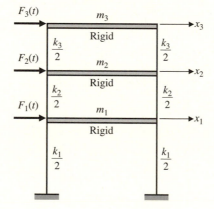

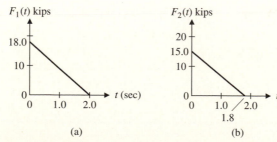

(b) The elastic limit displacement limits are

$$(x_{\text{el}})_1 = \frac{R_{m1}}{k_1} = \frac{40}{30.5} = 1.31115 \text{ in}$$

$$(x_{\text{el}})_2 = (x_{\text{el}})_3 = \frac{R_{m2}}{k_2} = \frac{R_{m3}}{k_3} = 0.58427 \text{ in}$$

(c) Mass and stiffness matrices

$$[m] = \begin{bmatrix} 0.141 & 0 & 0 \\ 0 & 0.132 & 0 \\ 0 & 0 & 0.066 \end{bmatrix} \text{kip-sec}^2/\text{in}$$

$$[k] = \begin{bmatrix} 75.0 & -44.5 & 0 \\ -44.5 & 74.17 & -29.67 \\ 0 & -29.67 & 29.67 \end{bmatrix} \text{kips/in}$$

(d) In a direct integration analysis, the damping matrix must be defined explicitly as discussed in Chapter 13. Thus the damping matrix is determined from Eq. (13.28) as

$$[c] = \sum_{r=1}^{n} \left(\frac{2\zeta_r \omega_r}{M_r} \right) [m]\{\Phi\}_r \left[[m]\{\Phi\}_r \right]^T \tag{1}$$

from which

$$[c] = \begin{bmatrix} 0.497 & -0.250 & 0.019 \\ -0.250 & 0.513 & -0.162 \\ 0.019 & -0.162 & 0.203 \end{bmatrix} \text{kip-sec/in}$$

(e) The integration constants for the central difference method are (refer to A.4 in Table 15.8)

$$a_0 = \frac{1}{(\Delta t)^2} = 10^6 \qquad a_1 = \frac{1}{2\Delta t} = 500$$

$$a_2 = 2a_0 = 2 \times 10^6 \qquad a_3 = \frac{1}{a_2} = 5 \times 10^{-7}$$

The BASIC computer program for the central difference method used to evaluate the responses is listed in Table 15.9. Displacement time histories for $x_1(t)$, $x_2(t)$, and $x_3(t)$ are presented in Figures 15.38, 15.39, and 15.40, respectively. From Figure 15.38 it is evident that the first level $x_1(t)$ experiences significant permanent deformation of approximately 2.75 in. However, the second level $x_2(t)$ exhibits minimal inelastic behavior as illustrated in Figure 15.39, and the top level $x_3(t)$ remains elastic throughout the entire response as indicated in Figure 15.40.

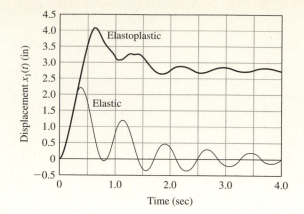

Figure 15.38

Displacement response histories for first story $x_1(t)$.

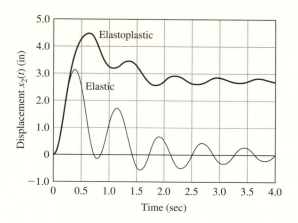

Figure 15.39

Displacement response histories for second story $x_2(t)$.

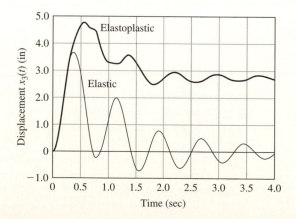

Figure 15.40

Displacement response histories for third story $x_3(t)$.

TABLE 15.9. **Listing of BASIC Computer Program Used for Example 15.5**

```
1       CLS
2       OPEN "XXX" FOR INPUT AS#2
3       INPUT #2,A:CLOSE#2:IF A=5*INT(A/5) THEN SAVE"center3":PRINT"saving please wait"
4       A=A+1
5       OPEN "XXX" FOR OUTPUT AS#2:PRINT#2,A:CLOSE#2
6       OPEN "p3p" FOR OUTPUT AS#1
10      REM This program calculates the response of an MDOF elastoplastic system
14      REM hg=1 for force hg=2 for acceleration
15      N=3:KK=386:DT=.001:TMAX=4:HG=1:N1=4:AN=2:REM an=1 linear an=2 elasto-plastic
16      DT2=DT*DT
21      PRINT "stiffness matrix
22      FOR I=1 TO N:FOR J=1 TO N:READ K(I,J):PRINT K(I,J),:NEXT J:PRINT :NEXT I
23      DATA 75,-44.5,0
24      DATA-44.5,74.17,-29.67
25      DATA 0,-29.67,29.67
30      REM read mass matrix
31      PRINT "mass matrix
32      FOR I=1 TO N:FOR J=1 TO N:READ M(I,J):PRINT M(I,J),:SK(I,J)=M(I,J):NEXT J:PRINT :NEXT I
33      DATA 0.141,0,0
34      DATA 0,.132,0
35      DATA 0,0,0.066
36      FOR I=1 TO N:ML(I)=1:NEXT I
40      REM damping matrix
41      E1=67.8:E2=490.3:E3=985:Z1=.07:Z2=.07
42      W1=SQR(E1):W2=SQR(E2):W3=SQR(E3):T3=2*3.14/W3:IF DT>(T3/10) THEN PRINT "change time step to less than ";T3/10:STOP
43      PRINT "W1";W1,"W2";W2,"Z1";Z1,"Z2";Z2
44      PRINT "PHI":FOR I=1 TO N:READ FI(I):PRINT FI(I):NEXT I:PRINT
45      DATA 1.2656,1.86087,2.19178
46      AK=2*Z2/W2:ZZ=Z1-Z2*W1/W2:AH=2*ZZ*W1
47      FOR I=1 TO N:DX(I)=M(I,I)*FI(I):NEXT I
48      FOR I=1 TO N:FOR J=1 TO N:FR(I,J)=DX(I)*DX(J):NEXT J:NEXT I
49      PRINT "DAMPING MATRIX"
50      FOR I=1 TO N:FOR J=1 TO N:C(I,J)=AK*K(I,J)+AH*FR(I,J)
51      PRINT USING"##.### ";C(I,J),:NEXT J:PRINT :NEXT I
52      PRINT "initial conditions
54      FOR I=1 TO N:READ X(I):PRINT X(I):NEXT I
56      DATA 0,0,0
57      FOR I=1 TO N:READ X1(I):PRINT X1(I):NEXT I
58      DATA 0,0,0
62      PRINT "time", "force"
63      FOR I=1 TO N:PRINT "force No";I:FOR J = 1 TO N1: READ T(I,J),F(I,J):IF TT>T(I,J) THEN 64 ELSE TT=T(I,J)
64      IF HG=1 THEN 65 ELSE F(I)=KK*F(I,J)*M(I,I)
65      PRINT T(I,J),F(I,J)
66      NEXT J,I
71      DATA 0,18,1.5,4.5,1.8,1.8,2,0
72      DATA 0,15,1.5,2.5,1.8,0,2,0
73      DATA 0,10,1.5,0,1.8,0,2,0
75      PRINT "initial force":FOR I=1 TO N:P(I)=F(I,1):PRINT I,P(I):NEXT I
76      PRINT "rm":FOR I=1 TO N:READ
        RM(I):KK(I)=(K(I,I)+K(I,I+1)):XEL(I)=RM(I)/KK(I)
77      RT(I)=RM(I):PRINT RM(I),XEL(I):NEXT I
78      DATA 40.0, 26.0, 26.0
80      GOSUB 1500
81      GOSUB 2000:PRINT "x11o"
84      FOR I=1 TO N:T(I,N1+1)=TMAX :NEXT I
85      N4=INT((TMAX-TT)/DT)
86      FOR I=1 TO N:FOR J=1 TO N:X11(I)=X11(I)+A(I,J)*FX(J):NEXT J:NEXT I
87      FOR I=1 TO N:PRINT X11(I):NEXT I
88      AO=1/DT2:A1=1/(2*DT):A2=2*AO:A3=1/A2
90      PRINT "mh":FOR I=1 TO N:FOR J=1 TO N:SK(I,J)=AO*M(I,J)+A1*C(I,J)
91      K1(I,J)=A2*M(I,J):K2(I,J)=AO*M(I,J)-A1*C(I,J)
```

(continued)

```
92    PRINT SK(I,J),:NEXT J:PRINT :NEXT I
93    PRINT "k1":FOR I=1 TO N:FOR J=1 TO N:PRINT K1(I,J),:NEXT J:PRINT :NEXT I
94    PRINT "k2":FOR I=1 TO N:FOR J=1 TO N:PRINT K2(I,J),:NEXT J:PRINT :NEXT I
101   PRINT"K−1":GOSUB 2000
102   PRINT "xp":FOR I=1 TO N:XP(I)=X(I)−DT*X1(I)+A3*X11(I):PRINT XP(I):NEXT I
140   PRINT#1, USING "###.### ";T1,X(1),X(2),X(3)
199   PRINT
200   PRINT "starting"
210   FOR J = 1 TO N1
230   IF J = N1 THEN GOTO 231 ELSE GOTO 235
231   FOR U=1 TO N1:DF(U)=0:NEXT U
232   AS1=N4:GOTO 260
235   FOR H=1 TO N
240   DF(H)=F(H,J+1)−F(H,J): DT1(H)= T(H,J+1) − T(H,J):N3=DT1(H)/DT
250   AS1=N3−1
255   PRINT N3,DF(H),DT1(H):NEXT H
260   FOR G = 0 TO AS1
280   T1=T1+DT
290   FOR H=1 TO N
291   IF T1>T(H,N1) OR F(H,J)=−999 THEN F1(H)=0:GOTO 310
300   F1(H) = F(H,J) + DF(H)/DT1(H)* (T1 − T(H,J))
310   PRINT T1,H,F1(H)
311   NEXT H
312   IF AN=1 THEN GOSUB 4000:GOTO 440
320   FOR Q=1 TO N
325   X=X(Q)−X(Q−1)
330   GOSUB 3000
335   PRINT USING "##.### ";T1,X,X1(Q),XX(Q),RM(Q),FS(Q),MN(Q),ML(Q)
340   NEXT Q
345   PRINT "ff final"
350   FOR R=1 TO N
360   FF(R)=FS(R)−FS(R+1)
365   PRINT USING "###.### ";R,FF(R),FS(R),FS(R+1),X(1),X(2)
370   NEXT R
440   GOSUB 1300:'PRINT "pe"
445   FOR Q=1 TO N
450   PE(Q)=F1(Q)−FF(Q)+K3(Q)−K4(Q)
451   'PRINT Q,PE(Q)
455   NEXT Q
465   FOR V1=1 TO N:XN(V1)=0:FOR V=1 TO N:XN(V1)=XN(V1)+A(V1,V)*PE(V):NEXT V,V1
467   FOR Q=1 TO N
470   X11(Q)=AO*(XP(Q)−2*X(Q)+XN(Q))
480   X1(Q)=A1*(−XP(Q)+XN(Q))
482   NEXT Q
485   'PRINT USING "##.### ";T1,XP(1),X(1),XN(1),XP(2),X(2),XN(2)
486   PRINT "displacement ";:PRINT USING "##.### ";T1,XN(1),XN(2),XN(3),X1(1),X1(2),X1(3)
487   PRINT#1, USING "###.### ";T1,X(1),X(2),X(3)
490   FOR Q=1 TO N:XP(Q)=X(Q):X(Q)=XN(Q):NEXT Q
500   NEXT G
510   NEXT J
800   CLOSE#1
900   STOP
1000  IF FS1(Q)<RT(Q) THEN RM(Q)=FS1(Q) ELSE RM(Q)=RT(Q)
1005  RETURN
1100  IF FS2(Q)>−RT(Q) THEN RM(Q)=FS2(Q) ELSE RM(Q)=RT(Q)
1105  RETURN
1200  IF XX(Q)<0 THEN FS(Q)=−RM(Q)−KK(Q)*(XX(Q)−X) ELSE FS(Q)=RM(Q)−KK(Q)*(XX(Q)−X)
1205  RETURN
1300  REM subroutine for k3,k4
1305  'PRINT "k3,k4"
1310  FOR E=1 TO N
1315  K3(E)=0:K4(E)=0
```

(continued)

```
1320   FOR R=1 TO N
1330   K3(E)=K3(E)+K1(E,R)*X(R):K4(E)=K4(E)+K2(E,R)*XP(R)
1331   'PRINT R,X(R),XP(R)
1340   NEXT R
1344   'PRINT K3(E),K4(E)
1345   NEXT E
1350   RETURN
1500   REM subroutine for y11
1510   FOR Q=1 TO N: M1(Q)=0:C1(Q)=0:FOR KP=1 TO N
1520   M1(Q)=M1(Q)+K(Q,KP)*X(KP)
1530   C1(Q)=C1(Q)+C(Q,KP)*X1(KP)
1540   NEXT KP,Q
1550   FOR Q=1 TO N :FX(Q)=P(Q)-M1(Q)-C1(Q):NEXT Q
1600   RETURN
2000   REM subroutine inverse
2005   FOR I=1 TO N:FOR J=1 TO N:A(I,J)=0:NEXT J,I
2020   FOR I=1 TO N
2060   A(I,I)=1
2070   NEXT I
2100   FOR I= 1 TO N
2105   TX=SK(I,I)
2110   FOR J=1 TO N
2120   SK(I,J)=SK(I,J)/TX:A(I,J)=A(I,J)/TX
2130   NEXT J
2150   FOR K= (I+1) TO N
2155   M=−SK(K,I)/SK(I,I)
2160   FOR J= 1 TO N
2180   SK(K,J)=SK(K,J)+SK(I,J)*M
2182   A(K,J)=A(K,J)+A(I,J)*M
2190   NEXT J,K,I
2400   REM back sub.
2410   FOR I=N TO 1 STEP −1
2420   FOR K= I−1 TO 1 STEP −1
2430   M=−SK(K,I)
2440   FOR J=N TO 1 STEP −1
2450   SK(K,J)=SK(K,J)+M*SK(I,J)
2455   A(K,J)=A(K,J)+M*A(I,J)
2499   NEXT J,K,I
2900   PRINT "inv":FOR Q=1 TO N:FOR R=1 TO N :PRINT A(Q,R),:NEXT R:PRINT:NEXT Q
2950   RETURN
3000   REM subroutine check x
3100   IF X < XEL(Q) AND S(Q)=0 THEN FS(Q)=KK(Q)*X:MN(Q)=1:XX(Q)=X:FS1(Q)=FS(Q):GOTO 3200
3110   IF X >XEL(Q) AND X1(Q)>0 AND S(Q)=0 THEN FS(Q)=RM(Q):MN(Q)=2:XX(Q)=X:FS1(Q)=FS(Q):GOTO 3200
3120   S(Q)=1
3130   IF X >(XX(Q)−2*XEL(Q)) AND X1(Q)<0 AND ML(Q)=1 THEN GOSUB 1000:FS(Q)=RM(Q)−KK(Q)*(XX(Q)−X):MN(Q)=3:ML(Q)=1:
       FS2(Q)=FS(Q):GOTO 3200
3140   IF (ML(Q)=1 OR ML(Q)=2) AND X1(Q)<0 THEN FS(Q)=−RM(Q):MN(Q)=4:XX(Q)=X:ML(Q)=2:FS2(Q)=−RT(Q):GOTO 3200
3150   IF X<(XX(Q)+2*XEL(Q)) AND X1(Q)>0 AND (ML(Q)=2 OR ML(Q)=3) THEN FS(Q)=−RM(Q)+KK(Q)*(X−XX(Q)):MN(Q)=5:ML(Q)=3:
       GOSUB 1100:GOTO 3200
3160   IF ML(Q)=3 AND X1(Q)>0 THEN FS(Q)=RM(Q):MN(Q)=6:XX(Q)=X:ML(Q)=1:GOTO 3200
3170   IF FS(Q)<RT(Q) THEN GOSUB 1200:MN(Q)=7:ML(Q)=4:GOTO 3200
3175   IF FS(Q)>=RT(Q) AND MN(Q)=7 THEN FS(Q)=RT(Q):MN(Q)=8:XX(Q)=X:ML(Q)=1:GOTO 3200
3180   PRINT "error":STOP
3200   RETURN
4000   REM
4100   FOR R=1 TO N:FF(R)=0:FOR W=1 TO N
4110   FF(R)=FF(R)+K(R,W)*X(W)
4120   NEXT W,R
4200   RETURN
```

▲

REFERENCES

1 Bathe, Klaus-Jürgen, *Finite Element Procedures,* Prentice Hall, Englewood Cliffs, NJ, 1996.

2 Bathe, K.J., Ramm, E., and Wilson, E.L., "Finite Element Formulations for Large Deformation Dynamic Analysis," *International Journal of Numerical Methods in Engineering,* Vol. 9, pp. 353–386, 1975.

3 Oden, J.T., *Finite Elements of Nonlinear Continua,* McGraw-Hill, New York, 1972.

4 Bathe, K.J., "An Assessment of Current Finite Element Analysis of Nonlinear Problems in Solid Mechanics," in *Numerical Solution of Partial Differential Equations—III,* B. Hubbard, ed., Academic Press, New York, 1976.

5 Wilson, E.L., Farhoomand, I., and Bathe, K.J., "Nonlinear Dynamic Analysis of Complex Structures," *Earthquake Engineering and Structural Dynamics,* Vol. 10, pp. 241–252, 1982.

6 Paz, M., *Structural Dynamics: Theory and Computation,* Van Nostrand, New York, 1980.

7 Biggs, J.M., *Introduction to Structural Dynamics,* McGraw-Hill, New York, 1964.

8 Humar, J.L., *Dynamics of Structures,* Prentice Hall, Englewood Cliffs, NJ, 1990.

9 Chopra, A.K., *Dynamics of Structures: Theory and Applications to Earthquake Engineering,* Prentice Hall, Englewood Cliffs, NJ, 1995.

10 ADINA, "A Finite Element Program for Automatic Dynamic Incremental Nonlinear Analysis," *System Theory and Modeling Guide,* Report ARD 87-7 ADINA R&D, Watertown, MA, 1987.

11 Thomson, W.T., *Theory of Vibration with Application,* 2nd ed., Prentice Hall, Englewood Cliffs, NJ, 1981.

12 Tse, F.S., Morse, I.E., and Hinkle, R.T., *Mechanical Vibrations: Theory and Applications,* Allyn and Bacon, Boston, 1978.

13 Abramson, H.N., "Nonlinear Vibration," in *Shock and Vibration Handbook,* 2nd ed., C.M. Harris and C.E. Crede, eds., McGraw-Hill, New York, 1976.

14 Timoshenko, S., Young, D.H., and Weaver, W., *Vibration Problems in Engineering,* 4th ed., Wiley, New York, 1974.

15 Disque, R.O., *Applied Plastic Design in Steel,* Robert E. Krieger, Huntington, NY, 1971.

16 Moy, S.S.J., *Plastic Methods for Steel and Concrete Structures,* Wiley, New York, 1981.

17 Beedle, L.S., *Plastic Design of Steel Frames,* Wiley, New York, 1958.

18 Cook, R.D., *Concepts and Applications of Finite Element Analysis,* 2nd ed., Wiley, New York, 1981.

19 Szuladzinski, G., *Dynamics of Structures and Machinery,* Wiley, New York, 1982.

20 Horne, M.R., *Plastic Theory of Structures,* MIT Press, Cambridge, MA, 1971.

NOTATION

a_0, a_1, a_2, a_3	integration constants used in central difference method for MDOF systems
A	cross-sectional area
c	viscous damping coefficient
C_c	critical damping constant
E	modulus of elasticity
E_T	tangent modulus
$\hat{F}$	effective force
F_D	damping force
F_I	inertia force
F_S	spring force or restoring force in system
F_y	yield stress
$F(t)$	externally applied force
$F_{\text{eff}}(t)$	effective load for ground acceleration
g	acceleration due to gravity
h	height

k	translational stiffness	$[\hat{k}]$	effective stiffness matrix
k_S	secant stiffness	$[K]$	generalized (modal) stiffness matrix
k_T	tangent stiffness	$[m]$	mass matrix
k_{T_0}	initial tangent stiffness	$[\hat{m}]$	effective mass matrix
$\hat{k}_T$	effective tangent stiffness	$[M]$	generalized (modal) mass matrix
K_r	modal stiffness for rth normal mode	$\{x\}$	displacement vector
L	length	$\{x_0\}$	initial displacement vector
m	mass	$\{\dot{x}\}$	velocity vector
$\hat{m}$	effective mass	$\{\dot{x}_0\}$	initial velocity vector
M_p	plastic moment	$\{\ddot{x}\}$	acceleration vector
M_r	modal mass for rth normal mode	$\{\ddot{x}_0\}$	initial acceleration vector
R_m	resistance force corresponding to a condition of full plastification	$\{\ddot{x}_t\}$	total (absolute) acceleration vector
		α	integration parameter used in Newmark method
R_{mT}	total plastic resistance for a structure	δ	axial elongation; also integration parameter for Newmark method
R_y	resistance force corresponding to a condition of first yielding	ΔF	incremental externally applied force
t	time	$\Delta \hat{F}$	effective incremental force
T	natural period for elastic structure	ΔF_D	incremental damping force
T_{ep}	natural period for elastoplastic structure	ΔF_I	incremental inertia force
T_i	initial axial tension	ΔF_S	incremental spring (restoring) force
T_n	smallest vibration period of a MDOF system	$\Delta P^{(k)}$	incremental force associated with $\Delta x^{(k)}$ for kth iteration
x	displacement	$\Delta R^{(k)}$	incremental residual force associated with $\Delta x^{(k)}$ for kth iteration
x_{el}	elastic limit displacement		
x_0	initial displacement	Δt	time increment
$\dot{x}$	velocity	Δt_{cr}	critical time step
$\dot{x}_0$	initial velocity	Δx	incremental displacement
$\ddot{x}$	acceleration	$\Delta x^{(k)}$	approximation of incremental displacement for kth iteration
$\ddot{x}_0$	initial acceleration		
$\ddot{x}_g(t)$	ground acceleration	$\Delta \dot{x}$	incremental velocity
Z	plastic section modulus	$\Delta \ddot{x}$	incremental acceleration
MDOF	multi-degree-of-freedom	ε	normal strain
MNO	materially nonlinear only	μ	constant indicating degree of spring hardening or softening
SDOF	single degree of freedom		
TL	total Lagrangian	σ	normal stress
TOL	specified convergence tolerance	σ_y	yield stress
UL	updated Lagrangian	θ	angle of rotation; also integration parameter for Wilson-θ method
$[c]$	viscous damping matrix		
$[C]$	modal damping matrix	ω_r	undamped natural circular frequency for rth normal mode
$\{\hat{F}\}$	effective force vector		
$\{F(t)\}$	vector of externally applied forces	ζ_r	damping factor for rth normal mode
$\{F_S(t)\}$	vector of restoring forces	$[\Phi]$	modal matrix
$[k]$	stiffness matrix	$\{\Phi\}_r$	modal vector for rth normal mode

PROBLEMS

15.1–15.4 For the systems shown in Figures P15.1 through P15.4, determine (a) the nonlinear equation of motion for large displacements (rotations) and (b) the linearized equation of motion for small displacements (rotations).

15.6 A cantilever beam having stiffness k supports a tip mass m as shown in Figure P15.6. If the vibration amplitude exceeds the gap distance b, the mass comes in contact with one of the springs also having stiffness k. Construct the resistance-deflection relationship for this system.

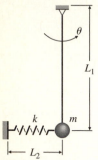

Figure P15.1

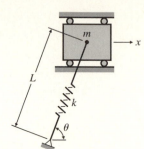

Figure P15.2

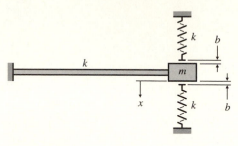

Figure P15.6

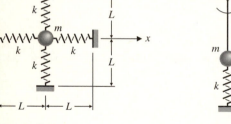

Figure P15.3

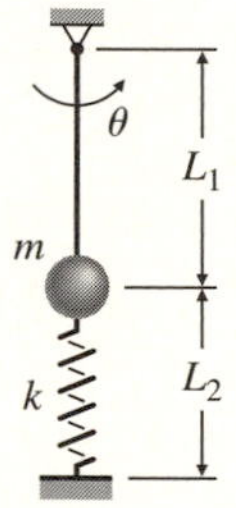

Figure P15.4

15.7 Evaluate the elastoplastic displacement response of the single-story shear frame building shown in Figure P15.7 subject to the triangular pulse load shown in Figure P15.8 by the Newmark method without equilibrium iterations. Use a time step $\Delta t = 0.001$ sec. Assume $m = 0.15$ kip-sec^2/in and $F_y = 36$ ksi. Perform the analysis for the cases (a) $\zeta = 0$ and (b) $\zeta = 0.05$. Compare the elastoplastic response with the elastic response.

15.5 When the mass m shown in Figure P15.5 is in the neutral position, the springs are unstrained and the springs are not attached to the mass. If the system is set into free vibration with an initial velocity to the mass of $\dot{x}_0$, determine (a) the maximum spring force, (b) the maximum deflection, and (c) the natural period of vibration. Assume $k_2 > k_1$.

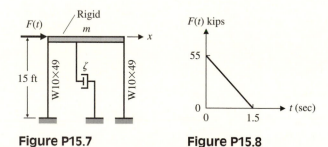

Figure P15.7

Figure P15.8

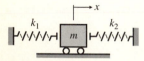

Figure P15.5

15.8 Evaluate the elastoplastic displacement response of the structure shown in Figure P15.7 subjected to the horizontal ground acceleration shown in Figure 15.30 by the Newmark method with equilibrium iterations. Use a time step $\Delta t = T/10$ and compare the result with the calculated response without equilibrium iterations using the same time step. Perform the analysis for the cases (a) $\zeta = 0$ and (b) $\zeta = 0.06$.

15.9 Evaluate the elastoplastic displacement response of the single-story shear frame building shown in Figure P15.9 subject to the ramp function shown in Figure P15.10 by the Wilson-θ method without equilibrium iterations. Use a time step $\Delta t = 0.001$ sec. Assume modulus of elasticity $E = 29,000$ ksi, moment of inertia $I = 125$ in^4, $m = 0.25$ kip-sec^2/in, maximum resistance $R_m = 17$ kips, and a damping factor $\zeta = 0.07$. Compare the elastoplastic response with the elastic response.

15.12 By the central difference method evaluate the elastoplastic displacement response of the simple beam supporting a dead load $W = 15$ kips at midspan as shown in Figure P15.12, subjected to the ideal step force shown in Figure P15.13. Compare the elastoplastic response to the elastic response. Assume modulus of elasticity $E = 29,000$ ksi, yield stress $F_y = 36$ ksi, and damping factor $\zeta = 0.065$. Select a time step $\Delta t < \Delta t_{cr}$.

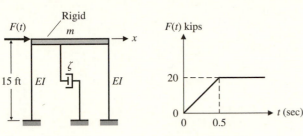

Figure P15.9

Figure P15.10

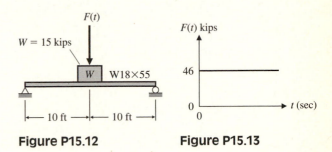

Figure P15.12

Figure P15.13

15.10 Evaluate the elastoplastic displacement response of the structure shown in Figure P15.9 subjected to the horizontal component of ground acceleration $\ddot{x}_g(t)$ shown in Figure 15.30 by the Wilson-θ method with equilibrium iterations. Use a time step $\Delta t = T/10$ and compare the result with the calculated response without equilibrium iterations using the same time step. Assume a damping factor $\zeta = 0.08$.

15.11 Repeat Problem 15.9 for the loading condition shown in Figure P15.11 by the central difference method. Select a time step $\Delta t < \Delta t_{cr}$.

15.13 By the central difference method evaluate the elastoplastic displacement response of the single-story shear frame structure shown in Figure P15.14 subjected to the dynamic force specified in Figure P15.15. Compare the elastoplastic response to the elastic response. Use a time step $\Delta t < \Delta t_{cr}$. Assume a damping factor $\zeta = 0.03$, modulus of elasticity $E = 29,000$ ksi, and $m = 0.2$ kip-sec^2/in.

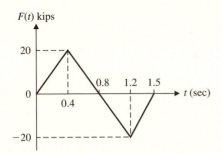

Figure P15.11

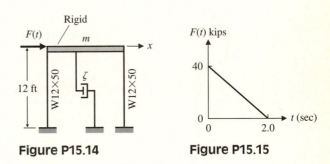

Figure P15.14

Figure P15.15

15.14 Repeat Problem 15.13 for the horizontal ground acceleration $\ddot{x}_g(t)$ shown in Figure 15.30, in lieu of the applied force.

15.15 Evaluate the elastoplastic displacement response of the three-story shear frame building shown in Figure P15.16 subjected to the dynamic forces specified in Figure P15.17 by the central difference method. Assume modulus of elasticity $E = 29{,}000$ ksi and yield stress $F_y = 36$ ksi. Select a time step $\Delta t < T_3/\pi$. Perform the analysis for the cases (a) for all three modal damping factors $\zeta = 0$ and (b) for $\zeta_1 = 0.03$, $\zeta_2 = 0.05$ and calculate ζ_3 on the assumption of Rayleigh damping. Compare the elastoplastic response with the elastic response.

15.16 Repeat Problem 15.15 for the horizontal ground acceleration shown in Figure P15.18, in lieu of applied force.

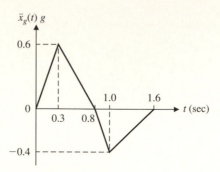

Figure P15.18

15.17 Repeat Problem 15.15 for the three-story shear frame structure shown in Figure P15.19 subjected to the dynamic forces specified in Figure P15.20.

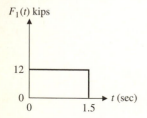

Figure P15.16

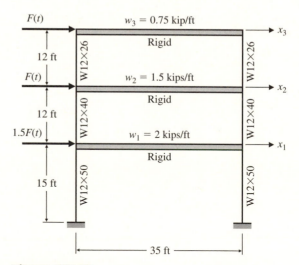

Figure P15.19

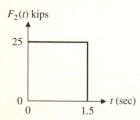

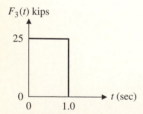

Figure P15.17

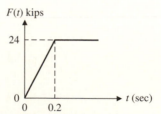

Figure P15.20

15.18 Repeat Problem 15.17 for the horizontal ground acceleration shown in Figure P15.18, in lieu of applied force.

PART V

Practical Applications

16 ▲ Elastic Wave Propagation in Solids

In the study of engineering mechanics, one encounters *statics,* that branch of mechanics dealing with the equilibrium of forces on bodies at rest or at a uniform velocity; *kinematics,* that branch of mechanics limited to the description of motion of bodies, with no attention to forces that precipitate the motion; and *solid-body dynamics,* that branch of mechanics dealing with forces that cause the motion of the body. In all these branches of mechanics it is assumed that if a force is applied to one point on the body, then every other point in the body responds to this force instantaneously. Also, if a force is applied at one point and an acceleration occurs, then every point in the body attains its own acceleration instantaneously. Even in basic *strength of materials* or classical *elasticity* theory the entire body is assumed to be in equilibrium with the applied forces, and all elastic deformations have reached their final values before any observations are made. These assumptions are correct if the time required to reach equilibrium is very short compared to the time necessary to make measurements or observations. However, stress-wave propagation must be considered when forces are applied for short time durations and when observations or measurements are made in very small intervals of time after application of the applied loads. For example, consider a hammer blow to one end of a steel bar 10 m in length. No motion at the distant end of the bar will be observed for a time equal to the length of the bar divided by the stress-wave speed of the bar. This characteristic time for the disturbance to travel the length of the bar is 2.0 msec (that is, 10 m ÷ 5000 m/sec), where 5000 m/sec is the approximate sound speed in steel. This implies that any measurements or observations made prior to 2.0 msec would indicate no displacement at the bar end opposite the hammer blow.

It is a well-established fact, as will be demonstrated throughout this chapter, that a finite wave speed in a material may be obtained directly from the equation of motion for these materials. This finite wave speed can be shown to be the square root of the ratio of material stiffness Q to the material mass density ρ (that is, $\sqrt{Q/\rho}$). Therefore, to clarify the discussion of wave propagation, it is important to review such general topics as stress at a point, constitutive relations, and the resulting equations of motion. These concepts are addressed in this chapter prior to the development of the basic theory of stress wave propagation.

16.1 STRESS AND STRAIN AT A POINT

Usually, the stress in a body varies from one point to another, and the general description of stress at a point cannot be described by a vector of three components. A full description of stress at a point requires a second-order tensor of nine components. However, it can be shown that this second-order stress tensor is symmetric and only six independent terms are required when there are no distributed body or surface couples. The text by Malvern [1] is recommended as a reference for stress and strain principles in a continuum.

The positive stress components used to described stress at a point are shown in a Cartesian axes system in Figure 16.1. These are the components of stress at the origin of the axes system as the cube is shrunk to the origin. Once these six components of stress are known at a point in the body, then a component of stress in any direction at this point can be determined by a rotation of the axes about any arbitrary axis passing

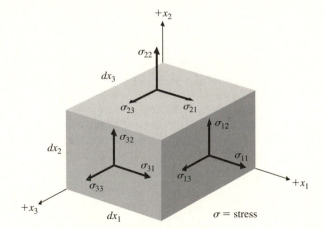

Figure 16.1

Stress at a point schematic showing positive stress directions.

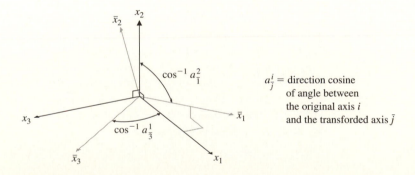

Figure 16.2

Tensor transformation by rotation of axes.

through the original point. The tensor transformation by rotation of axes (refer to Figure 16.2) is expressed as

$$\bar{\sigma}_{ip} = a_i^j a_p^q \sigma_{jq} \tag{16.1}$$

where a_i^j = direction cosine of the angle between the j axis in the original system and the i axis in the transformed system

a_p^q = direction cosine between the q axis in the original system and the p axis in the transformed system

$\bar{\sigma}_{ip}$ = stress components in the transformed axis system

σ_{jq} = components of stress in the original axis system

The preceding discussion is not essential to the understanding of wave propagation phenomena, but it does serve to provide a clearer understanding of the meaning of stress at a point. The definition of three-dimensional small strain tensor is also very basic to the process of obtaining the wave speed from the equation of motion. Consider the three simple cases associated with two-dimensional strain shown in Figure 16.3. In each pure strain case illustrated in Figure 16.3, the definitions of the strain terms approach the well-known partial derivatives as $\Delta x \to 0$ and $\Delta y \to 0$. The two-dimensional small strain tensor terms are defined as

Extensional x strain: $\quad \varepsilon_{xx} = \dfrac{\partial u_x}{\partial x}$

Extensional y strain: $\quad \varepsilon_{yy} = \dfrac{\partial u_y}{\partial y}$ $\qquad$ (16.2)

Shear strain: $\quad \varepsilon_{xy} = \dfrac{1}{2}\left(\dfrac{\partial u_x}{\partial y} + \dfrac{\partial u_y}{\partial x}\right)$

where u_x and u_y are the analytical continuous displacement functions of x and y, respectively.

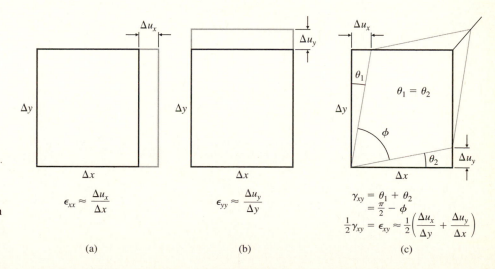

Figure 16.3

Two-dimensional small strain schematic: (a) extensional x strain; (b) extensional y strain; (c) shear strain.

$$\epsilon_{xx} \approx \frac{\Delta u_x}{\Delta x}$$

(a)

$$\epsilon_{yy} \approx \frac{\Delta u_y}{\Delta y}$$

(b)

$$\gamma_{xy} = \theta_1 + \theta_2$$
$$= \frac{\pi}{2} - \phi$$
$$\frac{1}{2}\gamma_{xy} = \epsilon_{xy} \approx \frac{1}{2}\left(\frac{\Delta u_x}{\Delta y} + \frac{\Delta u_y}{\Delta x}\right)$$

(c)

EXAMPLE 16.1 ▼

For the given two-dimensional displacements u_1 and u_2, show that for $x_1 = x_2$ the shear strain is equal to one-half the sum of the extensional strains, or $\varepsilon_{12} = 1/2(\varepsilon_{11} + \varepsilon_{22})$:

$$u_1 = A(x_1^2 + x_2^2)$$

$$u_2 = B(x_1^2 + x_2^2)$$

Solution

From Eqs. (16.2)

$$\varepsilon_{11} = \frac{\partial u_1}{\partial x_1} = 2Ax_1 \tag{1a}$$

$$\varepsilon_{22} = \frac{\partial u_2}{\partial x_2} = 2Bx_2 \tag{1b}$$

$$\varepsilon_{12} = \frac{1}{2}\left(\frac{\partial u_1}{\partial x_2} + \frac{\partial u_2}{\partial x_1}\right) \tag{1c}$$

For $x_1 = x_2$, Eqs. (1) become

$$\varepsilon_{11} = 2Ax_1 \tag{2a}$$

$$\varepsilon_{22} = 2Bx_1 \tag{2b}$$

$$\varepsilon_{12} = (A + B)x_1 \tag{2c}$$

Thus, substituting Eqs. (2a) and (2b) into Eq. (2c) yields

$$\varepsilon_{12} = \frac{1}{2}(\varepsilon_{11} + \varepsilon_{22}) \tag{3}$$

▲

In a similar manner, by incorporating the z direction into the analysis, the three-dimensional strain state may be derived. The three-dimensional small (linear) strain tensor $\boldsymbol{\varepsilon}$ may be written in a (1, 2, 3) Cartesian axis system as

$$\boldsymbol{\varepsilon} = \begin{bmatrix} \dfrac{\partial u_1}{\partial x_1} & \dfrac{1}{2}\left(\dfrac{\partial u_1}{\partial x_2} + \dfrac{\partial u_2}{\partial x_1}\right) & \dfrac{1}{2}\left(\dfrac{\partial u_1}{\partial x_3} + \dfrac{\partial u_3}{\partial x_1}\right) \\[4mm] & \dfrac{\partial u_2}{\partial x_2} & \dfrac{1}{2}\left(\dfrac{\partial u_2}{\partial x_3} + \dfrac{\partial u_3}{\partial x_2}\right) \\[4mm] \text{(Symmetric)} & & \dfrac{\partial u_3}{\partial x_3} \end{bmatrix} \tag{16.3}$$

or more succinctly,

$$\boldsymbol{\varepsilon} = \begin{bmatrix} \varepsilon_{11} & \varepsilon_{12} & \varepsilon_{13} \\ & \varepsilon_{22} & \varepsilon_{23} \\ \text{(Symmetric)} & & \varepsilon_{33} \end{bmatrix} \tag{16.4}$$

From Eq. (16.3) a general strain equation may be written as

$$\varepsilon_{rs} = \frac{1}{2}\left(\frac{\partial u_r}{\partial x_s} + \frac{\partial u_s}{\partial x_r}\right) \tag{16.5}$$

Equations (16.3) and (16.4) define the strain tensor as symmetric. The symmetry of this tensor may be illustrated by expanding the Jacobian matrix of displacement into a symmetric matrix, defined as the strain matrix, and a skew symmetric matrix, denoted as the rotation matrix. With this in mind the linear strain tensor is defined to be symmetric from the outset.

At this point it is important to note that the shear-strain tensor terms may be expressed in terms of the engineering shear strain as

$$\varepsilon_{12} = \frac{\gamma_{12}}{2} \qquad \varepsilon_{13} = \frac{\gamma_{13}}{2} \qquad \varepsilon_{23} = \frac{\gamma_{23}}{2} \tag{16.6}$$

where γ represents the engineering shear strain, which is proportional to the shear stress through the shear modulus. Some problems can arise as a result of the original definition of the engineering shear strain as a nontensorial term. The major difficulty occurs when the stress-strain relations must be transformed by an axis rotation. To overcome this difficulty, modifications must be made to account for the nontensorial shear strain terms.

16.2 CONSTITUTIVE RELATIONS

Load displacement relations or stress-strain relations that describe the reaction of the material to applied loads are called *constitutive relations* since they describe macroscopic behavior as a result of the internal constitution of the material. An ideal elastic solid that is commonly chosen for stress analysis is described by a linear relation between stress and strain. It is assumed to obey Hooke's law, which in a uniaxial stress condition relates the stress to strain by Young's modulus. For small strains, which are sufficiently small compared to unity, and stresses, which are based on the underformed coordinate system, the classical elastic constitutive generalized Hooke's law may be written as

$$\sigma_{ij} = A_{ijrs}\varepsilon_{rs} \tag{16.7}$$

where σ_{ij} = engineering stresses based on underformed coordinate areas
ε_{rs} = small strain terms
A_{ijrs} = elastic stiffnesses

For the generalized Hooke's law, the fourth-order tensor A_{ijrs} [Eq. (16.7)] exhibits 81 terms for unsymmetric stress and strain. A_{ijrs} transforms the strain, a second-order tensor, into stress, another second-order tensor. For symmetric stress and strain, the number of independent terms in each second-order tensor is reduced to 6, and in turn, the number of independent elastic constants (stiffnesses) are then reduced to 36. This yields a stiffness matrix symmetric in $i, j,$ and r, s, where

$$A_{ijrs} = A_{jisr} \qquad (16.8)$$

However, this does not yield a completely symmetric elastic constant matrix.

A completely symmetric elastic stiffness matrix may be obtained by the use of a strain energy function of thermodynamic origin. Assuming the existence of a homogeneous quadratic strain energy function of strain V such that

$$V = \frac{1}{2}\sigma_{ij}\varepsilon_{ij} \qquad (16.9)$$

and

$$\sigma_{ij} = \frac{\partial V}{\partial \varepsilon_{ij}} \qquad (16.10)$$

it may be shown that

$$A_{ijrs} = A_{rsij} \qquad (16.11)$$

The strain energy term V is based on strain energy per unit of underformed area, and the assumption of small strain. With symmetry in stress, strain, and the elastic constants, the number of independent elastic stiffness terms is reduced to 21 for a generally anisotropic material. It is important to note that the inverse of Eq. (16.7) may be expressed in terms of a compliance term S_{ijrs} such that

$$\varepsilon_{ij} = S_{ijrs}\sigma_{rs} \qquad (16.12)$$

where S_{ijrs} is the inverse of the stiffness A_{ijrs} and

$$S_{ijrs} = S_{rsij} = S_{jisr} \qquad (16.13)$$

In consideration of the symmetries discussed in the foregoing, a contracted notation has been adapted in the literature. To illustrate this notation expand Eq. (16.7) for $i = j = 1$, resulting in

$$\begin{aligned}
\sigma_{11} &= A_{1111}\varepsilon_{11} + A_{1122}\varepsilon_{22} + A_{1133}\varepsilon_{33} + A_{1123}\varepsilon_{23} + A_{1132}\varepsilon_{32} \\
&\quad + A_{1113}\varepsilon_{13} + A_{1131}\varepsilon_{31} + A_{1112}\varepsilon_{12} + A_{1121}\varepsilon_{21} \\
&= A_{1111}\varepsilon_{11} + A_{1122}\varepsilon_{22} + A_{1133}\varepsilon_{33} + 2A_{1123}\varepsilon_{23} \\
&\quad + 2A_{1113}\varepsilon_{13} + 2A_{1112}\varepsilon_{12}
\end{aligned} \qquad (16.14)$$

Expanding the additional stress terms results in the shear-strain tensor terms being multiplied by 2, whereas all the stress terms and the remaining strain terms are not prefixed by the multiplier 2. Therefore, the contracted notation is written as

$$\sigma_q = Q_{qr}\varepsilon_r \tag{16.15}$$

where stress and strain are expressed as vectors and the stiffnesses Q_{qr} are defined as a 6×6 symmetric matrix. Equation (16.15) is then rewritten as

$$
\begin{Bmatrix} \sigma_{11} = \sigma_1 \\ \sigma_{22} = \sigma_2 \\ \sigma_{33} = \sigma_3 \\ \sigma_{23} = \sigma_4 \\ \sigma_{13} = \sigma_5 \\ \sigma_{12} = \sigma_6 \end{Bmatrix} = \begin{bmatrix} Q_{11} & Q_{12} & Q_{13} & Q_{14} & Q_{15} & Q_{16} \\ & Q_{22} & Q_{23} & Q_{24} & Q_{25} & Q_{26} \\ & & Q_{33} & Q_{34} & Q_{35} & Q_{36} \\ & & & Q_{44} & Q_{45} & Q_{46} \\ \text{(Symmetric)} & & & & Q_{55} & Q_{56} \\ & & & & & Q_{66} \end{bmatrix} \begin{Bmatrix} \varepsilon_1 = \varepsilon_{11} \\ \varepsilon_2 = \varepsilon_{22} \\ \varepsilon_3 = \varepsilon_{33} \\ \varepsilon_4 = 2\varepsilon_{23} \\ \varepsilon_5 = 2\varepsilon_{13} \\ \varepsilon_6 = 2\varepsilon_{12} \end{Bmatrix} \tag{16.16}
$$

Comparison of Eqs. (16.14) and (16.16) yields

$$Q_{1111} = Q_{11} \qquad Q_{1122} = Q_{12} \qquad A_{1133} = Q_{13}$$
$$A_{1123} = A_{1132} = Q_{14}$$
$$A_{1113} = A_{1131} = Q_{15} \tag{16.17}$$
$$A_{1112} = A_{1121} = Q_{16}$$

It can be shown that further reduction of the number of independent elastic constants is possible when planes of elastic symmetry [1, 2] exist in the material. An important class of manmade materials, fibrous composite materials, for example, possesses three mutually perpendicular planes of elastic symmetry and are referred to as *orthotropic materials.* For these materials the number of independent elastic constants is nine. Materials for which the stiffness terms are independent of orientation, designated as *isotropic,* have only two independent elastic constants.

Since expressions for wave speed in orthotropic materials are developed in Section 16.4, a detailed description of this class of materials is presented. The stress-strain relations for a specially orthotropic material, where the reference axes are aligned with the principal material axes, are given as

$$
\begin{Bmatrix} \sigma_1 \\ \sigma_2 \\ \sigma_3 \\ \sigma_4 \\ \sigma_5 \\ \sigma_6 \end{Bmatrix} = \begin{bmatrix} Q_{11} & Q_{12} & Q_{13} & 0 & 0 & 0 \\ Q_{12} & Q_{22} & Q_{23} & 0 & 0 & 0 \\ Q_{13} & Q_{23} & Q_{33} & 0 & 0 & 0 \\ 0 & 0 & 0 & Q_{44} & 0 & 0 \\ 0 & 0 & 0 & 0 & Q_{55} & 0 \\ 0 & 0 & 0 & 0 & 0 & Q_{66} \end{bmatrix} \begin{Bmatrix} \varepsilon_1 \\ \varepsilon_2 \\ \varepsilon_3 \\ \varepsilon_4 \\ \varepsilon_5 \\ \varepsilon_6 \end{Bmatrix} \tag{16.18}
$$

Based on Eqs. (16.12) and (16.13) the inverse of Eq. (16.18) is expressed as

$$\begin{Bmatrix} \varepsilon_1 \\ \varepsilon_2 \\ \varepsilon_3 \\ \varepsilon_4 \\ \varepsilon_5 \\ \varepsilon_6 \end{Bmatrix} = \begin{bmatrix} S_{11} & S_{12} & S_{13} & 0 & 0 & 0 \\ S_{12} & S_{22} & S_{23} & 0 & 0 & 0 \\ S_{13} & S_{23} & S_{33} & 0 & 0 & 0 \\ 0 & 0 & 0 & S_{44} & 0 & 0 \\ 0 & 0 & 0 & 0 & S_{55} & 0 \\ 0 & 0 & 0 & 0 & 0 & S_{66} \end{bmatrix} \begin{Bmatrix} \sigma_1 \\ \sigma_2 \\ \sigma_3 \\ \sigma_4 \\ \sigma_5 \\ \sigma_6 \end{Bmatrix} \tag{16.19}$$

In matrix notation, Eqs. (16.18) and (16.19) are given, respectively, as

$$\{\sigma\} = [Q]\{\varepsilon\} \tag{16.20a}$$

and

$$\{\varepsilon\} = [S]\{\sigma\} \tag{16.20b}$$

With use of the notations $[Q]^{-1}$ and $[S]^{-1}$ as the inverse of $[Q]$ and $[S]$, respectively, the relation between $[Q]$ and $[S]$ becomes

$$[Q]^{-1} = [S] \tag{16.21a}$$

and

$$[S]^{-1} = [Q] \tag{16.21b}$$

where $[Q]$ and $[S]$ are called the stiffness and compliance matrices, respectively.

When an engineering notation generally reserved for three-dimensional orthotropic materials [2, 3] is employed, there are three Young's moduli, E_{11}, E_{22}, and E_{33}; three shear moduli, G_{23}, G_{13}, and G_{12}; and six Poisson's ratios, ν_{32}, ν_{31}, ν_{21}, ν_{23}, ν_{13}, and ν_{12}. If the three Young's moduli and the three shear moduli are chosen as independent constants, then only three of the Poisson's ratios may be considered as independent. Using the notation ν_{ij} as the Poisson's ratio for transverse strain in the j direction when stressed in the i direction, then symmetry of the compliance matrix requires

$$\frac{\nu_{ij}}{E_{ii}} = \frac{\nu_{ji}}{E_{jj}} \tag{16.22}$$

where the repeated index merely implies a repeated subscript and not the usual summation convention of indicial notation.

For an elastic solid it is assumed that the constitutive relation is a linear function, and thus the application of multiple loads produces stresses and strains that are the sums of the individual components. With this assumption, and the knowledge that normal stress produces no shear strain and shear stress produces no extensional stress, the compliances can be formulated more expeditiously than the stiffnesses. By use of a simple strength of materials approach, with known elastic properties, the stresses may be applied individually and the resulting strains in the various directions determined. As demonstrated in Table 16.1, if the stress σ_1 is applied, the results are a strain in the 1 direction equal to σ_1/E_{11}, a strain in the 2 direction equal to $-\nu_{21}\sigma_1/E_{11}$ (attributed to the Poisson effect for transverse strain in the 2 direction when a stress is applied in the 1 direction), and in a similar manner a strain produced in the 3 direction is equal to $-\nu_{31}\sigma_1/E_{11}$.

TABLE 16.1.   Strain in a Specially Orthotropic Material

Resulting Strain	Applied Stress					
	σ_1	σ_2	σ_3	σ_4	σ_5	σ_6
ε_1	$\dfrac{\sigma_1}{E_{11}}$	$-\dfrac{\nu_{21}\sigma_2}{E_{22}}$	$-\dfrac{\nu_{31}\sigma_3}{E_{33}}$	0	0	0
ε_2	$-\dfrac{\nu_{12}\sigma_1}{E_{11}}$	$\dfrac{\sigma_2}{E_{22}}$	$-\dfrac{\nu_{32}\sigma_3}{E_{33}}$	0	0	0
ε_3	$-\dfrac{\nu_{13}\sigma_1}{E_{11}}$	$-\dfrac{\nu_{23}\sigma_2}{E_{22}}$	$\dfrac{\sigma_3}{E_{33}}$	0	0	0
ε_4	0	0	0	$\dfrac{\sigma_4}{G_{23}}$	0	0
ε_5	0	0	0	0	$\dfrac{\sigma_5}{G_{13}}$	0
ε_6	0	0	0	0	0	$\dfrac{\sigma_6}{G_{21}}$

Summing the strain terms, and placing them and the stresses in columns, allows Eq. (16.19) to be rewritten. The resulting Eq. (16.23) represents the stress-strain relations or constitutive relations in terms of the compliances for a specially orthotropic material. The term "specially orthotropic" refers to the special case where the stress axes coincide with the principal material axes:

$$\begin{Bmatrix} \varepsilon_1 \\ \varepsilon_2 \\ \varepsilon_3 \\ \varepsilon_4 \\ \varepsilon_5 \\ \varepsilon_6 \end{Bmatrix} = \begin{bmatrix} \dfrac{1}{E_{11}} & \dfrac{-\nu_{21}}{E_{22}} & \dfrac{-\nu_{31}}{E_{33}} & 0 & 0 & 0 \\[2mm] \dfrac{-\nu_{12}}{E_{11}} & \dfrac{1}{E_{22}} & \dfrac{-\nu_{32}}{E_{33}} & 0 & 0 & 0 \\[2mm] \dfrac{-\nu_{13}}{E_{11}} & \dfrac{-\nu_{23}}{E_{22}} & \dfrac{1}{E_{33}} & 0 & 0 & 0 \\[2mm] 0 & 0 & 0 & \dfrac{1}{G_{23}} & 0 & 0 \\[2mm] 0 & 0 & 0 & 0 & \dfrac{1}{G_{13}} & 0 \\[2mm] 0 & 0 & 0 & 0 & 0 & \dfrac{1}{G_{21}} \end{bmatrix} \begin{Bmatrix} \sigma_1 \\ \sigma_2 \\ \sigma_3 \\ \sigma_4 \\ \sigma_5 \\ \sigma_6 \end{Bmatrix} \qquad (16.23)$$

The following simplifications in Eq. (16.23) are noted by referring to Eq. (16.22):

$$S_{12} = S_{21} \qquad \frac{\nu_{21}}{E_{22}} = \frac{\nu_{12}}{E_{11}}$$

$$S_{13} = S_{31} \qquad \frac{\nu_{31}}{E_{33}} = \frac{\nu_{13}}{E_{11}} \tag{16.24}$$

$$S_{23} = S_{32} \qquad \frac{\nu_{23}}{E_{22}} = \frac{\nu_{32}}{E_{33}}$$

As previously indicated by Eq. (16.21), $[Q] = [S]^{-1}$, and after an extended inverse calculation it can be shown that the stiffnesses defined by Eq. (16.18) are given by

$$Q_{11} = \frac{1 - \nu_{23}\nu_{32}}{E_{22}E_{33}\,\Delta} \qquad Q_{44} = G_{23}$$

$$Q_{22} = \frac{1 - \nu_{13}\nu_{31}}{E_{11}E_{33}\,\Delta} \qquad Q_{55} = G_{13}$$

$$Q_{33} = \frac{1 - \nu_{12}\nu_{21}}{E_{11}E_{22}\,\Delta} \qquad Q_{66} = G_{12}$$

$$Q_{12} = \frac{\nu_{21} + \nu_{31}\nu_{23}}{E_{22}E_{33}\,\Delta} = \frac{\nu_{12} + \nu_{32}\nu_{13}}{E_{11}E_{33}\,\Delta} \tag{16.25}$$

$$Q_{23} = \frac{\nu_{32} + \nu_{12}\nu_{31}}{E_{11}E_{33}\,\Delta} = \frac{\nu_{23} + \nu_{21}\nu_{13}}{E_{11}E_{22}\,\Delta}$$

$$Q_{13} = \frac{\nu_{31} + \nu_{21}\nu_{32}}{E_{22}E_{33}\,\Delta} = \frac{\nu_{13} + \nu_{12}\nu_{23}}{E_{11}E_{22}\,\Delta}$$

$$\text{where } \Delta = \frac{1 - \nu_{12}\nu_{21} - \nu_{23}\nu_{32} - \nu_{31}\nu_{13} - 2\nu_{21}\nu_{32}\nu_{13}}{E_{11}E_{22}E_{33}}$$

It is noteworthy that the stiffness and compliance matrices have many zero terms, or in matrix terminology, the matrices are not fully populated. If Eq. (16.1) was employed to determine stresses associated with axes not aligned with the principal material axes, the resulting elastic constant matrices would be fully populated. These fully populated matrices contain the nine independent elastic constants and the direction cosines of the tensor transformation as given in Eq. (16.1). This tensor transformation is described in detail in [2] and [3].

When the elastic constants are independent of the axes of orientation or they are the same in all directions, the material is isotropic and there are only two independent elastic constants. The engineering elastic constants given previously in Eqs. (16.23) and (16.25) for an orthotropic material are reduced to

$$E_{11} = E_{22} = E_{33} = E$$
$$\nu_{12} = \nu_{13} = \nu_{23} = \nu \tag{16.26}$$
$$G_{23} = G_{13} = G_{12} = G$$

for an isotropic material. Although there are three terms presented in Eqs. (16.26), G can be expressed in terms of E and ν as

$$G = \frac{E}{2(1 + \nu)} \tag{16.27}$$

An alternative description of an isotropic material that proves the existence of only two independent elastic constants may be described in terms of Lamé's constants λ and μ. By use of these constants the stiffness matrix $[Q]$ is given by

$$[Q] = \begin{bmatrix} \lambda + 2\mu & \lambda & \lambda & 0 & 0 & 0 \\ \lambda & \lambda + 2\mu & \lambda & 0 & 0 & 0 \\ \lambda & \lambda & \lambda + 2\mu & 0 & 0 & 0 \\ 0 & 0 & 0 & \lambda & 0 & 0 \\ 0 & 0 & 0 & 0 & \lambda & 0 \\ 0 & 0 & 0 & 0 & 0 & \lambda \end{bmatrix} \tag{16.28}$$

The Lamé's constants are related to the engineering constants by

$$\lambda = \frac{\nu E}{(1 + \nu)(1 - 2\nu)} \tag{16.29}$$
$$\mu = G$$

In reality this description renders a less complicated short-hand description than that obtained by the use of Eq. (16.26) in the expression of Eq. (16.25).

Be aware that all the equations developed previously for the more general material descriptions also apply to isotropic materials. The stress, strain, and elastic constants are all symmetric for isotropic materials and operate relative to each other as previously indicated. In the following section the equations of motion are presented. These equations are applicable to all material types presented.

EXAMPLE 16.2 ▼

Plane strain is defined as

$$\varepsilon_3 = \varepsilon_{33} = 0$$

$$\varepsilon_4 = \varepsilon_{32} = 0$$

$$\varepsilon_5 = \varepsilon_{13} = 0$$

Using this definition, derive the stress equations for plane strain of an isotropic material.

Solution

For an isotropic material Eqs. (16.26) yield

$$\begin{aligned} E_{11} &= E_{22} = E_{33} = E \\ \nu_{11} &= \nu_{23} = \nu_{12} = \nu \\ G_{13} &= G_{23} = G_{12} = G \end{aligned} \tag{1}$$

Substituting Eqs. (1) into Eqs. (16.23) results in

$$\varepsilon_1 = \frac{\sigma_1}{E} - \frac{\nu}{E}(\sigma_2 + \sigma_3) \tag{2}$$

$$\varepsilon_2 = \frac{\sigma_2}{E} - \frac{\nu}{E}(\sigma_1 + \sigma_3) \tag{3}$$

$$\varepsilon_3 = 0 = \frac{\sigma_3}{E} - \frac{\nu}{E}(\sigma_1 + \sigma_2) \tag{4}$$

$$\varepsilon_6 = \frac{\sigma_6}{G} \tag{5}$$

and also note that

$$\sigma_4 = \sigma_5 = 0 \tag{6}$$

From Eq. (4)

$$\sigma_3 = \nu(\sigma_1 + \sigma_2) \tag{7}$$

and substituting Eq. (7) into Eqs. (2) and (3) yields

$$\varepsilon_1 E = \sigma_1(1 - \nu) - \sigma_2\nu(1 + \nu) \tag{8}$$

$$\varepsilon_2 E = -\sigma_1\nu(1 + \nu) - \sigma_2(1 - \nu) \tag{9}$$

Solving Eqs. (8) and (9) simultaneously results in the following plane strain-stress equations:

$$\sigma_1 = \frac{E(1 - \nu)}{(1 + \nu)(1 - 2\nu)}\varepsilon_1 + \frac{\nu E}{(1 + \nu)(1 - 2\nu)}\varepsilon_2 \tag{10}$$

$$\sigma_2 = \frac{\nu E}{(1 + \nu)(1 - 2\nu)}\varepsilon_1 + \frac{(1 - \nu)E}{(1 + \nu)(1 - 2\nu)}\varepsilon_2 \tag{11}$$

$$\sigma_3 = \frac{\nu E(\varepsilon_1 + \varepsilon_2)}{(1 + \nu)(1 - 2\nu)} \tag{12}$$

$$\sigma_6 = G\varepsilon_6 \tag{13}$$

Alternatively, by use of Lamé's constants given in Eqs. (16.29), the plane strain-stress equations may be expressed as

$$\sigma_1 = (\lambda + 2\mu)\varepsilon_1 + \lambda\varepsilon_2 \tag{14}$$

$$\sigma_2 = \lambda\varepsilon_1 + (\lambda + 2\mu)\varepsilon_2 \tag{15}$$

$$\sigma_3 = \lambda(\varepsilon_1 + \varepsilon_2) \tag{16}$$

$$\sigma_6 = \mu\varepsilon_6 \tag{17}$$

▲

16.3 EQUATIONS OF MOTION

The general Cauchy equations of motion are based on Newton's second law and the premise that the time rate of change of the momentum of a collection of particles is equal to the sum of the externally applied forces. Further, it is assumed that Newton's third law applies and that the action-reaction principle is imposed upon the individual particles. If this collection of particles is a solid body of some given constant mass whose motion is restricted to a simple translation, the equation of motion may be simply stated as: mass times acceleration equals applied force.

Cauchy's equations of motion are derived in great detail by Malvern [1], and for zero body forces the general equation may be reduced to

$$\frac{\partial \sigma_{ij}}{\partial x_j} = \rho \frac{\partial^2 u_i}{\partial t^2} \tag{16.30}$$

where $\sigma_{ij} =$ stresses
$x_j =$ right-handed coordinate axes
$u_i =$ displacements
$\rho =$ mass density
$t =$ time

The indices i, j assume the integer values 1, 2, 3, and i is the free index indicating that there are three equations. Application of the general strain relation represented by Eq. (16.5) and the general constitutive relation given by Eq. (16.7) to Eq. (16.30) gives the resulting equation of motion

$$A_{ijrs} \frac{\partial^2 u_r}{\partial x_j \, \partial x_s} = \rho \frac{\partial^2 u_i}{\partial t^2} \tag{16.31}$$

Again, the only free index is i, which assumes the values of 1, 2, 3, thus yielding three equations. Equation (16.31) is generally referred to as the *wave equation*.

16.4 STRESS WAVE PROPAGATION

Before attempting further development of the wave equation, it is important to note that waves may be described in terms of bounded or unbounded media when discussing wave propagation in solids. For the unbounded medium it is assumed that the material extends unbounded (indefinitely) in all directions. For this type of medium it has been shown that waves of all frequencies travel without dispersion, and the wave velocity is independent of frequency. Dispersion is used in this context to mean the phenomenon of changes in wave speed as the frequency of transmission changes. For bounded media, such as rods and plates, the effects of dispersion must be taken into account. In this chapter the emphasis is directed primarily toward the unbounded media wave propagation, and only a limited discussion on the wave propagation in long rods (a common example of bounded media) is presented. For complete details on wave propagation in bounded media, texts by Kolsky [4], Redwood [5], and Achenbach [6] are recommended.

For wave propagation in an unbounded medium, assuming a plane wave solution, Hearmon [7] proposed the following solution:

$$(u_1, u_2, u_3) = (U_1, U_2, U_3)\exp\frac{2\pi i}{\lambda}(n_k x_k - c_n t) \qquad (16.32)$$

where U_1, U_2, and U_3 = amplitudes of the displacements
$\qquad\qquad \lambda$ = wave length
$\qquad\qquad n_k$ = direction cosines of the normal to the wave front
$\qquad\qquad c_n$ = phase velocity
$\qquad\qquad i$ = imaginary number $\sqrt{-1}$

This solution may be used to determine expressions for the phase velocities in terms of the elastic constants and the medium density. Substituting the expression given by Eq. (16.32) into Eq. (16.31) and collecting all nonzero terms on the left-hand side (LHS) results in

$$\begin{bmatrix} \Gamma_{11} - \rho c^2 & \Gamma_{12} & \Gamma_{13} \\ \Gamma_{12} & \Gamma_{22} - \rho c^2 & \Gamma_{23} \\ \Gamma_{13} & \Gamma_{23} & \Gamma_{33} - \rho c^2 \end{bmatrix} \begin{Bmatrix} U_1 \\ U_2 \\ U_3 \end{Bmatrix} = \begin{Bmatrix} 0 \\ 0 \\ 0 \end{Bmatrix} \qquad (16.33)$$

where Γ_{ir} are the Christoffel stiffnesses [7, 8], which are given by

$$\Gamma_{ir} = n_j n_s A_{ijrs} \qquad (16.34)$$

The condition for nonzero or nontrivial solution requires that the determinant of the square symmetric matrix on the LHS of Eq. (16.33) must vanish. This condition yields a cubic equation in ρc^2, and in general, there are three velocities corresponding to the given normal (n_1, n_2, n_3). This same general approach has been employed for the dynamic response of composite materials [9]. With use of the two-indices subscripts for the elastic stiffness, as demonstrated in Eqs. (16.16) and (16.17), the Christoffel stiffnesses of Eq. (16.34) are expanded as

$$\begin{aligned} \Gamma_{11} &= n_1^2 Q_{11} + n_2^2 Q_{66} + n_3^2 Q_{55} + 2n_1 n_2 Q_{16} \\ &\quad + 2n_1 n_3 Q_{15} + 2n_2 n_3 Q_{56} \\[4pt] \Gamma_{12} &= n_1^2 Q_{16} + n_2^2 Q_{26} + n_3^2 Q_{45} + n_1 n_2 (Q_{12} + Q_{66}) \\ &\quad + n_1 n_3 (Q_{56} + Q_{14}) + n_2 n_3 (Q_{46} + Q_{25}) \\[4pt] \Gamma_{13} &= n_1^2 Q_{15} + n_2^2 Q_{46} + n_3^2 Q_{35} + n_1 n_2 (Q_{56} + Q_{14}) \\ &\quad + n_1 n_3 (Q_{13} + Q_{55}) + n_2 n_3 (Q_{35} + Q_{36}) \\[4pt] \Gamma_{22} &= n_1^2 Q_{66} + n_2^2 Q_{22} + n_3^2 Q_{44} + 2n_1 n_2 Q_{26} \\ &\quad + 2n_1 n_3 Q_{46} + 2n_2 n_3 Q_{24} \\[4pt] \Gamma_{23} &= n_1^2 Q_{56} + n_2^2 Q_{24} + n_3^2 Q_{34} + n_1 n_2 (Q_{25} + Q_{46}) \\ &\quad + n_1 n_3 (Q_{36} + Q_{45}) + n_2 n_3 (Q_{23} + Q_{44}) \\[4pt] \Gamma_{33} &= n_1^2 Q_{55} + n_2^2 Q_{44} + n_3^2 Q_{33} + 2n_1 n_2 Q_{45} \\ &\quad + 2n_1 n_3 Q_{35} + 2n_2 n_3 Q_{34} \end{aligned} \qquad (16.35)$$

Equation (16.35) is representative of a general anisotropic material expressed by Eq. (16.16). For an orthotropic material, such as given by Eq. (16.18) where there are only nine independent elastic constants, Eq. (16.35) reduces to

$$\Gamma_{11} = n_1^2 Q_{11} + n_2^2 Q_{66} + n_3^2 Q_{55}$$
$$\Gamma_{12} = n_1 n_2 (Q_{12} + Q_{66})$$
$$\Gamma_{13} = n_1 n_3 (Q_{13} + Q_{55})$$
$$\Gamma_{22} = n_1^2 Q_{66} + n_2^2 Q_{22} + n_3^2 Q_{44} \qquad (16.36)$$
$$\Gamma_{23} = n_2 n_3 (Q_{23} + Q_{44})$$
$$\Gamma_{33} = n_1^2 Q_{55} + n_2^2 Q_{44} + n_3^2 Q_{33}$$

At this point all the necessary equations to predict the theoretical wave propagation velocities for a specified unbounded medium are in place. For an orthotropic material with the stiffnesses as defined in Eq. (16.25) and assuming all moduli and Poisson's ratios are known, the propagation velocities may be determined by use of Eq. (16.33). As the elastic wavefront propagates in an unbounded medium, the particles affected by the wavefront are translated in one of the three mutually orthogonal displacement directions that are aligned in the wavefront as shown in Figure 16.4. A disturbance that occurs at a boundary and generates stress waves into a body may excite the three particle directions, but they will separate and propagate as separate waves through the body.

To illustrate the phenomenon, consider a plane wave traveling in an unbounded medium in the x_2 direction that excites the material particles in the x_2 direction. The wave normal (n_1, n_2, n_3) for this case becomes $(0, 1, 0)$. Thus from Eq. (16.36)

$$\Gamma_{11} = Q_{66} \qquad \Gamma_{22} = Q_{22} \qquad \Gamma_{33} = Q_{44}$$
$$\Gamma_{12} = \Gamma_{13} = \Gamma_{23} = 0 \qquad (16.37)$$

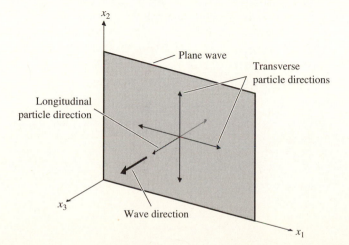

Figure 16.4

Plane wave propagation schematic showing longitudinal and transverse waves.

If the particle motion is assumed to be in the direction of x_2, then displacement amplitudes corresponding to $U_2 \neq 0$ and $U_1 = U_3 = 0$. From Eq. (16.33)

$$(\Gamma_{22} - \rho c^2)U_2 = 0 \tag{16.38}$$

and since U_2 is arbitrary or nonzero, then from Eqs. (16.38) and (16.37)

$$\rho c^2 = \Gamma_{22} = Q_{22} \tag{16.39}$$

and the wave velocity is expressed as

$$c = \left(\frac{Q_{22}}{\rho}\right)^{1/2} \tag{16.40}$$

EXAMPLE 16.3 ▼

A boron-epoxy composite material has material stiffness $Q_{11} = 35.32 \times 10^6$ psi, $Q_{22} = 3.532 \times 10^6$ psi, $Q_{12} = 1.06 \times 10^6$ psi, and $Q_{66} = 1.5 \times 10^6$ psi. Determine the wave speed in the x_2 direction if the mass density of the material ρ is 0.207×10^{-3} lb-sec^2/in^4.

Solution

From Eq. (16.40), the wave speed in the x_2 direction is represented as

$$c = \left(\frac{Q_{22}}{\rho}\right)^{1/2} \tag{1}$$

Thus

$$c = \left(\frac{3.532 \times 10^6}{0.207 \times 10^{-3}}\right)^{1/2} = 1.306 \times 10^5 \text{ in/sec}$$

▲

For the particular case examined in Example 16.3, the wave motion and particle motion are in the same direction, and the wave is referred to as a *longitudinal* wave. When the wave motion and particle motion are perpendicular to each other, the resulting wave is called a *transverse*, or *shear*, wave. An example of a shear wave, using the same wave direction of $n_k = (0, 1, 0)$, occurs when the amplitude of the particle motion U_1 or $U_3 \neq 0$. If $U_1 \neq 0$, then Eq. (16.33) becomes

$$(\Gamma_{11} - \rho c^2)U_1 = 0 \tag{16.41}$$

and since $U_1 \neq 0$, then

$$\rho c^2 = \Gamma_{11} = Q_{66} \tag{16.42}$$

and

$$c = \left(\frac{Q_{66}}{\rho}\right)^{1/2} \tag{16.43}$$

With similar reasoning for wave motion in the x_2 direction and particle motion in the x_3 direction, the wave velocity becomes

$$c = \left(\frac{Q_{44}}{\rho}\right)^{1/2} \tag{16.44}$$

It is noteworthy that the longitudinal wave velocities are greater than the transverse velocities, although it is not readily apparent. This condition will be discussed a bit later when it is applied to an isotropic material.

EXAMPLE 16.4 ▼

The longitudinal wave speed in the $n_k = (0, 0, 1)$ direction of a uniaxial fibrous composite is given in Table 16.2 as $c = \sqrt{Q_{33}/\rho}$. Determine the numerical value of the longitudinal wave velocity if the composite has the following material properties:

$E_1 = 20 \times 10^6$ psi $\qquad G_{13} = G_{12} = 0.81 \times 10^6$ psi

$E_2 = E_3 = 1.5 \times 10^6$ psi $\qquad G_{23} = 0.1 \times 10^6$ psi

$\nu_{13} = \nu_{12} = 0.12 \qquad \rho = 0.155 \times 10^{-3}$ lb-sec^2/in^4

$\nu_{23} = 0.3$

Solution

From Eqs. (16.25), the material stiffness Q_{33} is expressed as

$$Q_{33} = \frac{1 - \nu_{12}\nu_{21}}{E_{11}E_{22}\Delta} \tag{1}$$

or

$$Q_{33} = \frac{(1 - \nu_{12}\nu_{21})E_{33}}{1 - \nu_{12}\nu_{21} - \nu_{23}\nu_{32} - \nu_{31}\nu_{13} - 2\nu_{21}\nu_{23}\nu_{13}} \tag{2}$$

From the given data and from Eqs. (16.25), the remaining Poisson's ratios are calculated as

$$\nu_{21} = \frac{E_{22}}{E_{11}}\nu_{12} = \frac{1.5}{20}(0.12) = 0.009 \tag{3}$$

$$\nu_{31} = \frac{E_{33}}{E_{11}}\nu_{13} = \frac{1.5}{20}(0.12) = 0.009 \tag{4}$$

$$\nu_{32} = \frac{E_{33}}{E_{22}}(\nu_{23}) = \frac{1.5}{1.5}(0.3) = 0.3 \tag{5}$$

TABLE 16.2. **Relations Between Elastic Constants and Wave Velocities for an Orthotropic Elastic Material**

Wave Normal	Particle Direction	Wave Type	Phase Velocity Relation
$n_1 = 1$ $n_2 = 0$ $n_3 = 0$	x_1 x_2 x_3	Long Trans Trans	$\rho c^2 = Q_{11}$ $\rho c^2 = Q_{66}$ $\rho c^2 = Q_{55}$
$n_1 = 0$ $n_2 = 1$ $n_3 = 0$	x_1 x_2 x_3	Trans Long Trans	$\rho c^2 = Q_{66}$ $\rho c^2 = Q_{22}$ $\rho c^2 = Q_{44}$
$n_1 = 0$ $n_2 = 0$ $n_3 = 1$	x_1 x_2 x_3	Trans Trans Long	$\rho c^2 = Q_{55}$ $\rho c^2 = Q_{44}$ $\rho c^2 = Q_{33}$
$n_1 = 0$ $n_2 \neq 0$ $n_3 \neq 0$	x_1	Trans	$\rho c^2 = n_2^2 Q_{66} + n_3^2 Q_{55}$
	$x_2 - x_3$ plane	Quasi-long Quasi-trans	$(n_2^2 Q_{22} + n_3^2 Q_{44} - \rho c^2)(n_2^2 Q_{44} + n_3^2 Q_{33} - \rho c^2)$ $= n_2^2 n_3^2 (Q_{23} + Q_{44})^2$
$n_2 = 0$ $n_1 \neq 0$ $n_3 \neq 0$	x_2	Trans	$\rho c^2 = n_1^2 Q_{66} + n_3^2 Q_{44}$
	$x_1 - x_3$ plane	Quasi-long Quasi-trans	$(n_1^2 Q_{11} + n_3^2 Q_{55} - \rho c^2)(n_1^2 Q_{55} + n_3^2 Q_{33} - \rho c^2)$ $= n_1^2 n_3^2 (Q_{13} + Q_{55})^2$
$n_3 = 0$ $n_1 \neq 0$ $n_2 \neq 0$	x_3	Trans	$\rho c^2 = n_1^2 Q_{55} + n_2^2 Q_{44}$
	$x_1 - x_2$ plane	Quasi-long Quasi-trans	$(n_1^2 Q_{11} + n_2^2 Q_{66} - \rho c^2)(n_1^2 Q_{66} + n_2^2 Q_{22} - \rho c^2)$ $= n_1^2 n_2^2 (Q_{12} + Q_{66})^2$

Substituting the values for the material constants into Eq. (2) yields the stiffness Q_{33}; thus

$$Q_{33} = \frac{[1 - (0.009)(0.12)](1.5 \times 10^6)}{\begin{array}{c} 1 - (0.009)(0.12) - (0.3)(0.3) \\ - (0.009)(0.12) - 2(0.009)(0.3)(0.12) \end{array}}$$

or

$$Q_{33} = 1.651 \times 10^6 \text{ psi}$$

From Table 16.2, the wave velocity c is expressed as

$$c = \sqrt{\frac{Q_{33}}{\rho}} \tag{6}$$

or

$$c = \sqrt{\frac{1.651 \times 10^6}{0.155 \times 10^{-3}}} = 1.03 \times 10^5 \text{ in/sec}$$

▲

Another case where it is ambiguous whether the wave should be designated as longitudinal or transverse is examined next. For this demonstration the wave motion is chosen such that $n_1 = 0$, $n_2 \neq 0$, $n_3 \neq 0$, and when the particle motion is such that $U_1 \neq 0$, $U_2 = U_3 = 0$, then for an orthotropic material Eqs. (16.36) and (16.33) yield

$$\Gamma_{11} = n_2^2 Q_{66} + n_3^2 Q_{55} \tag{16.45}$$

and

$$(n_2^2 Q_{66} + n_3^2 Q_{55} - \rho c^2)U_1 = 0 \tag{16.46}$$

Again, since $U_1 \neq 0$, then Eq. (16.46) gives

$$\rho c^2 = n_2^2 Q_{66} + n_3^2 Q_{55} \tag{16.47}$$

which is a transverse wave because the wave motion and the particle motion are perpendicular. Similarly, for the same wave motion of $n_k = (0, n_2, n_3)$ and a particle motion in the x_2-x_3 plane such that $U_1 = 0$, $U_2 \neq 0$, $U_3 \neq 0$, Eq. (16.33) then reduces to

$$\begin{bmatrix} \Gamma_{22} - \rho c^2 & \Gamma_{23} \\ \Gamma_{23} & \Gamma_{33} - \rho c^2 \end{bmatrix} \begin{Bmatrix} U_2 \\ U_3 \end{Bmatrix} = \begin{Bmatrix} 0 \\ 0 \end{Bmatrix} \tag{16.48}$$

If both U_1 and U_2 are nonzero, Eq. (16.48) becomes

$$(\Gamma_{22} - \rho c^2)(\Gamma_{33} - \rho c^2) - \Gamma_{23}^2 = 0 \tag{16.49}$$

and from Eq. (16.36)

$$\Gamma_{22} = n_2^2 Q_{22} + n_3^2 Q_{44}$$
$$\Gamma_{23} = n_2 n_3 (Q_{23} + Q_{44}) \tag{16.50}$$
$$\Gamma_{33} = n_2^2 Q_{44} + n_3^2 Q_{33}$$

Substituting Eq. (16.50) into Eq. (16.49) results in

$$\begin{aligned} (n_2^2 Q_{22} + n_3^2 Q_{44} - \rho c^2)(n_2^2 Q_{44} + n_3^2 Q_{33} - \rho c^2) \\ = n_2^2 n_3^2 (Q_{23} + Q_{44})^2 \end{aligned} \tag{16.51}$$

which will yield two real velocities, one called quasi-longitudinal and one called quasi-transverse.

If the previous exercises are performed for the various combinations of wave motion and particle motion, then a series of wave velocities in unbounded media for an orthotropic material can be obtained. The results are summarized in Table 16.2.

In the case of an isotropic material, all elastic moduli and Poisson's ratios are equal and are the same in all directions. This implies that there are two wave velocities for a given elastic isotropic material, a longitudinal wave and a shear wave. In the isotropic solid unbounded body, if the wave motion direction and the particle motion direction coincide, a longitudinal wave is propagated and the wave velocity is given as

$$c_l = \left(\frac{\lambda + 2\mu}{\rho}\right)^{1/2} \tag{16.52}$$

where $\lambda + 2\mu = E(1-\nu)/(1+\nu)(1-2\nu)$.

When the wave motion direction and the particle motion direction are perpendicular, a shear wave is propagated. In the isotropic material case, particle motion may be in either of the other two orthogonal directions, and both propagate with a velocity of

$$c_t = \left(\frac{\mu}{\rho}\right)^{1/2} \tag{16.53}$$

For a bounded medium these two transverse waves are quite different in the bounding surface of the medium. The discussion of waves propagating in a bounding surface or plane is beyond the scope of this chapter. However, for the two transverse waves propagating to a bounding surface, one will excite the particles in the plane of the surface and the other will excite the particles normal to the surface. In Chapter 17 the wave motions in the plane of the surface are called Love waves and the wave motions normal to the surface are called Rayleigh waves.

The ratio of the wave speeds is denoted as κ, given by

$$\kappa = \frac{c_l}{c_t} = \left(\frac{\lambda + 2\mu}{\nu}\right)^{1/2} \tag{16.54}$$

The mechanical properties and associated unbounded medium wave velocities for some common isotropic materials are summarized in Table 16.3.

TABLE 16.3. **Approximate Unbounded Wave Velocities for Some Common Materials**

Material	Young's Modulus[*] E, Mpsi (GPa)	Shear modulus G,[*] Mpsi (GPa)	Poisson's[*] Ratio ν	Spec. Wt.,[*] lb/in³ Density (g/cm³)	Long. Vel. c_l, ft/sec (m/sec)	Trans. Vel. c_t, ft/sec (m/sec)	κ c_l/c_t
2024 aluminum	10.0 (68.97)	3.85 (26.55)	0.33	0.098 (2.70)	20,140 (6,141)	10,270 (3,130)	1.96
Cast iron	20.0 (137.93)	8.0 (55.18)	0.20	0.260 (7.21)	15,140 (4,617)	9,086 (2,770)	1.67
Tough pitch copper	17.0 (17.24)	7.0 (48.28)	0.33	0.322 (8.90)	14,490 (4,417)	7,638 (2,329)	1.90
70-30 brass	15.9 (109.66)	6.0 (41.38)	0.33	0.308 (8.53)	14,330 (4,368)	7,230 (2,204)	1.98
AISI 1030 steel	29.0 (200.01)	11.5 (79.32)	0.287	0.283 (7.85)	18,910 (5,765)	10,440 (3,184)	1.81
Ti-Al-V titanium	16.5 (113.80)	6.20 (42.76)	0.330	0.164 (4.54)	20,000 (6,098)	10,070 (3,071)	1.99
Med. str. concrete	4.5 (31.04)	1.84 (12.69)	0.22	0.086 (2.34)	12,660 (3,860)	7,577 (2,310)	1.67
Glass	9.60 (66.21)	4.10 (28.28)	0.17	0.079 (2.18)	18,720 (5,708)	11,800 (3,598)	1.59

*Refs. [10, 11].

16.5 APPLICATIONS

16.5.1 Stress Wave Velocities in Rods

For long isotropic material rods, where the rod length is much longer than the wavelength of a traveling wave, the wave velocity can be shown [4, 6, 12] to be

$$c_0 = \sqrt{\frac{E}{\rho}} \tag{16.55}$$

where E is the modulus of elasticity and ρ is the material mass density. This wave velocity c_0 is called a longitudinal rod, or bar, velocity. For the conditions discussed above, this velocity is nondispersive. Based on the properties given in Table 16.3, the approximate corresponding rod velocities for several elastic isotropic materials are presented in Table 16.4.

16.5.2 Reflected and Transmitted Plane Waves

Some interesting and important phenomena occur at connected interfaces between two dissimilar materials: the reflection and the transmission of an incident wave that impinges on the interface. These reflections and transmissions will occur at inclined and normal interfaces relative to the incident wave in both unbounded and rod (bounded) wavefronts. However, to clarify the discussion it is restricted to a normal interface in a rod as shown in Figure 16.5. The basic assumptions for this condition are that the inci-

TABLE 16.4. Bar Velocities for Materials Given in Table 16.3

Material	c_0	
	ft/sec	m/sec
Aluminum	16,550	(5045)
Cast iron	14,370	(4380)
Copper	11,900	(3629)
Brass	11,770	(3588)
Steel	16,580	(5056)
Titanium	16,430	(5009)
Concrete	11,850	(3613)
Glass	18,060	(5505)

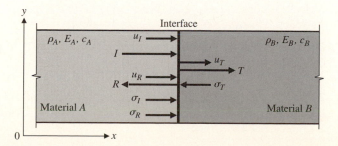

Figure 16.5

Plane longitudinal wave in rod impinging on normal interface between dissimilar isotropic materials A and B.

dent wave is traveling in the positive x direction, the rods have identical cross-sectional areas, and there is normal stress and displacement continuity in the x direction across the interface. These assumptions ensure that the interfaces remain in contact, and a secondary condition of equality of velocity, across the interface, may be utilized. From Figure 16.5 the following equations represent the aforementioned assumptions:

$$\sigma_I + \sigma_R = \sigma_T$$
$$\dot{u}_I + \dot{u}_R = \dot{u}_T \qquad (16.56)$$
$$u_I + u_R = u_T$$

where σ = normal stress
u = normal displacement
$\dot{u}$ = normal velocity
$I,\ R,\ T$ = incident, reflected, and transmitted waves, respectively

The concept of particle velocity and its relation to stress and an impedance term ρc is discussed in detail in [4, 6, 12]. This relation is given as

$$\sigma = \rho c \dot{u} \qquad (16.57)$$

where σ = normal stress in the direction of the particle velocity $\dot{u}$
ρc = characteristic impedance
ρ = mass density
c = wave velocity

In the discussion of stress wave motion it is invariably assumed that the wave normal and particle motion are in the same direction. The velocity calculation is unaffected if the wave normal and particle motion are in opposite directions. However, the sign of the stress is dependent upon the direction of the particle motion. If the wave normal and particle motion are in the same direction the stress is compressive; if they are opposite in direction the stress is tensile.

With this relation the stresses defined by Eq. (16.56) can be expressed as

$$\sigma_I = -(\rho c)_I \dot{u}_I$$
$$\sigma_T = -(\rho c)_T \dot{u}_T \qquad (16.58)$$
$$\sigma_R = +(\rho c)_R \dot{u}_R$$

where the negative sign indicates a compressive stress if all particle velocities are assumed positive in the x direction. From Figure 16.5, the characteristic impedances may be written as

$$(\rho c)_I = (\rho c)_R = (\rho c)_A$$

$$(\rho c)_T = (\rho c)_B \qquad (16.59)$$

where $(\rho c)_A$ and $(\rho c)_B$ are determined from the isotropic elastic properties of materials A and B, respectively.

From the second equation of Eqs. (16.56) and Eqs. (16.58) and (16.59), the following relation may be derived:

$$\frac{\sigma_I}{(\rho c)_A} - \frac{\sigma_R}{(\rho c)_A} = \frac{\sigma_T}{(\rho c)_B} \tag{16.60}$$

From Eq. (16.60) and the first equation of Eqs. (16.56), it is recognized that there are two equations and three unknowns of σ_I, σ_R, and σ_T. This implies that only the following stress ratios can be solved for:

$$\frac{\sigma_R}{\sigma_I} = \frac{(\rho c)_B - (\rho c)_A}{(\rho c)_A + (\rho c)_B}$$

$$\frac{\sigma_T}{\sigma_I} = \frac{2(\rho c)_B}{(\rho c)_A + (\rho c)_B} \tag{16.61}$$

It may also be shown that the displacement and velocity ratios become

$$\frac{\dot{u}_R}{\dot{u}_I} = \frac{u_R}{u_I} = \frac{(\rho c)_B - (\rho c)_A}{(\rho c)_A + (\rho c)_B}$$

$$\frac{\dot{u}_T}{\dot{u}_I} = \frac{u_T}{u_I} = \frac{2(\rho c)_A}{(\rho c)_A + (\rho c)_B} \tag{16.62}$$

Equations (16.61) and (16.62) demonstrate that the ratios of reflected stress and reflected displacements, relative to their respective components of the incident wave, are identical, but the transmitted ratios relative to the incident wave are different.

By application of Eqs. (16.61) and (16.62) to a wide range of characteristic impedances, $(\rho c)_A$ and $(\rho c)_B$, several important observations can be made:

1. If $(\rho c)_A = (\rho c)_B$, there is no reflected stress, displacement, or velocity since no material boundary exists.

2. If $(\rho c)_A > (\rho c)_B$, the reflected stress magnitude is less than full intensity and is opposite of the sign of the incident wave components. (A compressive incident wave is reflected as a tensile wave.) The transmitted wave has the same sign as the incident wave and its magnitude is greater than the reflected wave.

3. If $(\rho c)_A < (\rho c)_B$, the reflected stress magnitude is less than full intensity and has the same sign as the incident wave. (A compression wave is reflected as a compression wave.) The transmitted wave has the same sign as the incident wave and its magnitude is greater than the reflected wave.

4. If $(\rho c)_B \to 0$ (free surface or zero stiffness), the entire wave is reflected at full intensity with a change in sign. No stress is transmitted but the free-surface displacement is twice that of the incident displacement. (A compressive wave is reflected as a tensile wave.) The reflection of a rectangular wave at a free surface is illustrated in Figure 16.6.

5. If $(\rho c)_B >> (\rho c)_A$ (rigid interface or fixed boundary), the wave is reflected almost at full intensity with no change in sign. (A compressive wave is reflected as a compressive wave.) The reflection of a rectangular wave at a fixed boundary is illustrated in Figure 16.7.

Figure 16.6

Reflection of a rectangular
wave from a free surface
$[(\rho c)_B = 0]$.

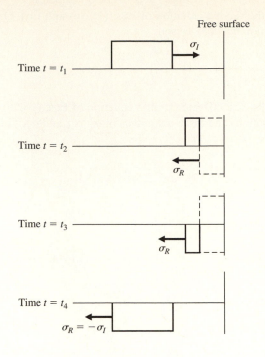

Figure 16.7

Reflection of a rectangular
wave at a fixed boundary
$[(\rho c)_B >> (\rho c)_A]$.

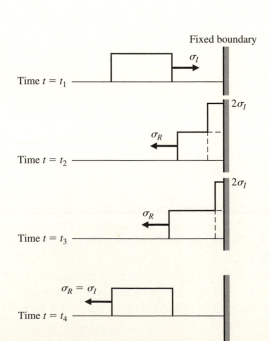

16.5.3 Colinear Impact of Bars

The colinear impact of bars having unequal cross-sectional areas and composed of different materials produces stress waves in each bar whose magnitudes are functions of the material properties and the cross-sectional areas of the bars. A schematic of the impact of two bars is shown in Figure 16.8. For impact to occur, V_1 must be greater than V_2 and both velocities must be in the same direction. Upon impact, a particle velocity of v_0 is established behind two separate wavefronts in each bar. These wavefronts, illustrated in Figure 16.8b, travel at the wave velocity associated with material of each bar. However, the particle velocity relative to a fixed coordinate, in each bar behind each respective wavefront, is equal to v_0. The forces in each bar at the impacted interface are equal, and in terms of the compressive stresses the equilibrium condition is expressed as

$$A_1 \sigma_1 = A_2 \sigma_2 \tag{16.63}$$

where σ_1 and σ_2 are the stresses in each bar, respectively. If the time of impact is chosen as $t = 0$, then the changes in momentum of each compressed volume at an incremental time of Δt after impact are equal, provided $0 < \Delta t < L_1/c_1$ and $0 < \Delta t < L_2/c_2$. The changes in momentum are expressed as

$$A_1(c_1 \Delta t)\rho_1(V_1 - v_0) = A_2(c_2 \Delta t)\rho_2(v_0 - V_2) \tag{16.64}$$

Canceling Δt and solving for v_0 in Eq. (16.64) yields

$$v_0 = \frac{V_2 + RV_1}{1 + R} \tag{16.65}$$

where $R = A_1\rho_1 c_1 / A_2\rho_2 c_2$. By application of Eqs. (16.63) through (16.65), the stresses may be determined as

$$\sigma_1 = \frac{\rho_1 c_1 V_1}{1 + R}\left(1 - \frac{V_2}{V_1}\right) \tag{16.66}$$

$$\sigma_2 = -R\left[\frac{\rho_2 c_2 V_2}{1 + R}\right]\left(1 - \frac{V_1}{V_2}\right) \tag{16.67}$$

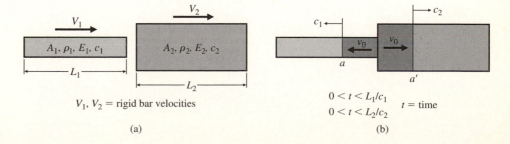

Figure 16.8

Impact of two colinear isotropic bars: (a) before impact; (b) after impact for $V_1 > V_2$.

V_1

V_2

A_1, ρ_1, E_1, c_1

A_2, ρ_2, E_2, c_2

L_1

L_2

V_1, V_2 = rigid bar velocities

(a)

c_1

c_2

v_0 v_0

a

a'

$0 < t < L_1/c_1$

$0 < t < L_2/c_2$

t = time

(b)

Consideration of two simple cases renders interesting results: (1) If $V_1 \leq V_2$, there is no impact; (2) if $V_1 > V_2$, $A_1 = A_2 = A$, $\rho_1 = \rho_2 = \rho$, and $c_1 = c_2 = c$. Then the stresses are equal and

$$\sigma_1 = \sigma_2 = \frac{\rho c}{2}(V_1 - V_2) \tag{16.68}$$

Upon examination of Eq. (16.68) and comparing it to Eq. (16.57), it is concluded that the particle velocity of Eq. (16.68) is $(V_1 - V_2)/2$, which is the same as $(V_1 - v_0)$ or $(v_0 - V_2)$ using Eq. (16.64). This implies that the normal stress given by Eq. (16.57) is equal to the product of the characteristic impedance and the change in the particle velocity across the wavefront.

A simple interpretation of this result can be attained if it is assumed that one is stationed at point a shown in Figure 16.8b; as the wavefront (moving at velocity V_1) moves past that position a change in particle velocity of $(V_1 - v_0)$ is experienced and the corresponding change in stress will be $\rho_1 c_1 (V_1 - v_0)$. For point a' in Figure 16.8b, the change in the particle velocity is then $(v_0 - V_2)$ and the assumed change in stress will be $\rho_2 c_2 (v_0 - V_2)$. If the bars are of the same material, Eq. (16.68) yields

$$(V_1 - v_0) = (v_0 - V_2) = \frac{(V_1 - V_2)}{2} \tag{16.69}$$

and the changes in the particle velocity for each bar are equal. This implies that the initial stress changes at the impacted ends of bars of the same material, having equal cross-sectional areas, are equal.

EXAMPLE 16.5 ▼

A 75-mm-diameter circular aluminum bar traveling at a velocity of 20 m/sec impacts a 50-mm-diameter circular steel bar traveling at a velocity of 10 m/sec in the same direction as illustrated in Figure 16.9. If the longitudinal wave velocity for steel and aluminum are considered equal at 5.05 km/sec, determine (a) the longitudinal stress in each bar and (b) the pulse length t_d of the stress wave in the steel bar.

Solution

(a) The pertinent physical characteristics for each bar are summarized in Table 16.5. The stress in the bars is determined from Eqs. (16.66) and (16.67) as

$$\sigma_{al} = \frac{\rho_{al} c_{al} V_{al}}{1 + R}\left(1 - \frac{V_{st}}{V_{al}}\right) \tag{1}$$

Figure 16.9

Impacting colinear bars of Example 16.5.

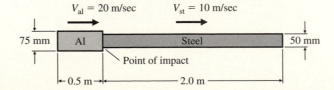

$V_{al} = 20$ m/sec $V_{st} = 10$ m/sec

75 mm   Al   Steel   50 mm

Point of impact

← 0.5 m →   2.0 m

and

$$\sigma_{st} = \frac{\rho_{st} c_{st} V_{st}}{1 + R}\left(1 - \frac{V_{al}}{V_{st}}\right) \tag{2}$$

in which R is calculated from

$$R = \frac{A_{al}\rho_{al}c_{al}}{A_{st}\rho_{st}c_{st}} = \frac{(44.18)(2.7)(5.05)}{(19.64)(7.85)(5.05)} = 0.774$$

Substituting this value for R and the pertinent material properties summarized in Table 16.5 into Eq. (1) yields the stress in the aluminum bar σ_{al} as

$$\sigma_{al} = \frac{2700\ \text{kg/m}^3 \times 5.05\ \text{m/sec} \times 20.0\ \text{m/sec}}{1.774}\left(1 - \frac{1}{2}\right)$$

$$= 76.86 \times 10^6\ \text{kg-m/sec}^2(1/\text{m}^2) = 78.86 \times 10^6\ \text{N/m}^2$$

$$= 76.86\ \text{MPa}$$

The stress in the steel bar σ_{st} can be determined from Eq. (1). However, since σ_{al} is known, σ_{st} may be more readily determined from Eq. (16.63) as

$$\sigma_{st} = \sigma_{al}\frac{A_{al}}{A_{st}} \tag{3}$$

thus

$$\sigma_{st} = 76.86\ \text{MPa}\frac{44.18\ \text{cm}^2}{19.64\ \text{cm}^2} = 172.9\ \text{MPa}$$

(b) The pulse length t_d is controlled by the aluminum bar. At impact, a compressive stress wave travels the length of the aluminum bar from the point of impact to its free end and reflects back from the free end as a tensile stress wave. The tensile stress wave then travels back to the impacted ends, thus causing the bars to separate. The pulse length t_d, in time, is then twice the transit time of the aluminum bar. Therefore

$$t_d = \frac{2L_{al}}{c_{al}} = \frac{2(0.5\ \text{m})}{5050\ \text{m/sec}} = 0.198\ \text{m}\cdot\text{sec}$$

TABLE 16.5.   Physical Characteristics for Impacting Bars

	Material	
Bar Characteristic	Aluminum	Steel
Mass density ρ (g/cm^3)	2.7	7.85
Wave speed c (km/sec)	5.05	5.05
Velocity V (m/sec)	20.0	10.0
Area A (cm^2)	44.18	19.64
Diameter d (mm)	75.0	50.0
Length L (m)	0.5	2.0

▲

16.5.4 Split Hopkinson Pressure Bar (SHPB)

An important and useful application of impacting bars is the device called a *split Hopkinson pressure bar* (SHPB). The original device was a single bar called simply a Hopkinson bar [13] that was used to determine the pressure generated from an explosive material when detonated at one end of a metal bar. At the opposite end of the bar a momentum trap (small metal disk of the same material as the bar) was used to measure the pressure in the wave by application of Eq. (16.57).

The Hopkinson bar was modified by Kolsky [14], and a material specimen was placed between two bars of equal cross-sectional area. A compression mode SHPB is shown schematically in Figure 16.10. This apparatus has been used primarily for obtaining tensile and compressive properties of metals at high strain rates [15, 16]. However, in recent years the SHPB has been used to study high strain-rate effects of soil and concrete [17].

The standard operating procedure for the compression mode SHPB is described with reference to Figure 16.10 as follows:

1. The striker bar is put in motion by the use of a gas gun or mechanical device, and the velocity of the striker bar is measured immediately before it impacts the incident bar. It can be shown using Eq. (16.68) with $V_2 = 0$ that the peak stress of the incident rectangular stress wave is determined from

$$\sigma_I = -(\rho c_0)_B \frac{V}{2} \tag{16.70}$$

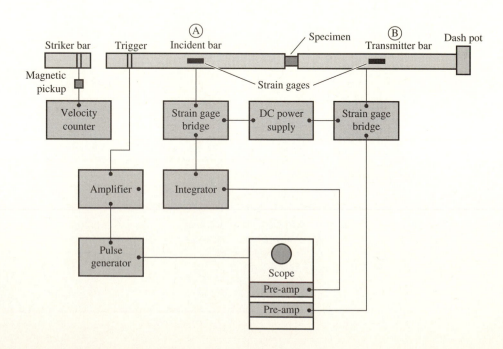

Figure 16.10

Schematic of SHPB illustrating compressive mode of operation.

where $(\rho c_0)_B$ is the characteristic impedance of the bar and V is the impact velocity of the striker bar. The negative sign indicates compression for a positive velocity. The wavelength (in time) of the pulse is equal to twice the wave transit time of the striker bar, or

$$t_I = \frac{2L_B}{c_0} \tag{16.71}$$

where t_I = pulse length of the incident wave
L_B = length of the striker bar
c_0 = wave speed of the bar (refer to Example 16.5)

2. The incident wave travels along the incident bar, crossing the strain gage where it is recorded. The strain and stress magnitude of the incident pulse are calibrated by use of Eq. (16.70). The calibration may also be determined by a shunt resistance method associated with the strain gage conditioner or by various experimental techniques [18].

3. The incident wave continues along the incident bar and impinges on the specimen. At the incident bar–specimen interface, a portion of the wave is reflected back to the incident bar and a portion is transmitted through the specimen into the transmitter bar. Similar calibration procedures must be performed on the transmitter strain gage circuit.

Rewriting Eq. (16.57) for a positive velocity $\dot{u}$ and solving for strain yields

$$\varepsilon = -\frac{\sigma}{E} = -\frac{\rho c}{E} \frac{du}{dt} \tag{16.72}$$

From Eq. (16.55) note that $c_0^2 = E/\rho$, which defines the strain as

$$\varepsilon = -\frac{1}{c_0} \frac{du}{dt} \tag{16.73}$$

Integration of Eq. (16.73) yields

$$u = -c_0 \int_0^{t'} \varepsilon \, dt \tag{16.74}$$

where t' is the pulse length of the incident wave. The time displacement functions at the incident and transmitter interfaces are given by

$$u_I(t) = -c_0 \int_0^{t'} [\varepsilon_I(t) - \varepsilon_R(t)] \, dt$$

$$u_T(t) = -c_0 \int_0^{t'} \varepsilon_T(t) \, dt \tag{16.75}$$

In Eqs. (16.75) the negative sign indicates positive displacements in a positive direction away from the striker bar for compression or negative strain. The nominal strain in the specimen is the difference between the two displacements defined by Eqs. (16.75) divided by the specimen length, and the strain rate is the derivative with respect to time of this nominal strain. Performance of these operations yields the nominal strain rate in the specimen given by

$$\dot{\varepsilon}_s = -\frac{1}{L_0}(\dot{u}_I - \dot{u}_T) = \frac{c_0}{L_0}(\varepsilon_I - \varepsilon_T - \varepsilon_R) \tag{16.76}$$

where $\dot{\varepsilon}_s$ is the strain rate of the specimen and L_0 is the length of the specimen.

The basic assumption in the operation of the SHPB is that the specimen is stressed uniformly along its length. The forces F at the interfaces of the specimen with the incident and transmitter bars are expressed as

$$F_I = EA(\varepsilon_I + \varepsilon_R) \tag{16.77a}$$

$$F_T = EA\varepsilon_T \tag{16.77b}$$

where E and A are the elastic modulus and cross-sectional area of the bars, respectively. Invoking the assumption of uniform stress along the specimen length results in

$$\varepsilon_T = \varepsilon_I + \varepsilon_R \tag{16.78}$$

Using Eq. (16.78) in (16.76) determines the strain rate of the specimen as

$$\dot{\varepsilon}_s = \frac{2c_0}{L_0}\varepsilon_R \tag{16.79}$$

and the strain in the specimen is determined by integrating Eq. (16.79) over the pulse length t' of the reflected wave. Thus

$$\varepsilon_s = -2\frac{c_0}{L_0}\int_0^{t'} \varepsilon_R(t)\, dt \tag{16.80}$$

where ε_R is positive or tensile strain (opposite sign of ε_I) because the specimen characteristic impedance is always less than the bar characteristic impedance, or the specimen is smaller in cross-sectional area than that of the bar.

From Eq. (16.77b)

$$F_T = \sigma_s A_S = EA\varepsilon_T \tag{16.81}$$

and then

$$\sigma_s = \frac{A}{A_S}E\varepsilon_T \tag{16.82}$$

where σ_s is the stress in the specimen and A_S is the specimen cross-sectional area.

The derivation of Eqs. (16.80) and (16.82) leads to several very important observations relative to the SHPB. That is, the strain in the SHPB specimen is proportional to the integral of the reflected strain, and the stress in the specimen is proportional to the transmitted strain. Thus, when Eq. (16.82) is plotted against Eq. (16.80), the result is a dynamic stress-strain curve. A typical set of strain pulses is shown in Figure 16.11, and the resulting dynamic stress-strain curve is shown in Figure 16.12. As shown in Figure 16.12 the peak dynamic stress is larger that the peak quasi-static stress. This general trend of increased stress at high strain rates (load rates) is exhibited by most materials.

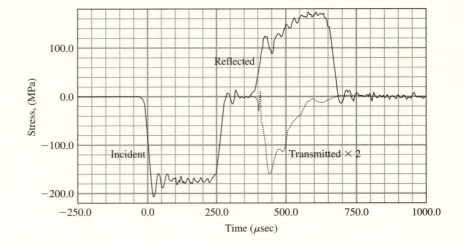

Figure 16.11

Typical set of SHPB strain signals for a direct compression test of a concrete specimen.

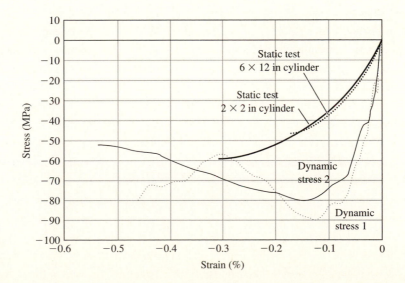

Figure 16.12

Comparison of dynamic stress-strain curve with static stress-strain curve for a concrete specimen in direct compression.

REFERENCES

1 Malvern, L.E., *Introduction to the Mechanics of a Continuous Medium,* Prentice Hall, Englewoods Cliffs, NJ, 1969.

2 Jones, R.M., *Mechanics of Composite Materials,* McGraw-Hill, New York, 1975.

3 Vinson, J.R. and Sierakowski, R.L., *The Behavior of Structures Composed of Composite Materials,* Martinus Nijhoff, Boston, 1986.

4 Kolsky, H., *Stress Waves in Solids,* Dover, New York, 1963.

5 Redwood, M., *Mechanical Waveguides,* Pergamon Press, New York, 1960.

6 Achenbach, J.D., *Wave Propagation in Elastic Solids,* North-Holland, Amsterdam, 1987.

7 Hearmon, R.F.S., *An Introduction to Applied Anisotropic Elasticity,* Oxford University Press, London, 1961.

8 Christoffel, E.B., *Ann. Mat. Pura. Appl.,* 8, 193, 1877.

9 Ross, C.A., Sierakowski, R.L., and Sun, C.T., *Dynamic Response of Composite Materials,* Soc. for Exp. Mech., 7 School St., Bethel, CT, 06081, 1985.

10 *Mark's Standard Handbook for Mechanical Engineers,* 8th ed., T. Baumeister, editor, McGraw-Hill, New York, 1978.

11 *Metals Handbook,* T. Lyman, editor, Amer. Soc. of Metals, Novelty, OH, 1948.

12 Rinehart, J.S., *Stress Transients in Solids,* Hyperdynamics, Santa Fe, NM, 1975.

13 Hopkinson, B., *Collected Scientific Papers,* Cambridge University Press, London, 1921.

14 Kolsky, H., "An Investigation of the Mechanical Properties of Materials at Very High Rates of Loading," *Proceedings of the Physics Society,* Ser. B., Vol. 6E, London, 1949.

15 Nicholas, T., "An Analysis of the Split Hopkinson Bar Technique for Strain-Rate-Dependent Material Behavior," *Journal of Applied Mechanics,* Vol. 40, Trans. ASME, Vol. 94, Series E, pp. 277–282, 1973.

16 Lindholm, U.S. and Bessey, R.J., "An Investigation of the Behavior of Materials Under High Rates of Deformation," AFML-TR-68-194, July 1968, Air Force Materials Lab. Tech Dept., Wright-Patterson AFB, OH.

17 Ross, C.A., "Split Hopkinson Pressure Bar Tests," Tech. Rept. ESL-TR-88-82, AFESC, Tyndall AFB, FL, Mar. 1989.

18 Dalley, J.W. and Riley, W.F., *Experimental Stress Analysis,* McGraw-Hill, New York, 1965.

NOTATION

a_i^j, a_p^q	direction cosines	Q	material stiffness
A	cross-sectional area	t	time
A_{ijrs}	elastic stiffness tensor	t'	pulse length of the reflected wave
A_S	cross-sectional area of material specimen	u_x	continuous displacement function of x
c	wave velocity (speed)	u_y	continuous displacement function of y
c_l	longitudinal wave velocity	u_I	normal incident displacement
c_n	phase velocity	$\dot{u}_I$	normal incident velocity
c_0	longitudinal bar velocity	u_R	normal reflected displacement
c_t	shear wave velocity	$\dot{u}_R$	normal reflected velocity
F_I	force at incident bar–specimen interface	u_T	normal transmitted displacement
F_T	force at transmitter bar–specimen interface	$\dot{u}_T$	normal transmitted velocity
G_{ij}	shear moduli	U_1, U_2, U_3	displacement amplitudes
i	imaginary number ($\sqrt{-1}$)	v_0	particle velocity at impact
L_B	length of striker bar	V	strain energy function
L_0	specimen length	V_1, V_2	velocities of impacting bars
n_k	direction cosines of normal to the wavefront	$[Q]$	elastic stiffness matrix

$[S]$	elastic compliance matrix	λ	Lamé's first elastic constant; also represents wave length
Δt	time increment		
ε_I	incident strain	μ	Lamé's second elastic constant
ε_R	reflected strain	ρ	mass density
$\dot{\varepsilon}_S$	strain rate in a material specimen	σ_{ij}	engineering stresses
ε_T	transmitted strain	σ_I	incident stress
ε_{ij}	terms in the three dimensional small strain tensor ε	σ_{jq}	stress components in a Cartesian axis system
ε_{xx}	extensional strain in x-direction	σ_R	reflected stress
ε_{xy}	shear strain in x-y plane	σ_S	stress in material specimen
ε_{yy}	extensional strain in y-direction	σ_T	transmitted stress
γ_{ij}	engineering shear strains	$\bar{\sigma}_{jp}$	stress components in a Cartesian axis system
Γ_{ij}	Christoffel stiffness	ν_{ij}	Poisson's ratios
κ	ratio of longitudinal wave velocity to shear wave velocity	ε	three-dimensional small strain tensor

PROBLEMS

16.1 The displacement field in a body is given by

$$u_1 = Ax_2x_3 \qquad u_2 = Ax_2x_3 \qquad u_3 = A(x_2 + x_3)$$

(a) Determine the strain components as a function of x_1, x_2, x_3 and show them in matrix form.

(b) For $A = 0.005$ and $x_1, x_2, x_3 = (1, 0, 1)$ show the strain components in matrix form.

16.2 Using Eqs. (16.20), (16.28), and (16.29) show that for plane stress ($\sigma_{33} = \sigma_{23} = \sigma_{13} = 0$) the constitutive equations for an isotropic material reduce to

$$\sigma_{11} = \sigma_1 = \frac{E}{1 - \nu^2}(\varepsilon_1 - \nu\varepsilon_2)$$

$$\sigma_{22} = \sigma_2 = \frac{E}{1 - \nu^2}(\varepsilon_2 - \nu\varepsilon_1)$$

$$\sigma_{12} = \sigma_6 = 2G\varepsilon_{12}$$

16.3 Given the wave normal of Table 16.2 as $n_1, n_2, n_3 = (0, 0, 1)$, derive the three wave velocity relations.

16.4 The properties of a uniaxial carbon fibrous composite are given as

$$E_1 = 20 \times 10^6 \text{ psi} \qquad \rho = 1.55 \times 10^{-4} \text{ lb-sec}^2/\text{in}^4$$

$$E_2 = E_3 = 1.5 \times 10^6 \text{ psi} \qquad G_{13} = G_{12} = 0.81 \times 10^6 \text{ psi}$$

$$\nu_{13} = \nu_{12} = 0.12, \nu_{23} = 0.3 \qquad G_{23} = 0.1 \times 10^6 \text{ psi}$$

Determine the numerical values for the longitudinal wave velocity in the 1 and 2 directions.

16.5 A copper rod of 10-mm diameter with a velocity of 40m/sec impacts a steel rod of 20-mm diameter with a velocity of 20 m/sec. Determine the stresses at the impact end of each rod. Assume elastic wave propagation.

16.6 Assume the copper rod of Problem 16.6 is 0.25 m long and the steel rod is 1.0 m long. (a) How long in time is the stress pulse in the steel rod? (b) Explain in terms of transmission, reflection, stress, particle velocity, strain, and displacement what happens when the wave in the steel bar encounters the free end.

16.7 In a steel split Hopkinson pressure bar, a rectangular incident compressive wave of 3.5 MPa impinges on a concrete sample of the same diameter. Use the properties of Tables 16.3 and 16.4 to determine the initial magnitudes of the stress and particle velocity for the stress wave reflected into the steel and transmitted into the concrete.

17 ▲ Earthquakes and Earthquake Ground Motion

As discussed in Chapter 1, vibrations in structural systems may result from a wide variety of sources. Some of the most common and significant of dynamic loads imposed on structures are those caused by environmental conditions such as wind, earthquakes, and water waves. Of these environmental sources, earthquakes loom as the most important in terms of their enormous potential for damage to structures and loss of human life.

Each year thousands of people die from earthquakes, approximately 10,000 people on average worldwide. In addition to this staggering loss of lives, earthquakes are responsible for hundreds of millions of dollars in property damage annually. Costs of tangible damage incurred by buildings and other types of civil engineering structures in the Loma Prieta earthquake of 1989 and the Northridge earthquake of 1994 have been estimated at $6.8 billion and $25 billion, respectively. For the 1995 Kobe earthquake, reported repair costs have exceeded $100 billion.

Thousands of earthquakes occur each year around the world; however, the majority are imperceptible to human sensitivity. Consequently, relatively few of these earthquakes cause damage to structures or loss of lives. Seismic disturbances that are of interest to the design engineer are those with potential to cause structural damage, referred to as *strong-motion earthquakes*. Knowledge of strong-motion earthquakes is advancing rapidly due in large part to the increase in appropriately situated strong-motion accelerographs in seismic regions of the world. There are an estimated 4000 strong-motion accelerographs at present in the world, approximately 1000 of which are located in the United States [1].

This chapter summarizes the fundamental concepts of earthquake engineering. Included in the discussion are causes of earthquakes, earthquake measuring scales, seismic activity or seismicity, characteristics of strong seismic ground motion, and earthquake damage mechanisms. An introduction to the response of structures to earthquake ground motion is presented in Chapter 18.

17.1 CAUSES OF EARTHQUAKES

Simply stated, earthquakes are vibrations of the earth's surface caused by waves emanating from a source of disturbance inside the earth. The nature of the disturbing source can vary from a volcanic eruption to an underground explosion. However, from an engineering standpoint, the most important earthquakes are those that are *tectonic* in origin, or those associated with large-scale strains in the earth's crust.

17.1.1 Tectonic Earthquakes

Over the years, many theories have been proposed to explain the causes of earthquakes. The plate tectonics theory is generally considered to be the most reliable. According to this theory, the earth's outer layer, referred to as the *lithosphere,* consists of approximately one dozen hard tectonic plates as shown in Figure 17.1. These plates, having an average thickness of 50 miles, sit on a comparatively soft *asthenosphere* and move as rigid bodies [1]. The plates interact with one another, and these interactions have been the primary cause of *orogeny* (i.e., formation of mountains) throughout geological history. Three basic types of plate interactions can occur. The first form of interaction involves two plates slipping apart. As this slippage occurs, hot mantle flows up toward the earth's surface and cools down, forming *midoceanic ridges.* The second type of interaction occurs when two plates slide horizontally, one over the other, and create a *transform fault.*

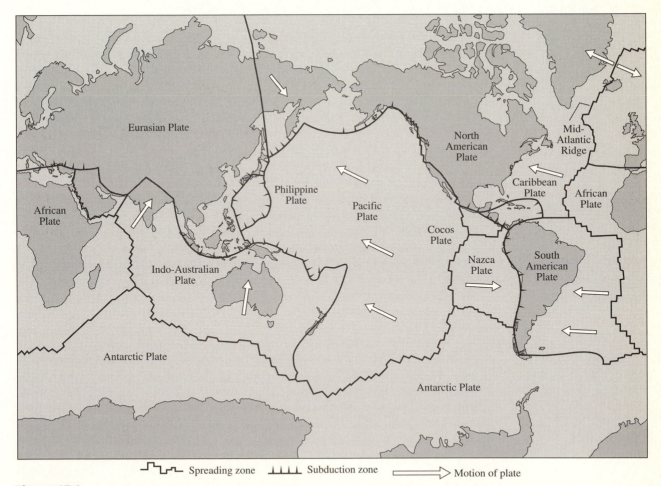

Figure 17.1

World map of tectonic plates. *Earthquakes* by Bruce A. Bolt, copyright © 1988. Reprinted by permission of W.H. Freeman and Co., New York, NY.

The third type of interaction occurs when a high-density oceanic plate *subducts* beneath a low-density continental plate, forming a *trench* and an *island arc.* An island arc is a chain of islands in the shape of an arc. An island arc is highly prone to seismic activity and generally includes volcanoes (e.g., the Aleutian Islands and the island chain of Japan).

Each type of plate interaction produces significant straining in crustal rocks. The strain is accumulated by the gradual shifting of the tectonic plates. The rocks become distorted but maintain their original positions because of continuity, mechanical bond, and friction. When accumulated stress finally exceeds the strength of the rocks, fracture occurs and the earth snaps back into an unstrained position. (This phenomenon is generally known as the *elastic rebound theory,* or Reid's theory.) The great release of energy associated with the rupture of the rocks creates shock waves that propagate through the earth's crust and cause an earthquake. The great (major) earthquakes usually are the transform-fault type and the subduction type.

17.2 FAULTS

Earthquakes generally originate on a plane of weakness in the earth's crust called a *fault.* Faults are formed when two crustal rock beds slip relative to each other. Faults are classified according to the direction of relative slippage. If the movement or slippage is primarily horizontal, the fault is called a *strike-slip* fault; whereas, if the slippage is vertical, the fault is termed a *dip-slip* fault. The strike-slip faults, which are essentially vertical fault planes, can be either *left lateral* or *right lateral* as illustrated in Figure 17.2a and b, respectively. Dip-slip faults may be either *normal faults* or *reverse faults* as shown in Figure 17.2c and d, respectively.

For a strike-slip fault, the strike is the direction of the fault line relative to north. With a *right-lateral fault* (Figure 17.2a), movement on the opposite side of the fault on which an observer is standing is to the right (typical of the San Andreas fault). Conversely, for a *left-lateral fault* (Figure 17.2b), movement on the opposite side of the fault on which an observer is standing is to the left. In a dip-slip fault, the dip of the fault is the angle the fault makes with a horizontal plane. If movement of the rock bed above the inclined fault surface (hanging wall) is downward, the fault is a *normal fault* (Figure 17.2c). In a *reverse fault* (Figure 17.2d), the hanging wall moves upward relative to the bedrock below the fault surface (footwall). In reality, fault slippage is generally a combination of strike-slip and dip-slip. Such faults are referred to as *oblique faults.*

A fault that reveals itself on the earth's surface due to past earthquake activity is called an earthquake fault. The well-noted San Andreas fault, for example, emerges to the surface between Point Arena and the Gulf of California and reveals its presence by a linear trough in the earth's surface approximately 190 miles long. The San Andreas fault occurs along the intersection of the Pacific Plate and the North American Plate, has a total length of approximately 600 miles, and extends almost vertically into the earth to a depth exceeding 20 miles.

Earthquakes often occur at *active faults.* Active faults are faults for which there is a past history of movement or deformation. Many of the individual faults in the San Andreas fault system are known to have been active during the past 200 years, and others are believed to have been active for thousands of years.

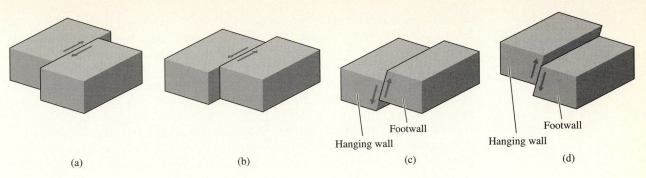

Figure 17.2

Types of faults: (a) right lateral fault; (b) left lateral fault; (c) normal fault; (d) reverse fault.

17.3   SEISMIC WAVES

At the first occurrence of an earthquake, seismic waves are generated that propagate through the earth's crust. The position on the fault plane where the seismic motion originates is called the *focus,* or *hypocenter,* of the earthquake as shown in Figure 17.3. The point on the earth's surface directly above the focus is the *epicenter.* The *focal distance* and the *epicentral distance* are the distances from the focus and the epicenter, respectively, to the point of observed ground motion. Depending upon the depth of the focus, the earthquake is classified as either shallow, intermediate, or deep [2] as illustrated in Table 17.1. The shallow-focus earthquakes are the most devastating. All known earthquakes to date in California have been the shallow-focus type.

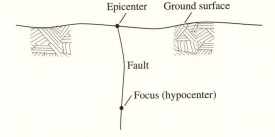

Figure 17.3

Definition sketch for earthquake fault, focus, and epicenter.

TABLE 17.1.   Classification of Earthquakes

Classification	Depth of Focus (mi)
Shallow	0–45
Intermediate	45–185
Deep	>185

Seismic waves fall into two categories [3]: *body waves* and *surface waves*. The body waves are of two types, the *P wave* and the *S wave*. The P wave (primary wave) is a dilatation wave and is often referred to as the longitudinal, or compressive, wave. It propagates in the same direction of its own vibration and creates a "push-pull" effect on the earth's material as it passes. It travels at a velocity of approximately 3.5 mi/sec and is the first to reach the surface. It propagates through both solid rock and liquid material.

The S wave (secondary wave) is also referred to as the shear, or transverse, wave. It propagates in a direction perpendicular to its vibration. It travels at approximately 2 mi/sec and causes the earth to move at right angles to the direction of propagation. This is comparable to snapping a rope like a whip: the wave moves along the length of the rope but the actual motion (vibration) is at right angles to the direction of propagation. Thus at the ground surface, S waves can produce both horizontal and vertical motion. The S waves, however, cannot propagate through the liquid parts of the earth [4].

The propagation velocities of the P waves, v_P, and S waves, v_S, in elastic materials are frequency independent and are expressed as

$$v_P = \left[\frac{E}{\rho} \frac{(1 + \nu)}{(1 + \nu)(1 - 2\nu)}\right]^{1/2} \tag{17.1}$$

and

$$v_S = \left(\frac{G}{\rho}\right)^{1/2} = \left[\frac{E}{\rho} \frac{1}{2(1 + \nu)}\right]^{1/2} \tag{17.2}$$

where E = modulus of elasticity
G = modulus of rigidity, or shear modulus
ρ = mass density
ν = Poisson's ratio

Equations (17.1) and (17.2) indicate that the actual velocities of P and S waves depend on the elastic properties of the rocks and soils through which they pass. In all materials $v_S < v_P$ [5]; therefore the P waves arrive at the surface first. Their effect is similar to that of a sonic boom that rattles windows in buildings. Seconds later, the S waves arrive bringing a significant component of side to side motion, thus inducing both vertical and horizontal ground shaking. Although S waves travel more slowly than P waves, they transmit more energy and are most effective in inflicting damage to structures.

The second general category of earthquake wave is the surface wave. Surface waves are most often detected in shallow earthquakes and their motion is restricted to near the ground surface. There are two types of surface waves, *Love waves* (or L waves) and *Rayleigh waves* (or R waves). The L wave vibrates in a plane parallel to the earth's surface and perpendicular to the direction of propagation. Their motion is essentially the same as that of an S wave without vertical displacement. The R wave vibrates in a plane perpendicular to the earth's surface, exhibiting both vertical and horizontal movement, in an elliptic motion. Surface waves travel more slowly than body waves, and R waves generally travel at a lower velocity than L waves.

The velocity differential between P waves and S waves can be used to locate the epicenter and focus of an earthquake. The time interval between the arrival of a P

Figure 17.4

Typical earthquake seismogram. *Seismic Design of Building Structures,* 6/e, by Michael R. Lindeburg, copyright © 1994. Reprinted by permission of Professional Publications, Belmont, CA.

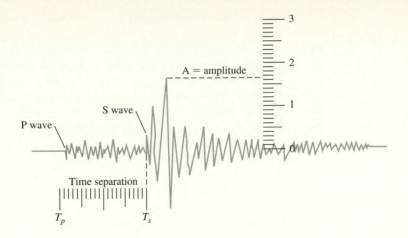

wave, T_P, and an S wave, T_S, to a seismograph station is called the *duration of preliminary tremors* and is expressed as

$$T_{SP} = \left(\frac{1}{v_S} - \frac{1}{v_P}\right)d \tag{17.3}$$

where

$$T_{SP} = T_S - T_P \tag{17.4}$$

and d is the distance traveled by the waves [6]. The quantity T_{SP} is determined from a seismogram, as illustrated in Figure 17.4, as the difference in the initial arrival times of the S wave and P wave. With the wave velocities v_S and v_P being approximately known from Eqs. (17.1) and (17.2), the focal distance d can be determined from Eq. (17.3). The locations of the focus and epicenter can then be ascertained if d is determined at three or more seismograph stations.

17.4 EARTHQUAKE INTENSITY

The *intensity* of an earthquake is used to denote its severity at a particular location as determined by human reaction to earth movement, observed damage to structures, and observation of other physical effects. Thus the intensity will vary with distance from the causative fault and with local ground conditions. Although the intensity of an earthquake is not assessed from scientific instrumentation and is highly subjective, intensities may be assigned to past major earthquakes based on historical records of observed damage and human responses.

The first intensity scale of note was the Rossi-Forel scale (1883), developed by Rossi of Italy, and Forel of Switzerland. The scale has 10 grades, I to X, where I is the least intense and X is the most intense. This scale is still used in some parts of Europe and was employed to assess the intensity of the 1906 San Francisco earthquake. More refined scales were developed in Italy by Mercalli (1902) and Cancani (1904). In 1931 Frank Neumann and H. O. Wood proposed a 12-grade Modified Mercalli (MM) scale

[7], which has been widely adopted in North America and other parts of the world. Observations are classified in 12 grades of intensity, ranging from I to XII. An abbreviated version of the Modified Mercalli Scale is presented in Table 17.2.

Other intensity scales in use today include the 12-grade Medvedev-Sponheuer-Karnik (MSK) scale and the 8-grade Japanese Meteorological Agency (JMA) scale. Because intensity scales are subjective and highly dependent on the construction practices and socioeconomic conditions of a country, and they bear no specific relation to the ground motion, correlation among the various intensity scales is not easily done.

The intensity of an earthquake is greatest in the vicinity of the causative fault and decreases with distance from the fault. Curves of equal intensity, called *isoseismals,* assume a bell-shaped pattern for small earthquakes. For "large" earthquakes, having a slipped length of fault several hundred miles, the idealized isoseismals become quite elongated in a direction parallel to the causative fault as illustrated in Figure 17.5. In actuality, however, the isoseismals are more complex, influenced by such factors as local site and geological conditions. The isoseismal map for the 1989 Loma Prieta earthquake [8] is shown in Figure 17.6.

TABLE 17.2.   **Abbreviated Modified Mercalli (MM) Earthquake Intensity Scale**

Intensity Value	Description
I	Not felt
II	Felt by persons at rest, especially by those on upper floors of buildings
III	Felt indoors. Hanging objects swing. Vibration similar to the passing of light trucks. Duration estimated
IV	Standing motor vehicles rock. Windows, dishes, doors rattle. Glasses clink. Wooden walls and frames creak. Vibrations similar to passing of heavy trucks
V	Felt noticeably outdoors. Sleepers awakened. Small unstable objects displaced or overturned. Some dishes and windows broken, pictures move. Pendulum clocks may stop
VI	Felt by all. Many frightened and run outdoors. Windows, dishes, glassware broken. Books fall off shelves and pictures off walls. Furniture moved or overturned. Weak plaster and weak masonry cracked
VII	Difficult to stand. Noticed by drivers of motor vehicles. Furniture broken. Damage to weak masonry. Weak chimneys broken at roof line
VIII	Steering of motor vehicles affected. Damage to ordinary quality masonry; partial collapse. Some damage to very good quality masonry; none to reinforced masonry. Fall of chimneys, monuments, towers, elevated tanks
IX	General panic. Weak masonry destroyed; ordinary masonry seriously damaged. General damage to foundations. Conspicuous cracks in ground. Underground pipes broken
X	Most masonry and frame structures destroyed along with their foundations. Some well-built wooden structures and bridges destroyed. Serious damage to dams, dikes, embankments. Rails bent slightly
XI	Rails bent greatly. Underground pipelines destroyed
XII	Damage total. Large rock masses displaced. Lines of sight and level distorted. Objects thrown into the air

Figure 17.5

Idealized isoseismals for a large earthquake.

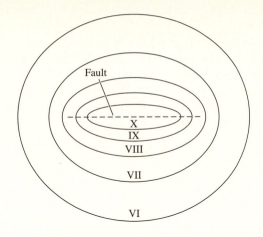

Figure 17.6

Isoseismal map showing the distribution of the Modified Mercalli intensity for the 1989 Loma Prieta earthquake. Intensity values for localities are given in Arabic numbers. Roman numerals represent the intensity level between isoseismal lines. "Lessons Learned from the Loma Prieta, California, Earthquake of October 17, 1989" by G. Plafker and J. Galloway, U.S. Geological Survey Circular 1045, 1989.

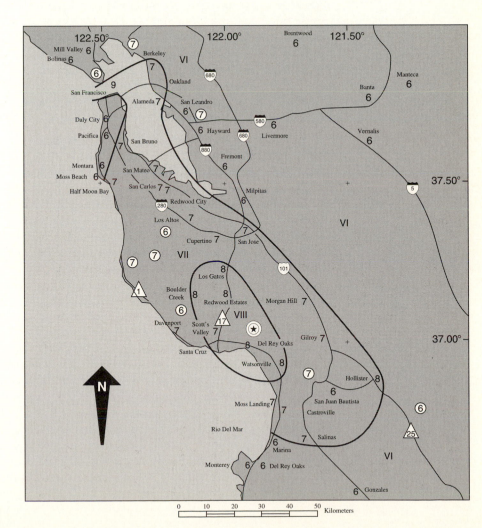

17.5 EARTHQUAKE MAGNITUDE

The *magnitude* of an earthquake is a measure of the amount of energy released. In 1935, Charles F. Richter developed the Richter magnitude scale to measure the strength or size of a local earthquake. The magnitude M is determined from the expression

$$M = \log_{10} A \qquad (17.5)$$

where A is the maximum seismic wave amplitude in microns (10^{-3} mm) recorded on a Wood-Anderson seismograph located at a distance of 100 km from the earthquake epicenter (see Figure 17.6). However, a standard seismograph is not always set at a distance of 100 km from the epicenter, in which case Eq. (17.5) can be modified and expressed as [9]

$$M = \log_{10} A - \log_{10} A_0 \qquad (17.6)$$

where A is the maximum seismic wave amplitude for the measured earthquake at a given epicentral distance and A_0 is the seismograph reading produced by a standard earthquake or calibration earthquake (A_0 is generally taken as 0.001 mm). A correlation between the amount of energy E_f released at the causative fault and the Richter magnitude M was developed by Gutenberg and Richter [10] and is expressed as

$$\log_{10} E_f = 4.8 + 1.5M \qquad (17.7)$$

Because the Richter magnitude is a logarithmic scale, an increase of unity in magnitude represents a 10-fold increase in the amplitude of the seismic waves (e.g., a reading of 6.0 represents a 10 times greater amplitude than a reading of 5.0). In general, earthquakes having a magnitude $M > 5$ generate ground motions sufficiently severe to be potentially damaging to structures. Earthquakes having $M < 5$ generate ground motions unlikely to be damaging because of their very short duration and moderate acceleration. An earthquake with a magnitude of 7.2 would be considered a *strong* earthquake. Earthquakes with magnitudes above 7.5 are referred to as *great earthquakes,* whereas earthquakes with magnitudes of 2.0 or less are known as *microearthquakes.* Earthquakes of larger magnitude occur less frequently than those of smaller magnitude. Several thousand earthquakes having magnitudes of 4.5 or greater occur each year; however, great earthquakes (e.g., 1906 San Francisco and 1964 Alaska) occur approximately once a year on average. Details of some significant earthquakes are presented in Table 17.3.

Richter's magnitude is actually a *local magnitude* and is often designated as M_L. At its inception, the local Richter magnitude was intended to measure shallow earthquakes having epicentral distances less than 600 km, and it therefore cannot be applied to earthquakes with a large epicentral distance. This has led to the emergence of new magnitude scales based on different formulas for epicentral distance and methods for choosing an appropriate seismic wave amplitude. Several of these alternative magnitude scales are the *surface wave magnitude* (M_s) and the *body-wave magnitude* (M_b). However, the best estimate of an earthquake's magnitude [10], especially for great earthquakes, is the *moment magnitude* (M_w). This scale evaluates the magnitude of an earthquake in terms of its seismic moment, M_0, that is directly related to the amount of energy released in the earthquake. The moment magnitude is expressed as [2]

$$M_w = \left(\log_{10}\frac{M_0}{1.5}\right) - 10.7 \qquad (17.8)$$

where

$$M_0 = \mu A_S D \qquad (17.9)$$

where μ = parameter characterizing the rigidity of the material surrounding the causative fault
A_S = slipped area
D = distance of slip

Attempts have been made in the past to correlate earthquake magnitude to earthquake intensity. One such effort, proposed by Richter [7], has correlated MM intensity

TABLE 17.3. **Magnitudes of Some Recent Significant Earthquakes**

Year	Location	Magnitude (Richter)
1906	San Francisco, CA	8.3
1908	Messina, Italy	7.5
1920	Kansu, China	8.5
1923	Kunto, Japan	7.9
1925	Yunnan, China	7.1
1931	Hawke's Bay, New Zealand	7.9
1933	Sunriku, Japan	8.3
1939	Erzincan, Turkey	8.0
1940	El Centro, CA	7.1
1943	Chile	7.9
1946	Nankaido, Japan	8.1
1949	Olympia, WA	7.1
1950	India	8.6
1952	Kern County, CA	7.7
1957	Mexico City, Mexico	7.9
1958	Kurile Islands	8.7
1960	Chile	8.3
1964	Prince William Sound, AK	8.4
1968	Iran	7.4
1970	Peru	7.6
1971	San Fernando, CA	6.5
1976	Tangshan, China	7.6
1978	Miyagiken-oki, Japan	7.4
1979	Montenegro, Yugoslavia	7.3
1985	Mexico City, Mexico	8.5
1989	Loma Prieta, CA	7.1
1990	Luzon, Philippines	7.7
1994	Northridge, CA	6.4
1995	Kobe, Japan	7.2

TABLE 17.4. **Correlation Between Earthquake Magnitude and Modified Mercalli Intensity**

Magnitude (Richter)	Intensity (MM)
2	I–II
3	III
4	V
5	VI–VII
6	VII–VIII
7	IX–X
8	X

with Richter magnitude and is presented in Table 17.4. This particular "rough correlation" was intended for ordinary ground conditions in metropolitan centers in California [11]. An empirical relationship between Richter magnitude M, Modified Mercalli intensity MM, and focal distance d (in kilometers) was suggested by Esteva and Rosenbleuth [12] and is expressed as

$$MM = 8.16 + 1.45M - 2.46 \ln (d) \tag{17.10}$$

In general, however, comparisons between earthquake magnitude and earthquake intensity are replete with difficulties, such as [13]: (1) intensity varies with distance from the epicenter; (2) a large earthquake may occur in a remote area causing little discernible damage; and (3) local ground conditions and construction practices have a significant effect on subjective assessments of damage.

17.6 SEISMICITY

Seismicity or seismic activity is usually defined by a *seismicity map,* a map indicating the locations of earthquake epicenters within a specified time interval. The map generally includes all earthquakes having a magnitude larger than a specified value. A typical *world seismicity map* is shown in Figure 17.7. It includes the epicenter locations of all earthquakes occurring worldwide during the period 1961 through 1967. Definite belts of seismic activity are apparent from the map, along with regions that appear to be completely devoid of seismic activity.

The most notorious belt of seismic activity, or *seismic zone,* is the Circum-Pacific seismic zone. This vast zone includes the Pacific sides of South America, Central America, and North America; the Aleutian Islands; the Kamchatka Peninsula; Japan; Indonesia; and New Zealand. Another important seismically active area is the Alpide belt or Eurasian seismic zone, extending from southeast Asia through the Middle East to the Mediterranean Sea. Other zones of high seismicity include parts of China and North America, and the midoceanic ridges along the center of the Atlantic and Indian Oceans.

Most moderate and large earthquakes are followed by numerous smaller earthquakes in the same vicinity called *aftershocks.* Aftershocks occur in the ensuing hours, and even over the next several months, of the main seismic event and can number in the hundreds for larger earthquakes. A few earthquakes are even preceded by smaller seismic disturbances called *foreshocks.*

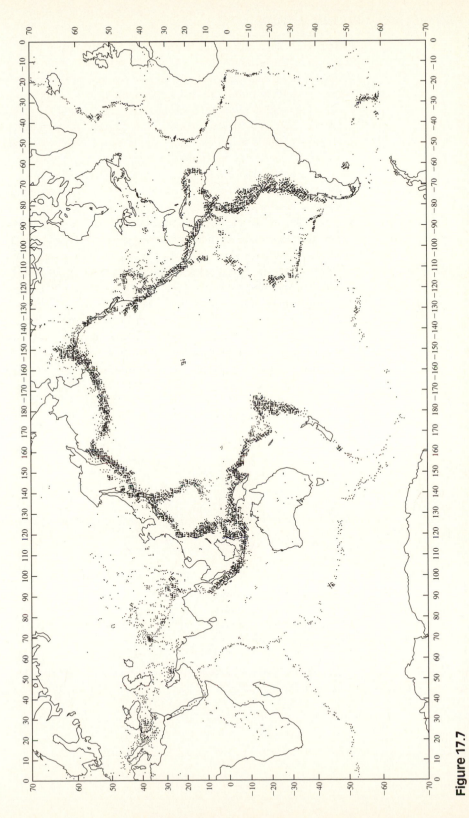

Figure 17.7

Seismicity of the world, 1961–1967; focal depths 0–450 mi. "World Seismicity Maps Compiled from ESSA, Coast and Geodetic Survey, Epicenter Data, 1961–1967" by M. Barazangi and J. Dorman, *Bulletin of the Seismological Society of America*, Vol. 59, No. 1, 1969.

17.7 EARTHQUAKE GROUND MOTION

Strong earthquake ground motion must be recorded for the purpose of seismic engineering. The ground motion is usually recorded with strong-motion accelerographs placed at various locations. The accelerographs are generally equipped with a triggering device that initiates recording when the ground acceleration exceeds a preset value. The acceleration record of a strong earthquake usually consists of two horizontal components and one vertical component. Generally, the two horizontal components are of equal magnitude and the vertical component is somewhat smaller. The accelerograph record frequently includes instrumentation errors, due to the frequency characteristics of the accelerograph and other inherent features, that must be corrected by filtering and other procedures [14]. The corrected accelerograms are then integrated to obtain the velocity and displacement time histories of ground motion. The N-S component of horizontal ground acceleration (in terms of the acceleration due to gravity, g) and the integrated velocity and displacement time history records for the 1994 Northridge, CA earthquake are shown in Figure 17.8. The Northridge accelerogram, extremely irregular and complex, is typical of earthquake accelerograms recorded on firm ground. This is due to the multiple reflections and refractions of the seismic waves as they traverse geologic interfaces. Alternatively, at the surface above soft strata, the earthquake ground motion assumes an almost harmonic nature resulting from filtering of the seismic waves as they travel through the soft strata and from successive reflections at the rock-soil interface and the ground surface.

Earthquake accelerograms are thus complex and can vary considerably from one another. They are significantly affected by local site conditions, distance from the caus-

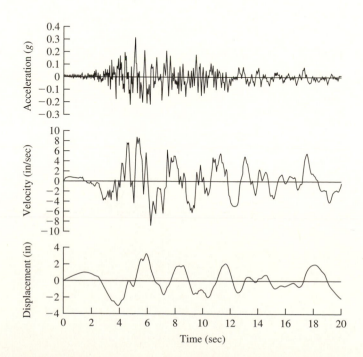

Figure 17.8

Northridge, CA earthquake of January 17, 1994, N-S component.

ative fault, and the transmission path of the seismic waves. Newmark and Rosenbleuth [3] classified earthquake ground motion into four groups, in accordance with their surface ground motion characteristics:

1. *Single-shock type.* This type of motion occurs only at close proximity to the epicenter, on firm strata, and for shallow earthquakes. A representative example of this type of earthquake is the Port Hueneme earthquake of 1957 as shown in Figure 17.9.

2. *A moderately long, extremely irregular motion.* This type of earthquake is associated with an intermediate focal depth and occurs only on firm ground. It is typical of earthquakes originating along the Circum-Pacific belt. The N-S component of the 1940 El Centro earthquake (Figure 17.10) is indicative of this type.

3. *A long ground motion exhibiting pronounced prevailing periods of vibration.* Motions of this type are recorded atop layers of soft strata, through which the seismic waves have been filtered and subjected to multiple reflections at the layer boundaries. The 1964 Mexico City earthquake exemplifies this behavior.

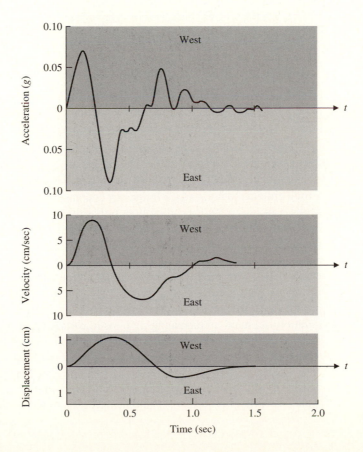

Figure 17.9

Port Hueneme, CA earthquake of March 18, 1957, E-W component. "The Port Hueneme Earthquake of March 18, 1957" by G.W. Housner and D.E. Hudson, *Bulletin of the Seismological Society of America,* Vol. 48, No. 2, 1958.

Figure 17.10

El Centro, CA earthquake of May 18, 1940, N-S component. *Design of Multistory Reinforced Concrete Buildings for Earthquake Motions* by J.A. Blume, N.M. Newmark, and L.H. Corning, Portland Cement Association, Skokie, IL, 1969.

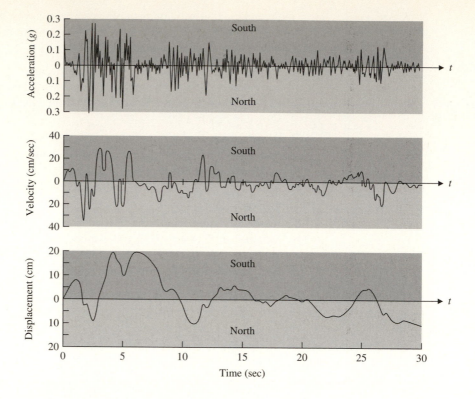

4. *A ground motion involving large-scale permanent deformation of the ground.* Earthquakes of this type may entail land slides or soil liquification. The Alaska and Niigata earthquakes of 1964 characterize this type of earthquake.

From the examination of ground motions of the basic types of earthquakes, three characteristics of ground motion that are important in earthquake engineering applications have been identified [15]: (1) peak (maximum) ground motion, (2) duration of strong ground motion, and (3) frequency content. Structural response is affected by each of these factors. Peak ground motion, primarily peak ground acceleration (PGA), influences the vibration amplitude and has been commonly employed to scale earthquake design spectra and acceleration time histories. The severity of ground shaking is significantly influenced by the duration of the strong ground motion. For example, an earthquake with a high peak acceleration poses a high hazard potential, but if it is sustained for only a short period of time it is unlikely to inflict significant damage to many types of structures. Conversely, an earthquake with a moderate peak acceleration and a long duration can build up damaging motions in certain types of structures. Finally, ground motion amplification in structures is most likely to occur when the frequency content of the ground motion is in close proximity to the natural frequencies of the structure.

Several attempts [6] have been made to correlate PGA with earthquake magnitude as well as with both earthquake magnitude and duration of strong ground shaking. A typical *correlation equation,* relating Richter magnitude M to PGA, is given by [9]

$$\log_{10} \text{PGA} = -2.1 + 0.81M - 0.027M^2 \tag{17.11}$$

TABLE 17.5. Peak Ground Accelerations and Durations of Strong-Phase Shaking

Magnitude (Richter)	PGA (g)	Duration (sec)
5.0	0.09	2
5.5	0.15	6
6.0	0.22	12
6.5	0.29	18
7.0	0.37	24
7.5	0.45	30
8.0	0.50	34
8.5	0.50	37

Correlation equations, such as Eq. (17-11), are very site-dependent. Although the PGA decreases with distance from the causative fault, the rate of decrease is relatively small over a distance comparable to the vertical dimension of the slipped fault [3]. Another commonly cited correlation between Richter magnitude and PGA, which also includes duration of strong-phase shaking, applicable in the vicinity of the epicenters of California earthquakes is presented in Table 17.5. These values are conservatively high, and most actual earthquakes exhibit somewhat smaller values of PGA. For example, the PGA recorded for the 1940 El Centro earthquake having a Richter magnitude $M = 7.1$ was approximately 0.33 g, whereas Table 17.5 would suggest a PGA in the vicinity of 0.39 g for the same magnitude.

17.8 EARTHQUAKE DAMAGE MECHANISMS

Earthquakes may damage structures in a variety of ways. Several of the most destructive damage mechanisms are [2, 16]: (1) large inertial forces developed in a structure due to earthquake ground motion could cause collapse; (2) landslides or other surficial movements instigated by the earthquake; (3) earthquake-induced soil consolidation or liquefaction beneath the foundation; (4) sudden fault displacement in close proximity to a structure (this mechanism is particularly hazardous for extended facilities such as pipelines, canals, and dams); (5) seismically induced water waves, such as tsunamis (sea waves), that may threaten coastal regions; and (6) earthquake-induced fires and explosions.

Severe ground shaking poses the most potentially hazardous damage mechanism to structures, and its effects may be widespread. Therefore, the response of structures to earthquake ground motion is addressed in Chapter 18.

REFERENCES

1 Okamoto, S., *Introduction to Earthquake Engineering,* 2nd ed., University of Tokyo Press, Tokyo, 1984.

2 Bolt, B.A., "The Nature of Earthquake Ground Motion," Chapter 1, *Seismic Design Handbook,* F. Naiem, editor, Van Nostrand Reinhold, New York, 1989.

3 Newmark, N.M. and Rosenbleuth, E., *Fundamentals of Earthquake Engineering,* Prentice Hall, Englewood Cliffs, NJ, 1971.

4 Wakabayashi, M., *Design of Earthquake Resistant Buildings,* McGraw Hill, New York, 1986.

5 Rosenbleuth, E., "Characteristics of Earthquakes," Chapter 1, *Design of Earthquake Resistant Structures,* E. Rosenbleuth, editor, Wiley, New York, 1980.

6 Housner, G.W., "Strong Ground Motion," Chapter 9, *Earthquake Engineering,* R.L. Wiegel, editor, Prentice Hall, Englewood Cliffs, NJ, 1970.

7 Richter, C.F., *Elementary Seismology,* W.F. Freeman, San Francisco, 1958.

8 "Loma Prieta Earthquake Reconnaissance Report," *Earthquake Spectra,* Supplement to Vol. 6., Earthquake Engineering Research Institute, El Cerrito, CA, May 1990.

9 Lindeburg, M.R., *Seismic Design of Building Structures,* Professional Publications, Belmont, CA, 1994.

10 Gutenberg, B. and Richter, C.F., "Earthquake Magnitude, Intensity, Energy and Acceleration," *Bulletin of the Seismological Society of America,* Vol. 46, No. 2, 1956, pp. 105–146.

11 Steinbrugge, K.V., "Earthquake Damage and Structural Performance in the United States," Chapter 9, *Earthquake Engineering,* R.L. Wiegel, editor, Prentice Hall, Englewood Cliffs, NJ, 1970.

12 Esteva, L. and Rosenbleuth, E., "Espectros de Temblores a Distancias Moderados y Grandes," *Bol. Soc. Mex. Ing. Sismica,* Vol. 2, 1964, pp. 1–18.

13 Smith, J.W., *Vibration of Structures, Applications in Civil Engineering,* Chapman and Hall, New York, 1988.

14 *Northridge Earthquake, Preliminary Reconnaissance Report,* J.F. Hall, editor, Earthquake Engineering Research Institute, Oakland, CA, 1994.

15 Hudson, D.E., *Reading and Interpreting Strong Motion Accelerograms,* Earthquake Engineering Research Institute, Berkeley, CA, 1979.

16 Mohraz, B. and Elghadamsi, F.E., "Earthquake Ground Motion and Response Spectra," Chapter 2, *Seismic Design Handbook,* F. Naiem, editor, Van Nostrand Reinhold, New York, 1989.

17 Hudson, D.E., *Ground Motion Measurements,* Chapter 6, *Earthquake Engineering,* R.L. Wiegel, editor, Prentice Hall, Englewood Cliffs, NJ, 1970.

18 Plafker, G. and Galloway, J., "Lessons Learned from the Loma Prieta, California Earthquake of October 17, 1989," U.S. Geological Survey Circular 1045, 1989.

19 Barazangi, M. and Dorman, J., "World Seismicity Maps Compiled From ESSA, Coast and Geodetic Survey, Epicenter Data, 1961–1967," *Bulletin of the Seismological Society of America,* Vol. 59, No. 1, pp. 369–80, 1969.

20 Housner, G.W. and Hudson, D.E., "The Port Hueneme Earthquake of March 18, 1957," *Bulletin of the Seismological Society of America,* Vol. 48, No. 2, pp. 163–168, 1958.

21 Bolt, B.A., *Earthquakes—A Primer,* W.H. Freeman, San Francisco, 1978.

NOTATION

A	maximum seismic wave amplitude (in microns) recorded on a seismograph		T_P	arrival time of P wave
A_S	slipped area of a fault		T_S	arrival time of S wave
d	focal distance		T_{SP}	duration of preliminary tremors
D	distance of slip along a fault		v_P	propagation velocity of P waves
E	modulus of elasticity		v_S	propagation velocity of S waves
E_f	energy released by an earthquake at the causative fault		JMA	Japanese Meteorological Agency earthquake intensity scale
g	acceleration due to gravity		MM	Modified Mercalli earthquake intensity scale
G	shear modulus or modulus of rigidity		MSK	Medvedev-Sponheuer-Karnik earthquake intensity scale
M	earthquake magnitude		N-S	north-south
M_b	body-wave earthquake magnitude		PGA	peak ground acceleration
M_s	surface-wave earthquake magnitude		μ	rigidity of earth material surrounding causative fault
M_L	local Richter earthquake magnitude		ν	Poisson's ratio
M_w	earthquake moment magnitude		ρ	mass density
M_0	seismic moment			

18 ▲ Earthquake Response of Structures

The earthquake excitation of structures must be defined in terms of ground motion. Indeed, defining the appropriate ground motion for a particular site has been the subject of much research. Since earthquake waves, which are responsible for the ground shaking, are initiated by irregular slippages along faults succeeded by multiple random reflections, refractions, and attenuations within the geological formation through which they pass, *stochastic* modeling of strong ground motion seems appropriate [1]. This has led to the development of procedures to generate synthetic accelerograms (i.e., artificially generated earthquake ground acceleration records) to reflect this stochastic behavior. In special instances (e.g., the seismic analysis of a nuclear power plant located in a zone of high seismicity), it may be desirable to conduct a stochastic seismic analysis that describes the response in probabilistic terms. However, the majority of practical seismic analyses performed are *deterministic*. Moreover, a great deal of knowledge pertaining to the seismic behavior of structures has been generated from deterministic methods of earthquake response.

This chapter discusses the deterministic earthquake response of structures, that is, the response of structures to prescribed earthquake loadings. There are two commonly used deterministic procedures for specifying seismic design loads: (1) *dynamic analysis* and (2) the *equivalent static force procedure* [2]. The dynamic analysis can be either a *response spectrum analysis* or a complete *time-history analysis*. The equivalent static force procedure, detailed in most seismic design codes, specifies the earthquake-induced inertial forces in structures as equivalent static forces by the use of empirical formulas. The type of seismic analysis to be conducted for a particular structure is generally a function of several parameters, such as (1) the type of structure and its social and economic importance, (2) the seismicity of the area, (3) the local soil and geological conditions, and (4) the dynamic characteristics of the structure.

This chapter provides an introduction to the deterministic earthquake response of structures. The two dynamic methods of earthquake analysis are discussed in detail and a general discussion of the provisions in seismic design codes is also presented. Both SDOF and MDOF systems are included in the discussion. The effects of soil-structure interaction, however, are not addressed in the dynamic analyses; that is, the foundation medium is assumed to be relatively stiff and the seismic motions applied to the structure's support locations are the same as the free-field earthquake motions at the site.

18.1 TIME-HISTORY ANALYSIS: BASIC CONCEPTS

The earthquake response problem is essentially a base excitation problem similar to those discussed in Sections 5.5 and 7.7 for SDOF systems and in Section 12.5 for MDOF systems. To clarify the discussion, consider the SDOF structure shown in Figure 18.1 subject to an arbitrary ground displacement $x_g(t)$. Newton's second law sets forth the equilibrium condition

$$F_I + F_D + F_S = F(t) \tag{18.1}$$

where

$$F_1 = \text{inertia force} = m\ddot{x}_t$$

$$F_D = \text{damping force} = c\dot{x}$$

$$F_S = \text{elastic restoring force} = kx$$

$$F(t) = \text{externally applied force} = 0$$

and $x(t)$ is the relative displacement of the mass and $x_t(t)$ is the total (absolute) displacement of the mass. Thus, the equilibrium condition given by Eq. (18.1) may be expressed as

$$m\ddot{x}_t + c\dot{x} + kx = 0 \tag{18.2}$$

The *effective load* that induces the dynamic response of this system is defined in recognition of the fact that the inertia force term (F_I) in Eq. (18.1) is proportional to the total (absolute) motion of the system, while the damping force (F_D) and elastic restoring force (F_S) are proportional to the relative motion of the system. Therefore, referring to Figure 18.1 and noting that

$$x_t = x_g + x \tag{18.3}$$

it follows that

$$\ddot{x}_t = \ddot{x}_g + \ddot{x} \tag{18.4}$$

Hence Eq. (18.2) may be rewritten as

$$m\ddot{x} + m\ddot{x}_g + c\dot{x} + kx = 0 \tag{18.5}$$

or

$$m\ddot{x} + c\dot{x} + kx = F_{\text{eff}}(t) \tag{18.6}$$

where $F_{\text{eff}}(t)$ is the effective load, attributed to the horizontal ground acceleration (see Figure 18.2), applied to the mass m and is given by

$$F_{\text{eff}}(t) = -m\ddot{x}_g(t) \tag{18.7}$$

Since $\omega^2 = k/m$ and $\zeta = c/2\sqrt{km}$, Eq. (18.6) becomes

$$\ddot{x} + 2\omega\zeta\dot{x} + \omega^2 x = -\ddot{x}_g(t) \tag{18.8}$$

Thus, for any arbitrary acceleration of the supports $\ddot{x}_g(t)$, the relative displacement of the mass $x(t)$ can be computed from the Duhamel integral expression, for zero initial conditions, as

$$x(t) = \frac{-1}{\omega\sqrt{1-\zeta^2}} \int_0^t \ddot{x}_g(\tau)e^{-\zeta\omega(t-\tau)} \sin \omega\sqrt{1-\zeta^2}(t-\tau)\,d\tau \tag{18.9}$$

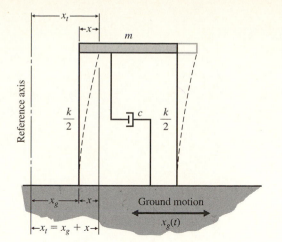

Figure 18.1

Rigid-base SDOF system subject to translational component of earthquake ground motion.

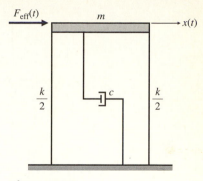

Figure 18.2

Effective load due to horizontal earthquake ground acceleration.

From Eq. (18.9) it is noted that the relative response of the structure is characterized by its natural frequency ω, the damping factor ζ, and the nature of the base excitation $\ddot{x}_g(t)$.

Generally, the undamped natural frequency is used in Eq. (18.9) in place of the damped frequency, and the negative sign is ignored (the sense of the response has little significance in earthquake analysis). Then Eq. (18.9) may be represented as

$$x(t) = \frac{1}{\omega} R(t) \tag{18.10}$$

where

$$R(t) = \int_0^t \ddot{x}_g(\tau) e^{-\zeta\omega(t-\tau)} \sin \omega(t - \tau)\, d\tau \tag{18.11}$$

and $R(t)$ is called the *earthquake response integral*. The relative displacement $x(t)$ is important in the earthquake analysis of structures because the strains (and stresses) in the structural members are directly proportional to the relative displacements. For example, the total shear force, or *base shear, V* transferred to the foundation by the elastic constraints (refer to Figure 18.3) is given by

$$V(t) = kx(t) \tag{18.12}$$

Notice that the base shear represented by Eq. (18.12) is equivalent to the elastic restoring force in the system, $F_S(t)$.

The *exact relative velocity $\dot{x}(t)$* of the mass is obtained by differentiating the relative displacement $x(t)$, given by Eq. (18.9), with respect to time, resulting in

$$\dot{x}(t) = -\int_0^t \ddot{x}_g(\tau) e^{-\zeta\omega(t-\tau)} \cos \omega\sqrt{1 - \zeta^2}(t - \tau)\, d\tau$$

$$+ \frac{\zeta}{\sqrt{1 - \zeta^2}} \int_0^t \ddot{x}_g(\tau) e^{-\zeta\omega(t-\tau)} \sin \omega\sqrt{1 - \zeta^2}(t - \tau)\, d\tau \tag{18.13}$$

Figure 18.3

Earthquake base shear.

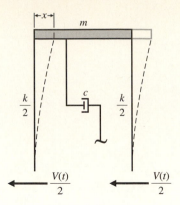

The *absolute acceleration* of the mass, $\ddot{x}_t(t)$, is obtained by differentiation of $\dot{x}(t)$, given by Eq. (18.13), with respect to time and noting that $\ddot{x}_t(t) = \ddot{x}(t) + \ddot{x}_g(t)$. This yields

$$\ddot{x}_t(t) = \frac{\omega(1 - 2\zeta^2)}{\sqrt{1 - \zeta^2}} \int_0^t \ddot{x}_g(\tau) e^{-\zeta\omega(t-\tau)} \sin \omega\sqrt{1 - \zeta^2}(t - \tau)\, d\tau$$

$$+ 2\omega\zeta \int_0^t \ddot{x}_g(\tau) e^{-\zeta\omega(t-\tau)} \cos \omega\sqrt{1 - \zeta^2}(t - \tau)\, d\tau$$

(18.14)

The absolute acceleration of the mass, $\ddot{x}_t(t)$, has several important applications. First, it is the quantity most easily measured experimentally during strong earthquake-induced vibrations. That is, an accelerograph located in a structure records a close approximation of $\ddot{x}_t(t)$ at that point. Second, the absolute acceleration is indicative of the force experienced by the mass m during the excitation.

Equations (18.9), (18.13), and (18.14) represent the *earthquake time-history* response for a SDOF structure. In practice, the time-history response is usually evaluated by direct numerical integration of Eq. (18.6) or Eq. (18.8). This is readily facilitated by employment of any of the numerical integration schemes discussed in Chapter 7.

At this point it is interesting to note that once the displacement response history $x(t)$ has been evaluated by dynamic analysis, the resulting internal forces that develop in the structure can be determined from static analysis of the structure at every instant of time. This procedure is based on the concept of the *effective earthquake force*. The effective earthquake force F_s is the equivalent static force that, when applied externally to the structure at any instant of time t, will produce the same displacement x in the stiffness component as that obtained from the dynamic analysis at the same instant of time. Therefore, the effective earthquake force and the elastic restoring force are equivalent at any instant of time. Thus, from Eq. (18.12)

$$F_s(t) = kx(t) = V(t)$$

(18.15)

Finally, if the stiffness term k is expressed in Eq. (18.15) in terms of the natural circular frequency of the system ω, then the effective earthquake force is equivalent to $m\omega^2 x(t)$. Notice here that the effective earthquake force is expressed as the product of the mass m and the *pseudoacceleration* $\omega^2 x(t)$, but not the absolute acceleration $\ddot{x}_t(t)$. The pseudoacceleration is discussed in detail in the following section.

EXAMPLE 18.1 ▼

An industrial building is shown in Figure 18.4. Idealize the structure as a SDOF system. Assume the structure acts as a braced frame in the E-W (having a total of six braced bays) direction and as an unbraced shear frame (with column bases pinned) in the N-S direction. Assume all columns bend about their strong axes in the N-S direction. The vertical cross bracings in the E-W direction are 1-in-diameter steel rods. The dead weight of the structure is 290 kips, which is concentrated at the base of the roof trusses. All columns are W10×39, all steel is ASTM A36, $\zeta = 0.05$, and story height $h = 14$ ft.

(a) Determine the natural period of the structure for the N-S and E-W directions.

(b) Conduct a time-history analysis of the structure in both directions. Use the N-S component of the Jan. 17, 1994, Northridge, CA, earthquake shown in Figure 18.5 as input. The digitized accelerogram for the Northridge earthquake is available on the authors' web site (refer to the Preface).

Solution

(a) Determine the natural period:

The mass of the structure is calculated as

$$m = \frac{290 \text{ kips}}{386.4 \text{ in/sec}^2} = 0.75 \text{ kip-sec}^2/\text{in}$$

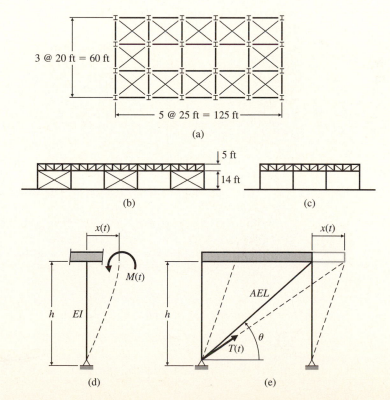

Figure 18.4

Industrial building of Example 18.1: (a) plan; (b) north and south elevation; (c) east and west elevation; (d) column moment; (e) axial force in bracing rod.

Figure 18.5

North-south component of horizontal ground acceleration for Northridge, CA, earthquake of January 17, 1994.

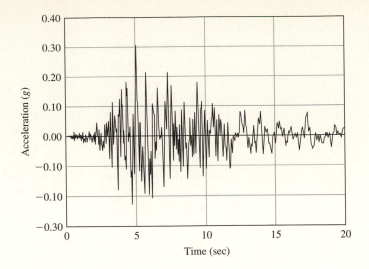

- Stiffness in the N-S direction (for all 24 columns):

$$k = 24\frac{3EI}{h^3} = \frac{72(29{,}000 \text{ ksi})(209 \text{ in}^4)}{(168 \text{ in})^3} = 92.03 \text{ kips/in}$$

- Stiffness in the E-W direction (for 6 braces acting in tension):

$$k = 6\frac{AE}{L}\cos^2\theta$$

where the cross-sectional area of the rod is

$$A = \frac{\pi d^2}{4} = \frac{\pi(1.0 \text{ in})^2}{4} = 0.785 \text{ in}^2$$

and its slope is

$$\theta = \tan^{-1}\frac{14 \text{ ft}}{25 \text{ ft}} = 29.24°$$

Therefore, the stiffness in the E-W direction is

$$k = \frac{(6)(0.785 \text{ in}^2)(29{,}000 \text{ ksi})(\cos 29.24°)^2}{343.8 \text{ in}} = 302.4 \text{ kips/in}$$

- Natural period in the N-S direction:

$$T = 2\pi\sqrt{\frac{m}{k}} = 2\pi\sqrt{\frac{0.75}{92.03}} = 0.567 \text{ sec}$$

- Natural period in the E-W direction:

$$T = 2\pi\sqrt{\frac{m}{k}} = 2\pi\sqrt{\frac{0.75}{302.4}} = 0.313 \text{ sec}$$

(b) Time-history response:

The pertinent parameters in the time-history response are the relative displacement $x(t)$, relative velocity $\dot{x}(t)$, and absolute acceleration $\ddot{x}_t(t)$. Also of interest are the base shear, $V(t)$, the bending moments in the columns $M(t)$, and the axial tension force in the steel rods $T(t)$.

(i) Response in the N-S direction:

The effective load applied to the mass is determined from Eq. (18.7) and is shown in Figure 18.6. The response for relative displacement $x(t)$ was determined from direct numerical integration of Eq. (18.8) using computer program NEWB on the authors' web site and is presented in Figure 18.7. The seismic base shear $V(t)$ is determined from Eq. (18.12) or Eq. (18.15) and is presented in Figure 18.8. The bending moment in each column is determined from the expression (refer to Figure 18.4d)

$$M(t) = \frac{3EIx(t)}{h^2}$$

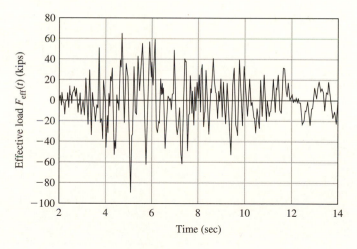

Figure 18.6

Effective load for earthquake horizontal ground acceleration $F_{\text{eff}}(t)$, N-S direction.

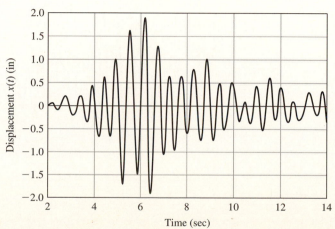

Figure 18.7

Relative displacement $x(t)$ in N-S direction.

The time-history response for bending moment $M(t)$ in each column is presented in Figure 18.9. The maximum responses for the N-S direction are summarized in Table 18.1.

(ii) Response in the E-W direction:

The relative displacement $x(t)$ for the E-W response is presented in Figure 18.10. The base shear time history $V(t)$ is presented in Figure 18.11. The axial force in each tension diagonal $T(t)$ is determined from the expression (refer to Figure 18.4e)

$$T(t) = \frac{kx(t)}{6 \cos \theta} = \frac{AE}{L}\cos \theta x(t)$$

The time-history response for $T(t)$ is presented in Figure 18.12. The maximum responses for the E-W direction are summarized in Table 18.2.

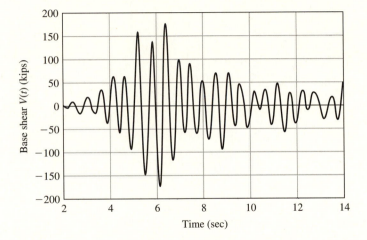

Figure 18.8

Total base shear $V(t)$ in N-S direction. "Loma Prieta Earthquake Reconnaissance Report" *Earthquake Spectra,* Supplement to Vol. 6, May 1990, Earthquake Engineering Research Institute (EERI, 499 14th St., Suite 320, Oakland, CA, 946-1934).

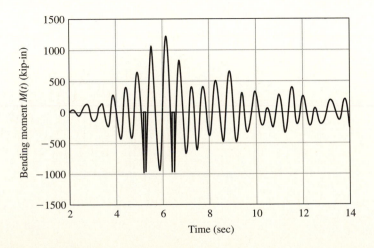

Figure 18.9

Bending moment $M(t)$ in each column; N-S direction.

TABLE 18.1.   Maximum Responses in the N-S Direction

Maximum relative displacement (in), x_{max}	1.928
Maximum relative velocity (in/sec), $\dot{x}_{max}$	21.957
Maximum absolute acceleration (in/sec^2), $(\ddot{x}_t)_{max}$	237.777
Maximum base shear (kips), V_{max}	178.333
Maximum bending moment in columns (in-kip), M_{max}	1230.096
Maximum bending stress in columns (ksi)	29.499

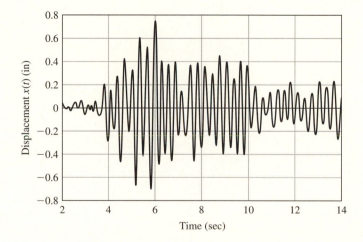

Figure 18.10

Relative displacement $x(t)$ in E-W direction.

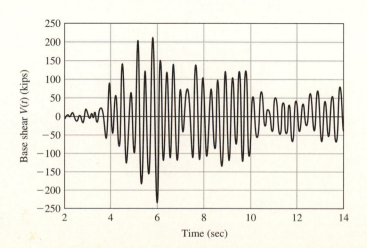

Figure 18.11

Total base shear $V(t)$ in E-W direction.

Figure 18.12

Axial tension force $T(t)$ in each bracing rod; E-W direction.

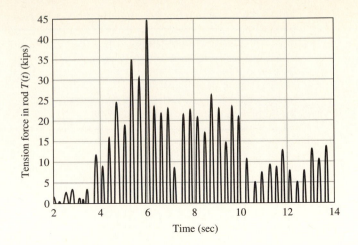

TABLE 18.2. **Maximum Responses in the E-W Direction**

Maximum relative displacement (in), x_{max}	0.775
Maximum relative velocity (in/sec), $\dot{x}_{max}$	13.546
Maximum absolute acceleration (in-sec²), $(\ddot{x}_t)_{max}$	313.951
Maximum base shear (kip), V_{max}	235.463
Maximum tension force in rod (kip), T_{max}	44.778
Maximum tension stress in rod (ksi)	57.042

18.2 EARTHQUAKE RESPONSE SPECTRA

As illustrated by Example 18.1, the evaluation of the dynamic response (i.e., displacements, accelerations, forces, etc.) at every instant of time during an earthquake time history can require significant computational effort, even for relatively simple structural systems [3]. However, for many engineering applications, only the maximum absolute values of the quantities $x(t)$, $\dot{x}(t)$, and $\ddot{x}_t(t)$ experienced by a structure during an earthquake are of interest. These quantities are commonly referred to as the *spectral displacement* S_d, the *spectral velocity* S_v, and the *spectral acceleration* S_a, respectively, and are defined by

$$S_d = |x(t)|_{max} \tag{18.16a}$$

$$S_v = |\dot{x}(t)|_{max} \tag{18.16b}$$

$$S_a = |\ddot{x}_t(t)|_{max} \tag{18.16c}$$

where $|x(t)|_{\max}$, $|\dot{x}(t)|_{\max}$, and $|\ddot{x}_t(t)|_{\max}$ are the maximum absolute values of relative displacement, relative velocity, and absolute acceleration determined from Eqs. (18.9), (18.13), and (18.14), respectively. However, these quantities are generally determined from numerical evaluation of Eq. (18.8).

Plots of S_d, S_v, and S_a versus the undamped natural period of vibration or natural frequency, for various damping factors, ζ are called earthquake response spectra. An earthquake response spectrum for a specific earthquake is constructed by considering a series of SDOF oscillators (or inverted pendulums) with varying periods of vibration attached to a moveable base as shown in Figure 18.13. The base is subjected to the

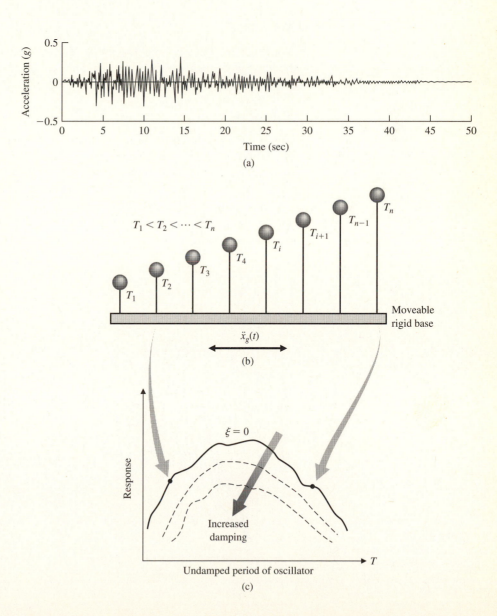

Figure 18.13

Construction of response spectrum: (a) earthquake accelerogram; (b) family of SDOF oscillators; (c) response spectrum.

same ground motion as that of a recorded earthquake and the pendulums respond differently to the ground motion. The maximum response (S_d, S_v, or S_a) for each pendulum is plotted against the natural period (or frequency) of the pendulums, resulting in a response curve or response spectrum. A family of such curves (for different values of damping) is referred to as *response spectra*. Such spectra are very useful in design because the designer can readily assess how structures of different natural periods would respond to a specific earthquake. The underlying concept of a response spectrum is that the maximum response of a linear SDOF system to any prescribed component of earthquake motion depends on the natural period of the system and the amount of damping.

EXAMPLE 18.2 ▼

Construct response spectra for S_d, S_v, and S_a for the N-S component of the Northridge earthquake. Consider damping factors of $\zeta = 0.02$, $\zeta = 0.05$, and $\zeta = 0.10$.

Solution

The spectral displacement S_d, spectral velocity S_v, and spectral acceleration S_a are determined from Eqs. (18.16a), (18.16b), and (18.16c), respectively. The responses $x(t)$, $\dot{x}(t)$, and $\ddot{x}_t(t)$ were evaluated numerically by direct integration of Eq. (18.8) for $\zeta = 0.02$, 0.05, and 0.10, from which the maximum response values $|x(t)|_{max}$, $|\dot{x}(t)|_{max}$, and $|\ddot{x}_t(t)|_{max}$ were determined. The response spectra for S_d, S_v, and S_a are presented in Figures 18.14, 18.15, and 18.16, respectively.

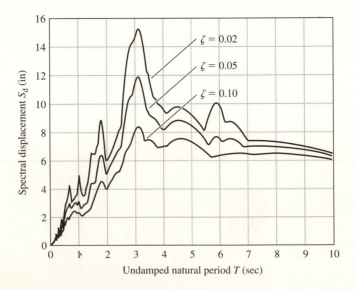

Figure 18.14

Displacement response spectrum for N-S component of Northridge, CA, earthquake.

Figure 18.15

Velocity response spectrum for N-S component of Northridge, CA, earthquake.

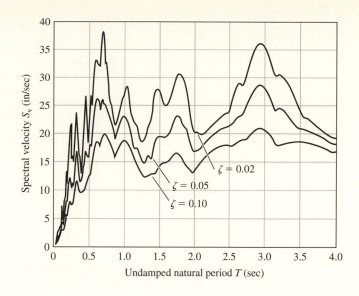

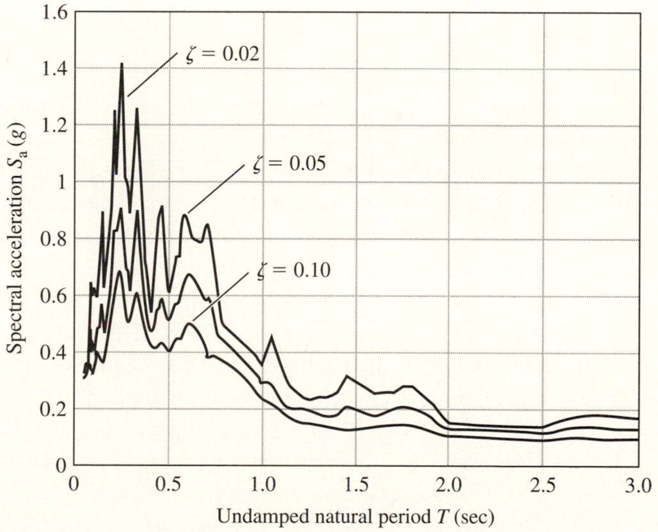

Figure 18.16

Acceleration response spectrum for N-S component of Northridge, CA, earthquake.

In typical engineering structures the percentage of critical damping is relatively small. It is approximately 2 to 8% for buildings and 5 to 10% for soil structures (refer to Table 13.1). Therefore the term $\sqrt{1-\zeta^2} \cong 1$ and the terms of order ζ and higher in Eqs. (18.13) and (18.14) may be neglected. Also, if the cosine term in Eq. (18.13) is replaced with a sine term, then $x(t)$, $\dot{x}(t)$, and $\ddot{x}_t(t)$ can be written as

$$x(t) = \frac{1}{\omega}\int_0^t \ddot{x}_g(\tau)e^{-\zeta\omega(t-\tau)}\sin\,\omega(t-\tau)\,\mathrm{d}(\tau) \tag{18.17}$$

$$\dot{x}(t) = \int_0^t \ddot{x}_g(\tau)e^{-\zeta\omega(t-\tau)}\sin\,\omega(t-\tau)d\tau = R(t) \tag{18.18}$$

$$\ddot{x}_t(t) = \omega\int_0^t \ddot{x}_g(\tau)e^{-\zeta\omega(t-\tau)}\sin\,\omega(t-\tau)d\tau \tag{18.19}$$

Then the following approximate relationships exist between the spectral quantities defined by Eqs. (18.16), and are represented as [4]

$$S_d \cong \frac{1}{\omega}S_v \tag{18.20}$$

and

$$S_a \cong \omega S_v \tag{18.21}$$

where ω is the natural circular frequency. For engineering applications, the following approximations are generally employed:

$$S_{pv} = \omega S_d \tag{18.22}$$

and

$$S_{pa} = \omega^2 S_d \tag{18.23}$$

where S_{pv} and S_{pa} are referred to as the *pseudospectral velocity* and *pseudospectral acceleration,* respectively.

The parameters S_{pv} and S_{pa} have certain characteristics that are of practical interest [5]. The pseudospectral velocity S_{pv} is close to the spectral velocity S_v for short-period structures and almost equal for intermediate periods, but is different for long-period structures. A comparison between S_{pv} and S_v for the N-S component of the Northridge earthquake for $\zeta = 0.05$ is illustrated in Figure 18.17. For zero damping, the pseudospectral acceleration S_{pa} is identical to the spectral acceleration S_a. However, for damping other than zero, the two are slightly different. Nevertheless, for damping levels encountered in most engineering applications, the two can be considered to be practically equal [6]. A comparison of S_{pa} and S_a for the N-S component of the Northridge earthquake for $\zeta = 0.05$ is presented in Figure 18.18.

The spectral relationships expressed by Eqs. (18.22) and (18.23) significantly expedite the construction of earthquake response spectra. Evaluation of the spectral displacement S_d by the use of Eq. (18.16a), after numerical integration on Eq. (18.8) to obtain the time-history response $x(t)$, the corresponding pseudospectral velocity S_{pv} and pseudospectral acceleration S_{pa} can readily be established from Eqs. (18.22) and (18.23) respectively. S_d, S_{pv}, and S_{pa} can then all be plotted as a single curve on four-way logarithmic paper. Therefore, at a given frequency or period, all three spectral quantities can be read simultaneously from the same tripartite plot. A tripartite plot of S_d, S_{pv}, and S_{pa} for the N-S component of the Northridge earthquake for damping factors of $\zeta = 0.02$, 0.05, and 0.10 is presented in Figure 18.19. The dashed line in the figure represents the maximum ground motion. The construction of response spectra such as illustrated in Figure 18.19 can be facilitated by use of the Fortran computer program GRESP, which is available on the authors' web site (refer to the Preface).

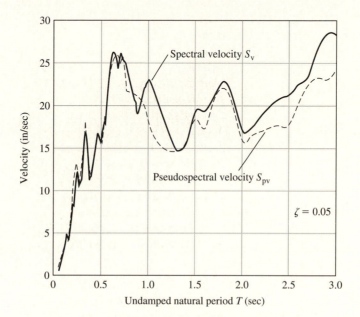

Figure 18.17

Comparison between spectral velocity and pseudospectral velocity for N-S component of Northridge, CA, earthquake ($\zeta = 0.05$).

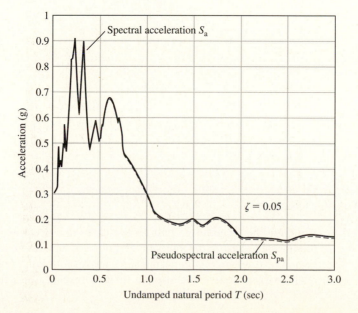

Figure 18.18

Comparison between spectral acceleration and pseudospectral acceleration for N-S component of Northridge, CA, earthquake ($\zeta = 0.05$).

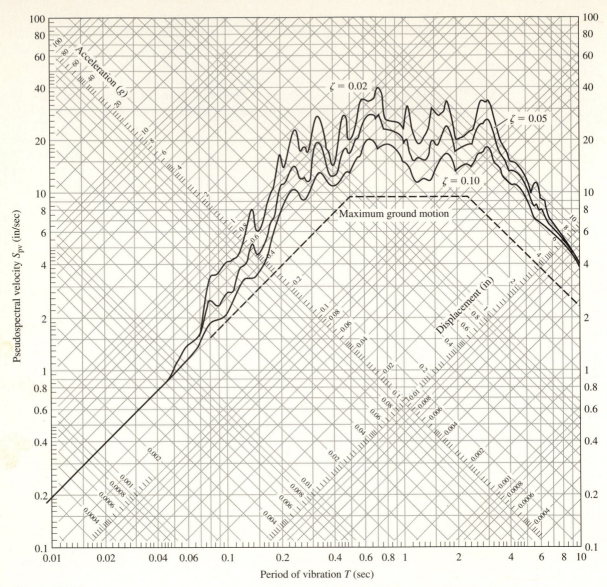

Figure 18.19
Combined response spectrum (S_d, S_{pv}, S_{pa}) for N-S component of Northridge, CA, earthquake ($\zeta = 0.02$, 0.05, and 0.10).

18.3   EARTHQUAKE DESIGN SPECTRA

Although the recorded ground accelerations and response spectra of past earthquakes provide a basis for the rational design of structures to resist earthquakes [7], they cannot be used directly in design since the response of a given structure to a past earthquake will invariably be different than its response to a future earthquake. However, certain similarities exist among earthquake ground motions recorded under similar conditions. These similarities are discussed in Section 18.3.1. On the basis of these similarities, response spectra from earthquakes with common characteristics have been averaged and "smoothed" to create *design spectra*.

It is important to note the basic conceptual difference between a calculated response spectrum and a specified design spectrum. A response spectrum is a convenient method of plotting the maximum response (e.g., displacement, velocity, etc.) of a family of SDOF oscillators, having different natural vibration periods and damping ratios, to a specific earthquake ground motion. A design spectrum is a specification of the seismic design force or displacement of a structure having a specified period of vibration and damping ratio [6, 7].

The first earthquake design spectrum was developed by George Housner [7, 8]. This spectrum was based on the two horizontal ground accelerograms of four earthquakes: (1) El Centro, CA, 1934; (2) El Centro, CA, 1940; (3) Olympia, WA, 1949; and (4) Taft, WA, 1952. This spectrum is presented in Figure 18.20 for 0, 0.5, 1, 2, 5, and 10% of critical damping, normalized to 0.2 *g* acceleration. To employ this spectrum for other ground accelerations, simply multiply the desired spectral quality by the ratio of the specified maximum ground acceleration to 0.2 *g*.

18.3.1 Trends in Earthquake Response Spectra

A systematic methodology for constructing design spectra can be developed by examining the common trends and characteristics of earthquake ground motion revealed by response spectra plotted on four-way logarithmic paper. Consider the response spectrum for the N-S component of the 1994 Northridge, CA, earthquake shown in Figure 18.19. Included on this spectrum is the maximum ground motion polygon (dashed line) that is composed of three bounds: (1) the inclined line on the left representing the maximum ground acceleration of 0.308 *g*, (2) the horizontal line representing the maximum ground velocity of 9.17 in/sec, and (3) the inclined line on the right representing the maximum ground displacement of 3.26 in. Such spectra exhibit approximately the same general characteristics [9]:

1. For extremely short periods (very high frequencies), the pseudospectral acceleration, S_{pa}, values approach magnitudes equal to the maximum ground acceleration $|\ddot{x}_g(t)|_{max}$.

2. For moderately short periods (0.1 to 0.3 sec), with damping factors ζ between 0.05 and 0.10, the pseudospectral accelerations are approximately twice the maximum ground accelerations.

3. For intermediate periods, the maximum pseudospectral velocity S_{pv} has a magnitude several times the maximum ground velocity $|\dot{x}_g(t)|_{max}$ for $\zeta = 0$, ranging down to about equal to the maximum ground velocity for $\zeta = 0.2$.

4. For very long periods (very low frequencies), the maximum spectral displacement S_d approaches the maximum ground displacement $|x_g(t)|_{max}$.

These broad generalizations for spectral values form the basis of a procedure for estimating earthquake design spectra.

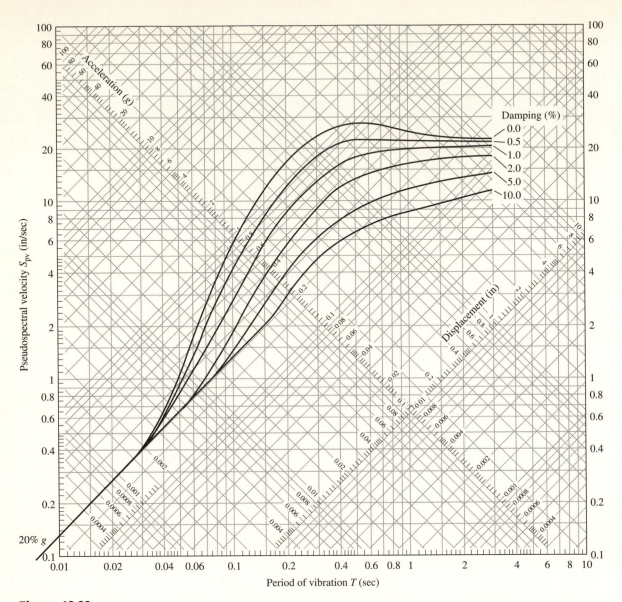

Figure 18.20

Housner design spectrum normalized to 0.2 *g* ground acceleration. "Design Spectrum" by G.W. Housner, Chapter 5 in *Earthquake Engineering,* R.L. Wiegel, editor, copyright © 1970. Reprinted by permission of Prentice Hall, Inc., Upper Saddle River, NJ.

18.3.2 Broad-Banded Design Spectra

A multilinear "broad-banded" (encompassing a broad frequency band) design spectrum can be constructed if the maximum ground motion inputs (acceleration, velocity, and displacement) can be estimated at a particular site by applying appropriate *amplification factors* to these ground motion maxima. Straight-line bounds are generally used for the delineation of the design spectrum. A popular method proposed by Newmark and Hall [10] for constructing broad-banded design spectra uses the anticipated maximum ground acceleration at the site as the primary input datum. The corresponding values for maximum ground velocity and maximum ground displacement are then proportioned relative to the maximum ground acceleration.

The standard Newmark-Hall design spectrum is normalized to a maximum ground acceleration of 1.0 g. The maximum ground velocity is specified as 48 in/sec and the maximum ground displacement is taken as 36 in. Three principal regions (acceleration, velocity, and displacement) of the design spectrum are identified in which the response is an approximately constant, amplified value. Amplification factors are then applied to the maximum ground motions in these regions to obtain the desired design spectrum. The procedure can be summarized as follows:

1. Plot the anticipated maximum ground motion polygon on four-way logarithmic paper.

2. Apply the appropriate amplification factors presented in Table 18.3 to the maximum ground motion to construct the design spectrum for specified damping values.

3. Draw the amplified displacement bound parallel to the maximum ground motion displacement.

4. Draw the amplified velocity bound parallel to the maximum ground velocity.

5. Draw the amplified acceleration region parallel to the maximum ground acceleration between 0.5 and 0.17 sec.

6. Below a period of 0.17 sec the amplified acceleration bound approaches the maximum ground acceleration. Draw a straight line from the amplified acceleration bound at 0.17 sec to the maximum ground acceleration line at 0.033 sec.

7. Below a period of 0.033 sec the acceleration bound is the same as the maximum ground acceleration line.

In general, the spectral intensities for vertical motion can be taken as approximately two-thirds the horizontal motion values when the fault motions are primarily horizontal [11]. Where fault motions are expected to involve large vertical components, the spectral intensities for vertical motion can be assumed equal to the horizontal.

TABLE 18.3. **Relative Values of Spectrum Amplification Factors**

Percent of Critical Damping	Amplification Factors		
	Displacement	Velocity	Acceleration
0.0	2.5	4.0	6.4
0.5	2.2	3.6	5.8
1.0	2.0	3.2	5.2
2.0	1.8	2.8	4.3
5.0	1.4	1.9	2.6
7.0	1.2	1.5	1.9
10.0	1.1	1.3	1.5
20.0	1.0	1.1	1.2

EXAMPLE 18.3   ▼

Construct a Newmark-Hall design spectrum for a maximum ground acceleration equal to that of the Northridge earthquake, 0.308 g, and for a damping factor $\zeta = 0.05$.

Solution

(a) Determine the maximum ground motion parameters:

$$\text{Maximum ground acceleration} = |\ddot{x}_g(t)|_{\max} = (0.308)(1g)$$
$$= 0.308 \ g$$

$$\text{Maximum ground velocity} = |\dot{x}_g(t)|_{\max} = (0.308)(48 \text{ in/sec})$$
$$= 14.78 \text{ in/sec}$$

$$\text{Maximum ground displacement} = |x_g(t)|_{\max} = (0.308)(36 \text{ in})$$
$$= 11.09 \text{ in}$$

(b) Determine the amplified response parameters:
- From Table 18.3 for 5% critical damping (or $\zeta = 0.05$):

$$\text{Amplified (pseudospectral) acceleration} = S_{pa} = 2.6(0.308 \ g)$$
$$= 0.801 \ g$$

$$\text{Amplified (pseudospectral) velocity} = S_{pv} = 1.9(14.78 \text{ in/sec})$$
$$= 28.08 \text{ in/sec}$$

$$\text{Amplified (spectral) displacement} = S_d = 1.4(11.09 \text{ in})$$
$$= 15.52 \text{ in}$$

(c) Construction of the design spectrum is illustrated in Figure 18.21:
- Draw the maximum ground motion polygon using $|\ddot{x}_g(t)|_{\max}$, $|\dot{x}_g(t)|_{\max}$, and $|x_g(t)|_{\max}$.
- Draw the amplified displacement S_d bound parallel to the maximum ground displacement line.
- Draw the amplified velocity S_{pv} bound parallel to the maximum ground velocity line. It intersects the amplified displacement bound to the right of the period $T_1 = 3.5$ sec. Extend the amplified velocity bound horizontally to the left.
- Draw the amplified acceleration S_{pa} bound parallel to the maximum ground acceleration line. It will intersect the amplified velocity bound at

$T_2 = 0.55$ sec. Extend the amplified acceleration bound downward left to the point corresponding to $T_3 = 0.17$ sec.

- Draw the amplified acceleration bound linearly from the point corresponding to $T_3 = 0.17$ sec so that it intersects the maximum ground acceleration line at a point corresponding to $T_4 = 0.033$ sec.

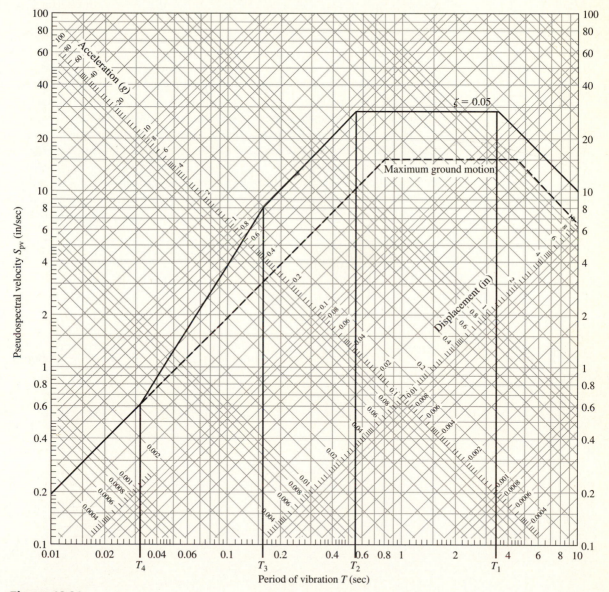

Figure 18.21

Construction of Newmark-Hall broad-banded design spectrum ($\zeta = 0.05$).

18.3.3 Influence of Soil Conditions on Design Spectra

Design spectra such as those presented in Figures 18.20 and 18.21 were based on earthquake records on alluvium and did not consider soil conditions as a parameter [6]. However, subsequent studies [12–15] on the effect of soil conditions on earthquake ground motions have indicated that soil conditions at the site significantly affect spectral amplifications and shapes as illustrated in Figure 18.22. Thus, the ground motions near the surface where a structure might be located are affected by the properties of the soil (e.g., stiffness, strength, and layering) and rock strata between the site and the source.

In general, motions in rock where the rock outcrops near the surface are very closely related to those in deep underlying strata. However, ground motions in soil can be greatly affected by transmission phenomena through the soil. Usually, the softer the soil through which the earthquake waves are propagated, the more the high-frequency waves are filtered out and the greater is the amplification of the wave motions that have frequencies corresponding to frequencies of the soil strata structure.

The available data suggest that there is a major difference between spectral amplification factors calculated for soft soils and those calculated in competent rock. In relatively soft soils, spectral amplifications vary with the frequency and intensity of the ground motion, and spectral velocities and accelerations may be twice those on competent rock. In extremely soft soils, the acceleration may decrease slightly, but spectral displacements and velocities may increase by a factor of 2 compared to those on rock.

Figure 18.22

Average acceleration spectra for different site conditions. "Site Dependent Spectra for Earthquake Resistant Design" by H.B. Seed, C. Ugas, and J. Lysmer, *Bulletin of the Seismological Society of America,* Vol. 66, No. 1, 1976.

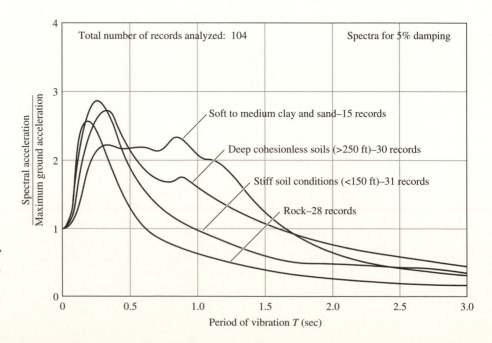

TABLE 18.4. Modification Factors for Spectral Amplification

Soil Condition	Modification Factor
Competent rock	0.67
Soft rock or firm sediment	1.0
Soft sediment	1.5

To account for variability in the soil conditions at the site in an approximate manner, modification factors for the spectral amplification factors presented in Table 18.3 have been suggested. These modification factors are presented in Table 18.4.

18.3.4 Estimating Ground Motion

To construct a design spectrum for a particular site, an estimate of the earthquake ground motions is necessary. This estimate must be based upon the seismic history in the vicinity of the site, the geology and fault activities of the region, and records of past earthquakes. Unfortunately, in many regions of the world this information is not available. Nevertheless, based upon the limited information available, seismic risk procedures and attenuation relationships for estimating the peak horizontal ground acceleration (PGA) at the site have been developed. This, in turn, has led to the development of isoseismal maps for peak ground accelerations and velocities [16, 17]. A contour map of the peak acceleration on rock having a 90% probability of not being exceeded in 50 years is illustrated in Figure 18.23.

With the availability of a large number of recorded earthquake ground motions, a number of statistical studies [14, 15, 18] were conducted to determine average values of ground velocity and ground displacement corresponding to a specified ground acceleration. In these studies it was recommended that the ratio of the peak velocity to peak acceleration (v/a) be used to estimate the peak ground velocity and the ratio of the peak acceleration–peak displacement product to the square of the peak velocity (ad/v^2) be used to estimate the peak displacement.

A lognormal distribution summary of v/a and ad/v^2 for records on four different soil conditions is presented in Table 18.5. The results for each soil category are presented in three groups: the L group includes the components with the larger of the two peak horizontal accelerations, the S group includes the group with the smaller of the two peak horizontal accelerations, and the V group includes the vertical components [6].

Therefore, Figure 18.23 can be used in conjunction with Table 18.5 to estimate the maximum ground motion at a site. The peak ground acceleration for the site can be determined from Figure 18.23, and the corresponding values for peak ground velocity and peak ground displacement can be estimated from Table 18.5. This information may then be used to construct a site specific design spectrum. The procedure is illustrated in the following example.

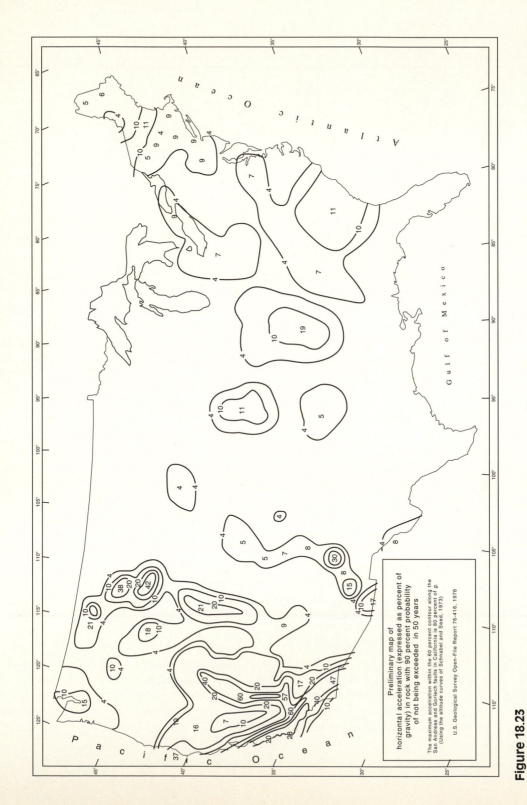

Figure 18.23

Algermissen and Perkins seismic risk map. "Tentative Provisions for the Development of Seismic Regulations for Buildings," ATC publication 3-06, 1978. Reprinted courtesy of National Institute of Standards and Technology, Technology Administration, U.S. Department of Commerce.

TABLE 18.5. Summary of v/a and ad/v^2 Ratios (Lognormal Distribution)

Soil Category	Group	v/a (in/sec)/g Percentile 50	v/a (in/sec)/g Percentile 84.1	ad/v^2 Percentile 50	ad/v^2 Percentile 84.1	d/a (in/g) Percentile 50
Rock	L	24	38	5.3	11.0	8
	S	27	44	5.2	11.2	10
	V	28	45	6.1	11.8	12
< 30 ft of alluvium underlain by rock	L	30	57	4.5	7.7	11
	S	39	62	4.2	8.2	17
	V	33	53	6.8	13.3	19
30–200 ft of alluvium underlain by rock	L	30	46	5.1	7.8	12
	S	36	58	3.8	6.4	13
	V	30	46	7.6	13.7	18
Alluvium	L	48	69	3.9	6.0	23
	S	57	85	3.5	4.9	29
	V	48	70	4.6	7.0	27

"A Study of Earthquake Response Spectra for Different Geological Conditions" by B. Mohraz, *Bulletin of the Seismological Society of America,* Vol. 66, No. 3, 1976.

EXAMPLE 18.4 ▼

Construct a Newmark-Hall broad-banded design spectrum for a site in San Diego, CA. Develop spectrum curves for damping factors of $\zeta = 0.02$, 0.05, and 0.10. Estimate the maximum horizontal ground acceleration from the Algermissen and Perkins seismic risk map shown in Figure 18.23. Determine the maximum ground velocity and maximum ground displacement from Table 18.5. Assume group L, 84.1 percentile on 150 ft of alluvium underlain by bedrock.

Solution

- Estimate maximum ground motion:
  From Figure 18.23, the maximum ground acceleration is given as

$$a = 0.4 \ g = 154.56 \ \text{in/sec}^2$$

From Table 18.5 the v/a and ad/v^2 ratios are determined as

$$\frac{v}{a} = 46 \ \frac{\text{in/sec}}{g} \quad \text{and} \quad \frac{ad}{v^2} = 7.8$$

Therefore, the maximum ground velocity is calculated as

$$v = \left(\frac{v}{a}\right)a = 46(0.4) = 18.4 \ \text{in/sec}$$

and the maximum ground displacement is determined by

$$d = \left(\frac{ad}{v^2}\right)(v^2)\left(\frac{1}{a}\right) = \frac{7.8(18.4)^2}{154.56} = 17.09 \ \text{in}$$

- Construct the maximum ground motion polygon with the values of *a*, *v*, and *d* calculated above as shown in Figure 18.24 as the dashed line.
- Apply the appropriate amplification factors given in Table 18.3 to the maximum ground motion components in a manner similar to that described in Example 18.3. The resulting design spectrum is shown in Figure 18.24.

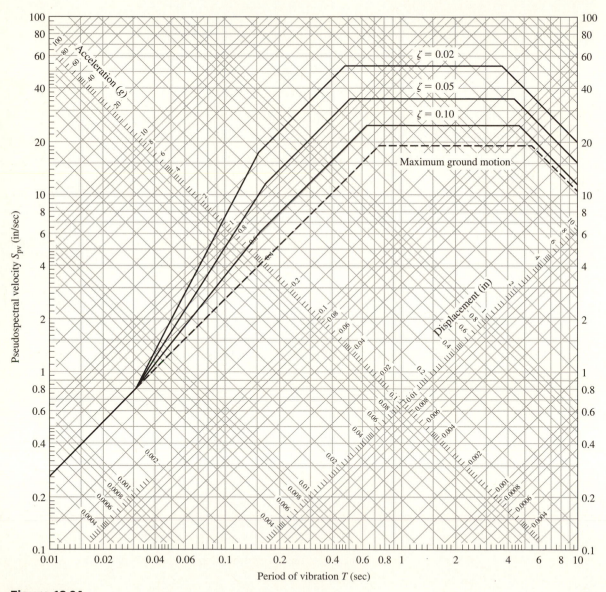

Figure 18.24

Design spectrum for 0.4 *g* ground acceleration ($\zeta = 0.02$, 0.05, and 0.10).

The use of the peak ground acceleration (PGA), as illustrated in Figure 18.23, for design has been deemed too conservative by several investigators. The use of the effective peak acceleration (EPA) and the effective peak velocity-related ground acceleration (EPV) in lieu of the PGA has been recommended as more appropriate for design. The EPA characterizes the short-period components of ground motion, and the EPV characterizes the longer period ground motion components. The EPA and EPV, respectively, are obtained by normalizing the spectral accelerations at periods between 0.1 and 0.5 sec and the spectral velocity at a period of approximately 1.0 sec with respect to the amplification factor of 2.5 for a 5% damping spectrum [6]. A contour map for the EPA of the United States is presented in Figure 18.25.

18.3.5 Spectral Response of SDOF Systems

Design spectra may be conveniently employed to estimate the maximum response of SDOF systems to a prescribed seismic input. Consider the shear frame structure shown in Figure 18.26a. The maximum relative displacement response $|x(t)|_{max}$ is equal to the spectral displacement S_d, as shown in Figure 18.26b. Similarly, the maximum relative velocity $|\dot{x}(t)|_{max}$ and the maximum absolute acceleration $|\ddot{x}_t(t)|_{max}$ are approximately equal to the pseudospectral velocity S_{pv} and the pseudospectral acceleration S_{pa}, respectively. Thus, from an appropriate design spectrum the maximum response may be established as

$$x_{max} = S_d(T, \zeta) \tag{18.1a}$$

$$\dot{x}_{max} \cong S_{pv}(T, \zeta) \tag{18.24b}$$

$$\ddot{x}_{t_{max}} \cong S_{pa}(T, \zeta) \tag{18.24c}$$

where T and ζ represent the natural period and damping factor, respectively.

The maximum seismic base shear V_{max} or maximum elastic force $(F_s)_{max}$ resisted by the structure can be calculated by

$$V_{max} = (F_s)_{max} = mS_{pa} = kS_d \tag{18.25}$$

As previously mentioned, in seismic analysis the elastic restoring force in the structure F_s is also referred to as the *effective earthquake force*. The calculation of the maximum bending moment in the column is illustrated in Figures 18.26c and 18.26d. For a column of height h fixed at its base, the maximum bending moment M_{max} is determined as

$$M_{max} = \frac{6EIS_d}{h^2} \tag{18.26}$$

and for a column pinned at its base

$$M_{max} = \frac{3EIS_d}{h^2} \tag{18.27}$$

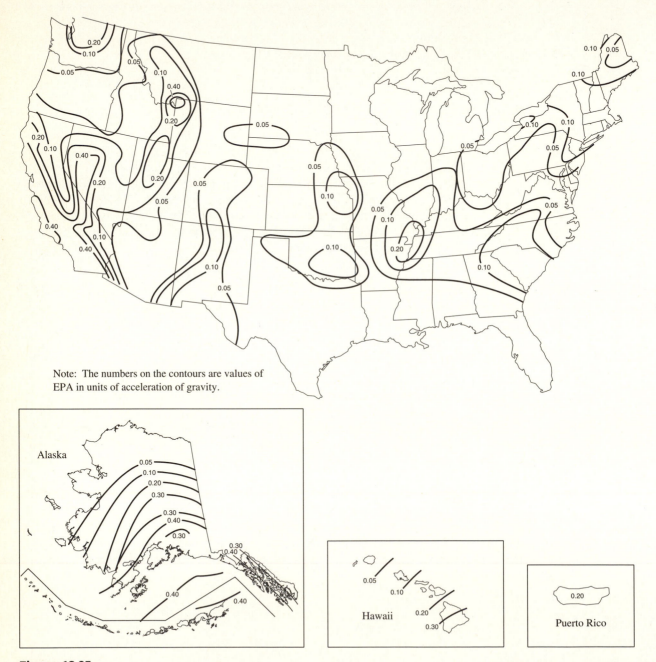

Note: The numbers on the contours are values of
EPA in units of acceleration of gravity.

Figure 18.25
Contour map for EPA for the United States. "Tentative Provisions for the Development of Seismic Regulations for Buildings,"
ATC publication 3-06, 1978. Reprinted courtesy of National Institute of Standards and Technology, Technology Administration,
U.S. Department of Commerce.

Figure 18.26

(a) SDOF shear frame building; (b) maximum displacement response; (c) moment in column, fixed base; (d) moment in column, pinned base.

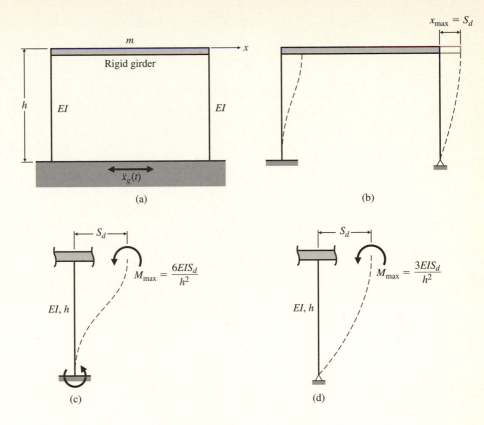

EXAMPLE 18.5 ▼

Estimate the maximum seismic response for the industrial building of Example 18.1 using the Newmark-Hall design spectra constructed in Example 18.3 for an anticipated maximum ground acceleration of 0.308 g and for a damping factor $\zeta = 0.05$. Compare the results with the maximum responses obtained from the time-history analysis (i.e., response spectrum values) conducted in Example 18.1.

Solution

(a) The maximum responses using the Northridge design spectra (Figure 18.21) for $\zeta = 0.05$:

(i) N-S direction, $T = 0.567$ sec:

• Spectral values

$$S_d = 2.5 \text{ in}$$

$$S_{pv} = 28 \text{ in/sec}$$

$$S_{pa} = 308.8 \text{ in/sec}^2$$

• Maximum base shear

$$V_{max} = mS_{pa} = (0.75 \text{ kip-sec}^2/\text{in})(308.8 \text{ in/sec}^2)$$
$$= 231.6 \text{ kips}$$

- Column bending moment

$$M_{max} = \frac{3EIS_d}{h^2} = \frac{3(29,000 \text{ ksi})(209 \text{ in}^4)(2.5 \text{ in})}{(168 \text{ in})^2} = 1610.6 \text{ in-kip}$$

(ii) E-W direction, $T = 0.313$ sec:
- Spectral values

$$S_d = 0.79 \text{ in}$$

$$S_{pv} = 15.5 \text{ in/sec}$$

$$S_{pa} = 308.8 \text{ in/sec}^2$$

- Maximum base shear

$$V_{max} = mS_{pa} = 231.6 \text{ kips}$$

- Axial force in rod

$$T_{max} = \frac{AE}{L}\cos\theta S_d = \frac{(0.785 \text{ in}^2)(29,000 \text{ ksi})}{343.8 \text{ in}}\cos 29.24° \, (0.79 \text{ in})$$

$$T_{max} = 55.52 \text{ kips}$$

(b) Comparisons of the maximum responses obtained from the time-history analysis (response spectrum) and the design spectrum analysis are presented in Tables 18.6 and 18.7 for the N-S and E-W directions, respectively.

It should be apparent that the maximum responses obtained from the time-history analysis are essentially the same as those that would be obtained from the response spectrum. The conceptual difference between the responses obtained from the response spectrum and the design spectrum are demonstrated in Figure 18.27. In this figure, notice that for the E-W response (i.e., for $T = 0.313$ sec) the ordinates of the response spectrum and design spectrum are approximately equal, thus accounting for the close correlation of results exhibited in Table 18.7. However, for the N-S response (i.e., for $T = 0.567$ sec), there is a considerable discrepancy between the results obtained from the response spectrum and those obtained from the design spectrum as illustrated in Table 18.6. In general, a response spectrum and a design spectrum do not yield the same results since the former represents the response to a specific earthquake, while the latter (in this particular instance) represents only the predicted response to any earthquake having the same PGA.

TABLE 18.6. **Comparison of Maximum Responses for the N-S Direction**

Response Quantity	Response Spectra	Design Spectra	% Difference
Maximum relative displacement (in), x_{max}	1.928	2.5	+ 29.69
Maximum relative velocity (in/sec), $\dot{x}_{max}$ (approximate)	21.957	28	+ 27.52
Maximum absolute acceleration (in/sec²), $(\ddot{x}_t)_{max}$ (approximate)	237.777	308.8	+ 29.87
Maximum base shear (kips), V_{max}	178.333	231.6	+ 29.87
Maximum bending moment in column (in-kip), M_{max}	1230.096	1610.6	+ 30.93
Maximum bending stress in column (ksi)	29.499	38.34	+ 29.97

TABLE 18.7. **Comparison of Maximum Responses for the E-W Direction**

Response Quantity	Response Spectra	Design Spectra	% Difference
Maximum relative displacement (in), x_{max}	0.775	0.79	+ 1.91
Maximum relative velocity (in/sec), $\dot{x}_{max}$ (approximate)	13.546	15.5	+ 14.43
Maximum absolute acceleration (in/sec^2), $(\ddot{x}_t)_{max}$ (approximate)	313.951	308.3	− 1.80
Maximum base shear (kips), V_{max}	235.463	231.6	− 1.64
Maximum tension force in rod (kips), T_{max}	44.778	45.52	+ 1.66
Maximum tension stress in rod (ksi)	57.042	57.96	+ 1.61

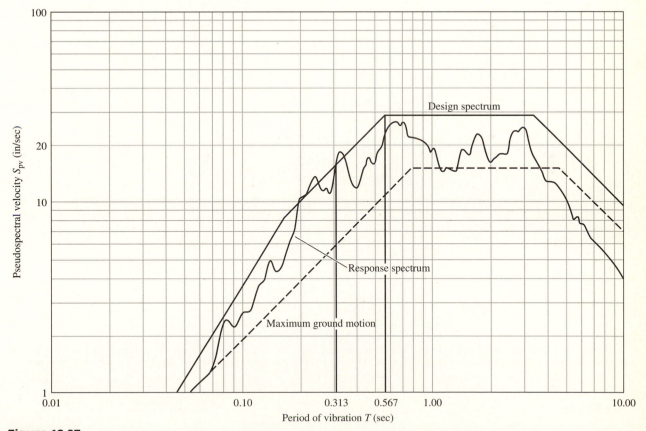

Figure 18.27

Comparison of Northridge, CA, earthquake response spectrum and broad-banded design spectrum ($\zeta = 0.05$).

18.4 RESPONSE OF MDOF SYSTEMS

The methods described in Chapters 12 and 13 for evaluating the dynamic response of MDOF systems can readily be applied to earthquake excitation problems. The time-history response of MDOF systems to a specified earthquake ground motion can be evaluated from either the mode superposition method discussed in Chapter 12 or by one of the direct integration procedures discussed in Chapter 13. The maximum response of MDOF systems may be calculated by the response spectrum method, in a manner similar to that employed for SDOF systems, by making some minor modifications to the modal analysis technique.

For purposes of developing a procedure for the dynamic earthquake response of discrete MDOF systems, consider the MDOF building model shown in Figure 18.28. In this shear frame building model, the masses m_i are localized at the floor levels and are interconnected with massless columns that have an equivalent spring constant k_i. Viscous dashpots, characterized by the damping constants c_i, represent the energy dissipation in the system. The horizontal shear forces acting at each level are designated by V_i, and V_n corresponds to the total base shear acting between the structure and the ground.

The equations of motion for the system in matrix form are expressed as

$$[m]\{\ddot{x}\} + [c]\{\dot{x}\} + [k]\{x\} = -\ddot{x}_g(t)[m]\{I\} \tag{18.28}$$

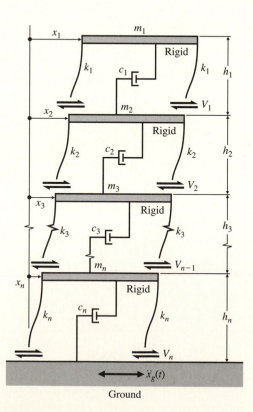

Figure 18.28

MDOF model for a typical multistory shear frame building.

where $[m]$ = mass matrix

$\quad[k]$ = stiffness matrix

$\quad[c]$ = damping matrix

$\quad\{I\}$ = unit vector

$\quad\ddot{x}_g(t)$ = ground acceleration

Analogous to the effective load $F_{\text{eff}}(t)$ defined for a SDOF system, an effective load vector $\{F_{\text{eff}}(t)\}$ can be defined for MDOF systems and is given by

$$\{F_{\text{eff}}(t)\} = -\ddot{x}_g(t)[m]\{I\} \tag{18.29}$$

In theory, the $[m]$, $[c]$, and $[k]$ matrices could have all elements different from zero (i.e., fully populated matrices with nonzero terms). However, for most MDOF structural systems these matrices have only the diagonal terms and a relatively small number of off-diagonal terms different from zero (i.e., these matrices have a narrow bandwidth, or are "well banded").

18.4.1 Time-History Response by Mode Superposition

If the complete dynamic response of the structure is required for a portion of or for the entire seismic event duration, a time-history analysis must be performed. For the special case in which the system damping matrix $[c]$ is a linear combination of the mass and/or stiffness matrices (proportional or classical damping; refer to Section 12.3), the mode superposition analysis technique may be employed. However, if the damping matrix is non-proportional and/or the response is nonlinear, a mode superposition analysis is precluded and a direct numerical analysis procedure must be implemented (refer to Chapter 13).

In a mode superposition analysis or a modal analysis, a set of normal (principal) coordinates is defined, such that, when expressed in those coordinates, the equations of motion become uncoupled. The normal or principal coordinates $\{q\}$, as discussed in Chapter 12, are related to the physical coordinates $\{x\}$ through the transformation

$$\{x\} = [\Phi]\{q\} \tag{18.30}$$

where $[\Phi]$ is the modal matrix. Substituting Eq. (18.30) and its time derivatives $\{\dot{x}\}$ and $\{\ddot{x}\}$ into Eq. (18.28) and premultiplying by $[\Phi]^T$ results in

$$[\Phi]^T[m][\Phi]\{\ddot{q}\} + [\Phi]^T[c][\Phi]\{\dot{q}\} + [\Phi]^T[k][\Phi]\{q\}$$
$$= (-\ddot{x}_g(t)[\Phi]^T[m]\{I\}) \tag{18.31}$$

or, more succinctly

$$[M]\{\ddot{q}\} + [C]\{\dot{q}\} + [K]\{q\} = \{P_{\text{eff}}(t)\} \tag{18.32}$$

where

$$[M] = [\Phi]^T[m][\Phi] \tag{18.33}$$

$$[C] = [\Phi]^T[c][\Phi] \tag{18.34}$$

$$[K] = [\Phi]^T[k][\Phi] \tag{18.35}$$

$$\{P_{\text{eff}}(t)\} = -\ddot{x}_g(t)[\Phi]^T[m]\{I\} \tag{18.36}$$

in which $[M]$, $[C]$, and $[K]$ are the diagonalized modal mass matrix, modal damping matrix, and modal stiffness matrix, respectively, and $\{P_{\text{eff}}(t)\}$ is the effective modal force vector.

Equation (18.31) represents a set of n uncoupled modal equations, in which each equation describes an independent SDOF system response for the rth mode in terms of the normal coordinate q_r, expressed as

$$\ddot{q}_r + 2\zeta_r\omega_r\dot{q}_r + \omega_r^2 q_r = \frac{1}{M_r}[P_{\text{eff}}(t)]_r, \qquad r = 1, 2, \ldots, n \qquad (18.37)$$

where ζ_r, ω_r, and M_r are the damping factor, natural frequency, and modal mass, respectively, for the rth mode. Equation (18.37) may be expressed alternatively as

$$\ddot{q}_r + 2\zeta_r\omega_r\dot{q}_r + \omega_r^2 q_r = -\ddot{x}_g(t)\Gamma_r \qquad (18.38)$$

where

$$\Gamma_r = \frac{\{\Phi\}_r^T[m]\{I\}}{\{\Phi\}_r^T[m]\{\Phi\}_r} = \frac{\{\Phi\}_r^T[m]\{I\}}{M_r} \qquad (18.39)$$

in which $\{\Phi\}_r$ is the modal vector for the rth mode and Γ_r is termed the *earthquake participation factor* for the rth mode.

The displacement response q_r in normal coordinates for the rth mode can be obtained by the Duhamel integral expression

$$q_r(t) =$$
$$-\frac{\Gamma_r}{\omega_r\sqrt{1 - \zeta_r^2}} \int_0^t \ddot{x}_g(\tau)e^{-\zeta_r\omega_r(t - \tau)}\sin \omega_r\sqrt{1 - \zeta_r^2}(t - \tau)d\tau \qquad (18.40)$$

Equation (18.40) can be solved by the methods discussed in Chapter 6 for SDOF systems. However, this is generally not feasible. It is highly recommended to apply one of the numerical integration schemes discussed in Chapter 7 directly to Eq. (18.38) to evaluate the response. With q_r having been evaluated for all modes of interest, the displacement in physical coordinates $\{x\}$ is calculated from the transformation expression given by Eq. (18.30), or

$$\{x(t)\} = \sum_{r = 1}^{n} \{\Phi\}_r q_r(t) \qquad (18.41)$$

As was discussed in Chapter 12, however, when conducting a modal analysis of large MDOF systems, it is generally necessary to consider only a relatively small number of modes p in the response calculations, such that $p << n$. Therefore, Eq. (18.41) can be expressed as

$$\{\hat{x}(t)\} = \sum_{r = 1}^{p} \{\Phi\}_r q_r(t) \qquad (18.42)$$

where $\{\hat{x}(t)\}$ represents the truncated response for $p < n$.

The *effective mass concept* can be useful in determining the number of modes p to be included in the modal analysis. The effective mass for the rth mode, M_{er}, is defined as

$$M_{er} = \{\Phi\}_r^T [m]\{I\}\Gamma_r \tag{18.43}$$

The sum of the effective masses for all modes ($r = 1, 2, \ldots, n$) is equal to the total mass of the structure. This results in a means of determining the number of individual modal responses necessary to accurately represent the structural response.

If the total response of the system is to be represented in terms of a small number of modes p (where $p << n$), and if the sum of the p effective masses is greater than a predefined percentage of the total mass of the structure, then the number of modes p considered in the analysis is adequate. However, if this is not the case, then additional modes must be considered. For example, the dynamic analysis procedure described in the SEAOC code [19] specifies that for the p modes considered in the analysis, at least 90% of the participating mass of the structure must be included in the response calculations for each principal horizontal direction.

When the relative displacements of the masses $\{x(t)\}$ have been established, the *effective earthquake forces* or the elastic restoring forces $F_{si}(t)$ acting at each mass m_i are determined from

$$\{F_s(t)\} = [k]\{x(t)\} \tag{18.44}$$

The story shears $V_i(t)$, referring to Figure 18.28, are calculated as

$$\{V(t)\} = [S][k]\{x(t)\} \tag{18.45}$$

where $[S]$ is the ($n \times n$) lower triangular matrix given by

$$[S] = \begin{bmatrix} 1 & 0 & 0 & . & . & . & & 0 \\ 1 & 1 & 0 & . & . & . & & 0 \\ 1 & 1 & 1 & . & . & . & & 0 \\ . & . & . & . & . & & . & . \\ 1 & 1 & 1 & . & . & . & . & 1 \end{bmatrix} \tag{18.46}$$

Similarly, the story moments $M_i(t)$ at each level of the structure are determined from the expression

$$\{M(t)\} = [H][S][k]\{x(t)\} = [H]\{V(t)\} \tag{18.47}$$

where $[H]$ is the ($n \times n$) lower triangular matrix given by

$$[H] = \begin{bmatrix} h_1 & 0 & 0 & . & . & . & & 0 \\ h_1 & h_2 & 0 & . & . & . & & 0 \\ h_1 & h_2 & h_3 & . & . & . & & 0 \\ . & . & . & . & . & . & . & . \\ h_1 & h_2 & h_3 & . & . & . & h_{n-1} & h_n \end{bmatrix} \tag{18.48}$$

Note that if $p << n$ modes are considered in the analysis, the dimensions of $[S]$ and $[H]$ are (nxp).

An important objective of the time-history analysis, from a structural engineering point of view, is to calculate the envelope of maximum responses, $\{x\}_{max}$, $\{F_s\}_{max}$, $\{V\}_{max}$, and $\{M\}_{max}$, that are to be used for design purposes. It should be noted that the maximum response envelopes do not define the response condition of the structure at any particular instant of time. For example, the maximum displacement response vector $\{x\}_{max}$ contains the maximum relative displacement experienced by each level of the structure during the entire time history. Frequently, the maximum responses of the individual stories occur at different times during the event.

EXAMPLE 18.6   ▼

Conduct a time-history analysis for a typical interior bent of a four-story shear frame building shown in Figure 18.29 for the N-S component of the Northridge earthquake shown in Figure 18.5. Compute the maximum responses for displacements $\{x\}_{max}$, effective earthquake forces $\{F_s\}_{max}$, story shears $\{V\}_{max}$, and story moments $\{M\}_{max}$. Assume $m = 1.5$ kip-sec^2/in, $k = 75$ kips/in, and $\zeta = 0.05$ for all modes.

Solution

The mass and stiffness matrices for the structure are

$$[m] = \begin{bmatrix} 1.5 & 0 & 0 & 0 \\ 0 & 1.5 & 0 & 0 \\ 0 & 0 & 3.0 & 0 \\ 0 & 0 & 0 & 3.0 \end{bmatrix} \text{kip-sec}^2/\text{in}$$

$$[k] = \begin{bmatrix} 75 & -75 & 0 & 0 \\ -75 & 150 & -75 & 0 \\ 0 & -75 & 225 & -150 \\ 0 & 0 & -150 & 300 \end{bmatrix} \text{kips/in}$$

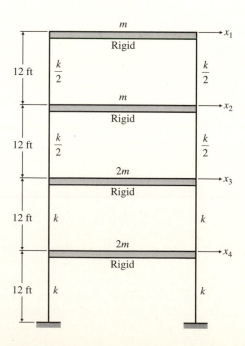

Figure 18.29

Four-story shear frame building of Example 18.6.

The natural frequencies and mode shapes for the system are determined from solutions of the system eigenproblem as discussed in Chapters 10 and 11. For this structure the eigensolution yields

$$\{\omega\} = \begin{Bmatrix} 2.88 \\ 6.32 \\ 10.95 \\ 12.51 \end{Bmatrix} \text{rad/sec}$$

$$[\Phi] = \begin{bmatrix} 0.525 & -0.472 & 0.341 & -0.229 \\ 0.437 & -0.095 & -0.477 & 0.488 \\ 0.277 & 0.358 & -0.145 & -0.325 \\ 0.151 & 0.298 & 0.373 & 0.287 \end{bmatrix}$$

The mode shapes are illustrated in Figure 18.30. The modal mass and modal stiffness matrices are determined from Eqs. (18.33) and (18.35), respectively, as

$$[M] = [\Phi]^T[m][\Phi] = \begin{bmatrix} 1.0 & 0 & 0 & 0 \\ 0 & 1.0 & 0 & 0 \\ 0 & 0 & 1.0 & 0 \\ 0 & 0 & 0 & 1.0 \end{bmatrix} \text{kip-sec}^2/\text{in}$$

(Notice that the modal matrix $[\Phi]$ has been normalized such that $[\Phi]^T[m][\Phi] = [I]$), and

$$[K] = [\Phi]^T[k][\Phi] = \begin{bmatrix} 8.3 & 0 & 0 & 0 \\ 0 & 39.47 & 0 & 0 \\ 0 & 0 & 120.0 & 0 \\ 0 & 0 & 0 & 156.7 \end{bmatrix} \text{kips/in}$$

The modal damping matrix is constructed from the expression

$$[C] = \text{diag}(2M_r\zeta_r\omega_r) = \begin{bmatrix} 0.288 & 0 & 0 & 0 \\ 0 & 0.632 & 0 & 0 \\ 0 & 0 & 1.045 & 0 \\ 0 & 0 & 0 & 1.251 \end{bmatrix}$$

The effective modal force vector is determined from Eq. (18.36) and is given by

$$\{P_{\text{eff}}(t)\} = -\ddot{x}_g(t)[\Phi]^T[m]\{I\} = -\ddot{x}_g(t) \begin{Bmatrix} 2.73 \\ 1.11 \\ 0.48 \\ 0.274 \end{Bmatrix} \tag{1}$$

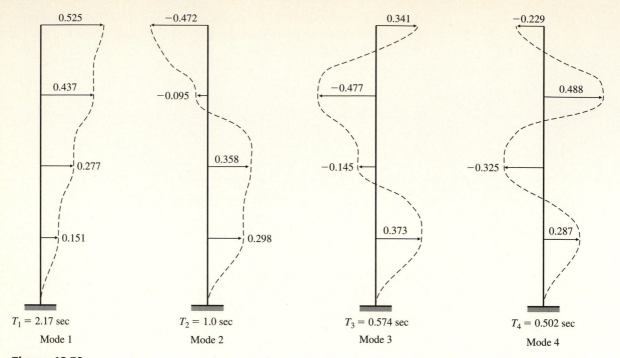

Figure 18.30

Natural vibration modes for four-story shear frame building.

The equations of motion in normal (or principal) coordinates are given by Eqs. (18.32) as

$$[M]\{q\} + [C]\{\dot{q}\} + [K]\{q\} = \{P_{\text{eff}}(t)\} \tag{2}$$

where $\{q\}$ is the vector of normal coordinates. For each mode r, the individual uncoupled equations, represented by Eq. (18.38), are expressed as

$$\ddot{q}_r + 2\zeta_r\omega_r q_r + \omega_r^2 q_r = -\ddot{x}_g(t)\Gamma_r \tag{3}$$

where Γ_r is the earthquake participation factor for the rth mode determined from Eq. (18.39) as

$$\Gamma_r = \frac{\{\Phi\}_r^T[m]\{I\}}{\{\Phi\}_r^T[m]\{\Phi\}_r} = \frac{\{\Phi\}_r^T[m]\{I\}}{M_r} \tag{4}$$

Since $M_r = 1.0$ for $r = 1, 2, 3, 4$, then from Eq. (4)

$$\Gamma_r = \{\Phi\}_r^T[m]\{I\} \tag{5}$$

and from Eq. (5) the earthquake participation factors for all four modes are

$$\{\Gamma\} = \begin{Bmatrix} 2.73 \\ 1.11 \\ 0.48 \\ 0.274 \end{Bmatrix}$$

Thus the uncoupled equations of motion in normal coordinates expressed by Eq. (3) are

$$\ddot{q}_1 + 0.288\dot{q}_1 + 8.3q_1 = -2.73\ddot{x}_g(t) \qquad (6a)$$

$$\ddot{q}_2 + 0.632\dot{q}_2 + 39.97q_2 = -1.11\ddot{x}_g(t) \qquad (6b)$$

$$\ddot{q}_3 + 1.095\dot{q}_3 + 120.0q_3 = -0.48\ddot{x}_g(t) \qquad (6c)$$

$$\ddot{q}_4 + 1.251\dot{q}_4 + 156.7q_4 = -0.274\ddot{x}_g(t) \qquad (6d)$$

Equations (6) can be evaluated by any of the numerical methods discussed in Chapter 7 and can be implemented as discussed in Chapter 12. After evaluating the response in normal coordinates q_r at each integration time step, the response in physical coordinates is obtained by applying the transformation and summation given by Eq. (18.41):

$$\{x(t)\} = \sum_{r=1}^{4} \{\Phi\}_r q_r(t) \qquad (7)$$

The time-history response for the top story relative displacement $x_1(t)$, the effective earthquake force at the bottom story $F_{S4}(t)$, the total base shear $V_4(t)$, and the overturning moment at the base $M_4(t)$ are presented in Figures 18.31, 18.32, 18.33, and 18.34, respectively. The maximum responses at each story are summarized in Table 18.8.

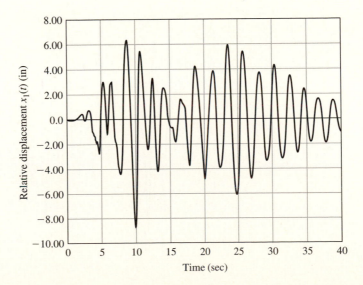

Figure 18.31

Relative displacement response history for top story $x_1(t)$.

Figure 18.32

Effective earthquake force
response history at bottom
story $F_{S4}(t)$.

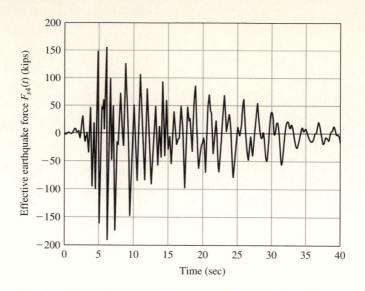

Figure 18.33

Total base shear response
history $V_4(t)$.

Figure 18.34

Base-overturning moment response history $M_4(t)$.

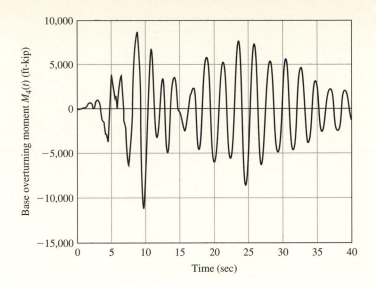

TABLE 18.8. **Maximum Response for Example 18.6**

Location	x_{max}(in)	$(F_S)_{max}$(kips)	V_{max}(kips)	M_{max}(ft-kip)
x_1	8.74	145.63	145.3	1747.62
x_2	7.07	119.8	251.24	4676.72
x_3	4.36	190.5	298.14	7943.32
x_4	2.44	192.17	356.57	11314.95

It was mentioned in Chapter 12 that a mode superposition analysis is effective for large MDOF systems having n DOF only when a relatively small number of vibration modes p (where $p << n$) need be considered for an accurate solution. It was discussed previously in this section that an accurate mode superposition analysis could be obtained by considering a number of modes p such that the sum of the effective masses for the p modes represented at least 90% of the total mass of the structure. The effective mass for the rth mode is determined from Eq. (18.43) and is expressed as

$$M_{er} = \{\Phi\}_r^T [m]\{I\}\Gamma_r \tag{8}$$

For this example the effective masses for all modes are summarized in Table 18.9. The results presented in Table 18.9 indicate that an accurate solution to this problem could be obtained by consideration of the first two modes only.

TABLE 18.9. Effective Modal Masses

Mode(r)	M_{er}	*$\sum M_{er}/m_{tot} \times 100$
1	7.44	82.67 %
2	1.24	96.44 %
3	0.24	99.11 %
4	0.08	100 %

$$* \sum_{r=1}^{4} M_{er} = m_{tot} = (9.0) \text{ kip-sec}^2/\text{in}$$

▲

18.4.2 Response Spectrum Analysis

The maximum response of a MDOF system to a prescribed seismic input can be determined in a manner analogous to that employed for a SDOF system, as illustrated in Section 18.3.5. When the equations of motion for MDOF systems are expressed in normal coordinates, they become uncoupled. Hence, the response of the system in any mode r is equivalent to that of an independent SDOF system. Therefore, the total response may be computed by summing the contributions of each individual mode. For example, the contribution of the rth mode to the total displacement of a MDOF system can be expressed as

$$\{x(t)\}_r =$$

$$-\frac{\{\Phi\}_r \Gamma_r}{\omega_r \sqrt{1 - \zeta_r^2}} \int_0^t \ddot{x}_g(\tau) e^{-\zeta_r \omega_r(t-\tau)} \sin \omega_r \sqrt{1 - \zeta_r^2}(t - \tau) d\tau \tag{18.49}$$

The contribution expressed by Eq. (18.49) is directly proportional to $1/\omega_r\sqrt{1 - \zeta_r^2}$ times the integral term. The maximum response will therefore depend on the integral's maximum absolute value. The maximum absolute value of the integral is generally given in terms of displacement or velocity spectra. Thus the maximum displacement response for the rth mode can be determined as

$$|\{x(t)\}_r|_{\max} = |\{\Phi\}_r \Gamma_r|(S_d)_r = |\{\Phi\}_r \Gamma_r|\frac{(S_{pv})_r}{\omega_r}$$

$$= |\{\Phi\}_r \Gamma_r|\frac{(S_{pa})_r}{\omega_r^2} \tag{18.50}$$

where the subscript r on S_d, S_{pv}, and S_{pa} indicates that the spectral values are computed for a natural period $T_r = \omega_r/2\pi$ and damping ζ_r, corresponding to the rth mode of vibration.

An upper bound to the maximum system response (for all modes $r = 1, 2, \ldots, n$) can be obtained by taking the sum of the absolute values of the responses in each mode. This may be expressed as

$$|\{x(t)\}_r|_{\max} = \sum_{r=1}^{n} |\{\Phi\}_r \Gamma_r|(S_d)_r \tag{18.51}$$

Equation (18.45) is neither a rational nor an accurate means for combining the maximum modal responses, because it assumes that the modal maxima occur at the same time and that they also have the same sign. This type of modal combination is generally too conservative for design purposes. A more reasonable method of combining modal maxima, which is based on probability theory, is the square-root-of-the-sum-of-the-squares (SRSS) method. The general expression for the SRSS combination of modal maxima is given by

$$R_{max} = \sqrt{\sum_{r=1}^{p} R_r^2} \tag{18.52}$$

where R_{max} is the representative maximum value for a particular response (e.g., displacement, story shear, etc.), R_r is the peak value of the particular response for the rth mode, and p is the number of significant modes considered in the modal response combination. The truncated maximum displacement response, for example, for a practical MDOF system is expressed as (for $p \ll n$)

$$|\{\hat{x}(t)\}_r|_{max} = \left\{ \sum_{r=1}^{p} \left[|\{\Phi\}_r \Gamma_r| (S_d)_r \right]^2 \right\}^{1/2} \tag{18.53}$$

The SRSS method for combining modal maxima has been shown to render accurate approximations for two-dimensional structural systems exhibiting well-separated vibration frequencies. For three-dimensional systems and/or systems with closely spaced modes, the complete-quadratic-combination (CQC) method [20] renders significant improvement in estimating the response. (Two consecutive modes are defined as closely spaced if their corresponding frequencies differ from each other by 10% or less of the lower frequency.) The CQC combination is expressed as

$$R_{max} = \sqrt{\sum_{r=1}^{p} \sum_{s=1}^{p} R_r P_{rs} R_s} \tag{18.54}$$

where R_r and R_s are the peak values of the particular response for the rth and sth modes, respectively. For constant modal damping (that is, $\zeta = \zeta_1 = \zeta_2 = \ldots = \zeta_p$) the expression for the parameter P_{rs} is given by

$$P_{rs} = \frac{8\zeta^2(1 + \lambda)\lambda^{3/2}}{(1 + \lambda^2)^2 + 4\zeta^2\lambda(1 + \lambda)^2} \tag{18.55}$$

in which $\lambda = \omega_s/\omega_r$.

EXAMPLE 18.7 ▼

Estimate the maximum responses for displacements $\{x\}_{max}$, effective earthquake forces $\{F_s\}_{max}$, story shears $\{V\}_{max}$, and story moments $\{M\}_{max}$ for the four-story shear frame building of Example 18.6 using the Newmark-Hall design spectrum constructed in Example 18.3 for $\zeta = 0.05$. Compare the results with the maximum responses obtained from the time-history analysis conducted in Example 18.6.

Solution

From the design spectrum of Figure 18.21 and the results of Example 18.6, the spectral values, building natural periods and earthquake participation facts are summarized in Table 18.10.

Let the maximum response component (e.g., displacement, story shear, etc.) for each individual mode be represented by R_{ir}, where the subscript r represents the mode and the subscript i represents the location on the structure as illustrated in Figure 18.35. Thus the individual modal displacements, for example, are expressed as

$$x_{ir} = \phi_{ir}\Gamma_r(S_d)_r \tag{1}$$

where ϕ_{ir} is the ith component of the modal vector for the rth mode and $(S_d)_r$ is the spectral displacement for the rth mode. Then the modal displacement vector for the rth mode is given by

$$\{x\}_r = \{\Phi\}_r\Gamma_r(S_d)_r \tag{2}$$

TABLE 18.10. **Summary of Spectral Responses**

Mode	Period T(sec)	S_d(in)	S_{pv}(in/sec)	S_{pa}(in/sec²)	Γ
1	2.17	10.0	28.8	82.94	2.72
2	1.0	4.56	28.8	182.13	1.11
3	0.574	2.63	28.8	315.34	0.48
4	0.502	2.0	25.02	313.0	0.274

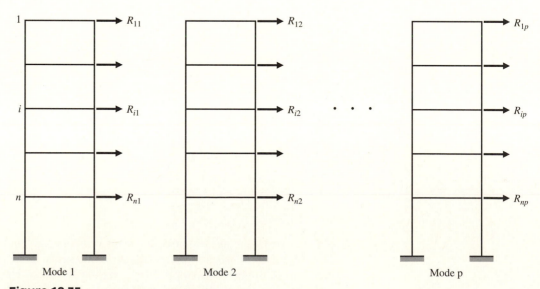

Figure 18.35
Maximum modal response components.

Similarly, the maximum effective earthquake forces for the rth mode are expressed as

$$\{F_S\}_r = [k]\{x\}_r \tag{3}$$

The maximum effective earthquake forces for each mode are illustrated in Figure 18.36. The modal story shears and story moments, respectively, are given by

$$\{V\}_r = [S]\{F_S\}_r \tag{4}$$

and

$$\{M\}_r = [H]\{V\}_r \tag{5}$$

where $[S]$ and $[H]$ are the lower triangular matrices defined by Eqs. (18.46) and (18.47), respectively.

The modal responses $\{R\}_r$ represent the maximum responses occurring in each mode r. Thus, since it is highly unlikely that the maximum response for each mode will occur simultaneously, a statistical combination of the modal maxima, such as SRSS, is generally recommended. The SRSS combination is expressed by Eq. (18.52) and is given as

$$R_{max} = \sqrt{\sum_{r=1}^{p} R_r^2} \tag{6}$$

The individual modal maxima and the corresponding SRSS combinations of modal maxima for all four modes (that is, $p = n = 4$) are summarized in Tables 18.11 through 18.14, respectively, for displacements, effective earthquake forces, story shears, and story moments.

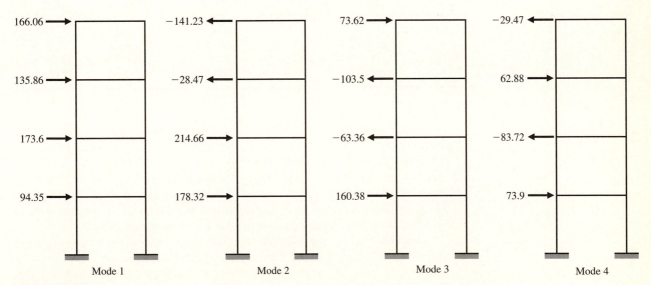

Figure 18.36
Maximum effective earthquake forces F_S (kips).

TABLE 18.11. **Relative Displacement Modal Maxima (in)**

Location	Mode 1	Mode 2	Mode 3	Mode 4	SRSS
x_1	13.21	−2.36	0.41	−0.13	13.42
x_2	10.99	−0.47	−0.57	0.27	11.02
x_3	6.97	1.79	−0.17	−0.18	7.20
x_4	3.80	1.49	0.45	0.16	4.11

TABLE 18.12. **Effective Earthquake Force Modal Maxima (kips)**

Location	Mode 1	Mode 2	Mode 3	Mode 4	SRSS
x_1	166.06	−141.23	73.62	−29.47	231.97
x_2	135.86	−28.47	−103.50	62.88	184.22
x_3	173.60	214.66	−63.36	−83.72	295.37
x_4	94.35	178.32	160.38	73.90	268.11

TABLE 18.13. **Story Shear Modal Maxima (kips)**

Location	Mode 1	Mode 2	Mode 3	Mode 4	SRSS
x_1	166.06	−141.23	73.62	−29.47	231.97
x_2	301.92	−169.71	−29.88	33.41	349.23
x_3	475.52	44.96	−93.24	−50.31	489.25
x_4	569.87	223.28	67.14	23.59	616.18

TABLE 18.14. **Story Moment Modal Maxima (ft-kip)**

Location	Mode 1	Mode 2	Mode 3	Mode 4	SRSS
x_1	1992.67	−1694.80	883.44	−353.62	2783.63
x_2	5615.71	−3731.27	524.88	47.35	6762.86
x_3	11322.00	−3191.81	−594.00	−556.33	11791.42
x_4	18160.49	−512.49	211.68	−273.23	18171.01

Comparisons between the maximum responses obtained from the time-history analysis (response spectrum) conducted in Example 18.6 and the design spectrum results are presented in Table 18.15. It is not surprising that the maximum responses are not in close agreement since the design spectrum is based upon a broad frequency band that is not intended to represent any particular earthquake. This is evident in Figure 18.27 in which the broad-banded design spectrum is compared with the response spectrum for the Northridge earthquake.

However, if one compares the maximum time-history responses with the maximum responses obtained using the Northridge response spectrum shown in Figure 18.27, then a close correlation of results is expected as illustrated in Table 18.16. This comparison verifies the accuracy of the SRSS combination of modal maxima.

TABLE 18.15. Comparison Between Time History and Design Spectra Maxima

Location	Relative Displacement (in)			Effective Earthquake Force (kips)			Story Shear (kips)			Story Moments (ft-kip)		
	Time History	Design Spectra	% difference	Time History	Design Spectra	% difference	Time History	Design Spectra	% difference	Time History	Design Spectra	% difference
x_1	8.74	13.42	53.64	145.63	231.97	59.28	145.63	231.97	59.28	1747.62	2783.63	59.28
x_2	7.07	11.02	55.98	119.80	184.22	53.77	251.24	349.23	39.00	4676.72	6762.86	44.61
x_3	4.36	7.20	65.00	190.50	295.37	55.05	298.14	489.25	64.10	7943.32	11791.42	48.44
x_4	2.44	4.11	68.09	192.17	268.11	39.52	366.57	616.18	68.09	11314.95	18171.01	60.59

TABLE 18.16. Comparison Between Time History and Response Spectra Maxima

Location	Relative Displacement (in)			Effective Earthquake Force (kips)			Story Shear (kips)			Story Moments (ft-kip)		
	Time History	Response Spectra	% difference	Time History	Response Spectra	% difference	Time History	Response Spectra	% difference	Time History	Response Spectra	% difference
x_1	8.74	8.57	−1.97	145.63	148.98	2.30	145.63	148.98	2.30	1747.62	1787.75	2.30
x_2	7.07	7.04	−0.44	119.80	120.99	0.99	251.24	222.59	−11.41	4676.72	4307.35	−7.90
x_3	4.36	4.59	5.24	190.50	188.52	−1.04	298.14	313.26	5.07	7943.32	7521.88	−5.31
x_4	2.44	2.62	7.27	192.17	176.16	8.33	366.57	393.22	7.27	11314.95	11595.69	2.48

18.5 GENERALIZED SDOF SYSTEMS

Any structural system of arbitrary configuration, regardless of its complexity, can be approximated by an equivalent or generalized SDOF system by assuming that its displacement field can be represented by a single shape function. This generalized coordinate approach can be used effectively in earthquake engineering, especially for a preliminary analysis. The success of the procedure, however, depends on an appropriate selection of a shape function $\psi(z)$. This is often a difficult task because the vibration shape depends on the type of loading as well as on the physical characteristics of the system. As discussed in Section 2.4.2, the minimum requirements for $\psi(z)$ are that it satisfy the geometric boundary conditions and that it be differentiable to the order specified in the strain energy expression.

18.5.1 Formulation of the Problem

Consider the arbitrary structure having distributed mass and stiffness properties shown in Figure 18.37. Let x_g represent the ground displacement, x_t represent the total (absolute) displacement of the structure, and x represent the relative displacement of the structure. The equilibrium of the system involves inertia F_I, damping F_D, and elastic restoring forces F_S, distributed along the vertical axis z, and can be expressed as

$$F_I(z, t) + F_D(z, t) + F_S(z, t) = 0 \tag{18.56}$$

The basic assumption underlying the generalized coordinate SDOF system approximation is that the displacements of the system are prescribed by the product of a single shape function $\psi(z)$ and a generalized coordinate amplitude $Y(t)$. Thus, the relative displacement $x(z,t)$ at any location on the structure at any given time t is specified by

$$x(z, t) = \psi(z)Y(t) \tag{18.57}$$

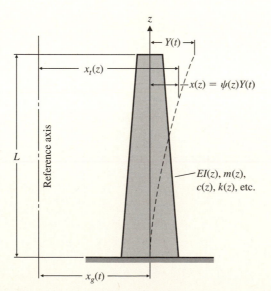

Figure 18.37

Definition sketch of generalized SDOF system.

When a virtual displacement of the form

$$\delta x(z) = \psi(z)\delta Y \tag{18.58}$$

is imparted to the structure, the principle of virtual work leads to the SDOF equilibrium relation given as

$$F_I^* \delta Y + F_D^* \delta Y + F_S^* \delta Y = 0 \tag{18.59}$$

where F_I^*, F_D^*, and F_S^* are the generalized inertia, generalized damping, and generalized elastic restoring forces, respectively, such that

$$F_I^* = \int F_I(z, t)\psi(z)\, dz \tag{18.60}$$

$$F_D^* = \int F_D(z, t)\psi(z)\, dz \tag{18.61}$$

$$F_S^* = \int F_S(z, t)\psi(z)\, dz \tag{18.62}$$

Since the distributed damping and elastic forces depend only on the relative motions of the system, the generalized damping and generalized elastic forces can be expressed as

$$F_D^* = C^* \dot{Y} \tag{18.63}$$

and

$$F_S^* = K^* Y \tag{18.64}$$

in which the generalized damping C^* and the generalized stiffness K^* are given by (referring to Figure 18.38)

$$C^* = \int c(z)[\psi(z)]^2\, dz + \sum_i c_i[\psi(h_i)]^2 \tag{18.65}$$

$$K^* = \int EI(z)[\psi''(z)]^2\, dz + \int k(z)[\psi(z)]^2\, dz + \sum_j k_j[\psi(h_j)]^2 \tag{18.66}$$

In Eq. (18.65), $c(z)$ represents distributed damping, c_i represents localized dashpots, and h_i represents the dashpot locations. In Eq. (18.66), $EI(z)$ represents distributed flexural rigidity, $k(z)$ represents distributed external stiffness, k_j represents localized external springs, and h_j represents the localized spring locations.

The inertia forces depend on the total (absolute) acceleration, $\ddot{x}_t$; thus

$$F_I(z, t) = m(z)\ddot{x}_t(z, t) \tag{18.67}$$

and since

$$\ddot{x}_t(z, t) = \ddot{x}(z, t) + \ddot{x}_g(t) = \psi(z)\ddot{Y}(t) + \ddot{x}_g(t) \tag{18.68}$$

Figure 18.38

Generalized SDOF system also having localized mass, stiffness, and damping.

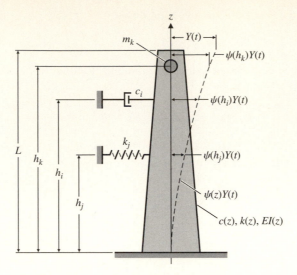

the *generalized inertia force* can be expressed as (refer to Figure 18.38)

$$F_I^* = \ddot{x}(t)\left\{\int m(z)[\psi(z)]^2\,dz + \sum_k m_k\left[\psi(h_k)\right]^2\right\}$$

$$+ \ddot{x}_g(t)\left\{\int m(z)[\psi(z)]^2\,dz + \sum_k m_k\left[\psi(h_k)\right]^2\right\}$$

(18.69)

Substituting the expressions for F_I^*, F_D^*, and F_S^* given by Eqs. (18.69), (18.53), and (18.64), respectively, into the virtual work equation [Eq. (18.59)], results in (noting that $\delta Y \neq 0$)

$$M^*\ddot{Y}(t) + C^*\dot{Y}(t) + K^*Y(t) = -\Gamma^*\ddot{x}_g(t)$$

(18.70)

where C^* and K^* have been defined by Eqs. (18.65) and (18.66), respectively. The generalized mass M^* is given by

$$M^* = \int m(z)[\psi(z)]^2\,dz + \sum_k m_k[\psi(h_k)]^2$$

(18.71)

The quantity Γ^* appearing on the right-hand side of Eq. (18.70) is termed the *earthquake response factor*. The earthquake response factor represents the extent to which the structure responds in the assumed shape $\psi(z)$, and is expressed as

$$\Gamma^* = \int m(z)\psi(z)dz + \sum_k m_k\psi(h_k)$$

(18.72)

In Eqs. (18.69), (18.71), and (18.72), $m(z)$ represents the distributed mass, m_k represents localized masses, and h_k represents the localized mass locations. Ignoring the negative sign in Eq. (18.70) and dividing through by the generalized mass M^* results in the expression

$$\ddot{Y}(t) + 2\zeta\omega\dot{Y}(t) + \omega^2 Y(t) = \frac{\Gamma^*\ddot{x}_g(t)}{M^*} \tag{18.73}$$

By analogy with the discrete SDOF system, the solution to Eq. (18.73) can be expressed as

$$Y(t) = \frac{\Gamma^* R(t)}{M^*\omega} \tag{18.74}$$

where $R(t)$ is the earthquake response integral given by

$$R(t) = \int_0^t \ddot{x}_g(\tau) e^{-\zeta\omega(t-\tau)} \sin \omega\sqrt{1-\zeta^2}(t-\tau) \, d\tau \tag{18.75}$$

Thus, the relative displacements are determined from Eq. (18.57) and are represented as

$$x(z,t) = \frac{\psi(z)\Gamma^*}{M^*\omega}R(t) = \psi(z)Y(t) \tag{18.76}$$

where

$$\omega = \sqrt{\frac{K^*}{M^*}} \tag{18.77}$$

and the system damping is given by

$$\zeta = \frac{C^*}{2M^*\omega} \tag{18.78}$$

18.5.2 Time-History Analysis

In addition to the displacement response given by Eq. (18.76), the time-history responses for the effective earthquake forces $F_S(z,t)$, the earthquake shears $V(z,t)$, and the earthquake moments $M(z,t)$ are also important for design purposes. Again note that the effective earthquake forces are tantamount to the elastic restoring forces that develop in the system, and thus, are essentially equal to the elastic forces specified in Eq. (18.56) that are also represented by the term $F_S(z,t)$. The effective earthquake forces are therefore expressed by (refer to Figure 18.39)

$$F_S(z,t) = m(z)\psi(z)\ddot{Y}_e(t) \tag{18.79}$$

where $\ddot{Y}_e(t)$ is the effective acceleration (or pseudoacceleration) expressed by

$$\ddot{Y}_e(t) = \omega^2 Y(t) \tag{18.80}$$

Substituting the expression for $Y(t)$ given by Eq. (18.74) into Eq. (18.80), Eq. (18.79) can be rewritten as

$$F_S(z,t) = m(z)\psi(z)\frac{\Gamma^*}{M^*}\omega R(t) \tag{18.81}$$

Figure 18.39

Effective earthquake forces for generalized SDOF system.

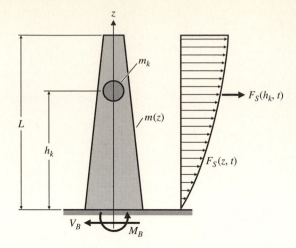

and for the case of a localized mass m_k, the effective earthquake force at that location is given by

$$F_S(h_k, t) = (M_k[\psi(h_k)])\frac{\Gamma^*}{M^*}\omega R(t) \tag{18.82}$$

where h_k represents the location of the localized mass m_k.

The total base shear $V_B(t)$, which is representative of the total seismic force acting on the structure, is obtained by summing the effective earthquake forces acting over the entire structure. Thus

$$\begin{aligned}
V_B(t) &= \int_0^L F_S(z, t)\ dz + \sum_k F_S(h_k, t) \\
&= \frac{\Gamma^*}{M^*}\omega R(t)\left\{\int_0^L m(z)\psi(z)\ dz + \sum_k m_k[\psi(h_k)]\right\} \tag{18.83} \\
&= \frac{(\Gamma^*)^2}{M^*}\omega R(t) = \omega^2 \Gamma^* Y(t)
\end{aligned}$$

Similarly, the base moment $M_B(t)$ is given by

$$\begin{aligned}
M_B(t) &= \int_0^L F_S(z, t)z\ dz + \sum_k F_S(h_k, t)h_k \\
&= \frac{\Gamma^*}{M^*}\omega R(t)\left[\int_0^L m(z)\psi(z)z\ dz + \sum_k m_k[\psi(h_k)]h_k\right] \tag{18.84} \\
&= \omega^2 Y(t)\left[\int_0^L m(z)\psi(z)z\ dz + \sum_k m_k[\psi(h_k)]h_k\right]
\end{aligned}$$

Figure 18.40

Effective earthquake forces acting at an arbitrary section.

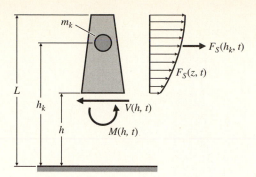

Expressions for the shear $V(h,t)$ and moment $M(h,t)$ at any arbitrary location $z = h$ on the structure (as illustrated in Figure 18.40) are given by

$$V(h, t) = \frac{\Gamma^*}{M^*}\omega R(t)\left[\int_h^L m(z)\psi(z)\ dz + \sum_k m_k\left[\psi(h_k)\right]\right]$$

$$= \omega^2 Y(t)\left[\int_h^L m(z)\psi(z)\ dz + \sum_k m_k\left[\psi(h_k)\right]\right]$$

(18.85)

and

$$M(h, t) = \frac{\Gamma^*}{M^*}\omega R(t)\left[\int_h^L m(z)\psi(z)z\ dz + \sum_k m_k\left[\psi(h_k)\right](h_k - h)\right]$$

$$= \omega^2 Y(t)\left[\int_h^L m(z)\psi(z)z\ dz + \sum_k m_k\left[\psi(h_k)\right](h_k - h)\right]$$

(18.86)

where $h_k > h$.

EXAMPLE 18.8 ▼

Conduct a time-history analysis of the tower structure shown in Figure 18.41 for the N-S component of the Northridge earthquake. The structure has distributed mass and stiffness properties defined by

$$m(z) = 0.2\left(2 - \frac{z}{h}\right)\ \text{kip-sec}^2/\text{ft}^2$$

$$EI(z) = 6 \times 10^7\left(1 - \frac{z}{h}\right)^2\ \text{kip/ft}^2$$

Use the generalized SDOF system approach and assume a shape function $\psi(z) = 1 - \cos(\pi z/2L)$. Also, assume length $L = 150$ ft, localized tip mass $m = 1.5$ kip-sec^2/ft, and $\zeta = 0.05$.

Figure 18.41
Tower structure of Example 18.8.

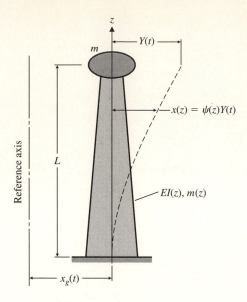

Solution

For the assumed shape function, the pertinent derivatives are

$$\psi'(z) = \frac{\pi}{2L}\sin\frac{\pi z}{2L} \tag{1}$$

$$\psi''(z) = \left(\frac{\pi}{2L}\right)^2 \cos\frac{\pi z}{2L} \tag{2}$$

The generalized stiffness K^* is given by Eq. (18.66) and is expressed as

$$K^* = \int_0^L EI(z)[\psi''(z)]^2 \, dz \tag{3}$$

$$K^* = (6 \times 10^7)\frac{\pi^4}{16L^4}\int_0^L\left(1 - \frac{2z}{L} + \frac{z^2}{L^2}\right)\cos^2\frac{\pi z}{2L}dz \tag{4}$$

$$K^* = 29.03 \text{ kips/ft} \tag{5}$$

The generalized mass M^* is described by Eq. (18.71) and is given by

$$M^* = \int_0^L m(z)[\psi(z)]^2 \, dz + m[\psi(L)] \tag{6}$$

$$M^* = 0.2\int_0^L\left(2 - \frac{z}{L}\right)\left(1 - 2\cos\frac{\pi z}{2L} + \cos^2\frac{\pi z}{2L}\right) dz + m(1.0) \tag{7}$$

$$M^* = 9.54 \text{ kip-sec}^2/\text{ft} \tag{8}$$

The natural circular frequency of the structure is determined from Eq. (18.77),

$$\omega = \sqrt{\frac{K^*}{M^*}} = \sqrt{\frac{29.03}{9.54}} = 1.74 \text{ rad/sec} \tag{9}$$

and the natural period of vibration is

$$T = \frac{2\pi}{\omega} = 3.6 \text{ sec} \tag{10}$$

The earthquake response factor Γ^* is given by Eq. (18.72) as

$$\Gamma^* = \int_0^L m(z)\psi(z) \, dz + m[\psi(L)] \tag{11}$$

or

$$\Gamma^* = 0.2\int_0^L \left(2 - \frac{z}{L}\right)\left(1 - \cos\frac{\pi z}{2L}\right) dz + m(1.0) \tag{12}$$

resulting in

$$\Gamma^* = 15.2 \text{ kip-sec}^2/\text{ft} \tag{13}$$

The equation of motion in terms of the generalized coordinate $Y(t)$ given by Eq. (18.73) is

$$\ddot{Y}(t) + 2\zeta\omega\dot{Y}(t) + \omega^2 Y(t) = \frac{\Gamma^*\ddot{x}_g(t)}{M^*} \tag{14}$$

Equation (14) can be evaluated by any of the numerical integration methods discussed in Chapter 7. Then the relative displacement response $x(z,t)$, at any location z on the structure, as determined from Eq. (18.76) is

$$x(z, t) = \psi(z)Y(t) \tag{15}$$

The relative displacement time history at the top of the structure, $x(L, t)$, is shown in Figure 18.42.

The base shear $V_B(t)$ is given by Eq. (18.83) as

$$V_B(t) = \omega^2\Gamma^* Y(t) \tag{16}$$

and the base moment $M_B(t)$ is given by Eq. (18.84) as

$$M_B(t) = \omega^2 Y(t)\left[\int_0^L m(z)\psi(z)z \, dz + M(\psi)(L)L\right] \tag{17}$$

The response histories for base shear $V_B(t)$ and base moment $M_B(t)$ are shown in Figures 18.43 and 18.44, respectively.

Figure 18.42
Relative displacement
response history at top of
structure $x(L, t)$.

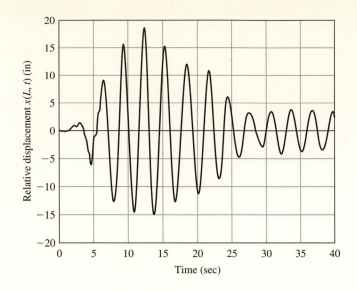

Figure 18.43
Base shear response history
$V_B(t)$.

Figure 18.44

Base moment response history $M_B(t)$.

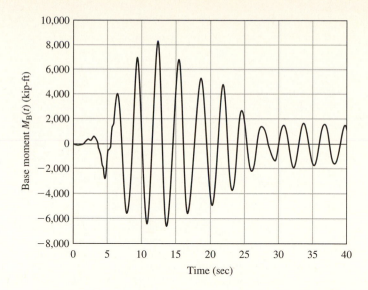

18.5.3 Response Spectrum Analysis

The response spectrum technique may be applied to a generalized SDOF system in a manner analogous to discrete SDOF systems. The maximum relative displacement $[x(z)]_{max}$ at any location on the structure is determined from

$$[x(z)]_{max} = \psi(z)Y_{max} \tag{18.87}$$

where

$$Y_{max} = \frac{\Gamma^*}{M^*\omega}|R(t)|_{max} \tag{18.88}$$

Since

$$|R(t)|_{max} \equiv S_{pv} \tag{18.89}$$

Eq. (18.87) can be expressed as

$$[x(z)]_{max} = \frac{\Gamma^*}{M^*\omega}\psi(z)S_{pv} = \frac{\Gamma^*}{M^*}\psi(z)S_d \tag{18.90}$$

The maximum effective earthquake force at any location on the structure is given by

$$[F_S(z)]_{max} = m(z)\psi(z)\frac{\Gamma^*}{M^*}\omega S_{pv} = m(z)\psi(z)\frac{\Gamma^*}{M^*}S_{pa} \tag{18.91}$$

or, for the case of a localized mass m_k located at h_k,

$$\left[F_S(h_k)\right]_{\max} = \left[m_k\psi(h_k)\right] = \frac{\Gamma^*}{M^*}S_{\text{pa}} \tag{18.92}$$

The maximum base shear force is given by

$$(V_B)_{\max} = \frac{(\Gamma^*)^2}{M^*}\omega S_{\text{pv}} = \frac{(\Gamma^*)^2}{M^*}S_{\text{pa}} \tag{18.93}$$

and the maximum base moment is defined by

$$(M_B)_{\max} = \frac{\Gamma^*}{M^*}S_{\text{pa}}\left[\int_0^L m(z)\psi(z)z \; dz + \sum_k m_k\psi(h_k)h_k\right] \tag{18.94}$$

The maximum shear force at any location $z = h$ on the structure is given by

$$(V_h)_{\max} = \frac{\Gamma^*}{M^*}S_{\text{pa}}\left[\int_h^L m(z)\psi(z) \; dz + \sum_k m_k\psi(h_k)\right] \tag{18.95}$$

and the maximum moment at the same location is determined from

$$(M_h)_{\max} = \frac{\Gamma^*}{M^*}S_{\text{pa}}\left[\int_h^L m(z)\psi(z)z \; dz + \sum_k m_k\psi(h_k)(h_k - h)\right] \tag{18.96}$$

where $h_k > h$.

EXAMPLE 18.9 ▼

Estimate the maximum relative displacement response $x_{\max}$, at $z = L/2$ and $z = L$, as well as $V_{\max}$ and $M_{\max}$, at $z = L/2$ and $z = 0$, for the generalized SDOF structure of Example 18.8 using the Newmark-Hall design spectrum constructed in Example 18.3 with $\zeta = 0.05$. Compare the results with the maximum responses obtained from the time-history analysis of Example 18.8.

Solution

From the design spectrum of Figure 18.21, for a natural period $T = 3.6$ sec, the appropriate spectral values are

$$S_d = 16.0 \text{ in} \qquad S_{\text{pv}} = 27.84 \text{ in/sec} \qquad S_{\text{pa}} = 48.44 \text{ in/sec}^2$$

The maximum relative displacement at the top of the structure is calculated from Eq. (18.90) for $z = L$ as

$$x(L) = \frac{\Gamma^*}{M^*}\psi(L)S_d = \frac{15.2 \text{ kip-sec}^2/\text{ft}}{9.54 \text{ kip-sec}^2/\text{ft}}(1.0)(16.0 \text{ in}) = 25.6 \text{ in} \tag{1}$$

Similarly, the maximum relative displacement at the structure's midheight ($z = L/2$) is

$$x\left(\frac{L}{2}\right) = \frac{\Gamma^*}{M^*}\psi\left(\frac{L}{2}\right)S_d = \frac{(15.2)(0.293)}{9.54}(16.0 \text{ in}) = 7.5 \text{ in} \tag{2}$$

The maximum base shear (i.e., at $z = 0$) is calculated from Eq. (18.93) as

$$V_B = \frac{(\Gamma^*)^2}{M^*}S_{pa} = \frac{(15.2 \text{ kip-sec}^2/\text{ft})^2}{9.54 \text{ kip-sec}^2/\text{ft}}\frac{48.44 \text{ in-sec}^2}{12 \text{ in-ft}} \tag{3}$$

$$= 97.76 \text{ kips}$$

and the maximum shear at midheight ($z = L/2$) is determined from Eq. (18.95),

$$V_{L/2} = \frac{\Gamma^*}{M^*}S_{pa}\left[\int_{L/2}^{L} m(z)\psi(z)\,dz + m\psi(L)\right] \tag{4}$$

$$= \frac{(15.2)(48.44)}{(9.54)(12)}[(0.2)(0.377)(150) + 1.5] = 82.39 \text{ kips}$$

The maximum base moment is expressed by Eq. (18.94) and is given as

$$M_B = \frac{\Gamma^*}{M^*}S_{pa}\left[\int_{h}^{L} m(z)\psi(z)z\,dz + m\psi(L)L\right] \tag{5}$$

$$= \frac{(15.2)(48.44)}{(9.54)(12)}[(0.2)(0.325)(150)^2 + 1.5(150)]$$

$$= 10,853.3 \text{ ft-kip}$$

and the maximum moment at midheight ($z = L/2$) is given by Eq. (18.96) as

$$M_{L/2} = \frac{\Gamma^*}{M^*}S_{pa}\left[\int_{L/2}^{L} m(z)\psi(z)z\,dz + m\psi\left(\frac{L}{2}\right)\left(L - \frac{L}{2}\right)\right] \tag{6}$$

$$= \frac{(15.2)(48.44)}{(9.54)(12)}\left[(0.2)(0.295)(150)^2 + 1.5\frac{150}{2}\right]$$

$$= 9261.5 \text{ ft-kip}$$

A comparison of the maximum responses obtained from the time-history analysis of Example 18.8 with the responses from the design spectrum analysis is presented in Table 18.17. The discrepancy in the results can be understood by noticing the difference in spectral values, between the design spectrum and the Northridge response spectrum illustrated in Figure 18.19, for a period $T = 3.6$ sec. However, if the response spectrum of Figure 18.19 is used ($S_d = 11.5$ in) to calculate the maximum responses, the correlation of results with the time-history analysis is very close (as should be expected), and is illustrated in Table 18.18.

TABLE 18.17. **Comparison of Time-History and Design Spectrum Results**

	Relative Displacement (in)		Shear (kips)		Moment (kip-ft)	
Location on structure	$z = L$	$z = L/2$	$z = 0$	$z = L/2$	$z = 0$	$z = L/2$
Time history	18.78	5.48	71.99	60.94	8026.89	6787.41
Design spectrum	25.6	7.5	97.76	82.39	10853.3	9261.5
Percent difference	36.0	36.0	36.0	35.0	35.0	36.0

TABLE 18.18. **Comparison of Time-History and Response Spectrum Results**

	Relative Displacement (in)		Shear (kips)		Moment (kip-ft)	
Location on structure	$z = L$	$z = L/2$	$z = 0$	$z = L/2$	$z = 0$	$z = L/2$
Time history	18.78	5.48	71.99	60.94	8026.89	6787.41
Response spectrum	18.40	5.39	70.27	59.22	7800.81	6656.7
Percent difference	2.02	1.64	2.4	2.82	2.82	1.93

▲

18.6   IN-BUILDING RESPONSE SPECTRUM

Experience from past earthquakes has demonstrated that mechanical and electrical equipment in buildings subject to earthquakes has often been severely damaged or destroyed, even though the structure itself exhibited very little or no structural damage. It should be realized that in some instances the equipment and structure are of equal importance [21]. This is especially true in the case of nuclear power plants and many industrial installations.

To clarify the discussion, consider a component piece of equipment located on an upper story of a multistory building structure as shown in Figure 18.45. When the base of the building is excited by earthquake motions, the equipment tends to move with the motion $\ddot{x}_{tk}$ of the floor at elevation h_k on which it is supported rather than the base motions of the building itself. That is, the equipment response is dependent on the *in-building* earthquake motions.

Many factors influence the seismic response of equipment in buildings. Several of the more important factors include the intensity and duration of strong ground motion, the type of building structure in which the equipment is housed, the type of connection between the equipment and the building structure (i.e., rigid, resilient, etc.), and the relative height of the equipment in the building. For example, the response of equipment located on very rigid structures is very similar to that of equipment located directly on ground, and is primarily a function of the frequency content of the postulated earthquake. However, the response of equipment located on relatively flexible structures is primarily a function of the structure's natural frequencies. In a flexible structure, the earthquake motions are filtered and amplified at the building's natural frequencies.

Figure 18.45

Equipment located in a multistory building subject to earthquake ground acceleration.

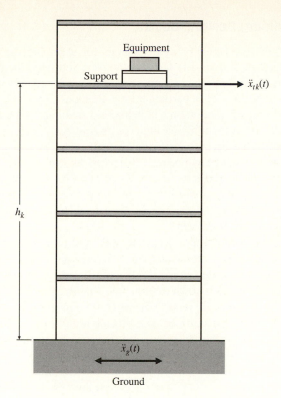

18.6.1 In-Building Motion Criteria

To develop earthquake design criteria for equipment, it is essential to establish the characteristics of potential earthquake-induced floor motions at all equipment locations within the building. The primary concern is for equipment and structural subsystems within buildings having fundamental frequencies less than approximately 5 Hz, because most of the energy in an earthquake accelerogram is confined to this band [22]. Generally, in-building earthquake motions of a very rigid structure having a fundamental frequency greater than such frequencies are approximately the same as the free-field ground motions. However, buildings having lower fundamental frequencies exhibit increased sensitivity to the ground motion and experience upper-story motions that may be several times larger than at the building base.

Equipment that is idealized as either a SDOF system or a MDOF system can be analyzed for seismic response by incorporating it into the main structural model. This approach requires no further analysis since the results of the analysis of the structure-equipment model also provide the equipment response. However, considering the multitude of equipment (for some installations) that must be included in the overall structure-equipment model, this approach is seldom employed because the computational effort is unjustified and/or the accuracy of the problem becomes intractable [23]. Moreover, for most equipment located in structures, the equipment-structure interaction effects can be neglected due to the large disparity in mass and stiffness between the equipment and the structure.

18.6.2 In-Building Design Spectra

One of the most widely used methods for assessing the dynamic seismic forces on equipment in buildings is the development and implementation of in-building design spectra. An in-building design spectrum is created in much the same manner as a design spectrum for a particular site. Response spectrum and then design spectrum curves for any elevation on a structure can be obtained by exciting the structure with a specified site-compatible earthquake ground accelerogram. The responses at selected locations in the structure can then be evaluated. The response is generally an acceleration response, which can be regarded as an "in-building" earthquake occurring at a specific location in the structure. This in-building response accelerogram can be used either directly as a time-history base excitation for the equipment at that location, or response spectrum curves may be developed for that location. The latter approach is most frequently employed.

A typical in-building response spectrum is presented in Figure 18.46. The sharp peak corresponds to the fundamental frequency of the supporting structure. The structure acts as a filter that significantly amplifies the response only at those frequencies in close proximity to its own natural frequencies. Since the frequencies of the structure cannot be determined precisely because of simplifying assumptions made in the formulation of the analytical model (e.g., localizing of masses, idealization of stiffness properties, etc.), the resulting in-building response spectrum curve should be modified by a shifting of the peak responses, as illustrated in Figure 18.47, to compensate for the inaccuracies in the frequency calculations. The amount of shift on either side of the original curve can be estimated from parametric studies conducted on the important variables (i.e., mass and stiffness properties) in the structural model.

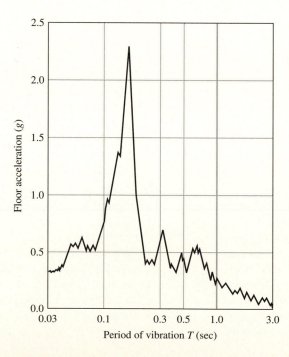

Figure 18.46

Typical in-building response spectrum. "Earthquake Forces on Equipment in Nuclear Power Plants" by A.H. Hadjian, ASCE *Journal of the Power Division,* Vol. 97, No. P03, 1971. Reprinted by permission of the American Society of Civil Engineers.

Figure 18.47

In-building response spectrum with horizontal shifting of peak response. "Earthquake Forces on Equipment in Nuclear Power Plants" by A.H. Hadjian, ASCE *Journal of the Power Division,* Vol. 97, No. P03, 1971. Reprinted by permission of the American Society of Civil Engineers.

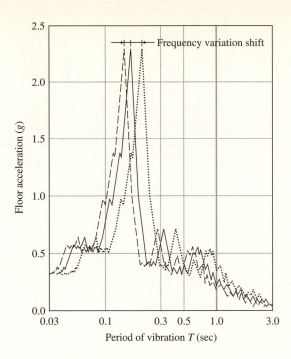

Design spectrum curves can subsequently be obtained from the augmented (peak-shifting) response spectrum curves. Indeed, the large peak amplifications exhibited by typical in-building design spectra should serve as a warning to equipment designers to avoid these frequency bands whenever possible.

EXAMPLE 18.10 ▼

A sensitive piece of equipment is to be supported on the third level of the four-story shear frame building of Example 18.6 shown in Figure 18.29. (a) Construct an in-structure response spectrum (for the Northridge earthquake) at the third level of the building for $\zeta = 0.02$. From this response spectrum construct the augmented design spectrum considering a 10% horizontal peak shift. (b) The equipment cannot withstand an acceleration greater than 0.8 g. Determine the acceptable range of stiffness for the equipment support k_s such that the acceleration limit is not violated. Assume equipment weights W_e of 5, 10, and 15 kips.

Solution

(a) In-building response spectrum:

The total acceleration response of the third level $\ddot{x}_{t2}(t)$ of the four-story building structure subjected to the Northridge earthquake is shown in Figure 18.48. Using this accelerogram as input, an acceleration response spectrum for $\zeta = 0.02$ can be generated by the procedure discussed in Section 18.2. The in-building acceleration response spectrum for the third level of the structure is shown in Figure 18.49. The augmented design spectrum with a 10% horizontal shift of the maximum response is shown in Figure 18.50.

Figure 18.48

Absolute (total) acceleration response history $\ddot{x}_{t2}(t)$ for third story of four-story shear frame building.

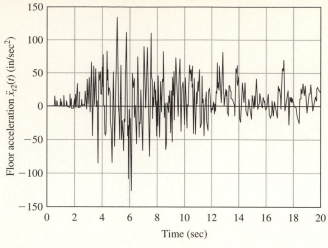

Figure 18.49

In-building acceleration response spectrum for third story of four-story building ($\zeta = 0.02$).

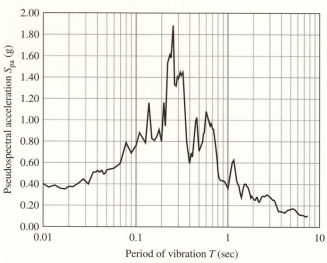

Figure 18.50

Augmented acceleration design spectrum for third story of four-story building ($\zeta = 0.02$).

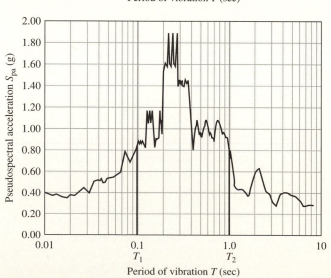

TABLE 18.19. Acceptable Ranges for Equipment Support Stiffness k_s

Equipment Weight W_e (kips)	k_s (kips/in)	
	Less Than	Greater Than
5	0.463	53.2
10	0.926	106.4
15	1.39	159.6

(b) Equipment response:

The natural period of the equipment-support system is given by

$$T = 2\pi\sqrt{\frac{W_e}{gk_s}} \qquad (1)$$

where W_e is the weight of the equipment and k_s is the stiffness of the equipment support. From Eq. (1), k_s may be expressed as

$$k_s = \frac{4\pi^2}{g}\frac{W_e}{T^2} \qquad (2)$$

To maintain the acceleration of the equipment under 0.8 g, the natural period of the equipment-support system must be less than T_1 (0.1 sec) or greater than T_2 (1.05 sec), as shown in Figure 18.50. With this information the appropriate ranges for k_s may be determined from Eq. (2). The results for k_s are summarized in Table 18.19.

▲

18.7 INELASTIC RESPONSE

Past experience has indicated that many structures subjected to severe earthquake ground shaking undergo deformations beyond their elastic limit. Indeed, many structures designed in accordance with the equivalent lateral force procedures specified in most seismic design codes are expected to exhibit limited inelastic behavior during an extreme seismic event. That is, the recommended static design forces in seismic design codes, which the structure must resist elastically, represent only a fraction of the effective (elastic) earthquake forces associated with the design earthquake. Therefore, the inelastic response of structures during severe ground shaking represents a critical aspect of earthquake engineering. For this reason, some basic concepts of inelastic earthquake response are discussed in this section.

18.7.1 Elastoplastic Response

The inelastic resistance-displacement relationship exhibited by most ductile structures is illustrated in Figure 18.51 by the dashed line. However, for convenience and to simplify the response calculations, the actual resistance-displacement relationship is often idealized as a linear elastic—perfectly plastic (or elastoplastic) representation as shown by the solid line in Figure 18.51. The elastoplastic approximation to the actual curvilinear resistance-displacement curve is drawn so that the areas under both curves are the same at the effective yield displacement x_y and at the selected value of the maximum permissible displacement x_m. With this type of force-displacement relation, the maximum restoring force exhibited by the system is $F_S = F_{Sy}$, corresponding to the yield strength. Yielding in the structure initially occurs when the displacement reaches x_y, and any increase in displacement beyond x_y occurs at a constant force F_{Sy}.

Due to the cyclic nature of earthquake ground motion, structures experience successive loading and unloading during a seismic event. The subsequent elastoplastic force-displacement relationship exhibits a sequence of loops known as *hysteresis loops*. A typical hysteresis loop for an idealized elastoplastic system is illustrated in Figure 18.52. A detailed discussion of the various stages of loading and unloading as defined by the hysteresis loop of Figure 18.52 is presented in Section 15.5. Most importantly, the hysteresis loop is a measure of the structure's energy dissipation capacity, which in turn is related to the structure's ductility.

The ductility μ of a structure is defined as the ratio of the maximum permissible displacement to the yield displacement. A comparison of the elastoplastic force-displacement relation to the purely linear elastic case is illustrated in Figure 18.53. The ductility factor for the elastoplastic system is thus expressed as

$$\mu = \frac{x_m}{x_y} \tag{18.97}$$

We also notice from this figure that the ductility factor represents the ratio of the force F_m that would have occurred under the specified ground motion, if the structure had displaced elastically to the same maximum value x_m to the force F_{Sy} that caused effec-

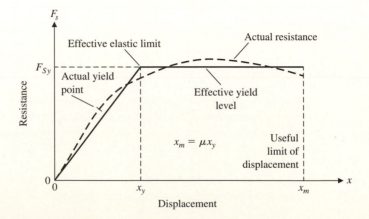

Figure 18.51

Inelastic resistance-displacement relationship.

tive yield. In most instances it would be uneconomical to design buildings to remain elastic when subjected to earthquakes of high intensity, but low probability, since it is well known that structures have considerable reserves of strength beyond the point at which first yield occurs. This is the reasoning behind the reduced elastic design forces recommended in seismic design codes, which are discussed in Section 18.8.

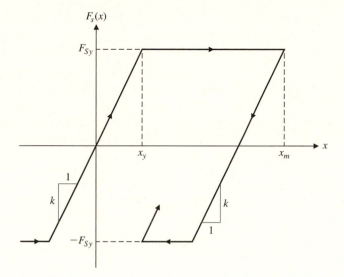

Figure 18.52

Idealized elastoplastic resistance-displacement relationship for cyclic loading.

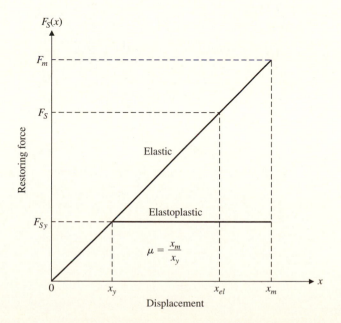

Figure 18.53

Comparison of elastic and elastoplastic restoring forces.

18.7.2 Inelastic Design Spectra

The analysis of the response of elastoplastic systems was discussed in Chapter 15. The governing equation of motion for a SDOF elastoplastic system excited by a ground acceleration $\ddot{x}_g(t)$ is given by

$$m\ddot{x} + c\dot{x} + F_s(x) = -m\ddot{x}_g(t) \tag{18.98}$$

or

$$\ddot{x} + 2\zeta\omega\dot{x} + \frac{F_s(x)}{m} = -\ddot{x}_g(t) \tag{18.99}$$

where $F_s(x)$ is the elastoplastic restoring force characterized by the resistance-displacement relationship shown in Figure 18.52. The time domain solution of this equation for any specified set of parameters and prescribed earthquake excitation can be obtained using the step-by-step integration procedures discussed in Chapter 15. However, for design purposes, the maximum system response is of primary interest, for which inelastic response spectra and design spectra are very useful and informative.

Studies conducted by Housner [24] and Blume [25] resulted in development of inelastic response spectra similar to the one shown in Figure 18.54. From observation of such response spectra, it was concluded [26] that: (1) the elastic and inelastic systems exhibit the same total displacement in the low-frequency (long-period) range; (2) in the intermediate-frequency (period) range, the elastic and inelastic systems absorb the same total energy; and (3) in the high-frequency (small-period) region, the elastic and inelastic systems have the same restoring force. In terms of the system parameters, these observations may be summarized as:

1. Low-frequency (long-period) range

$$x_{el} \cong x_{ep} \tag{18.100}$$

$$F_{Sy} \cong \frac{F_S}{\mu} \tag{18.101}$$

2. Intermediate-frequency (period) range

$$x_{ep} = \frac{\mu}{\sqrt{2\mu-1}} x_{el} \tag{18.102}$$

$$F_{Sy} = \frac{1}{\sqrt{2\mu-1}} F_S \tag{18.103}$$

3. High-frequency (small-period) range

$$F_{Sy} \cong F_S \tag{18.104}$$

$$x_{ep} \cong \mu x_{el} \tag{18.105}$$

where x_{el} = elastic displacement
x_{ep} = elastoplastic displacement
F_S = elastic restoring force
F_{Sy} = elastoplastic or yield restoring force

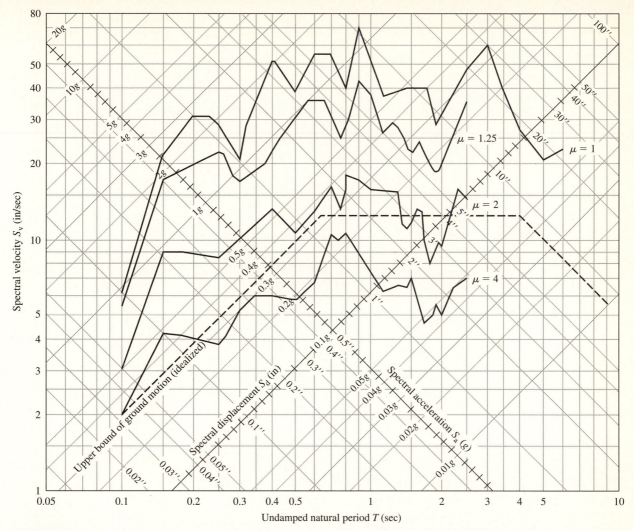

Figure 18.54

Comparison of elastic and elastoplastic restoring forces. *Design of Multistory Reinforced Concrete Buildings for Earthquake Motions* by J.A. Blume, N.M. Newmark, and L.H. Corning, Portland Cement Association, Skokie, IL 1969.

Based upon these conclusions, Newmark [26] recommended a procedure for constructing inelastic design spectra from elastic spectra by factoring the ordinates of the elastic spectra by appropriate coefficients depending on the ductility μ. Referring to Figure 18.55, the procedure is summarized as follows:

1. For the specified ground motion maxima and damping value, construct the elastic broad-band spectra as discussed in Section 18.3.2. The elastic spectrum is represented by the line A_0QPAVD, where A_0 is the maximum ground acceleration.

Figure 18.55

Construction of inelastic acceleration and inelastic total displacement spectra from the elastic spectrum.

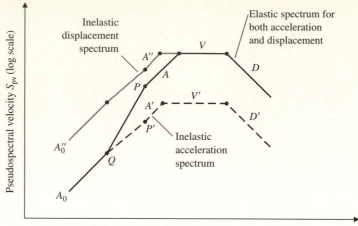

2. Draw the lines D' and V' parallel to the lines D and V by dividing the ordinates of lines D and V by μ.

3. Divide the ordinate of the point P by $\sqrt{2\mu - 1}$ to locate the point P'.

4. From the point P', draw a line A' at 45° until it intersects the line V'.

5. Join points P' and Q to complete the spectrum for inelastic accelerations. The line $A_0A'V'D'$ is the bound for the elastoplastic spectrum for acceleration.

6. Draw the line segments A'' and A_0'' obtained by multiplying the ordinates of the lines A' and A_0, respectively, by μ. The boundary $A_0''\ A''\ VD$ represents the elastoplastic spectrum for displacement.

EXAMPLE 18.11 ▼

(a) Construct an inelastic design spectrum for an earthquake having a maximum ground acceleration of 0.308 g (corresponding to the elastic design spectrum of Example 18.3) for a ductility factor of $\mu = 4$ and damping factor $\zeta = 0.05$. (b) For a structure having a natural period $T = 0.2$ sec and mass m, determine the maximum displacement of the elastic and inelastic systems, and the elastic and inelastic design forces.

Solution

(a) Following the procedure previously outlined in this section and referring to Figure 18.56:
   (i)   Construct the elastic spectrum as illustrated in Figure 18.56.

Figure 18.56

Construction of inelastic design spectrum for Example 18.11 ($\zeta = 0.05$, $\mu = 4$).

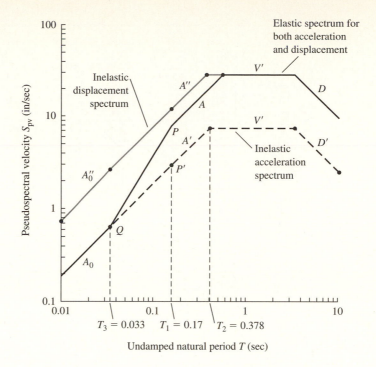

(ii) Establish the inelastic bounds D' and V' by dividing the elastic bounds D and V by $\mu = 4$.

(iii) Locate point P' by dividing the ordinate of P (8.19 in/sec) on the elastic spectrum (at $T_1 = 0.17$ sec) by $\sqrt{2\mu - 1}$. The resulting ordinate is 3.09 in/sec.

(iv) From point P' draw a line upward to the right at 45° until it intersects the horizontal line V' at $T_2 = 0.378$ sec.

(v) Then draw a straight line A' from point P' downward to the left until it intersects point A_0 at $T_3 = 0.033$ sec.

(vi) Complete the inelastic spectrum by drawing the line segments A'' and A_0'', which are determined by multiplying the ordinates of lines A' and A_0, respectively, by μ.

The combined elastic and inelastic spectra are plotted on four-way logarithmic paper in Figure 18.57.

(b) For $T = 0.2$ sec, the appropriate spectral values from the elastic and inelastic spectra are summarized in Table 18.20. The maximum displacements for the systems are defined by the spectral displacement S_d, and the design seismic force is equal to the product of the structure's mass and spectral acceleration (that is, $F_S = mS_{pa}$). Therefore, although the inelastic structure experiences a two-thirds increase in displacement over the elastic system, the maximum design force for the inelastic system is approximately one-third that for the elastic system.

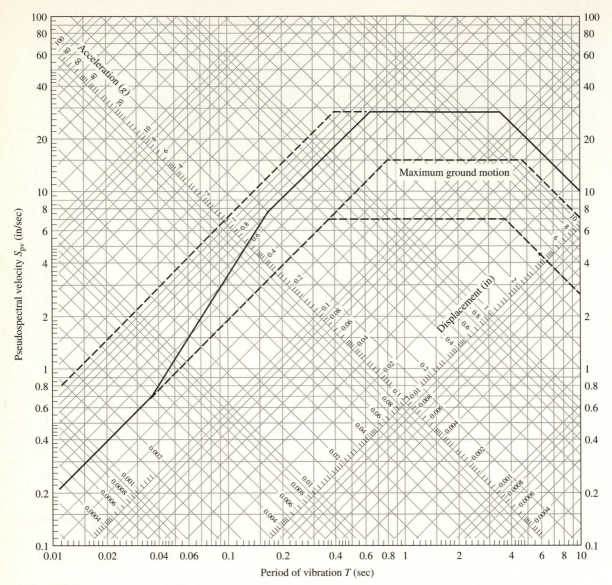

Figure 18.57
Inelastic design spectrum ($\zeta = 0.05$, $\mu = 4$).

TABLE 18.20. **Comparison of Elastic and Inelastic Structures**

System	S_d (in)	S_{pa} (g)	$F_S = mS_{pa}$
Elastic	0.315	0.8	0.8 mg
Inelastic	0.525	0.3	0.3 mg

▲

The use of inelastic spectra for design and analysis purposes has been limited primarily to systems that can be modeled as SDOF systems [6]. However, such analysis of simplified SDOF models can lend valuable insight to understanding the inelastic response of more complicated structures during an extreme seismic event.

18.8 SEISMIC DESIGN CODES

The seismic analysis and design of most "regular" building structures is done in accordance with seismic design procedures prescribed in various building codes. All seismic design codes offer the equivalent-static-lateral-force (ELF) procedure, but some also prescribe a dynamic analysis procedure. The dynamic analysis procedures described in seismic codes are, in essence, very similar to the methodologies discussed in the previous sections of this chapter.

Seismic design procedures appear in a multiplicity of codes and standards, which can be grouped into three basic categories:

1. Model building codes
2. Industry standards
3. Miscellaneous codes and standards

Of these groups, the model building codes are the most widely used or referenced. Except for some of the largest cities, drafting of building codes in the United States is done by the model code organizations, with the model codes intended for adoption by local governments. In larger jurisdictions, the model codes usually serve as a basis for a completely independent code. At the present time there are four recognized model codes that affect structural design:

1. The Uniform Building Code [27]
2. The Standard Building Code [28]
3. The BOCA Basic Building Code [29]
4. The National Building Code [30]

The Uniform Building Code (UBC) contains the most comprehensive seismic design provisions of the model codes. Moreover, the seismic design provisions contained in the UBC are adopted from the SEAOC Code [19] with some minor modifications.

18.8.1 ELF Procedure in the UBC

The concept of equivalent static lateral force (ELF) is somewhat misleading and often misunderstood [31]. The static lateral design forces specified in the seismic design codes do not represent the maximum dynamic forces expected to be exerted on a structure during a major earthquake. These design seismic forces are "equivalent" to the actual earthquake forces in the context that a structure designed to resist such forces (elastically) should be able to:

1. Resist minor earthquakes without damage (neither structural nor nonstructural)
2. Resist moderate earthquakes without structural damage, but possibly experience some nonstructural damage
3. Resist a major earthquake without collapse, but possibly with some structural damage as well as nonstructural damage

The code-specified design forces are based on the assumption that a significant amount of inelastic behavior may occur in the structure during a major earthquake. Thus the code design forces and associated elastic deformations are much lower than those that would occur if the structure were to remain elastic. Indeed, the design should account for stress reversals, provide adequate member ductility, and provide connections of sufficient strength and resilience.

The determination of the seismic design forces by the ELF procedure consists of two basic steps: (1) calculation of the total base shear and (2) distribution of the total seismic (base shear) force along the height of the structure. The ELF procedure exhibits features drawn from structural dynamics theory and also includes consideration of regional seismicity, geology and soil conditions, and the anticipated importance of the building and its occupancy. However, it must be noted that use of the ELF procedure is restricted primarily to "regular" structures. A regular structure is one that exhibits reasonably uniform distribution of mass and stiffness, both in height and in plan. Structures not meeting this criterion are "irregular," for which a dynamic analysis must be conducted, but for a few exceptions.

18.8.2 Dynamic Lateral Force (DLF) Procedures

The distribution of seismic forces prescribed by the ELF procedure is based on assumptions that are applicable only to regular structures. Dynamic effects resulting from unusual mass or stiffness distributions in a building are not accounted for in the ELF procedure. Therefore, for irregular structures and regular structures exceeding 240 ft in height, a DLF procedure is required by the UBC.

The notion of structural irregularity is divided into two categories: (1) plan structural irregularities and (2) vertical structural irregularities. Plan irregular buildings include those that undergo substantial torsion when subject to seismic loads, or that have reentrant corners, discontinuities in floor diaphragms, discontinuity in the lateral force path, or lateral load resisting elements that are not parallel. Vertical irregularities include soft or weak stories, large changes in mass between adjacent floors, and large discontinuities in the dimensions or in-plane locations of lateral load-resisting elements from story to story.

The two DLF procedures outlined in the UBC are: (1) response spectrum analysis and (2) time-history analysis. These procedures incorporate dynamic aspects of seismic response into the design process. The DLF procedures presented in the UBC were developed to address the fact that the distribution of seismic forces in some structures is often considerably different from that defined by the ELF procedure. The guidelines for the DLF procedures provide the engineer with a method for dealing with structures that violate the assumptions inherent in the use of the ELF procedures. Implementation of either of the UBC DLF procedures will be similar to the methodologies described in Sections 18.1, 18.4.1, and 18.5.2 for a time-history analysis or Sections 18.3.5, 18.4.2, and 18.5.3 for a response spectrum analysis.

It is anticipated that the majority of DLF analyses conducted will be response spectrum analyses. The time-history-analysis procedure is recommended for use in situations where it is important to represent inelastic response characteristics or to incorporate time-dependent effects when evaluating the dynamic response of a structure.

REFERENCES

1 Clough, R.W. and Penzien, J., *Dynamics of Structures,* 2nd ed., McGraw-Hill, New York, 1993.

2 Di Julio, R.M., "Static Lateral-Force Procedure," Chapter 4, *The Seismic Design Handbook,* F. Naeim, editor, Van Nostrand Reinhold, New York, 1989.

3 Anderson, J.C., "Dynamic Response of Buildings," Chapter 3, *The Seismic Design Handbook,* F. Naeim, editor, Van Nostrand Reinhold, New York, 1989.

4 Hudson, D.E., "Some Problems in the Application of Spectrum Techniques to Strong Motion Earthquake Analysis," *Bulletin of the Seismological Society of America,* Vol. 52, 1962, pp. 417–430.

5 Newmark, N.M. and Hall, W.J., *Earthquake Spectra and Design,* Earthquake Engineering Research Institute, Berkeley, CA, 1982.

6 Mohraz, B. and Elghadamsi, F.E., "Earthquake Ground Motion and Response Spectra," Chapter 2, *The Seismic Design Handbook,* F. Naeim, editor, Van Nostrand Reinhold, New York, 1989.

7 Housner, G.W., "Design Spectrum," Chapter 5, *Earthquake Engineering,* R.L. Weigel, editor, Prentice Hall, Englewood Cliffs, NJ, 1970.

8 Housner, G.W., "Behavior of Structures During Earthquakes," *Journal of the Engineering Mechanics Division,* ASCE, Vol. 85, No. EM4, 1959, pp. 109–129.

9 Blume, J.A., Newmark, N.M., and Corning, L.H., *Design of Multistory Reinforced Concrete Buildings for Earthquake Motions,* Portland Cement Association, Old Orchard Rd. Skokie, IL, 1961.

10 Newmark, N.M. and Hall, W.J., "Procedures and Criteria for Earthquake Resistant Design," *Building Practices for Disaster Mitigation,* U.S. Department of Commerce, Building Research Series, 46, 209–236, 1973.

11 Newmark, N.M. and Hall, W.J., "Seismic Design Criteria for Nuclear Reactor Facilities," *Proceedings of the 4th World Conference on Earthquake Engineering,* B4, Santiago, Chile, 1969, pp. 37–50.

12 Hagashi, S., Tsuchida, H., and Kurata, E., "Average Response Spectra for Various Subsoil Conditions," *Third Joint Meeting, U.S.–Japan Panel on Wind and Seismic Effects,* UNJR, Tokyo, 1971.

13 Kuribayashi, E., Iwasaki, T., Iida, Yi, and Tuji, K., "Effects of Seismic and Subsoil Conditions on Earthquake Response Spectra," *Proceeding of the International Conference on Microzonation,* Seattle, WA, 1972, pp. 499–512.

14 Mohraz, B., Hall, W.J., and Newmark, N.M., *A Study of Vertical and Horizontal Earthquake Spectra,* Nathan N. Newmark Consulting Engineering Services, Urbana, IL, AEC Report WASH-1255, 1972.

15 Hall, W.J., Mohraz, B., and Newmark, N.M., *Statistical Studies of Vertical and Horizontal Earthquake Spectra,* Nathan N. Newmark Consulting Engineering Services, Urbana, IL, 1975.

16 Algermissen, S.T. and Perkins, D.M., "A Technique for Seismic Risk Zoning, General Considerations and Parameters," *Proceedings of the Microzonation Conference,* Seattle, WA, 1972, pp. 865–877.

17 Algermissen, S.T. and Perkins, D.M., *A Probabilistic Estimate of Maximum Acceleration in Rock in Contiguous United States,* USGS Open File Report, 76–416, 1976.

18 Mohraz, B., "A Study of Earthquake Response Spectra for Different Geological Conditions," *Bulletin of the Seismological Society of America,* Vol. 66, No. 3, 1976, pp. 915–935.

19 *Recommended Lateral Force Requirements and Commentary,* Seismology Committee of the Structural Engineers Association of California (SEAOC), Sacramento, CA, 1990.

20 Wilson, E.L., Der Kiureghian, A., and Bayo, E.P., "A Replacement for the SRSS Method in Seismic Analysis," *Earthquake Engineering and Structural Dynamics,* Vol. 9, 1981.

21 Jordan, C.H., "Seismic Restraint of Equipment in Buildings," *Journal of the Structural Division,* ASCE, Vol. 104, No. ST5, May 1978, pp. 829–839.

22 Liu, S.C., Fagel, L.W., and Dougherty, M.C., "Earthquake-Induced-In-Building Motion Criteria," *Journal of the Structural Division,* ASCE, Vol. 103, No. ST1, January 1977, pp. 133–152.

23 Hadjian, A.H., "Earthquake Forces on Equipment in Nuclear Power Plants," *Journal of the Power Division,* ASCE, Vol. 97, No. PO3, July 1971, pp. 649–665.

24 Housner, G.W., "Limit Design of Structures to Resist Earthquakes," *Proceedings of the First World Conference on Earthquake Engineering,* Berkeley, CA, 1956, pp. 5-1 to 5-13.

25 Blume, J.A., "A Reserve Energy Technique for the Earthquake Design and Rating of Structures," *Proceedings of the Second World Conference on Earthquake Engineering,* Vol. II, Tokyo, 1960, pp. 1061–1084.

26 Newmark, N.M., "Current Trends in the Seismic Analysis and Design of High-Rise Structures," Chapter 16, *Earthquake Engineering,* R.L. Wiegel, editor, Prentice Hall, Englewood Cliffs, NJ, 1970.

27 *Uniform Building Code,* International Conference of Building Officials, Whittier, CA, 1997.

28 *Standard Building Code,* Southern Building Code Congress, Birmingham, AL, 1994.

29 *BOCA Basic Building Code,* Building Officials and Code Administrators International, Howewood, IL, 1994.

30 *National Building Code,* American Insurance Associates, New York, 1994.

31 Berg, G.Y., *Seismic Design Codes and Procedures,* Earthquake Engineering Research Institute, Berkeley, CA, 1982.

32 Seed, H.B., Ugas, C., and Lysmer, J., "Site-Dependent Spectra for Earthquake-Resistance Design," *Bulletin of the Seismological Society of America,* Vol. 66, No. 1, 1976, pp. 22–243.

33 Applied Technology Council, National Bureau of Standards and National Science Foundation, "Tentative Provisions for the Development of Seismic Regulations for Buildings," ATC Publication 3-06, NBS Publication 510, NSF Publication 78-8, 1978.

NOTATION

A	cross-sectional area		k	stiffness
c	viscous damping coefficient		K^*	generalized stiffness
C^*	generalized damping		K_r	modal stiffness for rth normal mode
C_t	coefficient for estimating natural period of buildings		k_s	stiffness of equipment support
E	modulus of elasticity		L	length
F_D	damping force		m	mass
F_D^*	generalized damping force		M	base moment, story moment, or bending moment
F_I	inertia force		M_{er}	effective mass for rth normal mode
F_I^*	generalized inertia force		M_p	plastic moment
F_S	elastic restoring force; also represents effective earthquake force		M_r	modal mass for rth normal mode
F_S^*	generalized elastic restoring force		M_x	design story moment at story x
F_{Sy}	elastoplastic (yield) restoring force		P_{rs}	parameter used in CQC method
$F_s(z,t)$	effective earthquake forces for generalized SDOF		$P_{eff}(t)_r$	effective modal force for rth normal mode
$F(t)$	externally applied force		q_r	modal displacement for rth normal mode
$F_{eff}(t)$	effective load for ground acceleration		$\dot{q}_r$	modal velocity for rth normal mode
g	acceleration due to gravity		$\ddot{q}_r$	modal acceleration for rth normal mode
h	height		R_{max}	maximum value of a particular response obtained by combining the corresponding peak modal responses R_r
h_k	height above base to story k		R_r	peak value of a particular response for rth normal mode
I	static moment of inertia of cross-section		$R(t)$	earthquake response integral

S_a	spectral acceleration		SDOF	Single-Degree-of-Freedom
S_d	spectral displacement		SEAOC	Structural Engineers Association of California
S_{pa}	pseudospectral acceleration		SRSS	square root of the sum of the squares
S_{pv}	pseudospectral velocity		UBC	Uniform Building Code
S_v	spectral velocity		$[c]$	viscous damping matrix
t	time		$[C]$	modal damping matrix
T	undamped natural period		$\{F_{eff}(t)\}$	effective load vector for ground acceleration
V	base shear or story shear		$\{F_S(t)\}$	effective earthquake force vector
W_e	weight of equipment		$[H]$	transformation matrix for story moments
x	relative displacement		$\{I\}$	unit vector
x_{el}	elastic limit displacement		$[k]$	stiffness matrix
x_{ep}	elastoplastic displacement		$[K]$	modal stiffness matrix
$x_g(t)$	ground displacement		$[m]$	mass matrix
x_m	maximum permissible displacement in an elastoplastic system		$[M]$	modal mass matrix
x_t	total (absolute) displacement		$\{M(t)\}$	story moment vector
x_y	effective yield displacement		$\{P_{eff}(t)\}$	effective modal force vector
$x(z,t)$	relative displacement for generalized SDOF		$\{q\}$	normal (or principal) coordinates vector
$\dot{x}$	relative velocity		$[S]$	transformation matrix for story shears
$\ddot{x}$	relative acceleration		$\{V(t)\}$	story shear vector
$\ddot{x}_g(t)$	ground acceleration		$\{x\}$	relative displacement vector
$\ddot{x}_t$	total (absolute) acceleration		$\{\dot{x}\}$	relative velocity vector
$\ddot{x}_{tk}$	absolute acceleration of kth story of a multistory building		$\{\ddot{x}\}$	relative acceleration vector
$Y(t)$	displacement amplitude of generalized coordinate $z(t)$		$\{\ddot{x}_t\}$	total (absolute) acceleration vector
$\ddot{Y}_e(t)$	effective acceleration (pseudoacceleration) of generalized coordinate $z(t)$		$\{\hat{x}(t)\}$	truncated modal displacement response for $p < n$
$z(t)$	generalized coordinate		δx	virtual displacement
BOCA	Building Officials Conference of America		δy	virtual displacement of generalized coordinate
CQC	complete quadratic combination		Γ_r	earthquake participation factor for rth normal mode
DLF	dynamic lateral force		Γ^*	earthquake response factor for generalized SDOF system
ELF	equivalent static lateral force			
EPA	effective peak ground acceleration		λ	ratio of two consecutive natural frequencies
EPV	effective peak velocity-related ground acceleration		μ	ductility ratio
			ω	undamped natural circular frequency
E-W	east-west		ζ	damping factor
MDOF	Multi-Degree-of-Freedom		ζ_r	damping factor for rth normal mode
PGA	peak ground acceleration		$\psi(z)$	shape function
N-S	north-south		$[\Phi]$	modal matrix
			$\{\Phi\}_r$	modal vector for rth normal mode

PROBLEMS

Note: For many of these problems the N-S ground acceleration component of the 1994 Northridge, CA, earthquake is the required input. The digitized accelerogram is available on the authors' web site cited in the Preface. The solution to many of these problems can be greatly facilitated by the use of the computer software available on the authors' web site (refer to the Preface).

18.1–18.4 Conduct a time-history analysis of the shear frame structure shown in Figures P18.1 through P18.4 for the N-S component of the Northridge earthquake. Plot the time-history responses for the relative displacement $x(t)$, base shear $V(t)$, and column bending moment $M(t)$. Assume all columns bend about their strong axis, $\zeta = 0.05$, and all steel is ASTM A36.

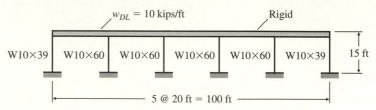

Figure P18.1

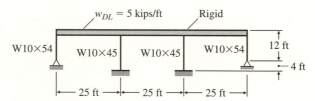

Figure P18.2

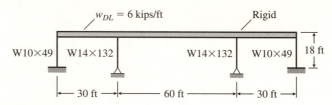

Figure P18.3

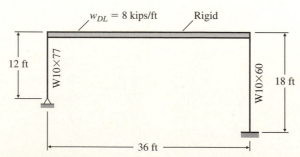

Figure P18.4

18.5–18.8 Conduct a time-history analysis of the braced frame structures shown in Figures P18.5 through P18.8 for the N-S component of the Northridge earthquake. Plot the time-history responses for relative displacement $x(t)$, base shear $V(t)$, and tension force in the cross bracing $T(t)$. Assume the cross braces are effective in tension only, $\zeta = 0.05$, and all steel is ASTM A36.

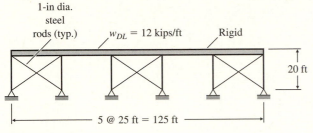

Figure P18.5

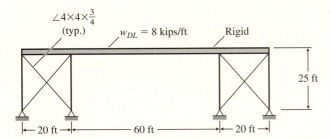

Figure P18.6

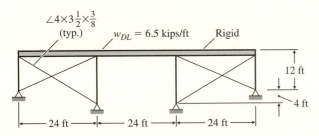

Figure P18.7

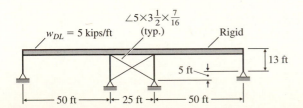

Figure P18.8

18.9 Construct a Newmark-Hall broad-banded design spectra for an earthquake having a peak ground acceleration (PGA) of 0.35 g for $\zeta = 0.02, 0.05$, and 0.10. Use the four-way logarithmic paper in Figure A.1 of Appendix A.

18.10 Repeat Problem 18.9 for a PGA of 0.45 g and for $\zeta = 0.03, 0.07$, and 0.12. Use the four-way logarithmic paper in Figure A.1 of Appendix A.

18.11–18.14 Estimate the maximum relative displacement x_{max}, maximum base shear V_{max}, and maximum column bending moments M_{max} for the shear frame structures shown in Figures P18.1 through P18.4 using the broad-banded design spectrum shown in Figure 18.21. Compare these results with the results obtained using the response spectrum for the Northridge earthquake, for $\zeta = 0.05$, shown in Figure 18.19. Explain any discrepancies in the results.

18.15–18.18 Estimate the maximum relative displacement x_{max}, maximum base shear V_{max}, and maximum tensile force in the cross braces for the braced frame structures shown in Figures P18.5 through P18.8 using the broad-banded design spectrum shown in Figure 18.21. Compare these results with the results obtained using the response spectrum for the Northridge earthquake, for $\zeta = 0.05$, shown in Figure 18.19. Explain any discrepancies in the results.

18.19–18.23 Conduct a time-history analysis of the multi-story shear frame buildings shown in Figures P18.9 through P18.13 for the N-S component of the Northridge earthquake. Compute the maximum responses for displacements $\{x\}_{max}$, effective earthquake forces $\{F_S\}_{max}$, story shears $\{V\}_{max}$, and story moments $\{M\}_{max}$. Assume all columns bend about their strong axis (where applicable) and $\zeta = 0.05$ for all modes.

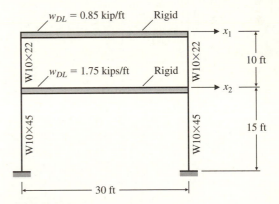

Figure P18.9

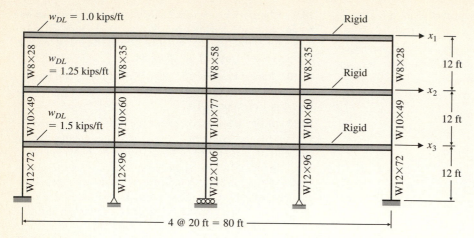

Figure P18.10

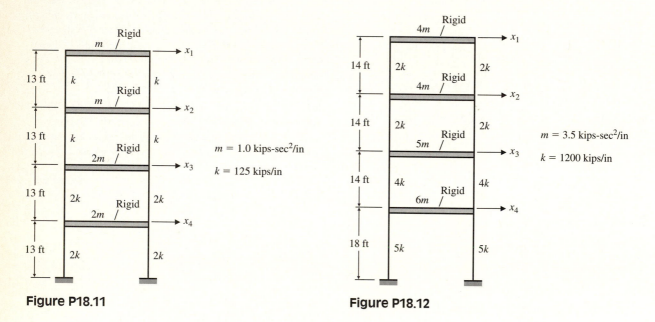

Figure P18.11

$m = 1.0$ kips-sec^2/in

$k = 125$ kips/in

$m = 3.5$ kips-sec^2/in

$k = 1200$ kips/in

Figure P18.12

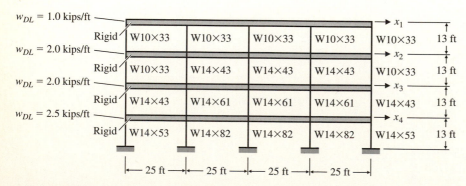

Figure P18.13

18.24–18.28 Estimate the maximum responses for relative displacement $\{x\}_{max}$, effective earthquake forces $\{F_S\}_{max}$, story shears $\{V\}_{max}$, and story moments $\{M\}_{max}$ for the multistory shear frame buildings shown in Figures P18.9 through P18.13 using the broad-banded design spectra shown in Figure 18.21. Determine the effective mass for each mode. Combine the modal maxima by the SRSS method. Compare these results to the results obtained using the response spectrum for the Northridge earthquake, for $\zeta = 0.05$, shown in Figure 18.19. Explain any discrepancies in the results.

18.29 Conduct a time-history analysis for the tower structure shown in Figure 18.41 for the N-S component of the Northridge earthquake. The structure has distributed mass and stiffness properties defined by

$$m(z) = 0.3\left(1.5 - \frac{z}{L}\right) \ \text{kip-sec}^2/\text{ft}^2$$

$$EI(z) = 10^8\left(1 - \frac{z}{L}\right) \ \text{kip-ft}^2$$

Assume a shape function $\psi(z) = (z/L)^2$, and using the generalized SDOF system approach plot the relative displacement time history $x(t)$ at $z = L$ and $z = L/2$. Also plot the shear time history $V(t)$ and moment time history $M(t)$ at $z = 0$ and $z = L/2$. Assume $L = 200$ ft, $m = 2.0$ kip-sec^2/ft, and $\zeta = 0.05$.

18.30 A building structure having uniformly distributed mass and stiffness properties is shown in Figure P18.14. Assume $m_z = 0.02$ kip-sec^2/ft per foot of building height and $EI = 2 \times 10^6$ kip-ft^2 is constant over the building height. Using the generalized SDOF system approach with an assumed shape function for the fundamental vibration mode of $\psi(z) = 1 - \cos(\pi z/2L)$, conduct a time-history analysis of the structure for the N-S component of the Northridge earthquake. Plot the relative displacement time history $x(t)$ at $z = L$ and $L/2$. Also plot the shear time history $V(t)$ and the moment time history $M(t)$ at $z = 0$ and $z = L/2$. Assume $L = 150$ ft and $\zeta = 0.05$.

18.31–18.32 For the structures of Problems 18.29 and 18.30, estimate the maximum relative displacement x_{max} at $z = L$ and $L/2$ and the maximum shear V_{max} and moment at $z = 0$ and $L/2$ using the broad-banded design spectrum shown in Figure 18.21.

18.33–18.36 Construct an in-building acceleration response spectrum ($\zeta = 0.03$) for the N-S component of the Northridge earthquake at the second story above ground level for the shear frame buildings shown in Figures P18.10 through P18.13. From the response spectrum, construct the augmented in-building design spectrum considering a 10% peak response shift. Determine the acceptable range of support stiffness k_s for equipment to be located on the second story of the structure if the equipment cannot withstand an acceleration greater than 0.75 g. Assume equipment weights of 6, 12, and 18 kips.

18.37 Using the broad-banded elastic design spectra constructed in Problem 18.9, construct corresponding inelastic design spectra for ductility factors $\mu = 3$ and $\mu = 5$. Use the four-way logarithmic paper in Figure A.1 of Appendix A.

18.38 A structure modeled as a SDOF system has a natural period of $T = 0.5$ sec and a mass m. Using the elastic design spectra of Example 18.9 and the inelastic design spectra of Example 18.37, estimate the elastic and inelastic displacement response and the elastic and inelastic design forces for the system for the following cases:

(a) $\zeta = 0.03$ and $\mu = 3$

(b) $\zeta = 0.05$ and $\mu = 5$

(c) $\zeta = 0.10$ and $\mu = 5$

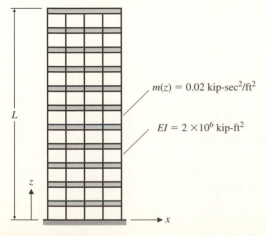

Figure P18.14

19 ⏶
Blast Loads on Structures

A very general definition of a blast is a violent gust of wind, or the effect or accompaniment of such a gust. In this chapter a blast represents the violent effect produced in the vicinity of an explosion. This violent effect consists of a shock accompanied by an instantaneous increase in ambient atmospheric pressure followed by a monotonic decrease in pressure below the local atmospheric pressure. Accompanying the dramatic changes in pressure associated with the shock wave or shock front are similarly dramatic changes in density and temperature. A blast is usually the result of an explosion defined as a sudden expansion of some energy source, with strict emphasis placed on the phrase "sudden expansion." Regardless of the causative energy source, which may vary from a mundane automobile tire "blowout" to an atomic device, the common essential feature in all explosions is that a very sudden expansion occurs.

This chapter discusses the basic concepts that define blast loads on structures and the corresponding structural response. Topics included in the discussion are sources of blast, shock waves, shock pressures, and material strain-rate effects. Procedures for estimating blast pressures on structures are developed, and an approximate solution technique for predicting the blast response of SDOF systems is presented.

19.1 SOURCES OF BLAST LOADS

Detonations of explosive materials such as dynamite or TNT are well known for their ability to generate blast in rock, soil, water, and air. However, energy in high-pressure gas cylinders, if released suddenly, can also cause an explosion and result in a blast. The muzzle blast of a gun or an electrical energy discharge are explosions that will cause blasts. For example, a device called a lithotriper, which employs an electrical discharge between two probes, is used to create a shock source to break up bladder stones in the human body. Lightning is an explosion in air, also resulting from an electrical discharge.

Generally, the most common sources of significant explosions are derived from chemical or nuclear materials. The measure of the effectiveness of chemical explosive materials is usually defined in terms of the peak pressure or specific impulse (area under pressure–time diagram). The effectiveness of an explosive substance is typically expressed as a TNT equivalent. A listing of several chemical explosive materials with their equivalent weight of TNT is presented in Table 19.1. A similar list based on the equivalent volume of TNT is given in Table 19.2.

For most military applications the explosive material generally comes from the list of explosives included in Tables 19.1 and 19.2. However, accidental explosions and resulting blast loads can be generated from such sources as grain dust, fuel/gas-air mixtures, an overpressurized steam boiler, or accidentally detonated ammonium nitrate fertilizer. A fairly large explosion of ammonium nitrate in a hold of a liberty ship in the port of Texas City (1949) is described in some detail by Kinney and Graham [4]. The recent (1995) intentional detonation of ammonium nitrate adjacent to the federal building in downtown Oklahoma City is an alarming example of the destructive use of this commonly available explosive material.

19.2 SHOCK WAVES

19.2.1 Introduction

Given the proper conditions, shock waves may form in any type of material, whether it be solid, liquid, or gas. In all materials, shock waves are associated with an instantaneous change of mechanical properties across an almost zero thickness wavefront separating the undisturbed medium from the shocked medium. In all media the shock is usually generated by an explosive device. However, under certain circumstances, high-velocity impact will initiate a shock front in solids or liquids. Shock waves associated with airplanes or missiles traveling in air are well known, and are very similar to shock waves or fronts that emanate from explosive sources in free air. In this chapter emphasis is restricted to shock waves in air.

For explosions in free air, which are not influenced by the ground plane, the blast is developed very quickly near the source. The characteristics of shock waves generated by strong air blasts are very different from the stress waves in elastic media discussed in Chapter 16. There, a linear system was described for stress waves or waves traveling at the speed of sound within the material. In this context, an equivalent sonic wave travels in air and produces very small pressure changes and practically no changes in particle velocity, and therefore does not "shock-up." However, in reality, air is a compressible gas having properties that will cause an initial disturbance or pressure rise to steepen, and form an almost discontinuous shock front.

TABLE 19.1. Relative Peak Pressure and Specific Impulse of Explosive Materials Compared to an Equivalent Volume of TNT

Explosive	Pressure	Specific Impulse
Torpex 30% Al	1.23	1.29
Torpex-2	1.22	1.22
Minol-3	1.18	1.20
DBX	1.16	1.18
HBX	1.15	1.18
Tritonal 75/25	1.13	1.17
Minol-2	1.15	1.16
Tritonal 8/20	1.13	1.15
Trialer	1.11	1.13
Barconal	1.09	1.09
Comp. B	1.09	1.06
Pentolite	1.07	1.03
Ednatol	1.02	1.01
TNT	1.00	1.00
Picratol	0.97	0.96
Amatex	0.96	0.90
Amatol	0.93	0.85

TABLE 19.2. Relative Peak Pressure and Specific Impulse of Explosive Compared to an Equivalent Weight of TNT

Explosive	Pressure	Specific Impulse
Comp. A-3	1.09	1.08
Comp. B	1.11	0.98
Comp. C-4	1.37	1.19
Cyclotol (70/30)	1.14	1.09
HBX-1	1.17	1.16
HBX-3	1.14	0.97
H-6	1.38	1.15
Minol-2	1.20	1.11
PETN	1.27	—
Pentolite	1.42	1.00
Picrotol	0.90	0.93
Tetryl	1.07	—
TNETB	1.36	1.10
TNT	1.00	1.00
TRITONAL	1.07	0.96
ANFO (94/6 Ammo. Nitrate/Fuel Oil)	0.82	—

19.2.2 Sound Speed and Mach Number

In linear elastic solids a three-dimensional, or volumetric, strain Δ, defined by the expression

$$\Delta = \varepsilon_{11} + \varepsilon_{22} + \varepsilon_{33} \tag{19.1}$$

is called the *dilatation,* in which the ε_{ii} represent the principal strains. The dilatation can be shown to be the linear portion of the change in volume divided by the original volume, when the original volume is subjected to a mean stress. The mean stress is defined as

$$\sigma_m = -p = \frac{1}{3}(\sigma_{11} + \sigma_{22} + \sigma_{33}) \tag{19.2}$$

where the σ_{ii} are the principal stresses and p is pressure, which is positive when directed toward the surface of the volume element. This sign convention for pressure is opposite to the standard sign convention used for stress, thus the negative sign in Eq. (19.2).

The constitutive relation between pressure p and the dilatation Δ may be obtained by summing the first three equations of Eq. (16.18) and then employing Eqs. (19.1) and (19.2). The result is an expression for Δ given by

$$\Delta = -p\left(\lambda + \frac{2}{3}G\right) \tag{19.3}$$

where G is the shear modulus defined by Eq. (16.27). The bulk modulus B, defined as the linear slope of the pressure-dilatation curve, is expressed as

$$B = \lambda + \frac{2}{3}G = -\frac{p}{\Delta} \tag{19.4}$$

where λ is Lamé's constant defined by Eq. (16.29). Using the very basic definition of wave speed discussed in Section 16.4, it is determined that the dilatation wave or spherical stress (hydrostatic) travels at a wave speed given by

$$c = \sqrt{\frac{B}{\rho}} \tag{19.5}$$

in which ρ is the mass density. It is also noted that for an elastic solid the slope of the pressure-dilatation curve is linear.

Assuming air to be an ideal gas, and also assuming an *isentropic process* (adiabatic and reversible), the relationship between pressure, density, and specific volume is defined by

$$pv^k = \frac{p}{\rho^k} = \text{constant} \tag{19.6}$$

where k is the ratio of specific heats of air and v is the specific volume, such that $v = 1/\rho$.

The counterpart of dilatation, termed the fraction change in volume, in air may be written as $(\partial v/v)_s$, where the subscript s denotes the isentropic process and entropy is held constant. The slope of the nonlinear curve relating pressure and the fractional volume (or bulk modulus) for air is not a constant, as in the case of elastic solids, but may be written as

$$B_s = -\left[\frac{\partial p}{(\partial v/v)}\right]_s \tag{19.7}$$

Using the definition of specific volume as the inverse of density yields

$$v = \frac{1}{\rho} = \rho^{-1} \tag{19.8}$$

and taking the logarithm of both sides of Eq. (19.8) gives

$$\log v = -\log \rho. \tag{19.9}$$

Differentiation of Eq. (19.9) results in

$$\frac{dv}{v} = -\frac{d\rho}{\rho} \tag{19.10}$$

and the bulk modulus of air, given by Eq. (19.7), becomes

$$B_s = \left[\frac{\partial p}{\partial \rho/\rho}\right]_s \tag{19.11}$$

Substituting Eq. (19.11) into Eq. (19.5) and defining the speed of sound by the letter a,

$$a = \left(\frac{\partial p}{\partial \rho}\right)_s^{1/2} \tag{19.12}$$

For an ideal gas it has been shown [2] that

$$\left(\frac{\partial p}{\partial \rho}\right)_s = kRT \tag{19.13}$$

where R is the universal gas constant and T is the absolute temperature. This then leads to the well-known equation for the speed of sound in air given by

$$a = \sqrt{kRT} \tag{19.14}$$

The Mach number, defined as the local velocity $\dot{u}$ divided by the local speed of sound, is written as

$$M = \frac{\dot{u}}{a} = \dot{u}(kRT)^{-1/2} \tag{19.15}$$

and, as will be shown, is a very useful term for describing the defining parameters of shock waves.

19.2.3 Shock Pressure

A full mathematical description of the principles and associated equations necessary to completely describe the shock front is beyond the scope of this chapter. However, since blast loads on structures are given in terms of pressure loads, some of the pertinent pressure terms will be described.

Based on developments presented by Baker [1] and Kinney and Graham [3], two pressures are important in predicting blast loads on structures: (1) the overpressure and (2) the reflected pressure. The pressure rise (overpressure) P_{so} of the ambient pressure across a normal shock front as it moves through the air, shown schematically in Figure 19.1a, is given as [2]

$$P_{so} = p_y - p_x = \left[\frac{2k}{k+1}(M_x^2 - 1)\right]p_x \tag{19.16}$$

where p_x = air pressure ahead of the shock front
p_y = air pressure behind the shock front
M_x = ratio of wave velocity $\dot{u}_x$ to sonic velocity a_x

Rearranging Eq. (19.16), the Mach number of the wavefront becomes

$$M_x = \sqrt{1 + \left(\frac{k+1}{2k}\right)\frac{P_{so}}{p_x}} \tag{19.17}$$

If the value for k is taken as 1.4, then Eqs. (19.16) and (19.17) become, respectively,

$$P_{so} = \frac{7(M_x^2 - 1)}{6}p_x \tag{19.18a}$$

and

$$M_x = \sqrt{1 + \frac{6}{7}\frac{P_{so}}{p_x}} \tag{19.18b}$$

Figure 19.1

Schematic of a shock encounter with a planar rigid surface and resulting normal reflection: (a) before reflection; (b) after reflection. *Explosions in Air* by W.E. Baker, copyright © 1973. Reprinted by permission of Wilfred Baker Engineering, San Antonio, TX.

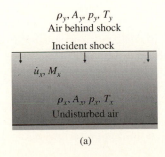

ρ_y, A_y, p_y, T_y
Air behind shock

Incident shock

$\dot{u}_x, M_x$

ρ_x, A_x, p_x, T_x
Undisturbed air

(a)

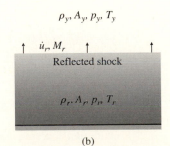

ρ_y, A_y, p_y, T_y

$\dot{u}_r, M_r$

Reflected shock

ρ_r, A_r, p_r, T_r

(b)

The pressure P_{so} is also called the side-on pressure and is measured experimentally using a pressure gage mounted normal to the direction of motion of the air blast. P_{so} is the pressure rise above the local atmospheric pressure P_0, when it is free of any interference with the ground or with any structural elements. Therefore P_{so} is often referred to as the *free-field pressure.*

As the shock wave encounters a rigid surface perpendicular or normal to the shock motion, a wave is reflected as shown in Figure 19.1b. In this description the gas particles are abruptly brought to rest at the rigid surface and then reflected back into the air. This pressure is then assumed to act on the reflecting surface as a "pressure loading," and is sometimes referred to as the transmitted, or blast, loading on the structure. The *reflected pressure* p_r behind the normally reflected shock is given as [2]

$$\frac{p_r}{p_x} = \frac{[(3k-1)(p_y/p_x) - (k-1)]}{(k-1)(p_y/px) + (k+1)}\left(\frac{p_y}{p_x}\right) \qquad (19.19)$$

and for $k = 1.4$, Eq. (19.19) becomes

$$\frac{p_r}{p_x} = \frac{p_y(8p_y - p_x)}{p_x(p_y + 6p_x)} \qquad (19.20)$$

This type of reflection is usually referred to as a *regular reflection*. The *reflection coefficient,* or *reflection factor,* C_r, defined as the ratio of the reflected overpressure ($p_r - p_x$), or P_r, to that of the overpressure in the incident shock ($p_y - p_x$), or P_{so}, may be written as [2]

$$C_r = \frac{p_r - p_x}{p_y - p_x} = \frac{P_r}{P_{so}} = \frac{(3k-1)(p_y/p_x) + (k+1)}{(k-1)(p_y/p_x) + (k+1)} \qquad (19.21)$$

and in terms of Mach number the reflection coefficient is expressed as

$$C_r = \frac{(3k-1)M_x^2 + (3-k)}{(k-1)M_x^2 + 2} \qquad (19.22)$$

Once again assuming a value of 1.4 for k, Eqs. (19.21) and (19.22) become, respectively,

$$C_r = \frac{8p_y + 6p_x}{p_y + 6p_x} \qquad (19.23)$$

and

$$C_r = \frac{8M_x^2 + 4}{M_x^2 + 5} \qquad (19.24)$$

For weak shocks ($M_x \cong 1.0$), where the shock velocity is equal to the wave speed, $C_r \cong 2.0$. This corresponds with the result for the transmitted stress described by Eq. (16.61), when the transmitted characteristic impedance is much larger than the incident characteristic impedance. For very strong shocks ($M_x \gg 1.0$), examination of Eq. (19.24) indicates that $C_r \rightarrow 8.0$, which is theoretically the upper limit of the normal reflection factor. Air shock characteristics such as Mach number, pressure ratios, etc., are listed as dimensionless parameters and are prescribed in tabular form in [1] and [2].

19.3 DETERMINATION OF BLAST LOADS

19.3.1 Defining Blast Loads

Explosions in air result in a shock wave, or blast, that propagates spherically away from the center of the explosion. In describing this explosion, several characteristic terms must be defined. The charge denotes the explosive, and charge weight W is the weight of the explosive in pounds or kilograms. The standoff R is the radial distance from the center of the explosive to some point in space, or to a particular location on a structure, measured in feet or meters. A normalized standoff Z is defined as the standoff R divided by the cube root of the charge weight $W^{1/3}$, or

$$Z = \frac{R}{W^{1/3}} \qquad (\text{ft/lb}^{1/3}, \text{m/kg}^{1/3}) \tag{19.25}$$

Figure 19.2a shows a schematic of an air blast with a series of rings denoting the advancing incident blast wave I. If the air blast encounters the ground plane, a regular reflection as described previously occurs until the angle between the incident wave and the ground exceeds a critical angle of approximately 40°. Above this critical angle, regular reflection yields to the formation of a Mach stem. The Mach stem propagates a wave normal to the surface and terminates in a triple point, at some distance above the ground. This triple point is the conjunction of the Mach stem, the reflected wave, and the incident wave, and is shown schematically in Figure 19.2b. The Mach stem is the shock that extends from the triple point to the ground. The Mach number M_s of the Mach stem is defined by

$$M_s = M_x \sin \alpha \tag{19.26}$$

where α is the incidence angle as shown in Figure 19.2b. Using this Mach number, the overpressures associated with the Mach stem can be determined and treated as any other shock wave. For an air blast, the importance of the Mach stem is paramount because it is this shock wave that will load structures at various distances from the blast along the earth's surface.

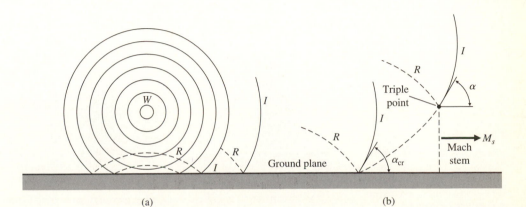

Figure 19.2

Schematic of an air blast reflecting from a ground plane and formation of a Mach stem.

(a)   (b)

An idealized pressure-time function for a blast front in free air that has not encountered any obstacles is shown in Figure 19.3. This type of loading will occur when the charge is located above the measuring point (Figure 19.4a), or the charge is positioned off to one side and the incident wave encounters the measuring point prior to the formation of the Mach stem as shown in Figure 19.4b.

Wherever parameters for free-air explosions are given in the literature, they are presented in graphical form and are based on data collected by Goodman [5], using Pentolite spherical charges, and on Brode's [6] theoretical prediction. An example of this data is shown in Figure 19.5 for spherical TNT explosions in free air. If the charge is a hemi-spherical-shaped TNT and is exploded on the ground surface, the pertinent blast parameters are different from the free-air data as illustrated in Figure 19.6. For the case of the charge positioned directly on the ground, the Mach stem does not form and the wave is assumed to travel along the ground as a plane wavefront. Extensive data for side-on pressure and specific impulse (area under the pressure-time curve) are given for several explosive materials in [4]. Multilinear or piecewise-linear equations of pressure and specific impulse, as functions of normalized standoff $R/W^{1/3}$ relations are given in [3 and 7].

The foregoing discussions on pressures are based on normal incidence of the shock with respect to the structure or measuring surface. When the shock front strikes the surface at incidence angles (Figure 19.7) then reflected waves occur, but the reflected angle

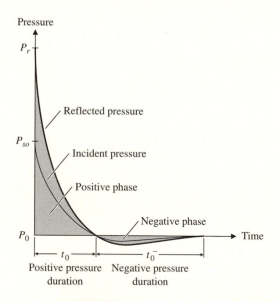

Figure 19.3

Idealized pressure-time curve for a free air explosion.

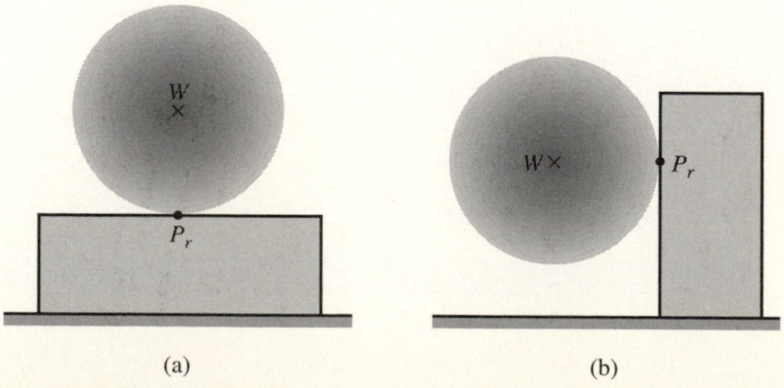

Figure 19.4

Conditions for measurement of parameters of a free air blast: (a) charge located above measuring point; (b) charge positioned side-on to measuring point.

is usually less than the incidence angle. This type of reflection is usually referred to as an oblique or regular reflection, and is defined by the general reflection principles up to a critical angle of approximately 40°. The effect of the angle of incidence on the reflected pressure is shown in Figure 19.8. This figure also demonstrates the effect of the local pressure rise or side on pressure on the reflection coefficient. For very low side-on pressures or weak waves, the reflection coefficient approaches the value of 2.0.

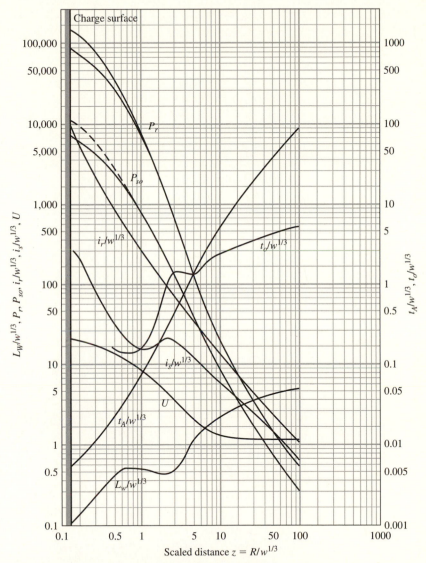

Figure 19.5

Parameters for spherical free air explosions of TNT at sea level [4]. "Structures to Resist the Effects of Accidental Explosions," Department of Defense Explosives Safety Board, 1990.

P_{so}	=	Peak positive incident pressure (psi)
P_r	=	Peak positive normal reflected pressure (psi)
$i_s/w^{1/3}$	=	Scaled unit positive incident impulse (psi·ms/lb$^{1/3}$)
$i_r/w^{1/3}$	=	Scaled unit positive normal reflected impulse (psi·ms/lb$^{1/3}$)
$t_A/w^{1/3}$	=	Scaled time of arrival of blast wave (ms/lb$^{1/3}$)
$t_o/w^{1/3}$	=	Scaled positive duration of positive phase (ms/lb$^{1/3}$)
U	=	Shock front velocity (ft/ms)
w	=	Charge weight (lb)
$L_w/w^{1/3}$	=	Scaled wavelength of positive phase (ft/lb$^{1/3}$)

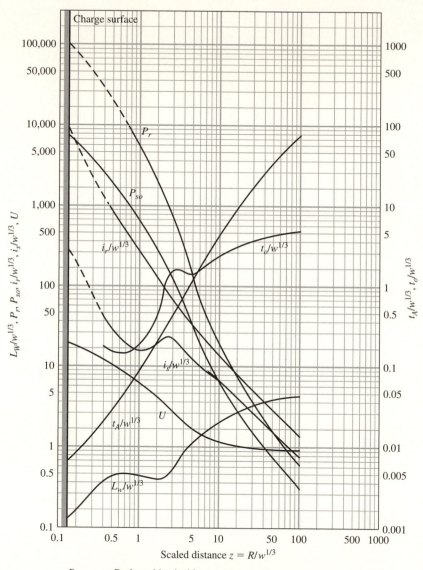

Figure 19.6

Parameters for hemispherical
TNT explosions on the ground
surface at sea level [4].
"Structures to Resist the Effects
of Accidental Explosions,"
Department of Defense
Explosives Safety Board, 1990.

P_{so}	=	Peak positive incident pressure (psi)
P_r	=	Peak positive normal reflected pressure (psi)
$i_s/w^{1/3}$	=	Scaled unit positive incident impulse (psi·ms/lb$^{1/3}$)
$i_r/w^{1/3}$	=	Scaled unit positive normal reflected impulse (psi·ms/lb$^{1/3}$)
$t_A/w^{1/3}$	=	Scaled time of arrival of blast wave (ms/lb$^{1/3}$)
$t_o/w^{1/3}$	=	Scaled positive duration of positive phase (ms/lb$^{1/3}$)
U	=	Shock front velocity (ft/ms)
w	=	Charge weight (lb)
$L_w/w^{1/3}$	=	Scaled wavelength of positive phase (ft/lb$^{1/3}$)

Figure 19.7

Schematic illustrating reflection of a shock wave having an angle of incidence with reflecting surface.

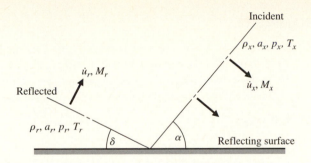

Figure 19.8

Reflected pressure factor versus angle of incidence. "Fundamentals of Protective Design for Conventional Weapons," Technical Manual TM 5-855-1, 1986. Reprinted courtesy of Headquarters, Department of the Army.

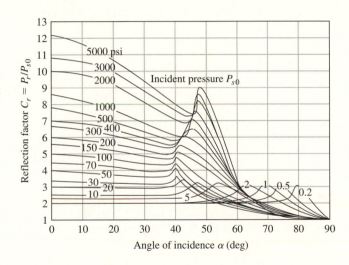

19.3.2 Structure Loading

For a given standoff and charge weight, Figure 19.5 may be used to determine the pressure-time relation or impulse-time relation for any location on a structure directly exposed to the shock front. However, these parameters are valid only for the pressure directed along a radial line from the explosive source normal to the point on the structure. At points other than the radial normal point on the structure, the pressure or impulse must be modified for effects of angle of incidence using Figure 19.8.

EXAMPLE 19.1 ▼

A spherical explosive charge weighing 27 lb (TNT equivalent weight) is positioned 15 ft above the midspan of a 20-ft-long simply supported beam as shown in Figure 19.9a. Determine the peak reflected pressure and the reflected specific impulse for the beam at the following locations: (a) at midspan and (b) at the beam ends.

Solution

(a) Blast parameters at midspan:

For a standoff distance $R = 15$ ft, the normalized standoff is determined from Eq. (19.25) as

$$Z = \frac{R}{W^{1/3}} = \frac{15 \text{ ft}}{(27 \text{ lb})^{1/3}} = 5.0 \text{ ft/lb}^{1/3} \tag{1}$$

From Figure 19.5, the peak positive incident overpressure P_{so} is determined as approximately 27 psi, and the peak reflected overpressure P_r is approximately 90 psi. The reflection factor for these pressures C_r is determined from Eq. (19.21) as

$$C_r = \frac{P_r}{P_{so}} = \frac{90}{27} = 3.33 \tag{2}$$

This result can be compared with the reflection factor determined from Figure 19.8 for an angle of incidence $\alpha = 0°$, which agrees quite favorably with 3.33.

Also from Figure 19.5, the reflected impulse term $i_r/W^{1/3}$ for $Z = 5.0$ is approximately 25 psi-msec/lb$^{1/3}$, which yields a reflected specific impulse i_r as

$$i_r = 25(27)^{1/3} = 75 \text{ psi-msec} \tag{3}$$

Most experts agree that the time of duration of the reflected pulse is approximately the same as the incident pulse duration. Therefore, assuming a triangular pressure-time pulse relation as illustrated in Figure 19.9b, the reflected positive pressure duration t_r may be estimated from the expression

$$t_r = \frac{2i_r}{P_r} \tag{19.27}$$

Thus for $i_r = 75$ psi-msec and $P_r = 90$ psi, the reflected pressure duration t_r as determined from Eq. (19.27) is 1.67 msec.

Figure 19.9

Set-up for Example 19.1: (a) schematic of explosive charge positioned over beam midspan; (b) simplified pressure-time curve.

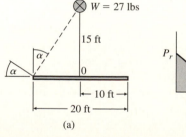

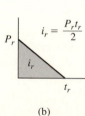

(a)   (b)

(b) Blast parameters at beam ends:

At the beam ends the standoff is the slant distance of $R = \sqrt{15^2 + 10^2} = 18.03$ ft as illustrated in Figure 19.9a. Thus the normalized standoff at the beam ends is given by Eq. (19.25) as

$$Z = \frac{R}{W^{1/3}} = 6.01 \tag{4}$$

From Figure 19.5, the peak side on pressure for this standoff is approximately 18 psi and the peak reflected pressure is approximately 53 psi. For these peak pressures the reflection coefficient C_r is approximately 2.95. This agrees with the reflection coefficient determined from Figure 19.9 for an incident angle $\alpha = \tan^{-1} 10/15 = 33.7°$. The specific impulse term for a normalized standoff of $Z = 6.01$, as determined from Figure 19.5, is approximately 20 psi-msec/lb$^{1/3}$, yielding a specific impulse $i_r = 60$ psi-msec. The duration of the reflected pressure pulse is again approximated by Eq. (19.27) as

$$t_r = \frac{2i_r}{P_r} = \frac{2(60)}{53} = 2.26 \text{ msec} \tag{5}$$

▲

The results of Example 19.1 pose a very interesting loading condition problem. The beam experiences a peak pressure load of approximately 90 psi at midspan, varying to approximately 53 psi at its ends. However, the reflected specific impulse exhibits significantly less variation over the same interval of length. The determination of an appropriate uniform loading condition for the beam can be obtained by taking an average of the values of pressure and duration at the beam midpoint and ends.

19.4 STRAIN-RATE EFFECTS

It is a well-established fact that most materials exhibit increases in strength as the loading rate or strain rate increases. There are a few exceptions, however, notably work-hardened aluminum alloys (2024-T3, 7075-T6) and high-strength steel such as rolled homogeneous armor. The majority of the evidence supporting the increase of strength with increasing strain rate is purely experimental, and therefore should be considered a material property in the same sense as the static strength and modulus of elasticity. The data is usually presented in the form of a plot of dynamic strength, or dynamic strength ratio relative to a quasi-static value, versus the logarithm to base 10 of strain rate. Data for several metals are given by Nicholas [10], and an example for copper and tungsten obtained by Ross [11] is shown in Figure 19.10. Strain-rate data for both the compressive and tensile strength of concrete [12] is shown in Figure 19.11, along with a general descriptor of some physical phenomena associated with certain strain rates. These general descriptors shown along the strain-rate axis are attributed to Kormeling et al. [13]. Notice in Figure 19.11 that concrete exhibits a *critical strain rate,* or *strain-rate threshold,*

Figure 19.10

Example of strain-rate data for several metals.

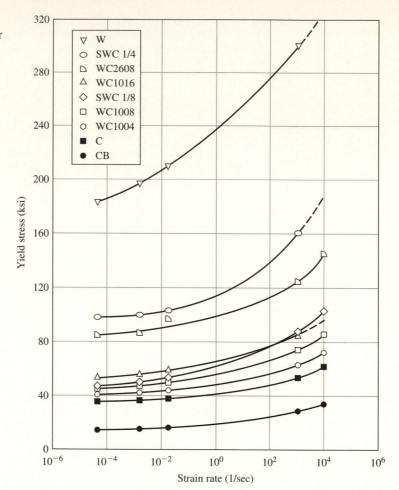

above which significant increases in strength are observed. This apparent strain-rate threshold range is approximately 1 to 10/sec for tension and 50 to 80/sec for compression (these values are shown as points *A* and *B,* respectively, in Figure 19.11).

It is rather difficult to predict a given load rate or strain rate based on a given impact velocity. However, if the time at which a particular stress or strain will occur can be calculated, then the stress rate or strain rate can be estimated by dividing that particular value by the time required to attain it. In the European scientific community it is generally the stress rate (psi/sec, MPa/sec) that is reported, and in the United States, strain rate (1/sec) is most commonly reported. For many materials it has been demonstrated that the elastic moduli are not particularly strain-rate sensitive; therefore the stress rate may be converted to strain rate by dividing the stress rate by the quasi-static Young's modulus.

The physical cause of strain-rate sensitivity is not completely understood. For brittle materials such as concrete, where the material failure occurs by fracture with little or no plastic deformation, a limited crack velocity phenomenon is assumed to occur. If an attempt is made to force the crack to propagate at velocities greater than the *limiting crack velocity,* then fracture will not occur and the local stress and deformation increases. This results in increased strength at an elevated strain rate. If the time to fracture of concrete is measured as a function of strain rate and plotted on a log-log scale as illustrated in Figure 19.12, it is noticed that a change in this slope occurs at approximately the same strain rate corresponding to the apparent strain rate sensitivity threshold illustrated in Figure 19.11.

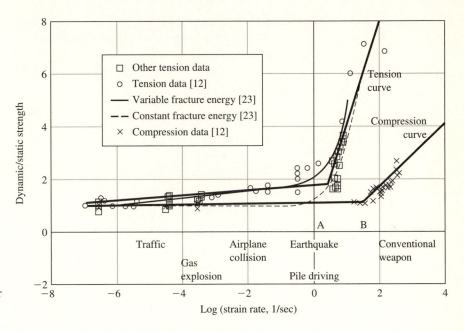

Figure 19.11

Example of strain-rate data for concrete.

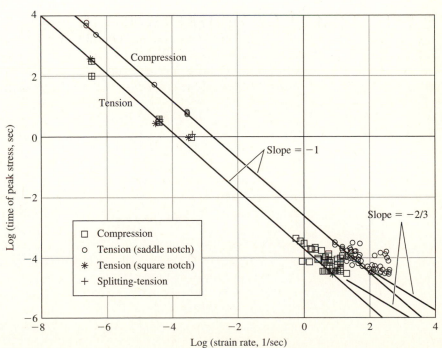

Figure 19.12

Time to fracture of concrete as a function of strain rate.

For materials such as ductile metals, where plastic deformation occurs by dislocation motion, a similar limiting velocity of dislocations may occur, resulting in increased strength. This phenomenon occurs at much higher strain rates in metal than in concrete. There are several strain-rate-related strength increase predictive techniques for metals worth noting. The classical equation developed by Malvern [14] is given as

$$\sigma = f(\varepsilon) + a \ln(1 + b\dot{\varepsilon}) \tag{19.28}$$

where σ = strength at strain rate
$f(\varepsilon)$ = quasi-static stress-strain relation
$\dot{\varepsilon}$ = plastic strain rate
a and b = experimental constants that describe the strain-rate sensitivity

A variation of Eq. (19.28), called the Johnson-Cook equation [15], gives the von Mises flow stress σ as

$$\sigma = (A + B\varepsilon^n)\left[1 + c\ln\left(\frac{\dot{\varepsilon}}{\dot{\varepsilon}_0}\right)\right](1 - T^{*m}) \tag{19.29a}$$

where

$$T^* = \frac{(T - T_{\text{room}})}{(T_{\text{melt}} - T_{\text{room}})} \tag{19.29b}$$

where ε = strain
$\dot{\varepsilon}$ = strain rate
$\dot{\varepsilon}_0$ = 1.0 sec
T = absolute temperature

and the empirical lookup parameters of A, B, m, n are specified in tabular form. Similar constitutive equations accounting for the strain-rate sensitivity of a variety of materials are given by Nicholas and Rajendran [16].

A very basic derivation of the tensile strength of brittle materials is given by Grady et al. [17, 18, 19]. A series of three equations for fracture strength σ_s, time of fracture t_s, and fracture size s are derived on the principle that the sum of the kinetic energy and the strain energy density of a given volume of material is equal to or greater than the fracture surface energy contained in the same volume. These basic equations are expressed as

$$\sigma_s = (3\rho c_0 K_{\text{IC}}^2 \dot{\varepsilon})^{1/3} \tag{19.30}$$

$$t_s = \frac{1}{c_0}\left(\frac{\sqrt{3}\, K_{\text{IC}}}{\rho c_0 \dot{\varepsilon}}\right)^{2/3} \tag{19.31}$$

$$s = 2\left(\frac{\sqrt{3}\, K_{\text{IC}}}{\rho c_0 \dot{\varepsilon}}\right)^{2/3} \tag{19.32}$$

where ρ = mass density
K_{IC} = mode I quasi-static fracture toughness
c_0 = dilatation wave speed
$\dot{\varepsilon}$ = strain rate
s = particle size

These three equations are valid for high strain rates, but no limiting value of strain rate is specified.

Close scrutiny of Eq. (19.30) indicates that the fracture tensile strength is proportional to the strain rate to the one-third power. Data to support this argument is given in [18]. The tensile strength data for concrete shown in Figure 19.11 also supports this criteria at strain rates above 1.0/sec.

EXAMPLE 19.2   ▼

Plot the fracture stress σ_s and particle size s as a function of strain rate for a material having the physical properties listed below. Consider strain rates $\dot{\varepsilon}$ from 0 to 10^6/sec.

Mass density	$\rho = 2247$ kg/m^3
Wave speed	$c_0 = 4000$ m/sec
Fracture toughness	$K_{IC} = 1.0 \times 10^6$ N/m$^{3/2}$

Solution

The fracture stress as a function of strain rate is determined from Eq. (19.30) as

$$\sigma_s(\dot{\varepsilon}) = (3\rho c_0 K_{IC}^2 \dot{\varepsilon})^{1/2} \times 10^6 \tag{1}$$

and the corresponding particle size is established from Eq. (19.32) as

$$s(\dot{\varepsilon}) = \left(\frac{\sqrt{3}\, K_{IC}}{\rho c_0 \dot{\varepsilon}}\right)^{2/3} \times 10^3 \tag{2}$$

The plots of the fracture stress as a function of linear strain rate and $\log_{10}$ strain rate are shown in Figure 19.13a and b, respectively, and the plots of particle size as a function of linear strain rate and $\log_{10}$ strain rate are illustrated in Figure 19.13c and d, respectively.

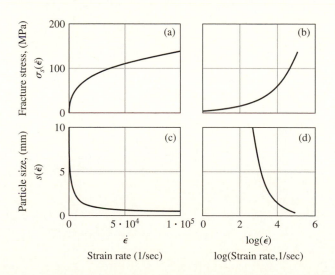

Figure 19.13

Results of Example 19.2: (a) fracture stress vs. strain rate; (b) fracture stress vs. log (strain rate); (c) particle size vs. strain rate; (d) particle size vs. log (strain rate).

Grady's equation, Eq. (19.30), was modified by Ross et al. [20] to account for the strain-rate dependence of fracture toughness and crack velocity in concrete. The resulting fracture strength is given as

$$\sigma_s^2 = \left\{ \frac{3K_{IA}^2 B\dot{\varepsilon}^{1-m}}{k\{1 - [k\dot{\varepsilon}^m/u_{cl}]^n\}^2} \right\}^{1/3}$$

(19.33)

where m is the slope and k is the intercept of the log (crack velocity) versus log (strain rate), K_{IA} is the arrest fracture toughness, B is the bulk modulus, u_{cl} is the limiting crack velocity, $\dot{\varepsilon}$ is strain rate, and n is an experimentally determined parameter for fracture toughness as a function of crack velocity. K_{IA} is the arrest fracture toughness when the crack is arrested or when the crack velocity goes to zero. This value for K_{IA} is not commonly measured and is not required if a ratio of fracture stresses is formed by dividing Eq. (19.33) with an expression for fracture stress at a quasi-static strain rate. This analytical ratio is shown in comparison to experimental values of normal weight concrete tensile strength in Figure 19.14. However, both lightweight and high-strength concrete exhibit slightly different strain-rate sensitivity characteristics from those illustrated in Figure 19.14.

High-intensity, short-duration impulse loads, which are the characteristic traits of a blast, impose high strain rates on structures. In this context, the "high" strain-rate region typically ranges from 10 to 10^4/sec. Therefore to accurately predict the response of structures to blast loading, it is important to ascertain the effects of these strain rates on structural materials, particularly concrete. Usually the strain-rate sensitivity data is used as a multiplier or dynamic increase factor (DIF) to the quasi-static strength data to obtain a fracture or yield strength at a higher strain rate. For computer calculations of structural and material response, the strain-rate sensitivity given by expressions such as Eqs. (19.28) to (19.33) are implemented directly into the material models employed by

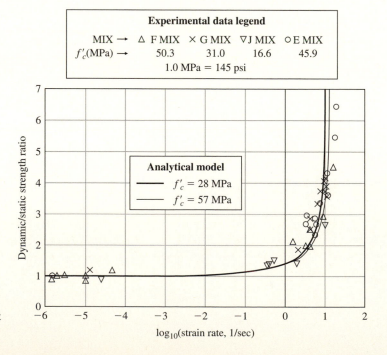

Figure 19.14

Fracture stress ratio vs. log (strain rate) for normal weight concrete.

computer programs. The strain-rate sensitivity of most materials is usually not a well-behaved phenomenon; however, some guidelines and practical applications are presented by Brubaker [21].

EXAMPLE 19.3 ▼

The tensile strength of concrete is often characterized by the modulus of rupture f_r. If a reinforced concrete beam is subjected to a dynamic bending moment reaching a maximum value of 50×10^6 in-lb in 10 msec, determine the dynamic modulus of rupture for the beam. Assume the beam has a section modulus S of 200 in^3 and the concrete has quasi-static values for modulus of rupture and modulus of elasticity of 500 psi and 5×10^6 psi, respectively.

Solution

The maximum tensile bending stress in the beam is calculated as

$$\sigma_b = \frac{M}{S} = \frac{50 \times 10^6 \text{ in-lb}}{200 \text{ in}^3} = 25 \times 10^4 \text{ psi}$$

and the stress rate is determined by

$$\dot{\sigma}_b = \frac{\sigma_b}{t} = \frac{25 \times 10^4 \text{ psi}}{10^{-2} \text{ sec}} = 2.5 \times 10^7 \text{ psi/sec}$$

The strain rate is then

$$\dot{\varepsilon} = \frac{\dot{\sigma}_b}{E} = \frac{2.5 \times 10^7 \text{ psi/sec}}{5 \times 10^6 \text{ psi}} = 5/\text{sec}$$

For a strain rate of 5/sec, the dynamic increase factor (DIF) for tension is estimated from Figure 19.11 as approximately 2.0. Therefore, the dynamic modulus of rupture f_{rd} is estimated as

$$f_{rd} = f_r \times (\text{DIF}) = 500(2) = 1000 \text{ psi}$$

▲

19.5 APPROXIMATE SOLUTION TECHNIQUE FOR SDOF SYSTEMS

For many preliminary design applications, an approximate solution for a blast-loaded structure is often desired or required. For near-field explosions where the explosive device is in close proximity to the structure, an almost ideal impulsive loading condition will exist. The definitive parameters for an impulsive load are: (1) the load duration and (2) the natural period of the structure. The natural period T is based on the fundamental flexural frequency of the structural element under load, and is defined as

$$T = \frac{2\pi}{\omega} = \frac{1}{f} \tag{19.34}$$

where ω (rad/sec) $= \sqrt{k/m}$ is the natural circular frequency for a SDOF system k is stiffness (N/m), m is mass (kg) and f (1/sec) is the natural frequency. When the duration of the loading pulse t_d (or t_r) is less than one-quarter of the natural period ($T/4$), the overall structural response is minimal, and the loading is termed impulsive.

For an impulsive loading, the effects of damping are negligible and therefore may be omitted in the response calculations. Thus the undamped system response can be evaluated analytically using the Duhamel integral procedure outlined in Section 6.3 or numerically by any of the numerical techniques discussed in Chapter 7. However, prior to conducting the actual analysis, the system must be reduced or transformed to an equivalent SDOF system.

The primary objective of the transformation is to reduce the system that exhibits distributed mass and stiffness characteristics to an equivalent SDOF system using an approach similar to the generalized coordinate approach previously discussed in Chapters 2 and 18. This transformation is based on an assumed deflected shape or an assumed mode for the structure. The deflected shape is usually normalized with respect to a unit displacement at a convenient location on the structure. For example, the elastic deflection equation for a simply supported beam of length L loaded at midspan with a concentrated load as illustrated in Figure 19.15a is expressed as

$$y(x) = \frac{P}{48EI}(3L^2x - 4x^3) \qquad 0 \le x \le \frac{L}{2} \tag{19.35}$$

where P = midspan load
E = Young's modulus
I = planar moment of inertia

The deflection at midspan ($x = L/2$) is then

$$y\left(\frac{L}{2}\right) = \frac{PL^3}{48EI} \tag{19.36}$$

Dividing Eq. (19.35) by Eq. (19.36), then the shape function $\psi(x)$ is written as

$$\psi(x) = \frac{1}{L^3}(3L^2x - 4x^3) \qquad 0 \le x \le \frac{L}{2} \tag{19.37}$$

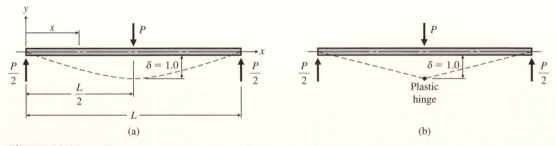

Figure 19.15
Schematic illustrating simply supported beam models with concentrated midspan loading: (a) elastic response; (b) plastic response.

If the beam forms a plastic hinge at the midspan, as shown in Figure 19.15b, then the shape function for this condition is expressed as

$$\psi(x) = \frac{2x}{L} \tag{19.38}$$

where the midspan deflection is unity. For a system with uniformly distributed mass and loading, an equivalent SDOF mass M_e and equivalent force F_e are defined as

$$M_e = \int \overline{m} \psi^2(x) \, dx \tag{19.39}$$

and

$$F_e = \int p(x)\psi(x) \, dx \tag{19.40}$$

where $\overline{m}$ = mass/length
$p(x)$ = force/length
$\psi(x)$ = shape function representative of a simply supported beam subjected to a uniformly distributed load

The load factor K_L and mass factor K_M are further defined as

$$K_L = \frac{F_e}{F_t} \tag{19.41}$$

and

$$K_M = \frac{M_e}{M_t} \tag{19.42}$$

where F_t is the total load equal to pL and M_t is total mass equal to $\overline{m}L$.

With these definitions the equation of motion for the equivalent SDOF system may be written as

$$K_M M_t \ddot{y} + K_L ky = K_L F(t) \tag{19.43}$$

Dividing through Eq. (19.43) by K_L yields

$$\frac{K_M}{K_L} M_t \ddot{y} + ky = F(t) \tag{19.44}$$

The ratio $(K_M/K_L = K_{LM})$ is called the load mass factor. This implies that only the mass requires a transformation, but this procedure will also yield an equivalent stiffness k_e and an equivalent period T_e defined as

$$T_e = 2\pi \sqrt{\frac{K_{LM} M_t}{k}} \tag{19.45}$$

To illustrate the effects of transforming the physical structure to an equivalent SDOF system, the case of a simply supported beam of length L subjected to an impulsive concentrated load at midspan is examined as illustrated in Figure 19.15a. The shape factor $\psi(x)$ is given by Eq. (19.37) and the load factor for concentrated loads is defined by

$$K_L = \frac{\sum F_r \psi(x_r)}{\sum F_r} \tag{19.46}$$

where F_r are the concentrated loads and $\psi(x_r)$ is the shape function evaluated at the position of the loads. Since there is only one load located at x equal to $L/2$, then $\Sigma F_r \psi(x_r)$ is equal to P multiplied by Eq. (19.37) evaluated at x equal to $L/2$. This results in $\psi(L/2)$ equal to 1, and therefore $\Sigma F_r \psi(x_r)$ is equal to P. The term ΣF_r is equal to P, and therefore $K_L = 1.0$. Since the mass is uniformly distributed, K_M is determined from Eqs. (19.39) and (19.42) as

$$K_M = \frac{\int_0^L \overline{m}\psi^2(x)\,dx}{\overline{m}L} = 2\int_0^{L/2}\left(\frac{3x}{L_2} - \frac{4x^3}{L^4}\right)^2 dx \tag{19.47}$$

where $\overline{m}$ is the mass per unit length. Equation (19.47) results in a value for $K_M = 0.49$, and the resulting K_{LM} also equal to 0.49. Using this value of K_{LM} in Eq. (19.45), the equivalent natural period T_e becomes

$$T_e = 2\pi\sqrt{K_{LM}\frac{M_t}{k}} \cong (0.7)2\pi\sqrt{\frac{M_t}{k}} \tag{19.48}$$

and the natural circular frequency ω is given as

$$\omega = \frac{2\pi}{T_e} \cong 1.43\sqrt{\frac{k}{M_t}} \tag{19.49}$$

Note that Eq. (19.49) yields a value for the natural frequency that is approximately 1.4 times greater than the frequency of a beam in which the mass is assumed to be concentrated at the midspan. This implies that a distributed mass system is stiffer than a system where the entire mass is concentrated at midspan, resulting in a higher frequency and in turn a lower period. These changes in stiffness and period will have a significant effect on the resulting response of the system. This underscores the importance of accounting for the effects of mass and load distribution when calculating the dynamic response of structural systems idealized as equivalent SDOF systems. Values for K_L, K_M, and K_{LM} for beams and slabs, for a variety of boundary conditions and for both elastic and elastoplastic responses, are presented in [22].

EXAMPLE 19.4 ▼

A simply supported S15 × 50 steel beam spanning 20 ft is subjected to an impulsive concentrated load at its midspan. Assuming the beam fails in flexure by forming a plastic hinge at midspan, calculate the effective natural period and natural frequency of the beam for (a) the mass uniformly distributed along the length and (b) the mass concentrated at midspan. Assume the total mass $M_t = 0.0833$ kip-sec^2/in and Young's modulus $E = 30,000$ ksi.

Solution

For the S15 × 50 the planar moment of inertia is given as $I_{xx} = 486$ in^4. The elastic stiffness k for a concentrated load at midspan is calculated as

$$k = \frac{48EI}{L^3} = \frac{48(30{,}000 \text{ ksi})(486 \text{ in}^4)}{(240 \text{ in})^3} = 50.6 \text{ kips/in} \tag{1}$$

and the shape function $\psi(x)$ for a simply supported beam forming a plastic hinge at midspan is given by Eq. (19.30) as

$$\psi(x) = \frac{2x}{L} \tag{2}$$

(a) The mass factor for the case of a uniformly distributed load is determined from Eq. (19.42). Noting that the mass per length $\bar{m} = M_t/L$, Eq. (19.42), used with Eq. (19.39), yields

$$K_M = \int_0^{L/2} \frac{\bar{m}\psi^2(x)\,dx}{\bar{m}L} = \frac{2}{L}\int_0^{L/2}\left(\frac{2x}{L}\right)^2 dx \tag{3}$$

or

$$K_M = 0.33 \tag{4}$$

(b) For concentrated masses M_i the mass factor is determined from the expression

$$K_M = \frac{\sum M_i \psi(x_i)}{\sum M_i} \tag{5}$$

For a single concentrated mass located at midspan ($x = L/2$), Eq. (5) results in

$$K_M = \frac{M\psi(L/2)}{M} \tag{6}$$

From Eq. (2), note that $\psi(L/2) = 1.0$; thus

$$K_M = 1.0 \tag{7}$$

(c) The load factor K_L for concentrated loads is given by Eq. (19.46). Since the mass distribution has no influence on K_L, Eq. (19.46) yields

$$K_L = \frac{\sum F_r \psi(x_r)}{\sum F_r} = \frac{\sum F_r \psi(L/2)}{\sum F_r} = 1.0 \tag{8}$$

(d) The load mass factor K_{LM} is given by

$$K_{LM} = K_L K_M \tag{9}$$

For the distributed mass case, Eq. (9) yields

$$K_{LM} = (1.0)(0.33) = 0.33 \tag{10}$$

and for the mass concentrated at midspan, Eq. (9) gives

$$K_{LM} = (1.0)(1.0) = 1.0 \tag{11}$$

(e) The equivalent natural period T_e and natural frequency f are determined from the expressions

$$T_e = 2\pi\sqrt{\frac{K_{LM}M}{k}} \tag{12}$$

and

$$\omega = \frac{2\pi}{T_e} \tag{13}$$

respectively. Note that Eq. (12) should be used with Eqs. (1) and (10) for the distributed mass case and with Eqs. (1) and (11) for the concentrated mass case. The results for both cases are summarized in Table 19.3.

The comparison presented in Table 19.3 illustrates the significant effect the mass distribution assumption has upon the dynamic characteristics of the beam.

TABLE 19.3. Effective Period and Natural Frequency

Case	Distributed Mass	Concentrated Mass
T_e (sec)	0.146	0.255
ω (rad/sec)	42.9	24.6

▲

REFERENCES

1 Baker, W.E., *Explosions in Air,* Winfred Baker Engineering, San Antonio, TX, 2nd Printing, 1983.

2 Kennedy, W.D., "Explosions and Explosions in Air," Chap. 2, Vol. 1, *Effects of Impact and Explosion,* Summary Tech. Rept. of Div. 2 NDRC, Washington, DC, 1946.

3 Drake, J.L., Twisdale, L.A., Frank, R.A., Dass, W.C., Rochefort, M.A., Walker, R.E., Britt, J.R., Murphy, C.E., Slawson, T.R., and Sues, R.H., "Protective Construction Design Manual: Airblast Effects (Section IV)," ESL-TR-87-57, Air Force Engineering & Services Laboratory, Tyndall AFB, FL, 32403, 1989.

4 Army TM 5-1300, Navy NAVFAC P-397, Air Force AFR-88-22, "Structures to Resist the Effects of Acciden-tal Explosions," Department of Defense Explosive Safety Board, Washington, DC, Nov. 1990.

5 Goodman, H.J., "Compiled Free-Air Blast Data on Bare Spherical Pentolite," BRL Report No. 1092, Aberdeen Proving Ground, MD, 1960.

6 Brode, H.L., "Numerical Solutions of Spherical Blast Wave," *Journal of Applied Physics,* Vol. 26, No. 6, pp. 766–775, 1955.

7 Army TB 700-2, Navy NAVORDCENINST 8020.2, Air Force T011A-1-47, Defense Logistics Agency DLAR 8220.1, "Department of Defense Ammunition and Explosive Hazard Classification Procedures," Depts. of the Army, the Navy, the Air Force, and the Defense Logistics Agency, Washington, DC, 1994.

8 "Fundamentals of Protective Design for Conventional Weapons," Dept. of the Army, Waterways Experiment Station, Vicksburg, MS, 1984.

9 Crawford, R.E., *Protection from Non-Nuclear Weapons,* AFWL-TR-70-127, Air Force Weapons Laboratories, Kirtland AFB, NM, Feb. 1971.

10 Nicholas, T., "Tensile Testing of Materials at High Rates of Strain," *Experimental Mechanics,* Vol. 21, No. 5, pp. 177–185, May 1981.

11 Ross, C.A., "Dynamic Compressive Properties of a Metal Matrix Composite Material," Ph.D. Dissertation, University of Florida, Gainesville, FL 1971.

12 Ross, C.A., "Fracture of Concrete at High Strain-Rate," *Toughening Mechanisms in Quasi-Brittle Materials,* S.P. Shah, editor, Kluwer Academic Publishing, Boston, pp. 577–596, 1991.

13 Kormeling, H.A., Zielinski, A.J., and Reinhardt, H.W., "Experiments on Concrete Under Single and Repeated Uniaxial Impact Tensile Loading," Rept. 5-80-3, Steven Lab., Delft University of Technology, Delft, The Netherlands, 1980.

14 Malvern, L.E., "The Propagation of Longitudinal Waves of Plastic Deformation in a Bar of Material Exhibiting a Strain-Rate Effect," *Journal of Applied Mechanics,* Trans. ASME, 18, pp. 203–208, 1951.

15 Johnson, G.R., and Cook, W.H., "A Constitutive Model and Data for Metals Subjected to Large Strains, High Strain Rates and High Temperatures," *Proceedings of the 7th International Symposium on Ballistics,* The Hague, The Netherlands, pp. 541–547, 1983.

16 Nicholas, T. and Rajendran, A.M., "Material Characterization at High Strain Rates," Chap. 3 of *High Velocity Impact Dynamics ,* J.A. Zakas, editor, Wiley, New York, pp. 127–319, 1990.

17 Grady, D.E., "The Mechanics of Fracture Under High Rate Stress Loading," Sandia National Lab., Rept. SAND-82-1148C, 1983.

18 Grady, D.E. and Lipkin, J., "Criteria for Impulsive Rock Fracture," *Geophysical Research Letters,* Vol. 7, No. 4, pp. 255–258, April 1980.

19 Grady, D.E., "Mechanisms of Dynamic Fragmentation: Factors Governing Fragment Size," Sandia National Laboratory Report SAND-84-2304C, 1983.

20 Ross, C.A., Jerome, D.M., Tedesco, J.W., and Hughes, M.L., "Moisture and Strain Rate Effects on Concrete Strength," *ACI Materials Journal,* Vol. 93, No. 3, pp. 293–300, May–June 1996.

21 Brubaker, D., "Modeling Strain-Rate-Dependent Tensile Failure in Concrete," Master's Thesis, University of Florida, Gainesville, FL 1995.

22 Biggs, J.M., *Introduction to Structural Dynamics,* McGraw-Hill, New York, 1964.

23 Weerheijm, J., "Concrete Under Impact, Tensile Loading, and Lateral Compression," Ph.D. Dissertation at Delft University of Technology, Delft, The Netherlands, 1992.

NOTATION

a	local speed of sound	I	static moment of inertia of cross section
a_r	sonic velocity of (reflected) wavefront	k	ratio of specific heats; also represents stiffness
a_x	sonic velocity of (incident) wavefront	k_e	equivalent stiffness
B	bulk modulus	K_{IA}	arrest fracture toughness
c	wave speed	K_{IC}	mode I quasi-static fracture toughness
c_0	dilatation wave speed	K_L	load factor
C_r	reflection coefficient (factor)	K_{LM}	load mass factor
E	Young's modulus	K_M	mass factor
f	undamped natural frequency (cps)	m	mass
f_r	modulus of rupture	$\bar{m}$	mass per unit length
f_{rd}	dynamic modulus of rupture	M	Mach number
F_e	equivalent force	M_e	equivalent mass
F_t	total force	M_s	Mach number of Mach stem
G	shear modulus	M_t	total mass
i_r	reflected specific impulse	M_x	Mach number of wavefront

p	pressure	v	specific volume
p_r	reflected pressure	W	charge weight
p_x	air pressure ahead of the shock front	Z	normalized standoff
p_y	air pressure behind the shock front	DIF	dynamic increase factor
P_r	reflected overpressure	TNT	trinitrotoluene
P_0	atmospheric pressure	α	incidence angle of shock front with target surface
P_{so}	pressure rise (overpressure) or side-on pressure	δ	reflected angle of wavefront with target surface
R	universal gas constant; also used to represent standoff	Δ	volumetric strain (dilatation)
t	time	ε_{ii}	principal strains
t_d	duration of loading pulse	$\dot{\varepsilon}$	strain rate
t_r	reflected positive pressure duration	λ	Lamé's constant
t_s	time of fracture	ρ	mass density
T	absolute temperature	σ_m	mean stress
T_e	equivalent period	σ_{ii}	principal stresses
u_{cl}	limiting crack velocity	σ_s	fracture strength
$\dot{u}$	local velocity	ω	undamped natural circular frequency (rad/sec)
$\dot{u}_r$	velocity of (reflected) wavefront	$\psi(x)$	shape function
$\dot{u}_x$	velocity of (incident) wavefront		

PROBLEMS

19.1   Assuming air is a perfect gas with a ratio of specific heats $k = 1.4$, plot the overpressure versus Mach number across a normal shock wave whose Mach number varies between 1.0 and 5.0. Note that atmospheric pressure is equal to 0.1 MPa.

19.2   For normal reflection determine and plot the reflection factor for the condition presented in Problem 19.1.

19.3   Assuming the critical incidence angle for regular reflection is 40°, determine the horizontal position on the ground where the Mach stem begins if the location of the blast is 10 m above the ground surface.

19.4   Assuming a spherical blast in free air of 122 lb of ANFO, determine the reflected pressure at a distance of 18 ft from the explosive. Assuming the effectiveness in terms of impulse is the same as for pressure, determine the reflected triangular-shaped pressure curve.

19.5   A spherical charge of TNT having a weight of 50 lb is detonated in free air 20 ft above the midpoint of a rectangular concrete slab measuring 25 ft by 30 ft in plan. Determine the peak reflected pressure and the reflected specific impulse at the center points, quarter points, and endpoints of the slab. Assuming a triangular pressure time history, determine the reflected positive pressure duration at each of these locations.

19.6   Repeat Problem 19.5 for a spherical charge of Amatol weighing 75 lb detonated 25 ft above the midpoint of the slab.

19.7   A spherical charge of tritonal 8/20 located 10 ft above the ground surface and weighing 500 lb is detonated 30 ft from a two-story concrete structure as shown in Figure P19.1a. The face of the building closest to the charge consists of four wall panels as shown in Figure P19.1b. Determine the peak reflected pressure and the reflected specific impulse at the center point of each of the wall panels. Assuming a triangular pressure time history, determine the reflected positive pressure duration at each of these locations.

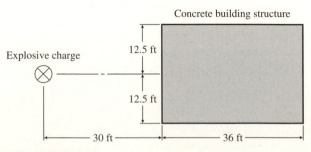

Figure P19.1(a)
Location of explosive charge with respect to building structure (plan view).

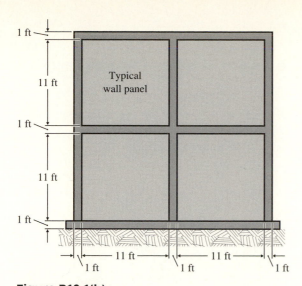

1 ft

11 ft

Typical
wall panel

1 ft

11 ft

1 ft

11 ft   11 ft
1 ft   1 ft   1 ft

Figure P19.1(b)
Elevation of building wall facing explosive charge.

19.8 Repeat Problem 19.7 for a hemispherical charge of Torpex-2 weighing 600 lb detonated at the same location on the ground surface.

19.9 A simply supported reinforced concrete beam spanning 100 in is loaded by a blast from 64 lb of TNT, centered 20 ft above the beam. The beam is 12 in wide and 10 in deep. The specific weight of the beam is 150 lb/ft^3 and has a Young's modulus of 5×10^6 psi. Assuming an average pressure loading with a triangular shape, use the solution of Example 6.4 to obtain the midspan displacement at the time corresponding to the end of the applied load. Use the equivalent solution technique discussed in Section 19.5.

20 ▲ Basic Concepts of Water Waves

Surface water waves can cause very large dynamic forces on structures. They can also generate currents, sediment transport, coastal erosion, harbor resonance, hazards to navigation, and many other influences. As a result, wind generated free surface waves have been studied in great detail. A full description of surface waves requires consideration of many physical aspects such as bottom conditions (geometry, roughness, permeability, deformation), surface conditions (atmospheric pressure, wind, surface tension, breaking), and flow and fluid conditions (currents, compressibility, viscous and turbulent dissipation). As a result, there is no single analytic solution that fully describes all wave conditions. Rather, there are a number of solutions that are applicable over certain ranges of conditions. The simplest (and most common) wave theory is linear wave theory, sometimes referred to as small amplitude theory or Airy wave theory after the nineteenth century astronomer George Airy.

20.1 LINEAR WAVE THEORY

The primary variables used to describe waves are the wave height H, which is the crest to trough vertical distance; the wave period T, which is the time between the passage of successive wave crests; the wave length L, which is the horizontal distance between wave crests; and the still water depth h, which is the vertical distance from the bottom to the free surface if no waves are present. The speed at which the wave crest moves is the wave celerity c. The free surface profile η is the vertical distance measured from the still water level (SWL). A definition sketch for these wave parameters is presented in Figure 20.1. In linear wave theory, the height of the crest a_c and the depth of the trough a_t are both equal to $H/2$ and are centered on the SWL. In other wave theories and often in nature, this is not necessarily the case and the crest height is greater than the trough depth, and the average position of the free surface is not the same as the still water level. It is important to note in Figure 20.1 the definition of the horizontal and vertical coordinates x and z, where z is defined positive up under the wave crest. This has an

Figure 20.1

Definition of surface waves.

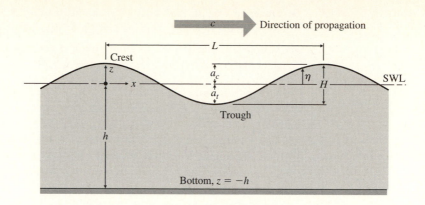

impact on how η is defined and specifies that water beneath the SWL corresponds to negative values of z. For convenience, two short-hand variables are often used to describe waves: the radian wave frequency $\omega = 2\pi/T$ and the wave number $k = 2\pi/L$. These are used because waves are periodic in both space x and time t. Because there is only one wave period T, these waves are termed monochromatic.

In linear wave theory, a number of simplifying assumptions are imposed: incompressible fluid; irrotational flow (i.e., no dissipation); horizontal, impermeable, rigid bottom; no surface stresses (pressure, wind, or surface tension); and no currents. For waves in intermediate to deep water depths over a silt or sand bottom, most of these assumptions are generally acceptable. However, an additional assumption is that the wave height is small with respect to the wave length. For linear wave theory to be valid, $H/L << 1$ or, in other words, the wave height must be nearly zero with respect to the wave length. This is a very severe assumption, but is invoked because it allows the development of a rather simple analytical description of the waves.

The irrotational flow assumption allows the introduction of a velocity potential ϕ, where the horizontal and vertical water particle velocities u and w are given by

$$(u, w) = \left[\frac{\partial \phi}{\partial x}, \frac{\partial \phi}{\partial z} \right] \tag{20.1}$$

For an incompressible fluid, conservation of mass yields

$$\nabla^2 \phi = 0 \tag{20.2}$$

which is the *Laplace equation*. This equation arises in many areas of engineering, such as groundwater flow and magnetic fields. The boundary conditions are as follows:

1. Bottom boundary condition (BBC): no flow into the horizontal, rigid, impermeable bottom.

2. Kinematic free surface boundary condition (KFSBC): the velocity of a water particle on the free surface is the same as the velocity of the free surface.

3. **Dynamic free surface boundary condition (DFSBC):** the assumption of zero stress on the free surface is used in the unsteady Bernoulli equation.

4. **Simple periodic:** the waves are periodic in time t and in space x (i.e., sinusoids) such that $f(x, z, t) = f(x, z, t + T)$ and $f(x, z, t) = f(x + L, z, t)$.

Two boundary conditions are required because the surface is a free boundary and will move if the forces do not balance. In a sense, one condition is to define the location of the boundary, and the other specifies the value of ϕ on the boundary. Figure 20.2 summarizes the boundary value problem for linear wave theory. In Figure 20.2, g is the acceleration due to gravity, a_x and $\mathbf{e}_x$ are the horizontal water particle acceleration and horizontal unit vector, and a_z and $\mathbf{e}_z$ are the vertical water particle acceleration and vertical unit vector.

Note that the free surface boundary conditions are applied at the SWL and not on the actual free surface η. This is because the boundary value problem with the boundary conditions applied on η is very difficult to solve. As an analytical convenience, the boundary is imposed at the SWL. This greatly simplifies the mathematics, but to have any validity, η must remain very close to the SWL. This requires that $H/L \ll 1$. Also note that in the DFSBC, which is based on the unsteady Bernoulli equation, the velocity squared term has not been included. The velocity is related to H/L, and if this is small, then $(H/L)^2$ is very small and can be neglected. These simplifications are the basis for the name, linear wave theory, since only linear terms in H/L are retained in the boundary value problem. Details of the solution technique are given in many coastal and ocean engineering texts [1, 2, 3]. Table 20.1 summarizes the solution for linear wave theory. These results are specific to the choice made for the coordinate system defined in Figure 20.1 and water particle kinematics expressed by Eq. (20.1). A FORTRAN computer program, LWT_AMP.FOR, that calculates the amplitudes of typical linear wave theory results is available on the authors' web site. The computer program LWT.FOR determines the time-dependent response at a given location and is also available on the authors' web site.

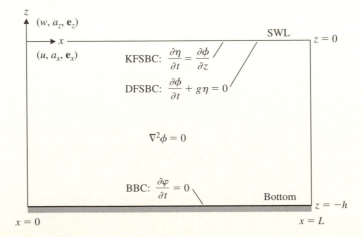

Figure 20.2

Summary of LWT boundary value problem.

TABLE 20.1. Summary of Linear Wave Theory (LWT)

Free surface	$\eta = \dfrac{H}{2}\cos(kx - \omega t)$
Velocity potential	$\phi = \dfrac{H}{2}\dfrac{g}{\omega}\dfrac{\cosh[k(h + z)]}{\cosh(kh)}\sin(kx - \omega t)$
Dispersion equation	$\omega^2 = gk\tanh(kh)$ or $L = \dfrac{gT^2}{2\pi}\tanh(kh)$
Celerity	$c = \dfrac{L}{T} = \dfrac{\omega}{k}$
Horizontal particle displacement	$\xi = -\dfrac{H}{2}\dfrac{\cosh[k(h + z)]}{\sinh(kh)}\sin(kx - \omega t)$
Vertical particle displacement	$\zeta = \dfrac{H}{2}\dfrac{\sinh[k(h + z)]}{\sinh(kh)}\cos(kx - \omega t)$
Horizontal water particle velocity	$u = \dfrac{H}{2}\dfrac{gk}{\omega}\dfrac{\cosh[k(h + z)]}{\cosh(kh)}\cos(kx - \omega t)$
Vertical water particle velocity	$w = \dfrac{H}{2}\dfrac{gk}{\omega}\dfrac{\sinh[k(h + z)]}{\cosh(kh)}\sin(kx - \omega t)$
Horizontal water particle acceleration	$a_x = \dfrac{H}{2}gk\dfrac{\cosh[k(h + z)]}{\cosh(kh)}\sin(kx - \omega t)$
Vertical water particle acceleration	$a_z = -\dfrac{H}{2}gk\dfrac{\sinh[k(h + z)]}{\cosh(kh)}\cos(kx - \omega t)$
Pressure	$p = \rho g\dfrac{H}{2}\dfrac{gk}{\omega}\dfrac{\cosh[k(h + z)]}{\cosh(kh)}\cos(kx - \omega t) - \rho g z$

20.1.1 Dispersion Equation

The *dispersion equation* defines the relationship among the wave frequency, wave number, and water depth. The dispersion equation may be written using either the wave number and wave frequency or the wave length and wave period:

$$\omega^2 = gk \tanh(kh) \quad \text{or} \quad L = \frac{gT^2}{2\pi} \tanh\left(\frac{2\pi h}{L}\right) \tag{20.3}$$

If the argument of the tanh term in Eq. (20.3) is large (which corresponds to h/L being large), then tanh $\rightarrow$ 1. This is referred to as deep water, for which the wave length is given by

$$L_0 = \frac{gT^2}{2\pi} \tag{20.4}$$

Since deep water wave conditions are frequently used as a reference value, they are denoted with a ( )$_0$ subscript. In Eq. (20.4), $L_0 = 1.56\,T^2$ for SI units (L in meters) and $L_0 = 5.12\,T^2$ in English units (L_0 in feet). Since deep water implies that h/L is large, then in terms of wave length, deep water is characterized by $h/L > 1/2$. At the other extreme is shallow water, which is usually taken as $h/L < 1/20$. For shallow water, the small argument approximations for the hyperbolic functions may be used. For shallow water, $\tanh(kh) \rightarrow kh$, so the shallow water wave length is $L = (gh)^{1/2}\,T$. Between these extreme conditions, the depth is intermediate and no simplifying approximations in the hyperbolic functions can be made.

For a specified wave period and water depth, the dispersion equation is transcendental and must be solved numerically for the wave length. This is a rather unfortunate circumstance because the dispersion equation must be routinely solved when using linear wave theory. However, there are several commonly employed alternative solution techniques: trial-and-error solutions; use of numerical solvers on calculators; numerical solutions on digital computers using techniques such as Newton-Raphson or half-interval; and graphical solutions, table look-ups, and analytical approximations. A graphical estimation to a solution to the dispersion equation is presented in Figure 20.3. The table look-up approach is given in appendix C of the *Shore Protection Manual* [4]. A portion of this appendix is presented in Table 20.2. A simple analytical approximation is given by

$$L = (2\pi h L_0)^{1/2}\left(1 - \frac{\pi h}{3L_0}\right) \tag{20.5}$$

This equation has a relative error of less than 2% when $h/L_0 < 0.3$. A second analytical approximation is given in Table 20.3, which is based on Pade approximates and is valid over a wide range of water depths. This FORTRAN subroutine may be easily implemented in numerical models. It is also available on the authors' web site.

As a wave propagates into shallow water, the wave length decreases. For example, a wave with a period of 10 sec has a deep water wave length of $L_0 = 156$ m. In depths of 50 m and 10 m, the lengths are $L = 151$ m and $L = 92$ m, respectively. Since the wave celerity is $c = L/T$, the wave slows down as the depth decreases. Note that in linear wave theory (and most other wave theories) the period is assumed to remain constant, which is consistent with observations.

Figure 20.3

Graphical solution of LWT dispersion equation.

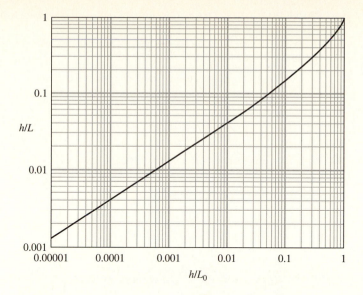

TABLE 20.2. Solutions to the Linear Wave Theory Dispersion Equation [4]

h/L_0	h/L	h/L_0	h/L	h/L_0	h/L
0.000100	0.003990	0.06500	0.1092	0.3400	0.3468
0.000500	0.008925	0.07000	0.1139	0.3500	0.3579
0.001000	0.01263	0.07500	0.1186	0.3600	0.3672
0.001500	0.01548	0.08000	0.1232	0.3700	0.3766
0.002000	0.01788	0.08500	0.1277	0.3800	0.3860
0.002500	0.02000	0.09000	0.1322	0.3900	0.3955
0.003000	0.02192	0.09500	0.1366	0.4000	0.4050
0.003500	0.02369	0.1000	0.1410	0.4100	0.4145
0.004000	0.02534	0.1100	0.1496	0.4200	0.4241
0.004500	0.02689	0.1200	0.1581	0.4300	0.4337
0.005000	0.02836	0.1300	0.1665	0.4400	0.4434
0.005500	0.02976	0.1400	0.1749	0.4500	0.4531
0.006000	0.03110	0.1500	0.1833	0.4600	0.4628
0.006500	0.03238	0.1600	0.1917	0.4700	0.4725
0.007000	0.03362	0.1700	0.2000	0.4800	0.4822
0.007500	0.03482	0.1800	0.2083	0.4900	0.4920
0.008000	0.03598	0.1900	0.2167	0.5000	0.5018
0.008500	0.03711	0.2000	0.2251	0.5100	0.5117
0.009000	0.03821	0.2100	0.2336	0.5200	0.5215
0.009500	0.03928	0.2200	0.2421	0.5300	0.5314
0.01000	0.04032	0.2300	0.2506	0.5400	0.5412
0.01500	0.04964	0.2400	0.2592	0.5500	0.5511
0.02000	0.05763	0.2500	0.2679	0.5600	0.5610
0.02500	0.06478	0.2600	0.2766	0.5700	0.5709
0.03000	0.07135	0.2700	0.2854	0.5800	0.5808
0.03500	0.07748	0.2800	0.2942	0.5900	0.5907
0.04000	0.08329	0.2900	0.3031	0.6000	0.6006
0.04500	0.08883	0.3000	0.3121	0.7000	0.7002
0.05000	0.09416	0.3100	0.3211	0.8000	0.8001
0.05500	0.09930	0.3200	0.3302	0.9000	0.9000
0.06000	0.1043	0.3300	0.3394	1.000	1.000

TABLE 20.3. **FORTRAN Subroutine for Pade Approximates to the Dispersion Equation [5]**

```
        SUBROUTINE PADE (DEPTH, PERIOD, GRAVITY, LENGTH)
C       This subroutine gives a solution to the linear wave theory
C       dispersion equation using Pade approximates.

C       DEPTH     STILL WATER DEPTH
C       PERIOD    WAVE PERIOD
C       GRAVITY   ACCELERATION DUE TO GRAVITY
C       LENGTH    WAVELENGTH

        REAL DEPTH, PERIOD, GRAVITY, LENGTH, C(6)

        DATA C/0.666, 0.355, 0.1608465608, 0.0632098765,
     1  0.0217540484, 0.0065407983/

        PI=4.*ATAN(1.)
        Y=DEPTH*(2.*PI/PERIOD)**2/GRAVITY
        SUM=0
        DO 100 I=1,6
    100 SUM=SUM+C(I)*Y**I
        LENGTH=2.*PI*DEPTH/SQRT(Y**2+Y/(1.+SUM))
        RETURN
        END
```

20.1.2 Water Particle Velocities

The water particle velocities exhibit an interesting behavior. Even though the wave crest propagates forward with celerity c, the water particles do not advance. Rather, they move in closed orbits. For the case where the water depth is very deep, the large argument approximations may be used for the hyperbolic functions. The amplitudes of the horizontal and vertical velocities are the same and both decrease exponentially with depth. Recalling that z is negative below the SWL, the two deep water velocity components are expressed as

$$u_0 = \frac{\pi H}{T} e^{kz} \cos(kx - \omega t) \tag{20.6a}$$

$$w_0 = \frac{\pi H}{T} e^{kz} \cos(kx - \omega t) \tag{20.6b}$$

and may be summed to yield the velocity vector given by

$$\boldsymbol{q} = u\boldsymbol{e}_x + w\boldsymbol{e}_z = \frac{\pi H}{T} e^{kz} [\cos(kx - \omega t)\boldsymbol{e}_x + \sin(kx - \omega t)\boldsymbol{e}_z] \tag{20.7}$$

where $\boldsymbol{e}_x$ and $\boldsymbol{e}_z$ are the unit vectors in the x and z directions. The term in brackets on the right-hand side of Eq. (20.7) is the equation for a unit circle. The paths that the water particles follow are simply closed circles with diameters that decrease with depth. In intermediate water depths the orbits are influenced by the bottom and are flattened into ellipses. Figure 20.4 illustrates wave particle orbits for different water depths.

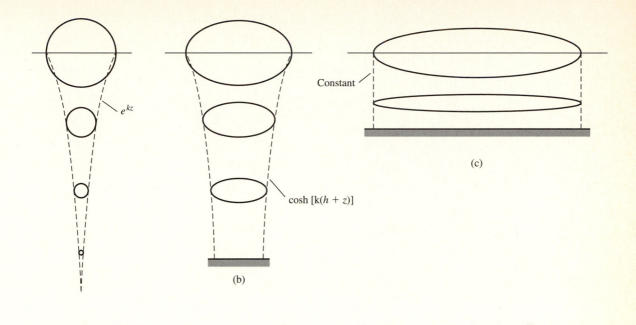

Figure 20.4
Wave orbits in (a) deep, (b) intermediate, and (c) shallow water depths. *Note:* The size of the orbits has been greatly exaggerated with respect to the water depth.

EXAMPLE 20.1 ▼

Determine the amplitudes of the horizontal velocity u and the vertical velocity w at mid-depth for a wave having height $H = 8$ ft and period $T = 10$ sec in a water depth $h = 60$ ft.

Solution

(a) First the wave length must be determined. The wave length will be estimated using three techniques previously discussed. The simple approximation in Eq. (20.5) gives

$$L_0 = \frac{gT^2}{2\pi} = 5.12T^2 = 5.12(10)^2 = 512 \text{ ft}$$

$$L = (2\pi hL_0)^{1/2}\left(1 - \frac{\pi h}{3L_0}\right) = [2\pi(60)(512)]^{1/2}\left[1 - \frac{\pi(60)}{3(512)}\right]$$

$$L = 385.4 \text{ ft}$$

Table 20.2 gives (using linear interpolation)

$$\frac{h}{L_0} = \frac{60}{512} = 0.1172$$

$$\frac{h}{L} = 0.1496 + \frac{0.1172 - 0.1100}{0.1200 - 0.1100}(0.1581 - 0.1496) = 0.1471$$

$$L = \frac{60}{0.1557} = 385.3 \text{ ft}$$

Using the Pade approximation gives

$$L = 385.3 \text{ ft}$$

(b) The amplitude of the horizontal velocity is given in Table 20.1 as

$$u = \frac{H}{2}\frac{gk}{\omega}\frac{\cosh[k(h+z)]}{\cosh(kh)}$$

where

$$z = -\frac{h}{2} = -30 \qquad \text{(recall } z \text{ is negative below the SWL)}$$

$$\omega = \frac{2\pi}{T} = \frac{2\pi}{10} = 0.6283 \text{ sec}^{-1}$$

$$k = \frac{2\pi}{L} = \frac{2\pi}{385.3} = 0.01631 \text{ ft}^{-1}$$

Thus

$$u = \frac{8}{2}\frac{32.18(0.01631)}{0.6283}\frac{\cosh[0.01631(60-30)]}{\cosh[0.01631(60)]} = 2.47 \text{ ft/sec}$$

(c) The amplitude of the vertical velocity is also specified in Table 20.1 and is

$$w = \frac{H}{2}\frac{gk}{\omega}\frac{\sinh[k(h+z)]}{\cosh(kh)} = 1.12 \text{ ft/sec}$$

▲

20.1.3 Wave Energy

The energy in linear waves is defined by the expression

$$E = \frac{1}{8}\rho g H^2 \tag{20.8}$$

where E is termed the wave energy density because it is the energy per unit area of horizontal surface. Wave energy flux, or wave power, is the rate at which energy is transmitted and is equal to the product of the energy density and the group velocity. The group velocity is defined as the speed at which a group of waves propagates, and is slower than the wave celerity until the waves reach shallow water. To clarify the concept, consider a group of waves propagating into still water. As a wave moves to the front of the group, it expends part of its kinetic energy to displace the still water. As this happens, the next wave advances to the front of the group and the process repeats. The ratio of the group velocity to the wave celerity is expressed as

$$\frac{c_g}{c} = \frac{1}{2}\left[1 + \frac{2kh}{\sinh(2kh)}\right] = n \tag{20.9}$$

20.2 NONLINEAR WAVES

As waves become steeper, the crests become higher and more peaked and the troughs become shallower and longer as shown in Figure 20.5. The amplitude of the crest may be much larger than the amplitude of the trough. As a result, the forces associated with the crest are much larger.

Unfortunately, in many situations linear wave theory does not provide sufficient accuracy for design purposes. There are several reasons for the inaccuracy: (1) it is often necessary to determine wave kinematics near the free surface, where LWT is least accurate, and (2) designs are usually considered for extreme wave conditions for which H/L is not small. Therefore, for these cases, more sophisticated (and complicated) wave theories must be employed. The two most common analytical wave theories are Stokes and cnoidal, and stream function is a commonly used numerical wave theory.

20.2.1 Stokes Wave Theory

In the derivation of LWT it was assumed that $H/L << 1$. Let this requirement be slightly relaxed and assume that $H/L = O(\varepsilon)$, where ε is small. The velocity potential for Stokes wave theory may be written

$$\phi(x, z, t) = \frac{c}{k}\sum_{n=1}^{N} \varepsilon^n \lambda_n \cosh[nk(h+z)]\sin[n(kx - \omega t)] \tag{20.10}$$

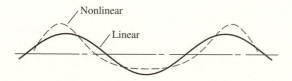

Figure 20.5

Comparison of linear and nonlinear free-surface profiles.

where N is the order of the solution and each λ_n is a constant that is a complicated function of kh. If ε is very small, then terms related to ε^2, ε^3, etc., are negligible and the solution reduces to the LWT problem. However, if ε is not small, it may be necessary to retain the ε^2 terms, that is, $N = 2$. This solution is referred to as *Stokes 2nd order*. To obtain this solution, the problem defined in ε must be solved first. The resulting terms from this solution then become the forcing terms in the ε^2 problem. As a result, the solution at ε^2 is rather lengthy. In a similar manner, to obtain Stokes 5th, this process is repeated for ε^3, ε^4, and ε^5. The resulting equations are very lengthy and are not presented here.

This solution technique is referred to as a perturbation method. Unfortunately, there are several different methodologies available to implement the perturbation. There is also an assumption of whether the mean velocity or momentum is zero. Therefore, several solutions exist in the literature [6, 7]. A FORTRAN computer program, based on the solution presented in [6], that calculates Stokes 5th wave theory results (STOKES_5.FOR) is available on the authors' web site.

20.2.2 Cnoidal Wave Theory

There are a number of other analytical wave theories available, one of which is cnoidal wave theory. Cnoidal wave theory is most applicable for steep waves in shallow water [8]. Like Stokes wave theory, higher order cnoidal wave theory requires a perturbation solution technique. Recall that for Stokes wave theory the perturbation parameter is $\varepsilon = H/L$, and an auxiliary parameter h/L (or kh) appears throughout the solution. In cnoidal theory, the perturbation parameter is $\varepsilon_c = H/h$, and the auxiliary parameter HL^2/h^3 appears. This auxiliary parameter is called the Ursell number U_r, which is the ratio $\varepsilon_c^3/\varepsilon^2$. If U_r is small, then the Stokes term is large. Conversely, if U_r is large, the cnoidal term dominates. As a general rule of practice,

$$U_r < 25 \qquad \text{Stokes wave theory is applicable}$$

$$\text{(20.11)}$$

$$U_r > 25 \qquad \text{cnoidal wave theory is applicable}$$

A solution for cnoidal waves is available through the U.S. Army Corps of Engineers [9]. A FORTRAN computer program, CNOIDAL.FOR, to calculate cnoidal wave theory results is included on the authors' web site.

Figure 20.6 illustrates LWT, Stokes 5th order, and 3rd-order cnoidal wave theory free surface profiles for $H = 1$ m, $T = 10$ sec, and $h = 6.5$ m ($U_r = 23.4$). Stokes theory overpredicts the wave amplitude. In fact, in shallow water, LWT may yield more accurate estimates than Stokes. Figure 20.7 indicates the region in which these three analytical theories provide the best estimate, based on errors in the dynamic free surface boundary condition. This curve is for reasonably steep waves with $H/H_B > 1/4$, where H_B is the breaking wave height. For shallow conditions, cnoidal wave theory provides the best estimates, and for deep water Stokes 5th yields the more accurate results. It is rather surprising, however, that over a range of intermediate water depths, LWT provides the best estimates.

Figure 20.6

Comparison of LWT, Stokes 5th order, and 3rd-order cnoidal wave theories for $H = 1$ m, $T = 10$ sec, and $h = 6.5$ m ($U_r = 23.2$).

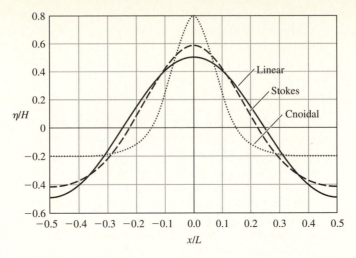

Figure 20.7

Periodic analytic wave theories providing the best fit to the dynamic free-surface boundary condition [10]. Adapted from *Evaluation and Development of Water Wave Theories for Engineering Applications* by R.G. Dean, courtesy of U.S. Corps of Engineers, Vols. I and II, 1974.

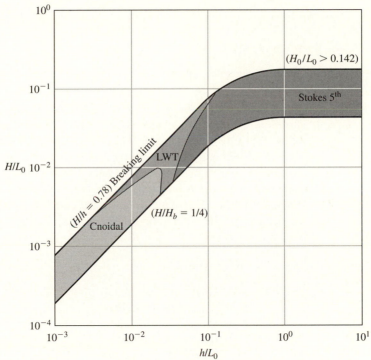

20.2.3   Stream Function Wave Theory

The development of Stokes 5th order wave theory is very laborious, and cnoidal theory involves special functions. To actually calculate a result from either of these theories requires a computer solution. An alternative is to simply seek a numerical solution. Stream function wave theory is one such solution. Since waves propagate at celerity, c, if we define a coordinate system that moves at this speed, then the wave form does not change. The problem can then be examined in terms of a stream function with the free surface being one stream line. For 2-D flow, the stream function has only one component and, therefore, may be treated as a scalar. In 2-D, the horizontal and vertical velocities are defined by

$$(u, w) = \left(\frac{\partial \Psi}{\partial z}, -\frac{\partial \Psi}{\partial x} \right) \qquad (20.12)$$

Stream function wave theory includes a rather clever aspect in the solution technique. In most wave theories, the greatest inaccuracies occur in predicting the kinematics at the free surface, which is often the location of interest in design. In stream function theory, the errors in free surface boundary conditions are numerically minimized in obtaining the solution. As a result, stream function estimates for kinematics near the free surface are generally more accurate than those predicted by Stokes 5th. However, since stream function wave theory is a numerical solution, application requires either a stream function program or the use of tables. Tabulated results are given by Dean [10] and a computer solution is given in [9].

20.3 WAVE TRANSFORMATIONS

As waves propagate toward the shoreline, they undergo a variety of transformations. These transformations include shoaling, refraction, diffraction, dissipation, reflection, and breaking. To date, no single theory has been developed that fully accounts for all of these physical processes. As a result, for practical considerations, some type of approximation is required to estimate these processes for design. Many of the simple methods commonly used in practice are discussed in publications such as the *Shore Protection Manual* [4]. The simplified approach presented assumes that each transformation is independent of the others. Under this assumption the wave height, before breaking occurs, is given by

$$H = K_S K_R K_D K_F H_0 \qquad (20.13)$$

where H = local wave height
H_0 = deep water wave height
K_S = shoaling coefficient
K_R = refraction coefficient
K_D = diffraction coefficient
K_F = dissipation coefficient

The examination of coupled wave transformations requires the use of numerical models.

20.3.1 Shoaling

Wave shoaling is simply a concept that requires the wave energy flux or wave power between two depths to remain constant. Recall that energy flux is the product of the wave energy density and group velocity. Since the group velocity decreases as the water depth decreases, the wave energy density must increase correspondingly to conserve energy flux. Stated another way, as waves propagate into shallower water, the wave length decreases, the wave celerity lessens, and the wave height increases. For linear wave theory, the shoaling coefficient K_s is given by

$$K_S = [2n \tanh(kh)]^{-1/2} \tag{20.14}$$

which is simply the square root of the deep water group velocity over the local velocity.

20.3.2 Refraction

Refraction is a very important wave transformation. It can cause wave energy to focus on certain locations of the shoreline, and it also determines the wave angle at which waves approach a structure. As waves propagate into shallow water, the crests tend to align themselves with the bottom contours. This occurs because the waves in deeper water travel faster than the waves in shallower water. It is more convenient to visualize the influence of refraction by examining wave rays. A *wave ray* is defined as the orthogonal to the local crest alignment pointing in the direction of wave propagation. Figure 20.8 shows wave rays for four different configurations. In Figure 20.8a, waves obliquely approach a shoreline with straight and parallel bottom contours. As the water depth decreases, the wave angle becomes more and more in alignment with the parallel bottom contours. In Figure 20.8b, the contours represent a shoal. It is seen in this case that the wave rays focus on the shoreline in the region of a shoal. In the region of an embayment, as illustrated in Figure 20.8c, the opposite behavior occurs, and the rays diverge. Figure 20.8d shows a natural shoreline with bays and headlands, and indicates

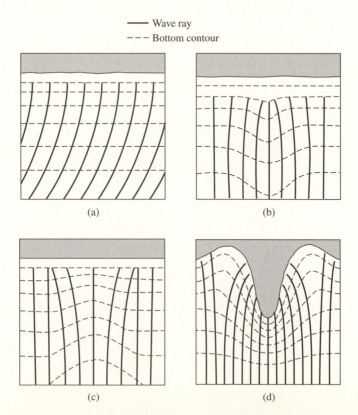

Figure 20.8

Wave rays for different shoreline configurations.

how waves are focused on the headlands and diverge in the bays. The wave energy flux between a pair of wave rays is constant. If the distance between wave rays increases as the waves propagate onshore, the wave height will decrease as a result of refraction. Conversely, if the distance between wave rays decreases, the wave height will increase.

If the bottom contours are straight and parallel, refraction may be estimated from Snell's law, given by

$$\frac{\sin \theta}{c} = \frac{\sin \theta_0}{c_0} \tag{20.15}$$

where θ_0 = deep water wave angle
c_0 = deep water celerity
θ = local wave angle
c = local celerity

This provides a straightforward estimate for refraction across the nearshore if the contours are nearly straight and parallel. Consider the energy flux between the two wave rays as illustrated in Figure 20.9. For this condition

$$(E_0 C_{g0})b_0 = (E c_g)b \tag{20.16}$$

where the subscript ( )$_0$ denotes deep water values. Substitution of the appropriate terms for the wave energies gives Eq. (20.16) as

$$\frac{1}{8}\rho g H_0^2 c_{g0} b_0 = \frac{1}{8}\rho g H^2 c_g b \tag{20.17}$$

and solving for the local wave height yields

$$H = \left(\frac{c_{g0}}{c_g}\right)^{1/2}\left(\frac{b_0}{b}\right)^{1/2} H_0 \tag{20.18}$$

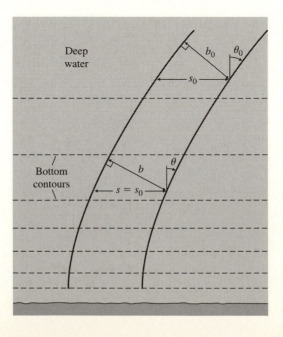

Figure 20.9

Wave rays for straight and parallel bottom contours.

The first term on the right-hand side (RHS) of Eq. (20.18) is K_S, the shoaling coefficient, and the second term is K_R, the refraction coefficient. Note that ray 2 is identical to ray 1; it is just displaced along the shore a distance s_0. It follows that $s = s_0$; therefore

$$b_0 = s_0 \cos \theta_0 \qquad b = s_0 \cos \theta \qquad (20.19\text{a, b})$$

and the refraction coefficient can be written as

$$K_R = \left(\frac{\cos \theta_0}{\cos \theta} \right)^{1/2} = \left(\frac{1 - \sin^2 \theta_0}{1 - \sin^2 \theta} \right)^{1/4} \qquad (20.20)$$

Substituting Eq. (20.15) into Eq. (20.20) results in

$$K_R = \left[\frac{1 - \sin^2 \theta_0}{1 - (c/c_0)^2 \sin^2 \theta_0} \right]^{1/4} \qquad (20.21)$$

Given the deep water wave angle, the wave period, and the local depth, the wave angle can be determined from Eq. (20.15) and the refraction coefficient from Eq. (20.21). A graphical solution for θ_0 and K_R is presented in Figure 20.10.

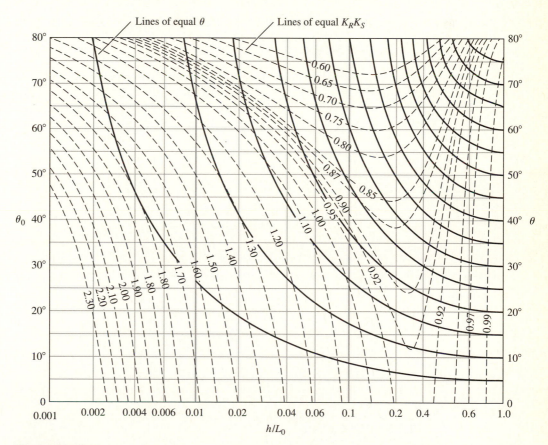

Figure 20.10

Change in wave height and direction due to shoaling and refraction [11]. Adapted from *Coastal Littoral Transport* by W.G. McDougal, N.C. Kraus, P.D. Komar, and J.D. Rosatti, courtesy of U.S. Army Corps of Engineers, EM1110-2-1502, 1992.

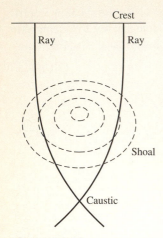

Figure 20.11

Caustic in the lee of a circular shoal.

If the bathymetry cannot be approximated by straight parallel contours, more sophisticated techniques must be used. One such technique is ray tracing. Ray tracing can be accomplished graphically by following procedures such as those detailed in the *Shore Protection Manual* [4]. These techniques are very laborious and are not recommended except for the most simple applications. Another technique is numerical wave ray tracing, which is straightforward to implement but often does not yield wave heights at the desired locations. This problem can be circumvented by implementing grid solution methods. While this is an improvement, the application of the technique, and of all the others as well, is subject to a serious common flaw: the crossing of two wave rays. At such a location, the wave height theoretically becomes infinite, and the method fails. This is because the wave ray separation distance b goes to zero, and this term appears in the denominator of Eq. (20.18). Locations where wave rays cross have been referred to as caustics. Figure 20.11 shows a simple case where a caustic might occur using a wave ray tracing technique. If a long-crested wave approaches an isolated circular shoal, then it is possible that the wave rays would cross in the lee of the shoal at the caustic shown. As a result of this problem, wave tracing techniques should be used with caution. However, they do work quite well with simple geometries that preclude this caustic formation.

EXAMPLE 20.2 ▼

An offshore storm generates waves with deep water height $H_0 = 3$ m and period $T = 12$ sec. The deep water wave direction θ_0 is 25° with respect to the bottom contours. If the contours are approximately straight and parallel, determine the local wave height H and direction θ in a water depth $h = 10$ m.

Solution

$$L_0 = 1.56T^2 = 224.6 \text{ m}$$

$$\frac{h}{L_0} = \frac{10}{224.6} = 0.04452$$

Using Figure 20.10 with $\theta_0 = 25°$,

$$\theta = 10°$$

$$K_R K_S = 1.1$$

$$H = K_R K_S H_0 = 1.1(3) = 3.3 \text{ m}$$

▲

20.3.3 Diffraction

Diffraction is the lateral transfer of wave energy due to a gradient in the wave height along the crest. Large gradients in the wave height along the crest can be developed as waves pass a fixed structure. A general illustration of slit diffraction is presented in Fig-

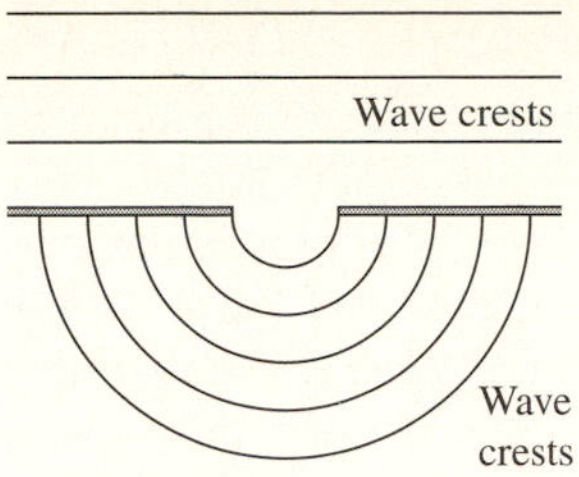

Figure 20.12
Slit diffraction.

ure 20.12. If plane waves propagate through a slit, which is narrow with respect to the wave length, then semicircular wave crests will form. Figure 20.13 shows wave diffraction occurring around the headlands of a coastal embayment. Analytical solutions have been developed for diffraction where the bottom is flat. A variety of these are presented in the literature [4, 12] for different wave angles and structure conditions.

20.3.4 Dissipation

Waves lose energy as they propagate, even though they are not breaking. This occurs through viscous dissipation in the water column and interactions with the sea bed. When waves travel great distances, viscous dissipation becomes important. However, if the travel distance is not great, this mechanism is generally negligible. Dissipation due to interactions with the bottom may result from friction, wave-induced flows in a permeable sea bed, and wave-induced displacements of a deformable sea bed. Again, if the travel distance over conditions where bottom friction and sea bed flows exist are not great, these interactions may generally be neglected. If there is a wide, shallow continental shelf, then these mechanisms should be included. Details for bottom friction and for permeable sea beds can be found in [13] and [14], respectively.

The deformation of the sea bed may lead to very substantial wave dissipation, even over short travel distances. This may occur at river mouths with large discharges of fine sediment. The sea bed is a very soft mud that undergoes large displacements due to the wave pressure on the bottom. If such conditions exist at a site, special theories are required to analyze the waves [15]. However, these conditions are the exception rather than the rule and generally do not represent a significant influence.

In summary, wave dissipation tends to decrease the wave height. Not including wave dissipation leads to conservative designs. Unless there are obvious reasons to do otherwise, neglecting dissipation is a prudent assumption when using the elementary techniques discussed in this chapter.

Figure 20.13

Diffraction around the headlands of a coastal embayment.

20.3.5   Breaking

Breaking waves are the primary forcing mechanism for most surf zone processes. Unfortunately, breaking waves have thus far eluded rigorous analytical treatment. As a result, various levels of empiricism are used to describe waves for engineering purposes. Several types of breaking waves are commonly recognized and are shown in Figure 20.14. Spilling breakers generally occur on low slope beaches. The breaker itself gradually dissipates energy by spilling off the face of the wave. For this type of wave, the height is often simply related to the local depth. Plunging breakers occur on moderately steep beaches. They represent the violent breaking that occurs when the wave curls over and crashes. Surging breakers peak up as they approach the beach, but rather than the crest breaking over, the base of the wave surges up the slope. Although breaking waves transition smoothly from one type to the other, they can generally be delineated by the surf similarity parameter ξ_I given by

$$\xi_{I.} = \frac{m}{[H_B/L_0]^{1/2}} \tag{20.22}$$

where m = beach slope
H_B = breaking wave height
L_0 = deep water wave length

This parameter can be thought of as a ratio between beach steepness and wave steepness. However, it is not a true wave steepness because the wave height is the breaking wave height, and the wave length is the deep water wave length. The breaker type as a function of the surf similarity parameter is given in Table 20.4.

The height of the breaking wave relative to the local breaking depth is the breaker index κ. This is shown in Figure 20.15 as a function of the surf similarity parameter. In this figure, the surf similarity parameter is defined somewhat differently using the breaking wave length rather than the deep water wave length. The value of κ is typi-

TABLE 20.4.   **The Breaker Type as a Function of Surf Similarity Parameter [17, 18]**

Breaker Type	Limiting ξ_I
Spilling	$\xi_I < 0.4$
Plunging	$0.4 < \xi_I < 2.0$
Surging	$2.0 < \xi_I$

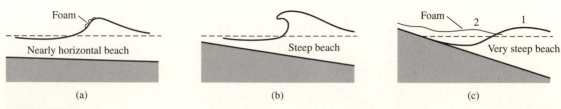

Figure 20.14

Common types of breaking waves: (a) spilling; (b) plunging; (c) surging [16]. Adapted from *Beach Processes and Sedimentation* by P.D. Komar, copyright © 1976 by permission of Prentice-Hall, Inc., Englewood Cliffs, NJ.

Figure 20.15

Breaker index k as a function of the surf similarity parameter z [17, 18]. Adapted from *Design and Construction of Mounds for Breakwater and Coastal Protection* by P. Brunn, Developments in Geotechnical Engineering, Elsevier, Amsterdam, 1985.

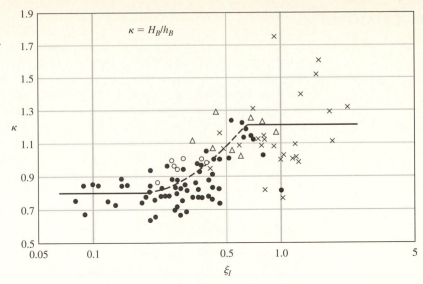

cally in the range 0.6 to 1.4 [4]. For mild sloped beaches, $\kappa = 0.8$ is a reasonable first estimate. The breaker index assumption that the local breaking wave height is proportional to the local water depth is used in many analytical and numerical models for surf zone processes. This may be a reasonable approximation for spilling breakers but is clearly in error for plunging and surging waves. If we equate the energy flux between the offshore wave and the breaking depth, assume deep water offshore and shallow water at breaking, and use a breaking index, the following equation may be written:

$$\frac{H_B}{H_0} = \left(\frac{15}{8\kappa}\right)^{1/5}\left(\frac{L_0}{H_0}\right)^{1/5} \tag{20.23}$$

where H_B = breaker height
H_0 = deep water wave height
L_0 = deep water wave length
κ = breaker index

Given the variability in κ, the constant in Eq. (20.23) ranges from 0.47 to 0.56. In field measurements it has been found that 0.56 fits the data [19]. This is at the upper limit of the range for the simple breaking equation and an indication that this is a very simplified approach.

Waves can also break in deep water. This occurs if the waves oversteepen. The limiting steepness is given by

$$\left(\frac{H_B}{L_0}\right)_{max} = 0.142 \approx \frac{1}{7} \tag{20.24}$$

In intermediate water depths, the limiting steepness is given [20] as

$$\left(\frac{H_B}{L}\right)_{max} = 0.142 \tanh{(kh)} \tag{20.25}$$

20.4 WAVE STATISTICS

The free surface profile of wind-generated water waves is rarely, if ever, the simple sinusoid or monochromatic wave assumed thus far. Actual wave observations include many different wave heights, periods, and directions that are superimposed to yield the local wave conditions (see Figure 20.16). However, real ocean waves are so complex that some idealization is required. In this section several techniques for representing waves are considered. The first technique is to simply pick a single wave that is in some way representative of the wave conditions. The second technique is to consider the statistics of the wave heights over a relatively short period of time (minutes to hours). Next, the distribution of wave heights with the wave period will be considered, and finally, the selection of an extreme wave for the design conditions will be discussed.

20.4.1 Significant Wave

One approach used to define the free surface profile of wind-generated waves is to represent all the waves by a single wave. That is, select a single wave height, period, and direction. This allows much easier analysis based on a single period (or monochromatic) technique. The wave height, period, and direction that are selected must in some way represent the combined effects of all the waves. The most commonly employed parameters for this technique are the significant wave height H_S, the zero crossing wave period T_z, and the direction corresponding to the dominant direction θ_c.

Traditionally, the significant wave height corresponded to the value that a trained observer would visually determine. However, this practice often resulted in the identification of a wave height greater than the average because an observer's visual estimate is biased toward the larger waves. Subsequently, it has been determined that a significant wave height actually represents an average of the highest one-third of the waves. Therefore, the notation $H_{1/3}$ is often used as the significant wave height.

The significant wave height may be determined in several ways from a measured free surface profile, such as illustrated in Figure 20.16. One common method for deter-

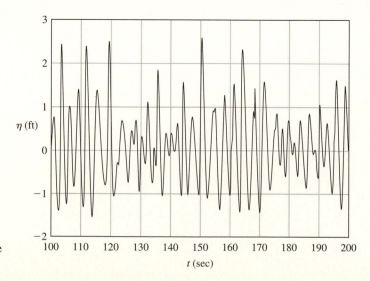

Figure 20.16

Measure of free surface profile for random waves.

mining the significant wave height is called the zero up-crossing technique. The first step in the procedure is to determine the mean water level, which is simply the average over the record length. Then subtract this mean from the signal, so that zero now represents the mean. Often this step has already been done through the calibration of the wave gage. Next, the location where the free surface crosses the zero line going up is designated as the start of the next wave. This divides the time series into a number of individual waves. The wave height for each of these waves is the vertical distance measured from the maximum free surface elevation in the interval to the minimum, and the wave period is the duration of the interval. Notice that by this definition, a bump on the free surface that does not pass through the mean water level is not considered a wave. Next the waves are ranked by wave height (i.e., largest to smallest) and the average of the largest one-third of the waves is taken. If there were 3000 waves in the time series, for example, then $H_{1/3}$ would correspond to the mean of the largest 1000 waves. The zero crossing wave period is the average of all of the measurements. This period is used rather than $T_{1/3}$, which could be either the average of the longest one-third of the wave periods or the period associated with $H_{1/3}$.

This technique for determining wave height can be extended to other waves. The most common are $H_{1/10}$ and $H_{1/100}$, which correspond to the average of the highest one-tenth and one-hundredth of the waves, respectively. Figure 20.16 presents a 100-sec sample of an 8-min wave record. Using the zero up-crossing method, $H_{1/3} = 3.32$ ft, $H_{1/10} = 3.83$ ft, and $H_{1/100} = 4.06$ ft. A FORTRAN computer program, UP_CROSS.FOR, to calculate $H_{1/n}$ from a measured time series is available on the authors' web site.

Wave conditions are sometimes described by the sea state. Wave conditions corresponding to the various sea states are given in Table 20.5. Also given in this table are storm conditions associated with the waves. Wave estimation from wind speeds should not be made using Table 20.5. Techniques for estimating wave conditions are discussed in Section 20.5. However, Table 20.5 does present a qualitative description of sea conditions under different sea states. In Table 20.5, U is the wind speed and T_p is the peak wave period, which corresponds to the wave period with the highest wave energy.

20.4.2 Short-Term Statistics

It has been observed that wave height probability in a single group of waves is given by the Rayleigh distribution. This result has also been determined analytically, assuming η is Gaussian. The Rayleigh distribution is shown in Figure 20.17, and a common definition is given by

$$P(H > \hat{H}) = \exp[-(\hat{H}/H_{\rm rms})^2] \tag{20.26}$$

$$f(\hat{H}) = \frac{2\hat{H}}{H_{\rm rms}^2}\exp[-(\hat{H}/H_{\rm rms})^2] \tag{20.27}$$

where $\hat{H}$ is an arbitrary height and $H_{\rm rms}$ is the root mean square (parameter of the distribution), given by

$$H_{\rm rms} = \left(\frac{1}{N}\sum_{n=1}^{N}H_n^2\right)^{1/2} \tag{20.28}$$

TABLE 20.5. Qualitative Sea-State Descriptions [21]

Sea State	Description	Beaufort Wind Force	Wind Description	U (kts)	H_S (ft)	T_p (sec)
0	Sea like a mirror	0	Calm	0	0	
1	Ripples with the appearance of scales are formed, but without foam crests.	1	Light airs	2.0	0.08	0.7
	Small wavelets, still short but more pronounced; crests have a glassy appearance, but do not break.	2	Light breeze	5.0	0.29	2.0
	Large wavelets, crests begin to break. Foam of glassy appearance. Perhaps scattered white horses.	3	Gentle breeze	8.5	1.0	3.4
2				10.0	1.4	4.0
				12.0	2.2	4.8
3	Small waves, becoming larger; fairly frequent white horses.	4	Moderate breeze	13.5	2.9	5.4
				14.0	3.3	5.6
				16.0	4.6	6.5
4	Moderate waves, taking a more pronounced long form; many white horses are formed (chance of some spray).	5	Fresh breeze	18.0	6.1	7.2
				19.0	6.9	7.7
				20.0	8.0	8.1
5	Large waves begin to form; the white foam crests are more extensive everywhere (probably some spray).	6	Strong breeze	24.0	12.0	9.7
6				24.5	13.0	9.9
				26.0	15.0	10.5
	Sea heaps up and white foam from breaking waves begins to be blown in streaks along the direction of the wind (spindrift begins to be seen).	7	Moderate gale	28.0	18.0	11.3
				30.0	22.0	12.1
				30.5	23.0	12.4
				32.0	26.0	12.9
7	Moderately high waves of greater length; edges of crests break into spindrift. The foam is blown in well-marked streaks along the direction of the wind. Spray affects visibility.	8	Fresh gale	34.0	30.0	13.6
				36.0	35.0	10.3
				37.0	37.0	14.9
				38.0	40.0	15.4
				40.0	45.0	16.1
8	High waves. Dense streaks of foam along the directions of the wind. Sea begins to roll. Visibility affected.	9	Strong gale	42.0	50.0	17.0
				44.0	58.0	17.7
				46.0	64.0	18.6
		10	Whole gale	48.0	71.0	19.4
	Very high waves with long overhanging crests. The resulting foam is in great patches and is blown in dense white streaks along the direction of the wind. On the whole, the surface of the sea takes a white appearance. The rolling of the sea becomes heavy and shocklike. Visibility affected.			50.0	78.0	20.2
				51.5	83.0	20.8
				52.0	87.0	21.0
				54.0	95.0	21.8
9	Exceptionally high waves (small and medium-sized ships might for a long time be lost to view behind the waves). The sea is completely covered with long white patches of foam lying along the direction of the wind. Everywhere the edges of the wave crests are blown in froth. Visibility affected.	11	Storm	56.0	103.0	22.6
				59.5	116.0	24.0
	Air filled with foam and spray. Sea completely white with driving spray; visibility very seriously affected.	12	Hurricane	>64.0	>128.0	(26)

Figure 20.17
Rayleigh distribution.

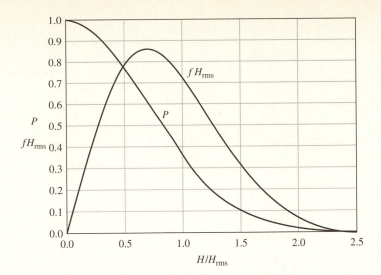

Equation (20.26) is the probability distribution or cumulative distribution, and Eq. (20.27) is the frequency distribution. The frequency distribution is simply the derivative of the cumulative distribution. Note that wave energy is proportional to H^2; therefore, $(H_{rms})^2$ is proportional to average wave energy. The relationship among wave heights for a Rayleigh distribution is given in Table 20.6. In general,

$$H_{1/n} = H_{rms} \left([\ln(n)]^{1/2} + \frac{n\pi^{1/2}}{2}\{1 - \text{erf}[\ln(n)]^{1/2}\} \right) \qquad (20.29)$$

where erf( ) is the error function. A computer model for estimating short-term wave statistics is given in [9].

The largest wave H_{max} in a wave record can be approximated by [12]

$$\frac{H_{max}}{H_S} = \left[\frac{\ln(N)}{2} \right]^{1/2} + e[8 \ln(N)]^{1/2} \qquad (20.30)$$

where N is the number of waves and e is Euler's constant ($e \approx 0.5722$).

TABLE 20.6. Statistically Representative Waves Based on Rayleigh Wave Height Relationships

Wave Height	Notation	H/H_{rms}
Mode	—	0.707
Median	—	0.833
Mean	$\overline{H} = H_1$	0.886
Root mean square	H_{rms}	1.000
Significant	$H_S = H_{1/3}$	1.416
Average of tenth-highest waves	$H_{1/10}$	1.800
Average of hundredth-highest waves	$H_{1/100}$	2.359

20.4.3 Wave Spectra

Fourier series and spectra were examined in Chapter 8. Since real waves involve many wave heights and periods, this is a useful technique for analyzing random waves. Recall that most reasonable functions may be approximated by a series of sine and cosine terms. This can be written in several ways:

$$F(t) = a_0 + \sum_{n=1}^{\infty} a_n \cos \omega_n t + \sum_{n=1}^{\infty} b_n \sin \omega_n t$$

$$= \sum_{n=0}^{\infty} A_n \cos (\omega_n t - \phi_n) \qquad (20.31)$$

$$= \sum_{n=0}^{\infty} B_n e^{-i\omega_n t}$$

where a_n, b_n, A_n, and B_n are amplitude coefficients and ϕ_n is the phase. In ocean engineering it is common to look at the magnitudes $(a_n^2 + b_n^2)$ or $(A_n)^2$ or $(B_n B_n^*)$ rather than just the amplitude coefficients. This is done for three reasons: (1) by examining the sum of the squares (always positive) phase information is removed; (2) since wave energy is proportional to H^2, $(a_n^2 + b_n^2)$ is proportional to the energy at frequency ω_n; and (3) the total energy is related to the variance, which is a statistic of the distribution and it is assumed, is invariant over the time series (stationary process). Figure 20.18 shows the wave spectrum for the wave time series presented in Figure 20.16. Note that Figure 20.16 shows only a portion of the total record on which the spectrum calculations are based. The energy density or power spectral density at each frequency is represented by $S_{\eta\eta}(\omega) = E(\omega)/\Delta\omega$. It is the wave energy (without ρg) in the frequency interval $\Delta\omega$. Since the spectrum is determined from the free surface profile squared, it is usually denoted as $S_{\eta\eta}$. A spectrum of the horizontal water particle velocity would be S_{uu}.

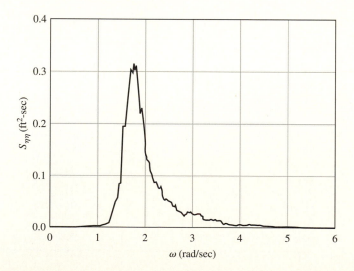

Figure 20.18

Wave spectrum for the time series in Figure 20.16 was extracted.

A wave spectrum may be determined from measurements of the free surface. However, it would be very useful from a design perspective if a general expression for the spectrum could be developed. From field measurements it has been observed that

$$S_{\eta\eta} \approx 0.0081 \; g^2 \omega^{-5} \qquad (20.32)$$

This corresponds to a fully arisen sea state. The terms fully arisen, fully developed, or saturated imply that the waves at each frequency contain their maximum amount of energy. If more energy is added, the waves will break and transfer the energy to other frequencies. At lower frequencies, the waves are much longer and, therefore, are probably not fully developed. As a result, two terms are used in a typical nonsaturated spectrum, one for the saturated high frequencies and a second to account for the lower unsaturated frequencies. The coefficients tend to be empirical and several of the more common one-sided spectra are given in Table 20.7. In Table 20.7, f_p is the peak frequency, $U_{19.5}$ is the wind speed 19.5 m above the surface, U_{10} is the wind speed 10 m above the surface, and F is the fetch length defined in Section 20.5.2. The term γ is a peak enhancement factor and causes the JONSWAP spectrum to have more energy near the peak frequency. The Pierson-Moskowitz spectrum has no fetch length dependency because it is for fully arisen seas. A fully arisen sea corresponds to the maximum waves that can be generated for a given wind speed and fetch length. Since the JONSWAP spectrum does include the fetch length, it is possible that the two techniques will give different estimates for the peak frequency f_p.

The relationship between U_{10} and $U_{19.5}$ is given approximately as $U_{10} = 0.91 \, U_{19.5}$ [4]. This allows a direct comparison of the two spectra. Spectra are plotted in Figure 20.19 for $F = 100$ km and $U_{10} = 100$ kph. For this case, the two estimates for f_p are very similar. When $\gamma = 1$, the two spectra are almost identical. However, for larger values of γ, the JONSWAP is more energetic and the energy is concentrated near the peak frequency. In wave spectra the frequency may be given as $\omega = 2\pi/T$ (rad/sec) or as $f = 1/T$ (Hz).

TABLE 20.7.   Several Common Wave Spectra

Pierson-Moskowitz [22]	$S_{\eta\eta}(f) = \dfrac{\alpha g^2}{(2\pi)^4 f^5} \exp\left[-\dfrac{5}{4}\left(\dfrac{f_p}{f}\right)^4\right]$	$\alpha = 0.0081$ $f_p = \dfrac{g}{1.14(2\pi)U_{19.5}}$
JONSWAP [23]	$S_{\eta\eta}(f) = \dfrac{\alpha g^2}{(2\pi)^4 f^5} \gamma^\delta \exp\left[-\dfrac{5}{4}\left(\dfrac{f_p}{f}\right)^4\right]$	$\alpha = 0.066\left(\dfrac{U_{10}^2}{gF}\right)^{0.22}$ $\delta = \exp\left[-2\sigma^2(f/f_p - 1)^2\right]$ $\gamma = 1 - 7 \text{ (mean} = 3.3)$ $f_p = 2.84\left(\dfrac{g^2}{U_{10}F}\right)^{0.33}$ $\sigma = \begin{cases} 0.07 & f \le f_p \\ 0.09 & f > f_p \end{cases}$

Figure 20.19

Pierson-Moskowitz and JONSWAP spectra.

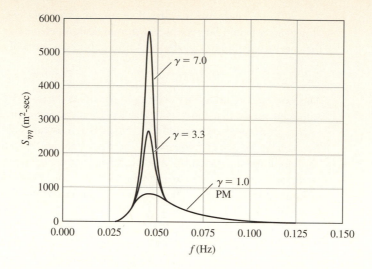

A discussion of wave spectra must address the effect of different wave directions as well as frequencies. If the wave number is considered a vector, the direction of travel may be expressed as the x and y components. Thus, the directional spectrum may be expressed as either

$$S_{\eta\eta} = f(k_x, k_y, \omega) \tag{20.33}$$

or

$$S_{\eta\eta} = f(\theta, \omega) \tag{20.34}$$

Unfortunately, the directional spectra expressed in these forms are difficult to employ and some type of simplification is usually made. The most common simplification involves the use of a directional spreading function, resulting in

$$S_{\eta\eta}(\theta, \omega) = G(\theta, \omega)f(\omega) \tag{20.35}$$

Equation (20.35) assumes that the direction can be decoupled from the energy spectrum. Two such spreading functions are given by

$$G(\theta) = \frac{2}{\pi}\cos^2(\theta - \theta_c) \qquad |\theta - \theta_c| < \frac{\pi}{2} \tag{20.36}$$

and

$$G\theta = \frac{1}{2\sqrt{\pi}} \frac{\Gamma(s+1)}{\Gamma\left(s + \frac{1}{2}\right)} \cos^{2s}(\theta - \theta_c) \qquad |\theta - \theta_c| < \frac{\pi}{2} \tag{20.37}$$

where θ_c = the direction about which the spectrum is centered

s = a spreading parameter

$\Gamma(\ )$ = the gamma function

Examples of these spreading functions are shown in Figure 20.20.

Figure 20.20
Spreading functions.

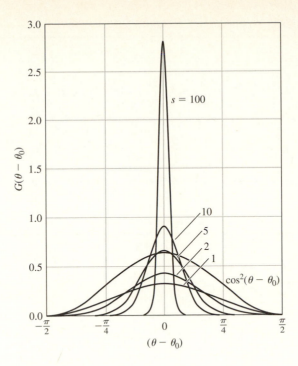

20.4.4 Long-Term Statistics

An important step in the design process is the estimation of the design wave relative to the design life of the structure and the degree of safety desired. On the basis of rather limited data and the use of an appropriate statistical distribution, the results must be extrapolated to extreme conditions. This process is similar to estimating the 50-year flood in hydrology. The counterpart in ocean engineering is the 50-year wave. Two common extreme value probability distributions are the lognormal and the Gumbel, which are summarized in Table 20.8. The values for the parameters α, θ, ε are estimated from data. Wave information at the design site is collected over several years or determined by hindcasting (Section 20.5). Data for as long an observation period as possible should be obtained because estimates of extreme conditions from small observation periods are not accurate. Estimates should not be extrapolated beyond 3 times the duration of the data. Usually, only the larger waves are examined. The waves can be defined by a wave height exceeding a specified threshold or by the largest wave that occurred each year. Using the waves exceeding a threshold provides more data to analyze and is the more common approach. These data are then plotted using a plotting formula that assigns a value of probability to each data point in the sample. To do this, the total number of observations are ordered from the largest to the smallest. A simple estimate of the exceedence is given by

$$Q(H) = 1 - P(H) = \frac{n}{N+1} \tag{20.38}$$

where $Q(H)$ = probability that the wave height is greater than H
$P(H)$ = probability that the wave height is less than H
n = number of wave heights greater than H
N = number of waves

TABLE 20.8. **Lognormal and Gumbel Extreme Value Probability Distributions**

Distribution	$P(H)$	Mean	Variance
Lognormal	$(2\pi)^{-1/2} \int_0^H \dfrac{1}{\alpha h} \exp\left[-\dfrac{1}{2}\left(\dfrac{\ln h - \theta}{\alpha} \right)^2 \right] dh$	$\exp\left(\theta + \dfrac{\alpha^2}{2} \right)$	$\exp(2\theta + \alpha^2)[\exp(\alpha^2) - 1]$
Gumbel	$\exp\left\{ -\exp\left[-\left(\dfrac{H - \varepsilon}{\theta} \right) \right] \right\}$	$\varepsilon + 0.58\theta$	$\dfrac{\pi^2}{6}\theta^2$

TABLE 20.9. **Scale Relationships for Lognormal and Gumbel Distributions**

Distribution	Abscissa Scale (x)	Ordinate Scale (y)	Slope (a)	Intercept (b)
Lognormal	$\ln H$	$(2\pi)^{-1/2} \int_0^y e^{-t^2/2} \, dt$	$\dfrac{1}{\alpha}$	$-\dfrac{\theta}{\alpha}$
Gumbel	H	$-\ln\{-\ln[P(H)]\}$	$\dfrac{1}{\theta}$	$-\dfrac{\varepsilon}{\theta}$

These points are now plotted on extreme value paper corresponding to the chosen probability density function. Prepared extreme value paper may be purchased, or constructed using the axis factors given in Table 20.9. The plotted points are then fit with a straight line, $ax + b$. The slope and intercept of this line specify the parameters of the distribution (see Table 20.9).

Once the probability distribution has been fit to the data, it may be used to select a design wave. The return period, or recurrence interval, T_R is the average time interval between successive occurrences of the design wave being equaled or exceeded. It is related to the probability of exceedence by the expression

$$T_R = \frac{r}{Q(H)} \tag{20.39}$$

where r is the recording interval associated with each data point. Since an expression for the exceedence from the data has already been developed, T_R may be added as a second ordinate to the plot.

It is often useful to know the probability that the design wave will be exceeded in the design life of the structure. This encounter probability E is given by

$$E = 1 - \left(1 - \frac{r}{T_r} \right)^{L/r} \tag{20.40}$$

where L is the design life. As with T_R, E may also be labeled as a third ordinate on the extreme value paper. A computer model for estimating long term statistics is given in [9].

EXAMPLE 20.3 ▼

Approximately 8.5 years of wave measurements are summarized in Table 20.10 in columns (1) and (2). Each observation is the significant wave height for a 6-hour interval. Determine the 10- and 20-year wave heights using a Gumbel distribution.

Solution

This problem can be solved using a spreadsheet. The given data are in columns (1) and (2) of Table 20.10 and have been ordered from largest to smallest. The running sum is given in column (3). Column (4) is the exceedence using Eq. (20.38). Column (5) is $-\ln[-\ln(Q)]$ as given in Table 20.9. This is fit with a straight line using linear regression. Column (1) is the independent variable and column (5) is the dependent variable. Only the data for $H \geq 10$ ft was used in the regression. The slope and intercept are found to be

$$a = -0.15815 \qquad b = 1.131305$$

Figure 20.21 shows a comparison of the data and the fit. It is seen that the fit is quite good for $H \geq 10$ ft ($r^2 = 0.993$). It is also seen that the larger waves have a behavior different from the smaller waves, and the break between the two is approximately $H = 10$ ft. Having determined a and b, the parameters of the distribution may be calculated using the equations in Table 20.9:

$$\theta = -6.32317 \qquad \varepsilon = 7.15343$$

With these, the probability of an arbitrary wave height can be determined using the equations in Table 20.8. However, the wave height that corresponds to specific

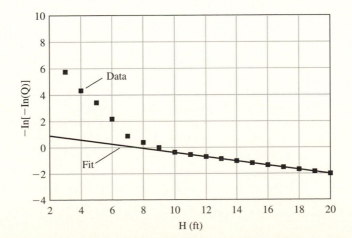

Figure 20.21

Comparison of measured data and Gumbel fit.

probabilities is desired. The probabilities associated with the 10- and 20-year return periods are calculated using Eq. (20.39) and with $r = 6$ hours. The equation in Table 20.8 is then solved for H. The results are

T_R (year)	Q	H(ft)
10	6.84932E-05	21.5
20	3.42466E-05	21.9

TABLE 20.10. **Design Wave Prediction Example**

(1)	(2)	(3)	(4)	(5)
H(ft)	Obs.	m	Q	$-\ln[-\ln(Q)]$
20	6	6	0.000	−2.030
19	16	22	0.002	−1.843
18	28	50	0.004	−1.704
17	68	118	0.010	−1.534
16	96	214	0.018	−1.396
15	132	346	0.028	−1.270
14	228	574	0.047	−1.117
13	260	834	0.069	−0.986
12	448	1,282	0.105	−0.811
11	628	1,910	0.157	−0.616
10	1,064	2,974	0.244	−0.343
9	1,692	4,666	0.383	0.042
8	1,428	6,094	0.501	0.369
7	1,780	7,874	0.647	0.831
6	2,916	10,790	0.887	2.117
5	992	11,782	0.968	3.427
4	232	12,014	0.987	4.344
3	120	12,134	0.997	5.794
2	36	12,170	1.000	9.407

▲

In Example 20.3, an important consideration was not explicitly addressed: the class intervals. The class intervals are the wave height ranges into which the data are sorted before the statistical analysis is conducted. If the data were plotted as a histogram, the class intervals are the width of the bars. These widths can be variable or all the bars can be the same width. In this example, the data were given in 1-ft bin widths. It happens that this is also a good class interval to analyze these data. In fact, when some data are given in bins, it is common to use the given bin width as the class interval. But, if there are raw wave height measurements, as opposed to binned data, the class interval must be selected. If the class interval is too wide, information in the data will be lost. In the present example, if a bin width of 20 ft was used, the histogram would have only one bar 20 ft wide that contained all the data points. On the other hand, if the class interval is too narrow, the histogram will be very noisy with large differences in the heights from one bar to the next. A rule of thumb is to take the class interval to be 40% of the standard deviation of the data. For the data in this example, this yields a bin width of 1.07 ft.

20.5 WAVE INFORMATION

Up to this point the analysis of measured waves and the determination of design statistics have been addressed. However, the actual task of obtaining wave data for these analyses has not been discussed. There are two approaches commonly employed to obtain the required data: (1) using measured data and (2) using hindcasts, the estimation of waves from historical storms.

20.5.1 Wave Measurements

There are a variety of techniques available for measuring the wave profile. Direct techniques actually measure the location of the free surface as a function of time. Wave heights can be determined from the changes in resistance or capacitance as the free surface moves up and down on a wave staff. The surface can also be monitored using sonic and ultrasonic systems similar to sonar. These systems measure the travel time for a signal to reach and reflect back from the free surface. Indirect techniques measure a wave parameter and then use a wave theory to convert it to the surface profile. Examples include surface accelerations, surface slope, water particle velocities, and wave pressure. The dependency on a wave theory (often LWT) is a weak aspect associated with these techniques, especially at higher wave frequencies.

There are a number of sources for obtaining wave data, such as:

LEO data. The Corps of Engineers program for collection of wave observations from shore is the Littoral Environment Observation (LEO) program. Volunteer observers obtain daily estimates that include the breaker height, wave period, direction of wave approach, wind speed, wind direction, current speed, and current direction.

NOAA buoy data. Since 1972 the National Oceanic and Atmospheric Administration (NOAA) has maintained a number of oceanographic buoys throughout United States coastal waters. Available information includes wind direction and speed, sea level pressure, air temperature, sea surface temperature, significant wave height, dominant wave period, and peak gust data.

Ship observations. Wave observations have been collected by observers aboard ships in passage for many areas of the world over many years. The observations include average wave height, period, and direction of the sea waves (locally generated) and the swell waves (generated elsewhere and propagated to the area).

A listing of sources that contain extensive summaries of oceanographic data is given in Table 20.11. In addition to wave and water level data, several sources listed include wind speed and direction, air and sea temperatures, and other information required for wave and water level studies. Information may also be available from the Coast Guard, port and harbor authorities, and local universities. Unfortunately, different wave measurement systems give slightly different results for the same conditions. This is due not only to differences in the measurement instruments but also to the data sampling rate, duration, and analysis technique. In most engineering designs, waves are not measured specifically for that particular project. This is because years of data are required to make extreme wave estimates for the design. Therefore, it is often necessary to use data obtained from previously conducted long-term wave measurement programs.

TABLE 20.11. Sources for Wave Data

Alaska Coastal Data Collection Program
  Plan Formulation Section
  U.S. Army Engineer District, Alaska
  Pouch 898
  Anchorage, AK   99506-0898
  (907) 753-2620
	Wind and wave data for coastal
	  Alaska

California Coastal Data Information Program
  Scripps Institute of Oceanography
  Mail Code A022
  University of California, San Diego
  LaJolla, CA   92093
  (619) 534-3033
	United States west coast gage
	  network and gage at CERC's FRF
	  in North Carolina

Coastal Engineering Information and Analysis Center
  USAEWES
  3909 Halls Ferry Rd.
  Vicksburg, MS   39180-6199
  (601) 634-2012
	Coastal Engineering Information
	  Management (CEIMS) LEO
	  Retrieval System, gage data from
	  the Corps Coastal Field Data
	  Collection Program and other
	  sources

Field Coastal Data Network
  Coastal & Oceanographic Engineering Department
  336 Weil Hall
  University of Florida
  Gainesville, FL   32611
  (352) 392-1051
	Coastal Florida wave gage network

National Oceanographic Data Center
  User Service (Code OC21)
  1825 Connecticut Ave., NW
  Washington, DC   20235
  (202) 673-5549
	Variety of oceanographic data

National Climate Data Center
  Federal Building
  Asheville, NC   28801
  (704) 259-0682
	Global, meteorological, and
	  oceanographic data and data
	  products

20.5.2 Hindcasts

Very often, measured wave data are not available for a particular design site. However, weather data have been collected for many years throughout the United States. From these sources, there is generally sufficient data to estimate the 50-year or 100-year storm. Using these historical storm events, estimates of the resulting wave conditions can be developed. This is referred to as hindcasting since it is based on historical storm conditions. The Corps of Engineers has conducted hindcasts for all coastal waters of the United States based on 20 years of weather conditions. Results are available in a series of Wave Information Studies (WIS) reports, available from U.S. Army Engineer Waterways Experiment Station (WES), Coastal Engineering Research Center.

If neither wave measurements nor WIS hindcasts are available, it is possible to make simple hindcast estimates. The wave hindcasting method outlined in this section follows the technique described in more detail in the *Shore Protection Manual* [4]. In this simplified technique, the significant wave height and period are estimated from the wind stress, storm duration, and fetch length. Unfortunately, there are a variety of locations (i.e., over land, over water, or different elevations) and techniques for presenting wind measurements. If measured wind speeds are provided, care must be taken to convert the appropriate wind speeds for wave hindcasting. These corrections are given in [4]. A common way to estimate the wind speed is from surface synoptic charts that are available from the U.S. Weather Service. An example synoptic chart is given in Figure 20.22. The pressure isobars are typically contoured at either 3 or 4 millibar (mb) inter-

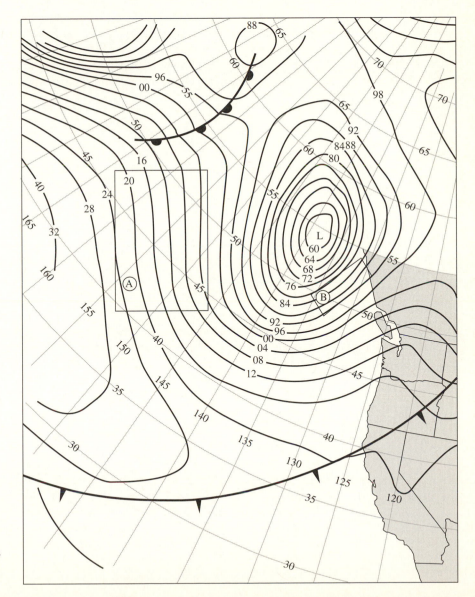

Figure 20.22
Simplified synoptic chart (pressure contours in millibars) [11].
Adapted from *Coastal Littoral Transport* by W.G. McDougal, N.C. Kraus, P.D. Komar, and J.D. Rosatti, courtesy of U.S. Army Corps of Engineers, EM 1110-2-1502, 1992.

vals, and this particular chart has a contour interval of 4 mb. Since the pressure is usually in the vicinity of 1000 mb, it is necessary to record only the last two digits of the pressure on the isobars. The pressure gradients are indicated by the isobar spacing. If they are closely spaced, the gradients are large. This pressure gradient is nearly in equilibrium with the Coriolis force produced by the rotation of the earth. The geostrophic wind is defined by assuming that an equilibrium or exact balance exists. The geostrophic wind blows approximately parallel to the isobars with low pressure to the left when looking in the direction of the wind in the northern hemisphere. In the southern hemisphere, the low pressure is on the right. The geostrophic wind is usually the best simple estimate of the wind speed. Figure 20.23 may be used to determine the geostrophic wind speed, which depends on the latitude, the average pressure gradient across the fetch, and the isobar spacing on the synoptic chart.

Once the geostrophic wind speed is known, several factors should be applied to adjust the wind speed. The first of these, R_T, accounts for the air-sea temperature difference. This correction is given in Figure 20.24. If no temperature data are available, then it is recommended that $R_T = 0.9$ for $T_a > T_s$, $R_T = 1.0$ for $T_a = T_s$, and $R_T = 1.1$ for $T_a < T_s$. Since the wave prediction curves are based on the wind speed measured at a 10-meter elevation, a correction must be applied to the geostrophic wind speed U_g to

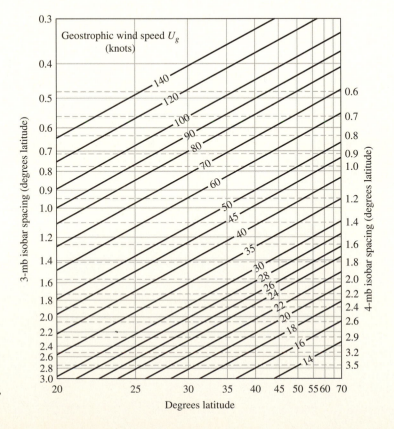

Figure 20.23

Geostrophic wind scale [4]. Adapted from *Shore Protection Manual,* courtesy of U.S. Army Coastal Engineering Research Center, Vicksburg, MS, 1984.

Figure 20.24

Correction for the air-sea temperature difference $(T_a - T_s)$, °C [4]. Adapted from *Shore Protection Manual,* courtesy of U.S. Army Coastal Engineering Research Center, Vicksburg, MS, 1984.

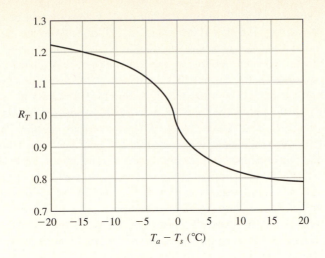

Figure 20.25

Correction factor to convert the geostrophic wind to the 10-m elevation wind speed [4]. Adapted from *Shore Protection Manual,* courtesy of U.S. Army Coastal Engineering Research Center, Vicksburg, MS, 1984.

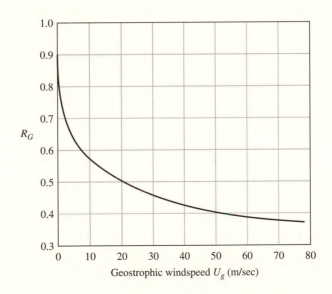

correct it to the 10-meter wind speed U. This correction factor, R_G, is given in Figure 20.25. The temperature corrected 10-meter wind speed is thus given by

$$U = R_T R_G U_g \tag{20.41}$$

Wave growth formula diagrams are expressed in terms of an adjusted wind speed or wind stress factor U_A, and wind speed is adjusted to wind stress factor by

$$U_A = 0.71 U^{1.23} \qquad (U \text{ in m/sec}) \tag{20.42}$$

Waves generated by a storm are primarily a function of three variables: (1) the wind speed or stress U_A, (2) the duration of the storm t, and (3) the fetch region over which the speed and duration are relatively constant. Best results are obtained when variation and direction of the wind speed (assumed parallel to the isobars) does not exceed ± 1.5°. Direction deviations of 30° should not be exceeded, and variations in the wind speed should not exceed ± 2.5 m/sec from the mean. Since the wind speed is related to the isobar spacing, this implies that the spacing should be nearly constant across the fetch. Following these guidelines, two fetches have been identified on the synoptic chart in Figure 20.22. Frequently, the discontinuity at a weather front will also limit a fetch. The fetch length is simply determined by measuring the length of the fetch and noting that 5° of latitude = 300 nautical miles (nm) = 555 km.

Estimates of the duration of the wind are also needed for wave prediction. Complete synoptic weather charts are prepared at 6-hour intervals. Thus, interpolation to determine the duration may be necessary. Linear interpolation is adequate in most cases.

With the estimates of the wind stress factor, wind duration, and fetch length available, the deep water significant wave height and peak spectral period may be determined from Figure 20.26, or with the Automated Coastal Engineering System (ACES) program "Wind-speed Adjustment and Wave Growth." For a given wind speed, the wave height can be limited by either the fetch length or the duration of the storm.

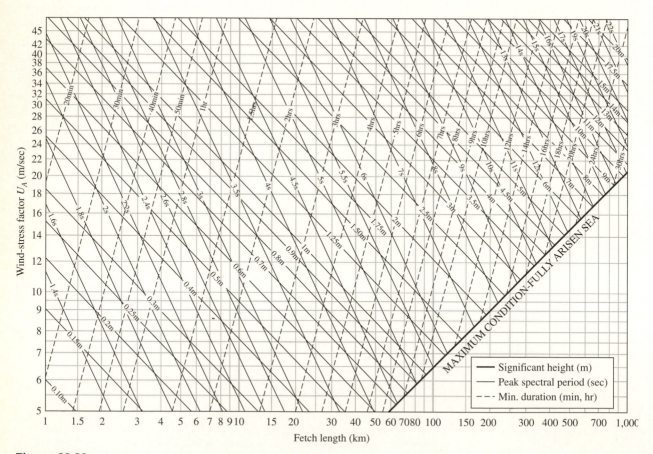

Figure 20.26
Nomograph of deep water significant wave prediction [4]. Adapted from *Shore Protection Manual,* courtesy of U.S. Army Coastal Engineering Research Center, Vicksburg, MS, 1984.

EXAMPLE 20.4 ▼

In a storm, the wind stress factor U_A is 20 m/sec, the fetch length F is 90 km, and the storm duration t is 5 hours. Determine the significant wave height H_S and peak spectral period T_p.

Solution

From Figure 20.26, two possible wave conditions can be estimated:
(a) $U_A = 20$ m/sec and $F = 90$-km yield

$$H_S = 3.0 \text{ m and } T_p = 7.6 \text{ sec}$$

(b) $U_A = 20$ m/sec and $t = 5$-hour yield

$$H_S = 2.5 \text{ m and } T_p = 6.6 \text{ sec}$$

The smaller wave of these two should be selected. Since the duration yields the smaller wave, this wave is termed duration limited. If the duration had been greater than 6.5 hours, then the wave would have been fetch limited. ▲

EXAMPLE 20.5 ▼

An examination of a series of synoptic charts indicated that the conditions described in Figure 20.22 persisted for 10 hours. The air and sea temperatures were reported at 9°C and 11°C, respectively. Estimate the significant wave height H_S and peak spectral period T_p generated by these weather conditions at approximately 54°N 130°W.

Solution

(a) Fetch. The appropriate fetch for this location is fetch B in Figure 20.22. Since 5° latitude = 555 km (300 nmi), the fetch length is

$$F = 600 \text{ km (328 nmi)}$$

(b) Geostrophic wind. The fetch width w and pressure change Δp are

$$w = 1.9° \text{ lat.} \quad \text{and} \quad \Delta p = 12 \text{ mb}$$

The isobar spacing s on this synoptic chart is

$$s = 4 \text{ mb}$$

The pressure gradient across the fetch is

$$p_g = \frac{w(°\text{lat.})}{\Delta p(\text{mb})} s(\text{mb}) = \frac{1.9}{12}4 = 0.63° \text{ lat.}$$

The center of the fetch is at 52°N; thus Figure 20.23 gives

$$U_g = 75 \text{ knots} = 38.6 \text{ m/sec}$$

(c) Wind stress. The air-sea temperature difference is

$$T_a - T_s = 9 - 11 = -2°C$$

and Figure 20.24 yields

$$R_T = 1.07$$

For $U_g = 38.6$ m/sec, Figure 20.25 gives

$$R_G = 0.44$$

The corrected wind speed is determined from Eq. (20.41) as

$$U = R_T R_G U_g = (1.07)(0.44)(38.6) = 18.2 \text{ m/sec}$$

This is converted to a wind stress using Eq. (20.43)

$$U_A = 0.71 U^{1.23} = 0.71(18.2)^{1.23} = 25.2 \text{ m/sec}$$

(d) Wave prediction. For $U_A = 25.2$ m/sec, $F = 600$ km and $t = 10$ hours, the significant wave height and peak periods are estimated from Figure 20.26 as

$$H_S = 5.4 \text{ m} \qquad T_p = 10.3 \text{ sec}$$

▲

REFERENCES

1 Ippen, A.T., *Estuary and Coastline Hydrodynamics,* McGraw-Hill, New York, 1966.

2 Sarpkaya, T. and Isaacson, M., *Mechanics of Wave Forces on Offshore Structures,* Van Nostrand Reinhold, New York, 1981.

3 Dean, R.G. and Dalrymple, R.A., *Water Wave Mechanics for Engineers and Scientists,* Prentice Hall, Englewood Cliffs, NJ, 1984.

4 *Shore Protection Manual,* U.S. Army Coastal Engineering Research Center, Vicksburg, MS, 1984.

5 Hunt, J.N., "Direct Solution Wave Dispersion Equation," *Journal of Waterways, Port, Coastal and Ocean Engineering Division,* ASCE, Vol. 105, 1979, pp. 457–459.

6 Skjelbreia, L. and Hendrickson, J.A., "Fifth Order Gravity Wave Theory," *Proceedings of the 7th International Conference on Coastal Engineering,* Vol. 1, The Hague, 1960, pp. 184–196.

7 Fenton, J.D., "A Fifth-Order Stokes Wave Theory for Steady Waves," *Journal of Waterways, Port, Coastal and Ocean Engineering Division,* ASCE, Vol. 111, 1985, pp. 216–234.

8 Laitone, E.V., "The Second Approximation of Cnoidal and Solitary Waves," *Journal of Fluid Mechanics,* Vol. 9, 1960, pp. 430–444.

9 Leenknecht, A.S., A. Szuwalski, and Sherock, A.R., *Automated Coastal Engineering System,* U.S. Army Corps of Engineers, Coastal Engineering Research Center, Vicksburg, MS, 1992.

10 Dean, R.G., *Evaluation and Development of Water Wave Theories for Engineering Applications,* U.S. Army Coastal Engineering Research Center, Fort Belvoir, VA, Vols. I & II, 1974.

11 McDougal, W.G., Kraus, N.C., Komar, P.D., and Rosatti, J.D., *Coastal Littoral Transport,* U.S. Army Corps of Engineers, EM 1110-2-1502, 1992.

12 Goda, Y., *Random Seas and Design of Maritime Structures,* University of Tokyo Press, Tokyo, 1985.

13 Kamphuis, J.W., "Attenuation of Gravity Waves by Bottom Friction," *Coastal Engineering,* Vol. 100, 1978, pp. 1725–1728.

14 Dalrymple, R.A., "Damping of Water Waves Over a Porous Bed," *Journal of Hydraulics Division,* ASCE, Vol. 100, 1974, pp. 1725–1728.

15 Dalrymple, R.A. and Liu, P.F., "Waves Over Soft Muds: A Two-Layer Fluid Model," *Journal of Physical Oceanography,* Vol. 100, 1978, pp. 1121–1131.

16 Komar, P.D., *Beach Processes and Sedimentation,* Prentice Hall, Englewood Cliffs, NJ, 1976.

17 Gunbak, A.R., *Rubble Mound Breakwaters,* Division of Port and Ocean Engineering Report, Technical University of Norway, Trondheim, 1977.

18 Brunn, P., *Design and Construction of Mounds for Breakwaters and Coastal Protection*, Developments in Geotechnical Engineering, Elsevier, Amsterdam, 1985.

19 Komar, P.D. and Gaughan, M.K., "Sine Wave Theory and Breaker Height Predictions," *Proceedings of the 13th International Conference on Coastal Engineering,* Vancouver, B.C.,1972.

20 Miche, R., "Undulatory Movements of the Sea in a Constant Decreasing Depth," *Annales des Ponts et Chaussees,* 1944.

21 Meyers, J.J., Holm, C.H., and McAllister, R.F., *Handbook of Ocean and Underwater Engineering,* McGraw-Hill, New York, 1969.

22 Moskowitz, L., "A Proposed Spectral Form for Fully Developed Wind Seas Based on the Similarity Laws of S.A. Kitaigorodskii," *Journal of Geophysical Research,* Vol. 69, 1964, pp. 5181–5190.

23 Hasselmann, K., et. al., "Measurements of Wind-Wave Growth and Swell Decay During the Joint North Sea Wave Project (JONSWAP)," *Deutsche Hydrographische Zeitschrift,* 1973.

24 Resio, D.T. and Vincent, C.L., "Estimation of Winds Over the Great Lakes," *Journal of Waterways, Port, Coastal and Ocean Engineering Division,* ASCE, Vol. 103, 1977, pp. 265–283.

25 Carson, W.D., et al., Atlantic Coast Hindcast, Phase II Wave Information, U.S. Army Coastal Engineering Research Center, Vicksburg, MS, 1982.

NOTATION

a_c	wave crest amplitude	H_{rms}	root-mean-square wave height
a_t	wave trough amplitude	H_S	significant wave height
a_x	horizontal water particle acceleration	$H_{1/3}$	average of the highest 1/3 wave heights
a_z	vertical water particle acceleration	$H_{1/10}$	average of the highest 1/10 wave heights
a_n, b_n, A_n, B_n	amplitude coefficients	$H_{1/100}$	average of the highest 1/100 wave heights
b	wave ray spacing	$\hat{H}$	arbitrary wave height
b_0	deep water wave ray spacing	$\overline{H}$	mean wave height
c	wave celerity	k	wave number
c_g	group velocity	k_x	x component of wave number
c_{g0}	deep water group velocity	k_y	y component of wave number
c_0	deep water wave celerity	K_D	diffraction coefficient
e	numerical constant (0.5722 . . .)	K_F	dissipation coefficient
E	wave energy density	K_R	refraction coefficient
E_0	deep water wave energy density	K_S	shoaling coefficient
f	wave frequency	L	wave length
f_p	peak frequency	L_0	deep water wave length
F	fetch length	m	beach slope
g	acceleration due to gravity	n	ratio of group velocity to wave celerity
G	wave direction spreading function	N	number of waves
h	still water depth	p	pressure
H	wave height	P	probability
H_B	breaking wave height	Q	exceedence probability
H_{max}	maximum wave height	R_G	geostrophic correction factor

R_T	temperature correction factor	z	vertical coordinate
s	wave-spreading parameter	α, θ, h	lognormal distribution coefficients
s	alongshore distance between wave rays	α, δ	wave spectrum coefficients
s_0	deep water alongshore distance between wave rays	ε	Stokes ordering term ($\varepsilon = H/L$)
		ε_c	cnoidal ordering term ($\varepsilon_c = H/h$)
$S_{\eta\eta}$	energy density or power spectral density of the free surface	ε, θ	Gumbel distribution coefficients
S_{uu}	horizontal water particle power spectral density	ϕ	velocity potential
		ϕ_n	phase
t	time; also storm duration	γ	JONSWAP peak enhancement coefficient
T	wave period	η	free surface displacement
T_a	air temperature	κ	breaker index
T_p	peak period	λ_n	coefficient in Stokes wave theory
T_s	water temperature	π	numerical constant (3.1415...)
T_z	zero crossing wave period	θ	wave angle
u	horizontal water particle velocity	θ_c	central or dominant wave direction
u_0	deep water horizontal water particle velocity	θ_0	deep water wave angle
U_A	adjusted wind speed or wind stress factor	ρ	water density
U_g	geostrophic wind speed	σ	wave spectrum coefficient
U_r	Ursell number	ω	radian wave frequency
U_{10}	wind speed 10 m above surface	ξ	horizontal water particle displacements
$U_{19.5}$	wind speed 19.5 m above surface	ξ_I	surf similarity parameter
w	vertical water particle velocity	ψ	stream function
w_0	deep water vertical water particle velocity	ζ	vertical water particle displacements
x, y	horizontal coordinates		

PROBLEMS

20.1 Show that the LWT velocity potential satisfies the conservation of mass equation, Eq. (20.2).

20.2 Determine the length of a 10-sec period wave in 50 ft of water.

20.3 Using the LWT dispersion equation, determine the deep and shallow water expressions for the wave celerity.

20.4 At a proposed platform site in the North Sea, the design conditions are: $T = 15$ sec, $h = 70$ m, and $H = 24$ m. Determine L, c, u, and w at $z = -20$ m using LWT.

20.5 An electromagnetic current meter records a maximum horizontal water particle speed of 6.0 ft/sec for a 10-sec wave measured at middepth in a total depth of 97 ft. Determine the wave height.

20.6 An observer on a 300-ft-long ship in the middle of the Pacific Ocean notices that: (1) when the bow is at a wave crest, the stern is in the trough; (2) the waves are propagating in the same direction as the ship; and (3) a different wave is at the bow every 20 sec. Determine the speed of the ship assuming the waves and ship are traveling in the same direction.

20.7 A wave with the following deep water characteristics is propagating toward the shore in an area where the bottom contours are all straight and parallel to the coastline: $H_0 = 20$ ft, $T = 15$ sec. The bottom is composed of sand of 0.1-mm diameter. If a water particle velocity of 1 ft/sec is required to initiate sediment motion, determine the greatest depth at which sediment motion can occur?

20.8 A satellite radar image of the mid-Pacific shows a group of 737-ft-long waves leading a group of 620-ft-long waves. The centers of the two wave groups are separated by a distance of 10 miles. Assuming that these waves were generated simultaneously by the same storm event, determine how far the groups have traveled from the originating storm, and how much time has elapsed since the waves left the generation area.

20.9 Determine the energy flux or power associated with a moderate swell of 6 ft and 10 sec. Ignoring refraction and energy loss, how many miles of coastline would be required to produce 1 MW?

20.10 A diver on the bottom in 30 ft of water is moved back and forth 12 ft ($\xi = 6$ ft) as each wave passes. By counting 11 waves and dividing by 10, the diver determines that the wave period is 8 sec. Assuming the diver is displaced with fluid particles, determine the wave height. Determine how fast the diver is moving when a wave crest passes.

20.11 Waves at location A with $H = 9$ ft and $T = 12$ sec approach the shore as shown in Figure P20.1. The inlet has a depth of 36 ft. Determine the wave height at point B.

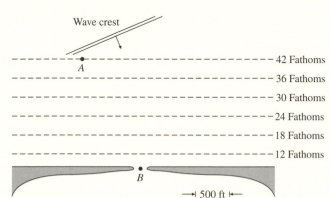

Wave crest

A

42 Fathoms
36 Fathoms
30 Fathoms
24 Fathoms
18 Fathoms
12 Fathoms

B

500 ft

Figure P20.1

20.12 Plot the horizontal water particle profile for $H = 10$ ft, $T = 10$ sec, and $h = 30$ ft for linear, Stokes 5th, and cnoidal wave theories.

20.13 For the time series measurement of the free surface given in Figure P20.2 determine $H_{1/3}$: (a) by zero up-crossing and (b) by assuming a Rayleigh distribution.

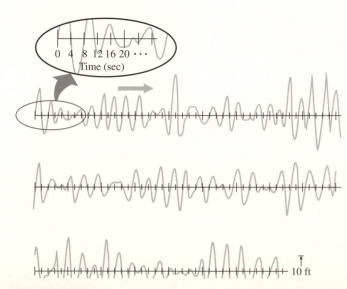

0 4 8 12 16 20 ···
Time (sec)

10 ft

Figure P20.2

20.14 The following waves were hindcast using 20 years of weather data at a site east of New York City [25]. Each estimate corresponds to the significant wave height for a 3-hour interval of time. Determine the 50-year wave height.

$H(m)$	Observations
0.5	9,285
1.0	20,342
1.5	12,028
2.0	6,781
2.5	3,312
3.0	2,819
3.5	1,897
4.0	981
4.5	430
5.0	204
5.0+	237

20.15 Estimate the wave height and period for $F = 200$ nm, $U_A = 40$ knots, and $t = 24$ hours. Estimate the wave condition for $t = 10$ hours.

20.16 The conditions in Figure P20.3 last for 16 hours. Estimate the deep water wave height, period, and direction along the northern California coastline.

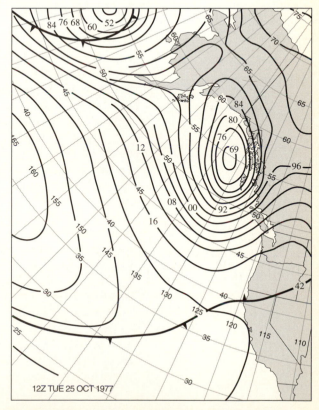

12Z TUE 25 OCT 1977

Figure P20.3. Courtesy of the U.S. Weather Service.

21 ▲ Response of Structures to Water Waves

The principal hydrodynamic forces acting on structures result from pressure and drag. The relative importance of these forces depends upon the wave conditions and structural geometry. As a result, there are several engineering methods to predict wave forces. The manner in which waves interact with a structure is significantly influenced by the size of the structure relative to the wavelength L. If the structure has a circular cross section, the characteristic dimension is the diameter D. If D/L is small, then wave diffraction is negligible and the Morison equation is used to estimate forces. If D/L is large, the modification of the wave field by the structure is important, and diffraction theory is used to calculate forces. If the drag forces are small and modifications to the wave field are not significant, the forces are inertia dominated for these conditions and the forces may be determined by the Froude-Krylov approach.

A physical interpretation of a small body is that the presence of the body has negligible influence on the wave field. For example, when typical ocean waves pass a pile, the modification of the wave field away from the pile is minimal. Just a wave length away from the pile (i.e., 50 to 100 pile diameters), it appears that the flow is completely unaffected by the pile. This absence of wave-structure interaction allows for an important simplification in the determination of wave forces on small bodies. The influence of the waves on the pile is calculated, but it is assumed that the pile has no influence on the waves. Wave forces are simply a function of the incident wave conditions at the location of the pile.

The behavior of a large body in the presence of a wave field is quite different. The structure is large enough that using the wave conditions at the center of the structure leads to significant errors, and the waves are strongly influenced by the presence of the structure. Consider a floating dock in small waves. Although some of the waves propagate around or under the dock, a significant portion of the incident wave is reflected back. Therefore, the wave field is highly modified, and this modification must be included in the wave force calculations.

21.1 MORISON EQUATION

As mentioned previously, in the context of wave loading, a small body is one that does not significantly modify the incident wave field. If the diameter of the structure is less than 5% of the wave length, this assumption tends to be valid. Examples of these types of structures are piles, structural members in oil platforms, pipelines, and moorings. Since these structures are common in the marine environment, significant effort has been directed toward the study of wave forces on small bodies. The most common technique for estimating the in-line wave force on small bodies is the Morison equation [1]. The Morison equation includes both inertia forces associated with pressure and drag forces associated with fluid friction and flow separation. However, it does not include hydrostatic forces. Because the Morison equation is so widely used and because it is partly empirical, a brief derivation is developed to demonstrate the underlying assumptions.

First, consider the horizontal wave-induced pressure component on a vertical cylinder as shown in Figure 21.1. A velocity potential for ideal flow about a circular cylinder in an unbounded fluid is given by [2]

$$\phi = u\left(1 + \frac{D^2}{4r^2}\right)r\cos\theta \tag{21.1}$$

where u is the water particle velocity and is a function of time only and r and θ are polar coordinates. On the surface of the cylinder, the normal, u_r, and tangential, u_θ, velocities are determined from the positive gradients of the velocity potential

$$u_r = 0 \qquad u_\theta = 2u\sin\theta \tag{21.2a, b}$$

From the Bernoulli equation, the fluid pressure is expressed as

$$p = -\rho\frac{\partial\phi}{\partial t} - \frac{\rho}{2}(u_r^2 + u_\theta^2) - \rho gz + C_B \tag{21.3}$$

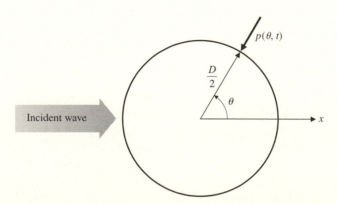

Figure 21.1

Pressure forces on a cylinder.

where C_B is the Bernoulli constant. Integrating the pressure over the cylinder surface gives the fluid-induced force on the structure. The horizontal force component per unit length of the cylinder in line with the wave direction is

$$f_{ix} = -\int_{-\pi}^{\pi} p\left(\frac{D}{2}, \theta, t\right) \cos\theta \, \frac{D}{2} d\theta \qquad (21.4)$$

It can be shown that the last three terms in Eq. (21.3) integrate to zero. Furthermore, on the surface of the cylinder,

$$\left.\frac{\partial\phi}{\partial t}\right|_{r = D/2} = D\frac{du}{dt} \cos\theta \qquad (21.5)$$

so the resulting horizontal force per unit length of the cylinder is

$$f_{ix} = \rho\frac{\pi D^2}{2}\frac{du}{dt} \qquad (21.6)$$

It is noted in Eq. (21.6) that this force is proportional to the acceleration of the fluid past the fixed cylinder. Hence, it is called the *inertia force*. Often, an inertia force is written as the product of mass and acceleration. Thus, the inertia force per unit length of cylinder is represented as

$$f = \rho\frac{\pi D^2}{4}\frac{du}{dt} \qquad (21.7)$$

It is seen that the quantity f_{ix} defined in Eq. (21.6) is twice this amount. This indicates that the inertia force is greater than the product of just the mass of the displaced fluid and its acceleration. Therefore the inertia force is expressed as

$$f_{ix} = (1 + C_a)\rho\frac{\pi D^2}{4}\frac{du}{dt} \qquad (21.8)$$

where C_a is termed the added mass coefficient. In solid, rigid-body dynamics, the inertia force is related simply to the mass of the body. However, in a fluid, the mass of the accelerated fluid is greater than just that displaced by the cylinder. This additional amount is called the added mass. For a vertical, circular cylinder, the analytical value of the added mass coefficient was shown above to be $C_a = 1.0$. The added mass coefficient can be analytically determined for a number of simple geometries and flow conditions. Often, the short-hand variable

$$C_m = 1 + C_a \qquad (21.9)$$

is used in Eq. (21.8). The term C_m is called the inertia coefficient.

The inertia force f_{ix} was determined assuming an ideal fluid. In a real fluid there will also be drag. There are two types of drag: (1) skin drag from the friction of the flow over the surface of the body and (2) form drag from the pressure differential across the body when the flow separates. These are shown in Figure 21.2. Both types of drag are proportional to the velocity squared, with empirical coefficients that are dependent on the Reynolds number, body shape, and roughness. The two types of

Figure 21.2

Example of friction and form drag.

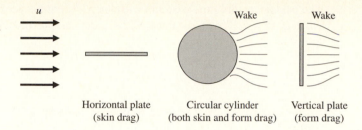

Horizontal plate
(skin drag)

Circular cylinder
(both skin and form drag)

Vertical plate
(form drag)

drag can be combined into one equation with a single drag force coefficient, and expressed as

$$f_{dx} = \frac{1}{2}\rho C_d D |u| u \tag{21.10}$$

where f_{dx} = in-line drag force per unit length of cylinder
C_d = drag force coefficient
u = water particle velocity in the direction of wave propagation

Since the fluid velocities in waves reverse direction during the flow cycle, the absolute value is required to ensure the proper sign on the drag force.

To arrive at the Morison equation, it is assumed that the drag and inertia forces can be added to define the total in-line force per unit length of cylinder. Thus

$$f_x = f_{dx} + f_{ix} = \frac{1}{2}\rho C_d D |u| u + \rho C_m \frac{\pi D^2}{4}\frac{du}{dt} \tag{21.11}$$

The velocity and acceleration are calculated at the center of the pile, assuming the pile is not present. If the body undergoes motion, then the drag force is based on the relative velocity between the fluid and the structure, and the inertia force is based on the relative acceleration. The Morison equation then becomes

$$f_x = \frac{1}{2}\rho C_d D |u - \dot{x}|(u - \dot{x}) + \rho\frac{\pi D^2}{4}C_m(\dot{u} - \ddot{x}) \tag{21.12}$$

where x is the horizontal structure displacement. Equation (21.12) is referred to as the relative motion Morison equation.

The Morison equation exhibits the interesting behavior that the maximum drag and inertia forces do not occur at the same time. If linear wave theory is used for the fluid velocity and acceleration with $x = 0$ at the center of the cylinder, the Morison equation is given by

$$f_x = \frac{1}{8}\rho C_d D H^2 \omega^2 \frac{\cosh^2[k(h + z)]}{\sinh^2(kh)}|\cos(-\omega t)|\cos(-\omega t)$$
$$+ \frac{\pi}{8}\rho C_m D^2 H \omega^2 \frac{\cosh[k(h + z)]}{\sinh(kh)}\sin(-\omega t) \tag{21.13}$$

which may be rewritten as

$$f_x = f_{d\max}|\cos \omega t| \cos \omega t - f_{i\max} \sin \omega t \tag{21.14}$$

The peak in the drag force occurs when $\omega t = 0$, and the peak in the inertia force occurs when $\omega t = \pi/2$. The phase at which the combined maximum total force occurs is determined as

$$\frac{\partial f_x}{\partial t} = 0 \rightarrow \omega t\big|_{max} \tag{21.15}$$

Figure 21.3 shows the two force components and the total force for $f_{dmax} = f_{imax} = 1.0$ lb/ft.

If $f_{dmax} \gg f_{imax}$, then the total force will be dominated by the drag component, and if $f_{imax} \gg f_{dmax}$, then the force is inertia dominated. The ratio of the two force components for LWT is

$$\frac{f_{dmax}}{f_{imax}} = \frac{C_d H}{\pi C_m D}\frac{\cosh[k(h+z)]}{\sinh(kh)} = \frac{1}{\pi^2}\frac{C_d}{C_m}\frac{u_m T}{D} \tag{21.16}$$

where u_m is the amplitude of the horizontal velocity and T is the wave period. This ratio determines the relative importance of the two force components. Since C_d and C_m do not have significant variations, the relative importance of drag and inertia is often related by Keulegan-Carpenter number K defined by

$$K = \frac{u_m T}{D} \tag{21.17}$$

This number can be considered as the distance traveled by a water particle, $u_m T$, relative to the pile diameter D. If K is large, the water particles move large distances relative to the cylinder diameter, resembling a condition similar to steady flow. In steady flow there is no inertia force; thus the total force is due to drag only. At the other extreme, if K is small, then the water particles move a small distance with respect to the cylinder diameter. In this case, the flow reversals are important and the total force is inertia dominated.

Figure 21.3

Wave force components.

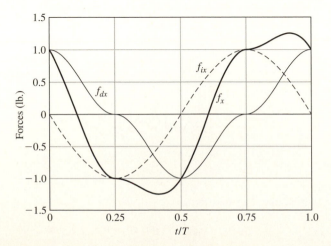

A common application of the Morison equation is to determine the force on a vertical member that extends over the full water depth. The total force is simply the integral of the depth-dependent force over the total depth (see Figure 21.4) expressed as

$$F_x = \int_{-h}^{\eta} f_x(z)\, dz \tag{21.18}$$

Note that uppercase F_x is used to denote the total horizontal force, while lowercase f_x is used for the force per unit length. The upper limit on the integral in Eq. (21.18) is a reminder of the limitations of LWT. In LWT, integration is performed only up to $z = 0$ and not to the actual water surface $z = \eta$. If LWT is used, then

$$F_x = F_{d\max}|\cos \omega t|\ \cos \omega t - F_{i\max} \sin \omega t \tag{21.19}$$

where

$$F_{d\max} = \frac{1}{16}\rho g C_d D H^2 \frac{\sinh(2kh) + 2kh}{\sinh(2kh)} \tag{21.20}$$

$$F_{i\max} = \frac{\pi}{8}\rho g C_m D^2 H \tanh(kh) \tag{21.21}$$

In Eq. (21.19), it is assumed that the drag and inertia coefficients are both time independent and do not vary with depth beneath the free surface. The expression delineating the moment at the mudline is

$$M = \int_{-h}^{\eta} (h + z) f_x(z)\, dz \tag{21.22}$$

For LWT, Eq. (21.22) becomes

$$M = M_{d\max}|\cos \omega t|\ \cos \omega t - M_{i\max} \sin \omega t \tag{21.23}$$

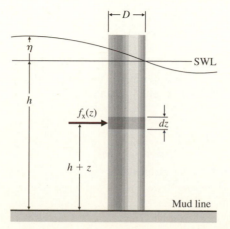

Figure 21.4

Definition sketch for force and moment on a vertical pile.

where

$$M_{d\max} = 4F_{d\max}h\left[1 - \frac{\sinh^2(kh)}{kh\sinh(kh) + (kh)^2}\right] \qquad (21.24)$$

$$M_{i\max} = F_{i\max}h\left[1 - \frac{\sinh(kh)}{kh\cosh(kh) + kh}\right] \qquad (21.25)$$

Integrating only to the still water level ($z = 0$) and not to the free surface ($z = \eta$) results in the forces and moments being underestimated. If the wave height is large, this error can be significant. For LWT, there are two general techniques for estimating kinematics up to the free surface: extrapolation and stretching.

There are three ways to extrapolate above the SWL: (1) constant, (2) linear, and (3) exponential. For a constant extrapolation, values above the SWL are taken as equal to those at the SWL. For linear extrapolation, values above the SWL are extended linearly using the same slope as at the SWL. Exponential extrapolation uses values of z above the SWL in the LWT equations. Constant and linear extrapolation lead to conservative estimates of kinematics. Exponential extrapolation leads to extremely conservative estimates and is not recommended.

Stretching redefines the water depth to be the sum of the still water depth h plus the free surface displacement. The stretched depth is defined by

$$d = h + \eta \qquad (21.26)$$

and z is in the range

$$-h < z < \eta \qquad (21.27)$$

It is assumed that LWT kinematics at the SWL (that is, $z = 0$) are now applied at $z = \eta$. The horizontal particle velocity and acceleration in stretch wave theory are, respectively,

$$u = \frac{Hgk}{2\omega}\frac{\cosh[k(h+z)]}{\cosh[k(h+\eta)]}\cos(kx - \omega t) \qquad (21.28a)$$

$$a_x = \frac{H}{2}gk\frac{\cosh[k(h+z)]}{\cosh[k(h+\eta)]}\sin(kx - \omega t) \qquad (21.28b)$$

Figures 21.5 and 21.6 show comparisons of the force and moment using stretch wave theory relative to LWT. In Figures 21.5 and 21.6, W indicates the relative importance of the inertia and drag terms and is defined as

$$W = \frac{C_mD}{C_dH} \qquad (21.29)$$

In shallow water, the influence is significant on both the force and the moment. This is because as the depth decreases, the wave height contribution to the total depth becomes more important. The mudline moments are strongly influenced by the force above the SWL because they have the longest moment arm.

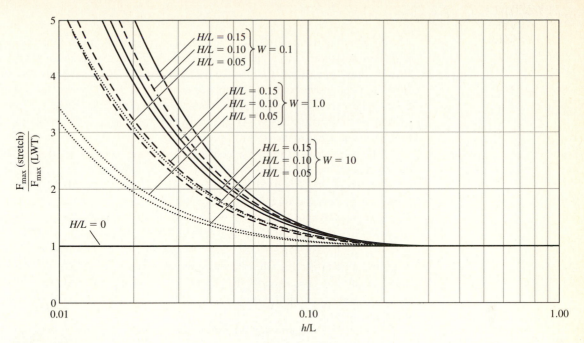

Figure 21.5

Ratio of maximum wave force using stretch wave theory to maximum wave force using LWT.

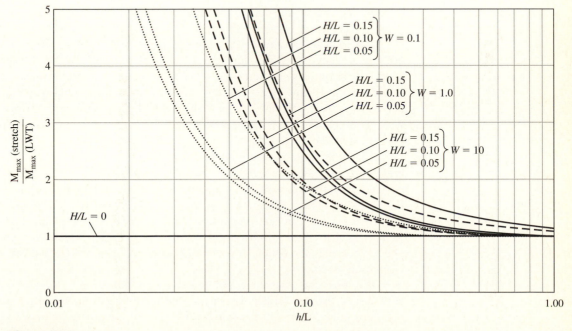

Figure 21.6

Ratio of maximum mudline moment using stretch wave theory to maximum mudline moment using LWT.

21.2   FORCE COEFFICIENTS

The drag and inertia coefficients C_d and C_m are empirically determined from laboratory and field measurements. Experimental coefficients are available for several classes of flow: steady flow, oscillatory flow, and wave flows. The water particles in oscillatory flows simply move back and forth horizontally, while in wave flows they move in orbits. There are also scale effects if the measurements were conducted in small experiments. In small scale tests it is generally not possible to achieve the high Reynolds numbers that occur in the field.

The estimation of the force coefficients from both laboratory and field measurements depends on the choice of wave theory used to analyze the data. The dependency on wave theory creates a difficulty when using linear wave theory. The largest wave forces occur near the wave crest, but linear wave theory extends only up to the SWL. Stretching can extend LWT to the water surface while retaining a linear dependency on H for statistical analyses. However, near the free surface, linear wave theory does not render accurate estimates for the wave kinematics. In this region, more sophisticated theories, such as Stokes 5th or stream function theory, must be used. Although linear wave theory has been used in this chapter to demonstrate the concepts of wave forces on small bodies, in engineering practice it is often necessary to use a more advanced nonlinear theory.

Figure 21.7 presents the drag and inertia coefficients for cylinders in oscillatory flow as a function of the Reynolds number and the relative roughness. The Reynolds number is defined as

$$R = \frac{u_m D}{\nu} \tag{21.30}$$

where u_m is the amplitude of the particle velocity and ν is the kinematic viscosity. The relative roughness is the ratio of the roughness height k to the cylinder diameter D. The values presented in Figure 21.7 correspond to a Keulegan-Carpenter number of $K = 20$. As the roughness increases, the drag coefficient increases and the inertia coefficient decreases. Above Reynolds numbers of 2×10^5, the coefficients are nearly independent of the Reynolds number. For $K = 100$, this cutoff is approximately 8×10^5. Drag and inertia coefficients for wave flows are a function of the Keulegan-Carpenter number and are given in Figure 21.8 for roughened cylinders. The drag coefficient increases with the roughness, but it has been observed in waves that the inertia coefficient is less dependent on the roughness.

It is common to develop marine growth on structural members. This growth, or biofouling, increases the weight on the members but does not increase the stiffness. It also increases the structure diameter, drag coefficient, and inertia coefficient. The fouling can be soft, such as seaweed and sponges, or hard, such as mussels and barnacles. In either case, the influence on the structure can be significant. Figure 21.9 shows typical soft and hard fouling in temperate waters. The growth is usually concentrated in the photic zone (i.e., the depths at which light penetrates the water), but can also exist at large water depths.

The force coefficients are also dependent on the proximity of other cylinders, the bottom, and the free surface. In general, for a cylinder in the vicinity of impermeable boundaries, the coefficients increase because the flows are more confined. If the structure pierces the free surface, wave slamming can occur, which may produce very large forces.

A thorough discussion of these topics is given in [1, 3]. With all the qualifiers given above, ranges of $0.6 < C_d < 1.2$ and $1.5 < C_m < 2.0$ are generally taken for smooth circular cylinders. Typical values are $C_d = 0.7$ and $C_m = 1.7$, except for $C_d = 1.0$ when using LWT.

The discussion of the Morison equation and the force coefficients has focused on circular cross sections. The results are also applicable to other shapes. In Eq. (21.11), D in the drag force term is replaced by the projected area per unit length of the cylinder. The $\pi D^2/4$ term in the inertia force is replaced by the displaced volume per unit length. Table 21.1 gives drag and inertia coefficients for common structural shapes.

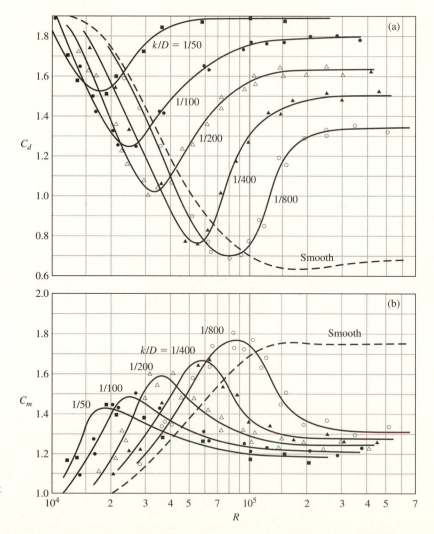

Figure 21.7

Drag and inertia coefficients for roughened cylinders in oscillatory flow, $K = 20$: (a) C_d; (b) C_m [3]. Adapted from *In-Line and Transverse Forces on Smooth and Sand-Roughened Cylinders in Oscillatory Flow at High Reynolds Numbers* by T. Sarpkaya, Report No. NDS-69SL76062, Naval Post Graduate School, Monterey, CA, 1976.

Figure 21.8

Drag and inertia coefficients for roughened cylinders in wave flows: (a) C_d; (b) C_m [4]. Adapted from *Hydrodynamics of Offshore Structures* by S.K. Chakrabarti, Springer-Verlag, Berlin, 1987.

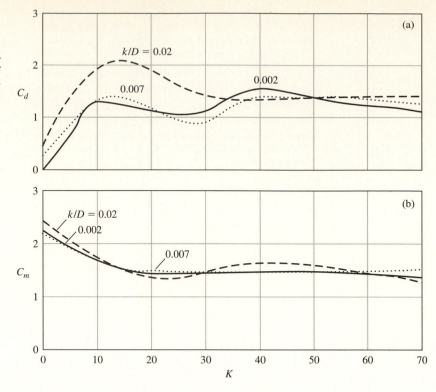

TABLE 21.1. Drag and inertia coefficients for common structural shapes

Section Shape	C_d	C_m
→ ■	2.0	2.5
→ ▢	0.6	2.5
→ ▷	2.0	2.3
→ ◁	1.3	2.3
→ ◇	1.5	2.2

Figure 21.9

Typical distribution of marine growth [5]. Adapted from *Dynamics of Marine Structures* by M.G. Hallman, N.J. Heaf, and L.R. Woolton, CIRIA Underwater Engineering Group, London, 1978.

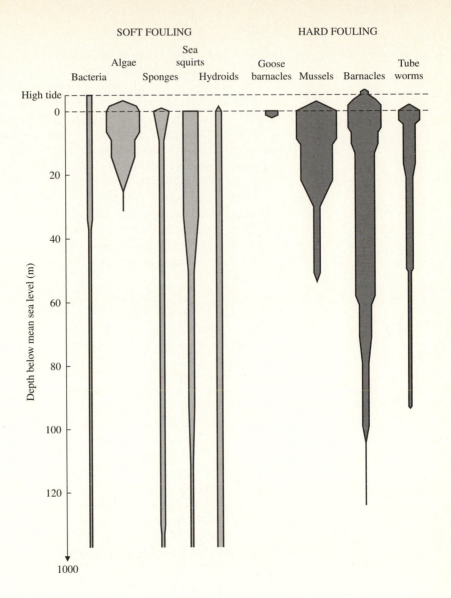

EXAMPLE 21.1 ▼

A discharge pipe 2 ft in diameter extends 30 ft vertically above the ocean bottom. The wave height is 10 ft, the wave period is 10 sec, and the water depth is 40 ft. Determine the maximum horizontal wave force and the maximum mudline moment using LWT, Stokes 5th, and cnoidal wave theories. Assume $C_d = 0.7$, $C_m = 1.7$, and water density $\rho = 1.99$ slugs/ft^3.

Solution

The force and moment may be determined by integrating Eq. (21.11). The steps are similar to the development of Eqs. (21.19) and (21.23), except the integration is not performed over the full depth. First consider the LWT problem. The integrations may be done analytically. The solution developed in Table 21.2 uses the symbolic mathematics package Maple.

A similar integration technique could also be developed for cnoidal and Stokes wave theories. An alternative is to numerically integrate the equations. Computer programs are provided on the authors' web site (see the Preface) to estimate u and a_x for each of these wave theories.

In this example, the 30-ft-high pipe is divided into three 10-ft elements. The force is calculated on each element, and the total force is the sum of the elements. Only three elements are used to demonstrate the solution technique. In practice, more elements may be required to provide better accuracy.

The drag and inertia forces on an element may be written as

$$f_d = \left[\frac{1}{2}\rho C_d D |u| u\right]\Delta L = \frac{1}{2}(1.99)(0.7)(2)|u|u(10) = 13.93|u|u$$

$$f_i = \left[\rho C_m \frac{\pi D^2}{4}a_x\right]\Delta L = 1.99(1.7)\frac{\pi(2)^2}{4}a_x(10) = 106.3a_x$$

Table 21.3 shows a spreadsheet solution for the force and moment using LWT. The horizontal velocity and acceleration (u and a_x) are calculated at the middle of each element using the FORTRAN program LWT.FOR, available on the authors' web site. The force on each element is estimated using the above equations. The moment is the product of the force and the appropriate moment arm.

In the last two columns of Table 21.3, the results from each element are summed. The maximum force and moment are the maximum values that occur over one wave period. In Table 21.3, the phases of u and a_x have been shifted by 180° to agree with the phase of the output from the cnoidal and Stokes 5th computer programs.

This same spreadsheet template is used to calculate the force and moment for cnoidal and Stokes 5th wave theories. Velocities and accelerations are determined using the computer programs CNOIDAL.FOR and STOKES.FOR, available on the authors' web site. Results are summarized in Table 21.4.

The variation of a wave cycle of the force and moment for each of the numerical computations is shown in Figure 21.10. The results are as anticipated; the nonlinear wave theories yield larger peak forces. In Table 21.4 it is seen that even though the elements were large the numerical results are reasonable. In Example 21.2, a FORTRAN program is used to solve for the forces, rather than a spreadsheet. This allows the elements to be arbitrarily small. The load, shear, and moment along the discharge pipe are shown in Figure 21.11. These are based on the analytical LWT solution in Table 21.2. The shear and moment are calculated by integrating from the top of the pipe down to different elevations to determine the curves.

A final consideration is an assessment of which of these wave theories is most applicable to this problem. For this case

$$\frac{h}{L_0} = \frac{40}{5.12(10)^2} = 0.0781$$

$$\frac{H}{L_0} = \frac{10}{5.12(10)^2} = 0.0195$$

From Figure 20.7, the best choice is Stokes 5th.

TABLE 21.2. **Maple Analytical Solution to Determine the LWT Force and Moment for Example 21.1**

Example 21.1
Force and Moment using LWT

First clear the variable names.
```
> t:= 't' : tmax:= 'tmax' : k:= 'k' : z:='z' : f:='f' : F:='F':
```

Enter the constants for the problem.
```
> Cd:=0.7 : Cm:=1.7 : d:=2 : H:=10 : T:=10 : h:=40 : z_top:=−10 :
> g:=32.2 : rho:=1.99 : omega:=2*Pi/T :
```

Solve the dispersion equation for the wave number (Table 20.1).
```
> k:=evalf(solve(omega^2−g*k*tanh(k*h)=0,k));
                    k:= .01907334318.
```

Calculate the amplitudes of the velocity and acceleration (Table 20.1).
```
> um:=evalf(g*k*H/(2*omega)*cosh(k*(h+z))/cosh(k*h));
> axm:=evalf(g*k*H/2*cosh(k*(h+z))/cosh(k*h));
          um:= 3.743864105 cosh(.7629337272 + .01907334318 z)
          axm:= 2.352339194 cosh(.7629337272 + .01907334318 z)
```

Calculate the magnitude of the drag force and moment (Eq. 21.10).
```
> fdmax:=evalf((1/2)*rho*Cd*d*um^2) :
> Fdmax:=int(fdmax,x=−h..z_top) ;
> Mdmax:=int((h+z)*fdmax,z=−h..z_top) ;
              Fdmax:= 653.9968364
              Mdmax:= 10332.76720
```

Calculate the magnitude of the inertia force and moment (Eq. 21.8).
```
> fimax:=evalf(rho*Cm*(Pi*d^2/4)*axm) :
> Fimax:=int(fimax,z=−h..z_top) ;
> Mimax:=int((h+z)*fimax,z=−h..z_top);
              Fimax:= 791.6234106
              Mimax:= 12188.06855
```

The total time dependent force and moment are the sums of the components.
```
> F:=evalf(Fdmax*abs(cos(omega*t))*cos(omega*t)−Fimax*sin(omega*t)) ;
> M:=evalf(Mdmax*abs(cos(omega*t))*cos(omega*t)−Mimax*sin(omega*t)) ;
F:=653.9968364 |cos(.6283185305 t)|cos(.6283185305 t)− 791.6234106 sin(.6283185305 t)
M:=10332.76720 |cos(.6283185305 t)|cos(.6283185305 t)− 12188.06855 sin(.6283185305 t)
```
Determine the time of the maximum force (Eq. 21.16)
and the corresponding maximum force and moment.
```
> tmax:=evalf((1/omega)*arcsin(Fimax/(2*Fdmax))) ;
> t:=tmax : Fmax:=F : Mmax=M ;
              tmax:= −1.034573499
              Fmax:= 893.5498393
              Mmax:= 13924.42798
```

TABLE 21.3. Spreadsheet Numerical Solution to Determine the LWT Force and Moment for Example 21.1

Linear Wave Theory Forces

fd const.=13.930
fi const.=106.280

max vals= 890.7 lb   13822 ft-lb
min vals= −890.7 lb −13822 ft-lb

time (sec)	z = −15 ft				z = −25 ft				z = −35 ft				F (lb)	M (ft-lb)
	u (ft/sec)	ax (ft/sec²)	f (lb/ft)	m (lb)	u (ft/sec)	ax (ft/sec²)	f (lb/ft)	m (lb)	u (ft/sec)	ax (ft/sec²)	f (lb/ft)	m (lb)		
0.0	−4.18	0.00	−243.39	−6084.76	−3.90	0.00	−211.88	−3178.13	−3.76	0.00	−196.94	−984.68	−652.2	−10247.6
0.1	−4.17	0.16	−225.22	−5630.56	−3.89	0.15	−194.85	−2922.72	−3.75	0.15	−179.95	−899.74	−600.0	−9453.0
0.2	−4.14	0.33	−203.68	−5092.06	−3.87	0.31	−175.68	−2635.22	−3.73	0.30	−161.92	−809.61	−541.3	−8536.9
0.3	−4.10	0.49	−182.09	−4552.15	−3.83	0.46	−155.45	−2331.73	−3.69	0.44	−142.91	−714.55	−480.4	−7598.4
0.4	−4.05	0.65	−159.4	−3985.12	−3.78	0.61	−134.21	−2013.10	−3.64	0.59	−121.86	−609.31	−415.5	−6607.5
0.5	−3.97	0.81	−133.46	−3336.56	−3.71	0.76	−110.96	−1664.42	−3.58	0.73	−100.95	−504.74	−345.4	−5505.7
0.6	−3.88	0.97	−106.62	−2665.40	−3.62	0.90	−86.89	−1303.38	−3.50	0.87	−78.18	−390.89	−271.7	−4359.7
.	.	.	.	.	.	.	.	.	.	.	.	.	.	.
.	.	.	.	.	.	.	.	.	.	.	.	.	.	.
9.8	−4.14	−0.33	−273.83	−6845.68	−3.87	−0.31	−241.58	−3623.63	−3.73	−0.30	−225.69	−1128.45	−741.1	−11597.8
9.9	−4.17	−0.16	−259.23	−6480.80	−3.89	−0.15	−226.73	−3400.98	−3.75	−0.15	−211.83	−1059.16	−697.8	−10941.0
10.0	−4.18	0.00	−243.39	−6084.76	−3.90	0.00	−211.88	−3178.13	−3.76	0.00	−193.94	−984.68	−652.2	−10247.6

TABLE 21.4. Summary of Force and Moment Calculations for Example 21.1

	F (lb)	M (ft-lb)
LWT (analytical)	894	13,925
LWT (numerical)	891	13,822
cnoidal (numerical)	1238	19,186
Stokes 5th (numerical)	1019	16,054

766

Figure 21.10

Calculated results using linear, Stokes 5th, and cnoidal wave theories: (a) forces; (b) moments.

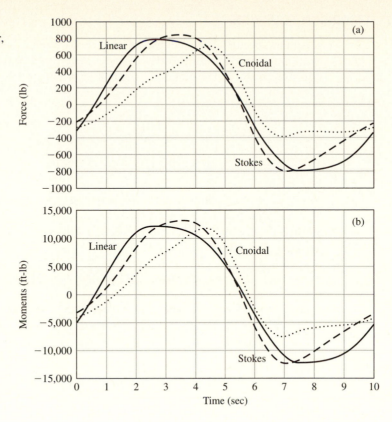

Figure 21.11

Analytical LWT force estimates: (a) load; (b) shear; (c) moment.

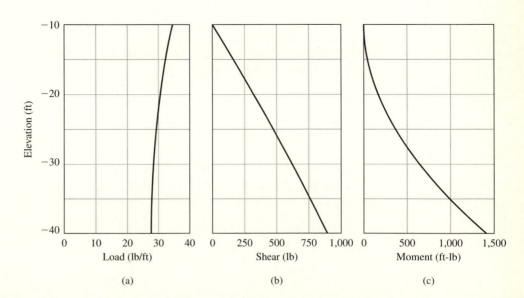

21.3 LINEARIZED MORISON EQUATION

Common spectral techniques use linear superposition. Even when using LWT, the drag term in the Morison equation is nonlinear. As a result, it is convenient to linearize the Morison equation. The drag force per unit length is given by

$$f_{dx} = \frac{1}{2}\rho C_d D |u| u \tag{21.31}$$

For simple periodic flows, this may be written as

$$f_{dx} = \frac{1}{2}\rho C_d D u_m^2 \,|\cos(\omega t)|\, \cos(\omega t) \tag{21.32}$$

where u_m is the amplitude of the velocity. One common method of linearizing is the Lorentz principle of equivalent work. In this technique the actual work per periodic cycle (or energy dissipation rate) is set equal to an equivalent linear term. The linearized drag is taken to be

$$f_{dL} = \frac{1}{2}\rho C_d D(\alpha u) = \frac{1}{2}\rho C_d D[\alpha u_m \cos(\omega t)] \tag{21.33}$$

where α is a linearizing coefficient. Work is the product of force and displacement and the dissipation rate is the product of force and velocity. Equating the power over one wave period results in

$$\int_t^{t+T} |u|u \cdot u \; dt = \int_t^{t+T} (\alpha u) \cdot u \; dt \tag{21.34}$$

Integrating Eq. (21.34) and solving for α gives

$$\alpha = \frac{8}{3\pi}u_m \tag{21.35}$$

Figure 21.12 shows a comparison of the linearized and nonlinear quadratic drag forces for the same wave and structural conditions. While the overall comparison is good, there are two significant differences: (1) the peak linearized force is about 15% less than the peak nonlinear force, and (2) the spectrum of the nonlinear force has higher frequency components while all the energy in the linear approximation is at $f = 1/T$. The linearized Morison equation may be written as

$$f_x = \frac{1}{2}\rho C_d D \frac{8u_m}{3\pi}u + \rho C_m \frac{\pi D^2}{4}\frac{\partial u}{\partial t} \tag{21.36}$$

In simple periodic waves, u_m is merely the amplitude of the horizontal particle velocity.

The appearance of higher harmonics in the quadratic force can be very important. It is possible that these harmonics occur in the vicinity of the natural period of the structure. The quadratic drag may be expanded in a Fourier series assuming simple periodic flow, resulting in

$$|u|u = u_m^2\,|\cos(\omega t)|\,\cos(\omega t)$$

$$= u_m^2 \frac{8}{3\pi}$$

$$\left[\cos(\omega t) + \frac{1}{5}\cos(3\omega t) - \frac{1}{35}\cos(5\omega t) + \frac{1}{105}\cos(7\omega t) + \cdots\right] \tag{21.37}$$

Figure 21.12

Comparison of quadratic and linearized drag forces.

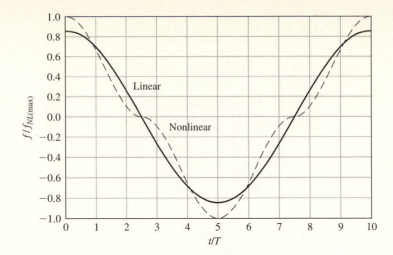

The appearance of the higher harmonics can give rise to a resonant behavior (say, at 3ω) for a structure subjected to wave loading for a wave of frequency ω. This physics is lost in the linearization. Note that the linear term in the Fourier series is the same as that obtained by the Lorentz principle of equivalent work.

Borgman [6] proposed a technique for linearizing the Morison equation for random waves. This technique requires the following change in Eq. (21.36):

$$\frac{8u_m}{3\pi} \rightarrow \left(\frac{8}{\pi}\right)^{1/2} u_{\text{rms}} \tag{21.38}$$

where u_{rms} is the root-mean-square velocity. Since the Morison equation (in this form) is linear, superposition may be used to determine the force spectrum corresponding to a wave spectrum. The steps are: (1) determine the velocity and acceleration spectra from the wave spectrum using the appropriate transfer functions, (2) determine u_{rms} from the velocity spectrum, and (3) use these results in the Morison equation. This approach is outlined below.

The horizontal velocity and acceleration, using linear wave theory, may be written as

$$u = \frac{H}{2}\omega\frac{\cosh[k(h+z)]}{\sinh(kh)}\cos(kx - \omega t) \tag{21.39}$$

$$a_x = \frac{H}{2}\omega^2\frac{\cosh[k(h+z)]}{\sinh(kh)}\sin(kx - \omega t) \tag{21.40}$$

From this, the velocity and acceleration spectra corresponding to a free surface spectrum are given as

$$S_{uu}(\omega) = \left\{\omega\frac{\cosh[k(h+z)]}{\sinh(kh)}\right\}^2 S_{\eta\eta}(\omega) \tag{21.41}$$

$$S_{a_x a_x}(\omega) = \left\{\omega^2\frac{\cosh[k(h+z)]}{\sinh(kh)}\right\}^2 S_{\eta\eta}(\omega) \tag{21.42}$$

in which $S_{\eta\eta}$ is the free surface spectrum. The velocity to linearize the drag, u_{rms}, is given by

$$u_{\mathrm{rms}}^2 = 8\int_0^\infty S_{uu}(\omega)\,d\omega \tag{21.43}$$

Finally, the one-sided spectral density of the force per unit length of the structure is

$$S_{ff}(\omega) = \left\{\omega\frac{\cosh[k(h+z)]}{\sinh(kh)}\right\}^2$$

$$\left[\frac{8}{\pi}\left(\frac{1}{2}\rho C_d D u_{\mathrm{rms}}\right)^2 + \left(\omega\rho C_m\frac{\pi D^2}{4}\right)^2\right]S_{\eta\eta} \tag{21.44}$$

Figure 21.13 shows a wave spectrum and the corresponding force spectrum. This figure represents the water line forces (that is, $z = 0$) in deep water with $C_d = 1.0$ and $C_m = 2.0$. The spectra are scaled by their respective peak values.

The relative motion Morison equation given in Eq. (21.12) is also nonlinear. The horizontal force on a vertical structure is the integral of the force per unit length over the depth. If the following notation for the depth-integrated water particle velocity and acceleration is adopted,

$$U = \int_z u(z,t)\,dz \tag{21.45}$$

$$\frac{dU}{dt} = \int_z a_x(z,t)\,dz \tag{21.46}$$

then the relative motion Morison equation may be written in terms of the total force as

$$F_x = \frac{1}{2}\rho C_d D|U - \dot{x}|(U - \dot{x}) + \rho C_m\frac{\pi D^2}{4}\left(\frac{dU}{dt} - \ddot{x}\right) \tag{21.47}$$

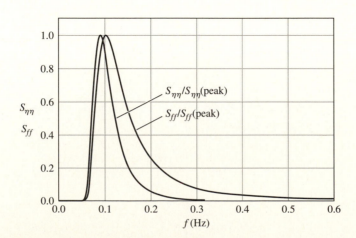

Figure 21.13

Wave spectrum and corresponding force spectrum.

Consider the vibration of a SDOF system due to wave forces expressed as

$$m\ddot{x} + c\dot{x} + kx = \frac{1}{2}\rho C_d D|U - \dot{x}|(U - \dot{x}) + \rho C_m \frac{\pi D^2}{4}\left(\frac{dU}{dt} - \ddot{x}\right) \quad (21.48)$$

where m, c, and k are the structure mass, damping, and stiffness, respectively. To linearize this equation, begin by writing the relative velocity

$$\dot{r} = \dot{x} - U \quad (21.49)$$

The equation of motion given in Eq. (21.48) may then be written

$$\left(m + \rho\frac{\pi D^2}{4}C_m\right)\ddot{r} + \left[c + \frac{1}{2}\rho C_d D|\dot{r}|\right]\dot{r} + kr$$

$$= -m\dot{u} - cu - k\int_t U \ dt \quad (21.50)$$

Notice that the term in the square brackets is nonlinear. As before, the Lorentz principle of equivalent work can be used to estimate a linear damping coefficient. First, define a linear damping term with a linear damping coefficient c^* that dissipates the same amount of energy over one wave period as the nonlinear damping as

$$\int_t^{t+T} c^*\left(\frac{1}{2}\rho C_d D\right)\dot{r} \ dt = \int_t^{t+T}\left(\frac{1}{2}\rho C_d D\right)|\dot{r}| \ \dot{r} \ dt \quad (21.51)$$

Next, consider the structure motion and wave velocity to be periodic, resulting in

$$x(t) = x_m \cos(\omega t - \phi) \quad (21.52)$$

$$U = U_m \cos(\omega t) \quad (21.53)$$

where x_m and U_m are the amplitudes of the motion and velocity and ϕ is the phase shift between the structure and wave motion. For a linear system it is assumed that the frequency of the steady-state response is the same as the forcing frequency. Defining the notational variable

$$S = \frac{\omega x_m}{U_m} \quad (21.54)$$

where S is the ratio of the structure velocity amplitude to the wave velocity amplitude, the linear damping coefficient is then given by

$$c^* = \left(\frac{8}{3\pi}U_m\right)G(S, \phi) \quad (21.55)$$

For very small structure displacements $G(S, \phi) = 1$, in which case c^* is the same as that for a fixed structure. As S increases, $G(S, \phi)$ increases, but is dependent on the phase

Figure 21.14

Linearization function for the relative motion Morison equation.

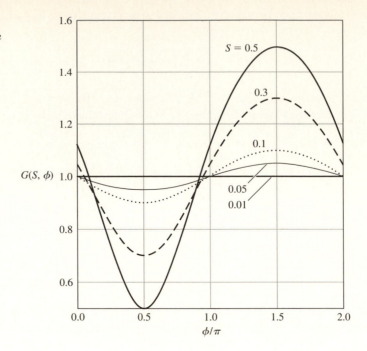

between the structure motion and waves. Figure 21.14 depicts $G(S, \phi)$ for a range of conditions. With this definition, the linearized equation of motion for a vertical cylinder is

$$\left(m + \rho\frac{\pi D^2}{4}C_m\right)\ddot{x} + \left[c + \frac{8}{3\pi}G\left(\frac{1}{2}\rho C_d D U_m\right)\right]\dot{x} + kx$$

$$= \left(\frac{8}{3\pi}GU_m\right)\left(\frac{1}{2}\rho C_d D\right)U + \rho C_m \frac{\pi D^2}{4}\frac{\partial U}{\partial t}$$

(21.56)

in which the acceleration term has also been linearized by neglecting the convective term.

21.4 INCLINED CYLINDERS

Thus far, the discussion of the Morison equation has been limited to vertical cylinders of arbitrary, but uniform, cross section. First consider a horizontal cylinder. Recall that the orbital velocities under waves decrease with depth. This means that the forward horizontal particle velocities are greater on the top of the cylinder than the returning velocities at the bottom. This leads to an asymmetry in the forcing. The complexity increases if the cylinder is inclined to the flow because there may also be a change in wave phase along the cylinder.

The forces can be represented by components that are normal to the axis of the cylinder. The normal forces are determined using the normal components of velocity and acceleration in the Morison equation. Assume the wave is propagating in the x

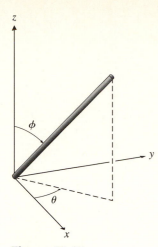

Figure 21.15

Definition sketch for arbitrary cylinder orientation.

direction and the cylinder is oriented as shown in Figure 21.15. The direction of the cylinder is defined by

$$\mathbf{c} = c_x\mathbf{i} + c_y\mathbf{j} + c_z\mathbf{k} \tag{21.57}$$

where

$$c_x = \sin\phi\cos\theta \qquad c_y = \sin\phi\sin\theta \qquad c_z = \cos\theta \tag{21.58a, b, c}$$

and the total velocity vector due to the horizontal and vertical wave particle velocities is

$$\mathbf{q} = u\mathbf{i} + w\mathbf{k} \tag{21.59}$$

The component of $\mathbf{q}$ that is normal to the cylinder axis is sought and is given by

$$\mathbf{v}_n = \mathbf{c} \times (\mathbf{q} \times \mathbf{c}) \tag{21.60}$$

Therefore the normal velocity is

$$\begin{aligned}\mathbf{v}_n &= [u - c_x(uc_x + wc_z)]\mathbf{i} \\ &\quad - c_y(uc_x + wc_z)\mathbf{j} + [w - c_z(uc_x + wc_z)]\mathbf{k}\end{aligned} \tag{21.61}$$

The Morison equation is then written in vector form as

$$\mathbf{f}_n = \frac{1}{2}\rho C_d D|\mathbf{v}_n|\mathbf{v}_n + \rho C_m\frac{\pi D^2}{4}\frac{d\mathbf{v}_n}{dt} \tag{21.62}$$

This derivation is based on the so-called independence principle in which the fluid velocity can be decomposed into components normal and tangential to the cylinder axis and that only the normal component induces a normal force. If the cylinder direction has a component in the x direction, then the wave phase changes along the cylinder. This phase change is simply kx. The Morison equation provides reasonable estimates for the forces on a vertical cylinder. But for inclined cylinders, there is much more scatter. In particular, the role of transverse lift forces, discussed in the next section, is unclear.

EXAMPLE 21.2 ▼

Calculate the wave forces on a subsea pipeline during the laying operation. Assume that the entire suspended pipe is at the same angle. The angle between the pipe and the bottom is 30° ($\phi = 60°$) and between the pipe and the wave direction is $\theta = 20°$. The circular pipe has a diameter of $D = 0.5$ m. The wave conditions are $H = 3$ m, $T = 10$ sec, and $h = 24$ m. Plot the force components for one wave period using LWT with $C_d = 1.0$, $C_m = 2.0$, and $\rho = 1025$ kg/m³.

Solution

A FORTRAN computer program that determines the forces on an inclined pipe is given in Table 21.5. The results are presented in Figure 21.16. The forces on the pipeline (and the large barge displacements) for these wave conditions would lead to rather difficult operational conditions. The typical operational conditions for lay barges correspond to a wave height of approximately 1 m. Generally, the operational wave height increases as the barge size increases and as the water depth increases.

Figure 21.16

Forces on the inclined pipeline in Example 21.2.

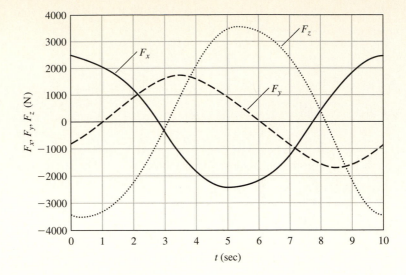

TABLE 21.5. **FORTRAN Program to Determine the Force on an Inclined Pipe in Example 21.2**

```
      Program Pipe

      real k
c     Read the input data
      write(*,*) ' Enter height, period, depth, and gravity:'
      read(*,*) height,period,depth,gravity
      write(*,*) ' Enter pipe diameter, theta, and phi:'
      read(*,*) diameter, theta, phi
      write(*,*) ' Enter drag, inertia coefficients, and water density '
     , '(cd,cm,rho):'
      read(*,*) cd,cm,rho

c     Write a header on the output file
      write(9,*)' height, period, depth, gravity'
      write(9,10)height,period,depth,gravity
      write(9,*)' diameter, theta, phi'
      write(9,10)diameter,theta,phi
      write(9,*)' cd, cm, rho'
      write(9,10)cd,cm,rho
10    format(1x,4f11.2)
      write(9,20)
20    format(1x,         time            fx            fy            fz            f')

c     Set number of time increments per wave period and number of
c     depth increments
      numtime=100
      numdepth=20

c     Determine the direction cosines using Eqs. (21.58)
      pi=4.0*atan(1.0)
      theta=theta*(pi/180.)
      phi=phi*(pi/180.)
      cx=sin(phi)*cos(theta)
      cy=sin(phi)*sin(theta)
      cz=cos(phi)
```

(continued)

```
c       Determine the pipe element length using numdepth elements in the vertical
        dL=(depth/numdepth)sqrt(cx**2+cy**2+1.0)

c       Determine wavenumber and frequency
        call PADE (depth,period,gravity,wavelen)
        k=2.0*pi/wavelen
        omega=2.0*pi/period

c       Set a couple force constants
c            These densities are for typical seawater.
             rho=1.99
             if(gravity .lt. 15.0)rho=1025.
        c1=0.5*rho*cd*diameter
        c2=rho*cm*pi*diameter**2/4.0

c       Start the time loop solving at numtime steps per wave period
        do 300 jtime=1,numtime+1
        t=(jtime-1)*(period/numtime)

c       Zero out the summing variables
        fx=0.
        fy=0.
        fz=0.
        tmax=0.
        fmax=0.

c       Start the depth loop to integrate the forces in depth
        do 100 jdepth=1,numdepth
        z=-(jdepth-0.5)*(depth/numdepth)
        x=cx*(depth+z)

c       Determine the velocities and accelerations
        u=(height*omega/2.0)*cosh(k*(depth+z))/sinh(k*depth))
    .    *cos(k*x-omega*t)
        w=(height*omega/2.0)*sinh(k*(depth+z))/sinh(k*depth))
    .    *sin(k*x-omega*t)
        ax=(height*omega**2/2.0)*cosh(k*(depth+z))/sinh(k*depth))
    .    *sin(k*x-omega*t)
        az=-(height*omega**2/2.0)*sinh(k*(depth+z))/sinh(k*depth))
    .    *cos(k*x-omega*t)

c       Determine the normal components of velocity and acceleration using Eq. (21.61)
        vnx=u-cx*(u*cx+w*cz)
        vny= -cy*(u*cx+w*cz)
        vnz=w-cz*(u*cx+w*cz)
        vn=sqrt(vnx**2+vny**2+vnz**2)

        anx=ax-cx*(ax*cx+az*cz)
        any= -cy*(ax*cx+az*cz)
        anz=az-cz*(ax*cx+az*cz)

c       Determine the force components using Eq. (21.62)
        fnx_z=c1*vn*vnx+c2*anx
        fny_z=c1*vn*vny+c2*any
        fnz_z=c1*vn*vnz+c2*anz

        fx=fnx_z*dL+fx
        fy=fny_z*dL+fy
        fz=fnz_z*dL+fz
        f=sqrt(fx**2+fy**2+fz**2)

100     continue
```

(continued)

```
c       Save the largest force
        if(f .gt. fmax)tmax=t
        if(f .gt. fmax)fmax=f
        write(9,200)t,fx,fy,fz,f
200     format(1x,5f11.2)
300     continue
        write(*,*) 'tmax, fmax', tmax,fmax
        end
```

▲

21.5 TRANSVERSE LIFT FORCES

In steady flow, vortices are shed in the lee of the body. This leads to the development of transverse lift forces. The shedding frequency of vortex pairs is characterized by the Strouhal number S, given by

$$S = \frac{f_v D}{v} \tag{21.63}$$

where f_v = vortex shedding frequency
D = cylinder diameter
v = mean flow velocity

Figure 21.17 shows the Strouhal number for steady flow. For oscillatory flow, the response is more complex. The velocity changes with time and it is common to use the maximum amplitude of the velocity as the velocity scale. As the flow begins to reverse, the formation of the present vortex will stop, developing something less than a full vortex. Then, as the flow continues to reverse, old vortices will be washed back across the structure. In spite of these difficulties, Figure 21.18 shows the dependency of the relative vortex shedding frequency f_r on the Keulegan-Carpenter number and the Reynolds number. The dimensionless relative vortex shedding frequency is the ratio of the shedding frequency and the wave frequency. For low K numbers, there are two or fewer vortices shed per wave. However, the number increases as K increases. Larger values of K are more representative of steady flow. In these cases, a number of vortices are shed before the flow reverses.

Figure 21.17

Strouhal number for slightly roughened circular cylinders [5]. Adapted from *Dynamics of Marine Structures* by M.G. Hallman, N.J. Heaf, and L.R. Woolton, CIRIA Underwater Engineering Group, London, 1978.

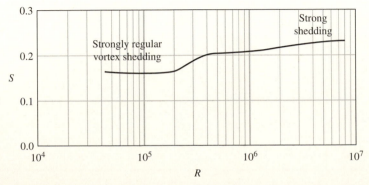

Figure 21.18

Relative frequency of vortex shedding [1]. Adapted from *Mechanics of Wave Forces on Offshore Structure* by T. Sarpkaya and M. Isaacson, copyright © 1981 by Van Nostrand Reinhold Company, New York, NY.

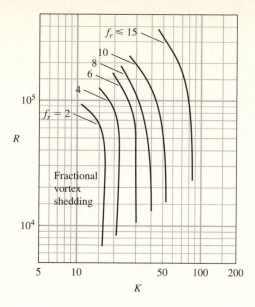

The lift force per unit length of cylinder is given approximately as

$$f_L = \frac{1}{2}\rho C_L D u_m^2 \cos(f_r \omega t + \varepsilon) \tag{21.64}$$

where u_m = amplitude of the wave particle velocity
f_r = relative vortex shedding frequency
ε = phase lag in the vortex shedding

The lift coefficient as a function of the Reynolds number and the Keulegan-Carpenter number is given in Figure 21.19. Note that the lift force is a transverse force and acts normal to the flow.

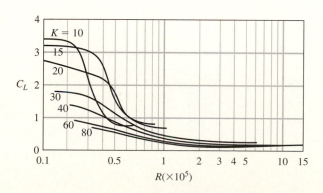

Figure 21.19

Lift force coefficients [3].

21.6 FROUDE-KRYLOV THEORY

In the preceding sections of this chapter, the diameter of the structure was considered small relative to the wave length. As a result, the wave kinematics could be estimated at the centerline of the structure, neglecting the influence of the structure on the waves. However, for large structures, diffraction effects (modifications of the wave field by the structure) are important.

Froude-Krylov theory is applicable to a range of structures that fall between the Morison equation and diffraction theory. The structure is large enough so that flow separation is not important, yet small enough so that diffraction effects are also small. In this rather narrow range of applications, the wave force may be estimated by integrating the pressure from the incident wave potential over the immersed surface of the body. In potential theory, there are no real fluid effects, and using only the incident potential neglects diffraction.

Consider the body with arbitrary, but uniform, cross section as shown in Figure 21.20. The force per unit length (in an ideal fluid) on the body is given by

$$\mathbf{f} = \int_{\Gamma} (-p\mathbf{n}) \ d\Gamma \tag{21.65}$$

where p is pressure and $\mathbf{n}$ is the outward unit normal to the structure boundary Γ. The moment per unit length is thus expressed as

$$\mathbf{m} = \int_{\Gamma} \mathbf{r}_c \times (-p\mathbf{n}) \ d\Gamma \tag{21.66}$$

From this discussion it follows that the forces and moments on a structure caused by waves are simply due to the pressure integrated over the structure surface. The pressure is obtained from the linearized Bernoulli equation of the form

$$p = -\rho \frac{\partial \phi_I}{\partial t} - \rho g z \tag{21.67}$$

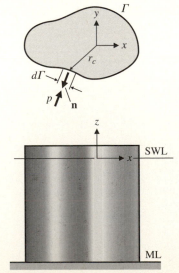

Figure 21.20

Definition sketch for a vertical cylinder of arbitrary, uniform cross section.

where ϕ_I is the incident wave velocity potential. In many cases, the force calculated by this technique may be reduced to a form

$$F_x = \rho V C_x \dot{u}_0 \qquad F_z = \rho V C_z \dot{w}_0 \qquad \text{(21.68a, b)}$$

where V = volume of the structure
$\dot{u}_0$ and $\dot{w}_0$ = water particle accelerations on the centerline of the structure
C_x and C_z = force coefficients

For a horizontal rectangular block $C_x = 1.5$, $C_z = 6.0$, and for a sphere $C_x = 1.5$, $C_z = 1.1$ [4].

21.7 DIFFRACTION THEORY: THE SCATTERING PROBLEM

When waves encounter a large fixed structure, a portion of the incident wave will be reflected or scattered. For this class of problems it is not appropriate to neglect this effect, since it has a significant influence on the development of the wave forces. Scattering problems are more complex than Froude-Krylov problems because forces due to both the incident and scattered wave fields must be determined.

Due to the complexity of this problem, it is common to neglect the real fluid effects of viscosity and separation. As a result, this approach is applicable only to large structures and low Keulegan-Carpenter numbers. This simplification allows the problem to be formulated in terms of a velocity potential. The wave force problem is then written as a boundary value problem that is similar to the linear wave theory problem. In the previously discussed wave problem, a significant simplification resulted by consideration of the linear problem. The same simplification can be manifested in the wave-structure problem. Figure 21.21 shows the regions in which the nonlinear terms become important. In general, the linear results presented in this chapter are applicable only to low Keulegan-Carpenter numbers (that is, $K < 3.0$) and low wave steepness ($H/L < 1/2\ H/L|_{\text{max}}$).

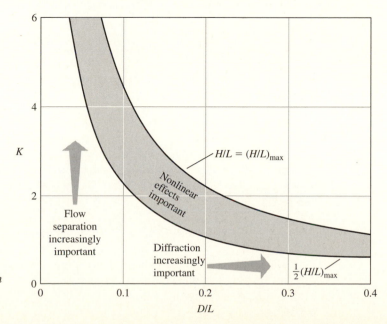

Figure 21.21

Wave force regimes [7]. Adapted from "Wave Forces Induced in the Diffraction Regime," *Mechanics of Waterways Induced Forced on Cylinders,* T.L. Shaw, editor, Pitman, London, 1979.

Because the solutions are linear, they can then be extended to random waves using linear superposition. Solutions to the wave-structure problem are obtained analytically for simple body geometries and numerically for complex geometrical configurations. The most common analytical approach is to use a separation of variables technique that leads to an eigenseries solution. The most common numerical techniques are the finite element method, source distribution, and the boundary element method.

As with the Froude-Krylov approach, the forces and moments on the structure caused by waves are simply the fluid pressure integrated over the immersed structural surface. The pressure is again obtained from the linearized Bernoulli equation. But now the velocity potential is considered to be the sum of incident and scattered wave components by

$$\phi = \phi_I + \phi_S \tag{21.69}$$

This is an implementation of linear superposition. The incident potential ϕ_I is known; thus only ϕ_S is unknown. On the surface of the impermeable fixed body, the velocity normal to the body must be zero, that is,

$$\mathbf{u} \cdot \mathbf{n} = 0 \quad \text{on } \Gamma \tag{21.70}$$

This is the kinematic boundary condition on the structure. It follows from $\mathbf{u} = \nabla \phi$ that

$$\nabla \phi_I \cdot \mathbf{n} + \nabla \phi_S \cdot \mathbf{n} = 0 \quad \text{on } \Gamma \tag{21.71}$$

In the LWT problem, the waves were assumed simple periodic in time and space. The waves are still assumed periodic in time, but unfortunately, due to the presence of the structure, they are no longer periodic in space. As a result, a boundary condition must be provided at a large distance from the structure. This is the radiation boundary condition. In the context of the eigenfunction expansion technique, the role of the radiation condition is to ensure the correct rate of decay of the scattered wave field at large distances from the structure. This also allows for a differentiation between inward (incident) and outward (scattered) propagating waves far from the structure.

In three dimensions the radiation condition takes the form

$$\lim_{r \to \infty} (\sqrt{r}) \left(\frac{\partial \phi_S}{\partial r} - ik\phi_S \right) = 0 \tag{21.72}$$

where r is the radial distance from the structure. If large distances from the structure, the velocity potential behaves as

$$\phi_S \approx e^{ikr} \tag{21.73}$$

then the radiation condition is satisfied and the result is an outward propagating wave. Consider incident waves reflecting from a large vertical circular cylinder. If it is assumed that the reflected waves radiate out as concentric rings, then conservation of energy flux gives

$$(2\pi r_1) \frac{1}{8} \rho g H_1^2 c_{g1} = (2\pi r_2) \frac{1}{8} \rho g H_2^2 c_{g2} \tag{21.74}$$

where r_1, r_2 = arbitrary radial distances from the structure
c_{g1}, c_{g2} = group velocities at r_1 and r_2
H_1, H_2 = wave heights

If the bottom is horizontal, the group velocities are the same; therefore

$$\sqrt{\frac{r_1}{r_2}} = \frac{H_2}{H_1} \sim \frac{\phi_2}{\phi_1} \tag{21.75}$$

The $\sqrt{r}$ in the radiation condition in Eq. (21.72) imposes the condition defined by Eq. (21.75). If the problem is two dimensional with one horizontal coordinate, i.e., in x and z, this $\sqrt{r}$ term is not included. An example of this condition is wave propagation in a long narrow channel with perfect reflections from the channel ends. Conservation of energy flux per unit width of channel gives

$$\frac{1}{8}\rho g H_1^2 c_{g1} = \frac{1}{8}\rho g H_2^2 c_{g2} \tag{21.76}$$

For a horizontal bottom, Eq. (21.76) reduces to

$$H_1 = H_2 \tag{21.77}$$

The reflected wave height does not decrease with distance from the ends of the channel, it is constant. Note that these are linear radiation conditions. For nonlinear formulations it is necessary to use approximations.

The other boundary conditions for the wave-structure problem are the same as for LWT. These are the kinematic and dynamic free surface boundary conditions and the bottom boundary condition. Note that ϕ_I satisfies the Laplace equation and these boundary conditions for the incident wave. Therefore, the problem is to find ϕ_S that satisfies these conditions and the structure conditions.

21.7.1 Wave Forces on a Vertical Wall

When waves interact with a fixed, impermeable structure, the kinematic boundary condition is that the normal component of wave particle velocity equals zero. Consider a full-depth, vertical, rigid impermeable wall as shown in Figure 21.22. The horizontal particle velocity must be zero at the wall. The incident horizontal water particle velocity is

$$u_I = \frac{H_I}{2}\frac{gk}{\omega}\frac{\cosh[k(h+z)]}{\cosh(kh)}\cos(kx - \omega t) \tag{21.78}$$

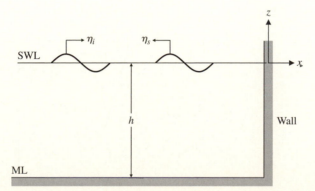

Figure 21.22

Definition sketch for a rigid impermeable wall.

If a scattered wave potential can be found such that when added to the incident wave potential gives zero velocity at the wall over the entire depth, the structure boundary condition will be satisfied. A scattered horizontal velocity that accomplishes this is

$$u_S = -\frac{H_I}{2}\frac{gk}{\omega}\frac{\cosh[k(h+z)]}{\cosh(kh)}\cos(-kx-\omega t) \tag{21.79}$$

Notice that this expression has the opposite sign of u_I, as expected, but it also corresponds to a wave moving in the negative x direction. This is usually called the reflected wave. The resulting total potential is

$$\phi = \phi_I + \phi_S = H_I\frac{g}{\omega}\frac{\cosh[k(h+z)]}{\cosh(kh)}\cos(kx)\sin(\omega t) \tag{21.80}$$

Equation (21.80) does not have the form of a propagating wave, but rather, is separable in x and t, and is known as a standing wave. Figure 21.23 shows the free surface profile in a standing wave system. The wave form does not appear to propagate; rather, the surface appears to move only vertically.

The force and mudline moment per unit width on the wall are determined from Eqs. (21.65) and (21.66) using the incident and scattered potentials; thus

$$f = \rho g H_1 h\frac{\tanh(kh)}{kh}\cos(\omega t) + \frac{1}{2}\rho g h^2 \tag{21.81}$$

$$m = \rho g H_1 h^2 \frac{kh\ \sinh(kh) - \cosh(kh) + 1}{(kh)^2\cosh(kh)}\cos(\omega t) + \frac{1}{6}\rho g h^3 \tag{21.82}$$

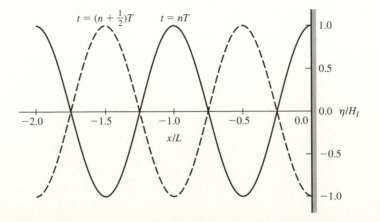

Figure 21.23

Standing wave at a vertical wall for perfect 100% reflection.

Figure 21.24

Linear wave theory dynamic forces on a vertical wall: (a) force; (b) moment.

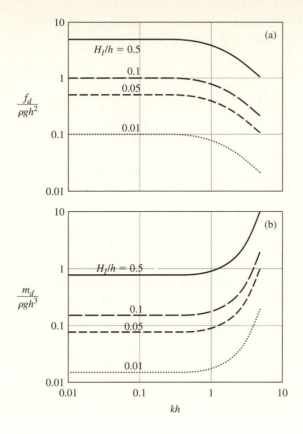

The dynamic components of these quantities are shown in Figure 21.24 for linear wave conditions.

For nonlinear waves, and especially breaking waves, the forces on the wall can be much larger than those predicted on the basis of Eqs. (21.73) and (21.74). When waves break on a wall there can be a shock pressure as the wave impacts the wall. The magnitude of these forces can be very large, but the durations are short. The magnitude of the pressure can be estimated using the *Minikin method* [8] given by

$$p_m = 101 \rho g \, \frac{H_B h_s}{L \, h_L} (h_L + h_s) \tag{21.83}$$

where p_m = maximum pressure
H_B = breaker height
h_s = depth at the structure
L = wave length at depth h_L
h_L = water depth at a distance L from the wall

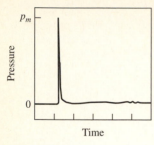

Figure 21.25

Measured breaking pressure on a wall [9]. Adapted from "Shock Pressure of Breaking Waves on Vertical Walls" by M.S. Kirkgoz, copyright © 1982 by permission of *Journal of Waterway, Port, Coastal and Ocean Engineering,* ASCE, 108: 81-95.

If the bottom is horizontal near the structure, then $h_L = h_s$. Otherwise, the determination of H_L and L is circular. The magnitudes of the resulting force and mudline moment per unit width of wall are

$$f = p_m \frac{H_B}{3} + \frac{\rho g}{2}\left(h_s + \frac{H_B}{2}\right)^2 \tag{21.84}$$

$$m = p_m \frac{H_B h_s}{3} + \frac{\rho g}{6}\left(h_s + \frac{H_B}{2}\right)^3 \tag{21.85}$$

Figure 21.25 shows the measured pressure on a vertical wall as a function of time. The rise time to the peak force is very short, and the peak impact pressure may be as large as $15\rho g H_B$. The largest pressures occur above the SWL.

21.7.2 Wave Forces on a Low Vertical Wall

Now consider the force on a thin, rigid, low vertical barrier, as shown in Figure 21.26. A common analysis technique is to determine the solutions on the left side of the barrier (region I) and on the right side of the barrier (region II) and require them to match at $x = 0$. In this case the incident and reflected waves cannot simply be added to attain no flow at the barrier. This is because the barrier does not extend over the full depth. On the barrier surface, the horizontal velocities on each side are zero. However, above the barrier, the velocity is continuous across the imaginary fluid interface between the two regions, that is, $u_I = u_{II}$. The pressure is also continuous across this interface; thus from the Bernoulli equation, $\phi_I = \phi_{II}$. This geometry results in a condition that cannot be satisfied by simply summing incident and reflected propagating waves. Near the structure, local waves develop that decay exponentially with distance away from the structure. These are called the *evanescent wave modes*. Beyond approximately three water depths from the structure, the evanescent waves become negligible. However, near the structure, they can be quite important. Interestingly, they exhibit oscillatory behavior with depth and exponential behavior in the horizontal. These evanescent modes sum to provide the required boundary conditions at the structure. In theory, there are an infinite number of evanescent modes, but in practice only the first 10 to 50

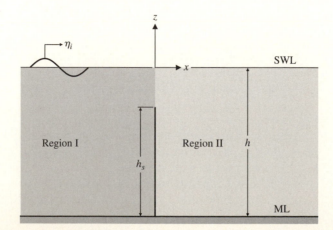

Figure 21.26

Definition sketch for a low vertical wall.

are used. Because of the number of modes that must be considered, the solution for even the very simple structure in Figure 21.26 is generally determined numerically. The amplitudes of the horizontal force and mudline moment per unit width of structure and the associated phase are shown in Figure 21.27.

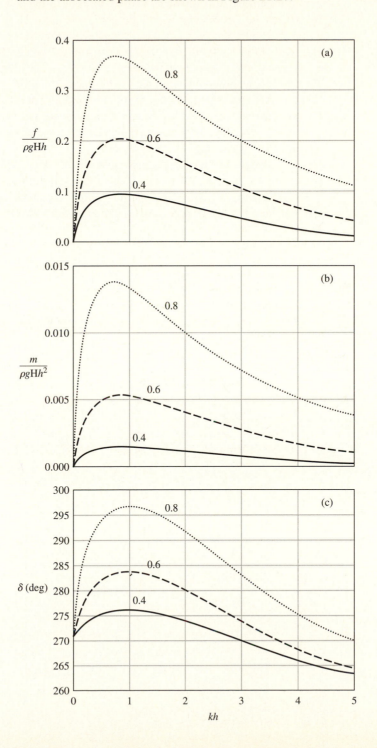

Figure 21.27

Forces on a low wall: (a) force; (b) mudline moment; (c) phase.

21.7.3 Wave Forces on a Rectangular Structure

Next consider the submerged structure shown in Figure 21.28, which has a uniform rectangular cross section. If the height of the structure is h_s and the length is b, the amplitudes of the horizontal and vertical forces and the overturning moment at the center of the base, all per unit width, are given in Figure 21.29. These results pertain to a structure height of $h_s/h = 0.5$. The dependency of the force on h_s is similar to that for a low wall, specifically, the higher the structure, the greater the force. The responses as a function of the structure width are more complex. They tend to be tuned to the ratio of the structure length to the wave length. For example, if the structure is one wave length long, the high pressure under the wave crest and low pressure under the through would integrate across the top of the structure to give zero net vertical dynamic force. Similarly, there would be a wave crest at each end of the structure so the horizontal forces would be equal and opposite, thus canceling.

This simple explanation assumes that it is not necessary to determine the scattered wave field to calculate the forces. If there is a substantial reflection, the waves across the structure are not symmetric and this simple argument fails. However, there is still the tendency for the responses to be tuned to the ratio of the structure length to the wave length.

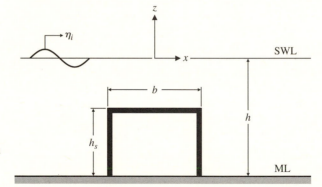

Figure 21.28

Definition sketch for a bottom founded, rigid rectangular structure.

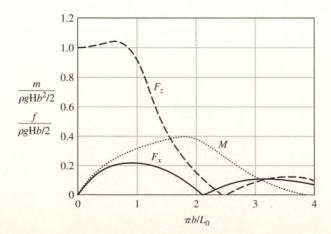

Figure 21.29

Force, moment, and phase on a bottom founded, rigid rectangular structure ($b/h = 4$, $hs/h = 0.5$) [10].

21.7.4 Wave Forces on a Vertical Circular Cylinder

A common engineering structure in marine applications is a bottom founded, surface piercing, vertical right circular cylinder. Consider the geometry shown in Figure 21.30. This problem is best addressed in cylindrical coordinates where $r^2 = x^2 + y^2$. In cylindrical coordinates, the waves are represented by Bessel functions rather than the cosines apparent in the Cartesian coordinate system. The classical solution to this problem is that of MacCamy and Fuchs [11], where the total force and mudline moment on a structure are determined as

$$F = \frac{1}{2}\rho g C_m \frac{\pi D^2}{4} H \tanh(kh) \cos(\omega t - \delta) \tag{21.86}$$

$$M = \frac{1}{2}\rho g C_m \frac{\pi D^2}{4} \frac{H}{k} \frac{kh \sin(kh) - \cosh(kh) + 1}{\cosh(kh)} \cos(\omega t - \delta) \tag{21.87}$$

where C_m is the mass coefficient and δ is the phase. In the present case,

$$C_m = \frac{8}{\pi} \frac{1}{\{[J'_1(kD/2)]^2 + [Y'_1(kD/2)]^2\}^{1/2}(kD)^2} \tag{21.88}$$

$$\delta = \tan^{-1}\left[\frac{Y'_1(kD/2)}{J'_1(kD/2)}\right] \tag{21.89}$$

where $J'_1(kD/2)$ is the derivative with respect to the argument of the Bessel function of the first kind of order 1 evaluated at $kD/2$ and $Y'_1(kD/2)$ is the analogous derivative but for a Bessel function of the second kind. There are two length scales in this problem, kD and kh. Figure 21.31 shows the amplitudes of the force and moment, and the phase for a range of wave and structural conditions.

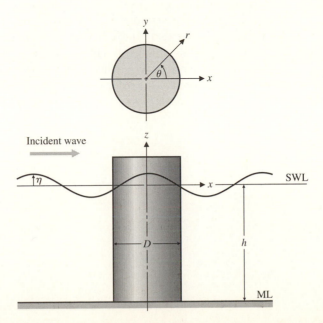

Figure 21.30

Definition sketch for diffraction by a vertical circular cylinder.

Figure 21.31

Horizontal force, mudline moment, and associated phase on a vertical circular cylinder.

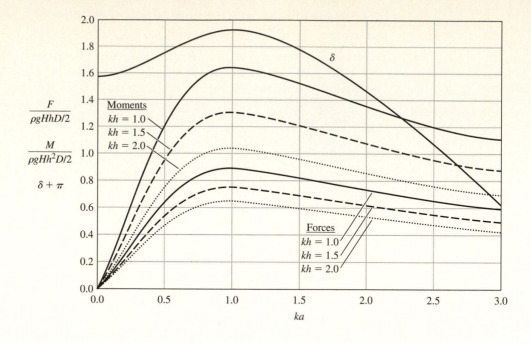

21.8 DIFFRACTION THEORY: THE RADIATION PROBLEM

To this point in the discussion, only fixed, rigid, large bodies have been considered. Many structures undergo motion in response to wave forces. These motions may be associated with rigid-body motions, such as floating bodies, or with structural deformations. In this section, only rigid floating bodies are considered.

A rigid floating body may exhibit as many as six DOF. The three translational DOF are surge, sway, and heave in the $x, y,$ and z directions. The three rotational DOF are roll, pitch, and yaw about the $x, y,$ and z axes. These six DOF are shown in Figure 21.32. If the structure experiences a steady velocity, as in the case of a ship, x is usually aligned in that direction. Most civil engineering marine structures do not exhibit a steady velocity. Typical examples of these types of structures are oil exploration and

Figure 21.32

Six DOF for a rigid floating body.

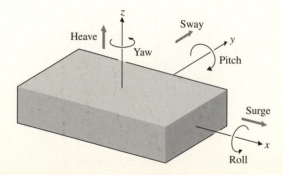

production platforms, piers, piles, and floating breakwaters. In these cases, it is common to align the coordinate system with the principal axis of the structure or with the wave approach direction.

For a fixed rigid body, the total velocity potential was expressed as the sum of the incident potential ϕ_I and the scattered potential ϕ_S. For a floating structure, the motion of the structure will also generate waves. Consider a block of wood floating in still water. If the block is moved back and forth in a horizontal direction, waves will be pushed off from the front and back of the block with each cycle of motion. These waves will radiate out into the still water. If the block is displaced cyclically in the vertical direction, a wave will radiate out for each cycle of motion. In this way, oscillation in each of the six DOF will produce waves in still water. These are called radiated waves. We will use the notation ϕ_{Rj} for these potentials, where $j = 1, 2, \ldots, 6$ correspond to each DOF. For a rigid structure that is floating or partially restrained, the total potential may be written as

$$\phi = \phi_I + \phi_S + \sum_{j=1}^{6} \phi_{Rj} \tag{21.90}$$

The kinematic boundary condition on the structure is that the normal component of the velocity equals the velocity of the structure,

$$\frac{\partial \phi_I}{\partial n} + \frac{\partial \phi_S}{\partial n} + \sum_{j=1}^{6} \frac{\partial \phi_{Rj}}{\partial n} = \frac{d\alpha_j}{dt} n_j \qquad \text{on } \Gamma \tag{21.91}$$

where $\alpha_j, j = 1, 2, \ldots, 6$ are the structure displacements in the six DOF, Γ is the surface of the structure, and repeated indices are summed. If the motions of the body are small, then the boundary condition may be applied at the mean position of the body surface Γ_0. A similar type of assumption was involved in the derivation of the LWT kinematic free-surface boundary condition. In that case, it was determined that H/L must be small if the approximations were to be meaningful. In the structure kinematic boundary condition, a comparable requirement is that $|\alpha|/L$ must be small. Therefore, the floating structure problem involves linearization of the wave and structural motions about the equilibrium free surface and the equilibrium structure surface locations, respectively.

Now the kinematic condition is applied on the fixed position Γ_0. From the diffraction problem, it is known that

$$\frac{\partial \phi_I}{\partial n} + \frac{\partial \phi_S}{\partial n} = 0 \tag{21.92}$$

on the fixed boundary. This condition leaves the radiation potentials to satisfy the motion. The unit normals for the linear displacements are the standard terms. However, for the rotation, the moment involves the cross products that give, for example,

$$\begin{aligned} n_1 &= n_x & n_2 &= n_y & n_3 &= n_z \\ n_4 &= zn_y - yn_z & n_5 &= xn_z - zn_x \\ n_6 &= yn_x - xn_y \end{aligned} \tag{21.93a, b, c, d, e, f}$$

Note that Eq. (21.91) involves 12 unknowns, the amplitudes of each DOF and the six radiated potentials. The additional conditions to determine these are provided by the structure equations of motion. In the LWT problem, the additional condition on the moving free surface was provided by the dynamic free surface boundary condition determined from the Bernoulli equation. In the case of the moving structure, the sum of the forces equals the product of the mass and the acceleration of the body (or mass moment of inertia and angular acceleration). The force is the integral of the pressure over the body surface as given in Eq. (21.65). The six equations of motion are thus represented as

$$m_{ij}\frac{d^2\alpha_j}{dt^2} = \int_{\Gamma_0} \rho\frac{\partial\phi}{\partial t}n_i\,ds - (h_{ij} + d_{ij})\alpha_j \qquad i = 1, 2, \ldots,6 \tag{21.94}$$

where m_{ij} = mass matrix components

$\qquad h_{ij}$ = hydrostatic stiffness matrix components

$\qquad d_{ij}$ = linearized mooring stiffness components

Wave forces are contained in the integral term. Hydrodynamic stiffness results for floating bodies. If a block of wood floating in still water is pushed down slightly, more of the block is submerged. As a result, the buoyancy force increases and tries to restore the block to its equilibrium position. For a body of uniform cross section, this force is proportional to the displacement, and the subsequent response can be described as the stiffness due to buoyancy. If the structure is always fully submerged, then the total buoyancy force does not change. However, if the body rotates, the relative position of the point through which the total buoyancy force acts (the center of pressure) and the center of gravity of the structure may change. These effects cause a moment on the structure. In the present development, the moorings are modeled as linear springs with forces proportional to structure motions.

Assume the motion in each of the six DOF is simple periodic in time, for which

$$\alpha_j = a_j e^{-i\omega t} \qquad j = 1, 2, \ldots,6 \tag{21.95}$$

where α_j is the time dependent motion, a_j is the amplitude of the motion (and may be complex to account for phase), $j = 1, 2, 3$ corresponds to displacements in x, y, and z, respectively, and $j = 4, 5, 6$ corresponds to rotations about the x, y, and z axes respectively. As before, the spatial and temporal components of the velocity potentials are separated assuming periodic flows, resulting in

$$\phi(x, y, z, t) = \Phi(x, y, z)e^{-i\omega t} \tag{21.96}$$

This equation applies to the incident, scattered, and radiated waves. The floating body problem, as presented herein, is completely linear. The governing equation (Laplace) is linear, the free surface boundary conditions have been linearized, the structure boundary conditions have been linearized, and the radiation condition is linear. In a linear system, if the input is doubled, then the output is doubled. In the present case, if the structure motion is doubled, the size of the radiated wave is doubled. Therefore, radiated potentials that are linear in the structure displacements are sought. The radiated velocity potentials can be written as

$$\phi_R = \Phi_{Rj}\alpha_j \tag{21.97}$$

where Φ_{Rj} are the scaled, spatial components of the radiated velocity potentials. The wave forces associated with the scattering problem are the same as examined previously. The forces associated with the motion of the body are denoted F_{Ri}, $i = 1, 2, \ldots, 6$. These radiated wave forces may be written as

$$F_{Ri} = \rho \int_{r_0} \frac{\partial \Phi_R}{\partial t} n_i \, d\Gamma = \rho \int_{\Gamma_0} \Phi_{Rj} \frac{d\alpha_j}{dt} n_i \, d\Gamma \qquad (21.98)$$

Decomposing Φ_{Rj} into real $R_e(\ )$ and imaginary $I_m(\ )$ parts gives

$$F_{Ri} = \rho \int_{\Gamma_0} \frac{d\alpha_j}{dt} \Big[\mathrm{Re}(\Phi_{Rj} n_i) + i\mathrm{Im}(\Phi_{Rj} n_i) \Big] d\Gamma \qquad (21.99)$$

Since

$$\frac{d^2 \alpha_j}{dt^2} = -i\omega \frac{d\alpha_j}{dt} \qquad (21.100)$$

Equation (21.99) may be written as

$$F_{Ri} = -\frac{\rho}{\omega} \int_{\Gamma_0} \frac{d^2 \alpha_j}{dt^2} \mathrm{Im}(\Phi_{Rj} n_i) d\Gamma + \int_{\Gamma_0} \rho \frac{d\alpha_j}{dt} \mathrm{Re}(\Phi_{Rj} n_i) d\Gamma \qquad (21.101)$$

Equation (21.101) may be simplified and expressed as

$$F_{Ri} = -\left(\mu_{ji} \frac{d^2 \alpha_j}{dt^2} + \lambda_{ji} \frac{d\alpha_j}{dt} \right) \qquad (21.102)$$

where

$$\mu_{ji} = \frac{\rho}{\omega} \int_{\Gamma_0} \mathrm{Im}(\Phi_{Rj} n_i) d\Gamma \qquad (21.103)$$

$$\lambda_{ji} = -\rho \int_{\Gamma_0} \mathrm{Re}(\Phi_{Rj} n_i) d\Gamma \qquad (21.104)$$

Equations (21.103) and (21.104) represent two important terms in defining the response of structures in waves. The term μ_{ji} is the amplitude of the force component in phase with the body acceleration and is known as the *added mass*. The term λ_{ji} is the coefficient of the force component related to the body velocity and is known as the *radiation damping*. Both terms are symmetric such that $\mu_{ij} = \mu_{ji}$ and $\lambda_{ij} = \lambda_{ji}$. The radiation damping is related to the energy in the waves resulting from body motions. The motions generate waves that propagate energy away from the body. Incorporating these results into the equation of motion gives

$$(m_{ij} + \mu_{ij})\frac{d^2\alpha_j}{dt^2} + \lambda_{ij}\frac{d\alpha_j}{dt} + (c_{ij} + d_{ij})\alpha_j = F_{Di} \tag{21.105}$$

where F_{Di} are the wave forces associated with the scattering problem (ϕ_I and ϕ_S).

Written in this way, it is seen that the scattering forces are in fact the exciting force on the structure. Therefore, the exciting forces (and moments) on a moving body are identical for the same fixed body. Furthermore, through the application of Green's theorem, it can be shown that it is not necessary to determine the scattered potential ϕ_S. Alternatively, numerical results for F_{Di} may be calculated based on the asymptotic radiated wave field far from the structure, making it unnecessary to compute the local evanescent modes. This rather surprising result was developed by Haskind [12]. This technique leads to a significant simplification in computational effort, if only forces are required in the analysis. A more thorough discussion is presented in [1] and [13].

The analysis of floating bodies with multiple DOF is mathematically laborious, even for simple structures. Figure 21.33 depicts a submerged buoyant structure with a uniform rectangular cross section. The structure is restrained by two vertical moorings with linear spring stiffness of k per unit length of structure. For normally incident waves, this problem is two dimensional in x-z. There are two translational DOF, surge and heave, and one rotational DOF, pitch. These may be denoted by the subscripts 1, 2, and 3, respectively. Figure 21.34 shows the added mass and radiation damping coefficients for a specific set of wave and structural conditions. The surge added mass and the surge radiation damping increase significantly at frequencies near the surge natural period. This strong surge response also influences the coupled surge-pitch coefficients. Note that the coupling term in the added mass, μ_{13}, is significant and leads to off-diagonal terms in the mass matrix.

Figure 21.35 shows the horizontal and vertical forces on the structure. Again, for this geometry, the response is dominated by surge.

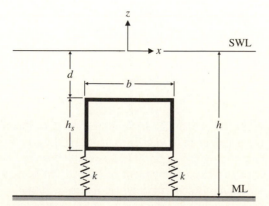

Figure 21.33

Definition sketch for a fully submerged, moored rectangular structure.

Figure 21.34

Force coefficients for a floating rectangular structure ($d/h = 0$, $b/h = 0.23$, $hs/h = 0.1$): (a) added mass; (b) radiation damping [10]. Adapted from "A Dynamic Submerged Breakwater" by A.N. Williams and W.G. McDougal, copyright © 1996 by permission of *Journal of Waterway, Port, Coastal and Ocean Engineering*, ASCE, 122: 702–713.

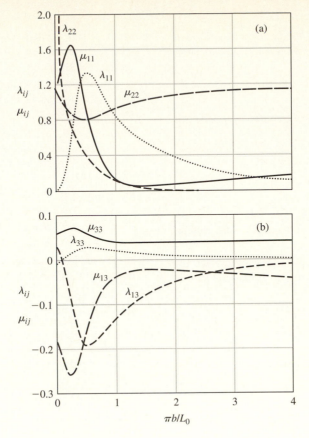

Figure 21.35

Horizontal and vertical force on a floating rectangular structure ($d/h = 0$, $b/h = 0.2$, $hs/h = 0.1$): (a) added mass; (b) radiation damping [10]. Adapted from "A Dynamic Submerged Breakwater" by A.N. Williams and W.G. McDougal, copyright © 1996 by permission of *Journal of Waterway, Port, Coastal and Ocean Engineering*, ASCE, 122: 702–713.

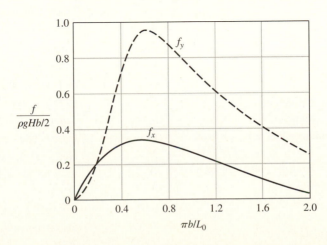

REFERENCES

1 Sarpkaya, T. and Isaacson, M., *Mechanics of Wave Forces on Offshore Structures,* Van Nostrand Reinhold, New York, 1981.

2 Lamb, H., *Hydrodynamics,* 6th ed., Dover Publications, New York, 1932.

3 Sarpkaya, T., *In-Line and Transverse Forces on Smooth and Sand-Roughened Cylinders in Oscillatory Flow at High Reynolds Numbers,* Report No. NDS-69SL76062, Naval Post Graduate School, Monterey, CA, 1976.

4 Chakrabarti, S.K., *Hydrodynamics of Offshore Structures,* Springer-Verlag, Berlin, 1987.

5 Hallam, M.G., Heaf, N.J., and Wootton, L.R., *Dynamics of Marine Structures,* CIR 1A Underwater Engineering Group, London, 1978.

6 Borgman, L.E., Spectral Analysis of Ocean Wave Forces on Piling, *Journal of Waterways and Harbors Division,* ASCE, Vol. 93, 1967, pp. 129–156.

7 Isaacson, M., Wave Induced Forces in the Diffraction Regime, *Mechanics of Wave Induced Forces on Cylinders.* T.L. Shaw, editor, Pitman, London, 1979.

8 Minikin, R.R., *Winds, Waves and Maritime Structures: Studies in Harbor Making and in the Protection of Coasts,* 2nd ed., Griffin, London, 1963.

9 Kirkgoz, M.S., Shock Pressure of Breaking Waves on Vertical Walls, *Journal of Waterway, Port, Coastal and Ocean Engineering,* ASCE, Vol. 108, 1982, pp. 81–95.

10 Williams, A.N. and McDougal, W.G., A Dynamic Submerged Breakwater, *Journal of Waterway, Port, Coastal and Ocean Engineering Division,* ASCE, Vol. 122, 1996, pp. 702–713.

11 MacCamy, R.C. and Fuchs, R.A., *Wave Forces on Piles: A Diffraction Theory,* U.S. Army Corps of Engineers, Beach Erosion Board, Technical Memo No. 69, 1954.

12 Haskind, M.D., *Oscillations of a Ship on a Calm Sea,* Society of Naval Architects and Marine Engineers, T & R Bulletin 1-12, 1953.

13 Newman, J.N., *Marine Hydrodynamics,* MIT Press, Cambridge, MA, 1977.

NOTATION

a_x	horizontal water particle acceleration
c	structural damping
C_a	added mass coefficient
c_B	Bernoulli constant
c_{g1}, c_{g2}	group velocity at locations r_1, r_2
C_m	inertia coefficient
C_d	drag coefficient
c_x, c_y, c_z	force component coefficients
c^*	linear damping coefficient
d	total depth $(h + \eta)$
D	diameter
f_{dmax}	maximum drag force per unit length of structure in the x direction
f_{dx}	drag force per unit length of structure in the x direction
f_{imax}	maximum inertia force per unit length of structure in the x direction
f_{ix}	inertia force per unit length of structure in the x direction
f_L	lift force per unit length of structure
f_r	relative vortex shedding frequency
f_v	vortex shedding frequency
f_x	total force per unit length of structure in the x direction
$\mathbf{f}_n$	fluid force normal to structure
F_x	total force in the x direction
F_{dmax}	maximum drag force
F_{imax}	maximum inertia force
F_{Di}	wave forces associated with scattering
F_{Ri}	radiated wave forces
g	acceleration due to gravity
G	damping coefficient parameter
h	still water depth
h_L	depth one wave length from structure
H	wave height
H_B	breaking wave height
H_1, H_2	wave heights at locations r_1, r_2
i	$\sqrt{-1}$
$\mathbf{i}, \mathbf{j}, \mathbf{k}$	unit vectors in Cartesian coordinates
J_1, Y_1	Bessel functions of the first and second kind of order 1

k	structural stiffness; also wave number ($2\pi/L$)
L	wave length
L_0	deep water wave length
m	structural mass
m	wave-induced moment per unit length of structure
M	total moment
$M_{d\max}$	maximum drag force moment
$M_{i\max}$	maximum inertia force moment
$\mathbf{n}$	outward unit normal vector
p	pressure
p_m	maximum pressure
$\mathbf{q}$	vector fluid velocity
r_c	moment arm on pressure force
$\dot{r}$	structure velocity relative to fluid velocity
r, θ	polar cooordinates
R	Reynolds number
S	ratio of structure velocity amplitude to water particle velocity amplitude; also Strouhal number
S_{ff}	force spectrum per unit length of structure
S_{uu}	horizontal water particle velocity spectrum
$S_{a_x a_x}$	horizontal water particle acceleration spectrum
$S_{\eta\eta}$	free-surface spectrum
t	time
u	water particle velocity
u_I	incident water particle velocity
u_m	velocity amplitude
u_{rms}	root-mean-square velocity

u_S	scattered water particle velocity
u_θ	tangential component of velocity
$\dot{u}_0, \dot{w}_0$	water particle accelerations at centerline of structure
U	depth integrated water particle velocity
$\dot{U}$	depth integrated water particle acceleration
$\mathbf{v}_n$	fluid velocity normal to structure
x	structure displacement
x_m	amplitude of structure motion
z	vertical coordinate
α	drag force linearizing coefficient
α_j	structure displacements; also horizontal coordinate
δ	phase
ε	phase of vortex shedding
λ_{ij}	radiation damping
Γ	structure surface
Γ_o	mean position of structure surface
η	free-surface elevation
κ	Keulegan-Carpenter number
μ_{ij}	added mass
π	numerical constant (3.1415...)
ρ	fluid density
ω	wave frequency ($2\pi/T$)
ϕ	velocity potential
Φ_I	incident wave potential
Φ_S	scattered wave potential
Φ_{Rj}	radiated wave velocity potential

PROBLEMS

21.1 A rigid vertical pile 4 ft in diameter supports a navigation aid at elevation +20 ft in a water depth of 40 ft. Determine the maximum horizontal force and mudline moment for a wave with $H = 7$ ft and $T = 10$ sec. Assume $C_d = 1.0$, $C_m = 1.7$, and $\rho = 1.99$ slugs/ft^3. Use (a) LWT and (b) stretch wave theory.

21.2 Repeat Problem 21.1 using Stokes 5th-order wave theory and assume $C_d = 0.7$ and $C_m = 1.7$.

21.3 A 2-m, 20-sec wave in 6 m of water passes a 2-m-diameter surface piercing pile that is encrusted with barnacles 4 cm high over its full length. Estimate the maximum mudline moment by the most appropriate method. Assume $\rho = 1025$ kg/m^3.

21.4 A structure has three vertical legs, each 8 ft in diameter, located at the corners of an equilateral triangle 60 ft on a side. Determine the maximum horizontal force using LWT for $H = 10$ ft, $T = 11$ sec, and $h = 40$ ft. Assume $\rho = 1.99$ slugs/ft^3.

21.5 For a Pierson-Moskowitz spectrum with $f_0 = 0.125$ hz, determine the rms force and mudline moment for a 2-ft-diameter pile in 60 ft of water. Assume $C_d = 1.0$, $C_m = 1.7$, and $\rho = 1.99$ slugs/ft^3.

21.6 Estimate the hydrodynamic load at each joint in the panel shown in Figure P21.1 for waves propagating in the y direction (i.e., into the plane of the page). The vertical members have an outside diameter of 1.5 m and the bracing has a diameter of 0.75 m. The wave has height $H = 4$ m, period $T = 9$ sec, and density $\rho = 1025$ kg/m³.

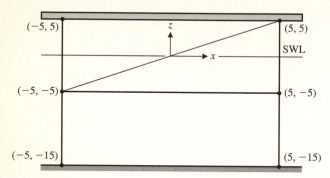

Figure P21.1

21.7 A circular cylinder is restricted to move in surge by only a horizontal carriage as shown in Figure P21.2. Attached to the cylinder are a spring and linear damping mechanism. The system has characteristics $m = 50$ kg, $D = 1$ m, $h_s = 1$ m, $h = 4$ m, $k = 200$ N/m, and $c = 10$ kg/sec. If the structure is subjected to linear waves of height 0.75 m and period 4 sec, calculate the amplitude and phase of the resulting response. Use the linearized, relative motion Morison equation for the hydrodynamic excitation with $C_d = 0.9$, $C_m = 2.0$, and $\rho = 1025$ kg/m³.

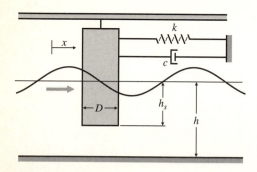

Figure P21.2

21.8 Figure P21.3 shows a slab-pile structure to be deployed in 40 m of water. The design wave for the intended location has a height of 6 m and period 10 sec. Assume that the presence of the structure does not interfere with the wave field. Each column has a diameter of 5 m. Assume that the presence of the columns decreases the force on the underside of the slab in accordance with the decrease in slab surface area presented to the waves. Calculate the maximum horizontal and vertical Froude-Krylov forces on the slab structure due to the design wave. Assume $\rho = 1025$ kg/m³.

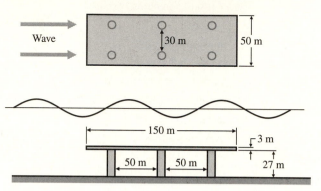

Figure P21.3

21.9 A rectangular barge of length 60 m, beam 15 m, and draft 6 m floats freely in a water depth of 10 m. Find the resonant wavelength for heave assuming $C_a = 1$ and neglecting damping. Assume $\rho = 1025$ kg/m³.

21.10 Determine the natural frequencies in surge, heave, and pitch for the submerged structure in Figure 21.31, for the conditions in Figure 21.33. Assume $h = 10$ m, $k = 100$ N/m, $\rho = 1025$ kg/m³, and the structure has a uniform mass distribution with a specific gravity of 0.3.

21.11 Determine the maximum force in the mooring lines for the structure in Figures 21.32 and 21.33 using Froude-Krylov to estimate forces and the condition in Problem 21.10.

APPENDIX A

Northridge Earthquake Ground Motion

The north-south component of ground motion recorded during the 17 January 1994 Northridge, CA, earthquake is shown in Figures 17.10 and 18.5. This earthquake record is used extensively throughout Chapter 18 and is required for the solution of many end-of-chapter problems in Chapter 18. Digitized values for the ground acceleration in units of g (acceleration due to gravity) are available on the authors' web site at www.Structural-Dynamics.com. The accelerogram includes 6000 data points at equal time intervals of 0.02 sec. The record is to be read row by row, with the first entry occurring at time $t = 0.0$ sec. Additionally, a sample of four-way logarithmic paper is presented in Figure A.1. This paper is necessary for the construction of earthquake response spectra and design spectra required by a number of end-of-chapter problems in Chapter 18.

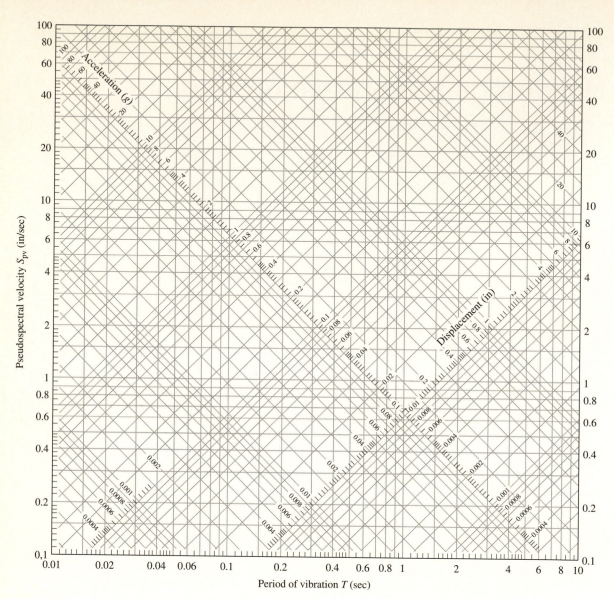

Figure A.1

Four-way logarithmic paper.

APPENDIX B

Computer Software

Many of the end-of-chapter problems in this text require a computer solution. A suite of computer programs has been developed for the solution of these problems. The programs are written for a PC in FORTRAN 77 for a Windows platform. Both the source code and an executable version of the program suite are available on the authors' web site at www.Structural-Dynamics.com.

Index